Natural Gas Processing from Midstream to Downstream

Natural Gas Processing from Midstream to Downstream

Edited by

Nimir O. Elbashir
Texas A&M University at Qatar, Qatar

Mahmoud M. El-Halwagi
Texas A&M University, USA

Ioannis G. Economou
Texas A&M University at Qatar, Qatar

Kenneth R. Hall
Bryan Research & Engineering
Texas, USA

WILEY

This edition first published 2019

Registered Offices
John Wiley & Sons, Inc., 111 River Street, Hoboken, NJ 07030, USA
John Wiley & Sons Ltd, The Atrium, Southern Gate, Chichester, West Sussex, PO19 8SQ, UK

Editorial Office
The Atrium, Southern Gate, Chichester, West Sussex, PO19 8SQ, UK

For details of our global editorial offices, customer services, and more information about Wiley products visit us at www.wiley.com.

Wiley also publishes its books in a variety of electronic formats and by print-on-demand. Some content that appears in standard print versions of this book may not be available in other formats.

Library of Congress Cataloging-in-Publication Data

Names: Elbashir, Nimir O. M., 1969– editor. | El-Halwagi, Mahmoud M., 1962– editor. | Economou, Ioannis G., editor. | Hall, K. R. (Kenneth R.), editor.
Title: Natural gas processing from midstream to downstream / edited by Nimir O. Elbashir, Mahmoud M. El-Halwagi, Ioannis G. Economou, Kenneth R. Hall.
Description: First edition. | Hoboken, NJ : John Wiley & Sons, 2019. | Includes bibliographical references and index. |
Identifiers: LCCN 2018027442 (print) | LCCN 2018042360 (ebook) | ISBN 9781119269632 (Adobe PDF) | ISBN 9781119269625 (ePub) | ISBN 9781119270256 (hardcover)
Subjects: LCSH: Natural gas. | Chemical processes.
Classification: LCC TP350 (ebook) | LCC TP350 .N36456 2018 (print) | DDC 665.7/3–dc23
LC record available at https://lccn.loc.gov/2018027442

Cover design by Wiley
Cover Images: Train Cars © B&M Noskowski/E+/Getty Images; Refinery © Brian Brown/Alamy Stock Photo

Set in 10/12pt WarnockPro by SPi Global, Chennai, India
Printed and bound in Singapore by Markono Print Media Pte Ltd

10 9 8 7 6 5 4 3 2 1

To our families and children

To the memory of our colleague Professor M. Sam Mannan.

Contents

List of Contributors

Noof Abdalla
Chemical Engineering Program
Texas A&M University at Qatar
Qatar

Shaik Afzal
Chemical Engineering Program
Texas A&M University at Qatar
Qatar

Monir Ahammad
Mary Kay O'Connor Process Safety Center
Artie McFerrin Department of Chemical Engineering
Texas A&M University
USA

Ahmad Al-Douri
Department of Chemical Engineering
Texas A&M University
USA

Ali Almansoori
Khalifa University
The Petroleum Institute
Abu Dhabi
United Arab Emirates

Saad A. Al-Sobhi
Department of Chemical Engineering
University of Waterloo
Waterloo
ON
Canada

and

Department of Chemical Engineering
Qatar University
Doha
Qatar

Andrew Avalos
Dwight Look College of Engineering
Texas A&M University
USA

Christos Boukouvalas
Laboratory of Thermodynamics and Transport Phenomena
School of Chemical Engineering
National Technical University of Athens
Athens
Greece

Joel G. Cantrell
Bryan Research & Engineering
LLC
Texas
USA

Nikolaos A. Diangelakis
Artie McFerrin Department of Chemical Engineering and
Texas A&M Energy Institute
Texas A&M University
USA

Ioannis G. Economou
Texas A&M University at Qatar
Qatar
TEES Gas and Fuels Research Center
Texas A&M Engineering Experiment Station
USA

Nimir O. Elbashir
Petroleum Engineering Program
Texas A&M University at Qatar
Qatar
TEES Gas and Fuels Research Center
Texas A&M Engineering Experiment Station
USA

Mahmoud M. El-Halwagi
TEES Gas and Fuels Research Center
Texas A&M Engineering Experiment
and Artie McFerrin Department of Chemical Engineering
Texas A&M University
USA

Ali Elkamel
Department of Chemical Engineering
University of Waterloo
Ontario
Canada

and

Khalifa University
The Petroleum Institute
Abu Dhabi
United Arab Emirates

Marwan El Wash
Chemical Engineering Program
Texas A&M University at Qatar
Qatar

Fatih S. Erenay
Department of Management Studies
University of Waterloo
ON
Canada

Michael Fowler
Department of Chemical Engineering
University of Waterloo
Ontario
Canada

Vinay Gadekar
EPCON Software™
Houston
USA

Kenneth R. Hall
Bryan Research & Engineering
LLC
Texas
USA

Rasha Hasaneen
Dwight Look College of Engineering
Texas A&M University
USA

Ahmed Farid Ibrahim
Petroleum Engineering Department
Texas A&M University
USA

Swarom R. Kanitkar
Cain Department of Chemical Engineering
Louisiana State University
Baton Rouge
USA

Eirini Karakatsani
Haldor Topsoe A/S
Nymøllevej 55, Kgs. Lyngby
Denmark

Tala Katbeh
Chemical Engineering Program
Texas A&M University at Qatar
Qatar

Georgios M. Kontogeorgis
Center for Energy Resources Engineering (CERE)
Department of Chemical and Biochemical Engineering
Technical University of Denmark
Denmark

Vasiliki Louli
Laboratory of Thermodynamics and Transport Phenomena
School of Chemical Engineering
National Technical University of Athens
Greece

M. Sam Mannan
Mary Kay O'Connor Process Safety Center
Artie McFerrin Department of Chemical Engineering
Texas A&M University
USA

Mozammel Mazumder
Dan F. Smith Department of Chemical Engineering
Lamar University
USA

Vasileios K. Michalis
Molecular Thermodynamics and Modeling of Materials Laboratory
Institute of Nanoscience and Nanotechnology
National Center for Scientific Research "Demokritos"
Greece

Nasr Mohamed
Petroleum Engineering Program
Texas A&M University at Qatar
Qatar

Rajib Mukherjee
TEES Gas and Fuels Research Center
Texas A&M Engineering Experiment Station
USA

Hisham A. Nasr-El-Din
Petroleum Engineering Department
Texas A&M University
USA

Paul A. Nelson
EPCON Software™
Houston
USA

Nefeli Novak
Laboratory of Thermodynamics and Transport Phenomena
School of Chemical Engineering
National Technical University of Athens
Greece

Emmanuel Ogbe
Department of Chemical Engineering
University of Waterloo
Ontario
Canada

and

Khalifa University
The Petroleum Institute
Abu Dhabi
United Arab Emirates

Gerald S. Ogumerem
Artie McFerrin Department of Chemical Engineering and
Texas A&M Energy Institute
Texas A&M University
USA

Ecem Özinan
Artie McFerrin Department of Chemical Engineering
Texas A&M University
USA

Eleni Panteli
Equinor ASA
Research & Technology Center
Trondheim
Norway

Konstantinos D. Papavasileiou
Molecular Thermodynamics and Modeling of Materials Laboratory
Institute of Nanoscience and Nanotechnology
National Center for Scientific Research "Demokritos"
Greece

Georgia Pappa
Laboratory of Thermodynamics and Transport Phenomena
School of Chemical Engineering
National Technical University of Athens
Greece

Loukas D. Peristeras
Molecular Thermodynamics and Modeling of Materials Laboratory
Institute of Nanoscience and Nanotechnology
National Center for Scientific Research "Demokritos"
Greece

Eirini Petropoulou
Laboratory of Thermodynamics and Transport Phenomena
School of Chemical Engineering
National Technical University of Athens
Greece

Efstratios N. Pistikopoulos
Artie McFerrin Department of Chemical Engineering and
Texas A&M Energy Institute
Texas A&M University
USA

Debalina Sengupta
TEES Gas and Fuels Research Center
Texas A&M Engineering Experiment Station
USA

Mostafa Shahin
ORYX GTL Company
Ras Laffan
Qatar

Munawar A. Shaik
Department of Chemical Engineering
Khalifa University
Abu Dhabi
UAE
Department of Chemical Engineering
Indian Institute of Technology (IIT)
New Delhi
India

Mohammed Shammaa
Dwight Look College of Engineering
Texas A&M University
USA

Nathan Sibley
Dwight Look College of Engineering
Texas A&M University
USA

Stathis Skouras
Equinor ASA
Research & Technology Center
Trondheim
Norway

Justin C. Slagle
Bryan Research & Engineering, Inc.
Texas
USA

James J. Spivey
Cain Department of Chemical Engineering
Louisiana State University
Baton Rouge
USA

Manolis Vasileiadis
Molecular Thermodynamics and Modeling of Materials Laboratory
Institute of Nanoscience and Nanotechnology
National Center for Scientific Research "Demokritos"
Greece

Epaminondas Voutsas
Laboratory of Thermodynamics and Transport Phenomena
School of Chemical Engineering
National Technical University of Athens
Greece

Ben R. Weber, Jr
Synfuels International
INC
Parkland Hall
USA

Moye Wicks III
EPCON Software™
USA

Todd J. Willman
EPCON Software™
Houston
USA

Qiang Xu
Dan F. Smith Department of Chemical Engineering
Lamar University
USA

About the Editors

Dr. Nimir O. Elbashir (editor)

Professor Elbashir holds a joint appointment as a professor in the Chemical Engineering Program and the Petroleum Engineering Program at Texas A&M University at Qatar and the chair of the Petroleum Engineering Program., He is the director of Texas A&M's Engineering Experiment Station Gas and Fuels Research Center, a major research center that involves 30 faculty members from both the Qatar and College Station campuses of Texas A&M University and the chair of the ORYX GTL Excellence Program in Gas-to-Liquid technology. He has extensive research and teaching experience from four different countries around the world, including his previous position as a researcher at BASF R&D Catalysts Center in Iselin, New Jersey. The focus of his research activities is the design of advanced reactors, catalysts, and conversion processes for natural gas, coal, and CO_2 to ultraclean fuels and value-added chemicals. He has established several unique global research collaboration models between academia and industry with research funds exceeding twelve million dollars during the past six years. He holds several U.S. and European patents and a large number of scientific publications in the form of peer-reviewed journals, conference papers, and technical industry reports as well as invited talks and conference presentations. The scholarship of his research activities has been recognized by awards from Qatar Foundation, BASF Corp., Texas A&M University Engineering Experiment Station, Texas A&M University Qatar, the American Institute of Chemical Engineers, and others

Dr. Mahmoud M. El-Halwagi (co-editor)

Professor El-Halwagi is the McFerrin Professor at Artie McFerrin Department of Chemical Engineering, Texas A&M University. He is also the managing director of the Texas A&M Gas and Fuels Research center. Dr. El-Halwagi's main areas of expertise are process integration, synthesis, design, operation, and optimization. Specifically, Dr. El-Halwagi's research focuses on sustainable design through the development of systematic and generally applicable approaches and tools. Dr. El-Halwagi is a fellow of the AIChE and is the recipient of several awards, including the AIChE Sustainable Engineering Forum Research Excellence Award and the National Science Foundation's National Young Investigator Award. Dr. El-Halwagi received his Ph.D. in chemical engineering from the University of California, Los Angeles, and his M.S. and B.S. from Cairo University.

Dr. Kenneth R. Hall (co-editor)

Professor Hall is a Senior Consulting Engineer at Bryan Research & Engineering. He formerly held positions as Professor of Chemical Engineering Program at Texas A&M University in College Station and Qatar and Regents Professor in the Texas A&M University System. He served in a large number of administrative positions at Texas A&M University, where he served as professor for over 42 years. Dr. Hall received numerous research and teaching awards. He graduated 34 doctorate students and 27 masters students. He received more than 25 million U.S. dollars of research funding during his tenure as professor. He has 16 U.S. patents and 271 peer-reviewed journal papers and 9 books. His primary emphasis in research is the thermodynamics of fluids and fluid mixtures with concentration upon the measurement and correlation of precise property data. Experimental efforts in this area include: PVT, VLE, enthalpy, vapor pressure, dew and bubble point apparatus. In each case, the emphasis is to gather state-of-the-art data. His theoretical efforts include equation-of-state studies and modern evaluation of properties in the vapor-liquid critical region focusing upon natural gas-type fluids and mixtures. A secondary interest is flow and energy measurements of compressible fluids.

Dr. Ioannis G. Economou (co-editor)

Dr. Economou is the Associate Dean for Academic Affairs and Professor of Chemical Engineering at Texas A&M University at Qatar. Prior to this, he was the Associate Provost for Graduate Studies and Professor of Chemical Engineering at the Petroleum Institute, Abu Dhabi (2009–12). From 1995 to 2009, he worked at the National Center for Scientific Research "Demokritos" in Athens, Greece. He holds a Diploma in Chemical Engineering from the National Technical University of Athens, Greece (1987) and a PhD also in Chemical Engineering from The Johns Hopkins University in Baltimore, Maryland, USA (1992). He worked as a post-doctoral researcher in Delft University of Technology in the Netherlands (1993–94) and in Exxon Research and Engineering Company, in New Jersey, USA (1994–95), as research fellow in University College London (1994–96) and Princeton University (2004 and 2015), and as visiting Professor in the Technical University of Denmark (2001 and 2006–07) and the American College of Greece (2007–09). He has consulted extensively for major oil and chemical companies in North America, Europe and Middle East. He has supervised 18 MSc students, 14 PhD students and 16 post-docs, he has published more than 190 peer-reviewed research papers in leading journals in Chemical Engineering, Physical Chemistry and Polymer Science, 10 book chapters and has given approximately 300 presentations in conferences, Universities and industrial research centers worldwide. His research interests are related to molecular thermodynamics, complex fluids, aqueous systems, CO_2 management, green solvents, and soft materials including polymers, ionic liquids, metal organic frameworks, etc. From 2007 to 2014, he was the Founding Chairman of the Working Party on Thermodynamics and Transport Properties of the European Federation of Chemical Engineering. He is Editor of *Fluid Phase Equilibria*, and member of the Editorial Boards in *Journal of Chemical and Engineering Data* and in *Journal of Supercritical Fluids.*

Preface

This book focuses on highlighting global experience in the production and monetization of shale gas and natural gas while providing techno-economic assessments of the midstream and downstream natural gas processing technologies. Natural gas has become a significant player in the global energy mix as a result of rising prosperity, which drives an increase in global energy demand that is expected to increase by more than 30% in a few decades. Natural gas, a fossil fuel, enjoys several environmental benefits over the other fossil fuels, such as coal and petroleum products. This book identifies the current status and the challenges facing the advancement of natural gas monetization looking at the entire process with emphasis upon midstream- and downstream-related topics from the fundamental to the applied. Particular focus is on the downstream processes including physical treatment (Liquefied Natural Gas or LNG) and the chemical conversion (both direct (e.g., natural gas to methanol) and indirect (e.g., gas-to-liquid or GTL).

The authors of this book are leading researchers from academia and industry, who provide a broad, but well-integrated coverage of several technologies and processes for treating and producing natural gas while highlighting the economics and risks facing the monetization technologies. Also, to offer a wide range of perspectives the book includes participants from different parts of the world, specifically from nations in which natural gas plays a critical role in the energy mix and the economy of the country. The book emphasizes success stories in natural gas monetization while identifying the challenges involved.

This book gives to the reader a broad picture of the current status of natural gas production, treatment, and monetization technologies and identifies challenges that these efforts face.

The book covers a broad range of topics interesting to various audiences, but it also can serve as a general reference for natural gas production/processing/use. It is useful for researchers and graduate students working with shale gas and natural gas production and monetization, while simultaneously being of interest to experts in the oil and gas industry who emphasize technical or economic aspects of the technology. Some chapters emphasize the importance of collaboration between academia and industry to advance existing commercial technologies and to develop innovative processes and techniques.

The initiative for this work emanates from the Texas A&M University Engineering Experiment Station Gas & Fuels Research Center (GFRC), which involves 30 professors from both Texas A&M University in College Station, Texas and the Qatar campus (http://gfrc.tamu.edu/).The GFRC is focusing upon advancing research and development activities that support growth in shale gas, specifically in the Gulf of Mexico region, which should bring to the United States chemical industry estimated incremental capital investments of $71.7 billion by 2020. In addition, the GFRC supports the State of Qatar in monetizing its natural gas wealth and sustaining its position as "The Gas Processing World Capital." GFRC works closely with industry and academia around the globe to provide solutions for several problems in the fields of natural gas exploration, production, and monetization and to work with industry to create skilled engineers and scientists capable of supporting and advancing this field.

This book has twenty-one chapters that are arranged in four sections:

Section 1: Introduction

This section highlights the importance of the global energy market and the different stages of natural gas exploration, production, handling, and monetization. Natural gas exploration and production appears in an oil and gas section while natural gas treatment to remove impurities and produce methane, and condensates, is in the midstream process section. Natural gas processing to final products in the form of fuels and chemicals is in the downstream process section. The section also briefly highlights the advantages and the constraints facing the different natural gas monetization techniques with emphasis on the United States, Russia, and Qatar.

Section 2: Upstream

Three book chapters address both fundamental and applied topics on upstream-related issues such as geologic sequestration of CO2, and its role in enhanced oil and gas recovery. A major challenge facing the environmental assessment of natural gas products is potential greenhouse gas emissions during the processing of natural gas to fuels. Also, this section includes a chapter that addresses thermodynamic modeling of natural gas and condensate properties that is essential in natural gas production and another chapter that covers the technical, microeconomic analyses and policy implications of environmental remediation techniques for shale gas wells in the Barnett Shales. Two other chapters, lying between upstream and midstream natural gas production, address fluid flow fundamentals and advanced techniques.

Section 3: Midstream

Seven chapters of this book address topics in midstream or in midstream-downstream parts of the natural gas monetization chart. The topics vary from the use of process simulators to design and optimize the operations of midstream and downstream plants to the optimization of natural gas network operations under uncertainty. This section also includes chapters that address the optimization of shale gas monetization supply chains and the identification of optimal operation for natural gas liquid recovery and products. Also, this section includes a chapter that covers fundamental topics such as thermodynamic modeling of natural gas processing units and another that focuses upon applied topics such as safety in midstream and downstream processing of natural gas.

Section 4: Downstream

The downstream and the monetization section has ten chapters that address several direct and indirect natural gas conversion routes to chemicals and fuels (including gas-to-liquid (GTL) technology) as well as the physical treatment of natural gas (liquefied natural gas processes (LNG)). Also, this section includes techno-economic assessment of the natural gas monetization processes as well as the characterization and the design of the GTL fuels and chemicals.

1

Introduction to Natural Gas Monetization

Nimir O. Elbashir

Petroleum Engineering Program, Texas A&M University at Qatar, Qatar
TEES Gas and Fuels Research Center, Texas A&M Engineering Experiment Station, USA

CHAPTER MENU

1.1 Introduction

Natural gas, mainly methane, has been known and utilized since the ancient Greek and Chinese civilizations. Natural gas began playing a prominent role in the energy market as early as the 1780s, during the start of the Industrial Revolution, where it was used in the United Kingdom as a source of lighting for homes and streets. Baltimore became the first city in the United States to light its streets using natural gas by the mid-1880s.

Currently, natural gas enjoys a significant share in the primary energy mix market compared to other fossil fuel sources (oil and coal) as well as renewables and other sources (hydro and nuclear). As shown in Figure 1.1 the contribution of natural gas as a primary energy source increased by almost 40% from 1995 to 2017, and as the fastest-growing fuel per annum, its share is expected to reach 30% by 2035 [1, 2]. Countries with the largest natural gas reserves are Russia (~1,688 trillion cubic feet (tcf)), Iran (~1,187 tcf), Qatar (~890 tcf), the United States of America (~388.8 tcf), Turkmenistan (~353 tcf), Saudi Arabia (~290 tcf), United Arab Emirates (~215 tcf), Venezuela (~195 tcf), Nigeria (~182 tcf), and Algeria (~159 tcf). These countries control almost 80% of the proven global natural gas reserves [3].

The global demand for natural gas is shown in Figure 1.2. The figure shows the apparent rise of natural gas demand in the United States and the rest of the world as a result of the significant enhancement in shale gas production, while the forecast shows a slight decrease in demand for the European nations. The world's largest consumers of natural gas are the United States, Russia, China, and Iran, while the most significant producers are Russia, the United States, Canada, Qatar, and Iran.

Qatar, a small country in the Middle East, is a good example of a success story in natural gas production and monetization since it is the fourth-largest producer of natural gas, globally [4].

Natural Gas Processing from Midstream to Downstream, First Edition.
Edited by Nimir O. Elbashir, Mahmoud M. El-Halwagi, Ioannis G. Economou, and Kenneth R. Hall.

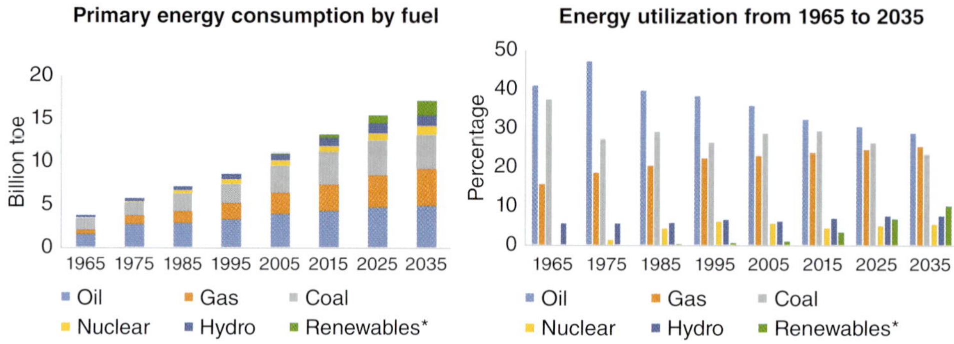

Figure 1.1 The global energy sources and their forecasted shares (*Renewables includes wind, solar, geothermal, biomass, and biofuels) [1].

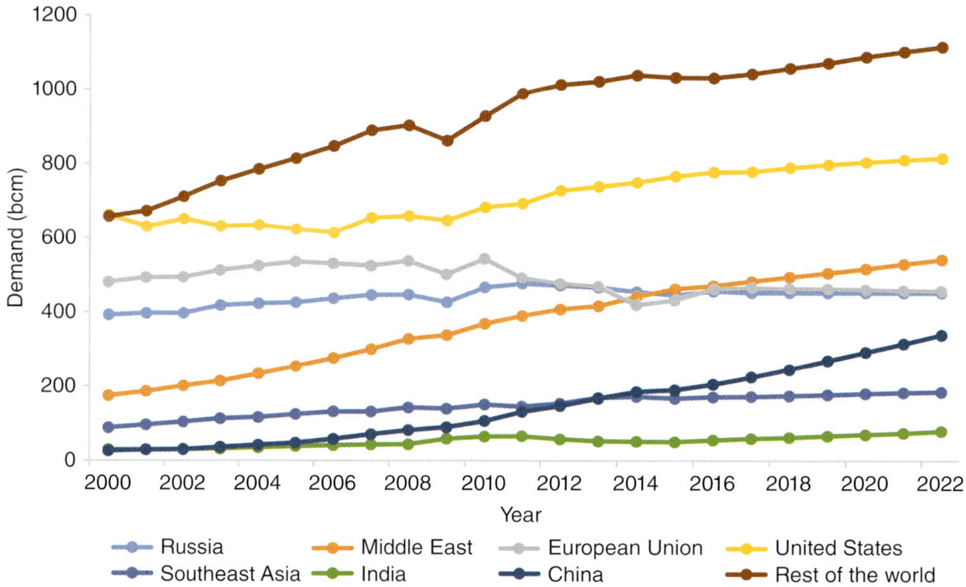

Figure 1.2 The past and the prospected demand of natural gas (data obtained from [2]).

At current reserves-to-production (R/P) rates, Qatar has more than 135 years' worth of natural gas [4]. Thus, natural gas will continue to be a major contributor to Qatar's economy for the foreseeable future. Qatar also aims to be at the forefront of developing innovative ways to monetize natural gas, not only in economic terms but also in environmental terms. This chapter sheds light on the differences in natural gas monetization pathways of major world players in this field, either as producers or as consumers, with a focus on Russia, the United States, and Qatar. The first section of this chapter will briefly highlight the differences between the significant monetization routes for natural gas while the second part will reflect the differences in natural gas monetization between Russia, the United States, and Qatar.

1.2 Natural Gas Chain

As shown in Figure 1.3, the "Upstream" part of natural gas chain starts with the exploration and the production of natural gas from either a conventional source (associated and non-associated

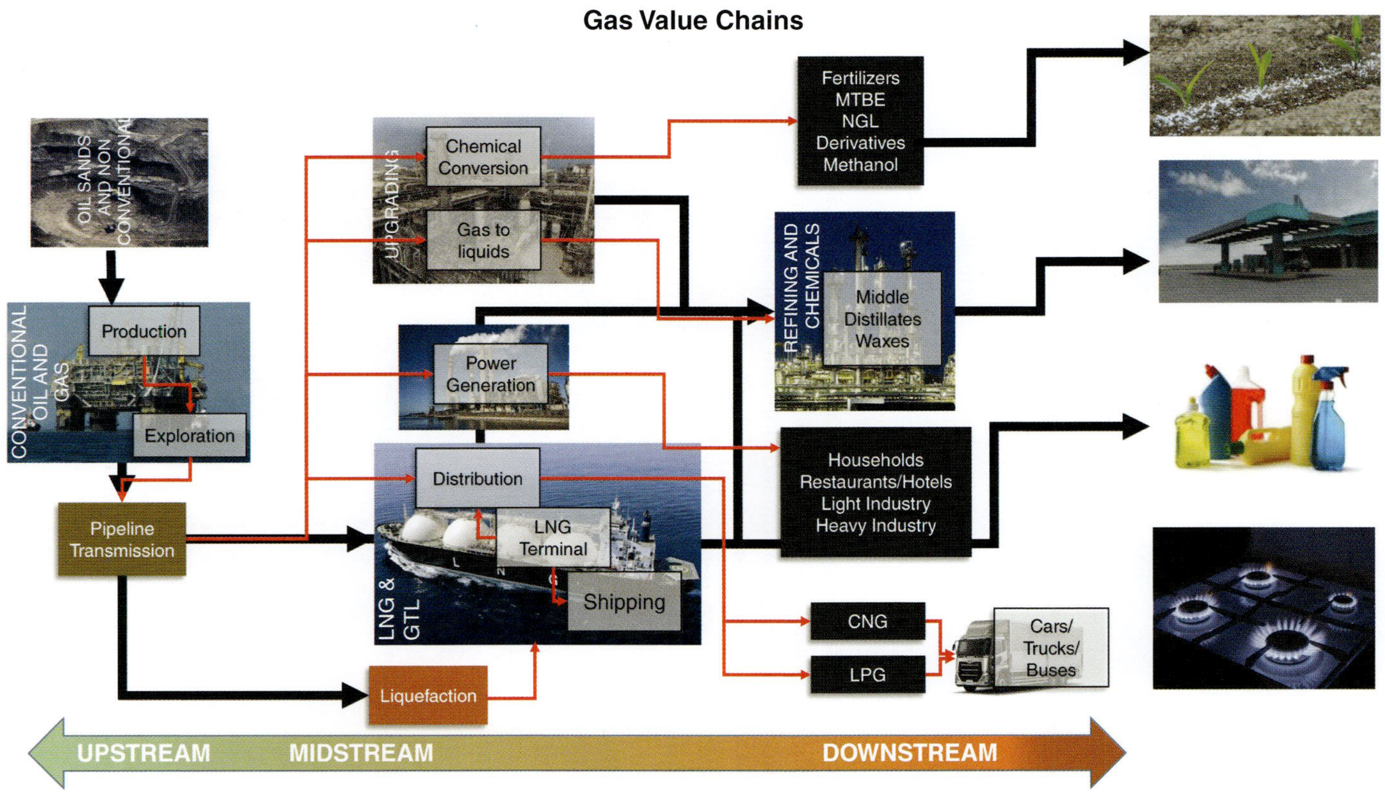

Figure 1.3 Natural gas chain from the upstream to the downstream.

Table 1.1 Typical composition of natural gas from the wellhead to the pipeline.

Component	Wellhead Gas, Mole%	Pipeline Gas, Mole%
Methane (CH_4)	70–98	95–98
Ethane (C_2H_6)	1–10	2–5
Propane (C_3H_8)	Trace–5	0.5–1.5
Butanes (C_4H_{10})	Trace–2	0.2–0.5
Pentanes (C_5H_{12})	Trace–1	Trace
Hexanes (C_6H_{14})	Trace–0.5	Trace
Heptanes & heavier (C_7H_{16}+)	Trace	Trace
Carbon Dioxide (CO_2)	Trace–3	0.5–2.0
Nitrogen (N_2)	Trace–15	0.5–1.5
Hydrogen Sulfide (H_2S)	Trace–2	<0.000004
Mercury (Hg)	*200 to 300 μg/m^3	*200 to 300 μg/m^3
Water (H_2O)	Trace–5	<0.0001

reservoirs) or a non-conventional source (shale, coalbed methane (CBM), oil sand, or tight gas reservoir). The different technologies that have been used to extract, process, transport, store, and distribute natural gas depend on the location and composition of the gas as well as the production location. The second part of the natural gas chain is the "Midstream," whereby the major treatment takes place, depends on the application of the gas and the specification required by downstream processes and the end users. A typical composition of natural gas from the wellhead to the pipeline is shown in Table 1.1. The purpose of the "Midstream" part is to remove components other than methane from natural gas in a series of separation processes that would combine different technologies and processes. Figure 1.4 shows a typical sequence of midstream natural gas processing plant. The "Downstream" part of the natural gas chain depends mainly on the end use of natural gas, and it could be composed of a physical treatment (e.g., liquefied natural gas (LNG)) or chemical treatment (e.g., gas-to-liquid (GTL)).

1.3 Monetization Routes for Natural Gas

1.3.1 Large Industries and Power Plants

The industry and the power plants sector account for the highest monetization of natural gas compared to others [5]. Specifically in the United States, coal began modestly in 2008 and dropped from 48.21% to 33.18% in 2015. Coal lost 15 % of the market, while natural gas increased 11% in the same period, as shown in Table 1.2. Renewable sources (not including solar and hydropower) increased 3.6% to 6.7% overall. The electricity sector is a major emitter of CO_2 in the United States, and it is assumed to be responsible of 29% of global warming emissions. Coal is the major source for these emissions, and therefore natural gas and renewables emerged to substitute for coal in this sector [6, 7]. That results in natural gas and renewables picking up 14.8% of the market (i.e., or ~99% of the market lost by coal). In 2016, natural gas become the major sources of electricity in the United States (~34%) followed by coal (30%), nuclear (~20%), and the renewables (~16%) [8].

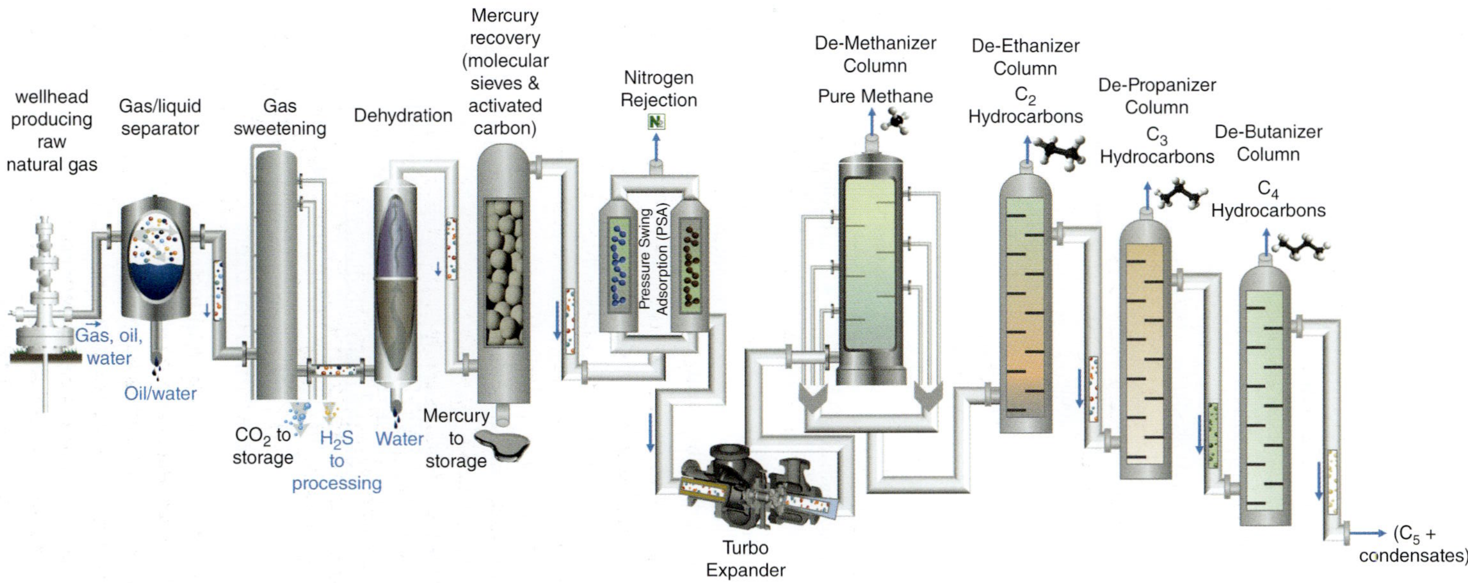

Figure 1.4 The processes of midstream natural gas plant.

Table 1.2 Role of natural gas in the United States electricity generation.

Annual Total	Coal	Natural Gas	Renewables
2006	48.97%	20.09%	2.36%
2007	48.51%	21.57%	2.52%
2008	48.21%	21.43%	3.04%
2009	44.45%	23.31%	3.63%
2010	44.78%	23.94%	4.02%
2011	42.28%	24.72%	4.69%
2012	37.40%	30.29%	5.29%
2013	38.89%	27.66%	6.01%
2014	38.64%	27.52%	6.39%
2015	33.13%	32.66%	6.65%
2016	30.4%	33.8%	13.10%

Table 1.3 Advantages and disadvantages for monetizing natural gas in industry and power plants.

Main advantages	Disadvantages and constraints
• Creates economies of scale (large individual offtakes) • Offers good load factor • Can avoid costly treatment facilities as gas quality is not usually critical • Provides basic gas infrastructure for subsequent expansion • Requires no storage on users' premises and avoids waste disposal problems • Uses conventional technology • More eco-friendly (no SO_x and less NO_x and CO_2 emissions, no particulates, etc.) than coal and most oil products	• Possibly lower value in competition with coal, fuel oil, hydro, etc. • Load factor may be low if the end consumer installs dual-fired capability or buys gas on an interruptible basis.

Table 1.3 lists the main advantages and disadvantages of monetizing natural gas in the large industry and power plant sectors. The major advantage is that natural gas doesn't require an expensive midstream treatment, while the major challenge is the low load factor due to the use of dual-fired generators, which is common practice in many places.

1.3.2 Small/Medium Industries and Commercial Users

The size and importance of the small/medium industries and the commercial user sector vary from country to country, and represent the use of natural gas in small machines like heated big/medium-sized washing machines that have dephosphatizing, rinsing, and drying treatments for fasteners and general small metal components. This sector has a strong presence in the developed nations but much lower contribution in developing nations. Table 1.4 shows the summary of the advantages and the disadvantages in monetizing natural gas in this sector. The major advantage in monetizing natural gas in this sector is that has high load factor offtakes and higher values in terms of heat compared to conventional steam. On the other hand, the

Table 1.4 Advantages and disadvantages for monetizing natural gas in small/medium industries and commercial users.

Main advantages	Disadvantages and constraints
• Higher value than steam-raising and similar basic heating uses • High load factor, unless space heating is involved • Requires no storage on users' premises and avoids waste disposal problems • Good volume potential, but slower build-up than for steam raising and similar uses • Ease of control/flexibility at point of use • More eco-friendly than coal and most oil products	• Need for treatment facilities, as gas quality is usually important • Cost and operation of a more complex distribution system • More sophisticated technology at point of use than steam raising • Need for service/maintenance back-up support

gas has to be treated in a more expensive midstream process compared to monetization in power plants.

1.3.3 Residential

One of the first priorities of "local" natural gas monetization is residential heating requirements. Gas is delivered to homes through pipelines or in tanks as CNG (compressed natural gas). This is the conventional use of natural gas to warm homes or for water heating. Also, it is also used in stoves, ovens, clothes dryers, lighting fixtures, and other appliances. Table 1.5 lists the major advantages in using natural gas in residential versus the disadvantages and the challenges facing the same.

1.3.4 Natural Gas Export

1.3.4.1 Pipeline Export

A major route of natural gas transportation both within state and out of the state is via pipelines. Russia is one of the world's largest producers of crude oil (including lease condensate) and the second-largest producer of dry natural gas, ~20 tcf in 2016. The majority of Russia's reserves are located in West Siberia, with the Yamburg, Urengoy, and Medvezhye fields accounting for a significant share of Russia's total natural gas reserves. Russia has built strong pipeline network

Table 1.5 Advantages and disadvantages for monetizing natural gas in small/medium industries and commercial users.

Main advantages	Disadvantages and constraints
• High gas value in free market economies • High load factor, unless space heating is involved • No storage on users' premises • Good volume potential if space heating is involved, but usually a slow build-up unless there is an existing town gas grid • Ease of control/flexibility at point of use • More eco-friendly than coal and most oil products	• Need for treatment facilities as gas quality is essential • High cost of gas distribution system unless there is an existing town gas grid • Substantial service/maintenance back-up support essential • Relatively low average offtakes per connection even if space heating is involved, in which case the load factor will be low

Table 1.6 Advantages and disadvantages for transporting natural gas via pipeline.

Main advantages	Disadvantages and constraints
• Where feasible/practicable, pipelines are usually quicker and easier to construct and operate than LNG projects • Conventional technology unless deep-water crossings are involved • Offtake prospects in other countries that may have to be crossed en route to the ultimate market • Substantial revenue-earning potential if volumes are large	• Large offtake volumes are usually necessary for exports to be economic • Potential supply security problems if several countries are involved • Maintenance/operation of the system in foreign countries may be difficult • Net-backs diminish with reducing volumes and/or increasing distances • Contractual and corporate structure complexities if several countries are involved

utilities (259,913 kilometers (km)) for local distribution of natural gas and for export of its natural gas to Europe. In 2014, almost 90% of Russia's 7.1 tcf of natural gas were exported to customers in Europe via pipelines, with Germany, Turkey, Italy, Belarus, and Ukraine receiving the bulk of these volumes. As a result, Russia's economy is highly dependent on its oil and gas, and hydrocarbon revenues, which account for more than 40% of the federal budget revenues. These pipelines that travel long distances should sustain high pressure in the pipeline anywhere from 200 to 1500 pounds per square inch (psi). The United States has the world largest pipeline infrastructure both for onshore and offshore lines, of 2,225,032 km of interstate and intrastate transmission pipelines, and distribution pipelines (1,984,321 km of this network is for natural gas transportation). Canada ranks number three after the United States and Russia with a total length of 100,000 km [9]. According to the U.S. Department of Transportation, pipelines are the safest, most environmentally friendly and most efficient and reliable mode of transporting natural gas [10].

Table 1.6 lists the advantages and the disadvantages for exporting natural gas via pipeline whether overland or subsea.

1.3.4.2 Liquefied Natural Gas (LNG)

Natural gas liquefaction is a physical treatment of natural gas to change methane from the gaseous phase to liquid phase to reduce its volume to 1/600 to ease its transportation and storage. The LNG process is a sophisticated technology. Only a very limited number of energy companies are capable of designing and operating such a process, which requires cooling methane to extremely low temperatures for the liquefaction process to take place. LNG is odorless, colorless, non-toxic and non-corrosive. It is classified as a flammable hazardous substance, specifically when vaporized into a gaseous state. The LNG process involves a gas treatment plant that is similar to the midstream plant shown in Figure 1.4. The gas is then cooled to separate the heavier hydrocarbons such as C_3, C_4, and C_{5+} components. These heavier components are fractionated to produce condensates (C_{5+} and liquefied petroleum gas (LPG) products). The lean gas is then liquefied in cryogenic exchangers at (−165 °C), and the liquefied LNG is then flashed to atmospheric pressures and stored in specialized atmospheric tanks prior to shipping. This technology allows the transportation of natural gas in the liquid phase form for thousands of miles under maximum pressure of 4 psi using special ships designed for this purpose.

Qatar is a peninsula that extends into the Arab Gulf followed by several islands and is connected with the eastern coast of the Arabian Peninsula only from the southern part of the country. The economy of Qatar before the oil and natural gas boom relied on pearl diving, fishing, agriculture, and handicrafts. Qatar's success in becoming a world-leading nation in natural

Table 1.7 Advantages and disadvantages for transporting natural gas via LNG technology.

Main advantages	Disadvantages and constraints
• Direct supply/contractual relationship between seller and buyer(s) • Greater local employment prospects than for pipeline gas exports • Technology transfer potential, which can be particularly important for developing countries • Expansion by increasing liquefaction capacity or ships is physically easier than expanding pipeline supplies. • Greater scope for producing countries (as commercial partners) to participate outside national boundaries, i.e., in shipping and/or trading	• Sophisticated technology • Capital intensive with long lead times • Need for regular, high offtakes to be economic • Some energy is lost in the liquefaction process • Net-backs diminish with reducing volumes and/or increasing distances

gas monetization is based on the discovery of the North Field, the world's largest non-associated gas reservoir with a reserve of ~900 tcf natural gas (14% of the total world's known reserves of natural gas). The North Field was discovered in 1971. It is a joint ownership with Iran's South Pars (6,000 square km is the area of the North Field that is in Qatari territorial waters while 3,700 square km (South Pars) is in Iranian territorial waters). Since 2012 Qatar has reached a natural gas production of around 70 million tons per year, and very recently in July 2017 Qatar Petroleum announced its plans to introduce new projects to increase this production to 100 million tons per year. The history of Qatar in building up the world's largest LNG facilities started with the creation of Qatargas in 1984 (a joint venture with Qatar Petroleum, the national energy company), which led several joint venture projects with world-leading energy corporations (e.g., ExxonMobil, Shell, and Total). In 1996 Qatargas sent its first LNG shipment to Japan, and in 2012 Qatar's national gas companies (Qatargas and Rasgas) became the world's largest producer of LNG, reaching a capacity of 42 million tons per annum. Qatar established Nakilat (Qatar Transport Company) in 2004 as the owner, operator, and manager of Qatar's LNG vessels and to provide shipping and marine-related services to a range of participants within the Qatari hydrocarbon sector. This company owns the world's largest LNG ships currently including 67 vessels. In January 2018, Qatar's two major LNG companies merged to become Qatargas [11].

Russia, a major producer of natural gas, has the Sakhalin II LNG project of 9.6 million tons of LNG per year. It sells its product to Japan, South Korea, the USA, and Mexico. The United States has only one operational LNG plant, the Cheniere plant at the Sabine Pass Facility, which has a current capacity of 2 billion cubic feet per day. It is expected that United States will become a major player in the LNG market by 2019, with five additional LNG projects under construction with a total capacity of about 7.5 billion cubic feet per day (see Figure 1.6).

Table 1.7 lists the major advantages and the constraints facing LNG technology, which has shown significant growth lately to reach 258 million tons in 2016 with more than 879 million tons per annum (MTPA) of proposed project development, concentrated in North America, East Africa and Asia Pacific [12].

1.4 Natural Gas Conversion to Chemicals and Fuels

Natural gas can be directly or indirectly converted to chemicals and fuels via different chemical catalytic routes (see Figure 1.5). The critical step in these chemistries is the breaking of the

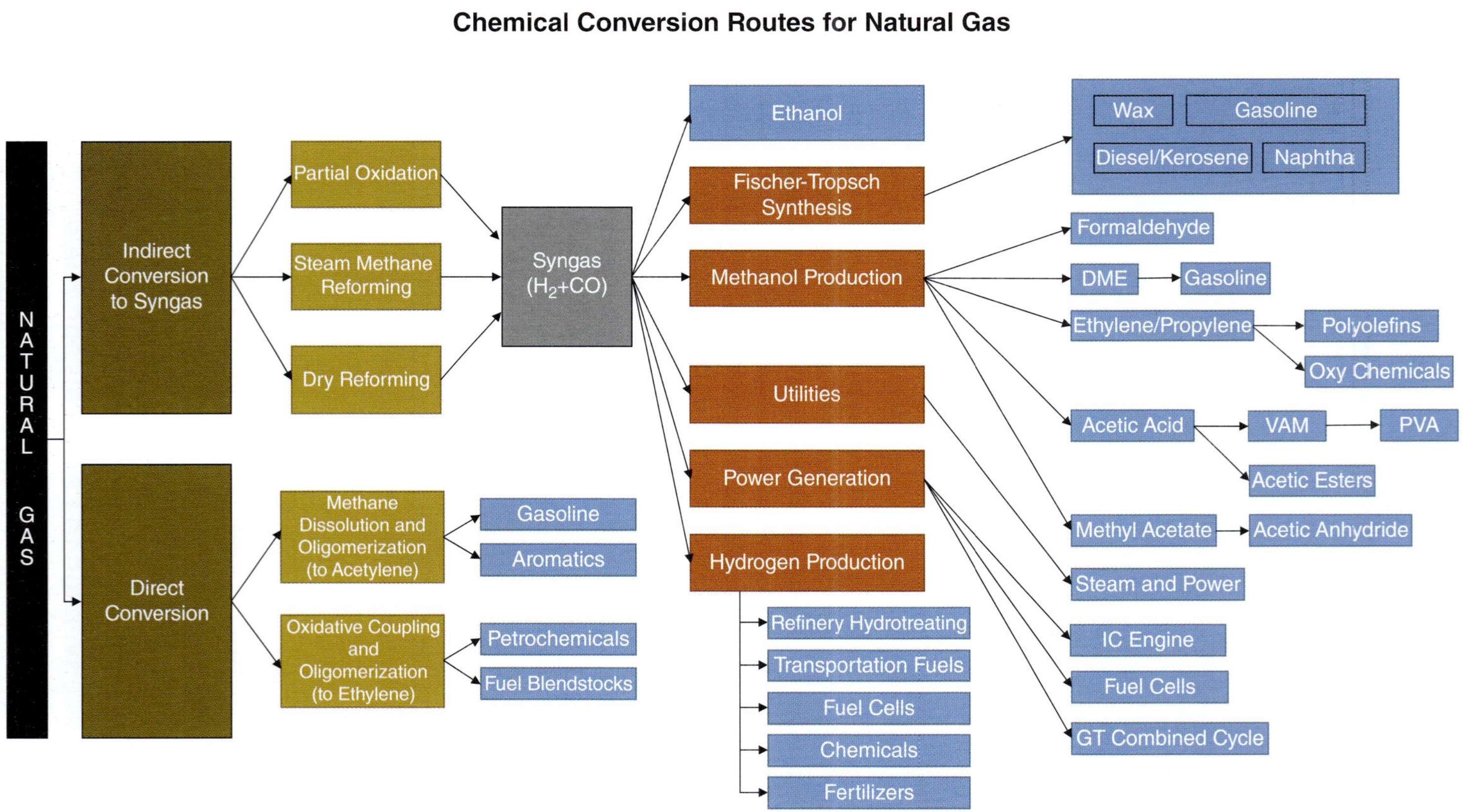

Figure 1.5 Natural gas chain with the different routes of natural gas monetization.

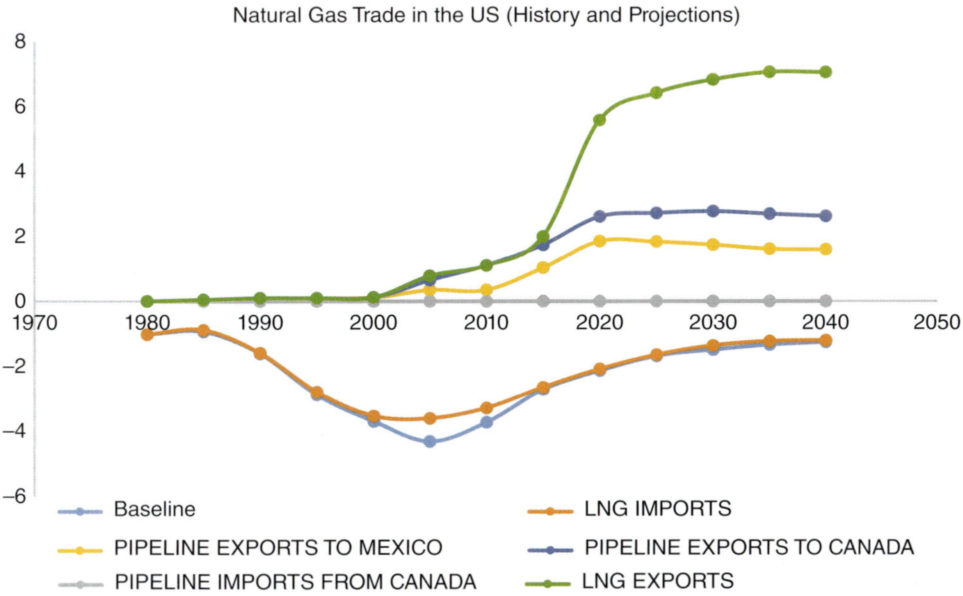

Figure 1.6 Historical, current and predicted trends in natural gas expansion in the United States. Y-axis represents US natural gas trade in trillion cubic feet.

carbon–hydrogen bond of methane, a step that requires a large amount of energy and normally takes place at relatively high temperatures. The indirect conversion routes of methane to synthesis gas or syngas (a mixture of carbon monoxide and hydrogen) takes place at temperatures above 800 °C. The gas-to-liquid (GTL) technology is one such indirect conversion route of natural gas and comprises three main stages. In the first one, methane is re-formed to form syngas (a high temperature process and the most expensive part of the GTL technology, which accounts for 65–70% of the total cost). The second stage involves the Fischer-Tropsch (FT) chemistry to convert syngas into condensates and liquid hydrocarbon over an iron-based or cobalt-based catalyst at lower temperatures than the reforming technologies. The FT reactor accounts for 20–24% of the cost of the GTL plant. The third stage is the refining stage to produce fuels and the other hydrocarbon products, which accounts for 9–11% of the GTL plant costs. The high cost of the first stage (the re-forming stage) is because of the high temperature requirements of the process and the need of pure oxygen in the plant, specifically for the partial oxidation re-former and the autothermal re-former. Despite the great potential of the GTL technology for the production of ultra-clean fuels (free of aromatics and sulfur), the energy-intensive technology faces three major hurdles:

1) High CO_2 emission especially in the re-forming stage that challenges its environmental benefit as an ultra-clean fuel
2) Economy-of-scale limitations of the FT stage to make the GTL process economically sound, requiring either a large-sized slurry reactor (Sasol technology), or multi-tubular reactors filled with catalyst particles (Shell technology)
3) GTL fuels and other products require special marketing campaigns as well as design blending with additives for them to compete as premium fuels with crude-oil refinery products. Typical GTL products compared are shown in Table 1.8.

Qatar is a global leader in the GTL field with two large-scale plants, the ORYX GTL plant – a joint venture between Qatar Petroleum (QP) (51%), and Sasol-Chevron (49%) that came online

Table 1.8 Typical composition of GTL products.

	Product	Carbon Number	Boiling Range
Petroleum gas	Methane	C1	−164°C
	Ethane	C2	−89°C
	Propane	C3	−44.5°C
	Butane	C4	−0.5°C
Light distillates	Naphtha	C5–C12	30–205°C
	Gasoline	C4–C12	35–220°C
Middle distillates	Kerosene	C8–C16	150–325°C
	Diesel/gas oil	C9–C16	180–380°C
Heavy distillates	Fuel oil	C9–C20	310–525°C
Residual fuels	Heavy fuel oil	C12–C70	425–600°C
Lubricants	Base oils	C20–C50	>400°C
Wax	Wax	C20–C60	520–620°C
Residuum	Coke, asphalt, tar	C50+	>600°C

in 2007 – and the world's largest GTL plant, the Pearl GTL Plant (a joint venture between Shell and QP). The ORYX GTL project uses about 330 million cubic feet per day (MMcf/d) of natural gas feedstock from the Al Khaleej field to produce 34,000 barrel/day (b/d) composed of 24,000 b/d diesel, 9,000 b/d naphtha, and 1,000 b/d of LP gas. The Pearl GTL project uses 1.6 billion cubic feet per day (Bcf/d) of natural gas feedstock to produce 140,000 b/d of GTL products as well as 120,000 b/d of natural gas liquids and LPG. Qatar is one of few countries having GTL plants along with South Africa, Nigeria (Escravos GTL plant of similar capacity and technology for the ORYX GTL Plant of Qatar) and Malaysia (the Bintulu GTL plant). Other major natural gas–producing countries, such as Russia and the United States, have yet to build large-scale GTL plants partially because of the challenges listed above as well as because of the strong market of natural gas on the local monetization in terms of electricity and other sectors as discussed above.

The major advantages and constraints of converting natural gas (mainly methane) to fuels and chemicals is listed in Table 1.9.

Table 1.9 Advantages and constraints in natural gas conversion to chemicals and fuels.

Main advantages	Constraints
• Volumes of gas utilized in conversion are modest compared with most pipeline gas and LNG projects • Low into-plant gas prices are usually needed to make conversion projects viable if the prices of alternative feedstock, notably oil products, are also relatively low • Capital intensive but come with long lead times and therefore create stable long-term markets	• Sophisticated technology is involved in most instances • Energy is "lost" in the conversion process • Some products may become over-supplied and therefore difficult to market • Export prices/net-backs will be influenced by competitors' activities and world prices • Project may need "under-writing" by government to succeed • Lower revenue-earning potential in absolute terms than pipeline gas or LNG

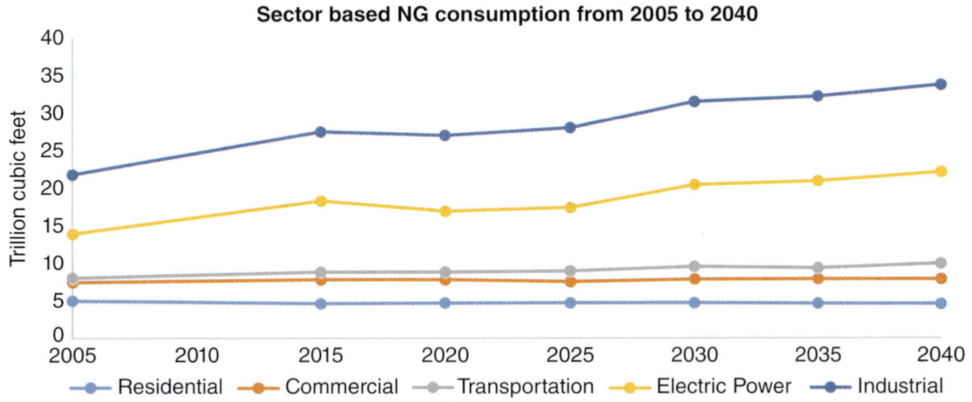

Figure 1.7 The United States natural gas historical, current and forecast natural monetization distribution [3].

1.5 Summary

This chapter summarizes the monetization routes of natural gas with focus on the experience of three countries that are considered among the major producers and/or exporters of natural gas and its products (the United States, Qatar, and Russia). Each of the aforementioned countries developed different natural gas monetization strategies depending on the local needs and the established marketing plans for its natural gas wealth. As shown in Figure 1.7, the United States natural gas monetization plant has not been changed in recent years despite the significant increase in the role of natural gas in its energy mix as a result of shale gas and tight gas production that reached almost 60% of total U.S. dry natural gas production in 2016 [13]. This book covers different topics related to the fundamentals and applied side of these aforementioned monetization technologies from the upstream to midstream to downstream.

Acknowledgment

Part of the material covered in this chapter is a result of research funded by Qatar National Research Fund (QNRF) in several National Priority Research Project (NPRP) that are focused on advancing natural gas processing in Qatar, specifically the Gas-to-Liquid Technology. Also, part of the assessment of natural gas monetization techniques is obtained from Shell workshops notes on natural gas processing. The author also would like to acknowledge ORYX GTL company for their support of their support of several projects in natural gas monetization summarized in this chapter.

References

1 BP Energy Outlook 2017 Report. Available from: https://www.bp.com/content/dam/bp/pdf/energy-economics/energy-outlook-2017/bp-energy-outlook-2017.pdf.
2 Key world energy statistics, 2016. International Energy Agency.
3 ExxonMobil Energy Outlook, 2016 report. ExxonMobil. Available from: http://corporate.exxonmobil.com/en/energy/energy-outlook/a-view-to-2040

4 Dudu P., The world's biggest natural gas reserves. In: Hydrocarbons technology. Available from: http://www.hydrocarbons-technology.com/features/feature-the-worlds-biggest-natural-gas-reserves/; November 2013.

5 International Energy Agency. Gas. market analysis and forecasts to 2022. Market report series; 2017. Available from: http://www.iea.org/bookshop/741-Market_Report_Series:_Gas_2017.

6 Environmental Protection Agency. Inventory of U.S. greenhouse gas emissions and sinks: 1990–2015; 2017.

7 Energy Information Agency (EIA). How much of the U.S. carbon dioxide emissions are associated with electricity generation? 2017.

8 U.S. Energy Information Agency. EIA—electricity data. Available at: https://www.eia.gov/electricity/; 2017.

9 The United States Central Intelligence Agency (CIA)The world fact book. Available from: https://www.cia.gov/library/publications/the-world-factbook/fields/2117.html.

10 U.S. Department of Transportation Pipeline and Hazardous Materials Safety Administration. Pipeline safety fact sheet. Available from: https://www.aga.org/research/fact-sheets/pipeline-safety/.

11 CNBC Energy News. Qatar announces huge rise in gas production amid diplomatic crisis. Available from: https://www.cnbc.com/2017/07/04/qatar-ratchets-up-gas-production-30-percent-despite-sanctions.html.

12 International Gas Union. 2017 world LNG report; 2017. Available from: https://www.igu.org/sites/default/files/103419-World_IGU_Report_no%20crops.pdf.

13 U.S. Energy Information Administration (EIA). How much shale gas is produced in the United States? 2016. Available from: https://www.eia.gov/tools/faqs/faq.php?id=907&t=8.

2

Techno-Economic Analyses and Policy Implications of Environmental Remediation of Shale Gas Wells in the Barnett Shales

Rasha Hasaneen, Andrew Avalos, Nathan Sibley, and Mohammed Shammaa

Dwight Look College of Engineering, Texas A&M University, USA

CHAPTER MENU

2.1 Introduction

2.1.1 Framing the Issues: The Energy and Environmental Equation

Global demand for energy as well the global carbon footprint is expected to rise consistently through 2050 (EIA 2017a). United States energy demand/consumption continues to be among the largest in the world, as does its carbon footprint, especially when compared to other global economies (EIA 2017a). The U.S. is expected to remain the largest consumer of energy and emitter of CO_2, after China, through 2050 as shown in Figure 2.1.

While it is true that both the energy and CO_2 intensities (per dollar of GDP) of the United States paint a much better picture (EIA 2017a), the issue still remains that the U.S. will continue to need more energy and emit more carbon dioxide than most other countries on the planet. Although it is difficult to accurately characterize energy reserves globally, it is clear that United States demand is expected to outstrip what can be produced domestically from traditional sources, across all sectors of the economy (EIA 2017a).

In addition, the current United States energy mix sways heavily toward coal and oil, totaling about 52% of total fuel consumption (EIA 2017b), which have a high environmental tax as compared to cleaner sources. Electricity generation sways more towards coal while transportation relies heavily on petroleum. Without clean, domestic alternatives to support this demand, the United States will continue to rely heavily on foreign energy sources and negatively impact the environment.

In recent years, shale gas has emerged as a potential alternative to this issue. Advances in horizontal drilling and hydraulic fracturing technologies have driven a rapid and widespread growth in natural gas production from shale formations in the United States. This development

Natural Gas Processing from Midstream to Downstream, First Edition.
Edited by Nimir O. Elbashir, Mahmoud M. El-Halwagi, Ioannis G. Economou, and Kenneth R. Hall.

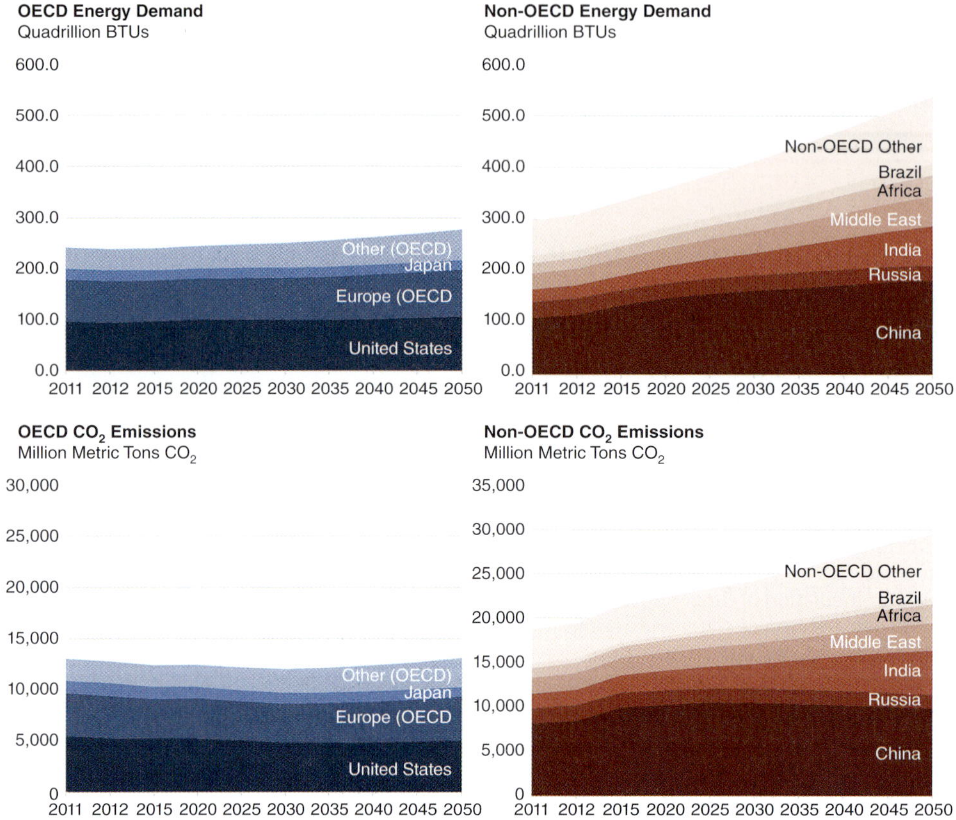

Figure 2.1 Projected global energy demand and CO_2 emissions.

has narrowed the production-consumption gap in U.S. energy and is expected to turn the United States into a net exporter of natural gas by 2020 (EIA 2013a). As natural gas is considered among the cleanest of conventional fuels, in terms of usage, this also promises a reduction of the carbon footprint of the United States (EIA 2012b), as many of the industries which are heavily reliant on coal and oil turn to natural gas as an alternative.(EIA 2012a; EPA 2013; EIA 2013a).

While natural gas itself is considered a clean fuel, the extraction process, attributed to natural gas derived from shale formations, has raised many debates as to the total lifecycle environmental footprint of the fuel (Armendariz 2009; Arthur et al. 2008; COGA 2012; Fontenot et al. 2013; McHugh et al. 2014; Fontenot et al. 2014; Skone et al. 2011; Jaramillo et al. 2007; EIA 2011). Many argue that the environmental impact of drilling and production processes employed in shale gas operations severely limit the environmental benefit of the use of shale gas as a clean fuel alternative (Howarth et al. 2010; Howarth et al. 2011; Cathes et al. 2012).

While research in this area is still developing, there are unique elements of shale gas drilling and production operations that inherently contribute to their environmental footprint. Shale reservoirs are massive, typically spanning multiple communities, many coming close to metropolitan areas and agricultural zones (EIA 2011). As a result, localized environmental impacts have less opportunity to dissipate and must be contained much more tightly. Also, as compared to vertical wells, in conventional oil and gas formations, unconventional wells tend to be more closely spaced and can take months to drill. They require horizontal drilling to access the formation and more fully drain the rock matrix and must employ hydraulic fracturing at extensive rates to stimulate production. The hydraulic fracturing process involves the injection of large amounts of chemical-laden water and mud into the well at high pressures, in order to fracture the rock and allow the gas to be produced (Kell 2009). Although hydraulic fracturing has been used for decades to stimulate traditional oil and gas wells, the main issue with shale is scale. The size and number of fractures required to release the gas from shale is much more significant than those previously employed by the industry (Clark et al. 2013).

As the current boom in shale production is relatively new, there is still fragmentation among operators in terms of drilling and production processes. Many of the producers typically operate using traditional techniques, which allow for minimal up-front capital investment. Also, since the breakeven gas/oil price for unconventional resources is much higher than those in conventional reservoirs, these operators are typically very cost sensitive. This has led to operations that have proven to be suboptimal from an environmental perspective (Fontenot et al. 2013; Howarth et al. 2010; Jenner and Lamadrid 2013).

These issues are becoming more pronounced as we discover and characterize more and more shale gas deposits. The situation has led to a negative perception of shale gas (Thomas et al. 2017) and growing opposition to the hydraulic fracturing process with several bans and moratoria on shale gas extraction, led by local and municipal governments, both in the United States and across the world (Hagstrom and Jackanich 2011; Metze 2017) Without alternatives that reduce this environmental impact, shale gas production may have significant adverse effects on both immediate environmental health and safety as well as on the broader environment.

The process it takes to develop a gas well requires an extensive use of diesel fuel, fresh water, contingencies to local water supply, roads, and reinjection of waste water into deep wells for disposal. The environmental impacts of these activities (Zoback et al. 2010; Annevelink et al. 2016) include:

- ***Land impacts*** – Number of wells that need to be drilled; acreage and clearing for well pads and impoundments; clearing land for building new roads
- ***Greenhouse gas emissions*** – Fugitive methane emissions and flaring; release of pollutants from diesel and gasoline engines used in the operation
- ***Impacts on water*** – Overconsumption of fresh water for fracturing (millions of gallons per well) (Kell, 2009) and waste water management required due to the contamination of water with fracturing chemicals and methane (local streams/rivers and well water); pond fires due to wastewater pond negligence (Tagliaferri et al. 2017)
- ***Seismic impacts*** – Increased seismic activity related to hydraulic fracturing (COGA 2012), natural gas extraction and waste water re-injection (Nicholson and Wesson 1987)
- ***Additional impacts*** – Release of VOCs from well installation and radioactive particles in waste water

While these issues are common among unconventional gas plays, the unique composition of the gas in each play could require additional processes, which would also need environmental remediation such as dewatering or CO_2 removal. As a result, looking at these issues can form a foundation for analysis of other plays, but cannot be applied directly without adaptation.

2.1.2 Well Lifecycle Analysis and Environmental Impacts

In order to focus the analysis, a specific shale gas play was evaluated. Depending on the underlying depositional system, different shale plays will require different remediation approaches, so limiting the analysis to a specific play enables a focused approach to the analysis. The Barnett Shale was chosen as it is among the most established and most mature shale gas plays in the

United States today, and it plays a critical role in the U.S. natural gas landscape. As a result, a robust data set collected from operations in the Barnett could be used to conduct the analyses.

In looking at the environmental impacts of shale gas, it was assumed that once the gas is produced and processed for transportation, its environmental footprint will be similar to that of natural gas from conventional sources. Therefore, the focus of this analysis is on the environmental footprint of the drilling and production processes associated with shale gas extraction, the "well" lifecycle, and not on the entire lifecycle of the shale gas itself.

As discussed previously, there are a number of environmental issues tied to shale gas development. Issues around greenhouse gas emissions, water consumption and disposal during hydraulic fracturing, and seismicity are the most consistent among shale gas plays and have the most direct impact on the immediate environment. In addition, these elements are among the most difficult to manage and mitigate.

As a result, the well lifecycle analysis across this chapter focuses on these three elements, and the proposed improvements to reduce the impact on the immediate environment were analyzed and discussed at length. In the first section, the analysis will focus on greenhouse gas emissions. Specifically, it focuses on emissions resulting from the burning of fuel to run drilling and production equipment rather than fugitive methane emissions, which can also be a source of greenhouse gas emissions in the operations. The second part of the analysis will turn its focus towards the impacts on water consumption and disposal and seismicity, which are inter-related.

2.2 Shale Gas Operations

To analyze the environmental impact of the drilling and production processes associated with shale gas, it is useful to understand the overall operation: how the gas is extracted and how it differs from conventional operations. This is the basis for the well lifecycle analysis.

2.2.1 Summary of Shale Gas Operations

After the initial exploration, permitting, and exploratory drilling stages are completed, an operator will begin drilling and production activities. Land is cleared for the operation and vertical wells are drilled into the shale formation through to the total formation depth. Conventional wells are typically drilled into sandstone, and the natural permeability allows the fluids to flow to the low-pressure wellbore; thus artificial stimulation, such as fracturing, is not necessary. In contrast, a wellbore in an unconventional reservoir may make contact with some natural fractures, but the hydrocarbons that could be produced from these are insufficient for economic production of gas. In order to bring enough fluid to the surface, the fracture network needs to be greatly expanded. Thus, a well needs to be designed in a way that will connect to the maximum number of fractures and allow induced fractures to expand as far as possible. This is achieved in the first two phases of the well's lifecycle: drilling and completion. The final phase of the lifecycle, the production phase, requires marginal effort, as it typically consists of a pumper making daily checks on the well or performing a workover, as required. Most new gas wells go through these three major phases in their life: drilling, completion/stimulation, and production. The drilling and completion phases are relatively brief, typically one to two months in duration, but require a tremendous amount of energy and capital. It is during these phases that much of the environmental impacts occur. The production phase starts when the well begins flowing and continues until it reaches the end of its economic life. Since many unconventional wells can live for decades, the marginal environmental impact of the production phase can sum up to a material amount.

Drilling: In the drilling phase, a well is drilled that descends into the targeted formation and bores through it, horizontally, increasing the contact area with the reservoir. Horizontal wells are drilled to a kick-off point (KOP), and the bit is driven in a soft angle until it reaches approximately 90° and will continue to bore to the desired measured depth. As a well is drilled, the hole will be encased with steel pipes, and cement will be pumped around it to improve isolation of the well from the subsurface zones. One of the advantages of horizontal wells is that multiple wells can be drilled into the targeted formation from a single well pad. This reduces the impact to the surface environment and reduces costs.

Completions: Once a well has been drilled, cased, and cemented, the stimulation or completion process begins. In the completion phase, the well bore is perforated to establish communication with the reservoir, and fractures are induced using hydraulic fracturing (Kell 2009). Hydraulic fracturing is a process that induces fractures into the rock, by pumping fluid down at very high rates. The fluid is often fresh-water-based and filled with proppant, a sandy or gel-like substance that fills in the newly created fractures and keeps them "propped" open to allow reservoir fluids to flow. This perforation and fluid pumping process is repeated several times in stages and connects the natural fractures within the reservoir. The combination of these fractures creates a large network of interconnected pores to the wellbore and allows for economical flow of oil and gas (Gale et al. 2007; Gottschling 2005).

Production: The gas is then produced by allowing fluid to flow to the surface. The gas is collected for processing and transportation.

The drilling and completion phases are relatively brief, typically one to two months in duration, but require a tremendous amount of energy and capital. It is during these phases that much of the environmental impacts occur. Specifically, it is during the completion phase that the well is hydraulically stimulated by fracturing the underlying matrix to release the gas. This hydraulic fracturing is the focus for much debate as it relates to water usage, disposal, and seismicity. The production phase starts when the well begins flowing and continues until it reaches the end of its economic life. During the production phase, much of the water that is used to fracture the well flows back and is "produced" with the natural gas. It must then be transported and disposed of. Since many unconventional wells can live for decades, the marginal environmental impact of the production phase can add up to a significant amount.

2.2.2 Hydraulic Fracturing and Water Impacts

Fracturing fluid is typically a slurry of water, proppant, and chemical additives, typically, 90% of the fluid is water and 9.5% is sand, with chemical additives accounting to about 0.5%.(Ciferno et al. 2012; API 2010) The quality of the water used is critical as impurities can reduce the efficiency of the additives used in the process. An average well requires 3 to 8 million U.S. gallons (11,000 to 30,000 m^3) of water over its lifetime (Ciferno et al. 2012), while the average well in the Barnett Shale formation uses from approximately 2.8 million (Nicot and Scanlon 2012) to 5 million (Nicot et al. 2014) gallons of water depending on horizontal fracture length and number of stages per well. Today, most water used in hydraulic fracturing comes from surface water sources such as lakes, rivers, and municipal supplies. However, groundwater can be used to augment surface water supplies where it is available in sufficient quantities.

While the cost of water also varies, water can behave as a public good, and water rights are typically granted with the land lease, along with mineral rights. In these cases, the cost of water represents the opportunity cost of selling that water on the market. In other cases, water is purchased from neighboring water authorities. This cost will vary dramatically based on the abundance of water in the region for alternatives such as farming, industrial operations, and urban usage.

The use of vast amounts of potable, fresh water has raised concerns in communities neighboring shale gas operations. This is heightened by the fact that shale reservoirs span such large territories, many across rural areas where potable water is in high demand (Grattan 2014). In addition, the transport of such large amounts of water requires the use of trucks, which has both cost and greenhouse gas implications. This transportation element becomes critical in arid areas or areas of high water stress that have large shale formations (Reig et al. 2014a).

Slickwater fracturing has additional issues in terms of the expected ultimate recovery of oil and gas from shale resources. As, typically, less than 50% of the water injected is produced over the life of the well, it can actually act as a barrier to optimal production by impeding the flow of oil and gas back through the network of fractures created (Burke et al. 2011; Reynolds et al. 2013).

Perhaps as critical as the use of fresh water, is the disposal of produced water once the fracturing operation is completed. The fluid that returns to the surface through the wellbore is not only the chemically treated fracture water, but also water from the rock formation that can contain salts, metals, and radionuclides. This water can contain hydrocarbons, high levels of total dissolved solids (TDS), suspended solids, and residual production chemicals. That wastewater must be captured and stored on site, and then is often shipped long distances for deep well injection. Produced water stored on the surface for long periods of time is subject to evaporation, which can further increase the salt concentration of the water. Additionally, flowback of produced water, generated during completion and production operations, could have catastrophic results on the underground drinking water supply, if done improperly. About 30% of injected water returns to the surface as "produced" water in the first month with another 20–30% returning over the life of the well. This implies that about 40–50% of the injected water remains in the rock matrix and in many cases impedes the flowback of natural gas through the well bore. A hydraulic fracture typically exhibits about 20% efficiency (Khuwaja et al. 2014).

2.2.2.1 Fresh Water Consumption

Although it is difficult to generalize on water usage by well, by focusing on the Barnett Shale play, averages can be used to estimate the potential benefits of environmental remediation. Nicot and Scanlon estimate that an average well in the Barnett Shale uses 2.8 million gallons of water for hydraulic fracturing (Nicot and Scanlon 2012). While the cost of water also varies, groundwater typically behaves as a public good, and water rights are typically granted with the land lease, along with mineral rights. Thus, the cost of water used in the simulator represents the publicly available price of water in the Barnett Shale area for "gas well class" water usage and represents cost of buying water from the various water authorities or the opportunity cost selling ground water on the market. This cost will vary dramatically based on the abundance of water in the region for alternatives such as farming, industrial operations, and urban usage. A 20% uplift was added to the price of fresh water to account for treatment of the water to produce "slickwater" for fracturing.

2.2.2.2 Transportation and Disposal of Produced Water

In looking at the disposal of produced water, rate of water production was taken into account as well as hauling and disposal costs per barrel of water produced. In addition, it is important to understand the cost of drilling a disposal well, which according to the EPA must be a Class II well as well as the maximum allowable injection rate into each well (EPA 2012). This will help to determine the number of Class II wells required for safe disposal of produced water and the baseline capital investment for current operations. It is this maximum allowable injection rate that minimizes the seismic impact of deep water injection.

2.2.3 Fuel Usage

Field data on heavy equipment fuel usage for the drilling and completion process, primarily for the operation of drilling rigs and pumper trucks, was taken into account in the model. Fuel usage costs for hauling and disposal are included in the overall cost of hauling and disposal assumptions discussed in the previous section.

2.2.4 Seismicity and Seismic Implications

Induced *seismicity* – seismic activity whose frequency is increased or triggered by human activity – is a controversial term as it relates to production of oil and gas. With respect to production of oil and gas from unconventional sources, induced seismicity can be attributed to three primary types of activities: hydraulic fracturing (COGA 2012), extraction of large amounts of fluids from the subsurface, and injection of large amounts of fluids (typically produced water) into the subsurface (Nicholson 2012).

Hydraulic fracturing is known to produce a seismic response, as it drives faults through the subsurface. However, this induced seismicity is more appropriately described as micro-seismicity, as it generally cannot be felt at the surface except by very sensitive instruments designed for that purpose to map the fracture propagation (COGA 2012). Many sources agree that a seismic event must generally be above a 3.0 on the Richter scale to have a chance of being felt on the surface, while others suggest a magnitude above 2.0 is sufficient. In any case, an event in this range (2.0–3.0) is not expected to cause damage on the surface. According to the Colorado Oil and Gas Association, there are only two known cases of seismic activity levels greater than 1.0 magnitude being induced by hydraulic fracturing; compared with the well over 1,000,000 fracture stimulations that have been done in the United States to date (COGA 2012).

Induced seismicity with a magnitude of greater than 3.0 is known to be caused by other hydrocarbon production–related activities, specifically massive extraction of fluid or fluid injection (Nicholson 2012). Although massive extraction of fluids has caused earthquakes in a few cases, it is not an issue specific to shale gas extraction. This is common to all operations that involve the extraction of large amounts of fluids. As a result, remediation of this cause will not be discussed here.

However, shale gas production does frequently necessitate the disposal of significant amounts of wastewater. It is this massive injection of fluid into the subsurface that is the primary driver of induced seismicity related to shale gas production. There are several cases of water-injection-induced seismicity associated with shale gas plays, including cases associated with the Barnett Shale. Wastewater injection has been linked to induced seismicity since the early 1960s, when injection at the Rocky Mountain Arsenal led to an increase in seismic activity (van Poollen and Hoover 1970).

Seismicity can occur when water injected into a formation causes a rise in pore pressure. If this pressure increase occurs along a fault, this increase can reduce the force required to trigger movement along that fault and cause an induced earthquake. The increase in pore pressure in a formation due to injection is related to several factors, both controllable and uncontrollable. The most important of the controllable factors is injection rate. Any injection well that operates near a fault has the potential to increase the likelihood of earthquakes if the pore pressure increase is large enough (Nicholson and Wesson 1987).

It is the injection of massive quantities of produced water that is believed to be the primary driver for this induced seismicity. In 1987, Wesson and Nicholson prepared a report for the EPA looking at seismic activity tied to deep well injection (Nicholson and Wesson 1987).

At that time, there were only two known cases of wastewater disposal wells triggering induced seismicity, while the cases of induced seismicity caused by injection wells used for secondary hydrocarbon recovery were much more numerous. With the recent shale gas boom in the U.S., the number of cases of wastewater-injection-induced seismicity has increased significantly. The United States Geologic Survey, shows a "hockey stick" style trend change in earthquakes of magnitude 3.0 or greater in the central and eastern United States starting in 2009 and continuing into 2015 (Survey 2015). In recent years there has been a spike in earthquakes in the Denton, Texas, area, which correlates with the well count in the Barnett Shale (Homefacts 2014; Browning et al. 2013a). This correlation suggests the presence of induced seismicity caused by injection wells disposing of the wastewater from Barnett Shale operations. Frohlich (2012) highlighted this correlation, concluding that injection well locations, when compared with earthquake epicenters, proved a definitive link between the injection wells and the earthquakes. Other surface areas overlying the Barnett Shale have recently seen even more drastic upticks in seismic activity. The area around Azle, Texas, experienced more than 30 earthquakes over a 5-month period from late 2013 to early 2014 (Grieder 2014). Additional studies have shown that areas with suspected anthropogenic earthquakes are more susceptible to earthquakes from natural transient stresses (van der Elst et al. 2013).

2.3 The Barnett Shale

In order to focus the analysis, a specific shale gas play was evaluated. Depending on the underlying depositional system, different shale plays will require different remediation approaches, so limiting the analysis to a specific play enables a focused approach to the analysis. The Barnett Shale was chosen as it is among the most established and most mature shale gas plays in the United States today, and it plays a critical role in the U.S. natural gas landscape. As a result, a robust data set collected from operations in the Barnett could be used to conduct the analyses.

The Barnett Shale is a geological formation, located in North Texas. It is estimated to extend 5000 square miles, across 25 counties with the core producing area located around Fort Worth (Armendariz 2009). The formation rests between 6,500 and 8,000 feet in depth, with an average thickness of 350 feet (Martineau 2007).

As of August 2014, there were over 17,500 wells, producing 4,856 million cubic feet per day of natural gas, according to the Texas Railroad Commission (2014). In addition, the Barnett Shale currently produces approximately 4,125 barrels per day of oil and 12,000 barrels per day of condensate, making it a considerable resource for Texas and placing it among the top five shale gas plays in the United States, with the success of horizontal drilling driving the success of the play. Today, horizontal well count is triple that of vertical wells in the formation, and horizontal well production dwarfs that of verticals wells in the play (Sieminski 2014; Dong 2012). Shale gas from the Barnett play does not require CO_2 and H_2O processing to make it usable and can therefore be used as a baseline for further analyses.

2.4 Environmental Remediation of Greenhouse Gas Emissions Using Natural Gas as a Fuel

To conduct the environmental impact analysis, a well-by-well approach was taken. Field data from Barnett Shale assets were obtained and compared with other public sources and research. The results were then extrapolated to encompass the entire play and develop the economic and environmental models, assuming that the shale gas wells across the play behaved and were produced in a similar manner. For each well, the step-by-step drilling and production processes were analyzed, and alternatives with reduced environmental footprints were proposed to address the three major environmental factors: greenhouse gas emissions, water consumption and disposal, and seismic impacts. These alternatives were then evaluated by looking at the environmental footprint improvement as well as the micro-economic implications to the operator. This environmental and economic model was then expanded to encompass the entire play, and conclusions were drawn regarding the economic viability of investing in the alternatives for each of the major environmental factors. This section discusses the greenhouse gas portion of the analysis, specifically focusing on greenhouse gas emissions resulting from the burning of fossil fuel for powering drilling and production equipment in the operation.

The process of drilling, completing, and producing a well consumes a large amount of energy. Power for the engines that operate the extensive number of rigs and trucks that make a gas well possible is provided primarily by diesel fuel supplemented by gasoline. There are three major categories of equipment used to drill, complete, and produce a shale gas well. Heavy-duty drilling rigs running primarily on diesel; heavy-duty diesel trucks used primarily for pumping; and light-duty transport vehicles, which are primarily standard gasoline-powered vehicles. Although diesel and gasoline offer many technical and logistical advantages, the resultant emissions pose a serious environmental issue. The EPA cites the development of natural gas and petroleum systems as the leading producer of greenhouse gas (EPA 2013).

The abundance of natural gas, from the development of shale resources, has driven the price down to levels where using natural gas could provide significant cost savings over diesel and gasoline. Furthermore, the environmental impact of burning diesel versus natural gas is vastly in favor of natural gas. Therefore, in looking at reducing the greenhouse gas footprint of shale operations, one key focus would be on replacing diesel and gasoline usage with natural gas throughout the lifecycle of a typical Barnett Shale gas well.

2.4.1 Single Fuel, Bi-Fuel, or Dual Fuel

Replacing diesel- and gasoline-based engines with natural gas–based ones has been examined for some time (Department of Energy 2001). Natural gas can be implemented in single-fuel, bi-fuel, and dual-fuel modes. Single-fuel mode involves an engine that only operates on natural gas. Bi-fuel engines have two independent engines, but can only be run one at a time, while dual-fuel engines are two independent engines that can be run simultaneously.

Dual-fuel rigs are seen as the best potential solution for natural gas utilization in the field, so an operator has the horsepower available with diesel, but can switch to natural gas as power requirements diminish (Hill et al. 2011). These alternatives are commercially available today, and their energy tax and required capital investment is known. Dual-fuel engines could also replace diesel engines used for trucking, hauling, and pumping. These engines offer improved economics over time, with current pricing spreads, and burn much cleaner (Seybold et al. 2013; Apache 2013; Associated-Press 2013).

For transport vehicles, single-mode natural gas systems are assumed, as conversion kits are readily available today and there are no specific power requirements that necessitate a bi-fuel or dual-fuel mode. The benefits of natural gas–based vehicles over gasoline vehicles are similar to their benefits over diesel.

2.4.2 Forms of Natural Gas

Natural gas as an engine fuel comes in three types: raw natural gas, compressed natural gas (CNG) and liquefied natural gas (LNG). CNG offers opportunity, as it can become compressed

Table 2.1 Emission Reductions from Natural Gas vs. Gasoline or Diesel Engines.

Carbon dioxide (CO_2)	10–20%
Carbon monoxide (CO)	80 90%
Hydrocarbons and volatile organic Compounds (VOCs)	50–70%
Nitrogen oxides (NOx)	30–40%
Sulfur dioxide (SO_2)	~70%.

on site for fuel use. LNG is a denser form and carries more value for heavy-engine use, but LNG compression trains are less portable and have more stringent requirements for compression that require preprocessing. Although the heating value of diesel is significantly higher than that of LNG, by a factor of 1.72 to 1.00, it is also about four times the cost, implying that LNG can be an economically favorable solution overall. Raw natural gas does not require any compression and therefore has no associated energy tax; however, not all natural gas engines can operate on raw natural gas without some sort of processing, and it must be piped from location to location so transportation becomes an issue.

2.4.3 Environmental Impact

In order to quantify the environmental impact of using natural gas in place of diesel or gasoline, each phase of the well lifecycle will be examined, and the opportunities present for natural gas substitution in operations determined. An assessment of the monetary and environmental impact of replacing traditional fuel–based engines with natural gas–based ones will then be made to determine the viability of the substitution. Compared to gasoline or diesel engines, natural gas engines typically demonstrate emissions reductions as shown in Table 2.1. Moreover, these vehicles produce no carbon smoke and particulate matter (PM) emissions, and harmful substances are reduced by 90% (Wang 2011).

The impact of these reductions will be evaluated using field data from the Barnett Shale operation and existing technologies for substitution.

2.5 Environmental Remediation of Water and Seismic Impacts

This section discusses the water and seismic impact portion of the analysis, specifically focusing on changes to the hydraulic fracturing process and the impact of those changes on both environmental footprint and economic viability of the fracturing process. Two alternatives to traditional "slickwater" hydraulic fracturing were explored. The first involves the use of alternative fracturing fluids that would eliminate the need for water, termed waterless fracturing. The second, more conservative alternative, involves modifying the existing hydraulic fracturing process to improve the environmental footprint of that process with respect to water and managing the seismic impact through an array of sensors.

2.5.1 Waterless Fracturing

The use of alternative fracturing fluids has been a practice in the oil and gas industry since 1962 (Reynolds et al. 2013). Although historically used as additives to "energize" water-based fracturing fluids, these fluids have shown promise as "slickwater" substitutes in the unconventional space due to their environmental and economic impact as well as their potential for improved oil and gas production (Burke et al. 2011). Due to their physical characteristics and environmental significance, two fluid alternatives were evaluated: liquefied petroleum gas and carbon dioxide.

2.5.1.1 Liquefied Petroleum Gas Fracturing

Fracturing with liquefied petroleum gas (LPG), known as gas fracturing, was developed in 2006 (Brino 2011), and over 1,300 wells have used this method for fracturing wells. The process involves the substitution of water with an LPG gel, mixing it with proppant and injecting it down the wellbore to induce fractures in the rock (Khuwaja et al. 2014). Once the gas-fracturing fluid is mixed, the fracturing process mimics that used in hydraulic fracturing.

Gas-fracturing fluid, as described by Tudor et al. (2009) and Lestz et al. (2007), is a mixture of LPG gases, including varying amounts of propane and butane, depending on the reservoir conditions. The LPG mixture components are designed to achieve the desired hydraulic fracturing and clean-up performance. Propane is the most commonly used LPG product in fracture treatments, but there are temperature limits for its use. Fracturing fluids made of pure propane can be used with formation temperatures of up to 96°C. When the formation temperature is above 96°C, commercial butane must be mixed with the propane in increasing amounts as the temperature increases. Pure butane allows fracturing formations with temperatures up to 150°C (LeBlanc et al. 2011). Because the gel retains sand better than water, it is possible to get the same results with a much smaller volume of fluid, and to pump at a slower rate. Field tests have resulting in successful fracturing with one-fourth to one-eighth of the amount of fluid used in traditional hydraulic fracturing (Khuwaja et al. 2014; Westenhaus 2012). When coupled with the lower specific gravity of propane, compared to water, the use of LPG would lead to additional benefits in transportation costs and the associated greenhouse gas implications (Khuwaja et al. 2014).

Field studies have also shown that, unlike water-based fracturing where gas was not observed until 5–20% of the load fluid had been recovered, propane-based fracturing yielded observable reservoir gas immediately after the well bore had been cleaned out (LeBlanc et al. 2011). In addition, propane-fractured wells demonstrate 100% recovery of propane within 12–15 days of production (LeBlanc et al. 2011) and an increase of 20–30% in expected ultimate recovery (EUR) of the well (Khuwaja et al. 2014).

2.5.1.2 Carbon Dioxide Fracturing

Liquid carbon dioxide (CO_2) fracturing systems were developed in 1981 (Reynolds et al. 2013) and have been used as fracturing substitutes to "slickwater" in limited applications in North America since the early 1990s (Gupta and Leshchyshyn 2005). The broader use of carbon dioxide for the recovery of shale resources is being explored in a number of situations; most of these are for specific conditions such as water-sensitive formations or partially depleted formations for enhanced gas recovery (Rogala et al. 2014; Khan et al. 2012b). As a fracturing fluid, carbon dioxide can be used in several forms, the most explored options include pumping CO_2 as a liquid or foam at high pressure (Burke et al. 2011) or, more recently, converting it to a supercritical fluid and pumping it at much lower pressures (Thomas 2015). This implies that with minor modifications to existing hydraulic fracturing equipment, supercritical CO_2 can be used in place of water, while using high-pressure liquid CO_2 would require some specialized equipment (He et al. 2014). However, this technology has a cost and energy tax that would render its use expensive in addition to its environmental impact. It is unclear whether its benefits over foamed CO_2 would warrant its use today, but the approach shows potential for future applications. On the other hand, carbon dioxide foam has shown in field tests that it not only reduces clean-up effort, but also improves flowback performance, by enabling gas to be produced at saleable rates within 48 hours of start of production, with complete recovery within two weeks (Reynolds et al. 2013). Similar to LPG, as carbon dioxide foam converts to gas in the formation, and as such does not contribute to formation damage or clay swelling, which can inhibit the flow of hydrocarbons to the surface. As a result, CO_2 fracturing has also demonstrated improvements in natural gas production by 15–40% (Burke et al. 2011). In addition, supercritical CO_2 has demonstrated superior fracturing network propagation in lab experiments and as a result, is expected to improve both production rates and expected ultimate recovery of reserves (Al-Adwani et al. 2008).

Finally, carbon dioxide can have two to three times the affinity to adsorption in shale as methane under lab conditions (Heller and Zoback 2014) and as much as five time the affinity under reservoir conditions (Gandossi 2013). As such, it is expected that a portion of the carbon dioxide used for fracturing would replace methane in the formation supporting the additional natural gas production while sequestering the carbon dioxide in the formation. Carbon dioxide can also react with formation elements, depending on the chemical characteristics of the shale play, to further sequester in the formation. As approximately 30% of gas in shale reservoirs is believed to be adsorbed in place (Mengal and Wattenberger 2011), a 30% sequestration rate for carbon dioxide fracturing will be assumed in this analysis. It is expected that should additional chemical sequestration occur it would render these estimates conservative.

2.5.2 Recycling Produced Water

While the use of large quantities of fresh water is a critical environmental issue, the use of fresh water for fracturing in comparison to other sectors represents less than 2% of water used in the United States (Maupin et al. 2014). In the counties covered by the Barnett Shale, water for shale gas production accounts for less than 1% of total consumption (Baker 2015). In addition, it was found that the use of purchased fresh water is not the major cost driver in the operation, given the low cost of water per gallon. While these costs can vary depending on the water scarcity level in the vicinity of the shale play, in general the cost of acquiring fresh water from water districts is a much smaller part of the operation.

The bigger environmental and economic issues arise with the management and disposal of produced water or wastewater. While the waterless fracturing techniques discussed above eliminate the need for both using and disposing of large quantities of water, which addresses water and seismic for the entire operation, there are established solutions for addressing the most critical of these issues through wastewater management. Two alternatives to standard storage and reinjection of the large quantities of wastewater produced were evaluated: fracturing with produced water and wastewater treatment.

2.5.2.1 Fracturing with Produced Water

The first alternative includes repurposing produced water for use in hydraulic fracturing operations. In addition to reducing the need for water disposal and reducing fresh water consumption during the completion process, using produced water for hydraulic fracturing has economic benefits, since reinjection of produced water into disposal wells can cost an average of USD 0.75 to USD 1.00 per barrel of water (LeBas et al. 2013).

Field experiments have shown that the use of high-TDS (>270,000 PPM) produced water can be recycled to create a cross-linked, gel-based hydraulic fracturing fluid. During bench scale experiments, cross-linked gels were formulated, and results showed that the gel was capable of successfully creating a fracture network and transporting sand into that network. When compared to a control well, the well fractured with the high-TDS treatment exhibited a comparable production profile, leading to a reduction in total water consumed per well and an economic benefit to the operator with no compromise in production levels (LeBas et al. 2013).

In the Barnett Shale, produced water only accounts for 5–10% of water used for hydraulic fracturing due to the quality of water produced, the limited amount of water produced in the time immediately after fracturing to be reused, and the ready availability of salt water disposal infrastructure in the area (Nicot et al. 2014; Mantell 2011a). In particular, the use of produced water does not negate the costs of hauling to and from treatment sites or reduce pumping costs, nor does it eliminate the need for the use of additional fresh water, as not all of the water used for hydraulic fracturing flows back to the surface for reuse. In addition, water quality produced from the Barnett formation requires significant filtering to enable reuse for fracturing, leading to an incremental cost for filtration and the disposal of solid waste, which could have its own environmental issues (Mantell 2011a). When coupled with the abundance of saltwater disposal wells and low-cost fresh water in the area, the economics of using produced water as a fracturing fluid are infeasible. Different economics exist for other plays where this option may be more feasible (Mantell 2011b).

2.5.2.2 Treating Wastewater

There are a number of methods that can be used to treat wastewater. They can be as simple as filters or conventional thermal evaporation processes, or as complicated as treatment systems, which use positively charged ions and bubbles to remove particles from the water at the fracturing site (Halliburton 2013).

In the last few years, membrane distillation (MD) has proven its importance as a powerful solution for water treatment. As it is a thermally driven separation process, this system has received attention as a possible water and wastewater treatment technology in applications such as water desalination, reuse, and recycling. Membrane distillation allows water to be reused without being diluted with freshwater (Dow et al. 2008). The main benefit of using membrane distillation is that it is a thermally driven process and not a pressure-driven process. It therefore does not need to overcome the high osmotic pressures that characterize produced waters in hydraulic fracturing operations. Also, since nonvolatile solutes cannot move through the membrane barrier in an MD system, it is capable of achieving near 100% rejection of dissolved salts and minerals (El-Bourawi et al. 2006).

There are a number of issues that make these solutions sub optimal for the treatment of wastewater from hydraulic fracturing (Estrada and Bhamidimarri 2016). For example, one of the issues is that some of the more traditional processes typically do not do a good job removing halides. Once this halide-rich water is treated for drinking purposes using conventional methods, such as chlorination or ozonation, toxic by-products can result (Parker et al. 2014). In addition, membrane distillation imposes an energy tax on the operation, as the availability of a heat source is a prerequisite for the process. This could increase the cost of the operation dramatically to maintain the required temperatures for the process. For these reasons, membrane distillation is a promising technology that has gained attention as a treatment alternative for high salinity source waters (El-Bourawi et al. 2006; Miller et al. 1997; NCTCOG 2012) around the world; but that needs to be further evaluated for shale gas operations. These options also produce highly concentrated waste as the water is separated from the dissolved solids. This water can contain high concentrations of toxic and potentially radioactive waste that require disposal.

Based on the literature reviewed, the uses of high-TDS water and wastewater treatment appear to be partial solutions to the water issue with hydraulic fracturing and in many cases create an incremental cost with little incremental benefit, with the potential to create additional environmental, health, and safety issues. As a result, this analysis will focus on waterless fracturing techniques and their potential use for both an economic and environmental gain.

2.6 Theoretical Calculations

2.6.1 Current Operations

As discussed, in order to determine the amount of energy a typical shale gas well operation requires, and therefore the amount of fuel to be consumed and greenhouse gases emitted, field data from Barnett Shale assets were obtained and compared with other public sources and research. The assumptions and equations detailed here were used to build a simulator that approximates the costs and greenhouse gas emissions of the operation. Greenhouse gas emissions were evaluated by looking at heavy equipment usage of diesel and light-duty truck usage of gasoline throughout the well lifecycle and then evaluating their conversion to fuel combinations with smaller environmental footprints.

2.6.1.1 Key Assumptions

- For a given well, there is a short-term phase, drilling and completion; and a long-term phase, production.
- The drilling and completion phase takes approximately 1 month for a Barnett Shale well (based on field data).
- The useful life of the equipment is uniform across equipment types and is assumed to be 5 years before a major overhaul, which would require an additional capital investment (at that point, the cycle will be assumed to repeat).
- Wells will be drilled continuously with no downtime between completing one and drilling the next well.
- Prices of fuel, water, and propane remain constant throughout the life of the operation.
- At one well/month, a given rig can drill and complete 12 wells/year, and this will constitute the basic "unit or cluster" of calculation.
- No produced water is reused or recycled, so each new well draws fresh water from the source.
- The life of one well is assumed to be 25 years (Browning et al., 2013b).
- Transport vehicles will be treated separately spanning the entire well lifecycle, and one vehicle will cover a bundle of 12 wells.
- All generators used in the drilling operation are assumed to be uniform in specifications and operating conditions.
- All drilling generators use fuel at the same rate and operate at 80% load.

2.6.1.2 Fuel Usage by Well

Heavy Equipment Usage: Drilling Rigs. From the field data, the average rig used in the drilling operation required a total of three 1250 horsepower diesel engines to deliver the necessary power for drilling a horizontal well of 3,500 feet. Over the 25 days, on average, required to drill the well, the engines each ran on average 351 hours at 80% load. To calculate the total amount of diesel fuel consumed per rig, for each well drilled, the total energy consumed by the three engines was determined using Equation 2.1. and the conversions in Table 2.2.

$$\text{Fuel Consumed per Well} = \frac{\text{(Engine Power)(Number of Engines)(Duration of Operation)}}{\text{Energy Density of Target Fuel}} \tag{2.1}$$

Heavy Equipment Usage: Pumper Trucks for Completions. Fuel usage during the completions phase is concentrated in gathering and pumping water and proppant down hole at rates powerful enough to overcome the friction of the wellbore and the small punctures at the end, and to break through long reaches of rock. Many pump trucks are typically gathered around the

Table 2.2 Energy consumption – conversions and constants (EIA, 2010).

Element	Conversion factor
1 horsepower (HP)	746 kilowatts (kW)
1 liter (L)	0.26417 gallons (G)
1250 HP @ 80% load	1000 horse power (HP)
Energy density of diesel	35.8 megajoules/liter (MJ/L)
1 Diesel Gallon Equivalent (DGE)	0.1395 million British Thermal Units (MM BTU) natural gas
1 DGE	143.94 standard cubic feet (SCF) natural gas
	0.14394 MCF natural gas (MCF= 1,000 cubic feet)
1 DGE	1.68 gallons liquefied natural gas

well site and work in series to add enough velocity to the water. The process of perforation and pumping of fracture fluid is repeated for all planned stages of the fracturing operation. From the field data, an average well covering 3,000 feet of lateral length is perforated in 12 stages, on average. Several pump trucks working for seven days intermittently consumed, on average, 31,000 gallons of diesel fuel to complete the operation.

Heavy Equipment Usage: Fresh and Waste Water Management. The hydraulic fracturing process uses several million gallons of water. That water must be hauled from fresh water sources. In addition, a portion of that water flows back to the surface as flowback or produced water. This water must then be hauled away from the site to a waste water disposal well. More detail on this topic is included in the second part of this chapter. Heavy-duty trucks running predominantly on diesel fuel are used to for this process, and pumper trucks typically dispose of the water into the disposal well. These trucks can carry approximately 8,000 gallons of water each (Worth 2015). To assess the amount of fuel used, it was assumed that the publicly available cost per barrel of water hauling and disposal was composed of the cost of fuel used. Those were used to determine the amount of diesel fuel used for these operations.

Transport Vehicles. During all major phases of the well lifecycle, transport vehicles are used for several purposes, making trips to and from the site to support the operation. In an effort to determine the environmental effects from truck mileage during the drilling, completion, and production phases, the North Central Texas Council of Governments (NCTCOG 2012), gathered data from operators, service companies, and regulatory authorities. Mileage assumptions were based on the expected number of vehicles utilized, the average number of trips taken, the average mileage per trip, and the average fuel efficiency. Table 2.3 summarizes those findings for each of the phases of the operation.

Table 2.3 Transport Vehicle Fuel Usage.

Phase	No. of trips per well	Distance/trip (miles)	Total mileage (miles)
Drilling	187	50	9,350
Completion	420	50	21,000
Production	3,613	0.5	1806.5
Total			32,157

Vehicles are assumed to be light-duty gasoline-powered trucks with a fuel efficiency of 20 miles/gallon. As the production phase can continue for many years, the transport vehicle calculations are annualized, using the 12-well assumption, to ensure consistency among the data.

2.6.1.3 Annual Fuel Usage and Costs

In looking at the greenhouse gas footprint of the shale gas operation and the resultant economic impact of the environmental remediation techniques, it is useful to look at annual fuel usage across the three stages and the various applications using current technologies. To determine the microeconomic impact of the environmental remediation recommendations, the annual operating costs of a diesel and gasoline based operation were determined and then compared to those of the alternative fuel options. Based on the field data, it takes approximately one month to drill and complete a single Barnett well. Thus, it can be assumed that a given "cluster" of equipment can be used to drill 12 consecutive wells per year.

Using the field data presented and applying the assumptions and equations listed here, the annual fuel consumption and resultant costs of the cluster of 12 wells was derived. To determine the annual operating costs of the equipment cluster, the price of fuel needed to be taken into account. In the case of diesel and gasoline, this is real cost to the operation as gasoline and diesel are not produced. In the case of natural gas, in its various forms, this represented the opportunity cost of selling that gas on the market rather than using it to drive the operation. The assumed prices are shown in Table 2.4 and changes in these market prices would impact the results of the analyses.

2.6.1.4 Greenhouse Gas Emissions from Fuel Burn

Of the many substances classified as greenhouse gases (GHG), perhaps the most discussed is carbon dioxide, due to its importance and prevalence in many industrial processes (Sikdar 2009). It has been estimated that more than 70% of the GHG increases over the past three decades are composed of CO_2 (Valbuena et al. 2012). While the combustion of gasoline and diesel fuel emits GHG across the entire spectrum, calculations are focused on CO_2, which will serve as an indicator of the benefits that can be derived from the use of alternative fuels in the shale gas production process. The broader spectrum of greenhouse gases will be discussed at a high level to demonstrate the expected magnitude of the benefits.

The U.S. Energy Information Administration estimates that about 19.64 pounds of CO_2 are produced from burning one gallon of gasoline that does not contain ethanol, about 22.38 pounds of CO_2 are produced from burning a gallon of diesel fuel (EIA 2015b). Table 2.5 summarizes the emissions factors from the technologies used in the field data. As these greenhouse gases can vary dramatically by technology the factors for the specific technologies studies are used here.

Table 2.4 Fuel Prices.

Fuel	Market price (USD)
Diesel	\$3.61/DGE (EIA, 2014a)
Gasoline	\$3.71/GGE (EIA, 2014a)
Natural gas	\$4.12/MM BTU (EIA, 2014b)
CNG	\$1.68/DGE (EIA, 2014c)
LNG	\$2.92/DGE (EIA, 2014c)

Table 2.5 Greenhouse Gas Emission Factors for Base Case Fuels.

Fuel type	NOx	CO	VOC	PM
Diesel (g/bhp-hr) (Stuver and Alonzo, 2014)	8.15	4.69	0.43	0.27
Gasoline (g/mi) (EPA, 2008a)	0.95	11.84	2.513	0.0094

Table 2.6 Baseline Water Assumptions.

Water consumption per well (Nicot et al. 2014)	4,600,000 gallons
Cost of water (City of Fort Worth Texas 2015a)	\$6.43 USD/ccf \$0.32 USD/barrel
Water recovery in Production Year 1 (Schaefer 2012)	30%
Annual water recovery in Production Years 2–25 (Browning et al. 2013b; Schaefer 2012)	1%
Cost of water hauling (Ecologix 2013)	\$1.60 USD/barrel
Cost of water disposal (Ecologix 2013)	\$0.90 USD/barrel
Cost of drilling a Class II injection well (City of Fort Worth Texas 2015b)	\$2,500,000 USD
Proppant consumption per well (McKeon 2013)	1,750 tons
Cost of proppant (Bleiwas 2015)	\$63 USD/ton
Amount of natural gas flared (Howarth et al. 2010)	159,087 MCF/yr
Water injection rate for each Class II well (EPA 2012)	~ 99,000,000 gallons/yr

Based on these figures, the annual GHG footprint of a "well cluster" can be calculated as a baseline for the environmental remediation process and can be extrapolated to express the broader environmental benefits.

2.6.1.5 Hydraulic Fracturing Impacts

The current hydraulic fracturing operation, using "slickwater" as a fracturing fluid, was analyzed to develop a baseline for comparison of various environmental remediation techniques. Table 2.6 summarizes the assumptions used to construct the baseline model for the economic simulator.

Environmental Impacts. The environmental implications of water consumption and disposal are inextricably linked with the greenhouse gas impacts has been discussed elsewhere. For baseline operations, a combination of diesel fuel for heavy equipment used for pumping and hauling water with gasoline for light-duty vehicles is used to develop the baseline environmental footprint of the hydraulic fracturing operation. The baseline fuel assumptions will be held constant for the analysis of alternative fracturing technologies to evaluate the true economic and environmental impact of the proposed remediation techniques.

2.6.2 Operations after Environmental Remediation of Greenhouse Gases

2.6.2.1 Conversion to Dual Fuel Systems

Utilizing a dual fuel system implies that a portion of the diesel fuel consumed with natural gas, and in this case 70% of the diesel fuel consumed per well, was replaced with natural gas in the form of CNG, LNG, or raw natural gas. During this process, the energy density of these replacement fuels was taken into account in both the cost and environmental impact equations (Howell and Harger 2013).

Table 2.2 includes the conversions used to determine these amounts. To determine the fuel consumption of natural gas, diesel, and gasoline across the well lifecycle, the simulator used the fuel consumption assumptions in the "current operations" base case and substituted 70% natural gas for diesel engines and 100% for gasoline. The energy density conversions were used to derive the absolute amount of natural gas consumed, and costs and greenhouse gas emissions were calculated off of that basis.

In addition, the energy burden of the CNG or LNG systems, to convert the reservoir gas to this form, must be taken into account as part of the overall calculation. This does not need to be taken into account when working directly with raw gas. This energy tax is accounted for in the price of these forms of natural gas, which can vary dramatically due to the processing requirements (Department of Energy 2014). By extrapolating the results and looking at the total number of wells drilled across the shale play, an expression for total annual cost and environmental footprint across the entire play can be calculated.

2.6.2.2 Environmental Improvements

To determine the improvements in greenhouse gas emissions from substituting natural gas for diesel and gasoline, the proposed configuration was analyzed and compared with the baseline configuration. Representative technologies were evaluated for each scenario. While the final results will vary depending on the technology implemented, the results discussed here will provide directional guidance for expected benefits from the use of natural gas as an alternative fuel.

Based on the estimates provided by the U.S. Energy Information Administration, 119.0 pounds of CO_2 are produced from burning one thousand cubic feet (MCF) of natural gas (EIA 2015b). In order to determine the reductions in other major pollutants specific technologies used in the field data and comparable alternative natural gas technologies were assumed. Table 2.7 summarizes these emissions for the considered fuel combinations

2.6.3 Operations after Environmental Remediation of Hydraulic Fracturing

2.6.3.1 Waterless Fracturing

The two environmental remediation options for waterless fracturing, LPG and CO_2 fracturing, were analyzed and compared against the base case representing current operations. Table 2.8 summarizes the assumptions used in evaluating the economic and environmental impact of the two alternatives. For each of these fluids, the key levers that drive both the environmental

Table 2.7 Emissions Factors for Technologies used in Environmental Remediation.

Technology	NOx	CO	VOC	PM
Heavy-duty diesel engine (g/bhp-hr) – used for 30% of heavy duty operations (Stuver and Alonzo 2014)	8.15	4.69	0.43	0.27
Heavy-duty natural gas engines (g/bhp-hr) – used for 70% of heavy-duty operations (Electric 2014; Center for Transport Analysis 2011)	0.5	1.4	5.88	0
Light-duty natural gas trucks (g/mi) – used for personal transport (Glover and Kremer 2001)	0.168	1.835	0.025	0
Heavy-duty diesel trucks (g/mi) – used for 30% of heavy duty hauling (EPA 2008b)	8.613	2.311	0.900	0.421

Table 2.8 Waterless Fracturing Assumptions.

Gas Fracturing	
Price of LPG (EIA 2015c)	\$0.44 USD/gallon
Volume of fluid reduction (Brino 2011)	75%
Increase in production (Khuwaja et al. 2014)	30%
LPG flowback/recovery (LeBlanc et al. 2011)	
5 days from start of production	55%
10 days from start of production	100%
Maximum	100%
Reduction in proppant	N/A
Amount of natural gas flared (Howarth et al. 2010):	
100% LPG flared	159,087 MCF
55% LPG flared	86,978 MCF
Carbon dioxide fracturing	
Price of liquid CO_2 (Lane 2014)	\$100 USD/ton
Volume of fluid reduction (Burke et al. 2011)	71%
Increase in production (Burke et al. 2011)	15%–40%
CO_2 flowback/recovery (Khan et al. 2012a; Khan et al. 2013; Rogala et al. 2014)	70%
Reduction in proppant (Burke et al. 2011)	34%
Amount of natural gas flared (Howarth et al. 2010)	159,087

footprint and micro-economic viability as they relate to the fracturing process were used to develop the simulation model.

Total Volume, Properties, and Cost of Alternative Fluid. In addition to having a direct impact on the economics of the project, these parameters also have a follow-on impact on the amount of fuel used for pumping and hauling as well as potential preprocessing costs unique to the fluid.

Potential Increase in Production. As an offset to the cost implications, this lever has the effect of offsetting potential cost increases in the model. Each of the alternative fracturing fluids has physical properties that can potentially result in an increase in well production. Estimated production increases based on field studies of other shale plays will be used in the simulator to determine the total economic impact of the substitution.

Flowback/Recovery of Fracturing Fluid. The management of fluids during the flowback period has a direct impact on costs as well as greenhouse gas emissions in several ways. This factor also impacts direct post-production clean-up costs and can have a potential impact on operating costs in the event fluids can be recycled and reused. The simulator assumes all flowback gasses will be flared to the legal limit of 10 days (Porter 2013) and not vented or recycled. The impact of recycling fracturing fluids will be addressed in the discussion.

In the case of LPG fracturing, two flaring options will be considered, flaring 100% of the flowback gas for the government allowed 10 day maximum. Based on available field data, it is expected that 100% of the propane would be flared during this period should the operator use this approach. This represents an extreme case, however, as gas-processing facilities accommodating natural gas streams with up to 45% LPG are typically part of natural gas production operations and would therefore be used to separate LPG from the natural gas. As a result, the second case, which involves flaring flowback gas until it falls below 45% LPG, will be developed with flow directed to a processing plant after this point. Based on similar production profiles, it is expected that this point will occur on the fifth day of production.

In the case of CO_2, flowback volumes are directly impacted by the amount of CO_2 that can be sequestered during the completion phase. In this case, it was assumed that 30% of the fracturing fluid would remain in the formation, adsorbed to the shale, based on the affinity of shale to CO_2 compared to methane. In addition, based on related field studies, it is expected that 100% of the CO_2 will flow back to the surface within the 10-day flaring limit for natural gas.

Reduction in Proppant Volume. An increase in efficiency in the deposition of proppant in the fracture has a direct impact on the cost of the operation, and an increase in proppant efficiency implies a reduction in the mining and use of sand, ceramics, and other materials and a reduced environmental footprint.

In determining the reductions in the above parameters, results from field studies in other plays were normalized and extrapolated to apply to the actual Barnett Shale operations. While this incorporates a level of uncertainty in the results, as compared to a full field study in the Barnett Shale, it provides a directional comparison on costs and environmental impact among the options. All reductions are in direct comparison to slickwater fracturing, and it is using that baseline that the comparisons are made.

Fuel consumption for the fracturing process is assumed to reduce linearly with fluid volume. This assumption could change depending on whether or not the fluid used has distinctive pumping characteristics. Literature implies that for both LPG and CO_2 these differences are minimal as they relate to the operation of the pumps, although specialized pumping equipment may be required, leading to an increase up front of the capital required.

The effect of substituting natural gas for diesel and gasoline in both heavy-duty and light-duty vehicles will initially be excluded from the calculation to provide a consistent platform to evaluate the impact of fracturing fluids exclusively. The integrated results of both greenhouse gas and water impacts will be discussed separately to demonstrate the potential impact of an integrated policy strategy that addresses all of these elements simultaneously. In performing this analysis, the results and assumptions used in the first part of this series will be used and extrapolated to account for the total reduction in fuel consumed in the alternative fracturing fluid scenario.

2.6.3.2 Environmental Improvements

Environmental improvements are classified into three categories, reduction in water consumption and disposal, reduction in proppant usage, and reduction in greenhouse gas emissions resulting from reductions in pumping, hauling, and disposal activities. In the case of CO_2 fracturing, the additional carbon sequestration resulting from the use of CO_2 as a fracturing fluid will also be taken into account. Table 2.9 summarizes the CO_2 emissions of the alternative fuel options.

2.6.4 Net Present Value and Expected Capital Outlay

The net present value (NPV) of the environmentally remediated operation was determined to identify the maximum capital outlay that would make the investment in environmental remediation economically viable for the operation. This was then compared against estimates for the up-front capital investment required for the technologies assumed in the environmentally improved case. Equation 2.2 was used to determine NPV.

$$NPV = -(CAPEX) + CF * \left[\frac{1-(1+i)^{-n}}{i}\right] \quad (2.2)$$

where:

CAPEX = the up-front capital investment
CF = annual net cash flows (or net savings)
i = the discount rate, and
n = the number of years before a major overhaul

Table 2.9 CO_2 Emissions for Technologies Used in Environmental Remediation of GHG.

Technology	CO_2 (lbs)
Diesel (per gallon) (EIA 2015a) (used for heavy duty operations)	22.38
Gasoline (per gallon) (EIA 2015a) (used for personal transport)	19.64
Natural gas (per MCF) (EIA 2015b) (flared during flowback)	119.9
Propane (per gallon) (EIA 2013b) (flared during flowback)	12.7

Depending on whether the investment is for a green field application or for a conversion of existing rigs, the up-front capital investment can vary. In a green field application, the difference between the base-case equipment and specialized equipment required for the environmentally remediated solutions would need to be taken into account. For a conversion project, the investment in existing technology would be considered a sunk cost, so the up-front capital investment would be the total cost of the specialized system without subtracting out the potential investment in existing technology. The discount rate was assumed to be 10%, and, based on the field data collected, equipment went through a major overhaul every five years.

2.7 Results and Discussion

2.7.1 Improved Operations with Environmental Remediation of Greenhouse Gas Emissions

With current operating conditions, the average well cluster costs approximately $6.9 million in fuel costs and emits 24 tons of CO_2 annually. This is the ongoing operating cost of using traditional equipment. The data, conversions, and equations highlighted in the previous section were used to build a simulator or "digital twin" of the relevant elements of the operation, and the results for the various fuel options were evaluated.

In looking at improving the greenhouse gas footprint of the shale gas operation, the three scenarios with 30% diesel and 70% natural gas (directly from the wellhead), CNG and LNG for the heavy-duty equipment were looked at along with conversion of light-duty vehicles to 100% compressed natural gas. Although LNG has the highest energy density of the three natural gas alternatives, then CNG, then natural gas directly from the wellhead, the environmental improvements are similar, while raw natural gas offers a much more attractive economic benefit annually. This is primarily due to higher opportunity cost of LNG and CNG compared to natural gas taken directly from the wellhead. Typically, the market price of CNG and LNG take into account the ongoing cost of gas processing and compression required to get the gas to those states. Once a company invests in these processes it is typically more profitable to sell those products than to reuse them for power generation. The price of raw natural gas (or associated petroleum gas (APG) in shale oil operations) excludes these mark-ups and as a result becomes a more cost-effective option for an operator. Table 2.10 summarizes the cost improvements per million BTU of natural gas produced assuming an average annual production rate of 172,000 MM BTU/well derived (Browning et al., 2013b).

The combined economic and environmental improvements for one well cluster (i.e., annual improvements) are shown graphically in Figure 2.2.

In the case of APG, natural gas extracted during shale oil operations is typically not sold and constitutes a cost to the operation, as gas must either be vented or flared. While this analysis

Table 2.10 Cost Improvements per Million BTU of Natural Gas Produced.

Fuel scenarios	Cost improvements per MM BTU (% of price @ $4.12/MM BTU))
Diesel LNG	$ 0.46 (11%)
Diesel CNG	$1.28 (31%)
Diesel Raw	$2.04 (49%)

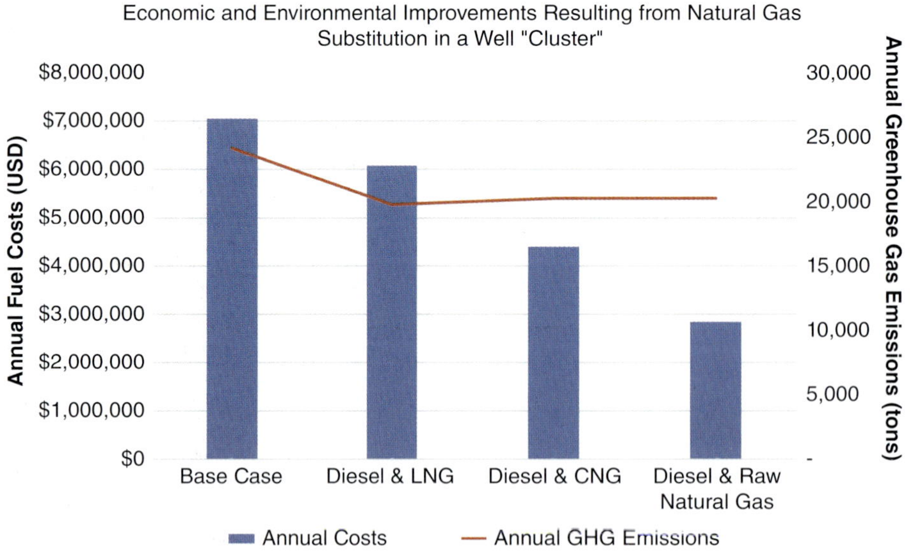

Figure 2.2 Economic and Environmental GHG Improvements for a Well "Cluster".

is out of scope for this chapter, it can be seen that both the environmental impact and economic impact would only be more significant than that for shale gas operations as it involves the recycling of process by-products to power the operation. In these cases the savings on diesel can be taken in their entirety rather than being offset by the opportunity cost of natural gas. In addition, the reduction of flared or vented greenhouse gases would have an additional positive impact on GHG footprint (Horwitt and Sumi 2014).

2.7.1.1 Capital Investment Analysis

While the improved annual operating costs are a benefit, an additional up-front capital investment will be required. This is a much more expensive proposition for converting existing rigs than for building new rigs, as the investment in traditional technologies is sunk already. For new rigs, only the difference between the natural gas–based technologies and the traditional diesel/gasoline-based technologies would be considered as an investment and would need to be made either way.

Based on the annual savings, the net present value calculations indicate the maximum up-front capital investment that can be made for the project to break even over the initial five years, as shown in Figure 2.3.

The five-year horizon is used in these analyses as all equipment considered would need a major overhaul, and repair typically after five years and an additional investment would need to be taken into account. This is a simplified assumption to enable comparison of existing technologies. A more sophisticated approach would be to understand the maintenance schedules of existing technologies over the useful life of the equipment and to compare total lifecycle costs of natural gas versus diesel/gasoline based equipment. As this field data is much more difficult to find, this simplified model illustrates the approach, which can be applied to the more sophisticated scenario in future work.

Based on the net present value, an operator can invest as much as \$15.9 million in up-front capital and break even if natural gas directly from the wellhead is used. With CNG, that number decreases to \$10.0 million, and with LNG that number drops significantly to \$3.7 million. To determine whether or not the required investment would be financially feasible, an evaluation

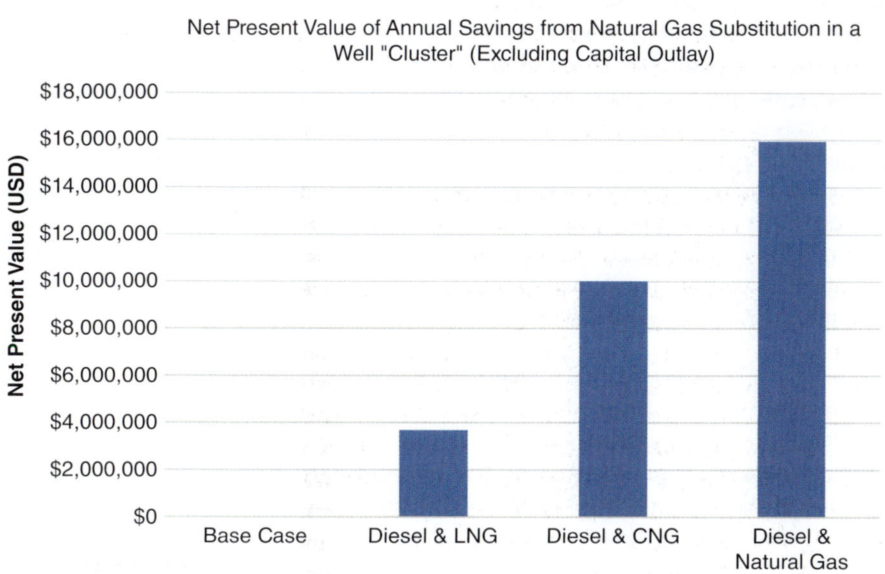

Figure 2.3 Net Present Value of a Well "Cluster" (Excluding Capital Outlay).

Table 2.11 Estimated Capital Outlays for Environmental Remediation Scenarios.

Scenario/case	Heavy equipment (drilling)	Heavy equipment (completion)	light-duty passenger vehicles	Total investment
Diesel and LNG	$2,490,000	$65,000	$66,500	$2,621,500
Diesel and CNG	$800,000	$65,000	$66,500	$931,500
Diesel and natural gas	$2,000,000	$65,000	$66,500	$2,131,500

of the cost of representative technologies was evaluated. Using data from equipment manufacturers, retailers, operators, and public information, several technology options were evaluated.

In looking at technologies, the major difference in technology among the three scenarios is in the drilling rigs. (GHS 2012; Howell and Harger 2013) The lowest up-front investment is in the diesel-CNG scenario, as the operator will need to process and compress the wellhead gas to pipeline quality for transport, so no additional investment in a CNG compression train is required specifically for this application. In addition, since compressed natural gas is high quality, a lower-cost engine can be used to drive the drilling rig. For raw natural gas direct from the wellhead, a more robust engine must be used. In this case, the cost of the engine would be significantly higher; however, only two such engines could be used, and in speaking with both the manufacturer and distributor of such engines, the resultant horsepower could potentially be used to displace all three diesel engines for most situations, so that the diesel engine is used for backup purposes only. For this analysis, however, a conservative scenario is used with the understanding that these engines would have a one-to-one replacement ratio, two diesel engines would be replaced with two natural gas engines, and the third engine would remain and operate one-third of the time. Finally, while the engines scoped for the CNG scenario could be used for LNG, additional investment in a small LNG compression train would be required to liquefy the natural gas. Typically, an operator does not need this level of compression to prepare natural gas for pipelines, and LNG trains require a much higher level of purity for liquefaction, so additional gas processing, over and above pipeline quality, would also be required. This is modeled in the simulator as an LNG uplift to the CNG investment to cover the up-front capital cost of these systems. LNG trains typically have a much higher throughput than that required for one cluster, so the train would be shared among five to six clusters depending on throughput and the cost divided among the clusters to determine the effective investment. Table 2.11 summarizes the capital outlays required for the various options.

The net present value analysis shown in Figure 2.4 indicates that the most favorable economic scenario is the use of raw natural gas directly from the wellhead. The higher up-front capital investment narrowed the gap between raw natural gas and CNG; however, the increased annual savings of the wellhead gas compensated for this over the five-year period. Also, given the high up-front investment in LNG, the diesel-LNG scenario actually yielded a negative net present value when the capital investment was taken into account.

2.7.1.2 Broader Economic and Environmental Benefits

Based on the well cluster analysis, it can be seen that the use of natural gas in place of diesel and gasoline can yield both an environmental and economic benefit for operators, provided either CNG or raw natural gas taken directly from the wellhead is used. Using the wellhead gas scenario and extrapolating these results to encompass the 1,000–1200 new wells to drilled annually across the entire Barnett Shale play (NGI 2014) and the ~3,700 new wells to be drilled annually across the United States (Hughs 2014), operators can reduce the environmental impact of natural gas extraction and save the industry money, as shown in Table 2.12.

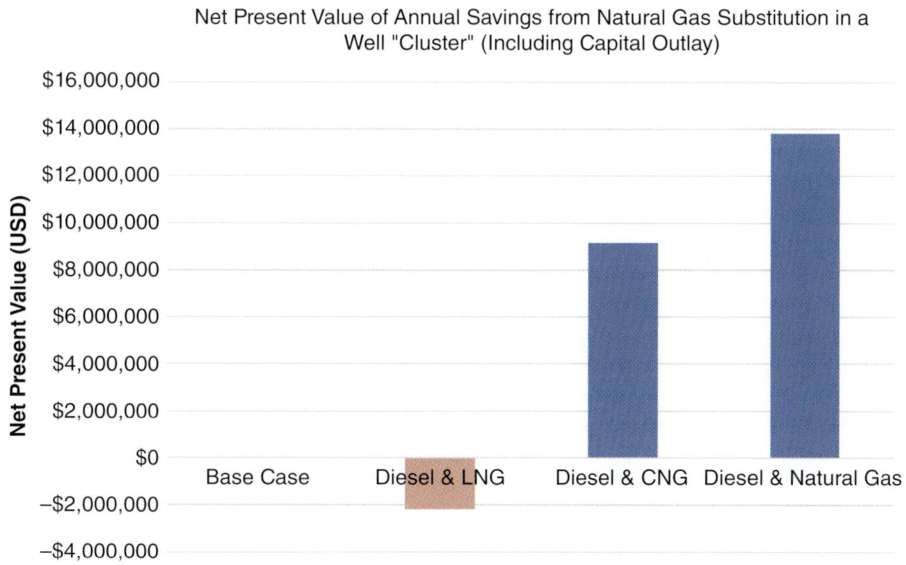

Figure 2.4 Net Present Value of a Well "Cluster" (including Capital Outlay).

Table 2.12 Extrapolation of Results across Barnett Shale and All U.S. Shale Plays.

	Impact across Barnett Shale play	Impact across all United States shale plays
Annual savings (USD)	$385,234,346	$1,295,788,256
CO_2 reductions (tons)	355,277	1,195,022
NOx reductions (tons)	19,944	67,086
CO reductions (tons)	11,487	38,638
VOC reductions (tons)	1,494	5,026
PM reductions (tons)	657	2,1211

While the long-term financial impact of converting to natural gas is significant, as discussed earlier, the shale gas industry is highly cost sensitive and as a result, would be hesitant to invest heavily up front in the operation.

2.7.2 Improved Operations with Alternative Fracturing Fluids

In evaluating both LPG and CO_2 as alternative fracturing fluids, little appreciable difference in annual operating costs, compared to slickwater, was observed, with the cost of the alternative fracturing fluids compensating for the additional water management expense of slickwater fracturing. This is shown in Figure 2.5.

It is important to point out that the Barnett Shale formation has high levels of formation water, so a well may produce more than 100% of the water used for fracturing (Nicot et al. 2014). This additional water production is not taken into consideration in the simulation as it is not clear whether or not this water would ultimately be produced in the same volumes with alternative fracturing fluids. These base results assume that the fracturing fluid is not recycled or resold.

It is interesting to note that, even though less diesel and gasoline are used in the process, the greenhouse gas footprint for LPG fracturing is actually higher than that of slickwater fracturing due to the large amount of natural gas and LPG that must be flared to get LPG concentrations

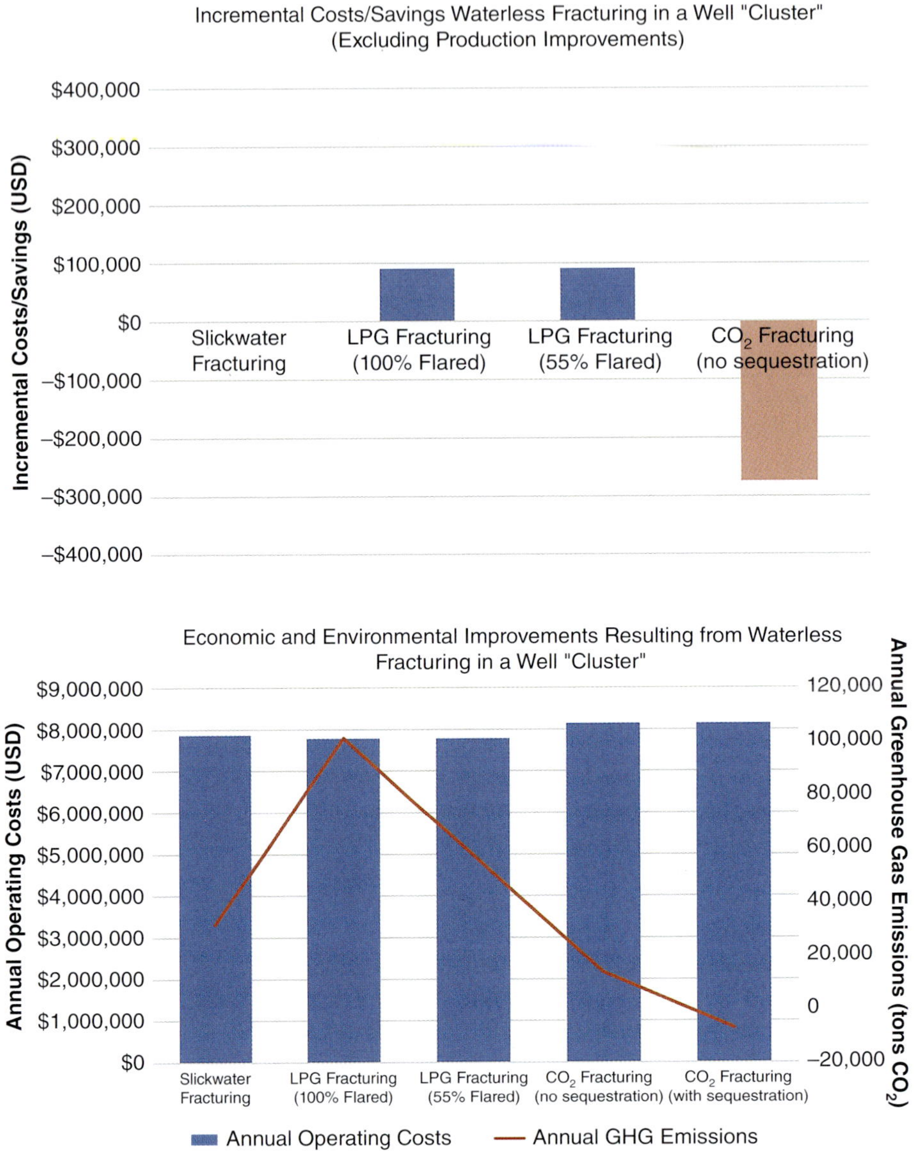

Figure 2.5 Impact of Waterless Fracturing on Annual Operating Costs and GHG Footprint.

below those required for transport and standard processing facilities. Although it represents the highest overall cost, CO_2 fracturing also has the lowest carbon footprint, and once carbon sequestration is taken into account, the operation becomes carbon negative. This also implies that, based on cost savings alone, there is little economic incentive for operators to move to alternative fracturing fluids that could represent a higher risk to production.

The picture changes when the impact on production, or revenue, is taken into account. Based on observed production increases in other shale plays, and their application to the Barnett Shale, a significant economic impact can be realized when production is enhanced by using alternative fracturing fluids. This impact is summarized in Table 2.13.

This implies that operators should be willing to invest between $10,000,000 and $25,000,000 in these new technologies provided production improvements can be guaranteed. In the case

Table 2.13 Economic Impact of Waterless Fracturing Options on a Well "Cluster".

Fracturing Fluid	Annual costs (USD/yr)	Annual savings (USD/yr)	Incremental production (USD/yr)	NPV of impact – life of well (USD)
Slickwater	$7,874,531	—	—	—
LPG (55% flared)	$7,783,580	$90,952 1.15%	$2,551,104 (32%)	$23,247,425 (33%)
LPG (100% flared)	$7,783,580	$90,952 1.15%	$1,988,235 (23%)	$18,138,240 (26%)
CO_2	$8,148,547	–$274,016 (–3.36%)	$1,275,552–$3,401,472 (16%–43%)	$11,304,222–$30,601,282 (16%–44%)

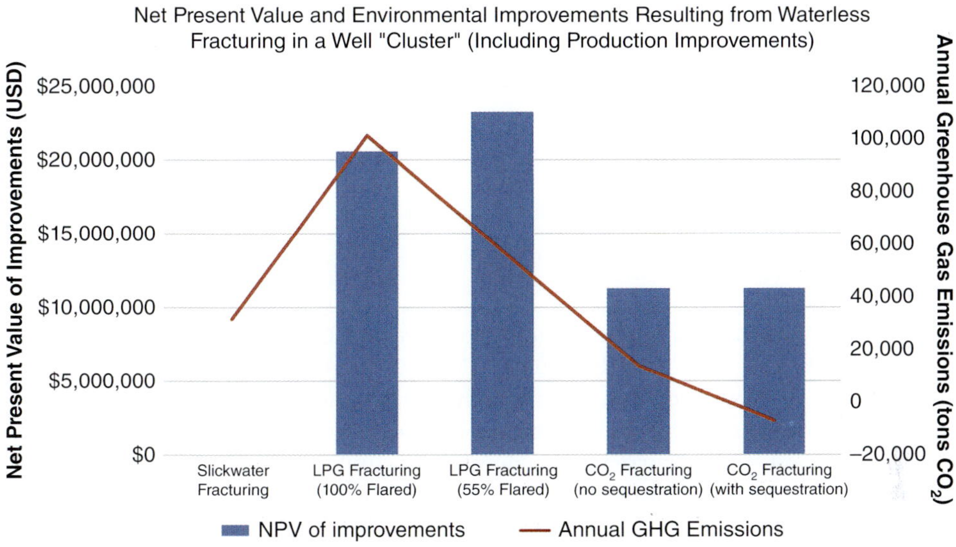

Figure 2.6 Impact of Production Improvements on the NPV of a Well "Cluster".

of CO_2 fracturing, this converts a negative return on investment to a positive one. Figure 2.6 demonstrates this impact on net present value.

As discussed previously, there are number of key factors influencing the outcome of this analysis; the most critical of these will be discussed here. While the remaining elements have an impact on the analysis, that impact is less pronounced and will not be discussed at length in this analysis.

2.7.2.1 Cost of Alternative Fracturing Fluids

Given the volumes of fluids used, a small change in the price of fracturing fluids has a significant impact on the economics of the operation. As the prices used in the simulation include transportation costs to the well site, their prices can be dependent on both the supply and demand of these fluids as well as the transportation costs of these fluids to the site. While the market dynamics of water and LPG pricing can be somewhat predictable and there is existing infrastructure to support the use of both of these fluids, the issue becomes more complex in the case of carbon dioxide.

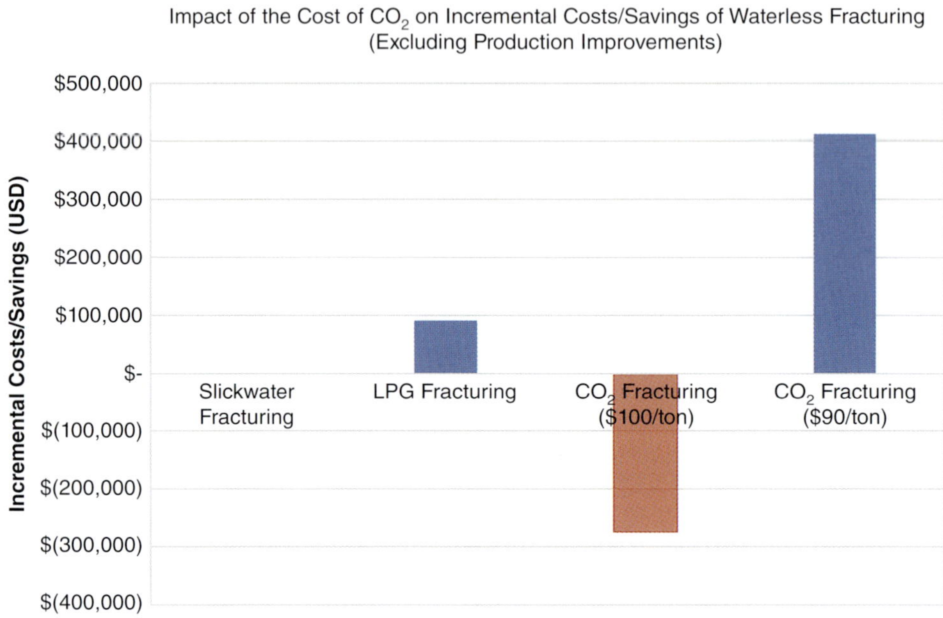

Figure 2.7 Impact of CO_2 Price on the Net Present Value of the Hydraulic Fracturing Operation (Excluding Production Improvements).

While carbon dioxide itself is abundant, the availability of captured, separated, and processed CO_2 can be limited. Also, the lack of infrastructure to transport CO_2 in large quantities makes its price vary quite dramatically across the country. Finally, most end consumers of liquid CO_2, such as the food and beverage industry, require high purity standards, thereby driving up the processing costs, and therefore the price, of available CO_2 supply. Therefore, carbon dioxide behaves as a public good, and its current price does not accurately reflect supply and demand dynamics. As there is a market failure in the case of CO_2, the impact of public policy can be profound. A small improvement in the price of carbon dioxide has a significant implication on the economics of the case and can make the option economically more favorable than both LPG and water. Figure 2.7 shows the significant impact on net present value of a 10% reduction in CO_2 price. This is a critical component of the analysis, as the CO_2 purity required for fracturing and other industrial processes is significantly lower than that for the other industries such as food and beverage and as a result could be less expensive the prevailing market prices assumed in the analysis and significantly impact the results.

2.7.2.2 Availability of Salt Water Disposal Sites

The cost of water management is heavily dependent on the costs of hauling and disposal. These costs are dependent the price for fuel (typically diesel) used for hauling and how far water must be hauled both to the well from water sources and from the well to disposal sites. The price of fuel, has an impact on all scenarios as its cost also impacts the cost for transporting all fracturing fluids. The distance from water sources and the availability of salt water disposal sites, does however have a significant impact on the "slickwater" operation significantly.

Extensive salt water disposal infrastructure has been developed within the Barnett Shale play with deep water injection into the Ellenberger formation. As a result, the cost of wastewater hauling and disposal is much lower than that of other shale plays, which can be between two and five times as expensive (Schaefer 2012; Burke et al. 2011). In addition, the drilling and

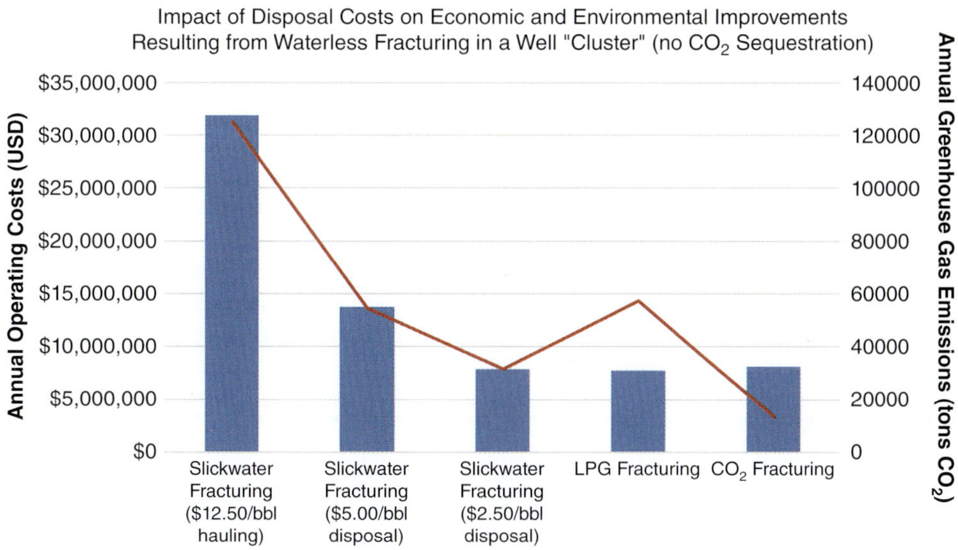

Figure 2.8 Impact of Hauling and Disposal Costs on the Benefits of Alternative Fracturing Fluids.

completion of a Class II salt water disposal well, provided there is a viable disposal formation or aquifer, can range anywhere from $400,000 (City of Fort Worth Texas, 2015b) to more than $2.5 million (Worth 2015), rendering the building of comparable infrastructure at other shale plays expensive for operators in those areas and in many cases prohibitive. The worse the scarcity of salt water disposal sites, the clearer the operating benefit of alternative fracturing fluids with or without production increases. Figure 2.8 shows the impact of these increases in hauling and disposal costs on the operation. Many of the fields where LPG and CO_2 fracturing have been employed, have had these higher cost profiles, rendering these alternative processes necessary for the profitability of the operations.

2.7.2.3 Fracturing with CO_2 vs. LPG

In looking purely at the economics of the operation, it can be inferred that while both waterless fracturing options offer economic benefits, LPG fracturing offers a larger economic incentive for operators.

Fracturing with liquefied petroleum gas has a number of benefits. Since the types of hydrocarbons in the fracturing gel are comparable to those in the formation, once exposed to the heat and pressures of the reservoir, the gel is broken down, and propane vaporizes and can simply be produced with the reservoir flow. This minimizes clean-up and maximizes well flowback. Unlike carbon dioxide fracturing, no specialized separation equipment is required; the natural gas processing plant used to produce dry gas from wellhead gas will ultimately remove propane with the stream where it can be reclaimed for reuse or resale. Also, since, LPG is electrically neutral and low in friction, many of the salts, heavy metals, and radioactive components that rise to the surface with water remain in the formation. In addition, the transportation infrastructure for propane is relatively extensive, and, as such, its transportation to the well site can be considered comparable to that of water.

However, there are several potential issues with the use of LPG as a fracturing fluid. The first is the obvious safety concern with substituting water with a significant volume of flammable gel and pumping it at high pressure down the wellbore. The substitution of the nonflammable base fluid, water, with a flammable one, propane, decreases the inherent safety of the operation, and therefore additional precautions must be taken to ensure the safety of the operation. In addition, every effort to run the operation remotely must be taken to minimize the impact of a potential fire hazard. The safety program for an operation that uses propane instead of water would be expected to be much more complex and prone to human error. Given the safety precautions taken to handle the flowback of natural gas, the extension of the safety system to include that fracturing fluid needs to be evaluated. This is especially true of storage, liquefaction, pumping, and transport units involved in the management of propane and/or butane before, during and after the fracturing operation.

Gas fracturing also requires highly specialized equipment(Loree and Mesher, 2007), which could result in a significant capital investment by the operator. While the expected net present value of the operation seems to support this investment, risk-averse, cost-sensitive operators will be reluctant to make the investment. Finally, this is truly a new technology not widely proven in the industry. Unlike hydraulic fracturing and CO_2 fracturing, which have been used in the oil and gas industry for decades and are used in shale gas extraction as a new application, gas fracturing is truly a new and relatively unproven technology. As the shale gas market is highly competitive, operators who have taken a risk on the technology have kept the benefits tightly controlled, thereby creating a state of information asymmetry in the marketplace and reducing technology adoption. More traditional, risk-averse operators, therefore, lack the data required to make informed decisions to adopt this technology.

In contrast, CO_2 fracturing has many of the same benefits of LPG fracturing in that it also does not damage the formation or dissolve salts, heavy metals, and radioactive components that rise to the surface with water, enabling them to remain in the formation. It also has comparable flowback characteristics and can drive similar increases in production. CO_2 has the added benefit of affinity to shale and its ability to displace methane in the shale formation, leading to enhanced gas recovery. This is especially helpful as shale wells are depleted and workover rigs are needed to sustain production. The use of CO_2 to prolong the production of the well would be a natural extension of the fracturing process. In addition, supercritical CO_2 has been shown to produce more extensive three-dimensional fractures, which is expected to further enhance production.(Zhou and Burbey 2014; Thomas 2015; Ishida et al. 2012; Middleton et al. 2015).

CO_2 is also inherently safer as a fracturing fluid than LPG, having similar flame-retardant characteristics to water. It is safe to transport, store, and pump and, therefore, does not introduce inherent safety risks to the operation. While creating foamed or supercritical CO_2 does impose an energy tax and, therefore a cost, on the operation, CO_2 in these forms does not require specialized pumping or monitoring equipment, nor does it require an extension of the safety management protocols in the operation. While flowback CO_2 can be vented to the atmosphere through a flaring process until concentrations reach pipeline levels, an incremental investment in a mobile CO_2 recovery system for the first two to three weeks after flowback is initiated would enable recovery and reuse/resale of CO_2.(Reynolds et al., 2013) For shale plays like the Barnett play, this would be an incremental investment to the operation, while for other plays where CO_2 is produced from the formations themselves, this would be an extension of the investment required to separate CO_2 from the formation gas. These costs, coupled with the higher market price of carbon dioxide make the alternative less attractive for operators than LPG or hydraulic fracturing.

Some of the most common drawbacks of CO_2 fracturing include the need to transport the fluid in a liquid state and store it in pressurized tanks, as well as the potential formation of ice on the hydraulic head and pipes during the fracturing operation should the wellhead pressure drop rapidly (Rogala et al. 2014).

From an environmental and safety perspective, it is clear that the use of CO_2 as a fracturing fluid could be much more attractive than any of the other options discussed here. In addition to the safety considerations and expected sequestration during hydraulic fracturing as well as the reduced amount of proppant required, the benefit of enhanced gas recovery and post-production sequestration in depleted shale wells with no incremental capital investment in equipment, render the potential environmental benefits much more appealing.

Several studies have evaluated the viability of depleted shale reservoirs for wide-scale carbon sequestration and have recommended their use (Khan et al. 2012a; Khan et al. 2013; Middleton et al. 2015). However, this long-term sequestration comes at a cost of \$36–\$73 per ton of CO_2, inclusive of enhanced gas recovery gains (Tayari et al. 2015a). In the absence of any external incentives, oil and gas operators would not be likely undertake the longer-term sequestration programs at this cost. In addition, the largest components of this cost are predicted to be transportation and injection costs with power supplied from an electric utility and pipeline-based transportation, which is not available in all areas. Thus, in the absence of public policy to drive these longer-term programs, this level of sequestration will not likely be realized.

In fact, while CO_2 fracturing is considered a more preferred solution from an environmental perspective, the economics could drive operators to remain with hydraulic fracturing or move towards LPG as an alternative. The implementation of short-term or temporary policy aimed at making the use of CO_2 more economically attractive could drive improvements in infrastructure that could sustain the alternative as economically viable, even after the policy instruments have expired. This type of environmental policy has proven successful in driving technology adoption in the past with renewables policy or even the unconventional fuels development policy, which spurred the current shale boom in the United States (Wang and Krupnick, 2013).

2.7.2.4 Flowback and Recycling of Fracturing Fluid

While the recycling of produced water may not be economically desirable in the Barnett Shale due to the abundance of water disposal infrastructure and the extended flowback periods, the recycling and reuse of alternative fracturing fluids are a different story. For LPG fracturing, field studies have shown that 100% of the propane used in fracturing is produced with the saleable natural gas within the first two weeks of production (LeBlanc et al. 2011). This implies that with separation and processing, a large portion of the LPG used in additional wells can be reused or resold. In which case, the majority of the fuel need only be purchased once, and the operator would just need to make additional purchases to replace process losses. A 10% loss rate is assumed for this analysis. Should recycling of propane be employed, flowback from the well could be processed on site or sent to the natural gas–processing facility and then transported back to site rather than flaring the flowback gas until the propane level is reduced. This additional cost would need to be taken into account in the analysis.

A similar rationale applies to CO_2 fracturing; again flowback of CO_2 is completed within weeks of the start of production(Zhenyun et al. 2015) with the exception that a portion of the CO_2 injected for fracturing will remain sequestered, so a 10% loss is assumed over and above the sequestered amount. The costs of CO_2 separation and fluid processing (either into foam or supercritical fluid) need to be taken into account as well (Tayari et al. 2015b). Figure 2.9 shows the impact of recycling on the economics and greenhouse gas footprint of the operation. The profitability of the well cluster was analyzed including separation costs, reduced operating costs due to fluid recycling, and the net present value of the incremental revenue from the excess natural gas recovered due to separation, rather than flaring.

Recycling has a threefold benefit for the operation. Firstly, it implies that less fracturing fluid is purchased for each subsequent operation after the first. Secondly, it reduces the amount of natural gas flared, thereby resulting in an increase in revenue. Finally, it reduces the overall greenhouse gas footprint of the operation significantly due to a reduction in flaring and fuel

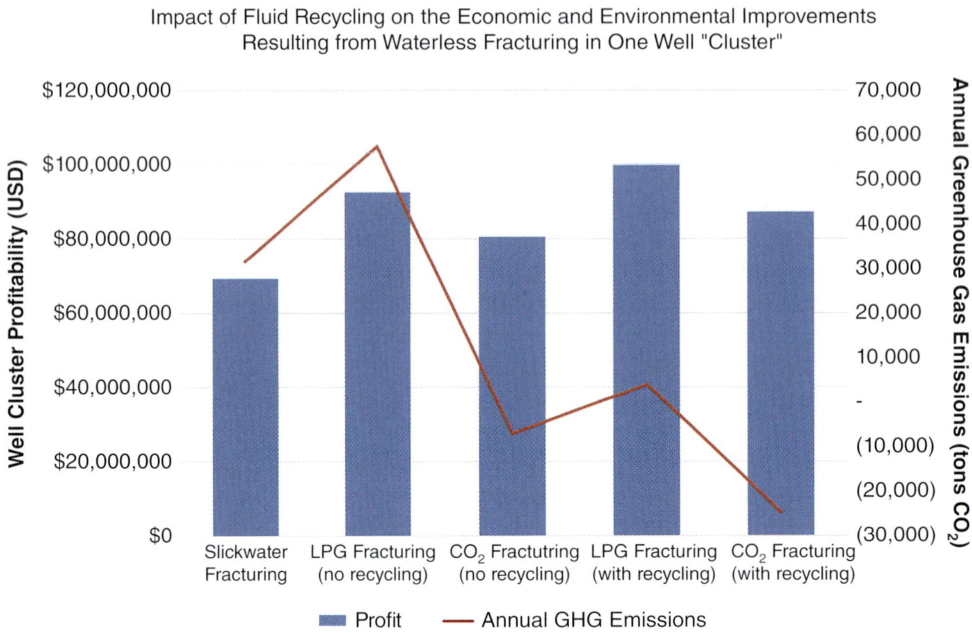

Figure 2.9 Impact of Recycling Fracturing Fluids on Profitability and Greenhouse Gas Footprint.

used for hauling fracturing fluid to the site. As a result of the sequestration of carbon dioxide in the formation, the economic impact of LPG recycling is much higher than that of CO_2. In this case, the more favorable economic solution is not the most favorable environmental solution due to the market failure attributed to CO_2 as a greenhouse gas.

2.7.2.5 Seismic Implications

One of the critical issues with hydraulic fracturing is the potential to trigger seismic activity during the re-injection of wastewater. As all fluids in the waterless fracturing process vaporize and flowback, reinjection is not required, and therefore the risk of seismic activity from re-injection is eliminated in the remediated processes. As there is no evidence of significant seismic activity resulting from the fracturing process itself, the elimination of water from the process addresses the seismicity issues directly.

2.7.2.6 Unlocking Arid and Water Sensitive Shales

In addition to improving the economics and environmental footprint of existing shale plays, an added benefit of waterless fracturing is unlocking additional shale plays that were previously infeasible either due to water scarcity or an undersaturation of the shale formation, rendering it water sensitive. This not only unlocks a significant resource within the United States, but also provides a platform for countries like China, Mexico and South Africa that have a significant, characterized resource in very arid or water-stressed parts of the country (Reig et al. 2014b). This could, in turn, have a significant impact on some of these developing nations' energy independence, or their dependence fuels such as coal and their greenhouse gas footprint. Again, for a country like China, that has a greenhouse gas footprint that is expected to surpass the United States this shift could significantly improve the global greenhouse gas profile and change the dynamics of the energy industry (EIA 2012b).

Table 2.14 Extrapolation of Results across Barnett Shale and All U.S. Shale Plays.

	Impact across Barnett Shale Play	Impact across all United States Shale Plays
CARBON DIOXIDE FRACTURING		
Annual savings (USD)	\$1,775,839,450	\$5,475,504,969
Annual CO_2 reductions (tons)	5,626,666	17,348,885
GAS FRACTURING		
Annual savings (USD)	\$3,040,524,683	\$9,374,951,107
CO_2 reductions (tons)	2,755,141	8,495,019

2.7.2.7 Broader Economic and Environmental Benefits

Based on the well cluster analysis, it can be seen that the use of either CO_2 or LPG in place of water as a fracturing fluid can yield both an environmental and economic benefit for operators, provided increases in production are taken into account, and especially when fluid recycling is employed. Using the scenario that includes both production impacts and fluid recycling and extrapolating these results to encompass the 1,000–1200 new wells to drilled annually across the entire Barnett Shale play (NGI 2014) and the ~3,700 new wells to be drilled annually across the United States (Hughs 2014), operators can reduce the environmental impact of natural gas extraction and save the industry money, as shown in Table 2.14.

As demonstrated by the analyses, the economic benefits tied to LPG outweigh those of CO_2 in the substitution of water as a fracturing fluid, while the environmental benefits of CO_2 outweigh those of LPG. While both options drive economic and environmental benefits, driving the industry towards the safer, more environmentally friendly alternative requires the institution of policy that will incentivize the development infrastructure to drive down the price of CO_2 and create a sustained carbon economy. With the right type and duration of policy actions, a carbon economy can be developed that far outlasts the policy itself and encourages the capture and sequestration of carbon for years to follow.

2.7.3 Environmental and Microeconomic Impacts of Combined Technology Alternatives

As much of the environmental impacts of the shale gas operation occurs during the drilling and completion phases of the well lifecycle, and it takes approximately one month to drill a well, a cluster of 12 wells was used as the unit of measure for the analysis. This represents an "annualized" cost model. The net present value of the remaining costs that run through the life of the well cluster (assumed to be 25 years), as well as the net present value of the lifetime revenue of the cluster, were then compared to this annualized cost to develop an expression for the lifetime profitability of the well. Figure 2.10 highlights the combined environmental and microeconomic impact of the proposed technology alternatives excluding production impacts.

While there is a reduction in operating costs resulting from each of the three alternatives, the largest cost reduction comes from the substitution of natural gas with diesel as a fuel, even though this option does not produce the most optimal greenhouse gas reduction. A different picture emerges when the increase in production resulting from the alternative fracturing fluids is taken into account, as shown in Figure 2.11.

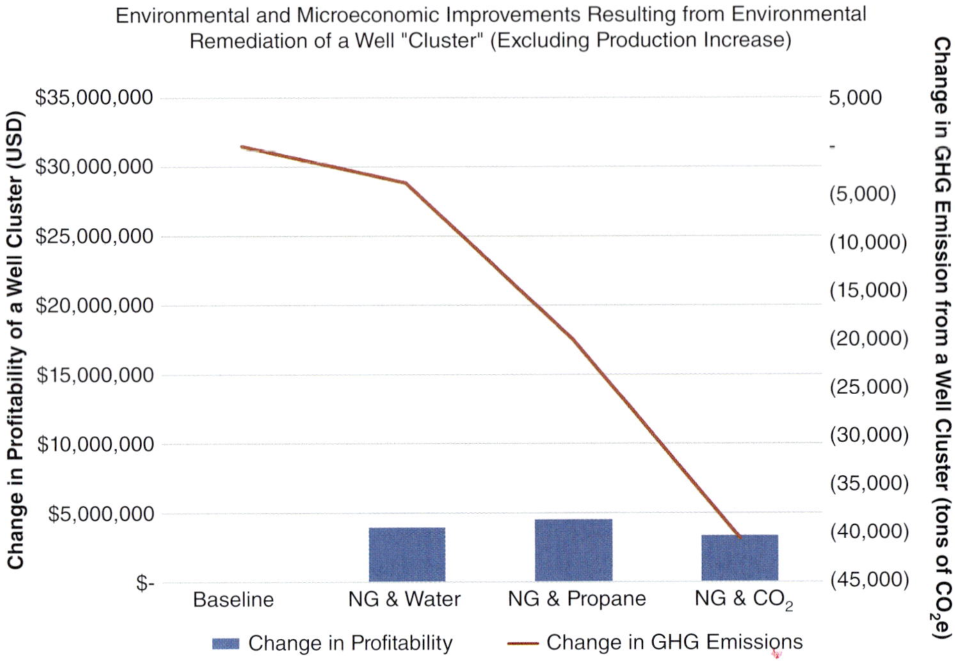

Figure 2.10 Environmental Improvements and Operations Savings of Technology Alternatives (Excluding Production Impacts).

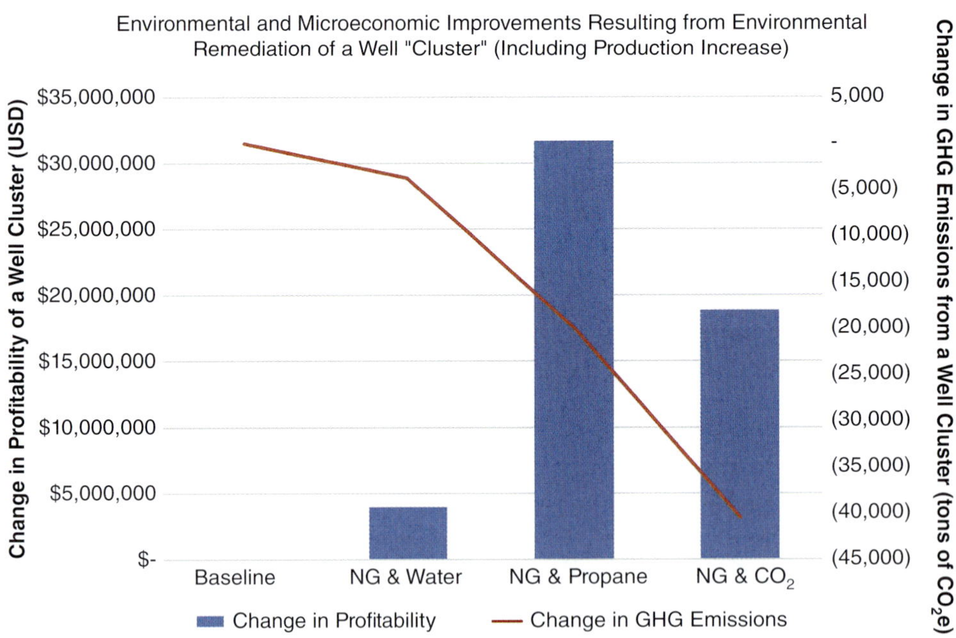

Figure 2.11 Environmental and Microeconomic Impacts of Technology Alternatives (including production impacts).

The improvements in the overall microeconomic footprint of the operation is shown by looking at the overall profitability of the well cluster over its expected life. The increases in profitability resulting from the alternative fracturing fluids far outweigh the cost savings involved. If only the economic view was taken, it can be surmised that the use of natural gas as a fuel coupled with the use of LPG or propane as a fracturing fluid would yield the best solution. This option does not, however, lead to the most effective environmental solution. This economic disparity results from a number of inherent issues, the most significant of which are:

- As more carbon dioxide is sequestered in the operation, a positive environmental outcome, more carbon dioxide must be purchased to compensate for the "loss."
- The cost of carbon dioxide is very high as there is a distinct lack of infrastructure to capture, process, and transport carbon dioxide to the site.
- In field studies, carbon dioxide foam did not increase production by as large a percentage as LPG gel, thereby leading to a higher boost in revenues tied to LPG.

As a result, even if operators were incentivized to invest in a more environmentally favorable solution to gain the long-term economic benefits, they would likely choose the gas-fracturing option as it has the most economically favorable outcome, instead of the CO_2 fracturing, which still provides economic value but has a much higher environmental impact. This case study vividly demonstrates the negative externalities associated with many environmental issues. As a result, it lends itself well to the institution of public policy to eliminate or reduce the dead-weight loss in the market and drive the desired economic and environmental outcomes.

2.8 Opportunities for Future Research

There are a number of opportunities for future research to validate and/or reinforce the conclusions of this chapter; these areas include:

- The energy tax, and the costs, of producing foamed or supercritical CO_2 compared to LPG gel should be evaluated and added to the analysis if a true representation of the environmental footprint is to be established. This chapter assumed that the cost of producing LPG would be lower than that of producing carbon dioxide, thereby increasing the economic distance between the two alternatives.
- For the Barnett Shale, much of the water produced in hydraulic fracturing is formation water; therefore this was excluded from the analysis to make it more comparable to other plays. The production of formation water should be further investigated.
- One of the most critical greenhouse gas issues with hydraulic fracturing is the leakage of methane directly into the atmosphere. Methane is known to have 25 times the greenhouse gas effect as carbon dioxide, and leakage can have a significant impact on the greenhouse gas footprint of the operation. Methane leakage was assumed to be consistent among all scenarios and as a result was excluded from the analysis. A similar analysis looking at the reduction of methane leakage could be conducted to supplement these results.
- In these analyses, the prices of fuel consumed and natural gas produced were held constant throughout the life of the well. Sensitivity analyses taking into account changes in fuel prices and the price of natural gas over time could be conducted and included in the calculations to provide a more accurate estimate of the economics over time.
- The amount of carbon dioxide sequestered during the CO_2 fracturing operation was estimated based on the amount of adsorbed methane expected in the formation. There are additional factors which impact the amount of CO_2 that can be sequestered including the

shut-in time, the amount of moisture in the formation, and potential chemical sequestration that can occur should the CO_2 react with formation elements. Field experiments to verify actual sequestration amounts and flowback curves of carbon dioxide should be conducted to verify those assumptions.

- Much of the literature highlights the use of binary foam (carbon dioxide mixed with another gas such as nitrogen) to offset some of the drawbacks of carbon dioxide. The economic, environmental and safety impacts of the use of binary foam instead of pure CO_2 should be analyzed to determine if they have a significant impact.

References

al-Adwani FA, Langlinais J, Hughes RG. 2008. Modeling of an underbalanced drilling operation utilizing supercritical carbon dioxide. In: Engineers, S. O. P., editor, SPE/IADC Managed Pressure Drilling and Underbalanced Operations Conference and Exhibition. Abu Dhabi, UAE: Society of Petroleum Engineers.

Annevelink MPJA, Meesters JAJ, Hendriks AJ. 2016. Environmental contamination due to shale gas development. Sci Total Environ, 550: 432–438.

Apache. 2013. Apache leads the way to new natural gas trend. Available from: http://www.apachecorp.com/News/Articles/View_Article.aspx?Article.ItemID=2957.

API 2010. Hydraulic Fracturing: Unlocking America's Natural Gas Resources. Freeing Up Energy. Washington D.C.: American Petroleum Institute.

Armendariz A. 2009. Emissions from natural gas production in the Barnett Shale area and opportunities for cost effective improvements. Southern Methodist University.

Arthur JD, Brian B, Jo CB, Mark L. 2008. Evaluating the environmental implications of hydraulic fracturing in shale gas reservoirs. ALL consulting.

Associated-Press. 2013. Fracking industry switching from diesel to natural gas. Pittsburgh Post-Gazette.

Baker MB. 2015. Barnett Shale wells rank high for water usage. Star Telegram, July 6, 2015.

Bleiwas D. 2015. Estimates of hydraulic fracturing sand production, consumption and reserves in the United States. Rock Products. U.S. Geological Survey. Available from: http://www.rockproducts.com/frac-sand/14403-estimates-of-hydraulic-fracturing-frac-sand-production-consumption-and-reserves-in-the-united-states.html#.Vig5YX6rSCc.

Brino A. 2011. Interview with Robert Lestz: Inventor of waterless fracking on why his method will be a game-changer. Available from: https://www.reuters.com/article/idUS243651410820111114.

Browning J, Ikonnikova S, Gulen G, Tinker SW. 2013a. Barnett Shale Production Outlook. Society of Petroleum Engineers.

Browning J, Tinker SW, Ikonnikova S, Gulen G, Potter E, Fu Q, Horvath S, Patzek T, Male F, Fisher W, Roberts F, Medlock III K. 2013b. Barnett study determines fill-field reserves, production forecast. OGJ, September 2.

Burke LH, Nevison GW, Peters WE. 2011. Improved Unconventional gas recovery with energized fracturing fluids: Montney Example. SPE Eastern Regional Meeting. Columbus, OH: Society of Petroleum Engineers.

Cathes LM, Brown L, Taam M, Hunter A. 2012. A commentary on "The greenhouse-gas footprint of natural gas in shale formations" by R.W. Howarth, R. Santoro, and Anthony Ingraffea. Climatic Change, 113: 525–535.

Center For Transport Analysis 2011. Biomass energy Data Book. In: Division, E. A. T. S. (ed.). Available from: http://cta.ornl.gov/bedb/appendix_a/Lower_and_Higher_Heating_Values_of_Gas_Liquid_and_Solid_Fuels.pdf.

Ciferno JP, Munson RK, Murphy JT. 2012. US DOE National Energy Technology Laboratory's post combustion carbon capture. Carbon Management Technology Conference. Orlando, FL: Society of Petroleum Engineers.

City of Fort Worth, Texas. 2015a. 2015 Water rates, Available from: http://fortworthtexas.gov/water/info/default.aspx?id=133048.

City of Fort Worth, Texas. 2015b. Salt Water disposal terms and data. http://fortworthtexas.gov/uploadedFiles/Gas_Wells/SWD_questions.pdf.

Clark CE, Horner RM, Harto CB. 2013. Life cycle Water consumption for shale gas and conventional natural gas. Environ Sci Technol, 47, 11829-11836.

COGA. 2012. Does Hydraulic fracking cause earthquakes? Facts on geo-seismic activity and natural resource devleopment. Colorado Oil and Gas Association. Available from: http://www.coga.org/pdfs_facts/SeismicActivity.pdf. :

Department of Energy 2001. A Dual fuel conversion system for diesel engines. Inventions and innovation. Washington D.C.: U.S. Department of Energy.

Department of Energy 2014. Clean cities alternative fuel price report. energy efficiency and renewable energy. Washington D.C.: U.S. Department of Energy.

Dong, Z. 2012. A New Global Unconventional Natural Gas Resource Assessment. [unpublished dissertation] College Station (TX): Texas A and M University.

Dow N, Zhang J, Duke M, Li J, Gray SR Ostarcevic E. 2008. Membrane distillation of brine wastes. CRC Water Quality and Treatement Research Report. Water Quality Research Australia.

Ecologix. 2013. What the Texas Railroad Commission's new ruling means for the future of frac water management. Ecologix Environ Syst. Available from: http://blog.ecologixsystems.com/what-the-texas-railroad-commissions-new-ruling-means-for-the-future-of-frac-water-management/.

EIA 2010. Annual energy outlook 2010 with projections to 2035. Department of Energy - Energy Information Association.

EIA. 2011. Shale and oil plays, North America. Available from: http://www.eia.gov/pub/oil_gas/natural_gas/analysis_publications/maps/maps.htm.

EIA 2012a. Annual energy outlook 2012 with projections to 2035. Washington, DC: Energy Information Administration.

EIA 2012b. International energy statistics database. November 2012 ed. Washington, DC: U.S. Energy Information Administration.

EIA 2013a. Annual energy outlook 2013 with projections to 2040. Washington, DC: U.S. Energy Information Administration.

EIA. 2013b. Carbon Dioxide Emissions Coefficients. U.S. Energy Information Administration. Available from: http://www.eia.gov/environment/emissions/co2_vol_mass.cfm.

EIA 2014a. Gasoline and diesel fuel update. U.S. Energy Information Administration. Available from: http://www.eia.gov/petroleum/gasdiesel/.

EIA 2014b. Henry Hub natural gas spot price. U.S. Energy Information Administration. Available from: http://www.eia.gov/dnav/ng/hist/rngwhhdm.htm.

EIA 2014c. Natural gas prices. U.S. Energy Information Administration. Available from: http://www.eia.gov/dnav/ng/ng_pri_sum_dcu_nus_m.htm.

EIA 2015a. FAQs: How much carbon dioxide is produced by burning gasoline and diesel fuel. In: Administration, U. S. E. I., ed. U.S. Energy Information Administration. Available from: http://www.eia.gov/tools/faqs/faq.cfm?id=307andt=11. p. 000–000.

EIA. 2015b. How much carbon dioxide is produced by burning gasoline and diesel fuel? U.S. Energy Information Association. Available from: http://www.eia.gov/tools/faqs/faq.cfm?id=307andt=11.

EIA. 2015c. Mont Belvieu, TX Propane spot price FOB. U.S. Energy Information Administration. Available from: http://www.eia.gov/dnav/pet/hist/LeafHandler.ashx?n=PETands=EER_EPLLPA_PF4_Y44MB_DPGandf=D.

EIA 2017a. Annual Energy Outlook 2017 With Projections to 2050. Washington, DC: U.S. Energy Information Administration.

EIA 2017b. Monthly energy review. In: Administration, U. S. E. I., ed. Available from: https://www.eia.gov/energyexplained/?page=us_energy_home.

El-Bourawi MS, Ding Z, Ma R, Khayet M. 2006. A framework for better understanding membrane distillatiion separation process. J Membr Sci, 285, 4–29.

Electric G. 2014. Waukesha gas engines VHP Series Four (L7044GSI)—specifications sheet. General Electric Company. Available from: https://www.ge-distributedpower.com/products/gas-compression-mechanical-drive/maxi-2-000-bhp/vhp-gp.

EPA 2008a. Average annual emissions and fuel consumption fof gasoline-fueled passenger cars and light trucks. United States Environmental Protection Agency.Available from: http://www.epa.gov/otaq/consumer/420f08024.pdf.

EPA 2008b. Average in-use emissions from heavy-duty trucks. U.S. Environmental Protection Agency. Available from: http://www3.epa.gov/otaq/consumer/420f08027.pdf.

EPA 2012. Class II wells - oil and gas related injection wells (class II). U.S. Environmental Protection Agency. Available from: http://water.epa.gov/type/groundwater/uic/class2/.

EPA 2013. Inventory of greenhouse gas emissions and sinks 1990–2011. Washington D.C.: Environmental Protection Agency.

Estrada JM, Bhamidimarri R. 2016. A review of the issues and treatment options for wastewater from shale gas extraction by hydraulic fracturing. Fuel, 182: 292–303.

Fontenot BE, Hildenbrand ZL, Carlton, DDJ, Walton, JL,Schug KA. 2014. Response to Comment on An evaluation of water quality in private drinking water wells near natural gas extraction sites in the Barnett Shale formation. Environ Sci Technol, 48: 3597–3599.

Fontenot BE, Hunt LR, Hildenbrand ZL, et al. 2013. An evaluation of water quality in private drinking water wells near natural gas extraction sites in the Barnett Shale formation. Environ Sci Technol, 47, 10032–10040.

Frohlich C. 2012. Two-year survey comparing earthquake activity and injection-well location in the Barnette Shale, Texas. PNAS, 109 13934–13938.

Gale JFW, Reed RM, Holder J. 2007. Natural fractures in the Barnett Shale and their importance for hydraulic fracture treatments. AAPG Bull, 91 603-622.

Gandossi, L. 2013. An overview of hydraulic fracturing and other formation stimulation technologies for shale production. In: Transport JRC-IFEA, ed. Luxembourg: Publications Office of the European Union.

GHS 2012. Thought of the day: Converting diesel powered drilling rigs to natural gas. Energy: exploration and production. Houston, TX: Global Hunter Securities.

Glover EL, Kremer JC. 2001. MOBILE6 Emissions factors for natural gas vehicles. Washington, DC: Environmental Protection Agency. Available from: http://www.epa.gov/otaq/models/mobile6/r01033.pdf.

Gottschling JC. Openhole vs. Cased-hole hydraulic fracturing. SPE Annual Technical Conference and Exhibition, 2005. Dallas, TX. Society of Petroleum Engineers.

Grattan, R. 2014. Water outreach should be a priority for drillers, report says. Shale, November 18.

Grieder, E. 2014. Shaken and stirred. Texas Monthly, March.

Gupta DVS, Leshchyshyn TT. 2005. CO2 energized hydrocarbon fracturing fluid: history and field application in tight gas wells. SPE Latin America and Caribbean Petroleum Engineering Conference, June 20–23, 2005. Society of Pertoleum Engineers. Rio de Janeiro, Brazil.

Hagstrom EL, Jackanich K. 2011. Hydraulic fracturing litigation is on the rise. Hydraulic Fracturing Digest. Sedgewick L.L.P.

Halliburton 2013. CleanWave Mini-Water Treatment Service. H010503.

He L, Feng W, Jin Z, Siwei M, Yongwei D. 2014. Fracturing with carbon dioxide: application status and development trend. Petroleum Exploration and Development, 41, 513–519.

Heller R, Zoback M. 2014. Adsorption of methance and carbon dioxide on gas shale and pure mineral samples. J Unconven Oil Gas Res, 8, 14–24.

Hill DG, Johnson J, Bell J, Mayer N, Giberson H, Haessler R, Matthews W. 2011. Natural gas as a fuel in drilling operations—analysis, testing and implementation. SPE/IADC Drilling Conference and Exhibition. Amsterdam, The Netherlands: Society of Petroleum Engineers.

Homefacts. 2014. Denton, TX Earthquake report. Available from: http://www.homefacts.com/earthquakes/Texas/Denton-County/Denton.html.

Horwitt D, Sumi L. 2014. Up in Flames: U.S. shale oil boom comes at expense of wasted natural gas, increased CO_2. Earthworks. Available from: http://www.earthworksaction.org/library/detail/up_in_flames#.VXdbmM9VhBc.

Howarth R, Ingraffea A, Engelder T. 2011. Natural gas: Should fracking stop? Nature, 477 (7364): 272–275.

Howarth R, Santoro R, Ingraffea, A. 2010. Methane and the greenhouse-gas footprint of natural gas from shale formations. Climatic Change, 106, 679–690.

Howell J, Harger J. 2013. CNG and LNG: what's best for your fleet? Westport and Clean Energy Webinar. Westport.

Hughs JD. 2014. Drilling deeper: a reality check on U.S. government forecasts for a lasting tight oil and shale gas boom. Post Carbon Institute. Available from: http://www.postcarbon.org/books_and_reports/.

Ishida T, Aoyagi K, Niwa T, et al. 2012. Acoustic emission of hydraulic fracturing laboratory experiment with supercritical and liquid CO_2. Geophy Res Lett, 39.

Jaramillo P, Griffin WM, Matthews, HS. 2007. Comparative life-cycle air emissions of coal, domestic natural gas, LNG and SNG for electricity generation. Environ Sci Technol, 41: 6290–6296.

Jenner S, Lamadrid AJ. 2013. Shale gas vs. coal: policy implications from environmental impact comparisons of shale gas, conventional gas, and coal on air, water, and land in the United States. Energ Policy, 53, 442–453.

Kell S. 2009. Modern shale gas development in the United States: A Primer. Washington, DC: U.S. Department of Energy—Office of Fossil Energy.

Khan C, Amin R, Madden G. 2012a. Economic modelling of CO_2 injeciton for enhanced gas recovery and stroage: a reservoir simulation study of operational parameters. Energ Environ Res, 2: 65–82.

Khan C, Amin R, Madden G. 2012b. Economic modelling of CO_2 injection for enhanced gas recovery and storage: a reservoir simulation study of operational parameters. Energ and Environ Res, 2: 65–82.

Khan C, Amin R, Madden G. 2013. Carbon dioxide injection for enhanced gas recovery and stroage (reservoir simulation). Egyptian J eum, 22, 225–240.

Khuwaja AW, Basit A, Haneef I. waterless fracking - a new way to frack using LPG. Society of Petroleum Engineers. SPE Annual Technical ConferenceIslamabad, Pakistan, November 25–26, 2014.

Lane J. 2014. Liquid CO2, or liquid gold? Maybe both, as Aemetis adds CO2 liquefaction at its Keyes, CA plant. Available from: http://www.biofuelsdigest.com/bdigest/2014/10/27/liquid-co2-or-liquid-gold-maybe-both-as-aemetis-adds-co2-liquefaction-at-its-keyes-ca-plant/.

Lebas RA, Lord P, Luna D, Shahan T. 2013. Development and use of high-TDS recycled produced water for crosslinked-gel-based hydraulic fracturing. Society of Petroleum Engineers. SPE Hydraulic Fracturing Technology Conference. The Woodlands, TX.

Leblanc D, Martel T, Graves D, Tudor EH, Lestz R. 2011. Application of propane (LPG) based hydraulic fracturing in the McCully gas field, New Brunswick, Canada. Society of Petroleum Engineers. SPE North American Unconventional Gas Conference and Exhibition. The Woodlands, TX.

Lestz RS, Wilson L, Taylor, RS, Funkhouser GP, Watkins H, Attaway D. 2007. Liquid petroleum gas fracturing fluids for unconventional gas reservoirs. J Canad Petrol Technol, 46.

Loree DN, Mesher ST. 2007. Liquefied petroleum gas fracturing system. United States patent application.

Mantell ME. 2011a. Produced water reuse and recycling challenges and opportunities across major shale plays. Environmental Protection Agency. EPA Hydraulic Fracturing Study Technical Workshop #4: Water Resources Management. Arlington, VA.

Mantell ME. 2011b. Produced water reuse and recycling challenges and opportunities across major shale plays. U.S. Environmental Protection Agency. EPA Hydraulic Fracturing Study Technical Workshop #4: Water Resources Management. Washington, DC.

Martineau DF. 2007. History of the Newark East field and the Barnett Shale as a gas reservoir. AAPG Bull, 91, 399–403.

Maupin MA, Kenny JF, Hutson SS, Lovelace JK, Barber NL, Linsey KS. 2014. Estimated use of water in the United States in 2010: U.S. Geological Survey Circular. Available from: http://pubs.usgs.gov/circ/1405/.

McHugh T, Molofsky L, Daus A, Connor J. 2014. Comment on "An evaluation of water quality in private drinking water wells near natural gas extraction sites in the Barnett Shale Formation. Environ Sci Technol, 48: 3595–3596.

Mckeon M. 2013. Horizontal fracturing in shale plays. Petroleum Technology Transfer Council: Eastern Workshop. Available from: http://www.thepttc.org/workshops/eastern_062111/eastern_062111_McKeon.pdf.

Mengal SA, Wattenberger RA. 2011. Accounting for adsorbed gas in shale gas reservoirs. Society of Petroleum Engineers. SPE Middle East Oil and Gas Show and Conference. Manama, Bahrain.

Metze T. 2017. Fracking the debate: frame shifts and boundary work in Dutch decision making on shale gas. J Environ Policy Plan, 19, 35–52.

Middleton RS, Carey JW, Currier RP, et al. 2015. Shale gas and non-acqueous fracturing fluids: opportunities and challenges for supercritical CO_2. App Enery, 147, 500–509.

Miller JA, Lawrence AW, Hickey RF, Hayes TD. 1997. Pilot plant treatment of natural gas produced waters to meeting beneficial use discharge requirements. SPE/EPA Exploratrion and Production Environmental Conference. Dallas, TX.

NCTCOG 2012. Development of oil and gas mobile source inventory in the Barnett Shale in the 12-county Dallas–Fort Worth area. Arlington, TX. North Central Texas Council of Governments.

NGI. 2014. Information on the Barnett Shale. Natural Gas Intelligence: NGI's Shale Daily.

Nicholson C. 2012. Earthquakes associated with deep well activities—comments and case histories. American Rock Mechanics Association.

Nicholson C, Wesson RL. 1987. Earthquake hazard associated with deep well injection. U.S. Environmental Protection Agency.

Nicot J-P, Scanlon BR. 2012. Water use for shale gas production in Texas. Environ Sci Technol, 46: 3580–3586.

Nicot J-P, Scanlon BR, Reedy RR, Costley RA. 2014. Source and fate of hydraulic fracturing water in the Barnett Shale: a historical Perspective. Environ Sci Technol, 48, 2464–2471.

Parker KM, Zeng T, Harkness J, Vengosh A, Mitch WA. 2014. Enhanced formation of disinfection byproducts in shale gas wastewater-impacted drinking water supplies. Environ Sci Technol, 48, 11162.11169.

Porter D. 2013. Eagle Ford Shale Task Force report. Railroad Commission of Texas. Available from: http://www.rrc.state.tx.us/media/8051/eagle_ford_task_force_report-0313.pdf.

Railroad Commission of Texas. 2014. Barnett Shale Well count. Available from: http://www.rrc.state.tx.us/.

Reig P, Luo T, Proctor JN. 2014a. Global shale gas development: water availability and business risks. World Resources Institute.

Reig P, Luo T, Proctor JN. 2014b. Global shale gas devleopment: water availability and business risk. World Resources Institute. Available from: http://www.wri.org/publication/global-shale-gas-development-water-availability-business-risks.

Reynolds MM, Ku RYS, Vertz JB, Stashko ZD. 2013. First field application in Canada of carbon dioxide separation for hydraulic fracture flow back operations. Society of Petroleum Engineers. SPE Unconventional Resources Conference. Canada. Calgary, Alberta.

Rogala A, Ksiezniak K, Krzysiek J, Hupka J. 2014. Carbon dioxide sequestration during shale gas recovery. Physiochem Probs Mineral Process: 50, 682–692.

Schaefer K. 2012. Fracking and water: a new way to profit from the industry's biggest problem. OILPRICE.COM. Available from: http://oilprice.com/Energy/Energy-General/Fracking-and-Water-A-New-Way-To-Profit-from-the-Industrys-Biggest-Problem.html.

Seybold E, Goswisk S, King GE, et al. 2013. Evolution of dual fuel pressure pumping for fracturing: methods, economics, field trial results and improvements in availability of fuel. Society of Petroleum Engineers. Annual Technical Conference and Exhibition. New Orleans, LA.

Sieminski A. 2014. Outlook for U.S. shale oil and gas. U.S. Energy Information Administration. Houston, TX: Argus Americas Crude Summit.

Sikdar SK. 2009. Quo vadis energy sustainability? Clean Technol Environ Policy, 11: 367–369.

Skone TJ, Littlefield J, Marriott J. 2011. Lifecycle greenhouse gas inventory of natural gas extraction, delivery and electricity production. National Energy Technology Laboratory.

Stuver S Alonzo, J. 2014. A comparison of air emission estimation methods for drilling rig emissions. Environmentally Friendly Drilling Systerms Program [Online].

Survey, U. S. G. 2015. Increasing rate of earthquakes beginning 2009. United States Geological Survey. Available from: http://earthquake.usgs.gov/research/induced/.

Tagliaferri C, Clift R, Lettieri P, Chapman C. 2017. Shale gas: a life-cycle perspective for UK production. Intl J Life Cycle Assess, 22: 919–937.

Tayari F, Blumsack S, Dilmore R, Mohaghegh SD. 2015a. Techno-economic assessment of industrial CO_2 storage in depleted shale gas reservoirs. J Unconvent Oil Gas Res, 11: 82–94.

Thomas M. 2015. Interview with Danua Miedzinska: new patented technology of CO_2 sequestratoin to unlock Polish shales. Shale Gas International. Available from: http://www.shalegas.international/2015/04/02/interview-new-patented-technology-of-co2-sequestration-to-unlock-polish-shales/.

Thomas M, Pidgeon N, Evensen D, et al. 2017. Public perceptions of hydraulic fracturing for shale gas and oil in the United States and Canada. Climate Change, 8, e450–n/a.

Tudor EH, Nevison GW, Allen S, Pike B. 2009. Case study of a novel hydraulic fracturing method that maximizes effective hydraulic fracture length. Society of Petroleum Engineers. SPE Annual Technical Conference and Exhibition. New Orleans, LA.

Valbuena E, Barrufet MA. Falcone G. Analytical estimation of CO_2 storage capacity in depleated oil and gas reservoirs based on thermodynamic state functions. SPE Latin America and Caribbean Petroleum Engineering Conference, 2012 Mexico City, Mexico.

Van Der Elst NJ, Savage HM, Keranen KM, Abers GA. 2013. Enhanced remote earthquake triggering at fluid-injection sites in the midwstern United States. Science, 341: 164–167.

Van Poollen HK, Hoover DB. 1970. Waste disposal and earthquakes at the Rocky Mointain Arsenal, Derby, Colorado. J Petrol Technol: 983–993.

Wang G. 2011. Advanced vehicles: costs, energy use, and macroeconomic impacts. J Power Sources, 196, 530–540.

Wang Z, Krupnick A. 2013. A retrospective review of shale gas development in the United States: What let to the boom? Resources for the Future. Available from: http://www.rff.org/files/sharepoint/WorkImages/Download/RFF-DP-13-12.pdf:

Westenhaus B. 2012. A New way to fracture oil and gas wells. Oilprice.com. Available from: http://oilprice.com/Energy/Energy-General/A-New-Way-To-Fracture-Oil-and-Gas-Wells.html.

Worth COF. 2015. Salt water disposal terms and data. County of Fort Worth. Available from: http://fortworthtexas.gov/uploadedFiles/Gas_Wells/SWD_questions.pdf.

Zhenyun S, Weidong S, Yanzeng Y, et al. 2015. An experimental study on the CO_2/sand dry-frac process. Nat Gas Indus B, 1, 192–196.

Zhou X, Burbey J. 2014. Fluid effect on hydraulic fracture propagation behavior: a comparison between water and supercritical CO_2-like fluid. Geofluids, 14: 174–188.

Zoback M, Kitasei S, Copithorne B. 2010. Addressing the environmental risks from shale gas development. Worldwatch Institute.

3

Thermodynamic Modeling of Natural Gas and Gas Condensate Mixtures

Epaminondas Voutsas[1], *Nefeli Novak*[1], *Vasiliki Louli*[1], *Georgia Pappa*[1], *Eirini Petropoulou*[1], *Christos Boukouvalas*[1], *Eleni Panteli*[2], *and Stathis Skouras*[2]

[1] *Laboratory of Thermodynamics and Transport Phenomena, School of Chemical Engineering, National Technical University of Athens, Greece*
[2] *Equinor ASA, Research & Technology Center, Trondheim, Norway*

CHAPTER MENU

3.1 Introduction

Knowledge of the physical properties and the thermodynamic behavior of fluids is important in designing, operating, and optimizing any chemical and physical process. Considering the compositional variations and their various applications, it would require infinite time and effort to measure all physical properties and study their phase behavior, while the additional cost of such practice could be unbearable for the oil and gas industry. It is no surprise that from early on scientists and engineers have used empirical or more advanced, first-principle models to predict fluid behavior of diverse petroleum mixtures over a wide range of conditions, based on a limited amount of experimental measurements [1]. The increase in computing power has given the opportunity of making numerous calculations within a small computational timeframe, using theoretical or semi-theoretical approaches to fully simulate a process. The oil and gas industry is one of the driving forces in this field combining the performance of measurements of fluid properties and their modelling with appropriate thermodynamic models.

A reservoir fluid, either oil, gas, or gas-condensate, contains a variety of substances that include mainly hydrocarbons, ranging from methane up to heavier hydrocarbons with several carbon atoms, as well as non-hydrocarbon molecules like water, nitrogen, carbon dioxide, hydrogen sulfide, and trace components such as mercury. Such a reservoir fluid needs to be processed so as to remove these components, which lower the quality of the oil or gas, so as to produce hydrocarbon products with required specifications, either for further transportation or for final sale to the market [2].

Natural Gas Processing from Midstream to Downstream, First Edition.
Edited by Nimir O. Elbashir, Mahmoud M. El-Halwagi, Ioannis G. Economou, and Kenneth R. Hall.

In the Norwegian Continental Shelf for example, an extensive pipeline network system exists to transport partly processed gas from offshore installations to onshore facilities. Offshore processing basically prepares the gas for further transport in the pipeline network and ensures that the gas will reach the onshore facilities in the dense phase. To do so, a cricondenbar pressure specification or a dew point pressure specification at high pressure, set by the network operators, should be fulfilled. In a similar way, onshore processing prepares fully processed gas to be transported in pipelines to the European Continent. The most used dew point specification in the European market is the cricondentherm temperature specification, or a dew point temperature specification at intermediate pressures. Dew point specifications, such as cricondenbar and cricondentherm, are related to the phase diagrams of natural gas and gas condensate mixtures. Accurate knowledge of the phase diagrams of these mixtures is consequently very important to the oil and gas industry, for the accurate design of processing facilities and for optimal operation of transportation pipelines.

Figure 3.1 shows a typical pressure-temperature (PT) phase diagram of a multicomponent hydrocarbon mixture such as a natural gas or a gas condensate mixture. PT diagrams are essentially used to classify reservoirs and to describe the phase behavior of the reservoir fluid. In order to understand the significance of the PT diagrams, it is necessary to define the following key points:

- *Phase envelope* (two-phase region) is the boundary of the region where gas and liquid coexist in equilibrium. Outside this envelope, the fluid exists in single phase (gas, liquid, or dense fluid).
- *Isoquality lines* describe the pressure and temperature conditions for equal percentage of liquid (on a molar or volume basis) produced from a vapor–liquid split.

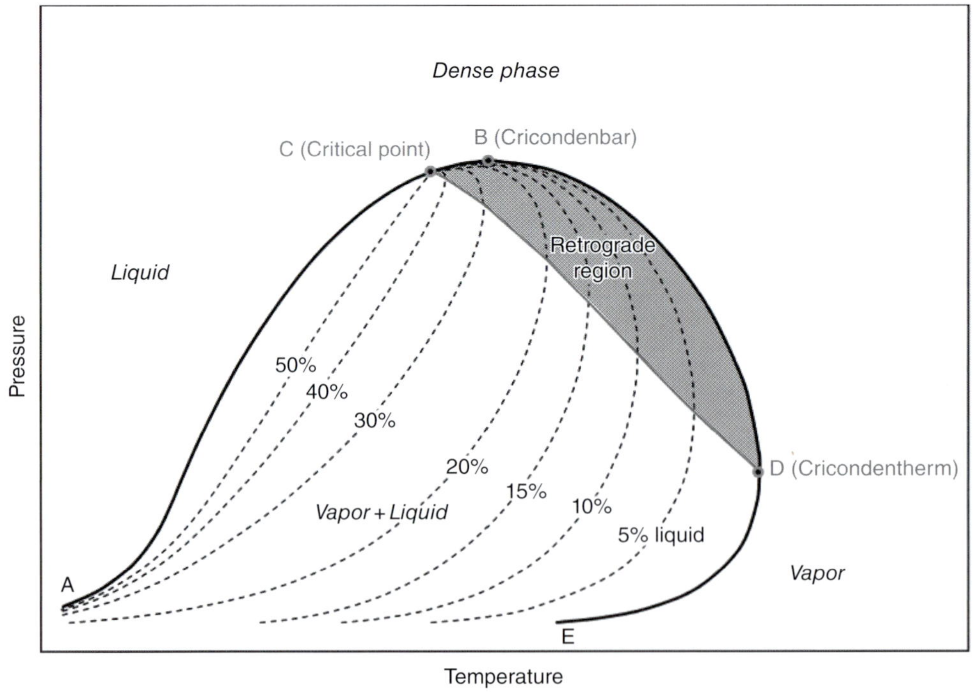

Figure 3.1 Typical pressure-temperature (PT) diagram of a multicomponent system such as natural gas or gas condensate with a specific overall composition.

- *Bubble-point curve* (line AC) is defined as the line separating the single liquid phase region from the two-phase region.
- *Dew-point curve* (line CBDE) is defined as the line separating the single vapor phase region from the two-phase region.
- *Cricondentherm temperature* (cricoT) is defined as the maximum temperature above which liquid cannot be formed regardless of pressure (point D).
- *Cricondenbar pressure* (cricoP) is defined as the maximum pressure above which no gas can be formed regardless of temperature (point B).
- *Critical point* is referred to as the state of pressure and temperature at which all intensive properties of the gas and liquid phases are equal (point C). As shown in Figure 3.1, the bubble-point and the dew point curves as well as all the isoquality lines converge at the critical point.
- *Retrograde region* is the region where liquid is formed as the pressure decreases below dew point pressure. It is called retrograde because gas condenses into a liquid under isothermal conditions instead of expanding or vaporizing when pressure is decreased.

The hydrocarbon dew point (HCDP) is the temperature at a given pressure, or equivalently the pressure at a given temperature, in which hydrocarbons will begin to condense; hence the expression "hydrocarbon liquid dropout". As already mentioned, accurate knowledge of the hydrocarbon dew point is of great importance for the design and operation of the oil and gas industry, both for gas processing and gas transport in the existing pipeline infrastructure. For a gas condensate reservoir, pressure drop caused by depletion leads to condensation when the pressure of the reservoir reaches the dew point, resulting in poorer gas quality and reduced well productivity. Also, during transportation of natural gas, special notice should be taken of the fact that in pipeline systems designed to transport single-phase gas, facilities may not exist to prevent, or accumulate and remove liquid dropout. Therefore, any portion of the gas condensed into liquid may not only cause operational or safety problems, but also result in loss of energy during transportation. Moreover, the liquid that is formed is always enriched to heavy hydrocarbons, those that have the higher calorific value and market price. To avoid hydrocarbon condensation or liquid dropout in gas pipelines, the HCDP needs to be monitored.

HCDP measurements can be conducted either with direct or indirect methods. Direct methods utilize apparatuses based on the chilled mirror approach either automatic or manual, while other methods are also available but they are not in common industrial practice [3–5]. However, due to the limited time for experimental measurements in industrial applications, the HCDP is usually determined by coupling a gas chromatograph (GC) for compositional analysis of the gas with a thermodynamic model able to perform dew point predictions (indirect method). The GC analysis is one of the most important factors that affect dew point predictions [6]. On the other hand, the availability of thermodynamic models that accurately predict the HCDP for a given gas composition is of vital importance.

Traditionally, the gas industry uses equations of state (EoSs), which are analytical expressions that relate pressure, temperature, and molar volume. The most commonly used EoSs are the cubic ones belonging to the van der Waals family of EoSs, such as Soave-Redlich-Kwong (SRK) [7, 8] and Peng-Robinson (PR) [9]. Cubic equations of state are simple models, available in every process simulation tool that makes their usage very direct. Many variations of the PR and SRK EoS have been proposed over the years to improve hydrocarbon phase equilibrium predictions. The most common modification of the cubic EoSs is targeting on pure compound predictions by modifying the temperature dependency of the attractive parameter of the EoS. The underlying reason for these modifications is the improvement of the description of the vapor pressures of the high-molecular-weight hydrocarbons, and the volumetric behavior of the supercritical components, mainly methane and nitrogen, which could lead to more accurate dew point predictions of natural gas mixtures [10–21]. Nasrifar et al. [22] evaluated various cubic EoS variations in dew point predictions of natural gas mixtures. Morch et al. [23], working with an EoS that belongs to the RK family, employed an optimization procedure to fit the coefficients of supercritical methane to both pure component fugacity and experimental dew points in order to achieve a good agreement between experimental dew point data and modelling results for natural gas mixtures. Modifications of the PR EoS have also been proposed for natural gas dew point predictions, as for example by Floter et al. [24], Saffari and Zahedi [16] and Haghtalab et al. [25]. Martinez and Hall [26] also compared a generalized three-parameter Redlich-Kwong/Peng-Robinson (RK-PR) EoS [27] with the three parameter Patel-Teja EoS [28] in dew point predictions.

Apart from pure component properties, an important role in the accurate dew point predictions has been played by the mixing rules employed. When the traditional van der Waals one fluid (vdW1f) mixing rules are used, utilization of binary interaction parameters (BIP) is expected to yield accurate results. This, however, is not always the case as shown by Nasrifar et al. [29] and Louli et al. [30], who demonstrated that the use of BIPs does not necessarily improve the accuracy of the EoS in dew point predictions of natural gases, especially at higher pressures. Apart from the vdw1f mixing rules, advanced mixing rules, like the LCVM (Linear Combination Vidal Michelsen) [31] or the UMR ones (Universal Mixing Rules) [32] coupled with cubic EoS have also been successfully applied to hydrocarbon dew point predictions. One very successful model that has proven to be an attractive alternative to cubic EoS is the Universal Mixing Rules-Peng Robinson UNIFAC (UMR-PRU) [30, 33], which couples the classical PR EoS with UNIFAC by utilizing the UMR mixing rules. UMR-PRU is ideal for industrial applications, as it is a purely predictive tool through the incorporation of UNIFAC and its group contribution approach.

Non-cubic EoSs have also been employed for hydrocarbon dew point predictions, mainly those belonging to the SAFT family [34, 35]. In the latest years, these models have gained attention from both academia and industry, with PC-SAFT being one of the most popular variations of SAFT that has been applied to dew point predictions of hydrocarbon mixtures [34–37].

The aim of this chapter is to present the capabilities of well-established thermodynamic models based on a cubic equation of state, namely PR and UMR-PRU, as well as a more sophisticated equation of state, like PC-SAFT, in the prediction of thermodynamic properties needed for the oil and gas industry, such as dew points and liquid dropouts of natural gas and gas condensate mixtures.

In order to illustrate, from an industrial point of view, the significance of the accurate knowledge of the hydrocarbon dew points such as cricondenbar, a simple example will be discussed. As shown in the work of Louli et al. [30] and Skylogianni et al. [38], the UMR-PRU model predicts the most accurate cricondenbar pressure as compared to the classical PR and SRK EoSs for synthetic and real natural gas mixtures. For an offshore process, a gas specification in terms of cricondenbar needs to be fulfilled. Offshore hydrocarbon dew point processing is accomplished by lowering the pressure of the reservoir fluid in a choke valve, separating gas, oil, and water in a three-phase inlet separator and then cooling and/or expanding the gas stream to remove the heavier hydrocarbons to match the cricondenbar specification. The gas is then recompressed to high pressures to enter a pipeline. Accurate knowledge of the cricondenbar has a direct impact on the design of the hydrocarbon dew point system and energy efficiency of the process. Taking into account the very high amount of gas transported as well as the length of the pipeline networks, a pressure reduction even by a few bars in terms of recompression, can lead to significant savings in energy needs and also in reduction of the CO_2 emissions.

The chapter is organized as follows: in section 3.2, the thermodynamic models considered are briefly presented; section 3.3 presents applications in synthetic and real natural gas mixtures; section 3.4 presents applications in synthetic and real gas condensates, and, finally, in section 3.5, a case study of a natural gas process simulation is presented by applying the considered thermodynamic models.

3.2 Thermodynamic Models

The Peng-Robinson equation of state [9], the UMR-PRU model [32, 39], and the PC-SAFT EoS [40, 41] are evaluated. These models are well documented in the literature, so they are only briefly discussed here.

3.2.1 Peng-Robinson EoS

The Peng-Robinson EoS [9] is described by the following equations:

$$P = \frac{RT}{v-b} - \frac{\mathrm{a}}{(v-b)(v+b)} \tag{3.1}$$

$$\mathrm{a} = \mathrm{a_c}\,\mathrm{a}(T) \tag{3.2}$$

$$\mathrm{a_c} = 0.45724\frac{(\mathrm{RT_c})^2}{\mathrm{P_c}} \tag{3.3}$$

$$\mathrm{a}(T) = [1 + m(1 - T_r^{0.5})]^2 \tag{3.4}$$

$$m = 0.37464 + 1.54226\omega - 0.26992\omega^2 \tag{3.5}$$

$$b = 0.07780\frac{RT_c}{P_c} \tag{3.6}$$

T_c and P_c are the compound critical temperature and pressure, ω is the acentric factor and T_r is the reduced temperature.

For the extension of the PR to mixtures, the classic van der Waals one-fluid mixing and combining rules are commonly used:

$$\mathrm{a} = \sum\sum \mathrm{x_i x_j a_{ij}} \tag{3.7}$$

$$\mathrm{b} = \sum \mathrm{x_i b_i} \tag{3.8}$$

$$\mathrm{a_{ij}} = (\mathrm{a_i a_j})^{1/2}(1 - \mathrm{k_{ij}}) \tag{3.9}$$

where x is the mole fraction, k_{ij} in Eq. 3.9 is a binary adjustable interaction parameter of components i and j, that is determined by fitting binary phase equilibrium data.

For the compounds investigated in this chapter, the critical properties (T_c and P_c) and the acentric factors (ω) are taken from DIPPR [42].

3.2.2 PC-SAFT EoS

One of the most successful attempts to correct the weaknesses of cubic equations of state is the statistical associating fluid theory (SAFT) family of equations of state. SAFT is based on the perturbation theory of Wertheim [43, 44] and was originally developed by Chapman et al. [45].

In this study, the Perturbed-Chain SAFT (PC-SAFT) proposed by Gross and Sadowski [40, 41] is considered, which is one of the most successful modifications of the SAFT theory.

The main difference between original SAFT and PC-SAFT is the reference fluid used. Specifically, PC-SAFT uses the hard-chain reference fluid to account for the dispersion interactions unlike SAFT, which uses the hard-sphere reference fluid. In the PC-SAFT equation, the residual Helmholtz energy a^{res}(T, V, N) is given as the summation of three terms, the hard-chain (a^{hc}), the dispersion (a^{disp}), and the association (a^{assoc}), each one representing contributions from the corresponding kind of intermolecular forces:

$$a^{res}(\mathrm{T,V,N}) = a(\mathrm{T,V,N}) - a^{ideal}(\mathrm{T,V,N}) = a^{hc} + a^{disp} + a^{assoc} \tag{3.10}$$

The hard-chain term (a^{hc}) is given by:

$$\frac{a^{hc}}{RT} = \overline{m}\frac{a^{hs}}{RT} - \sum_i x_i(m_i - 1)lng_{ii}^{hs}(\sigma_{ii}) \tag{3.11}$$

where m_i is the number of segments in a chain of component i, $\overline{m}$ is the mean segment number in the mixture ($\overline{m} = \sum_i x_i m_i$).

The hard-sphere contribution (a^{hs}) is given by:

$$\frac{a^{hs}}{RT} = \frac{1}{\zeta_0}\left[\frac{3\zeta_1\zeta_2}{1-\zeta_3} + \frac{\zeta_2{}^3}{\zeta_3(1-\zeta_3)^2} + \left(\frac{\zeta_2{}^3}{\zeta_3{}^2} - \zeta_0\right) ln(1-\zeta_3)\right] \tag{3.12}$$

$$\zeta_n = \frac{\pi}{6}\rho\sum_i x_i m_i d_i^n \tag{3.13}$$

$$d_i = \sigma_i\left[1 - 0.12exp\left(-3\frac{\varepsilon_i}{kT}\right)\right] \tag{3.14}$$

where σ_i is the segment diameter, and ε_i is the depth of the potential.

The radial distribution function of the hard sphere fluid (g_{ii}^{hs}) is calculated as follows:

$$g_{ii}^{hs} = \frac{1}{1-\zeta_3} + \left(\frac{d_i d_j}{d_i + d_j}\right)\frac{3\zeta_2}{(1-\zeta_3)^2} + \left(\frac{d_i d_j}{d_i + d_j}\right)^2\frac{2\zeta_2{}^2}{(1-\zeta_3)^3} \tag{3.15}$$

The dispersion term is modeled using a second order perturbation theory on chain molecules rather than hard spheres, using the following expression:

$$\frac{a^{disp}}{RT} = -2\pi\rho I_1(n,\overline{m})\overline{m^2\varepsilon\sigma^3} - \pi\rho\overline{m}C_1(n,\overline{m})I_2(n,\overline{m})\overline{m^2\varepsilon^2\sigma^3} \tag{3.16}$$

where the integrals I_1 and I_2 and the compressibility expression C_1 are functions of $\overline{m}$ and the packing fraction n (or ζ_3), while n is the reduced density.

The dispersion term is extended to mixtures by assuming the van der Waals one-fluid theory approximation:

$$\overline{m^2\varepsilon\sigma^3} = \sum_i\sum_j x_i x_j m_i m_j\left(\frac{\varepsilon_{ij}}{kT}\right)\sigma_{ij}^3 \tag{3.17}$$

$$\overline{m^2\varepsilon^2\sigma^3} = \sum_i\sum_j x_i x_j m_i m_j\left(\frac{\varepsilon_{ij}}{kT}\right)^2\sigma_{ij}^3 \tag{3.18}$$

The parameters for a pair of unlike segments are obtained by the conventional Berthelot-Lorentz combining rules:

$$\sigma_{ij} = \frac{1}{2}(\sigma_i + \sigma_j) \tag{3.19}$$

$$\varepsilon_{ij} = \sqrt{\varepsilon_i\varepsilon_j}(1 - k_{ij}) \tag{3.20}$$

The binary interaction parameter, k_{ij}, is introduced to correct the segment-segment interactions of unlike chains. Actually, it is a binary adjustable parameter that is determined by fitting binary phase equilibrium data, similarly to cubic EoS.

The association contribution (a^{assoc}) is calculated from the following expressions:

$$\frac{a^{assoc}}{RT} = \sum_{A=1}^{M}(lnX^A - 0.5X^A) + 0.5M \tag{3.21}$$

where M is the number of association sites per molecule and X^A is the mole fraction of molecules not bonded at site A. In Eq. 3.21, the summation is taken over all association sites of the molecule. The quantity X^A is calculated from the expression:

$$X^A = \left(1 + N_{Av}\sum_{B}\rho X^B \Delta^{AB}\right)^{-1} \tag{3.22}$$

The summation in the above equation is taken over all sites. N_{Av} is Avogadro's number, ρ is the molar density and Δ^{AB} is the association strength that is calculated from:

$$\Delta^{AB} = \frac{1 - 0.5n}{(1-n)^3}\left(exp\frac{\varepsilon^{AB}}{kT} - 1\right)(\sigma^3\kappa^{AB}) \tag{3.23}$$

where A and B are association sites on the molecule.

The extension of the associating term to mixtures is straightforward, and no mixing rules are required; yet, for the cross-associating compounds the following combining rules proposed by Wolbach and Sandler [46] are used:

$$\varepsilon^{A_iB_j} = \frac{1}{2}(\varepsilon^{A_iB_i} + \varepsilon^{A_jB_j}) \tag{3.24}$$

$$\kappa^{A_iB_j} = \sqrt{\kappa^{A_iB_i}\kappa^{A_jB_j}}\left(\frac{\sqrt{\sigma_{ii}\sigma_{jj}}}{\frac{1}{2}(\sigma_{ii} + \sigma_{jj})}\right)^3 \tag{3.25}$$

PC-SAFT requires three parameters for pure non-associating compounds: the segment number (m), the segment diameter (σ), and the segment energy parameter (ε/k). For pure associating compounds, like water, alcohols, etc., two more pure-component parameters are required: the association energy ε^{AB} and the effective associating volume κ^{AB}. These three or five parameters for PC-SAFT are usually determined by simultaneously fitting pure compound vapor pressure and saturated liquid volume experimental data. The pure component parameters for the non-associating components were taken from Ting et al. [35] and Gross and Sadowski [40, 41]. For water and methanol, the two-site association scheme has been adopted for both molecules as originally proposed by Gross and Sadowski using the pure compound parameters given in the pertinent publication [47].

3.2.3 UMR-PRU

The UMR-PRU model [39] belongs to the EoS/G^E class, and couples the PR EoS [15] with UNIFAC [48]. The temperature-dependent part of the attractive term parameter for pure non-polar compounds is calculated from Eqs. 3.4 and 3.5, while for polar compounds by the Mathias-Copeman [14] modification of the PR EoS:

$$a(T) = [1 + c_1(1 - T_r^{0.5}) + c_2(1 - T_r^{0.5})^2 + c_3(1 - T_r^{0.5})^3]^2 \quad \text{for} \quad T \leq T_r \tag{3.26}$$

$$a(T) = [1 + c_1(1 - T_r^{0.5})]^2 \quad \text{for} \quad T > T_r$$

The parameters c_1, c_2, and c_3 in Eq. 3.26 are pure-component specific adjustable parameters, which are determined by fitting pure-component vapor pressure data.

For the extension to mixtures, the following universal mixing rules (UMR) proposed by Voutsas et al. [32] are applied:

$$\frac{a}{bRT} = \frac{1}{-0.53}\frac{G_{AC}^{E,SG} + G_{AC}^{E,res}}{RT} + \sum_i x_i \frac{a_i}{b_i RT} \tag{3.27}$$

$$b = \sum_i \sum_j x_i x_j b_{ij} \text{ with } b_{ij} = \left(\frac{b_i^{1/2} + b_j^{1/2}}{2}\right)^2 \tag{3.28}$$

The Staverman-Guggenheim term of the combinatorial part ($G_{AC}^{E,SG}$) and the residual part ($G_{AC}^{E,res}$) of the excess Gibbs energy (G^E) respectively, are calculated from Eqs. 3.29 – 3.32.

$$\frac{G_{AC}^{E,SG}}{RT} = 5\sum_i x_i q_i \ln\frac{\theta_i}{\varphi_i}, \quad \frac{G_{AC}^{E,res}}{RT} = \sum_i x_i v_k^i (\ln\Gamma_k - \ln\Gamma_k^i) \tag{3.29}$$

$$\ln\Gamma_k = Q_k\left[1 - \ln\left(\sum_m \theta_m \Psi_{mk}\right) - \sum_m \frac{\theta_m \Psi_{mk}}{\sum_n \theta_n \Psi_{nm}}\right] \tag{3.30}$$

$$\text{For compound i:} \quad \varphi_i = \frac{x_i r_i}{\sum_j x_j r_j} \qquad \theta_i = \frac{x_i q_i}{\sum_j x_j q_j}, \tag{3.31}$$

$$\text{For group m:} \quad \theta_m = \frac{X_m Q_m}{\sum_n X_n Q_n}, \ X_m = \frac{\sum_j \nu_m^{(j)} x_j}{\sum_j \sum_n \nu_n^{(j)} x_j}, \tag{3.32}$$

The interaction parameter Ψ_{mk} between groups m and k (Eq. 3.30) is a function of temperature, and it is calculated by Eq. 3.33:

$$\Psi_{mk} = \exp\left[-\frac{A_{mk} + B_{mk}(T - 298.15) + C_{mk}(T - 298.15)^2}{T}\right] \tag{3.33}$$

where A_{mk}, B_{mk} and C_{mk} are binary adjustable interaction parameters, which are determined by fitting binary phase equilibrium data.

The UMR-PRU model has been successfully applied to the prediction of low- and high-pressure vapor-liquid, liquid-liquid, solid-gas, and solid-liquid-gas equilibria as well as heats of mixing, in binary and multicomponent systems of various degrees of non-ideality and asymmetry with respect to the size of the molecules involved, up to those observed in polymer/solvent systems [32, 39, 49]. Furthermore, UMR-PRU has been proven to be a very reliable model for thermodynamic property calculations in natural gas mixtures [30, 33, 38].

The critical properties (T_c and P_c) and the acentric factors (ω) are taken from DIPPR [42] as in PR, while the Mathias-Copeman parameters c_1, c_2 and c_3 for the polar components are taken by Boukouvalas et al. [50].

3.3 Prediction of Natural Gas Dew Points

Natural gas is the lightest of the petroleum mixtures; generally, it has the highest amount of methane and the smallest amount of heavy hydrocarbons, while its composition varies depending on the reservoir. It can be produced either from a natural gas reservoir or as a by-product of

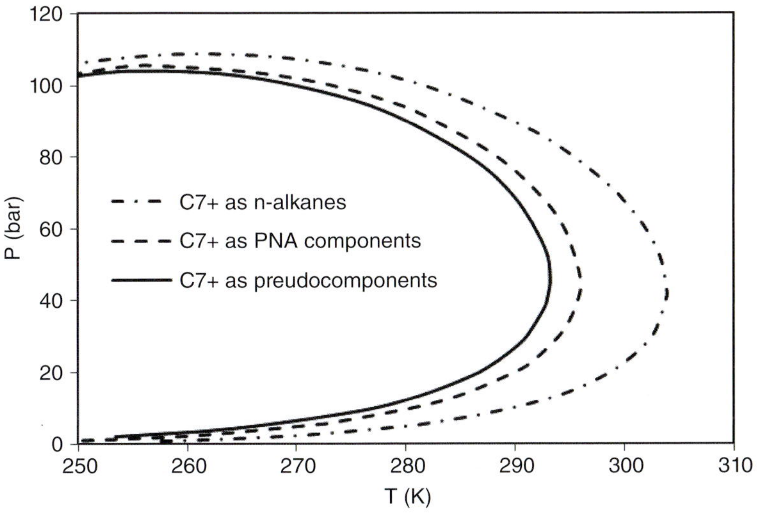

Figure 3.2 Effect of the characterization of the C7+ fraction of the natural gas (Calculations with a cubic EoS coupled with three different characterization schemes).

crude oil production (associated gas). As a petroleum fluid, real natural gas consists of several compounds (hydrocarbons and non-hydrocarbons) that cannot be fully determined by analytical methods. Usually, accurate compositional analysis is available up to normal hexane, while the rest of the components are either grouped into several fractions (i.e., C7, C8, C9 etc.) according to their boiling point and/or carbon number, or grouped in one single plus fraction, usually accompanied by its molecular weight and specific gravity, the latter being the common industrial practice. The plus fraction can be treated continuously through a distribution function, or split to subfractions by various methods. For natural gas, it is common to split an unidentified fraction into paraffinic (P), naphthenic (N) and aromatic (A) subgroups [51]. Since the heavy fraction is a very important factor that influences the dew point curve [52–54], its characterization affects also the dew point predictions obtained by a thermodynamic model and has to be appropriate in order to describe the mixture as accurately as possible. An example is shown in Figure 3.2 that presents dew point predictions calculated by a cubic EoS for a real natural gas, where its C7+ fraction has been characterized with three different schemes: (a) with three n-alkanes; (b) with three PNA components for each of the C7, C8, and C9 subfractions; and (c) with three pseudocomponents. Deviations up to 12°C at the cricondentherm temperature and up to 5 bar at the cricondenbar pressure are obtained with the three different schemes.

3.3.1 Synthetic Natural Gases

The term "synthetic natural gases" (SNGs) refers to mixtures with well-defined components (hydrocarbons and non-hydrocarbons that occur in natural gases) and compositions, and they are used to simulate real natural gas mixtures without the uncertainty of the unidentified heavy components. Thus, they are commonly used for the validation of thermodynamic models.

Louli et al. [33] using a database of 23 SNG mixtures concluded that the UMR-PRU model is a sufficiently accurate model for dew point predictions, especially at pressures near the cricondenbar pressure. This is confirmed by the typical dew point predictions presented in Figure 3.3. The corresponding compositions of the SNGs are given in Table 3.1. For PR and PC-SAFT zero binary interaction parameters (k_{ij}'s) have been used since they have a negligible effect on the

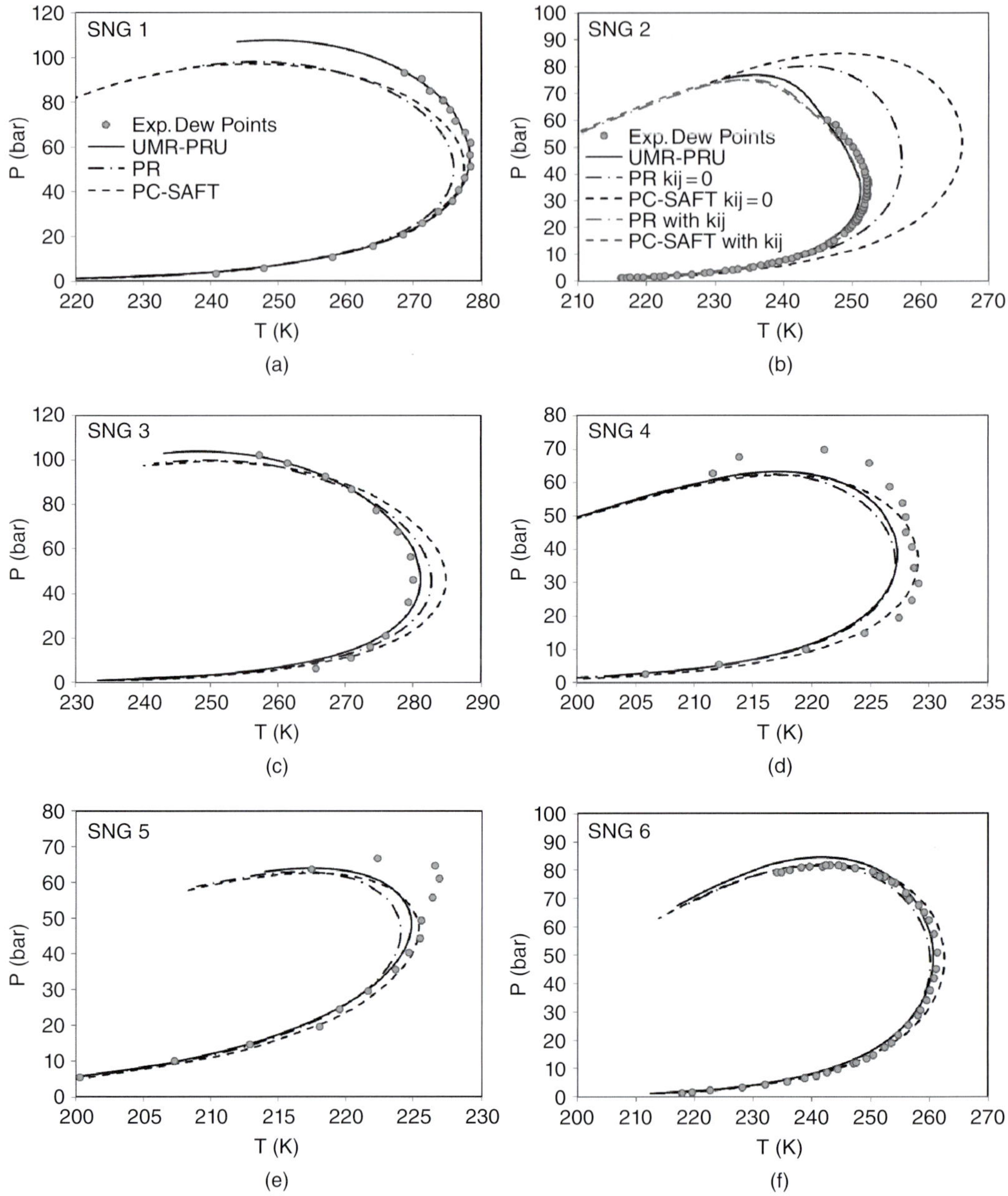

Figure 3.3 Experimental and predicted dew points for the SNGs presented in Table 3.1.

dew point predictions, except from SNGs that contain high amounts of carbon dioxide (CO_2), greater than about 10% molar [22, 33]. The presence of CO_2 increases the solubility of hydrocarbons in the vapor phase and, consequently, reduces the two-phase (vapor–liquid) region. In such cases, the use of k_{ij} significantly improves the accuracy of PR and PC-SAFT, as it is shown in Figure 3.3(b) for SNG2 that contains about 26% molar CO_2.

Natural gas is saturated with water at reservoir conditions, and its presence may lead to corrosive effects, especially in the case of sour gases. Water needs to be removed from the gas to fulfill a water dew point specification. At low temperature and high pressure conditions, such

Table 3.1 Compositions (mol %) of the SNG mixtures.

Components	SNG1 [23]	SNG2 [58]	SNG3 [30]	SNG4 [55]	SNG5 [56]	SNG6 [58]
Nitrogen	–	1.56	–	0.31	0.67	0.77
CO_2	–	25.91	–	0.20	–	1.70
Methane	93.60	69.11	93.12	90.42	89.96	84.45
Ethane	2.63	2.62	3.05	8.04	8.22	8.68
Propane	–	0.42	0.99	0.80	0.90	3.30
i-Butane	1.49	0.11	1.03	0.08	0.11	0.29
n-Butane	1.49	0.10	1.51	0.12	0.13	0.59
i-Pentane	–	0.03	–	0.01	0.01	0.08
n-Pentane	0.80	0.02	–	0.01	0.003	0.09
n-Hexane	–	0.11	–	0.31	–	0.05
n-Heptane	–	–	–	0.20	–	–
Benzene	–	–	0.29	–	–	–

as those occurring in subsea transportation pipelines, hydrates may form that affect significantly the equipment as well as the product quality. Modeling of phase equilibria of natural gas mixtures that contain water is a difficult and challenging task, since such systems show extremely non-ideal behavior due to the hydrogen-bonding interactions developed by the water molecules. The strong hydrogen-bonding interactions result in a limited miscibility of water and hydrocarbons over a broad range of conditions. In order to prevent hydrate formation, a certain amount of hydrate inhibitors, such as methanol or glycols, is injected into the pipeline aiming to shift the hydrate formation curve outside the pipeline operating conditions.

To investigate the influence of water and methanol on the phase equilibrium of natural gases, four of the SNGs presented previously in Table 3.1, namely SNG2, SNG4, SNG5, and SNG6, have been considered. UMR-PRU, PR, and PC-SAFT dew point predictions for SNGs that contain water or both water and methanol are shown in Figure 3.4. For the SNGs that contain only water, all models yield good predictions at the low-pressure region that deteriorate at higher pressures, where associating effects are more pronounced due to the higher densities. PR strongly underestimates the water and methanol effect on the dew point, yielding a more ideal phase behavior than the experimentally observed one. On the other hand, the advanced mixing rules employed in UMR-PRU improve substantially the dew point predictions of PR. Although PC-SAFT includes a specific term that accounts for hydrogen bonding, it presents higher deviations than UMR-PRU. This may be attributed to the implementation of the two-site association scheme considered in the original PC-SAFT for water, while the use of a more physically sound scheme as the four-site model [58] would lead to better results. However, this kind of investigation is outside the scope of this chapter. When methanol is added to the mixture, a different behavior is observed; for the mixtures that contain a small amount of water, PC-SAFT yields the best prediction results, while for the mixtures with higher water content, UMR-PRU dominates, as was the case for the SNG mixtures containing only water.

3.3.2 Real Natural Gases

For real natural gas (RG) mixtures an appropriate characterization method of the heavy end fraction should be implemented. For the RGs considered here, the individual components in

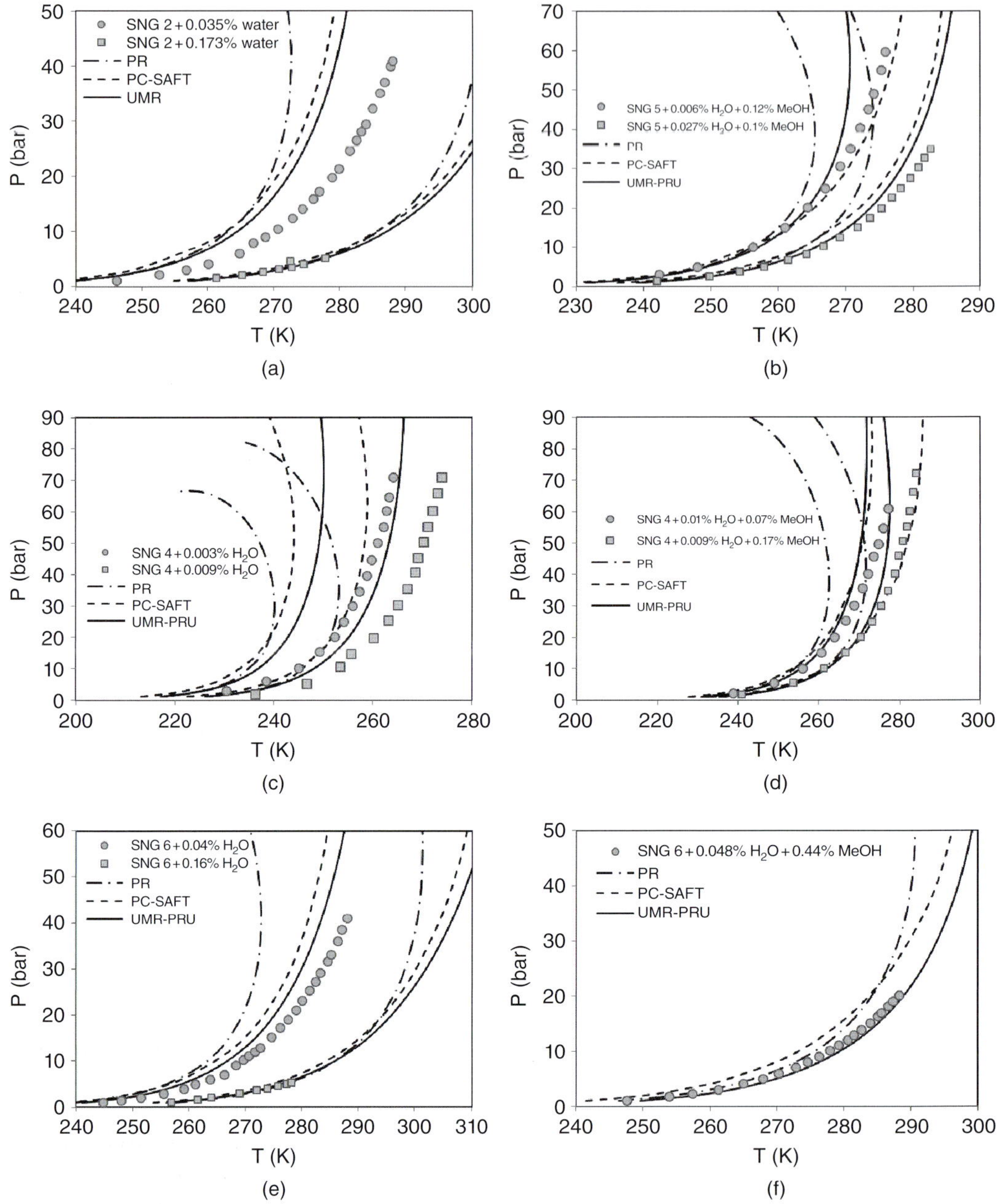

Figure 3.4 Dew point predictions for wet SNG mixtures and SNGs containing water and methanol (the experimental dew point data are taken from [55, 57, 59, 60]).

the C7, C8, and C9 fractions are grouped into paraffinic (P), naphthenic (N) and aromatic (A) subgroups, which is a common practice in the literature [51]. These fractions are acquired by gas chromatography, and the limits between each cut are straight-chain paraffins. The C7 fraction, for example, contains all components with boiling points higher than that of normal hexane and up to that of normal heptane. Therefore, in the C7 fraction all paraffinic compounds are grouped

Table 3.2 Compositions (mol %) of the examined RGs [38].

Components		RG1	RG2	RG3	RG4
Carbon dioxide		1.69920	1.78914	1.67151	2.41947
Nitrogen		0.62462	0.50552	0.62276	0.72766
Methane		88.39031	85.44976	88.16604	83.01782
Ethane		5.34631	7.08992	5.48476	7.79299
Propane		2.22515	2.93022	2.31347	3.79525
i-Butane		0.38658	0.55347	0.36385	0.52174
n-Butane		0.61403	0.84967	0.62520	0.98078
2,2-DM-propane		0.01341	0.01730	0.01129	0.02551
i-Pentane		0.17903	0.23616	0.18616	0.21128
n-Pentane		0.16064	0.21810	0.17430	0.21808
2,2- DM-butane		0.00491	0.00883	0.00519	0.00305
Cyclopentane		0.01480	0.01522	0.01665	0.01310
2,3-DM-butane		0.00865	0.01719	0.00896	0.00626
2-M-pentane		0.03813	0.04735	0.04210	0.04106
3-M-pentane		0.02105	0.02498	0.02291	0.02128
n-Hexane		0.04976	0.05355	0.05531	0.05400
C7	P: n-heptane	0.03293	0.03923	0.03304	0.03244
	N: cyclohexane	0.08713	0.09285	0.09323	0.05629
	A: benzene	0.03558	0.00834	0.03993	0.01629
C8	P: n-octane	0.00495	0.00718	0.00408	0.00566
	N: cycloheptane	0.03560	0.03547	0.03388	0.02422
	A: toluene	0.02286	0.00480	0.02199	0.01023
C9	P: n-nonane	0.00072	0.00164	0.00051	0.00121
	N: cyclooctane	0.00098	0.00242	0.00075	0.00169
	A: m-xylene	0.00267	0.00117	0.00213	0.00210
n-Decane		–	0.00052	–	0.00053

as C7P and are treated as n-heptane, all naphthenic compounds are grouped as C7N and are treated as cyclo-hexane, while all aromatic compounds are grouped as C7A and are treated as benzene. Similarly, C8P is treated as n-octane, C8N as cyclo-heptane, C8A as toluene, C9P as n-nonane, C9N as cyclo-octane, and C9A is treated as m-xylene. Finally, the C10 fraction is treated as n-decane. The compositions of the four real natural gases considered here, along with the PNA characterization for the C7, C8 and C9 subfractions, are given in Table 3.2.

Figure 3.5 presents the models' dew point predictions along with the experimental data. PR and PC-SAFT have been applied with non-zero binary interaction parameters, yet the use of zero binary interaction parameters leads to very similar results [23, 38]. UMR-PRU yields the lowest deviations in dew point predictions, while PC-SAFT the highest ones. UMR-PRU is also capable of predicting accurately the dew points at the higher pressures, and especially the cricondenbar one, yielding superior results over the other models. The models' behavior at the pressure region around the cricondenbar is of particular interest, since natural gas is usually transported in the dense phase region, above cricondenbar pressure.

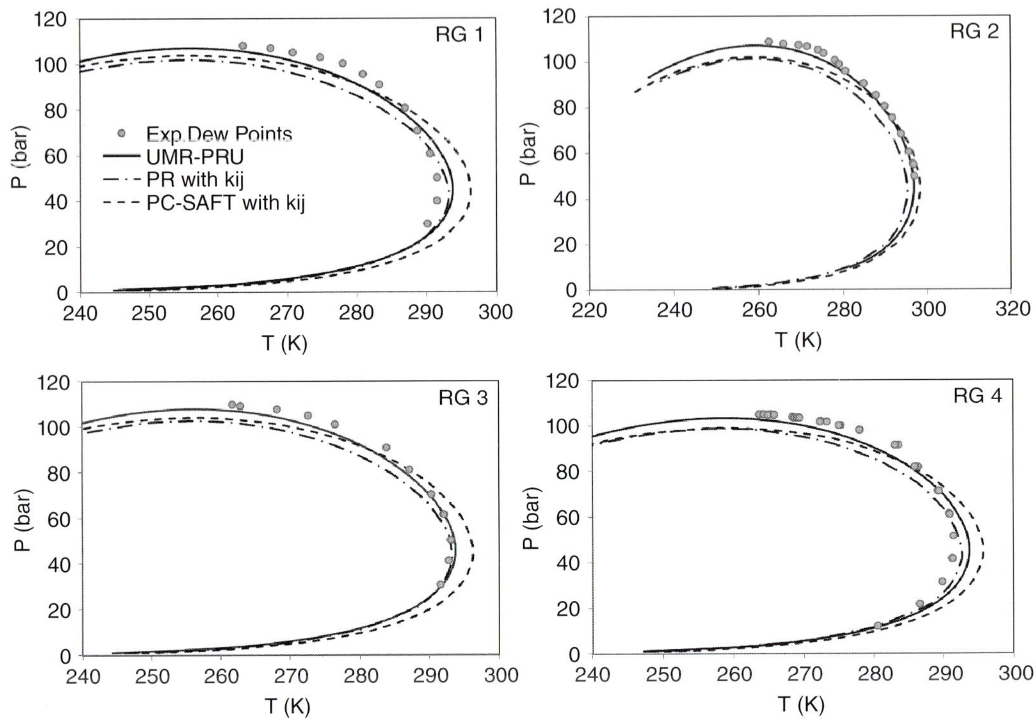

Figure 3.5 Experimental and predicted dew points for the RGs presented in Table 3.2.

3.4 Prediction of Dew Points and Liquid Dropout in Gas Condensates

Gas condensates are usually defined as petroleum mixtures heavier than natural gases. The main difference between natural gas and gas condensate wells is the condensation behavior with pressure drop. Natural gas wells continue to produce gas streams throughout the life span of the reservoir, unlike gas condensates where isothermal pressure drop in the reservoir due to production leads to partial condensation of a liquid phase rich in heavy ends. Similarly to natural gases, gas condensates are composed of known hydrocarbon compounds from C1 to C6 and at least one undefined fraction, containing hydrocarbons heavier than normal hexane (C7+ fraction) [61]. Accurate handling of the C7+ fraction is even more important for gas condensates than for natural gases, since it is found in higher molar fractions and contains much heavier hydrocarbons. Adding to that the inherently inaccurate results of EoS in handling high molecular weight hydrocarbons, the high reservoir pressures (up to 1000 bar), and the high gas/oil ratio of such fluids, the handling of gas condensates is obviously a challenge.

Therefore, the availability of accurate thermodynamic models combined with a proper characterization method of the C7+ fraction, which significantly affects the phase equilibrium simulations [51–54, 62], is of great importance in practice for reservoir and process engineers.

The most representative method to measure the C7+ fraction is the true boiling point analysis that splits the C7+ fraction into narrow boiling point range subfractions and provides information for the true boiling point, molecular weight, and density / specific gravity of each subfraction. The common industrial practice, however, combines experimental information for the C7+ fraction, e.g., specific gravity / density and molecular weight, and empirical methods that simulate the true boiling point analysis and split the C7+ fraction into single carbon number (SCN) fractions. The most popular methods in the oil and gas industry are those of Whitson et al. [63] and Pedersen et al. [64–66]. In order to handle the SCN fractions with an equation of state, properties of these fractions, such as the critical temperature, the critical pressure, etc. are required, for which various correlations are available [51]. Finally, a lumping procedure takes place in order to reduce the number of SCN fractions and facilitate the computational process.

3.4.1 Synthetic Gas Condensates

Synthetic gas condensates, like synthetic natural gases, are mixtures with defined components and compositions that are used to simulate real gas condensates. This way, the problem of the heavy end fraction characterization, which affects significantly the predictions of phase behavior, is avoided in the evaluation of a model's accuracy.

The molar compositions of the SGC mixtures examined here are presented in Table 3.3. Regarding PR and PC-SAFT, the use of binary interaction parameters is necessary due to the presence of heavy compounds that increase the mixture's asymmetry. These parameters were taken either from the literature [35, 37, 67, 68] or by the ASPEN HYSYS databank [69].

Typical dew point predictions are given graphically in Figure 3.6. It is shown that for such asymmetric mixtures, a cubic EoS requires advanced mixing rules such as those utilized by UMR-PRU. PC-SAFT, which explicitly accounts for the chain-length dependences of the interactions, leads also to satisfactory results.

Besides the phase equilibrium curve, the amount of liquid that condenses as the pressure declines is also very important and of major practical interest, especially during gas production

Table 3.3 Molar compositions of SGC mixtures.

Component	SGC 1 [70]	SGC 2 [71]	SGC 3 [72]	SGC 4 [73]	SGC 5 [73]
N_2	–	–	–	–	0.1519
CO_2	–	–	–	–	0.0019
Methane	0.7905	0.7472	0.8205	0.8159	0.6388
Ethane	–	0.1097	–	0.0595	0.0422
Propane	–	–	0.0895	0.0298	0.0308
n-Butane	–	0.0442	–	–	–
i-Pentane	0.0023	–	–	–	–
n-Pentane	0.1999	–	0.0500	0.0339	0.0309
n-Hexane	0.0023	–	–	–	–
n-Heptane	–		–	0.0249	0.0213
n-Octane	–	0.0387	–	–	–
Toluene	–	0.0272	–	0.0187	0.0664
n-Decane	–	–	0.0199	0.0173	0.0159
n-Hexadecane	–	0.0330	0.0201	–	–
n-Tetracosane	0.0020	–	–	–	–
Naphthalene	0.0030	–	–	–	–

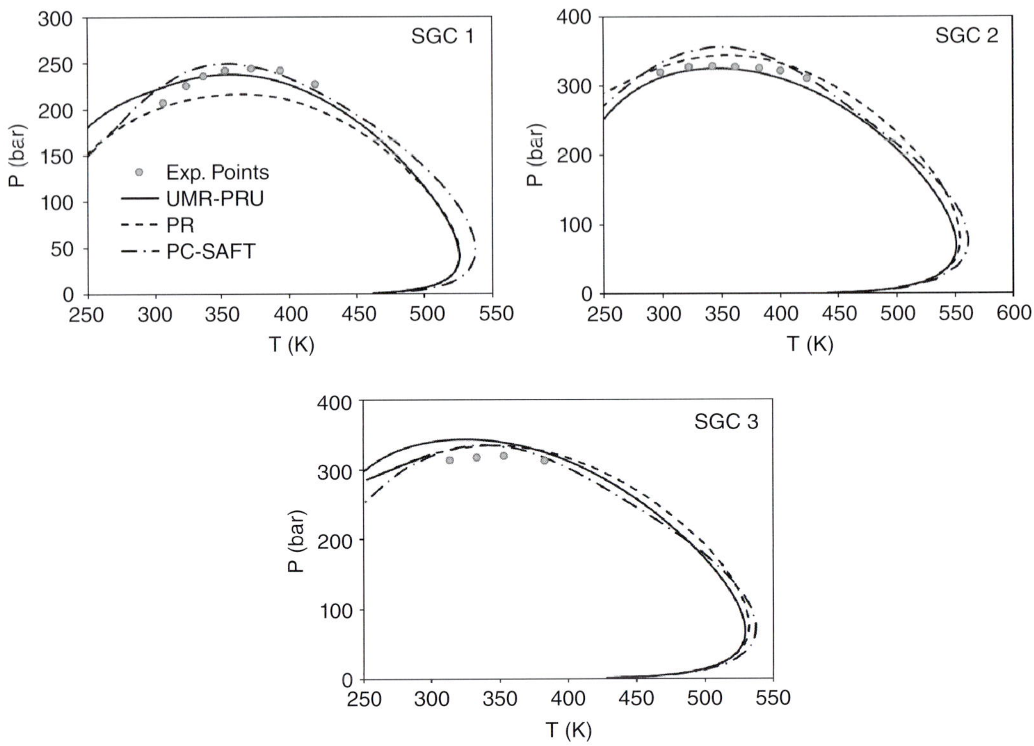

Figure 3.6 Dew point predictions for SGCs 1 to 3.

from the well. This is described by the liquid dropout, which is defined, in a volume basis, by Eq. 3.34.

$$VpL \quad \% = \frac{100 \cdot V_{liq}}{V_{liq} + V_{gas}} \tag{3.34}$$

where V_{liq} is the volume of the liquid phase and V_{gas} the volume of the gas phase. *VpL* includes information about the volumetric behavior of the fluids, and given the limited accuracy of most models in the calculation of liquid densities, *VpL* prediction is a good test for the models' performance.

Indicative liquid dropout predictions are presented in Figure 3.7. All models give similar results at the left part of the liquid dropout curve. However, PR and UMR-PRU predict higher values for the maximum liquid dropout than PC-SAFT, which is in agreement with the experimental data. PC-SAFT on the other hand predicts better than PR and UMR-PRU the right part of the liquid dropout curve, which corresponds to the retrograde region.

Conclusively, the available thermodynamic models can satisfactorily reproduce the behavior of synthetic gas condensates provided that the system asymmetries are taken into account.

3.4.2 Real Gas Condensates

Given its accuracy in SGC mixtures, the UMR-PRU model is further applied to predict the phase behavior of real gas condensate (RGC) mixtures. Similar to the case of real natural gas mixtures, the thermodynamic model has to be coupled with a proper characterization of the

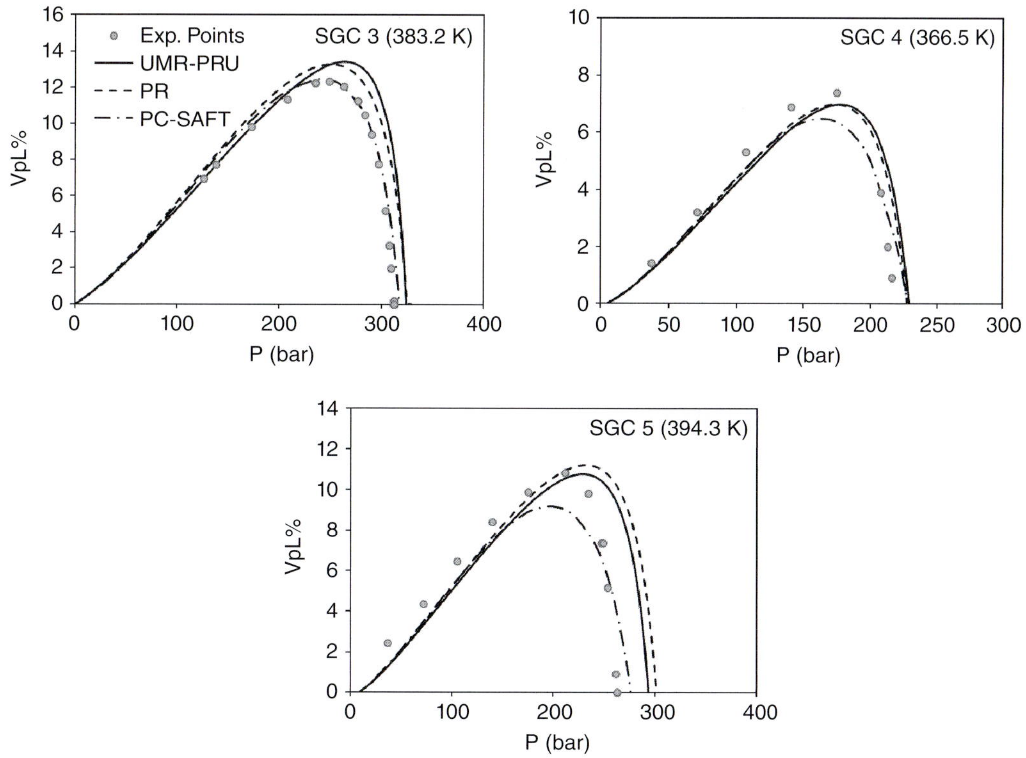

Figure 3.7 Liquid dropout predictions for SGCs 3 to 5 at various temperatures.

undefined heavy fraction of the RGC. The characterization method developed for this purpose is presented in the next section.

The composition and the properties of the RGC systems examined here are given in Table 3.4. They involve a wide range of methane concentrations and C7+ fraction properties: i.e. concentration, molecular weight and density.

3.4.2.1 Characterization of the Plus Fraction

The UMR-PRU model is applied to RGC systems combined with the characterization method proposed by Pedersen et al. [64–66, 78], properly modulated to fulfill the requirement of the model for pseudocomponents of known chemical structure.

The required input information is the total composition, the molecular weight and the density of the plus fraction. The basic steps of the characterization procedure are **splitting** and **lumping**, while for the case of UMR-PRU a third step is also employed, referred to as **assign UNIFAC structure**.

More specifically, the procedure is the following:

First Step (Splitting). The plus fraction is split into a number of single carbon number (SCN) pseudocomponents, from C7 up to maximum C80. In this procedure, components with molar fractions smaller than 10^{-8} are ignored.

The SCN pseudocomponents are obtained by solving Eqs 3.35 to 3.40.

$$C_{SCN} = A + B \cdot \ln z_{SCN} \tag{3.35}$$

Table 3.4 Molar compositions of RGC systems.

Component	RGC 1 [74]	RGC 2 [75]	RGC 3 [76]	RGC 4 [77]
CO_2	0.0060	0.0217	0.0057	0.0244
N_2	0.0556	0.0034	0.0037	0.0008
Methane	0.8519	0.7064	0.5939	0.8210
Ethane	0.0357	0.1076	0.1378	0.0578
Propane	0.0154	0.0494	0.0758	0.0287
i-Butane	0.0037	–	0.0079	0.0056
n-Butane	0.0057	0.0302	0.0328	0.0123
i-Pentane	0.0027	–	0.0084	0.0052
n-Pentane	0.0023	0.0135	0.0131	0.0060
n-Hexane	–	0.0090	0.0180	0.0072
C7+	0.0210	0.0588	0.1029	0.0310
MW(C+) (g/mol)	129.81	153.00	177.10	132.00
Density 60°F (kg/m^3)	772.78	809.27	800.20	773.38

$$\rho_{SCN} = C + D \cdot \ln C_{SCN} \tag{3.36}$$

$$MW_{SCN} = 14^{*}C_{SCN}\text{-}4 \tag{3.37}$$

$$Z_{C+} = \sum_{SCN=7}^{80} z_{SCN} \tag{3.38}$$

$$MW_{C+} = \frac{\sum_{SCN=7}^{80} z_{SCN} \cdot MW_{SCN}}{\sum_{SCN=7}^{80} z_{SCN}} \tag{3.39}$$

$$\rho_{C+} = \frac{\sum_{SCN=7}^{80} z_{SCN}\ MW_{SCN}}{\sum_{SCN=7}^{80} \frac{z_{SCN}\ MW_{SCN}}{\rho_{SCN}}} \tag{3.40}$$

where z is the mole fraction, C the carbon number, ρ the density and MW the molecular weight of each SCN fraction. The constants A and B in Eq. 3.35 are determined by solving Eqs. 3.37 to 3.38, while the constants C and D in Eq. 3.36 are determined by solving Eq. 3.40 and assuming that the density of the defined C6 fraction is equal to 685 kg/m^3 [79].

The critical properties and acentric factor of each SCN fraction are calculated using the equations developed by Retzekas et al. [80] and Ambrose-Walton [81] respectively.

Second Step (Lumping). The lumping procedure, i.e. the grouping of SCN fractions, is necessary to reduce the number of pseudocomponents and facilitate the calculations with any thermodynamic model for real-time applications. Pedersen et al. [66] have recommended a weight-based grouping where each lumped pseudocomponent contains approximately the same weight amount.

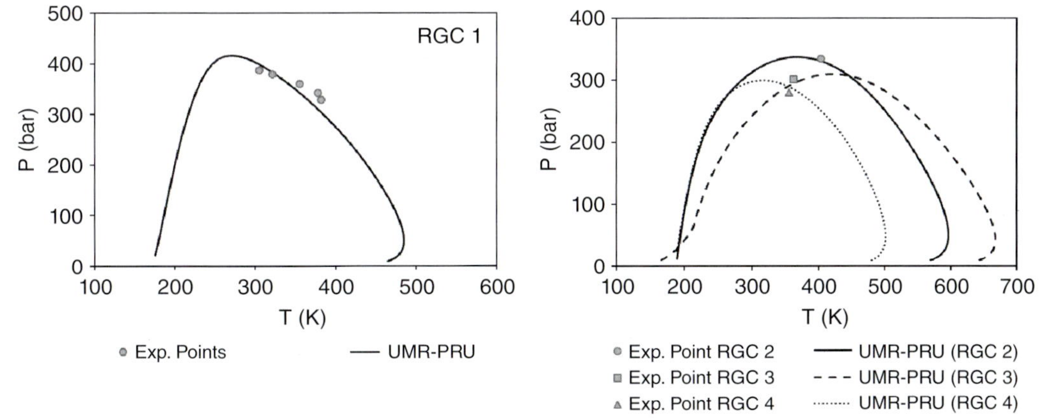

Figure 3.8 Dew point predictions for RGC mixtures using UMR-PRU model coupled with the adapted Pedersen characterization method and a lumping scheme of 12 pseudocomponents.

3rd Step (Assigning UNIFAC structure). In order to combine the UMR-PRU model with this characterization method, pseudocomponents of known chemical structure, as already mentioned, are required due to the incorporation of UNIFAC. To this purpose, each lumped pseudocomponent is assigned to the normal alkane with the closest MW in terms of chemical structure, while the properties for each lump are calculated as a molar mean average of the SCN fraction properties.

The number of pseudocomponents also affects the description of the RGC behavior. However, it has been shown by various researchers, e.g. Elsharkawy [82] and Nasrifar et al. [29], that a scheme of 10–12 pseudocomponents is sufficient for such systems. Thus, a scheme of 12 pseudocomponents is adopted in this case.

3.4.2.2 Dew Point Predictions

The performance of the UMR-PRU model coupled with the aforementioned characterization method, using 12 pseudocomponents, is illustrated in Figure 3.8, in which dew point predictions are plotted along with the experimental data. It should be noted that experimental dew point data for RGCs are very scarce, and in most cases only one saturation point is reported. As shown UMR-PRU is successfully extended to RGC systems leading to dew point predictions that are in good agreement with the experimental data.

3.5 Case Study: Simulation of a Topside Offshore Process

Process simulators are fundamental tools for the design, development, analysis, and optimization of industrial processes, as for example for natural gas processing. Process simulators can be employed in order to understand how inlet gas composition and process operating variables affect the quality and quantity of the final products as well as to estimate the energy requirements. For midstream natural gas processing, process simulations are useful tools for flow assurance, i.e. to ensure the safe and economical flow of the natural gas stream by avoiding two phase formation in one-phase pipelines, to avoid hydrate and scale formation by simultaneously satisfying product quality specifications. Comparison of simulation results to real process data is always necessary to validate the suitability of various property models for certain applications or to provide the possibility of tuning the model parameters if deemed necessary.

One of the main strengths of modern commercial process simulators is their flexibility and ability to make different kinds of thermodynamic calculations. Since 2001, the CO-LaN (CAPE-OPEN Laboratories Network) organization has promoted the creation of the CAPE-OPEN standards [83], where a set of free-to-use non-proprietary rules and interfaces are defined, that allow CAPE (Computer-Aided Process Engineering) applications or components to interoperate. Currently, most process simulators are somehow compliant with this standard.

Here, Honeywell's UniSim® software [84] is employed to simulate a typical platform operation encountered in midstream natural gas processing that treats a gas condensate well stream for the production of rich gas and stabilised condensate. Two thermodynamic models are compared: the commonly used in gas industry PR EoS, which is incorporated in UniSim as a standard fluid package, and the UMR-PRU model, which is implemented in UniSim through the CAPE-OPEN standard protocol.

The simplified process flow diagram is illustrated in Figure 3.9. It involves a number of separations aiming to remove the heavier compounds from the natural gas in order to achieve the required quality for the export gas and condensate. In this topside process, the well stream enters the first separator at 87 bar, aiming to split the stream to a heavy condensate and a light gas fraction. The condensate is further stabilized in two separation stages so as to remove the volatile light hydrocarbons, in order to meet a vapor pressure specification, usually defined as a true vapor pressure (TVP) or Reid vapor pressure (RVP). The flash gas from each of the condensate stabilisation stages will be then recompressed in three stages to meet the pressure and temperature of the main gas stream from the first separator. The resulting gas stream is then cooled down by sea water to meet the hydrocarbon dew point specification. The export gas product is finally compressed at 210 bar.

In gas processing facilities aiming to create pipeline-quality natural gas, other steps are additionally used, which have been omitted in this case study for simplification purposes. For example, a dehydration process is needed to eliminate water from the produced natural gas, since water may cause the formation of hydrates. Dehydration is accomplished by several methods. Among these the most common is the use of a glycol dehydration unit, e.g. with triethylene glycol (TEG). Since a dehydration process has not been involved, no water in the well fluid has been considered, and the first-stage separator that in the real process is a three-phase separator has been replaced with a two-phase separator. Also, if necessary, removal of contaminates may be involved that includes the elimination of acid gases, mainly hydrogen sulphide and carbon dioxide. The most commonly used technique is absorption in amine solutions, such as monoethanolamine (MEA).

A typical dry feed gas was employed as a case study fluid. The heavy end fraction of the fluid (C7+ fraction) has been characterized with six pseudocomponents. The composition of the feed gas is given in Table 3.5.

In the process simulation case study, the PR EoS and the UMR-PRU models are compared with respect to the product flow rates and the distribution of the compounds in the product streams, as well as the energy consumption throughout the process.

UMR-PRU and PR predict similar production rates for the export gas (Table 3.6). Figure 3.10 shows the recovery of the compounds in the export gas. Similarly to methane, acid gases (CO_2 and H_2S), as well as ethane and propane, are practically recovered in the export gas. On the other hand, only about 3% of the heavy end fraction goes to the export gas, while the intermediate hydrocarbons are distributed in both products. Figure 3.11 presents the export gas and condensate compositions predicted with UMR-PRU and PR, and shows that PR predicts

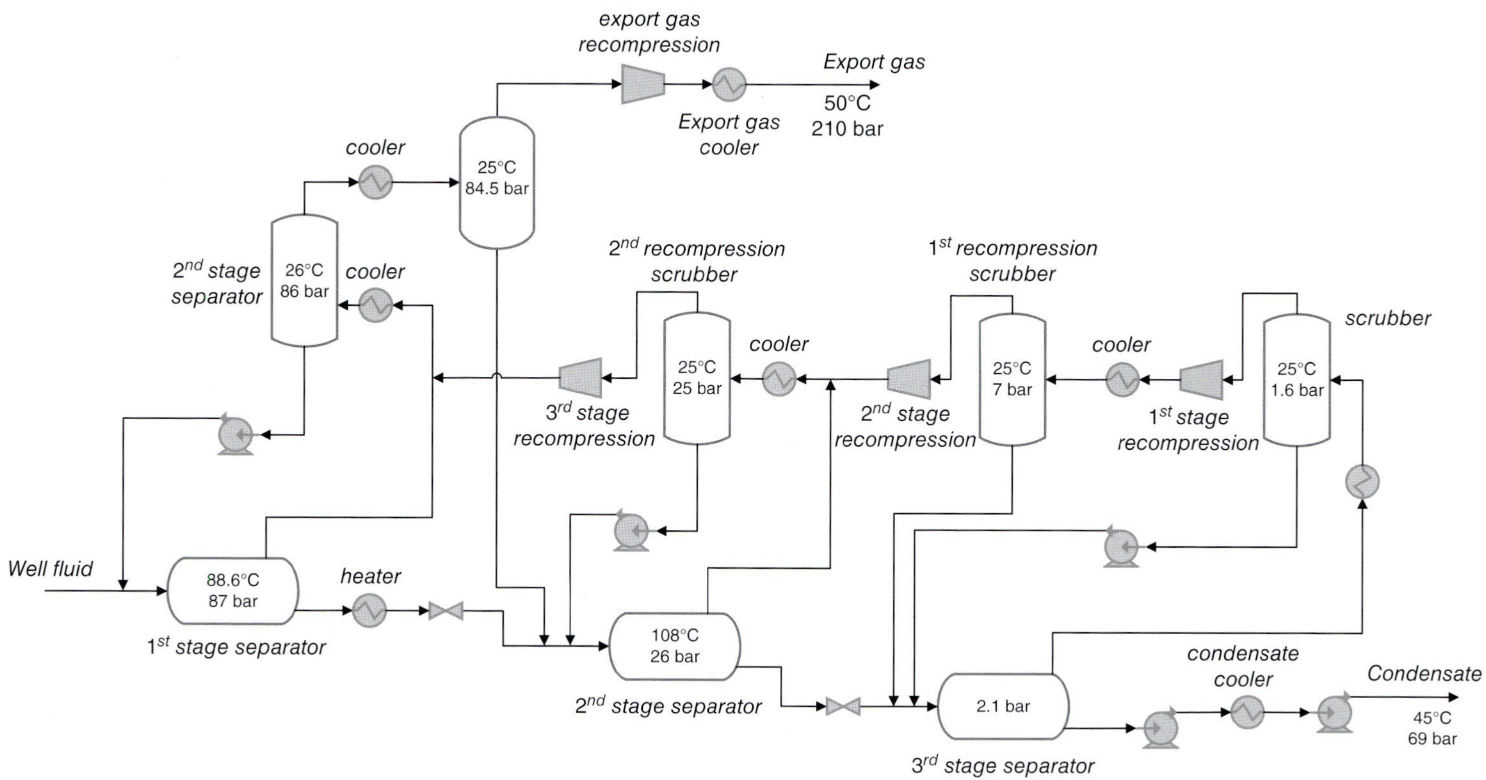

Figure 3.9 Simplified process flow diagram.

Table 3.5 Composition of the feed fluid.

Compound	Concentration (% mol)
N_2	0.26
CO_2	3.16
H_2S	0.0023
C1	73.98
C2	8.24
C3	4.00
iC4	0.78
nC4	1.41
iC5	0.51
nC5	0.51
nC6	0.76
C7+	6.39
Pseudo-components	
nC7*	1.409
nC8*	1.534
nC9*	1.023
nC11*	0.239
nC18*	1.670
nC30*	0.511

Table 3.6 Feed and products flow rates (kmol/hr).

Stream	PR	UMR-PRU
Feed	10000	10000
Export gas	9213	9199
Condensate	787	801

slightly higher compositions of the compounds that are heavier than C4. Figure 3.12 shows the enrichment of the export gas in lighter compounds compared to the feed, as predicted with the UMR-PRU model, while Figure 3.13 illustrates the corresponding effect in terms of the phase envelopes of the products compared to the feed fluid. The process results to a reduction in the cricondenbar pressure from about 375 to lower than 110 bar and a corresponding reduction in the cricondentherm temperature from about 335 to lower than 35°C.

Figures 3.14 and 3.15 present the phase envelopes of the export gas and export condensate streams respectively. Although the two models predict similar composition of the export gas (Figure 3.11), UMR-PRU predicts a higher cricondenbar pressure than PR, which as discussed in previous sections is closer to what is experimentally observed. For the very high amount of gas transported and length of the pipeline networks, better cricondenbar pressure

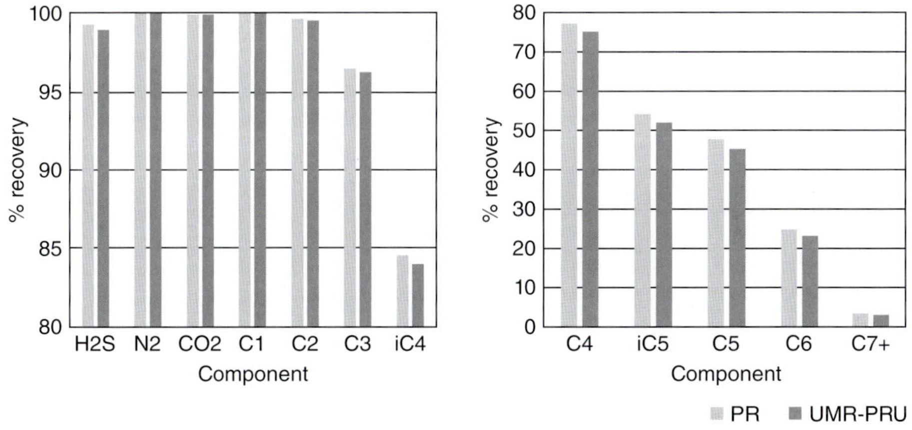

Figure 3.10 Component recovery in the export gas as predicted with UMR-PRU and PR.

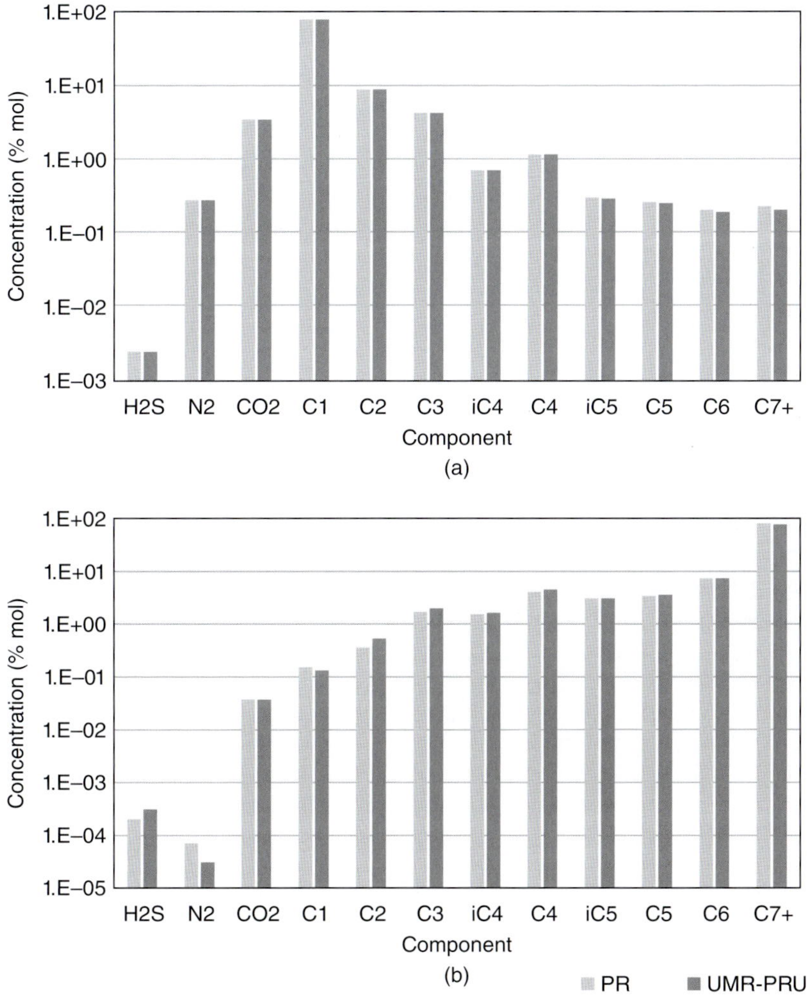

Figure 3.11 (a) Export gas and (b) condensate compositions predicted with UMR-PRU and PR.

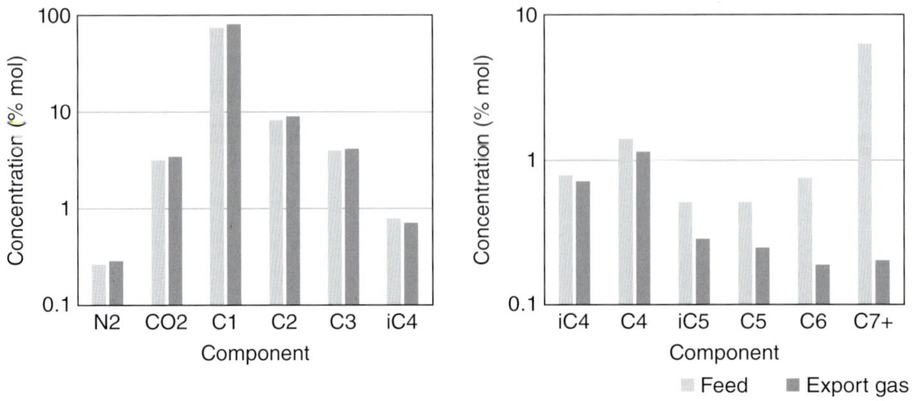

Figure 3.12 Export gas composition compared to the feed fluid predicted with the UMR-PRU model.

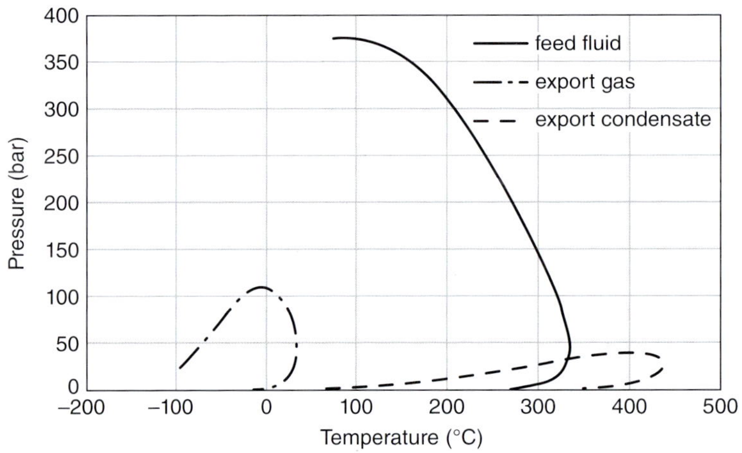

Figure 3.13 Phase envelopes of the process feed fluid and products predicted with the UMR-PRU model.

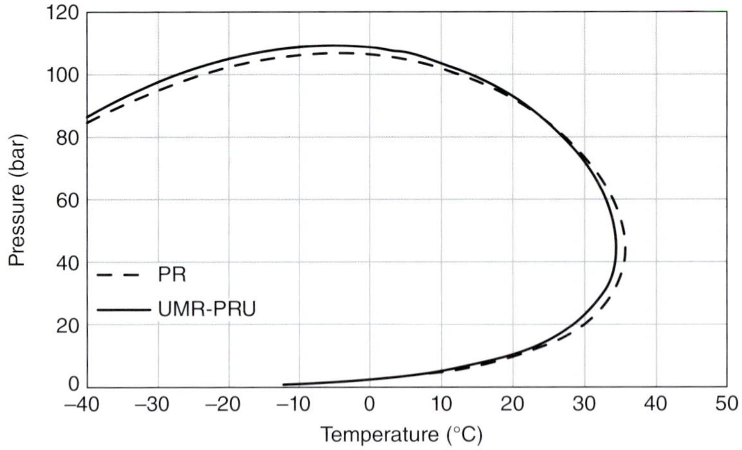

Figure 3.14 Export gas phase envelope obtained with UMR-PRU and PR.

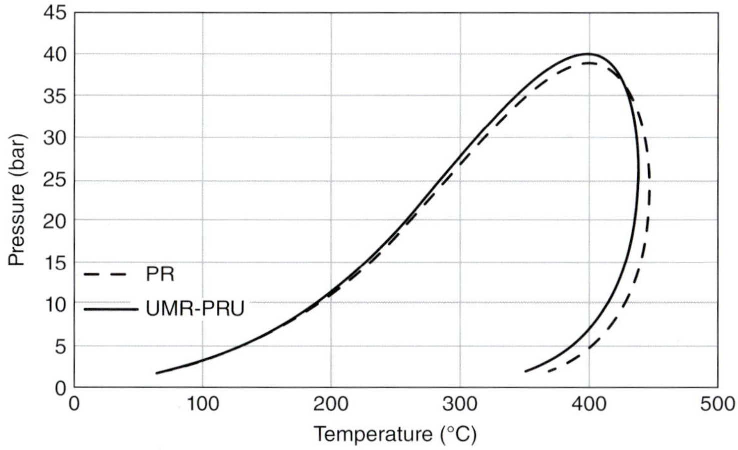

Figure 3.15 Condensate phase envelope obtained with UMR-PRU and PR.

Table 3.7 Energy requirements in the process.

Process	Power consumption (kW)	
	PR	UMR-PRU
Compression	7942	7924
Heating	2806	2908
Pumping	427	470

monitoring, even by a few bars, can lead to significant savings in energy needs for compression and also reduction of the CO_2 emissions. On the other hand, PR predicts higher cricondentherm temperatures than UMR-PRU for the export gas, due to the higher concentrations of C7+ components that it yields.

Table 3.7 compares the energy requirements of the process as predicted by UMR-PRU and PR. As shown, most of the energy is consumed at the recompression stages. Furthermore, a cooling duty of 25.1 MW with UMR-PRU and 25.4 MW with PR is calculated. Cooling is usually achieved using sea water and therefore the real cost for cooling is reduced to the cost for pumping the necessary sea water.

3.6 Concluding Remarks

The thermodynamic properties of petroleum fluids have been successfully described for many years with several empirical, semi-empirical, and more sophisticated models. However, a continuous effort for improving this modeling is necessary because of the more stringent requirements of the processes, as well as the introduction of new processes that require more accurate data.

This chapter has investigated the application of PR, UMR-PRU, and PC-SAFT in the prediction of very important thermodynamic properties of natural gas and gas condensate mixtures, such as dew points and liquid dropouts. Accurate knowledge of these mixture properties is significant to the oil and gas industry, for the optimal gas processing and the operation of the transportation pipelines.

For dry natural gas mixtures, the UMR-PRU model is shown to be a sufficiently accurate model, yielding better dew point predictions than PR and PC-SAFT, especially at the higher pressures and the cricondenbar pressure. For natural gases containing water and methanol, UMR-PRU gives fairly good dew point predictions, which are better than those obtained by classical Peng-Robinson, while PC-SAFT, although it comprises a specific term that accounts for hydrogen bonding interactions, presents overall higher deviations than UMR-PRU. For synthetic gas condensates, UMR-PRU and PC-SAFT give better dew point predictions than PR. For liquid dropout predictions in synthetic gas condensates, PR and UMR-PRU predict overall better than PC-SAFT the left part of the liquid dropout curve up to the maximum liquid dropout, while PC-SAFT shows better performance than PR and UMR-PRU at the right part of the curve which corresponds to the retrograde region. For real gas condensates, UMR-PRU combined with an adapted characterization method proposed by Pedersen et al. [64–66] yields very satisfactory dew point predictions. Moreover, the UMR-PRU model has been successfully implemented in commercial simulation tools through the CAPE-OPEN standards. The model has been applied for the simulation of a topside offshore operation that process a well-fluid to produce rich gas and stabilized condensate and has been compared with PR. UMR-PRU yields generally similar simulation results with the PR EoS, that is usually used for routine simulations by the engineers in the gas industry. However, UMR-PRU predicts higher cricondenbar pressure of the export gas than PR, which is in agreement with real natural gas dew point experiments. Also, UMR-PRU predicts slightly lower compositions of the heavier components in the export gas and condensate than PR, which results in lower predicted cricondentherm temperatures.

Apart from their predictive accuracy, computational times for models of different complexity are also of interest. Thus, in order to compute vapor-liquid equilibrium points, PC-SAFT requires approximately 10 times more CPU time than PR with conventional van der Waals one-fluid mixing rules, due to the need for molar volume solver. On the other hand, the UMR-PRU model that involves explicit mixing rules for both attractive and co-volume parameters and is cubic in terms of molar volume, requires only about 30% more time than PR.

As a closing remark, the UMR-PRU model has been proven to be a sufficiently accurate model for thermodynamic property predictions of natural gas and gas condensate mixtures, and it is an attractive alternative to the classical cubic equations of state traditionally used for natural gas process simulations.

References

1 Pedersen KS, Fredenslund A, Thomassen P. Properties of oils and natural gases. Houston: Gulf Pub Co; 1989.

2 Pedersen KS, Christensen PL, Shaikh JA. Phase behavior of petroleum reservoir fluids. Boca Raton, FL: CRC press; 2012.

3 Panneman H, ed. A traceable calibration procedure for hydrocarbon dew point meters. 23rd World Gas Conference, Amsterdam Gasunie Engineering & Technology; 2006, 5–9 June 2006.

4 Rahman S, Barrufet MA. A new technique for simultaneous measurement of PVT and phase equilibria properties of fluids at high temperatures and pressures. J Petrol Sci Eng. 1995;14(1–2):25–34.

5 Zhou J, Patil P, Ejaz S, Atilhan M, Holste JC, Hall KR. (p, V m, T) and phase equilibrium measurements for a natural gas-like mixture using an automated isochoric apparatus. J Chem Thermodynam. 2006;38(11):1489–1494.

6 Brown AS, Milton MJ, Vargha GM, et al. Measurement of the hydrocarbon dew point of real and synthetic natural gas mixtures by direct and indirect methods. Energ Fuels. 2009;23(3):1640–1650.

7 Redlich O, Kwong J. On the thermodynamics of solutions. V. An equation of state. fugacities of gaseous solutions. Chem Rev. 1949;44(1):233–244.

8 Soave G. Equilibrium constants from a modified Redlich-Kwong equation of state. Chem Eng Sci. 1972;27(6):1197–203.

9 Peng D-Y, Robinson DB. A new two-constant equation of state. Ind Eng Chem Fund. 1976;15(1):59–64.

10 Boston J, Mathias P, eds. Phase equilibria in a third-generation process simulator. Proceedings of the 2nd International Conference on Phase Equilibria and Fluid Properties in the Chemical Process Industries, West Berlin; 17–21 March 1980

11 Gasem K, Gao W, Pan Z, Robinson R. A modified temperature dependence for the Peng-Robinson equation of state. Fluid Phase Equilib. 2001;181(1):113–125.

12 Graboski MS, Daubert TE. A modified Soave equation of state for phase equilibrium calculations. 1. Hydrocarbon systems. Ind & Eng Chem Process Design Develop. 1978;17(4):443–8.

13 Magoulas K, Tassios D. Thermophysical properties of n-alkanes from C1 to C20 and their prediction for higher ones. Fluid Phase Equilib. 1990;56:119–40.

14 Mathias PM, Copeman TW. Extension of the Peng-Robinson equation of state to complex mixtures: evaluation of the various forms of the local composition concept. Fluid Phase Equilib. 1983;13:91–108.

15 Robinson DB, Peng D-Y. The characterization of the heptanes and heavier fractions for the GPA Peng-Robinson programs. Tulsa, OK: Gas Processors Association; 1978.

16 Saffari H, Zahedi A. A new alpha-function for the Peng-Robinson equation of state: application to natural gas. Chin J Chem Eng. 2013;21(10):1155–1161.

17 Soave G. Improving the treatment of heavy hydrocarbons by the SRK EOS. Fluid Phase Equilib. 1993;84:339–342.

18 Souahi F, Sator S, Albane SA, Kies FK, Chitour CE. Development of a new form for the alpha function of the Redlich-Kwong cubic equation of state. Fluid Phase Equilib. 1998;153(1):73–80.

19 Stryjek R, Vera J. PRSV: An improved Peng-Robinson equation of state for pure compounds and mixtures. Can J Chem Eng. 1986;64(2):323–333.

20 Twu CH, Coon JE, Cunningham JR. A new generalized alpha function for a cubic equation of state Part 2. Redlich-Kwong equation. Fluid Phase Equilib. 1995;105(1):61–69.

21 Twu CH, Coon JE, Cunningham JR. A new generalized alpha function for a cubic equation of state Part 1. Peng-Robinson equation. Fluid Phase Equilib. 1995;105(1):49–59.

22 Nasrifar K, Bolland O. Prediction of thermodynamic properties of natural gas mixtures using 10 equations of state including a new cubic two-constant equation of state. J Petroleum Sci Eng. 2006;51(3–4):253–366.

23 Mørch Ø, Nasrifar K, Bolland O, Solbraa E, Fredheim A, Gjertsen L. Measurement and modeling of hydrocarbon dew points for five synthetic natural gas mixtures. Fluid Phase Equilib. 2006;239(2):138–145.

24 Flöter E, de Loos TW, de Swaan Arons J. Improved modeling of the phase behavior of asymmetric hydrocarbon mixtures with the Peng-Robinson equation of state using a different temperature dependency of the parameter a. Ind Eng Chem Res. 1998;37(5):1651–1662.

25 Haghtalab A, Mahmoodi P, Mazloumi SH. A modified Peng-Robinson equation of state for phase equilibrium calculation of liquefied, synthetic natural gas, and gas condensate mixtures. Can J Chem Eng. 2011;89(6):1376–1387.

26 Martinez SA, Hall KR. Thermodynamic properties of light synthetic natural gas mixtures using the RK-PR Cubic Equation Of State. Ind Eng Chem Res. 2006;45(10):3684–92.

27 Cismondi M, Mollerup J. Development and application of a three-parameter RK-PR equation of state. Fluid Phase Equilib. 2005;232(1):74–89.

28 Patel NC, Teja AS. A new cubic equation of state for fluids and fluid mixtures. Chem Eng Sci. 1982;37(3):463–473.

29 Nasrifar K, Bolland O, Moshfeghian M. Predicting natural gas dew points from 15 equations of state. Energ Fuels. 2005;19(2):561–572.

30 Louli V, Pappa G, Boukouvalas C, et al. Measurement and prediction of dew point curves of natural gas mixtures. Fluid Phase Equilib. 2012;334(0):1–9.

31 Boukouvalas CJ, Magoulas KG, Stamataki SK, Tassios DP. Prediction of vapor–liquid equilibria with the lcvm model: systems containing light gases with medium and high molecular weight compounds. Ind Eng Chem Res. 1997;36(12):5454–5460.

32 Voutsas E, Magoulas K, Tassios D. Universal mixing rule for cubic equations of state applicable to symmetric and asymmetric systems: Results with the Peng-Robinson equation of state. Ind Eng Chem Res. 2004;43(19):6238–6246.

33 Louli V, Boukouvalas C, Voutsas E, Magoulas K, Tassios D. Application of the UMR-PRU model to multicomponent systems: Prediction of the phase behavior of synthetic natural gas and oil systems. Fluid Phase Equilib. 2007;261(1–2):351–358.

34 Sun L, Zhao H, McCabe C. Predicting the phase equilibria of petroleum fluids with the SAFT-VR approach. AIChE J. 2007;53(3):720–731.

35 Ting PD, Joyce PC, Jog PK, Chapman WG, Thies MC. Phase equilibrium modeling of mixtures of long-chain and short-chain alkanes using Peng–Robinson and SAFT. Fluid phase Equilib. 2003;206(1):267–286.

36 Alfradique M, Castier M. Calculation of phase equilibrium of natural gases with the Peng-Robinson and PC-SAFT equations of state. Oil Gas Sci and TechnolRevue de l'IFP. 2007;62(5):707–714.

37 Voutsas EC, Pappa GD, Magoulas K, Tassios DP. Vapor–liquid equilibrium modeling of alkane systems with Equations of State:"Simplicity versus complexity." Fluid Phase Equilib. 2006;240(2):127–39.

38 Skylogianni E, Novak N, Louli V, et al. Measurement and prediction of dew points of six natural gases. Fluid Phase Equilib. 2016;424:8–15.

39 Voutsas E, Louli V, Boukouvalas C, Magoulas K, Tassios D. Thermodynamic property calculations with the universal mixing rule for EoS/G E models: results with the Peng–Robinson EoS and a UNIFAC model. Fluid Phase Equilib. 2006;241(1):216–28.

40 Gross J, Sadowski G. Application of perturbation theory to a hard-chain reference fluid: an equation of state for square-well chains. Fluid Phase Equilib. 2000;168(2):183–199.

41 Gross J, Sadowski G. Perturbed-chain SAFT: an equation of state based on a perturbation theory for chain molecules. Ind Eng Chem Res. 2001;40(4):1244–1260.

42 Daubert TE. Physical and thermodynamic properties of pure chemicals: data compilation. Design Institute for Physacal Property Data (DIPPR). 1989.

43 Wertheim M. Fluids with highly directional attractive forces. II. Thermodynamic perturbation theory and integral equations. J Statist Physics. 1984;35(1–2):35–47.

44 Wertheim M. Fluids with highly directional attractive forces. I. Statistical thermodynamics. J Statist Physics. 1984;35(1–2):19–34.

45 Chapman WG, Gubbins KE, Jackson G, Radosz M. New reference equation of state for associating liquids. Ind Eng Chem Res. 1990;29(8):1709–1721.

46 Wolbach JP, Sandler SI. Using molecular orbital calculations to describe the phase behavior of cross-associating mixtures. Ind Eng Chem Res. 1998;37(8):2917–2928.

47 Gross J, Sadowski G. Application of the perturbed-chain SAFT equation of state to associating systems. Ind Eng Chem Res. 2002;41(22):5510–5515.

48 Fredenslund A, Jones RL, Prausnitz JM. Group-contribution estimation of activity coefficients in nonideal liquid mixtures. AIChE J. 1975;21(6):1086–1099.

49 Pappa GD, Louli V, Dedousi K, Voutsas EC. Phase equilibria of mixtures containing CO_2 and organic acids using the UMR-PRU model. J Supercrit Fluids. 2011;58(3):321–329.

50 Boukouvalas C, Spiliotis N, Coutsikos P, Tzouvaras N, Tassios D. Prediction of vapor–liquid equilibrium with the LCVM model: a linear combination of the Vidal and Michelsen mixing rules coupled with the original UNIF. Fluid Phase Equilib. 1994;92:75–106.

51 Riazi M. Characterization and properties of petroleum fractions: Philadelphia: ASTM international; 2005.

52 Aladwani H, Riazi M. Some guidelines for choosing a characterization method for petroleum fractions in process simulators. Chem Eng Res Design. 2005;83(2):160–166.

53 Naji HS. Feasible C7+ splitting methods: an object-oriented approach. King Abdulaziz University, Jeddah, Saudi Arabia, Intl J Eng Technol. 2010;10(1), 43–60.

54 Riazi M, Al-Adwani H, Bishara A. The impact of characterization methods on properties of reservoir fluids and crude oils: options and restrictions. J Petrol Sci Eng. 2004;42(2):195–207.

55 Jarne C, Avila S, Blanco ST, Rauzy E, Otín S, Velasco I. Thermodynamic properties of synthetic natural gases. 5. Dew point curves of synthetic natural gases and their mixtures with water and with water and methanol: measurement and correlation. Ind Eng Chem Res. 2003;43(1):209–217.

56 Avila S, Blanco ST, Velasco I, Rauzy E, Otín S. Thermodynamic properties of synthetic natural gases. 1. Dew-Point Curves Of Synthetic Natural Gases And Their Mixtures With Water And Methanol. Measurement And Correlation. Ind Eng Chem Res. 2002;41(15):3714–3721.

57 Avila S, Blanco ST, Velasco I, Rauzy E, Otín S. Thermodynamic properties of synthetic natural gases. 2. Dew point curves of synthetic natural gases and their mixtures with water and methanol. Measurement and Correlation. Energ Fuels. 2002;16(4):928–934.

58 Grenner A, Schmelzer J, von Solms N, Kontogeorgis GM. Comparison of two association models (Elliott-Suresh-Donohue and simplified PC-SAFT) for complex phase equilibria of hydrocarbon-water and amine-containing mixtures. Ind Eng Chem Res. 2006;45(24):8170–8179.

59 Avila S, Blanco ST, Velasco I, Rauzy E, Otın S. Thermodynamic properties of synthetic natural gases: Part 4. Dew point curves of synthetic natural gases and their mixtures with water: measurement and correlation. Fluid Phase Equilib. 2002;202(2):399–412.

60 Avila S, Blanco ST, Velasco I, Rauzy E, Otín S. Thermodynamic properties of synthetic natural gases. Part 3. Dew point curves of synthetic natural gases and their mixtures with water. Measurement and correlation. Energy Fuels. 2003;17(2):338–343.

61 Pedersen KS, Christensen PL, Shaikh JA. Phase behavior of petroleum reservoir fluids: Boca Raton, FL: CRC press; 2014.

62 Whitson CH. Effect of C7+ properties on equation-of-state predictions. SPE J. 1984;24(06):685–696.

63 Whitson CH. Characterizing hydrocarbon plus fractions. SPE J. 1983;23(04):683–94.

64 Pedersen KS, Thomassen P, Fredenslund A. Thermodynamics of petroleum mixtures containing heavy hydrocarbons. 1. Phase envelope calculations by use of the Soave-Redlich-Kwong equation of state. Ind Eng Chem Proc Design Develop. 1984;23(1):163–170.
65 Pedersen KS, Thomassen P, Fredenslund A. Thermodynamics of petroleum mixtures containing heavy hydrocarbons. 2. Flash and PVT calculations with the SRK equation of state. Ind Eng Chem Proc Design Develop. 1984;23(3):566–573.
66 Pedersen KS, Thomassen P, Fredenslund A. Thermodynamics of petroleum mixtures containing heavy hydrocarbons. 3. Efficient flash calculation procedures using the SRK equation of state. Ind Eng Chem Proc Design Develop. 1985;24(4):948–954.
67 Arce P, Aznar M. Modeling of critical lines and regions for binary and ternary mixtures using non-cubic and cubic equations of state. The J Supercrit Fluids. 2007;42(1):1–26.
68 Justo-García DN, García-Sánchez F, Díaz-Ramírez NL, Romero-Martínez A. Calculation of critical points for multicomponent mixtures containing hydrocarbon and nonhydrocarbon components with the PC-SAFT equation of state. Fluid Phase Equilib. 2008;265(1):192–204.
69 Hysys A. Version 8.0. Aspen Technology Inc. 2013.
70 Jensen M, Ungerer P, De Weert B, Behar E. Crystallisation of heavy hydrocarbons from three synthetic condensate gases at high pressure. Fluid Phase Equilib. 2003;208(1):247–260.
71 Ungerer P, Faissat B, Leibovici C, et al. High pressure–high temperature reservoir fluids: investigation of synthetic condensate gases containing a solid hydrocarbon. Fluid Phase Equilib. 1995;111(2):287–311.
72 Gozalpour F, Danesh A, Todd A, Tohidi B. Vapour-liquid equilibrium compositional data for a model fluid at elevated temperatures and pressures. Fluid Phase Equilib. 2003;208(1):303–313.
73 Yarborough L. Vapor–liquid equilibrium data for multicomponent mixtures containing hydrocarbon and nonhydrocarbon components. J Chem Eng Data. 1972;17(2):129–133.
74 Bonyadi M, Esmaeilzadeh F. Prediction of gas condensate properties by Esmaeilzadeh-Roshanfekr equation of state. Fluid Phase Equilib. 2007;260(2):326–334.
75 Coats K, Smart G. Application of a regression-based EOS PVT program to laboratory data. SPE Reservoir Eng. 1986;1(03):277–299.
76 Donohoe CW, Buchanan Jr RD. Economic evaluation of cycling gas-condensate reservoirs with nitrogen. J Petrol Technol. 1981;33(02):263–270.
77 Firoozabadi A, Hekim Y, Katz DL. Reservoir depletion calculations for gas condensates using extended analyses in the Peng-Robinson equation of state. Can J Chem Eng. 1978;56(5):610–615.
78 Pedersen KS, Blilie AL, Meisingset KK. PVT calculations on petroleum reservoir fluids using measured and estimated compositional data for the plus fraction. Ind Eng Chem Res. 1992;31(5):1378–1384.
79 Katz D, Firoozabadi A. Predicting phase behavior of condensate/crude-oil systems using methane interaction coefficients. J Petrol Technol. 1978;30(11): 1649–1655
80 Retzekas E, Voutsas E, Magoulas K, Tassios D. Prediction of physical properties of hydrocarbons, petroleum, and coal liquid fractions. Ind Eng Chem Res. 2002;41(6):1695–1702.
81 Poling BE, Prausnitz JM, O'Connell JP. The properties of gases and liquids: New York: Mcgraw-hill; 2001.

82 Elsharkawy AM, ed., Characterization of the plus fraction and prediction of the dewpoint pressure for gas condensate reservoirs. SPE Western Regional Meeting; 2001: Society of Petroleum Engineers, 2630 March 2001.
83 Belaud J-P, Pons M. Open software architecture for process simulation: the current status of cape-open standard. Computer Aided Chem Eng. 2002;10:847–852.
84 Solutions HP. UniSim® Design Suite, Houston: 2010, UniSim® is a registred trademark of Honeywell Process Solutions.

4

CO_2 Injection in Coal Formations for Enhanced Coalbed Methane and CO_2 Sequestration

Ahmed Farid Ibrahim and Hisham A. Nasr-El-Din

Petroleum Engineering Department, Texas A&M University, USA

CHAPTER MENU

Since the beginning of the Industrial Revolution in 1750, human activities have produced a 40% increase concentration of carbon dioxide in the atmosphere, from 280 ppm in 1750 to 406 ppm in early 2017. U.S. energy-related carbon dioxide (CO_2) emissions totaled 2.53 billion metric tons in the first six months of 2016. The main human activity that emits CO_2 is the combustion of fossil fuels (coal, natural gas, and oil) for energy and transportation, although certain industrial processes and land-use changes also emit CO_2. CO_2 sequestration is one of the effective ways to decrease CO_2 emissions to reduce the effect of global warming. Geologic sequestration is permanently storing CO_2 in subsurface structures such as oil reservoirs, natural gas deposits, deep saline formations, and unmineable coal seams. CO_2 sequestration in coal seams combines CO_2 storage with enhancing methane recovery.

This chapter investigates the ability of CO_2 sequestration in coal formations, and the effect of different parameters on the process behavior. It presents experimental results for adsorption isotherm, wettability behavior, and injectivity in coal formations, and then provides conclusions based on those results. The following sections of this chapter present the characterization of the coal formations, the CO_2, N_2, and CH_4 adsorption, wettability coal surface, CO_2 injectivity, and a different pilot field test for enhanced coalbed methane (ECBM) process.

4.1 Coalbed Characteristics

Coal is a complex organic rock of variable purity and moisture content deposited in multiple geological ages and is often classified by rank. Coal rank depends on the thermal maturity, defined by (American Society for Testing and Materials) ASTM D388-05. The ranks of coal (from the higher to lower rank) are as follows: anthracite, bituminous coal, subbituminous coal,

Natural Gas Processing from Midstream to Downstream, First Edition.
Edited by Nimir O. Elbashir, Mahmoud M. El-Halwagi, Ioannis G. Economou, and Kenneth R. Hall.

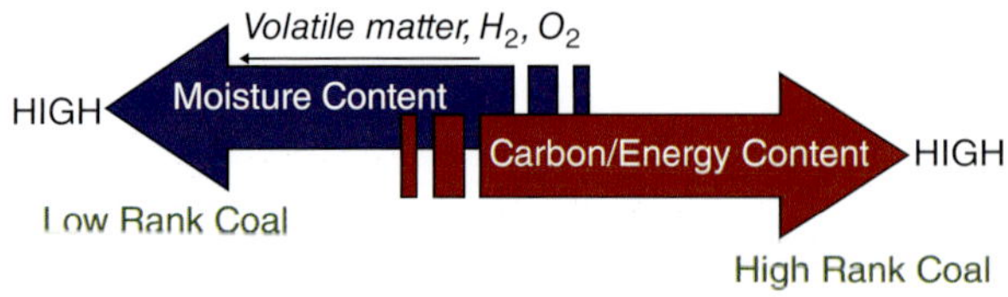

Figure 4.1 Schematic for the coal composition as a function of coal rank.

and lignite. Anthracite coal is a dense, hard rock with a jet-black color and metallic luster. It contains between 86 and 98 wt% carbon by weight. Bituminous coal contains between 69 and 86 wt% carbon. Subbituminous coal contains less carbon and more water. Lignite coal, or brown coal, is a very soft coal that contains up to 70 wt% water (Figure 4.1).

According to the U.S. Energy Information Administration (EIA), coalbed methane production in the U.S. in 2013 was 1.5 Tscf, nearly 5% of U.S. gas production that year (EIA 2013). The coalbed methane (CBM) is considered an unconventional resource where the coal is the source rock and the reservoir rock for methane (Gray 1987). The coalbed formations are characterized by their dual porosity: primary (micropores and macropores) and secondary (cleats network) (Laubach et al. 1998). The primary porosity system contains the majority of the gas-in-place as adsorbed gas in the coal matrix while the cleat network system is usually full of water and provides the path for mass transfer to the wellbore (Laubach et al. 1998; Shi and Durucan 2004). Coal-seam water usually has low salinity, which varies between 800 to 28,000 ppm (Table 4.1) (Hamawand, Yusaf, and Hamawand 2013).

The gas is stored in coal as three different ways: free gas within the cleats network system, adsorbed molecules on the organic surface of coal, and gas dissolved in groundwater within the coal.

The production of CBM formations contains three stages (Gray 1987; Seidle 2011; Ahmed and Meehan 2012). In the dewatering stage, the water flows from the formation and the pressure in the cleat network decreases, which allows gas to desorb from the coal matrix (Ibrahim and Nasr-El-Din 2015a). Once the gas saturation in the cleat network becomes higher than the critical gas saturation, it begins to flow through the cleat network to the producing wells. As the gas desorption from the matrix continues, the gas flow increases dominantly and reaches its maximum value (peak gas stage). Finally, in the decline stage, the gas flow decreases and the CMB behavior becomes similar to the conventional gas reservoirs. As the gas desorbs from the coal surface, the matrix shrinks. Matrix shrinkage increases cleat width, and the permeability increases (Harpalani and Schraufnagel 1990) (Figure 4.2).

CO_2 has been sequestrated in coal seams to enhance the coalbed methane production (ECBM) in addition to CO_2 storage (Seidle 2000; Mavor et al. 2002; Busch, Gensterblum, and Krooss 2003; Garduno et al. 2003; Schroeder, Ozdemir, and Morsi 2002; Shi and Durucan

Table 4.1 Salinity of coal seam associated water (Hamawan, Hamawand, Yusaf 2013).

Location			TDS mg/L
Australia	Bowen Basin	Durham Ranch	5,968
		Fairview	1,201
		Upper seam	6,207
		Lower seam	6,528
US		Black Warrior Basin	4,402
		San Juan Basin	28,783
		Uinta Basin	11,876
		Maramarua	837

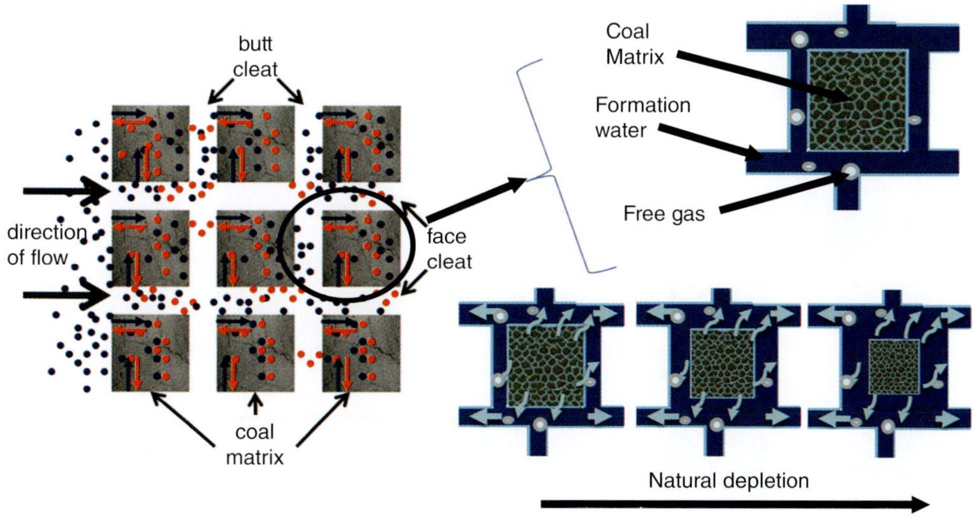

Figure 4.2 Schematic for gas storage in coal formations and natural depletion with matrix shrinkage.

2005). In the ECBM process, nitrogen, CO_2, or their binary mixture is injected into coalbeds to enhance CBM production and ultimate methane recovery. As CO_2 is injected into the coal formation, it replaces the methane on the coal surface due to its higher affinity to the coal matrix than methane (Busch, Gensterblum, and Krooss 2003). Busch, Gensterblum, and Krooss (2003) reported that the ratios of the final sorption capacities for pure CO_2 and methane (in molar units) on the five coal samples varied between 1.15 and 3.16. Battistutta et al. (2010) determined the adsorption isotherms of CO_2, CH_4, and N_2 and found that the adsorption ratio between the maximum in the excess sorption N_2:CH_4:CO_2 was 1:1.5:2.6. Both CO_2 adsorption and CH_4 desorption affect the absolute coal permeability. CH_4 desorption shrinks the coal matrix and increases cleat width, which in turn increases the absolute permeability. Opposite effects can be expected when CO_2 is adsorbed by coal. Previous research work has shown that the coal swells when exposed to carbon dioxide (Figure 4.3) (Harpalani and Schraufnagel 1990; Palmer and Mansoori 1998; Battistutta et al. 2010; Liu, Harpalani, and Pillalamarry 2012; Liu and Harpalani 2013).

4.2 Adsorption Isotherm Behavior

Adsorbed gas phase accounts for more than 90% of gas-in-place. Several studies report CO_2 and CH_4 sorption on the same coal sample under different moisture conditions (Figure 4.4) (Guo et al. 2015) to obtain information on the selectivity of the coal for either gas species. For all mentioned data sets (Clarkson and Bustin 2000; Busch, Gensterblum, and Krooss 2003; Reeves 2003; Busch et al. 2006; Harpalani, Prusty, and Dutta 2006; Kelemen and Kwiatek 2009; Pini, Storti, and Mazzotti 2011) higher CO_2 compared to CH_4 sorption was observed while the CO_2/CH_4 sorption ratio varies from 1.1 to 9.1 (Figure 4.5). Interestingly, a relative decrease in sorption ratio with coal rank can be observed for moisturized coal samples having a ratio of 9 at low coal rank and decreases to 1.2–1.5 for anthracite-rank coals. The high CO_2/CH_4 sorption ratios at low coal maturity can (partly) be due to a much higher solubility of CO_2 in water as compared to CH_4 considering the large water contents for low-rank coal (Guo et al. 2015). For single gas measurements, higher sorption of CO_2 compared to CH_4 was always observed for the same coal sample under the same pressure and temperature (Figure 4.6) (Harpalani, Prusty,

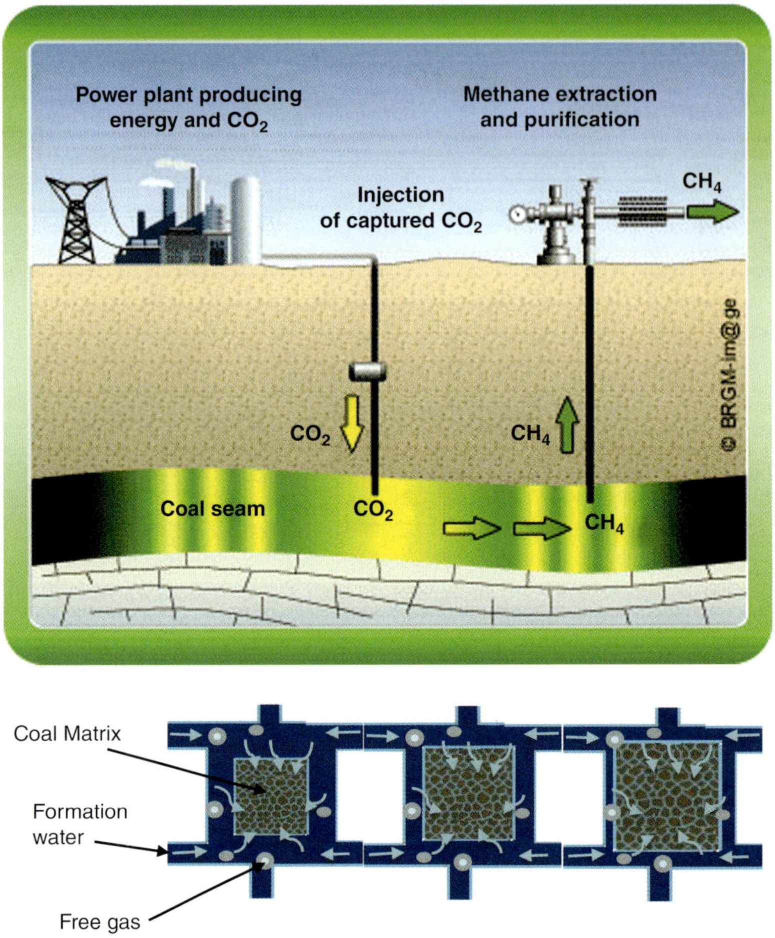

Figure 4.3 Schematic for gas storage in coal formation during CO_2 injection with coal matrix swelling effect (modified after CCUS 2018).

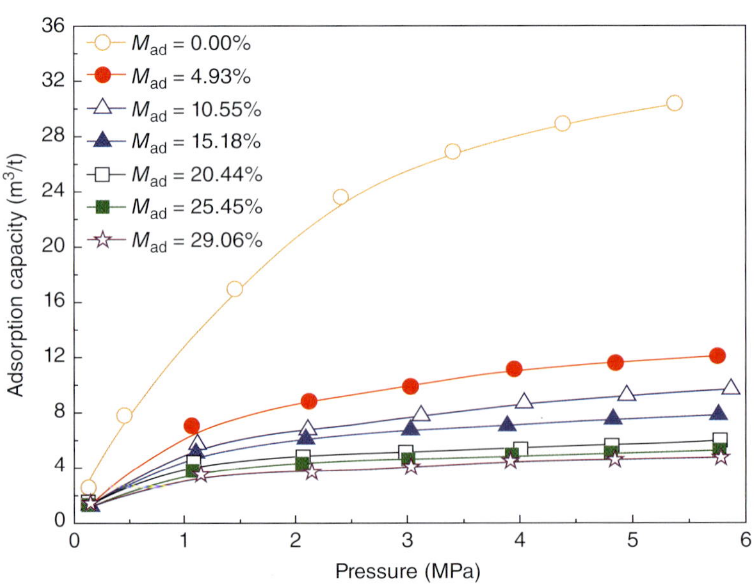

Figure 4.4 Change in methane adsorption isotherm in lignite as a function moisture at 30°C (Guo et al. 2015).

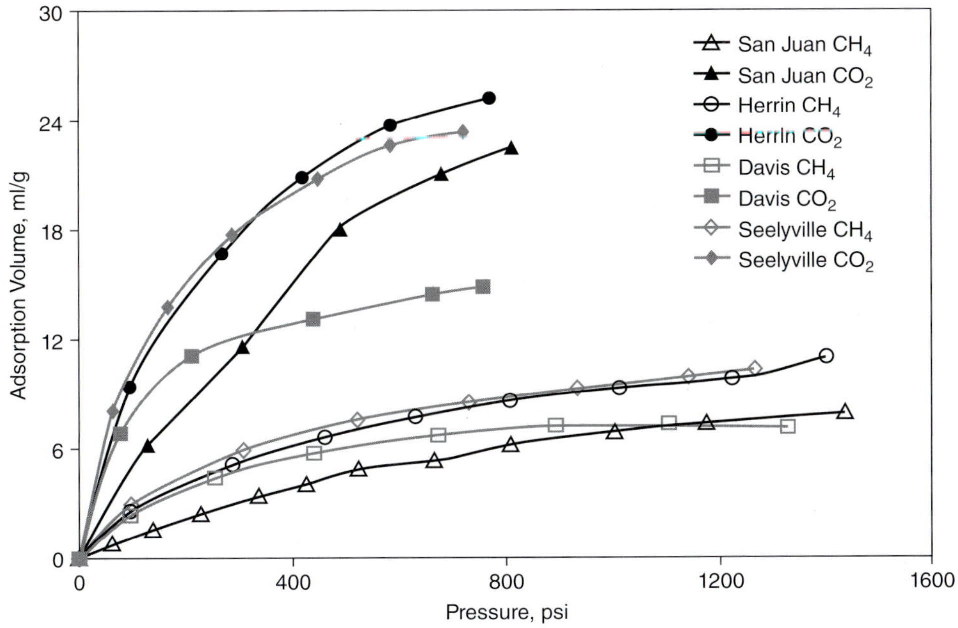

Figure 4.5 CO_2 versus CH_4 adsorption isotherm for different coal samples ((Harpalani, Prusty, B. K., and Dutta 2006).

Coal samples	Moisture (%)	Ash (%)	Volatility (%)	True density (cm^3/g)	Apparent density (cm^3/g)	Porosity (%)
Changcui coal	1.06	16.31	8.62	1.385	1.314	5.13
Qinan coal	0.86	11.60	33.16	1.521	1.452	4.54

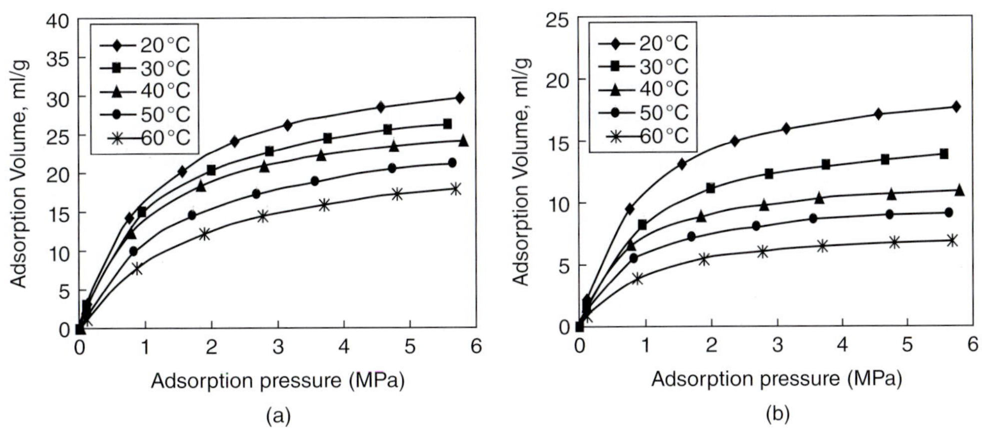

Figure 4.6 The effect of temperature on CH_4 adsorption isotherm in (a) Changcui coal; (b) Qinan coal (Chen, Jin, and Chen. 2011).

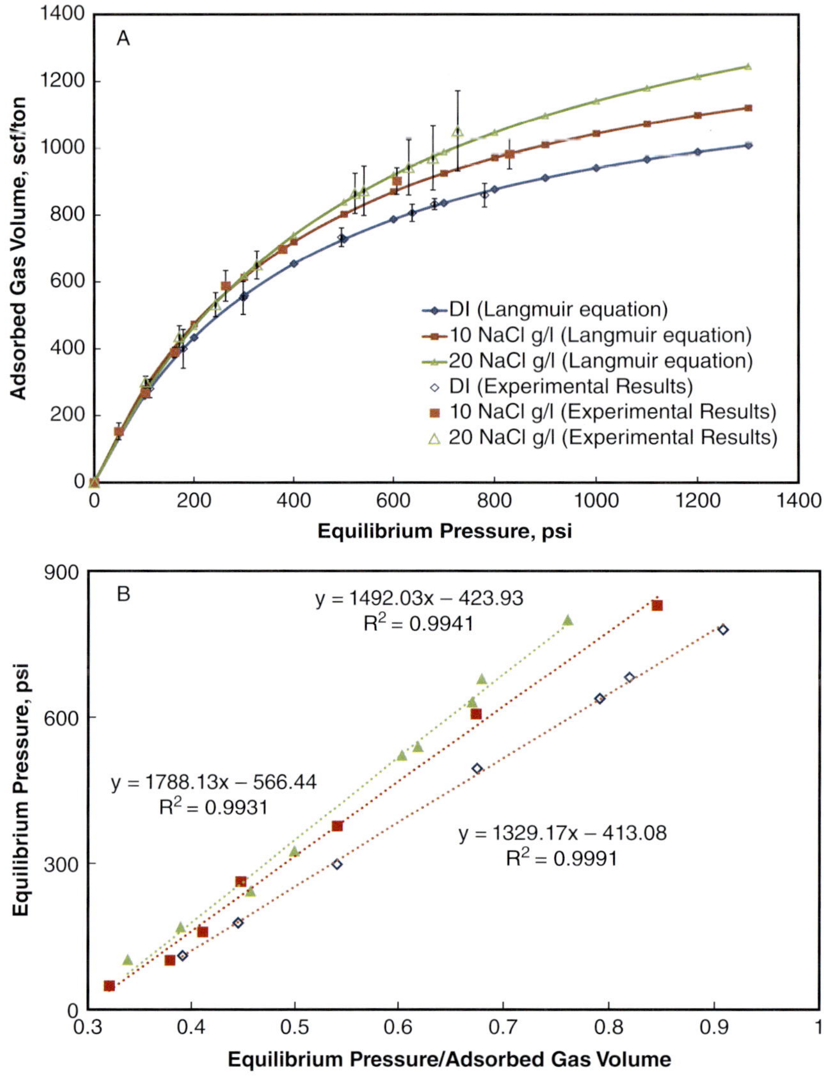

Figure 4.7 CO_2 adsorption isotherm in high volatile bitumen A coal; (A) as a function of salt concentrations, (B) fitted with Langmuir equation to get the Langmuir volume and pressure constants (Ibrahim and Nasr-El-Din 2016).

and Dutta 2006; Chen, Jin, and Chen 2011). Temperature has a negative effect on adsorption capability, which decreases as the temperature increases (Chen, Jin, and Chen 2011) (Figure 4.7).

Ibrahim and Nasr-El-Din (2016) investigated the effect of water salinity on the CO_2 adsorption on bitumen coal samples. It was demonstrated that the CO_2 adsorption increased as the NaCl concentration increased. The Langmuir adsorption volume changed from 1,330 scf/ton in deionized water (DI) water case to 1,788 scf/ton in the 20 g/l NaCl brine (Figure 4.7).

4.3 Coal Wettability

The efficiency of the ECBM and CO_2 sequestration process is dependent on the wettability behavior of the coal-water–CO_2 system. It depends specifically on the CO_2 diffusion rate from the cleat network, through the micro-cleats, to the surface of the coal matrix. If the coal is gas-wet (hydrophobic) then the gas will fill the smaller pores (Mazumder, Plug and Bruining 2003; Plug, Mazumder, and Bruining 2008; Saghafi, Pinetown, and Javanmard 2014), which leads to a faster diffusion of the injected gas to the coal surface (diffusion coefficient of CO_2 = 1.7×10^{-7} m^2/s at 100 bar and 300 K) (Plug, Mazumder, and Bruining 2008). If the coal is water-wet (hydrophilic), then water will fill the smaller pores, which decreases the CO_2 diffusion (from the main cleats) to the coal surface through the water (diffusion coefficient of CO_2 = 2×10^{-9} m^2/s at 100 bar and 300 K) (Plug, Mazumder, and Bruining 2008).

Wettability studies usually use the contact angle measurements as a simple and reliable method to characterize the degree of wetting when a solid and liquid interact (Kwok and Neumann 1999). The wettability behavior of coal was investigated in the literature based on contact angle measurements for coal-water–air systems (Siemons et al. 2006; Sakurovs and Lavrencic 2011; Shojai et al. 2011; Shojai et al. 2012). It was found that the coal rank highly affects the coal wettability behavior. These ranks have different oxygen-containing polar groups such as carboxyl (COOH), hydroxyl (OH), and methoxyl (OCH_3) groups. It was found that the coal becomes more water-wet (decreasing the hydrophobicity) with decreasing rank, carbon content and with increasing oxygen-containing groups. In other words, the lower the coal rank (lower carbon content and higher oxygen-containing groups), the lower the contact angle (lower hydrophobicity). The wettability behavior of coal depends on the pressure, and its effect is a function of the coal rank (Siemons et al. 2006; Sakurovs and Lavrencic 2011; Shojai et al. 2012). Sakurovs and Lavrencic (2011) and Shojai et al. (2012) investigated the wettability behavior for different coal ranks at different pressures up to 1,500 psi. For high-ranking coal, the wettability was modulated from water-wet to gas-wet with CO_2 injection at higher pressures. Siemons et al. (2006) indicated that coal was water-wet at atmospheric pressure and then turned into gas-wet at pressures greater than 375 psi. Sakurovs and Lavrencic (2011) and Shojai et al. (2012) showed that the wettability behavior of CO_2 in wet coal was dependent on the CO_2 adsorption behavior instead of other CO_2 properties, such as solubility and density. They found that the CO_2-coal wetting behavior was affected by the adsorption of CO_2 on the coal surface and that the surface tension between water and coal decreased due to adsorption.

Figure 4.8 shows the contact angle values for Siemons et al. (2006) are generally higher than the values in the study by Ibrahim and Nasr-El-Din (2016). Two reasons may explain this behavior. First, the contact angle measurements by Siemons et al. were conducted in a non-equilibrium system where the water was not pre-equilibrated with CO_2. Consequently, the mass transfer of CO_2 and water might have affected the behavior. Second, the contact angle measurements by Siemons et al. were conducted using anthracite coal, which is a higher rank than the high-volatile A bitumen (hvAb) coal sample used in the Ibrahim and Nasr-El-Din (2016) study. According to Sakurovs and Lavrencic (2011) and Shojai et al. (2012), the coal hydrophobicity increases as coal ranking decreases, which affects the contact angle measurements. Similar results were found by Shojai et al. (2012), especially for high-volatile B bitumen (hvBb) coal sample, and the difference between the two measurements is due to the coal-rank effect. Figure 4.8 shows a comparison with contact-angle measurements by Arif et al. (2016) for semi-anthracite, medium volatile bituminous, of lignite coal samples at

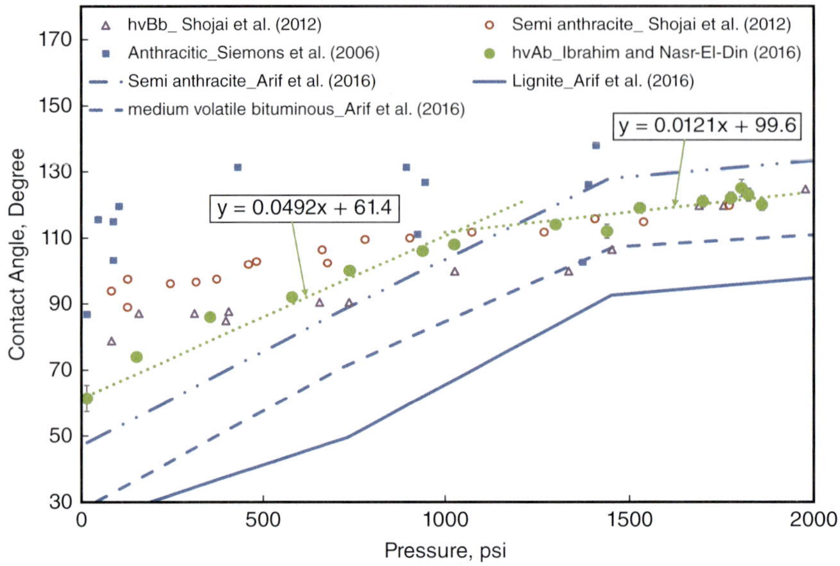

Figure 4.8 Comparison of contact-angle values from the literature (Siemons et al. 2006 (DI water, and 45°C); Shojai et al. 2012 (DI water, and 40°C); Arif et al. 2016 (DI Water, and 49°C); and Ibrahim and Nasr-El-Din 2016 (DI water, and 40°C).

49°C. Lower values were found in the case of medium volatile bituminous, and lignite coal samples compared to those of Ibrahim and Nasr-El-Din work. Lower coal rank and higher temperatures may account for these differences in the contact angle values. Contact angle measurements for semi-anthracite bituminous coal should be higher than those found by Ibrahim and Nasr-El-Din (2016): similar to the results obtained by Siemons et al. (2006), but the high temperature negates the high-rank effect.

Two factors affected the wettability behavior of the coal-water–CO_2 system during the pressure increase. First, the solubility of CO_2 in water increased and CO_2 adsorption on the coal surface increases at high pressures. As the solubility of CO_2 in water increased, the pH decreases (the CO_2 in water forms a carbonic acid that decreases the pH). Farokhpoor et al. (2013) described the pH variation with the CO_2 solubility as follows:

The carbonic acid was produced from the reaction of dissolved CO_2 with water according to Eq. 4.1:

$$\mathbf{CO_2 + H_2O \Longleftrightarrow H_2CO_3} \tag{4.1}$$

The reaction equilibrium constant can be defined as $\mathbf{K}_1 = \frac{[\mathbf{H_2CO_3}]}{[\mathbf{CO_2}]}$.
H_2CO_3 dissociates in water according to Eq. 4.2:

$$H_2CO_3 \Longleftrightarrow HCO_3^- + H^+ \tag{4.2}$$

The reaction equilibrium constant can be defined as $K_2 = [HCO_3^-] \times [H^+]/[H_2CO_3]$.

Millero and Roy (1997) reported that the equilibrium constants for Eqs. 3 and 4 are $K_1 = 2.612 \times 10^{-2}$ and $K_2 = 9.7 \times 10^{-7}$.
$HCO_3^-$and H_2O dissociate according to Eqs. 3 and 4:

$$\mathbf{HCO_3^- \Longleftrightarrow CO_3^{-2} + H^+ \quad K_c} = 4.69 \times 10^{-11} \text{ at } 25^\circ\text{C} \tag{4.3}$$

$$H_2O \Longleftrightarrow OH^- + H^+ \quad K_w = 1.8 \times 10^{-16} \text{ at } 25^\circ\text{C} \tag{4.4}$$

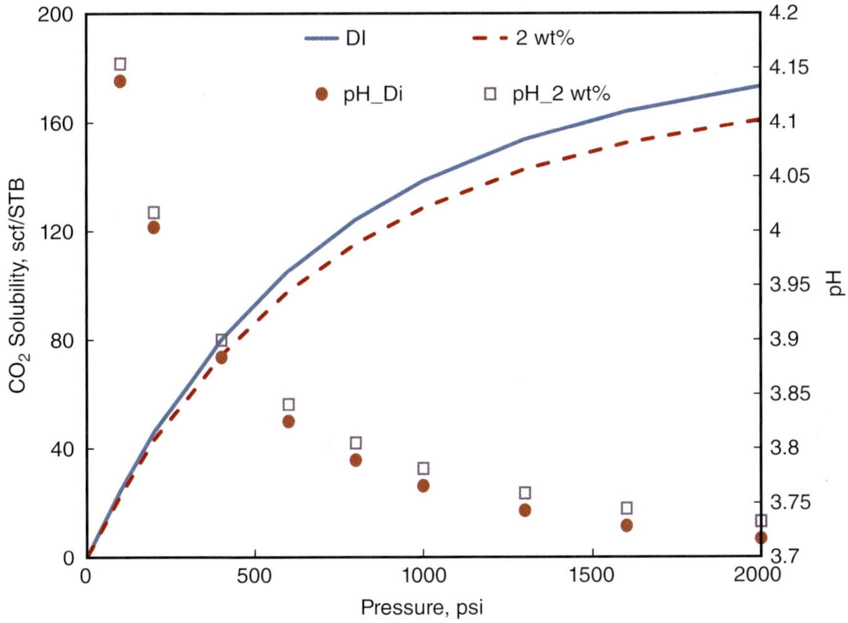

Figure 4.9 pH and CO_2 solubility as a function of salinity and pressure.

Eqs. 4.1–4.4 describe the CO_2-H_2O reaction system. Conducting a charge balance on the system yields the following equation.

$$[H^+] = [HCO_3^-] + 2[CO_3^{-2}] + [OH^-] \quad (4.5)$$

The coefficient (2) in Eq. 4.5 is present because each mole of CO_3^{-2} requires 2 moles of H^+ to neutralize.The dissociation of H_2O and HCO_3^- is very small so the concentration of $[CO_3^{-2}]$ + $[OH^-]$ can be ignored. Therefore Eq. 4.5 can be reduced to Eq. 4.6.

$$[H^+] = [HCO_3^-] \quad (4.6)$$

The pH can be calculated with $pH = -\log_{10}[H^+]$.

Figure 4.9 shows the solubility of CO_2 in DI water and 20 gm/L NaCl brine and pH with saturation pressure. In both brine and DI water, pH decreased rapidly with saturation pressure at the beginning, then slightly decreased. Decreasing pH reduces the density of the negative charges at the water/CO_2 and solid/water interfaces (Chiquet, Broseta, and Thibeau 2007). Hence, the repulsive force between the two interfaces becomes less effective at high pressures, which destabilizes the water film on coal surface, and the coal becomes more hydrophobic.

Adsorption-isotherm measurements show that the CO_2 adsorption onto the coal surface increased along with pressure. Hence, the gas interfacial tension to the solid decreased significantly (Espinoza and Santamarina 2010), which increased the contact angle with pressure.

As the salt concentration increases, it slightly decreases the solubility of CO_2 in water and the pH (Figure 4.9). For example, at 2,000 psi, a 175 scf of CO_2 dissolved in stock tank barrel of DI water was needed to decrease the pH to 3.72. Compared to 2 wt% NaCl brine, CO_2 solubility is 165 scf/STB. The solubility difference is nearly 10 at which it changes the pH by only 0.03 (from 3.71 to 3.74) as it is a logarithmic scale difference. As a result, the effect of salt on the pH can be neglected. As salt concentration increases, it compresses and destabilizes the hydrated layers (double layer) surrounding the coal surface, causing a reduction in the absolute value of the zeta potential and making the coal surface more hydrophobic. The destabilization of the hydrated

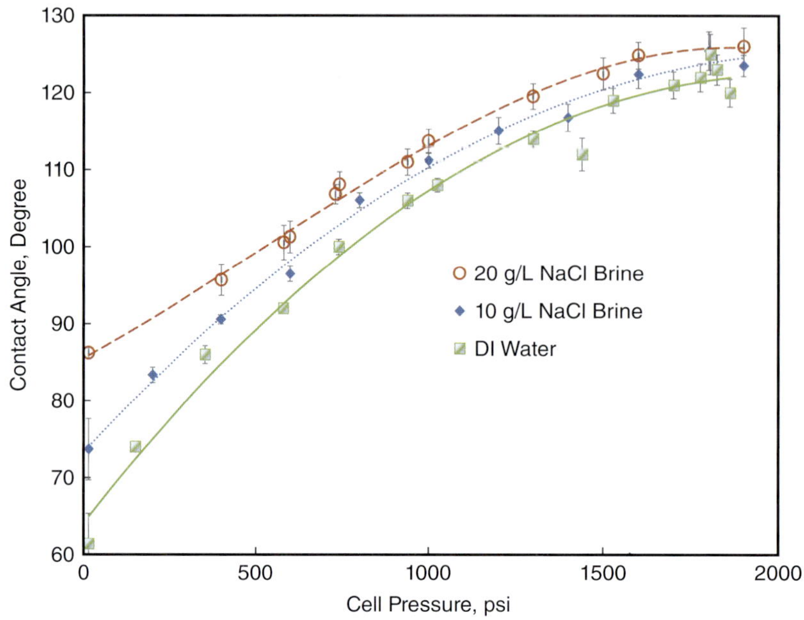

Figure 4.10 Contact angle as a function of cell pressure in the coal-water–CO_2 system at different NaCl concentrations.

layer improves the contact of the CO_2 with the coal surface resulting in a faster diffusion of the CO_2 from the gas phase to the coal surface, which enhances the CO_2 adsorption (as appears in adsorption isotherm measurements). All of these parameters enhance the coal to CO_2 wettability and increase its hydrophobicity (Figure 4.10). Similar results were observed by Nasralla, Bataweel, and Nasr-El-Din (2013) in the salinity effect on the wettability of oil-brine-sandstone system. These researchers found that the sandstone became less water-wet with salinity increasing as a result of the reduction of the negative charge on the brine/rock interface, which shrinks and destabilizes the water film surrounding the rock. Similar results were observed by Arif et al. (2016), where the contact angle increases as the NaCl concentration increased in different coal ranks (Figure 4.11).

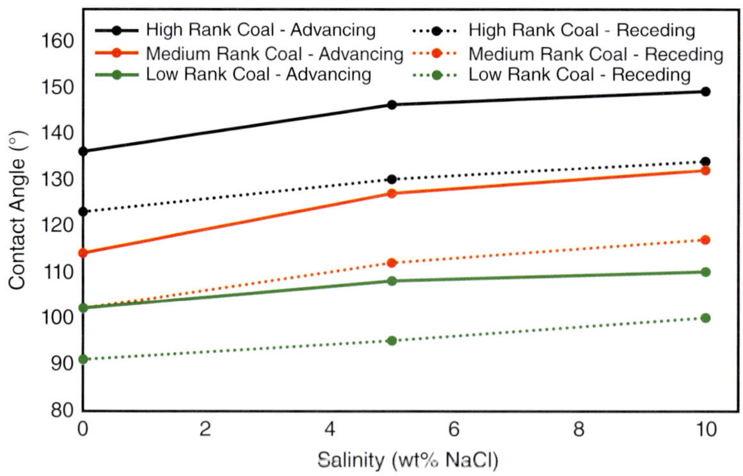

Figure 4.11 Effect of salinity on CO_2 wettability of coals at 49°C and 2000 psi (Arif et al. 2016).

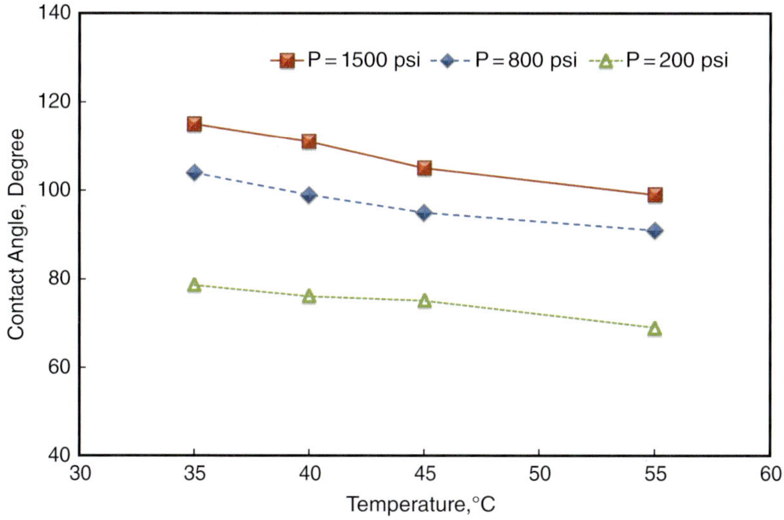

Figure 4.12 The contact-angle measurements as a function of temperature (Ibrahim and Nasr-El-Din, 2016).

Temperature has a negative effect on CO_2 wettability to coal. Figure 4.12 shows the effect of temperature on the contact-angle values after Ibrahim and Nasr-El-Din (2016). As the temperature increases, the contact angle decreases. Similar behavior was found by Arif et al. (2016) (Figure 4.13). As the temperature increased, the CO_2 adsorption to the coal surface decreased. As a result, the gas-coal interfacial tension increased (Espinoza and Santamarina 2010), which decreased the contact angle with temperature.

The composition of injected gas also affected the wettability behavior. Shojai et al. (2011) investigated the wetting behavior of flue gasses (73 mol% N_2, 11 mol% CO_2, and impurities). They found that the contact angle values of flue gasses on the coal surface were generally smaller than those of CO_2. Shojai et al. (2012) showed experimental results for flue gas (80 mol% N_2, 20 mol% CO_2) coal wettability measurements where the wettability of coal was intermediate-wet at all pressures, and the contact angle only slightly increased with increasing pressure (Figure 4.14).

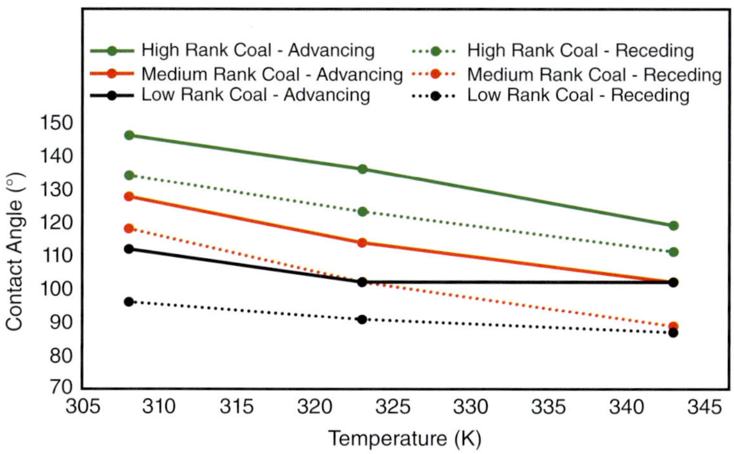

Figure 4.13 Effect of temperature on CO_2 wettability at 2100 psi for different coal samples used by Arif et al. (2016).

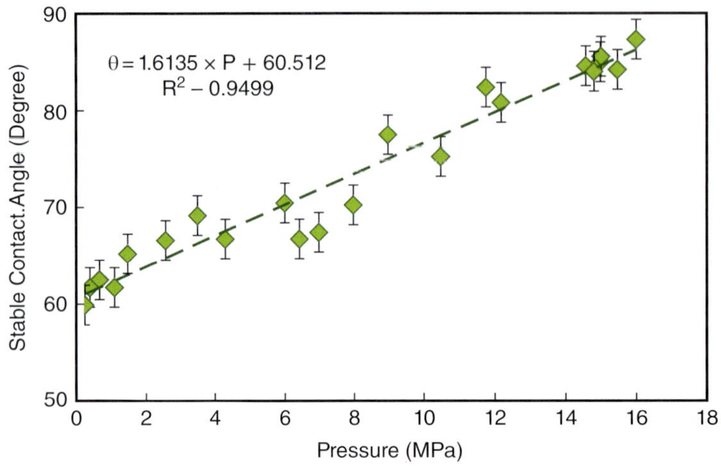

Figure 4.14 Contact angle as a function of pressure at 49°C for synthetic flue gas-wet coal system after Shojai et al. 2011.

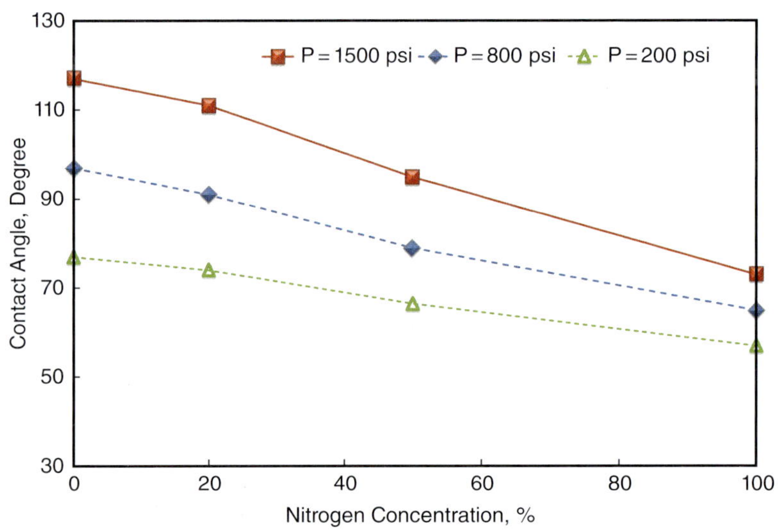

Figure 4.15 The contact-angle measurements as a function of nitrogen concentration of the injected gas (Ibrahim and Nasr-El-Din, 2016).

Figure 4.15 shows the contact angle measurements as a function of nitrogen concentration by Ibrahim and Nasr-El-Din (2016). The contact-angle values decreased as nitrogen concentration increased, and the corresponding slope increased with pressure. At 100% N_2, the coal remained water-wet, even at high pressures (contact angle was 75 at 1,500 psi). These changes occur because of the N_2 adsorption rate and the capacity of the coal surface is lower than that of CO_2. Chow et al. (2016) show that the gas/liquid interfacial tension increased as nitrogen concentration increased in the gas phase, which contributed to the reduction in the contact-angle values.

4.4 CO_2 Injectivity

In the ECBM process, nitrogen, CO_2, or their binary mixtures are injected into coalbeds to be adsorbed on the coal surface and replace the CH_4. CO_2, or N_2 adsorption, and CH_4 desorption affect the coal permeability (Palmer and Mansoori 1998; Harpalani, Prusty, and Dutta 2006; Battistutta et al. 2010; Liu, Harpalani, and Pillalamarry 2012; Ibrahim and Nasr-El-Din 2015a). During primary recovery, formation pressure reduction tends to decrease the formation permeability. However, CH_4 desorption shrinks the coal matrix and increases cleat width, which increases the absolute permeability. In the ECBM process, opposite effects can be expected when CO_2 is adsorbed by coal. The change in formation permeability will be controlled by matrix shrinkage or swelling during CH_4 desorption and CO_2 adsorption, respectively. Moreover, multiphase flow occurs at reservoir conditions for both primary and enhanced coalbed

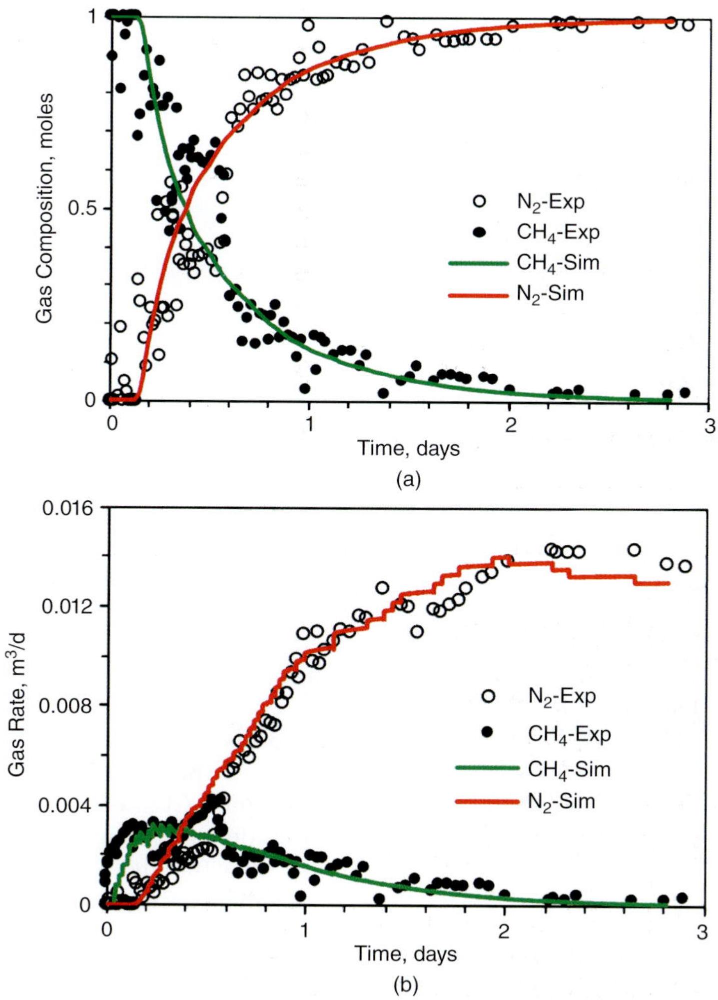

Figure 4.16 ECBM test after (Zhou et al. 2013). (a) Gas compositions of N_2 and CH_4 during the N_2-ECBM test, (b) gas production rates of CH_4 and N_2 during N_2-ECBM, (c) Gas compositions of CO_2 and CH_4 during the CO_2-ECBM test and (d) gas production rates of CH_4 and CO_2 during the CO_2-ECBM test.

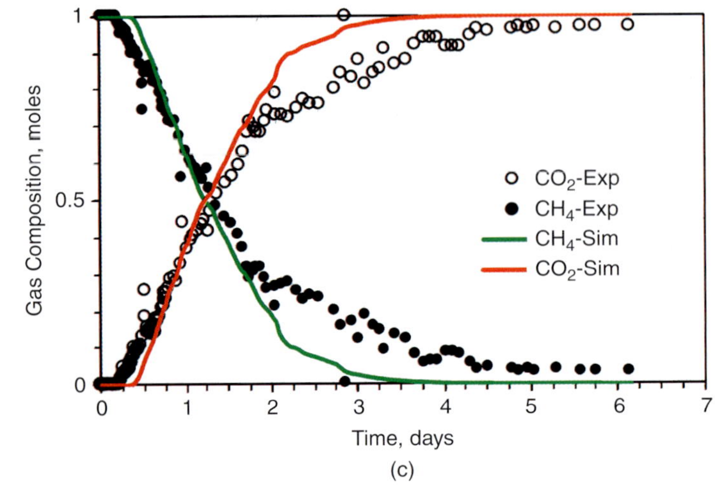

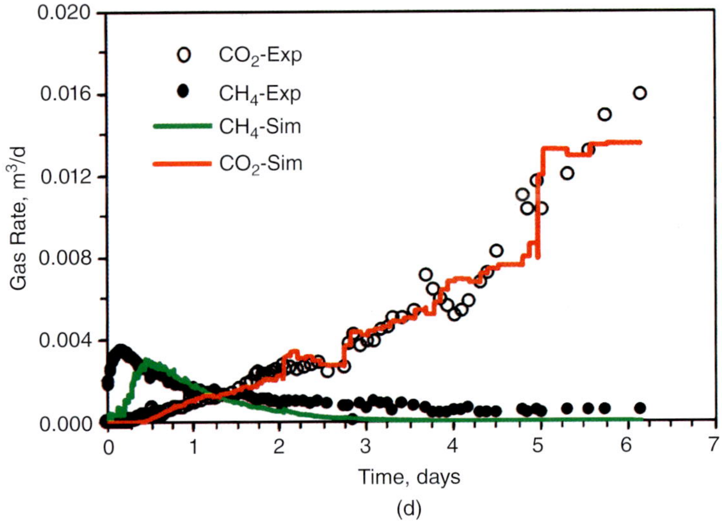

Figure 4.16 (*Continued*)

methane recovery processes. Hence, the formation injectivity is controlled by the water and gas relative permeabilities in addition to the swelling/shrinkage effects.

In comparing N_2-ECBM to CO_2-ECBM the N_2 breakthrough time occurs earlier than that of CO_2, and the recovery of CH_4 for N_2-ECBM is lower than that of CO_2-ECBM. The N_2 injection causes moderate increases in coal permeability, whereas the injection of CO_2 reduces coal permeability significantly (Zhou et al. 2013). Figure 4.16 shows the variation with time of recovery and composition of CH_4 for N_2- and CO_2-ECBM tests. Results show that the N_2-ECBM gives slightly quicker CH_4 recovery than CO_2-ECBM. However, the ultimate recovery of the N_2-ECBM test is similar to that of the CO_2-ECBM test. The CH_4 composition in a produced gas stream of the N_2-ECBM test reduces faster than that of the CO_2-ECBM test. Figure 4.16 shows the recovery against CH_4 composition in the produced gas stream for both N_2- and CO_2-ECBM tests. The recovery of the N_2-ECBM test increases slowly initially and then quickly compared to the CO_2-ECBM test. Eventually, both tests give the same ultimate recovery of about 92%. However, it is important to note that economics dictate when the project should be

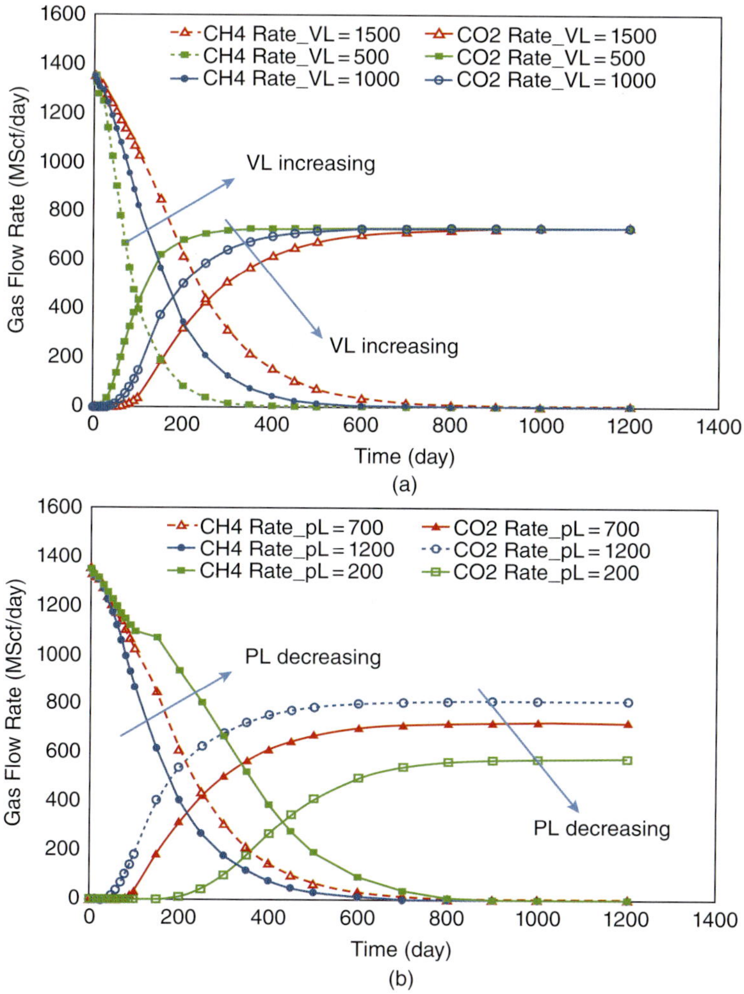

Figure 4.17 Effect of injected gas Langmuir volume and pressure on CO_2 and CH_4 production behavior after Ibrahim and Nasr-El-Din (2015b).

abandoned. This normally corresponds to a certain percentage of CH_4 in the produced gases. In this case, the recovery factor of CH_4 is 71% for the N_2-ECBM and 86% for the CO_2-ECBM project (Zhou, Hussain, and Cinar 2013).

Ibrahim and Nasr-El-Din (2015b) investigated the effect of injected gas on the ECBM performance. Figure 4.17 shows that the Langmuir pressure or the volume doesn't affect the total flow rate; however, the total flow rate decreased with time as Langmuir pressure decreased, due to permeability reduction, as a result of matrix swelling. The decreasing of Langmuir pressure of the injected gas delays the CO_2 breakthrough. The methane production rate was improved with the increasing Langmuir volume or decreasing Langmuir pressure of the injected gas.

As the flow in the ECMB is multiphase flow at reservoir conditions, the formation injectivity is controlled by the water and gas relative permeabilities in addition to the swelling/shrinkage effects (Ibrahim and Nasr-El-Din 2017). Figure 4.18 shows the relative permeability curves as a function of water salinity, N_2 concentration in the injected gas, and back pressure. As the

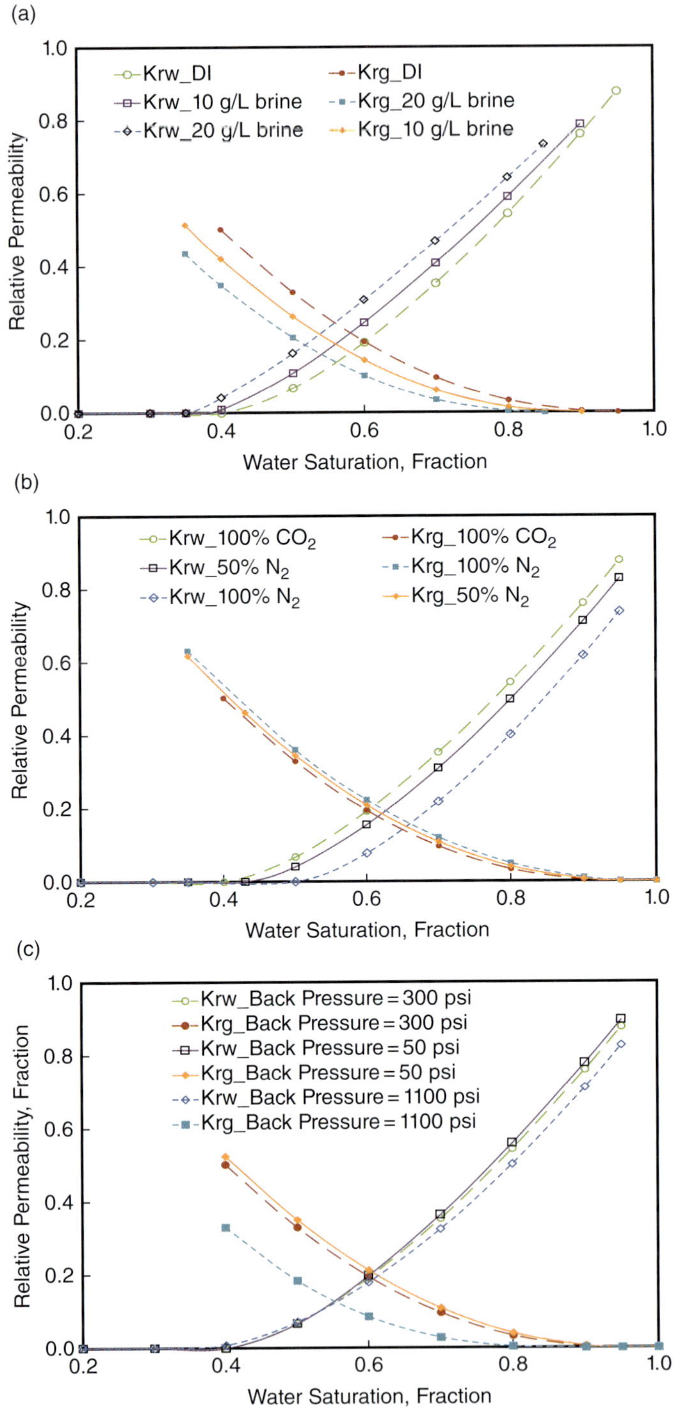

Figure 4.18 Relative permeability curves at 1 cm^3/min as a function of (A) salt concentration, (B) N_2 concentration, and (C) back pressure (after Ibrahim and Nasr-El-Din 2017).

water salinity increased, the coal became more CO_2-wet. As a result, the residual gas saturation increased, and the end-point and cross-point water saturation decreased. The overall gas relative permeability decreased, and the water relative permeability increased. A similar behavior was found as a function of N_2 concentration. Low displacement efficiency was found as N_2 concentration increased and the coal became more water-wet. Figure 18c shows a reduction in both water and gas relative permeabilities as the back pressure increased, but k_{rg} reduction was higher than k_{rw} reduction. This behavior was due to the Klinkenberg (slippage) effect that can increase the effective gas permeability at low pressures.

Ma et al. (2017) investigated the coal softening, decrease of Young's modulus and an increase of Poisson's ratio, effect due to CO_2 adsorption in coal matrix on the injectivity performance (Figure 4.19). The injectivity was significantly affected by the changes in Young's modulus and Poisson's ratio. At early stages of injection, an increase of Poisson's ratio promotes an increase of permeability and injection efficiency while the reduction of Young's modulus has an opposite effect that decreases injection efficiency. With continued injecting, a decrease of Young's modulus has a dominant impact on reducing the rebound pressure required to

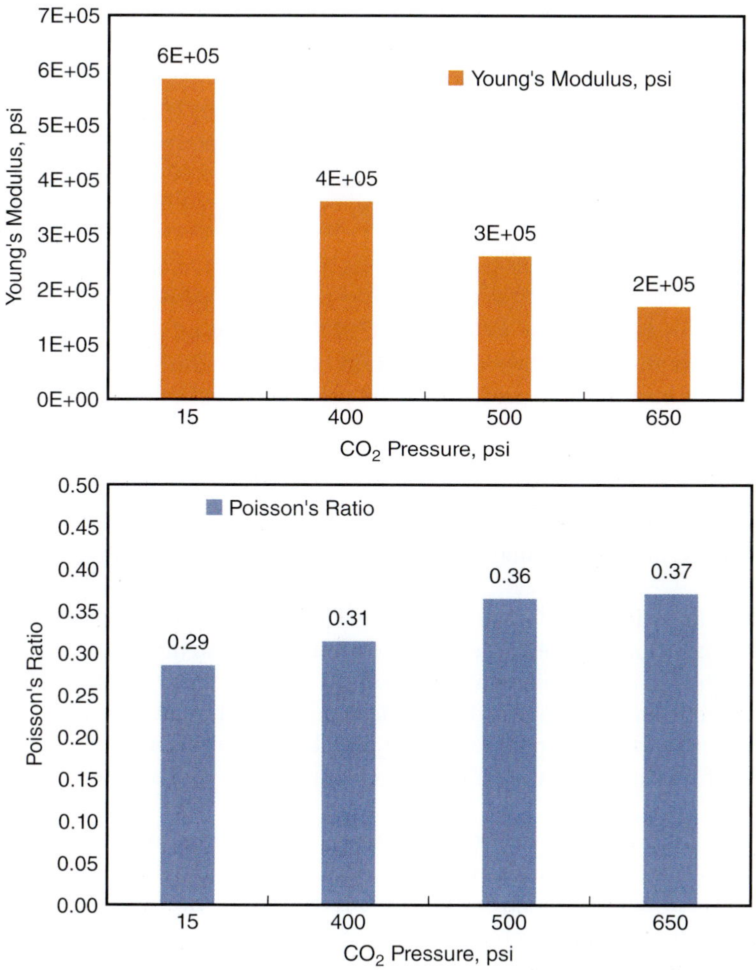

Figure 4.19 The effect of coal softening due to CO_2 adsorption on Young's modulus and Poisson's ratio of the coal formations, after (Ma et al. 2017).

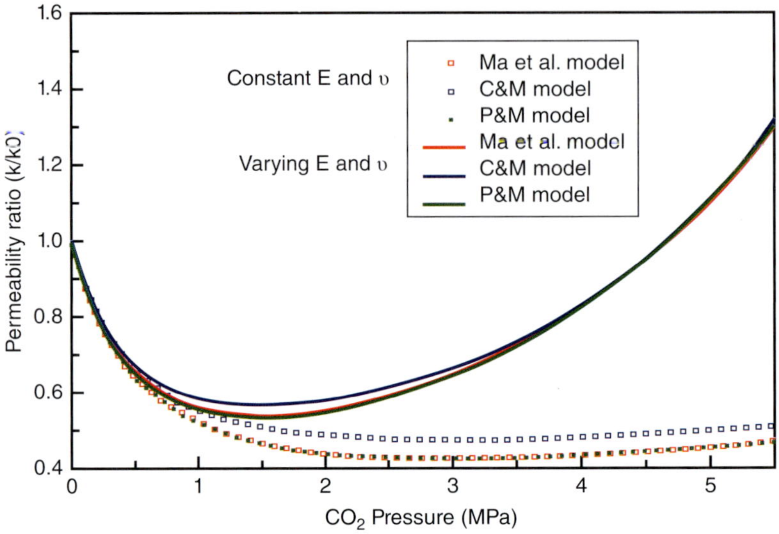

Figure 4.20 The effect of considering the coal softening on the formation absolute permeability using Ma et al. (2017), Cui and Bustin (2005), and Palmer and Mansoori (1998) models.

restore permeability, which significantly improves the longer-term CO2 injectivity and storage efficiency (Figure 4.20).

4.5 Pilot Field Tests

The Allison Unit in the San Juan Basin is the first CO_2-ECBM recovery pilot (Reeves and Oudinot 2005). Figure 4.21 shows the production and injection history for this unit. The CO_2 injection was applied at constant bottom-hole pressure (2,300 psi). The injection flow rate was started at 50 MMscf/month, then decreased with time to 20 MMscf/month. The reduction in the well injectivity was due to the swelling effect of the coal matrix. The increase in the injectivity of CO_2 (300 days after a decline) was due to the reduction in the reservoir pressure (the produced gas volume is larger than the injected gas volume). As a result, the CO_2 started to desorb from the coal surface and led to matrix shrinkage, increasing the absolute coal permeability.

To compare flue gas and N_2 injection for ECBM, the Tiffany Unit in the San Juan Basin was investigated for ECBM by N_2 injection. Figure 4.22 shows the production and injection rates in the Tiffany Unit. Almost no change in the injectivity was found. An earlier gas breakthrough was observed with an N_2 concentration of 16% in the produced gas, which related to low displacement efficiency. In comparison to CO_2 injection, a CO_2 breakthrough was minimal in the Allison Unit during the life of the injection with CO_2 at a concentration of 0.6% in the produced gas.

A series of CO_2 and N_2 injection cycles was conducted previously on the Yobari Test site in North Japan (Fujioka, Yamaguchi, and Nako 2010). A micro-pilot test was carried out from May 2004 to October 2007. Preliminary results from several injection tests indicated that CO_2 injectivity in fully water-saturated coal seams increased as the water saturation near the injector decreased during the injection of CO_2. The preliminary results of the pilot test showed that coal swelling and injectivity reduction were observed and the injection of N_2 counteracted the swelling effect.

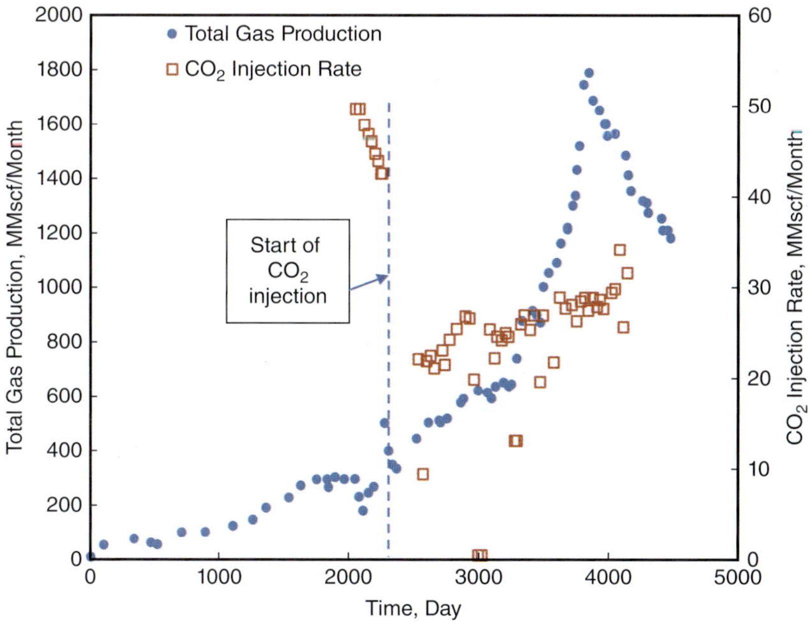

Figure 4.21 Gas injection and production flow in Allison Unit in the San Juan Basin.

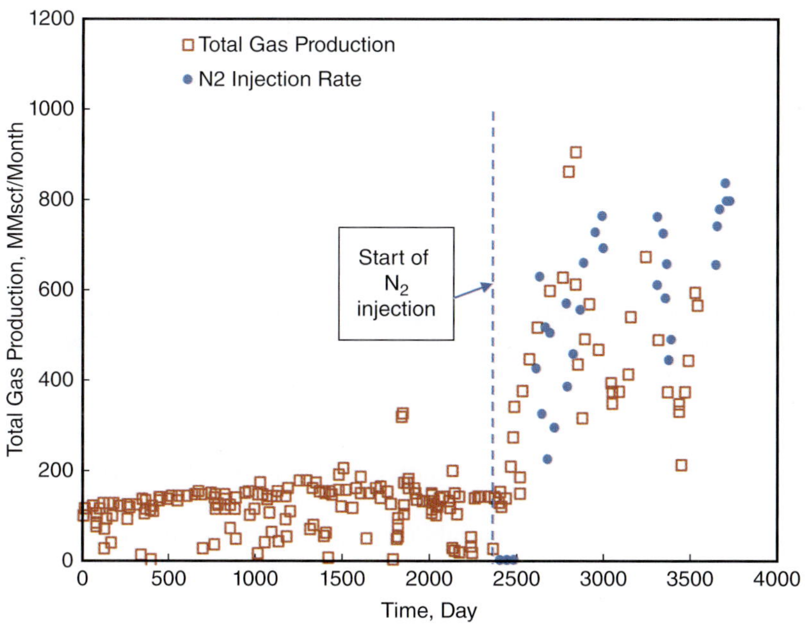

Figure 4.22 Gas injection and production flow in the Tiffany Unit in the San Juan Basin.

4.6 Conclusions

CO_2 can be sequestrated in coal seams to enhance the coalbed methane production (ECBM) in addition to CO_2 storage. The efficiency of this process is greatly affected by the coal formation properties and the operation conditions as following;

- As the coal rank increased from lignite to anthracite, gas adsorption capacity increases, and the coal becomes more hydrophobic.
- Adsorption capability decreases, and the coal becomes more water-wet as the formation temperature increases, or the formation pressure decreases.
- As the formation water salinity increases, the coal becomes more gas-wet, and the adsorption isotherm increases.
- During ECBM process, CO_2 injectivity increases initially as gas relative permeability near the well increases, then the injectivity decreases as a result of matrix swelling and absolute permeability reduction.
- As the water salinity in coal formations increases, the overall gas relative permeability decreases, and the water relative permeability increases. Similar behavior was found in the presence of N_2 in the injected gas.
- The injection flow rate controls the contact time of CO_2 on the coal surface, the CO_2 adsorption onto the coal matrix, and slightly affects the coal permeability.
- The presence of N_2 in the injected gas decreases the displacement efficiency, and reduces the swelling effect.

For carbon sequestration and ECBM application, the storage capacity of CO_2 in coal increased as the formation water salinity, coal rank, and the formation pressure increased. The gas storage as free gas increases, where the displacement efficiency increases as the coal become more CO_2 wet. Also, the adsorbed gas phase increases due to the CO_2 adsorption isotherm increase. Injectivity could be a special issue in the low-permeability and high-pressure formations; however, adding nitrogen to the injected CO_2 could reduce the matrix swelling effect and the formation injectivity.

References

Arif M, Barifcani A, Lebedev M, et al. 2016. CO_2-wettability of low to high rank coal seams: implications for carbon sequestration and enhanced methane recovery. Fuel, 181: 680–689. DOI: http://dx.doi.org/10.1016/j.fuel.2016.05.053.

ASTM D388-15. 2015. Standard classification of coals by rank. DOI: 10.1520/D0388-15.

Battistutta E, Van Hemert P, Lutynski M, et al. 2010. Swelling and sorption experiments on methane, nitrogen and carbon dioxide on dry selar Cornish Coal. Intl J Coal Geol, 84 (1): 39–48. DOI: http://dx.doi.org/10.1016/j.coal.2010.08.002.

Busch A, Gensterblum Y, Krooss BM. 2003. Methane and CO_2 sorption and desorption measurements on dry argonne premium coals: pure components and mixtures. Intl J Coal Geol, 55 (2–4): 205-224. DOI: http://dx.doi.org/10.1016/S0166-5162(03)00113-7.

Busch A, Gensterblum Y, Krooss BM, et al. 2006. Investigation of high-pressure selective adsorption/desorption behaviour of CO_2 and Ch4 on coals: an experimental study. Intl J Coal Geol, 66 (1–2): 53–68. DOI: http://dx.doi.org/10.1016/j.coal.2005.07.003

Busch A, Krooss BM, Gensterblum Y, et al. 2003. High-pressure adsorption of methane, carbon dioxide and their mixtures on coals with a special focus on the preferential sorption behaviour. J Geochem Explor, 78–79 (0): 671–674. DOI: http://dx.doi.org/10.1016/S0375-6742(03)00122-5.

Carbon Capture, Use and Storage (CCUS). 2017. CO_2 capture, storage and re-use: a solution for the future. Available from: http://www.captage-stockage-valorisation-co2.fr/en/coal-seams#!prettyPhoto.

Chen S, Jin L, Chen X. 2011. The effect and prediction of temperature on adsorption capability of coal/CH_4. Procedia Engg, 26 (Supplement C): 126–131. DOI: https://doi.org/10.1016/j.proeng.2011.11.2149.

Chiquet P, Broseta D, Thibeau S. 2007. Wettability alteration of caprock minerals by carbon dioxide. Geofluids 7 (2): 112–122. DOI: 10.1111/j.1468-8123.2007.00168.x.

Chow YT, Maitland GC, Trusler JP. 2016. Interfacial tensions of the (CO_2 + N_2 + H_2O) system at temperatures of (298 to 448) K and pressures up to 40 Mpa. J Chem Thermodynam, 93: 392–403. DOI: http://dx.doi.org/10.1016/j.jct.2015.08.006.

Clarkson CR, Bustin RM. 2000. Binary gas adsorption/desorption isotherms: effect of moisture and coal composition upon carbon dioxide selectivity over methane. Intl J Coal Geol, 42 (4): 241–271. DOI: http://dx.doi.org/10.1016/S0166-5162(99)00032-4,

Cui X, Bustin RM. 2005. Volumetric strain associated with methane desorption and its impact on coalbed gas production from deep coal seams. AAPG Bull, 89(9):1181–1202.

Espinoza DN, Santamarina JC. 2010. Water-CO_2-mineral systems: interfacial tension, contact angle, and diffusion—implications to CO_2 geological storage. Water Resour Res, 46 (7): n/a-n/a. DOI: 10.1029/2009WR008634.

Farokhpoor R, Bjørkvik BJ, Lindeberg E, et al. 2013. Wettability behaviour of CO_2 at storage conditions. Intl J Greenhouse Gas Control 12 (0): 18–25. DOI: http://dx.doi.org/10.1016/j.ijggc.2012.11.003.

Fujioka M, Yamaguchi S, Nako M. 2010. CO2-ECBM Field Tests in the Ishikari Coal Basin of Japan. Intl J Coal Geol, 82 (3): 287–298. DOI: https://doi.org/10.1016/j.coal.2010.01.004.

Garduno JL, Morand H, Saugier L, et al. 2003. CO_2 sequestration potential of Texas Low-Rank Coals. Presented at SPE AnnuAvailable from: al Technical Conference and Exhibition. Denver, Colorado, 5-8 October. SPE-84154-MS. https://doi.org/10.2118/84154-MS.

Gray I, 1987. Reservoir engineering in coal seams: Part 1—The physical process of gas storage and movement in coal seams. SPE Reserv Eng, 2(01): 28–34. Available from: http://dx.doi.org/10.2118/12514-PA.

Guo H, Cheng Y, Wang L, et al. 2015. Experimental study on the effect of moisture on low-rank coal adsorption characteristics. J Nat Gas Sci Engg, 24 (Supplement C): 245–251. DOI: https://doi.org/10.1016/j.jngse.2015.03.037.

Hamawand I, Yusaf T, Hamawand SG. 2013. Coal seam gas and associated water: a review paper. Renew Sustain Energ Revs, 22 (0): 550–560. DOI: http://dx.doi.org/10.1016/j.rser.2013.02.030.

Harpalani S, Prusty BK, Dutta P. 2006. Methane/CO_2 sorption modeling for coalbed methane production and CO_2 sequestration. Energ Fuels, 20 (4): 1591–1599. DOI: 10.1021/ef050434l.

Harpalani S, Schraufnagel RA. 1990. Shrinkage of coal matrix with release of gas and its impact on permeability of coal. Fuel, 69 (5): 551–556. DOI: http://dx.doi.org/10.1016/0016-2361(90)90137-F.

Ibrahim AF, Nasr-El-Din HA. 2015a. A comprehensive model to history match and predict gas/water production from coal seams. Intl J Coal Geol, 146 (Supplement C): 79–90. DOI: https://doi.org/10.1016/j.coal.2015.05.004.

Ibrahim AF, Nasr-El-Din HA. 2015b. History matching and predicting gas production during Ecbm. Paper presented at the Unconventional Resources Technology Conference, San Antonio, Texas, USA, 20-22 July. URTEC-2147766-MS. DOI: 10.15530/URTEC-2015-2147766.

Ibrahim AF, Nasr-El-Din HA. 2016. Effects of formation-water salinity, formation pressure, gas composition, and gas-flow rate on carbon dioxide sequestration in coal formations. SPE J 22 (5): 1530–1541. DOI: https://doi.org/10.2118/185949-PA.

Ibrahim AF, Nasr-El-Din HA. 2017. Effect of water salinity on coal wettability during CO_2 sequestration in coal seams. Energ Fuels 30 (9): 7532–7542. DOI: 10.1021/acs.energyfuels.6b01205.

Kelemen SR, Kwiatek LM. 2009. Physical properties of selected block argonne premium bituminous coal related to CO_2, CH_4, and N_2 adsorption. Intl J Coal Geol, 77 (1–2): 2–9. DOI: http://dx.doi.org/10.1016/j.coal.2008.05.020.

Kwok D, Neumann A. 1999. Contact angle measurement and contact angle interpretation. Advances Colloid Interface Sci, 81 (3): 167–249.

Laubach S, Marrett R, Olson J, Scott A. 1998. Characteristics and origins of coal cleat: a review. Int. J. Coal Geol. 35(01): 175–207. DOI: http://dx.doi.org/10.1016/S0166- 5162(97)00012-8.

Liu S, Harpalani S. 2013. Permeability prediction of coalbed methane reservoirs during primary depletion. Intl J Coal Geol, 113 (0): 1–10. DOI: http://dx.doi.org/10.1016/j.coal.2013.03.010.

Liu S, Harpalani S, Pillalamarry M. 2012. Laboratory measurement and modeling of coal permeability with continued methane production: part 1 – modeling results. Fuel 94 (0): 117–124. DOI: http://dx.doi.org/10.1016/j.fuel.2011.10.053.

Ma T, Rutqvist J, Liu W, et al. 2017. Modeling of CO_2 sequestration in coal seams: role of CO_2-induced coal softening on injectivity, storage efficiency and caprock deformation. Greenhouse Gas Sci Technol, 7(0): 562–578. http://dx.doi.org/10.1002/ghg.1664.

Mavor MJ, Gunter WD, Robinson JR, et al. 2002. Testing for CO_2 sequestration and enhanced methane production from coal. Presented at SPE Gas Technology Symposium, Calgary, Alberta, Canada, 30 April-2 May. SPE-75683-MS. https://doi.org/10.2118/75683-MS.

Mazumder S, Plug WJ, Bruining H. 2003. Capillary pressure and wettability behavior of coal–water–carbon dioxide system. Presented at the SPE Annual Technical Conference and Exhibition, Denver, Colorado, 5-8 October. SPE-84339-MS. DOI: 10.2118/84339-MS.

Millero F, Roy R. 1997. A chemical equilibrium model for the carbonate system in natural waters. Croatica Chemica Acta, 70 (1): 1–38.

Nasralla RA, Bataweel MA, and Nasr-El-Din HA. 2013. Investigation of wettability alteration and oil-recovery improvement by low-salinity water in sandstone rock. J Can. Petrol Tech, 52 (02): 144–154.

Palmer I, Mansoori J. 1998. How permeability depends on stress and pore pressure in coalbeds: a new model. SPE Res Eval Engg, 1 (6): 539–544. DOI: 10.2118/52607-pa.

Pini R, Storti G, Mazzotti M. 2011. A model for enhanced coal bed methane recovery aimed at carbon dioxide storage. Adsorption, 17 (5): 889–900. DOI: 10.1007/s10450-011-9357-z.

Plug W. J, Mazumder S, and Bruining J. 2008. Capillary Pressure and Wettability Behavior of CO_2 Sequestration in Coal at Elevated Pressures. SPE J. 13 (04): 455–464.

Reeves SR. 2003. Enhanced CBM recovery, coalbed CO_2 sequestration assessed. Oil Gas J, 101 (27): 49–53.

Reeves, S, Oudinot, A. 2005. The Allison Unit CO_2-ECBM Pilot–A Reservoir and Economic Analysis, 2005 International Coalbed Methane Symposium, pp. 16–20.

Saghafi A, Pinetown K, Javanmard H. 2014. Gas wettability of coal and implications for gas desorption and drainage. Paper presented at the 14th Coal Operators' Conference, University of Wollongong, The Australasian Institute of Mining and Metallurgy & Mine Managers Association of Australia, 2014, 266–273.

Sakurovs R, Lavrencic S. 2011. Contact angles in CO_2–water–coal systems at elevated pressures. Intl J Coal Geol, 87 (1): 26–32. DOI: http://dx.doi.org/10.1016/j.coal.2011.04.005

Schroeder K, Ozdemir E, Morsi B. 2004. Sequestration of carbon dioxide in coal seams. Gas 2: 2.

Seidle JP. 2000. Reservoir-engineering aspects of CO_2 sequestration in coals. J Petrol Tech, 52 (4): 64–66.

Shi JQ, Durucan S. 2005. CO_2 storage in deep unminable coal seams. Oil Gas Sci Tech Rev, IFP 60 (3): 547–558.

Shi JQ, Durucan S. 2004. Drawdown induced changes in permeability of coalbeds: a new interpretation of the reservoir response to primary recovery. Trans Porous Med 56(1):1–16.

shojai k. n, rudolph es, wolf kh, et al. 2011. wettability determination by contact angle measurements: hvbb coal–water system with injection of synthetic flue gas and CO_2. J Colloid Interface Sci, 364 (1): 237–247. DOI: http://dx.doi.org/10.1016/j.jcis.2011.07.091.

Shojai KN, Wolf KH, Ashrafizadeh SN, et al. 2012. Effect of coal petrology and pressure on wetting properties of wet coal for CO_2 and flue gas storage. Intl J Greenhouse Gas Control 11, Supplement (0): S91–S101. DOI: http://dx.doi.org/10.1016/j.ijggc.2012.09.009.

Siemons N, Bruining H, Castelijns H, et al. 2006. Pressure dependence of the contact angle in a CO_2–H_2O–coal system. J Colloid Interface Sci, 297 (2): 755–761. DOI: http://dx.doi.org/10.1016/j.jcis.2005.11.047.

U.S. Energy Information Administration. 2013. Coalbed methane production.

Zhou F, Hussain F, Cinar Y. 2013. Injecting pure N_2 and CO_2 to coal for enhanced coalbed methane: experimental observations and numerical simulation. Intl J Coal Geol, 116 (Supplement C) 53–62. DOI: https://doi.org/10.1016/j.coal.2013.06.004

5

Fluid Flow: Basics

Paul A. Nelson, Todd J. Willman, and Vinay Gadekar

**EPCON Software™, Houston, USA*

CHAPTER MENU

5.1 Introduction

Understanding the mechanics of fluid flow is an important aspect of designing and optimizing process and utility systems. For process systems, the throughput can be significantly increased by identifying and resolving high pressure drop and two-phase flow conditions. For water systems (i.e., steam, condensate, boiler feedwater, cooling water, process water), flow simulation modeling can identify opportunities for savings in fuel, electricity, and chemical treatment costs while reducing water usage and increasing overall capacity. Detailed flow simulation of fire water, flare, and pressure relief valve systems also improves process safety.

This chapter summarizes the relevant background information for performing fluid mechanical calculations. It is intended for use by anyone without formal training in fluid mechanics, and should serve as a useful refresher for those who have taken fluid mechanics course work. The next chapter is a companion piece, and presents some material that is normally not covered in an academic curriculum.

The authors are affiliated with EPCON Software (www.epcon.com), a company that has been providing practical process engineering software and services to the process industries since 1981. The fluid mechanics calculation methods described herein are performed in the EPCON Process Innovator® simulation software used by thousands of engineers worldwide.

The field of fluid mechanics is vast, and no attempt has been made to cover it in depth. We focus our attention on one-dimensional subsonic flow in pipes and fittings. This chapter is no substitute for a good text on fluid mechanics for someone who wants a more comprehensive understanding of the field. Many such texts are available, and reference will be made to the ones known best to the author, although these tend to be dated (See [1, 2].)

*www.epcon.com

Natural Gas Processing from Midstream to Downstream, First Edition.
Edited by Nimir O. Elbashir, Mahmoud M. El-Halwagi, Ioannis G. Economou, and Kenneth R. Hall.

There is no escaping the use of mathematics to describe the flow of fluids. The mathematics is not overly complex, and should be readily understood by a reader who has a working knowledge of calculus. The reader without that arrow in his/her quiver will probably not be able to follow all of the derivations, but should still be able to use the results that are presented.

Units of Measure

Consistency in the use of units of measure is vital if correct results are to be obtained. The common textbook practice is to introduce fluid mechanical concepts without much reference to units of measure, and then to sort it out later. In this chapter, generally speaking, we will use the English system of units. However, we will avoid the use of slugs and poundals for mass and force respectively, as these units are unfamiliar to many. Instead we will use pounds mass (lb_m) and pounds force (lb_f). It is critical that we carefully distinguish between these two very different usages of the generic term "pound." We will always have to specify one or the other, and never use the word "pound" alone. This means that we will have to introduce the gravitational conversion constant, g_c, which is often a source of confusion. Hopefully, by introducing g_c early in context and using it consistently we will be able to demystify it.

It is common for engineers to use both inches and feet for units of length, e.g., pressure in psia, and pipe diameter and roughness in inches, with other units of length in feet. In this chapter, we will depart from that practice and consistently use feet as our unit of length, and psfa as our unit of pressure. The material is heavily mathematical, and will be hard enough to understand without throwing in the additional complication of factors of 12 and 144 scattered throughout. Furthermore, the equations derived here will appear pretty much as they do in textbooks and literature references, and there is an advantage to that.

An exception to the use of English units is the dynamic viscosity. The use of centipoise (cp) for viscosity is essentially universal, no doubt because of the convenient fact that the viscosity of water at room temperature is about 1 cp. The English unit for viscosity is $lb_m/(ft\ sec)$, and we will have to convert to this unit from cp when viscosity enters an equation, e.g., in the calculation of the Reynolds number. The conversion factor is $1\ cp = 6.719688 \times 10^{-4}\ lb_m/(ft\ sec)$.

The units used for energy will be ft lb_f. The common English unit for thermal energy, as in enthalpy and heat capacity, is the British Thermal Unit (BTU), defined as the amount of energy required to raise the temperature of one pound mass of water by one degree Fahrenheit. The conversion factor (also known as "Joule's Constant") is $1\ BTU = 778.1938\ ft\ lb_f$.

Notation

A	= Cross-sectional area, ft^2
a	= Acceleration, ft/sec^2
	= Parameter in Eq. (5.30) and Eq. (5.31), $(lb_f\ ft^4)/lbmole^2$
b	= Parameter in Eq. (5.30) and Eq. (5.31), $ft^3/lbmole$
C	= Circumference, ft
C_n	= Generic notation for a constant
c	= Speed of sound, ft/sec
c_p	= Heat capacity at constant pressure, $(ft\ lb_f)/(lb_m\ R)$
c_v	= Heat capacity at constant volume, $(ft\ lb_f)/(lb_m\ R)$
D	= Pipe diameter, ft

F = Force, lb_f

f = Darcy friction factor, dimensionless

f_F = Fanning Friction factor, dimensionless

g = Acceleration of gravity = 32.174 ft/sec^2

g_c = Gravitational conversion constant = 32.174 (lb_m ft)/(sec^2 lb_f)

H = Enthalpy, (ft lb_f)/lb_m

h = Fluid head, (ft lb_f)/lb_m (note that this is *not* in units of ft)

h_e = Head loss in flow through equipment associated with a pipe, (ft lb_f)/lb_m

h_f = Frictional head loss, (ft lb_f)/lb_m

h_p = Head gain from a pump, (ft lb_f)/lb_m

K = Resistance factor for pipe fitting (K-factor), dimensionless

k = Polytropic exponent, dimensionless

L = Pipe length, ft

L_{max} = Maximum pipe length in compressible flow, ft

ℓ = Element of length in the direction of a streamline, ft

M = Mach number, dimensionless

M_w = Molecular weight, lb_m/lbmole

m = Mass, slug (= (lb_f sec^2)/ft) – See discussion of Eq. (5.37)

N_{Re} = Reynolds number, dimensionless

P = Pressure, lb_f/ft^2 (psfa)

Q = Heat added to (or removed from) a system, (ft lb_f)/lb_m

= Volumetric flow rate, ft^3/sec

q = Mass flow rate, lb_m/sec

R = Universal gas constant = 1545.349 (ft lb_f)/(lbmole R)

S = Entropy, (ft lb_f)/(lb_m R)

s = Safety factor (or fouling factor), dimensionless

T = Temperature, degrees Rankine (R)

t = Time, sec

U = Internal energy, (ft lb_f)/lb_m

V = Specific volume, ft^3/lb_m

$\mathcal{V}$ = Molar volume, ft^3/lbmole

v = Velocity, ft/sec

W = Work done by a system on its environment, (ft lb_f)/lb_m

W_s = Shaft work done by a system on its environment, (ft lb_f)/lb_m

Z = Gas compressibility, dimensionless

z = Elevation, ft

$\alpha(T)$ = Function of temperature in SRK and P-R equations of state, Eqs. (5.30) and (5.31)

γ = Ratio of heat capacities (isentropic exponent) = c_p/c_v, dimensionless

ε = Pipe roughness, ft

θ = Angle of inclination of pipe from vertical, radians

μ = Dynamic viscosity, cP (or in English units, lb_f-sec/ft^2)

ρ = Density, lb_m/ft^3

τ = Shear stress, lb_f/ft^2

5.2 Thermodynamics of Fluids

We will not need to concern ourselves with much thermodynamics until we come to the study of compressible fluids. However, we will need to understand the First Law of Thermodynamics and the definition of enthalpy when we derive the total energy balance in the next section. So, it makes sense to just go ahead and introduce as much thermodynamics as we will need at this point. The field of thermodynamics is vast, and we will present only as much as is needed to understand the impact of compressibility on fluid flow. There are many fine textbooks on thermodynamics which will provide a much more comprehensive introduction to the field with Ref. [3] being highly recommended. Accurate thermophysical properties are vital to process simulation, fluid flow simulation, process equipment sizing, and process safety analysis. Various industry groups have addressed this need. In particular, the American Petroleum Institute has produced the API Technical Data Book [4]. See Figure 5.1 for a screenshot of a typical inquiry to the API Technical Data Book.

In the text that follows we frequently refer to a "system." By "system" we simply mean a defined volume of fluid. If we are referring to a "closed system," we mean by that a certain collection of fluid particles which does not exchange mass with its environment. If we are referring to an

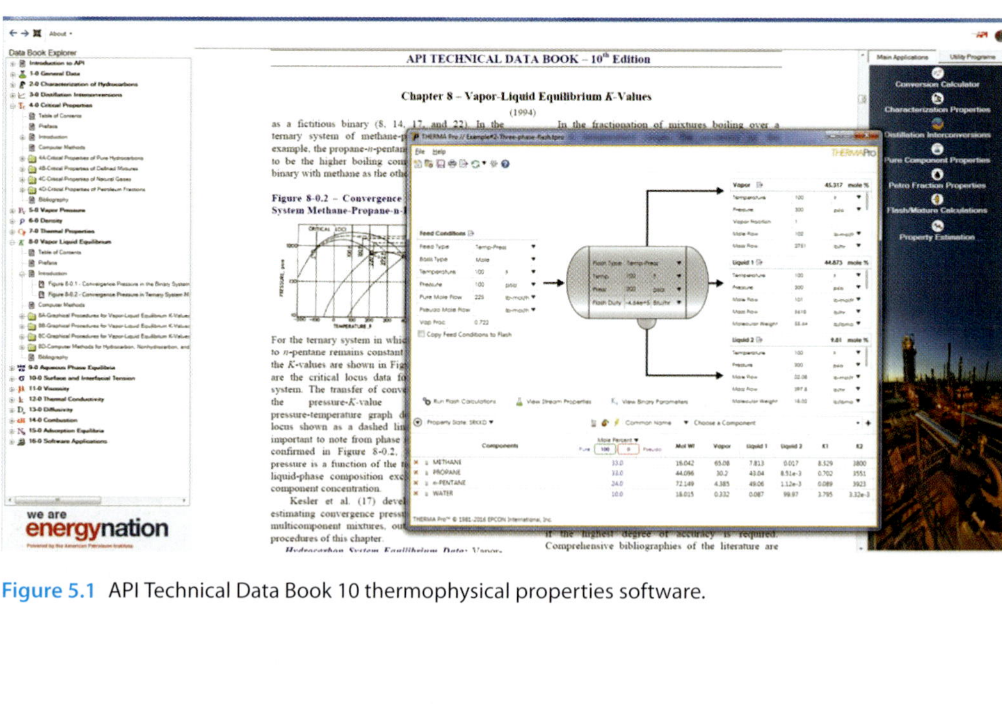

Figure 5.1 API Technical Data Book 10 thermophysical properties software.

"open system," we mean by that the fluid contained within a defined volume of space, through which it may be flowing.

5.2.1 First Law of Thermodynamics

The First Law of Thermodynamics states that energy is neither created nor destroyed; it is conserved. More specifically, it states that in a closed system the change in internal energy (dU) is equal to the heat added (dQ) minus any work (dW) that the system does on its environment. Mathematically, this is stated as:

$$dU = dQ - dW \tag{5.1}$$

Now for the moment we confine our attention to a "reversible" process, in which the thermal and mechanical forces driving the exchange of energy and work between the system and its environment differ only infinitesimally from a state of equilibrium. We introduce a quantity which we will call "entropy" (S). We do not address the absolute value of S, but only how its value changes during an exchange of heat:

$$dS = \frac{dQ}{T} \tag{5.2}$$

If the only mechanism for exchange of work between the system and its environment is an infinitesimal change of volume at pressure P, then we can write:

$$dW = PdV \tag{5.3}$$

If we substitute Eqs. (5.2) and (5.3) into Eq. (5.1) we obtain.

$$dU = TdS - PdV \tag{5.4}$$

In thermodynamic textbooks (such as Ref. [3]), it is proven that S is a state variable, and we will take that as given here. With that we note that every variable in Eq. (5.4) is a state variable, meaning that it is dependent only on the state of the system, and not on the process(es) by which the system attained that state. This means that Eq. (5.4) is true even if the process by which the change of energy content takes place is not reversible, i.e., even if Eqs. (5.2) and (5.3) are not valid. We can take Eq. (5.4) to be a fundamental statement of the First Law of Thermodynamics.

When we come to the consideration of the work exchanged in an open system, we need to make a distinction between "shaft work" (W_s) added to or withdrawn from the fluid, and the work which is performed simply by virtue of the fact that fluid may enter the defined space at one pressure and density, and leave at another. We call the latter "pressure-volume work."

Consider a volume fixed in space through which a quantity, q (in mass flow units), of fluid flows at steady state. The fluid enters across a surface of area A_1 with internal energy U_1, pressure P_1, and specific volume V_1. It leaves across a surface of area A_2, which need not be the same as A_1, with internal energy U_2, pressure P_2, and specific volume V_2. The fluid entering the system during an infinitesimal interval of time dt, acts with a force P_1A_1, and moves a distance $d\ell_1 = v_1 dt = \frac{qV_1}{A_1}dt$. So, the work that the entering fluid does on the system is $(P_1A_1)\left(\frac{qV_1}{A_1}dt\right) = P_1V_1qdt$. Likewise, the work that the leaving fluid does on the environment is P_2V_2qdt. The mass of fluid flowing through the system during this time interval is qdt. So, the net work done by the system upon its environment per lb_m of fluid is $(P_2V_2 - P_1V_1)$.

Now consider Eq. (5.1) integrated between sections 1 and 2:

$$U_2 - U_1 = Q - W \tag{5.5}$$

If we divide the work term into two parts, shaft work and pressure-volume work, we have:

$$U_2 - U_1 = Q - W_s - P_2V_2 + P_1V_1 \tag{5.6}$$

Rearranging Eq. (5.6):

$$(U_2 + P_2V_2) - (U_1 + P_1V_1) = Q - W_s \tag{5.7}$$

The combination of terms $U + PV$ is so important for the study of flowing systems that it has been given its own name ("enthalpy"), and symbol (H), defined as:

$$H = U + PV \tag{5.8}$$

For an open system, the enthalpy has the same significance as does the internal energy for a closed system. For an open system, the expression of the First Law of Thermodynamics equivalent to Eq. (5.1) is:

$$dH = dQ - dW_s \tag{5.9}$$

If we write Eq. (5.8) in differential form and substitute for dU using Eq. (5.4), we find that the expression of the First Law of Thermodynamics for an open system equivalent to Eq. (5.4) is:

$$dH = TdS + VdP \tag{5.10}$$

5.2.2 Second Law of Thermodynamics

The Second Law of Thermodynamics states that in a closed system the entropy cannot decrease.

$$dS \geq 0 \tag{5.11}$$

A complete proof is beyond the scope of this chapter, but a simple example will illustrate the principle. Consider a closed system consisting of two fluid volumes at different temperatures, T_1 and T_2, where $T_1 > T_2$, separated by a wall through which heat but not fluid can pass. If an infinitesimal amount of heat, dQ, passes from volume 1 to volume 2 the entropy of volume 1 decreases by an amount $\frac{dQ}{T_1}$, Eq. (5.2), and the entropy of volume 2 increases by $\frac{dQ}{T_2}$. The net change of entropy of the system is:

$$dS = -\frac{dQ}{T_1} + \frac{dQ}{T_2} > 0 \tag{5.12}$$

If the heat were to transfer in the opposite direction, from T_2, the lower temperature, to T_1, the entropy change would be negative. The First Law of Thermodynamics would not be violated. However, it is a simple fact of experience that this is never observed to happen. Thus, the Second Law of Thermodynamics puts our everyday experience into mathematical form.

5.2.3 Heat Capacity

The amount of energy that must be added to a quantity of fluid to raise its temperature one degree is known as the heat capacity of the fluid. To examine this, we rearrange Eq. (5.1) and take its derivative with respect to temperature:

$$\frac{dQ}{dT} = \frac{dU}{dT} + \frac{dW}{dT} \tag{5.13}$$

We consider two different measures of heat capacity. If the fluid is held at constant volume, no work is done by the fluid on its environment, and the second term on the right-hand side is zero. We obtain the "heat capacity at constant volume":

$$c_v = \left(\frac{\partial Q}{\partial T}\right)_V = \left(\frac{\partial U}{\partial T}\right)_V \tag{5.14}$$

If the fluid is held at constant pressure, pressure-volume work in the amount $dW = PdV$ will be done as the fluid expands against its environment. We obtain the "heat capacity at constant pressure":

$$c_p = \left(\frac{\partial Q}{\partial T}\right)_P = \left(\frac{\partial U}{\partial T}\right)_P + P\left(\frac{\partial V}{\partial T}\right)_P = \left(\frac{\partial H}{\partial T}\right)_P \tag{5.15}$$

where we have brought in Eq. (5.8).

The ratio between the two heat capacities is an important parameter in its own right, known as the "isentropic exponent":

$$\gamma = \frac{c_p}{c_v} \tag{5.16}$$

The heat capacity at constant pressure is always greater than the heat capacity at constant volume because of the work term. Therefore γ is always greater than unity, and it can be shown from statistical mechanics that the upper limit is $\frac{5}{3}$. This applies to the simplest (i.e., monatomic) molecules. The larger and more complex the molecule, the lower the value of γ. For air, γ is about 1.4.

5.2.4 Properties of a Perfect Gas

The concept of a perfect gas is an idealization; no gas is quite perfect. However, in many applications a real gas will behave very much like a perfect gas. When that is the case, the use of the perfect gas assumption often produces an analytical solution where otherwise none would be possible.

For a perfect gas, pressure, specific volume, and temperature are related by the "ideal gas law":

$$PV = \frac{P}{\rho} = \frac{RT}{M_w} \tag{5.17}$$

where R is known as the "universal gas constant," and is experimentally determined to be 1545.349 (ft lb$_f$)/(lbmole R). Thermodynamics textbooks generally start with a different definition of a perfect gas, and show that Eq. (5.17) is a consequence of that definition, which also includes the finding that the internal energy of a perfect gas, U, is a function of temperature only. Since by Eq. (5.17) the product PV is a function of temperature only, it follows that the enthalpy, defined in Eq. (5.8), is also a function of temperature only. This means that the partial derivatives in Eqs. (5.14) and (5.15) can be replaced by ordinary derivatives:

$$c_v = \frac{dU}{dT} \tag{5.18}$$

$$c_p = \frac{dH}{dT} \tag{5.19}$$

From the definition of enthalpy, Eq. (5.8), and from Eqs. (5.18) and (5.19) with Eq. (5.17), it is apparent that:

$$c_p = c_v + \frac{R}{M_w} \tag{5.20}$$

Eliminating c_v between Eqs. (5.16) and (5.20), we have a relationship that we will find useful:

$$c_p = \frac{\gamma}{\gamma - 1}\frac{R}{M_w} \tag{5.21}$$

Likewise, eliminating c_p between the same two equations yields:

$$c_v = \frac{1}{\gamma - 1}\frac{R}{M_w} \tag{5.22}$$

Let us return to the First Law of Thermodynamics in the form of Eq. (5.4) and consider a process where the entropy does not change, i.e., $dS = 0$. The first term on the right-hand side drops out, and we can substitute for dU on the left-hand side using Eq. (5.18) to yield

$$c_v dT = -PdV \tag{5.23}$$

Dividing through by T, and using Eqs. (5.17) and (5.22):

$$\frac{dT}{T} = -(\gamma - 1)\frac{dV}{V} \tag{5.24}$$

which we can integrate to:

$$\ln T = -(\gamma - 1)\ln V + C_1 \tag{5.25}$$

Eq. (5.25) can be rearranged to the following important relationship for an isentropic process:

$$TV^{\gamma-1} = e^{C_1} = C_2 \tag{5.26}$$

If we substitute for T in Eq. (5.26) using Eq. (5.17) we find:

$$PV^{\gamma} = \frac{P}{\rho^{\gamma}} = \frac{R}{M_w}C_2 = C_3 \tag{5.27}$$

If we substitute for V in Eq. (5.26) using Eq. (5.17), we find:

$$\frac{T^{\frac{\gamma}{\gamma-1}}}{P} = \frac{M_w}{R}C_2^{\frac{1}{\gamma-1}} = C_4 \tag{5.28}$$

Note that we have given each "constant" in Eqs. (5.25) through (5.28) a unique label to emphasize the fact that they are not all the same constant.

5.2.5 Equations of State

Eq. (5.17) is the simplest example of an "equation of state," i.e., an equation that expresses a relationship between the three primary properties of a system: temperature, pressure, and volume. The equation applies pretty well at relatively high temperatures and at relatively low pressures. It is normally a good approximation at atmospheric conditions. However, as pressure increases and/or temperature decreases, the relationship between these three properties departs from the perfect gas law. A measure of the degree of departure is the "compressibility," defined as:

$$Z = \frac{PVM_w}{RT} = \frac{PM_w}{\rho RT} \tag{5.29}$$

When the perfect gas law applies, $Z = 1$.

A great deal of work has gone into the search for more realistic relationships between T, P, and V. No equation has been found that is satisfactory in every respect, but many are useful within defined regions of applicability. Arguably the most popular equation of state for process industry work is the Soave-Redlich-Kwong (SRK) equation of state:[1]

$$P = \frac{RT}{\mathcal{V} - b} - \frac{a\alpha(T)}{\mathcal{V}(\mathcal{V} + b)} \tag{5.30}$$

where a and b are constants to be determined, and $\alpha(T)$ is a function of temperature which was proposed by Soave. Note that this equation, as is customary, is written in terms of the molar volume, which we denote by $\mathcal{V}$. The relationship between molar volume and specific volume is $\mathcal{V} = M_w V$. The SRK equation of state is pressure-explicit. That is, the pressure is expressed explicitly in terms of temperature and molar volume. It is also a "cubic equation." While not obvious from the form of Eq. (5.30), it turns out that with some manipulation, at constant temperature, the equation can be written as a cubic equation in terms of the compressibility. This means that the equation has some capacity to represent the existence of multiple roots (i.e., vapor, liquid, and an intermediate, unstable, root) over some regions of P-$\mathcal{V}$-T space. It also means that the equation can be solved directly using Cardan's solution.

Another equation of state, popular in the natural gas–processing industry, is the Peng-Robinson (P-R) equation of state:[2]

$$P = \frac{RT}{\mathcal{V} - b} - \frac{a\alpha(T)}{\mathcal{V}(\mathcal{V} + b) + b(\mathcal{V} - b)} \tag{5.31}$$

The benefits of a good equation of state go far beyond the relationship between P, $\mathcal{V}$, and T. It can be shown that all of the thermodynamic functions can be derived from an equation of state. Further discussion of equations of state is beyond the scope of this report, but can be found elsewhere (see Ref. [5]).

5.3 Fundamental Equations of Fluid Mechanics

5.3.1 Continuity Equation

Density and velocity can change with distance in pipeline flow, but the one property that remains constant is the total mass rate of flow. As normally derived in a fluid mechanics textbook, the equation of continuity is a partial differential equation relating fluid density to velocity in three dimensions. For our purposes, we can use a much simpler relationship applicable to one-dimensional flow:

$$q = \rho v A \tag{5.32}$$

If we consider the flow through two different cross-sections, $q_1 = q_2$, and Eq. (5.32) leads us to:

$$\rho_1 v_1 A_1 = \rho_2 v_2 A_2 \tag{5.33}$$

When the flow channel has a constant cross-sectional area, as is normally the case in pipe flow:

$$\rho_1 v_1 = \rho_2 v_2 \tag{5.34}$$

1 Soave, G. "Equilibrium Constants from a Modified Redlich-Kwong Equation of State, *Chem. Eng. Sci.*, **27**, 1197–1203 (1972)

2 Peng, D. -Y. and Robinson, D. B., "A New Two-Constant Equation of State", Ind. Eng. Chem Fundam., **15**, 59–64 (1976)

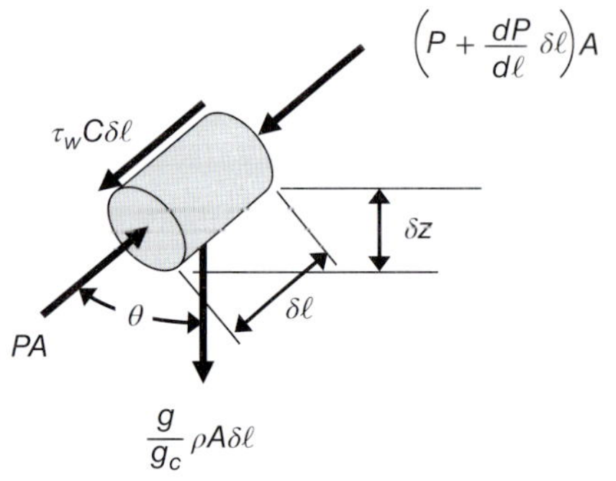

Figure 5.2 Forces acting on the fluid within an element of pipe.

5.3.2 Momentum Balance

We can develop an equation of motion for a fluid element in a pipe by considering all of the forces acting on the element, and using Newton's Second Law to calculate the change in momentum of a fluid element as a result of the net force.

Refer to Figure 5.2 for a diagram of the forces acting on the fluid within an element of pipe with constant cross-sectional area A, circumference C, and length $\delta\ell$. (Flow is from lower left to upper right.) The pressure force on the upstream face of the element is PA. The pressure force on the downstream face of the element acts in the opposite direction with magnitude $\left(P + \frac{dP}{d\ell}\delta\ell\right)A$. The effect of gravity is a downward-acting force of magnitude $\frac{g}{g_c}\rho A\delta\ell$. The component of this force acting in the direction of the flow is $-\frac{g}{g_c}\rho A\cos\theta\delta\ell$. A comment is necessary about the inclusion of the factor $\frac{g}{g_c}$. It has numerical value unity, and units of lb_f/lb_m. The effect of gravity is a force of 1 lb_f for each unit of mass in lb_m. To maintain consistency in units of measure, we must include this factor even though its numerical value is unity. The motion of the fluid is opposed by a force arising from wall friction of magnitude $\tau_w C\delta\ell$ acting in the direction opposite to the flow.

We now apply Newton's Second Law to the forces and acceleration acting on the fluid element contained in the control volume, $A\delta\ell$. The familiar form of Newton's Second Law is:

$$F = ma \tag{5.35}$$

In accordance with our preferred units of measure, F is in units of lb_f, and a is in units of ft/sec^2. But what is the value, and what are the units of m? We must make a clear distinction between weight and mass, and then establish the relationship between them. Consider a cubic foot of material with density ρ acted upon by the force of gravity. At rest, supported by a restraint, the pull of gravity upon this mass results in a force of $\frac{g}{g_c}\rho$ lb_f in the downward direction, which is its weight. If the restraint is removed, and the mass is allowed to fall freely under the pull of gravity, it will accelerate at a rate of $g = 32.174$ ft/sec^2. Now Eq. (5.35) becomes:

$$\frac{g}{g_c}\rho = mg \tag{5.36}$$

Solving Eq. (5.36) for m, we see that the relationship between mass and weight is:

$$m = \frac{\rho}{g_c} \tag{5.37}$$

In a consistent set of English units, the unit of mass is the slug. One slug is the mass that is accelerated at a rate of 1 ft/sec^2 by a force of 1 lb_f. The use of this unit for mass enables us to dispense with g_c, but introduces an unfamiliar unit of measure. We have chosen to stay with familiar units and to introduce g_c at this point.

Per the above, the mass of the fluid to be entered into Newton's Second Law is $\frac{\rho}{g_c}A\delta\ell$, and this mass experiences an acceleration $\frac{dv}{dt}$.

Plugging all of these terms into Eq. (5.35) we obtain:

$$PA - \left(P + \frac{dP}{d\ell}\delta\ell\right)A - \frac{g}{g_c}\rho A\cos\theta\delta\ell - \tau_w C\delta\ell = \frac{\rho}{g_c}A\delta\ell\frac{dv}{dt} \tag{5.38}$$

Simplifying Eq. (5.38) and dividing through by ρA:

$$\frac{1}{\rho}\frac{dP}{d\ell}\delta\ell + \frac{g}{g_c}\cos\theta\delta\ell + \frac{\tau_w C}{\rho A}\delta\ell + \frac{1}{g_c}\frac{dv}{dt}\delta\ell = 0 \tag{5.39}$$

Now noting that $\cos\theta = \frac{dz}{d\ell}$, $\frac{dv}{dt} = \frac{dv}{d\ell}\frac{d\ell}{dt} = v\frac{dv}{d\ell}$, $C = \pi D$, and $A = \frac{\pi D^2}{4}$, Eq. (5.39) becomes:

$$\frac{1}{\rho}dP + \frac{g}{g_c}dz + \frac{4\tau_w}{\rho D}\delta\ell + \frac{v}{g_c}dv = 0 \tag{5.40}$$

To proceed further we need a functional form for the wall shear stress, τ_w. It has been found that a good fit to data is obtained by representing the wall shear stress as a quadratic function of fluid velocity:

$$\tau_w = f_F\frac{\rho v^2}{2g_c} \tag{5.41}$$

The parameter f_F is called the "Fanning friction factor." Presently we will introduce the "Darcy friction factor," f, and will prove a very simple relationship between the two, namely that $f = 4f_F$. For now, we carry along the Fanning friction factor as being the one that represents the wall shear stress. Substituting Eq. (5.41) into Eq.(5.40):

$$\frac{1}{\rho}dP + \frac{g}{g_c}dz + \frac{4f_F}{D}\frac{v^2}{2g_c}d\ell + \frac{v}{g_c}dv = 0 \tag{5.42}$$

5.3.3 Bernoulli's Equation

The analysis of Figure 5.2 can also be performed along a streamline where wall shear stress is not a factor. Instead of Eq. (5.42) we then obtain:

$$\frac{1}{\rho}dP + \frac{g}{g_c}dz + \frac{v}{g_c}dv = 0 \tag{5.43}$$

For the case of incompressible flow, where ρ is not a function of P, we can integrate Eq. (5.43) to obtain:

$$\frac{1}{\rho}P + \frac{g}{g_c}z + \frac{v^2}{2g_c} = C_1 \tag{5.44}$$

If we consider the flow through two different cross sections, we have:

$$\frac{P_1}{\rho} + \frac{g}{g_c}z_1 + \frac{{v_1}^2}{2g_c} = \frac{P_2}{\rho} + \frac{g}{g_c}z_2 + \frac{{v_2}^2}{2g_c} \tag{5.45}$$

Eq. (5.45) is Bernoulli's equation in the form it is most commonly encountered for the solution of fluid flow problems. In textbooks, the second term on each side normally appears without the factor $\frac{g}{g_c}$. This apparent inconsistency is waved off with the assertion that a unit of "ft lb/lb" is equivalent to a unit of "ft." This overlooks the fact that the "lb" (= lb_f) in the numerator of "ft lb/lb" is not at all the same thing as the "lb" (= lb_m) in the denominator. No numerical error is made by this oversight, but it leads to the misconception that the elevation head, and by extension other fluid head terms, are measured in units of length when, in fact, they are measures of energy content per unit of mass.

Eq. (5.45) can be rearranged in a variety of ways, and may appear in different forms in different sources. But the form given in Eq. (5.45) is preferable, in that each term is a specific contribution to the mechanical energy per unit of mass.

Though derived for flow along a streamline, Bernoulli's equation is frequently applied to flow in a pipeline. When this is done, it is necessary to introduce some additional terms for practical applications. First of all, we introduce a head loss term, h_f, into Eq. (5.45) to account for frictional energy loss. A given pipe section may also contain a pump that increases the head, and to represent this we incorporate a head gain term, h_p. As will be seen in a moment, the term h_f, will be taken to be proportional to the square of the fluid velocity. This provides a good fit, not only to head loss as a result of wall friction, but also to loss due to fittings and some pieces of process equipment. However, in the event that a piece of equipment is included in the pipe which does not lend itself to representation in terms of the square of the velocity we include an additional head loss term due to equipment, h_e,. Eq. (5.45) thus becomes:

$$\frac{P_1}{\rho} + \frac{g}{g_c}z_1 + \frac{{v_1}^2}{2g_c} + h_p = \frac{P_2}{\rho} + \frac{g}{g_c}z_2 + \frac{{v_2}^2}{2g_c} + h_f + h_e \tag{5.46}$$

The terms h_p and h_e represent individual pieces of equipment, and their functional form in terms of fluid velocity must be individually determined. It is common to represent these terms as polynomials in some variable related to fluid velocity (e.g., volumetric flow rate).

For the frictional loss term, h_f, it has been found that a good fit to data can be obtained by representing the head loss as a quadratic function of velocity:

$$h_f = \frac{fL}{D}\frac{v^2}{2g_c} \tag{5.47}$$

Eq. (5.47) is called the Darcy Equation, and the parameter f is called the Darcy friction factor. It does have some dependence upon flow rate, and is generally charted as a function of the pipe Reynolds number, N_{Re}, (to be introduced in Subsection 5.4.1) with pipe roughness as a parameter. We will deal more with this friction factor in Subsection 5.4.2. Substituting Eq. (5.47) into Eq. (5.46), and neglecting for the moment h_p and h_e, we have:

$$\frac{P_1}{\rho} + \frac{g}{g_c}z_1 + \frac{{v_1}^2}{2g_c} = \frac{P_2}{\rho} + \frac{g}{g_c}z_2 + \frac{{v_2}^2}{2g_c} + \frac{fL}{D}\frac{v^2}{2g_c} \tag{5.48}$$

5.3.4 Mechanical Energy Balance

We know from the First Law of Thermodynamics that energy is conserved. In fluid flow problems we normally encounter energy in two forms, mechanical and thermal. It is actually the sum of the two that is conserved. A mechanical energy balance deals with only one side of this conservation relationship, and Eq. (5.43) is such a mechanical energy balance. Let us consider the units of measure in Eq. (5.43), and the significance of each term. The first and third terms clearly have units of ft lb_f/lb_m. As such they represent energy in the form of ft lb_f per unit of mass. But for the inclusion of the factor $\frac{g}{g_c}$, the second term would have units of feet. By including this factor, even though its numerical value is unity, we see that this term also represents energy in the form of ft lb_f/lb_m. The first term represents potential energy in the form of capacity to drive fluid flow. The second term represents potential energy in the form of work that has been performed upon the fluid in order to raise it to its elevation. And the third term represents kinetic energy. Together, these three terms represent the mechanical energy of the fluid that is conserved as the fluid moves along a streamline unless external forces act to increase or decrease its mechanical energy.

When there is conversion of mechanical energy into thermal energy, e.g., through the effect of wall friction, it is still possible to derive an equation for the mechanical energy balance by incorporating a frictional loss term, and that is how we got to Eq. (5.48). If we apply Eq. (5.48) to two cross-sections separated by a differential distance $d\ell$, we have:

$$\frac{1}{\rho}dP + \frac{g}{g_c}dz + \frac{1}{2g_c}dv^2 + \frac{f}{D}\frac{v^2}{2g_c}d\ell = 0 \tag{5.49}$$

Comparing Eq. (5.49) with Eq. (5.42) we see that they are the same, provided that $f = 4f_F$ as we previously suggested. From now on we will concern ourselves only with the Darcy friction factor as used in Eq. (5.49).

5.3.5 Total Energy Balance

By "total energy" we include thermal as well as mechanical energy. To get an equation for total energy we simply add Eqs. (5.43) and (5.1) to obtain:

$$dU + \frac{1}{\rho}d\text{P} + \frac{g}{g_c}dz + \frac{v}{g_c}dv = dQ - dW \tag{5.50}$$

We distinguish between shaft work, dW_s, and pressure-volume work, PdV. Breaking the work term in Eq. (5.50) into these two forms, moving the pressure-volume term to the left-hand side, and noting that $\frac{1}{\rho} = V$, we obtain:

$$dU + PdV + Vd\text{P} + \frac{g}{g_c}dz + \frac{v}{g_c}dv = dQ - dW_s \tag{5.51}$$

Using the definition of enthalpy in Eq. (5.8), our equation for total energy balance becomes:

$$dH + \frac{g}{g_c}dz + \frac{v}{g_c}dv = dQ - dW_s \tag{5.52}$$

5.3.6 Speed of Sound

The speed of sound plays a major role in the study of compressible flow. We can derive an expression for the speed of sound as follows. We consider a pressure front traveling through a duct of cross-sectional area A with velocity c. We imagine an observer traveling along with the pressure front. To this observer, fluid approaches at velocity c with pressure P and density ρ, and leaves with pressure $P+dP$, density $\rho+d\rho$, and velocity c-dv. We consider shear forces to be negligible, and consider the pressure forces acting on a control volume containing the pressure front. The net pressure force, acting in the direction opposite the direction of flow, is $A[(P + dP) - P] = AdP$. The change in momentum flux across the faces of the control volume is $\frac{q}{g_c}[c - (c - dv)] = \frac{q}{g_c}dv = \frac{\rho Ac}{g_c}dv$. Inserting these terms into Newton's Second Law, $F = ma$, we have

$$dP = \frac{\rho}{g_c}cdv \tag{5.53}$$

Now we apply Eq. (5.32) to the flow on both sides of the pressure front, noting that q and A are the same on both sides.

$$\rho cA = (\rho + d\rho)(c - dv)A \tag{5.54}$$

which we can rearrange to:

$$\frac{\rho}{d\rho} = \frac{c}{dv} \tag{5.55}$$

Eliminating dv between Eqs. (5.53) and (5.55):

$$c^2 = g_c \frac{dP}{d\rho} = g_c \left(\frac{\partial P}{\partial \rho} \right)_S \tag{5.56}$$

In Eq. (5.56) we have written the ratio $\frac{dP}{d\rho}$ as a partial derivative at constant entropy because the variations in pressure and density are vanishingly small, so the process is nearly reversible. Further, the rapidity of the process and the smallness of the temperature variation render the process nearly adiabatic. In the limit for an infinitesimal pressure wave, the process may be considered reversible and adiabatic, hence isentropic. So the equation for the speed of sound is:

$$c = \sqrt{g_c \left(\frac{\partial P}{\partial \rho} \right)_S} \tag{5.57}$$

The ratio of fluid velocity to the speed of sound is called the Mach number.

$$M = \frac{v}{c} \tag{5.58}$$

We can use Eq. (5.27) to evaluate the speed of sound in a perfect gas. Focusing on the equality between the second and fourth terms:

$$P = C_3 \rho^{\gamma} \tag{5.59}$$

Since Eq. (5.59) applies to an isentropic process, the derivative of P with respect to ρ is really a partial derivative at constant entropy. Then using Eq. (5.17):

$$\left(\frac{\partial P}{\partial \rho} \right)_S = C_3 \gamma \rho^{\gamma - 1} = \gamma \frac{P}{\rho} = \gamma \frac{RT}{M_w} \tag{5.60}$$

Substituting Eq. (5.60) into Eq. (5.57):

$$c = \sqrt{\frac{\gamma R T g_c}{M_w}} \tag{5.61}$$

5.4 Incompressible Pipeline Flow

5.4.1 Reynolds Number

For incompressible flow in a pipeline we will use Bernoulli's equation, Eq. (5.48). We need to evaluate the Darcy friction factor, f, which is a function of the Reynolds number. The Reynolds number is a dimensionless parameter that is fundamental for the study of fluid flow. For flow in a pipe it is defined as:

$$N_{Re} = \frac{vD\rho}{\mu} \frac{1}{6.719688 \times 10^{-4}} \tag{5.62}$$

where the numerical coefficient converts the units of measure of viscosity (centipoise) to those consistent with the other variables, such that all dimensions cancel. In effect, the Reynolds number represents a ratio between the rates of kinetic and viscous energy dissipation in a flow. For $N_{Re} > 4,000$, as is usually encountered in industrial applications, the flow is turbulent. For $N_{Re} < 2,000$ the flow is usually laminar. The middle range, $2,000 < N_{Re} < 4,000$, is called the "transition region" and the nature of the flow in this range is generally unpredictable.

5.4.2 Friction Factor

In Eq. (5.47) we introduced the Darcy friction factor, f. Experimental findings for f as a function of N_{RE} with pipe roughness (ε) as a parameter are plotted in widely available charts known as "Moody friction factor charts," based on the work of Professor Lewis Ferry Moody of Princeton University.[3] The group $\frac{v^2}{2g_c}$ is commonly known as the "velocity head." So the coefficient $\frac{fL}{D}$ gives the frictional head loss in terms of the number of velocity heads.

We also introduced the Fanning friction factor in Eq. (5.41). While Professor Moody's work dealt specifically with the Darcy friction factor, both the Darcy and Fanning friction factors are presented in charts following Moody's design, often simply labeled "friction factor" without specifying whether it is the Darcy or Fanning factor that is plotted. So how can one tell which is being plotted? The answer is that at low Reynolds numbers, where the flow is laminar, the friction factor is inversely proportional to the Reynolds number and independent of pipe roughness. There will be a single diagonal line at the far left side of the chart, which gives the friction factor for laminar flow. If this line is labeled $\frac{64}{N_{Re}}$ the plot is of the Darcy friction factor; if it is labeled $\frac{16}{N_{Re}}$ the plot is of the Fanning friction factor. (Note that this laminar flow limit for the Darcy friction factor is proven in Section 5.5.)

A number of mathematical correlations have been proposed to represent friction factor data in equation form. The most widely accepted is the Colebrook-White equation, which was actually the one used by Professor Moody for the turbulent flow region of his chart:[4]

$$\frac{1}{\sqrt{f}} = -2\log_{10}\left(\frac{\varepsilon}{3.7D} + \frac{2.51}{N_{Re}\sqrt{f}}\right) \tag{5.63}$$

This is an implicit equation for f, which must be solved using an iterative procedure. The author has found that a Newton-Raphson algorithm solving for $\frac{1}{\sqrt{f}}$ with a starting value of 7 converges quite rapidly.

5.4.3 K-Factors for Fittings

There are often a number of fittings in a pipe that add resistance to flow. These include elbows, tees, valves, etc. Rather than attempt a detailed analysis of flow in these obstructions, we represent the energy loss in a fitting in terms of velocity heads. The number of velocity heads lost is called the "K-factor" of the fitting. When there is more than one fitting, the K-factors for all of them can be summed to give a K-factor for the whole pipe. The K-factor can then be added to the group $\frac{fL}{D}$ in Eq. (5.47) to give an expression for frictional head loss which includes the fittings:

$$h_f = \left(\frac{fL}{D} + K\right)\frac{v^2}{2g_c} \tag{5.64}$$

Tabulations of values for the K-factors can be found in various references. The most useful and widely known to engineers would be Ref. [6].

The use of K-factors is not limited to simple pipe fittings. Any piece of equipment for which the head loss can be represented to an acceptable degree of accuracy as a quadratic function of the velocity of the fluid can be assigned a K-factor and handled as above.

3 Moody, LF. Friction factors for pipe flow. *T ASME*. 1944; **66** (8): 671–684.

4 Colebrook, Cyril Frank "Turbulent Flow in Pipes, with particular reference to the Transition Region between the Smooth and Rough Pipe Laws", *Journal of the Institution of Civil Engineers* (London), 11 (4), 133–156 (1939).

5.4.4 Fouling Factor

The friction factor obtained from a Moody chart or from solution of the Colebrook-White equation assumes that the pipe is clean and has its original diameter. In some services, over the course of time, the interior surface of a pipe becomes coated with deposits that decrease its diameter and increase its roughness. In such cases, the measured pressure drop through the pipe will not agree with a prediction based on the assumption of a clean pipe. What is needed is an adjustable parameter that can be used to bring the prediction in line with the actual observation. Such a parameter is the "fouling factor," which we denote here as "s" because it is sometimes called the "safety factor" (and because we have already assigned other meanings to the symbols "F" and "f"). In a design calculation, the safety factor can be used when the pipe is being sized to assure that it will have the needed carrying capacity even when fouled to a specified degree. The fouling factor is applied as a multiplier to the friction factor in Eq. (5.64). Our final expression for the frictional head loss is thus:

$$h_f = \left(\frac{fL}{D}s + K\right)\frac{v^2}{2g_c} \tag{5.65}$$

5.4.5 Other Head Loss and Gain Terms

In practical applications there may be other sources of energy loss and/or gain. For example, the purpose of a pump is to add energy to the fluid by raising its pressure. We will add a term, h_p, to Eq. (5.48) to represent the energy gain due to a pump in the line. There is not a standard form for this term as we had for h_f. The relationship between head gain and flow rate is known as the "pump curve." It must be individually determined for each pump. The pump curve is usually supplied by the pump manufacturer in the form of a polynomial relationship between head increase and volumetric flow of fluid in gallons per minute (gpm).

Likewise, there may be pieces of equipment in a pipe that do not lend themselves to representation in terms of a K-factor. For these we include an equipment head loss term, h_e. As for the pump, the relationship between head loss and flow rate must be individually determined for each such piece of equipment, and will normally be represented as a polynomial.

In summary the equation that we use for analyzing the flow of an incompressible fluid in a pipe with fittings is the Bernoulli equation, Eq.(5.45), with terms added as discussed above to represent energy loss and gain:

$$\frac{P_1}{\rho} + \frac{g}{g_c}z_1 + \frac{{v_1}^2}{2g_c} + h_p = \frac{P_2}{\rho} + \frac{g}{g_c}z_2 + \frac{{v_2}^2}{2g_c} + \left(\frac{fL}{D}s + K\right)\frac{v^2}{2g_c} + h_e \tag{5.66}$$

A comment about Eq. (5.66) is in order. For a truly incompressible fluid and a pipe of constant diameter, $v_1 = v_2$, the third terms on the left- and right-hand sides (the velocity head terms) would cancel each other and drop out. In this case it is also obvious what value of v is to be used in the frictional head loss term. As it turns out, the Bernoulli equation is quite powerful, and also describes reasonably well the flow of compressible fluids as long as the change in pressure from inlet to outlet is not too extreme, say less than 20% of the inlet pressure. In this case $v_1 \neq v_2$, and the third terms do not cancel. Moreover, we need to come up with an intermediate value of v to use in the frictional term. The way this is handled is to express all velocities in terms of the mass flow rate using Eq. (5.32). This introduces the density. For v_1 and v_2 we use the density at the inlet and outlet respectively. We use known or assumed values for the pressure to calculate the density. For v in the frictional loss term, we simply use an arithmetic average of the inlet and outlet densities.

There is another circumstance where it is necessary to retain the velocity head terms. Sometimes there is a change in diameter in a piping system. Normally we would treat pipes of different diameter as different pipes, and analyze them separately. However, where they join there is a velocity change. The change in diameter, and hence change in velocity, can be associated with either the end of the upstream pipe or the beginning of the downstream pipe. The choice is arbitrary, but it must be made so that the effect of the change in velocity head can be accounted for.

As will be discussed in Chapter 6, when Bernoulli's equation is applied using stagnation pressure rather than static pressure the velocity head terms drop out, and change in diameter can be handled at the nodes without resort to the strategy introduced in the preceding paragraph.

5.4.6 Example Application

Some of the most profitable systems to model hydraulically are cooling water systems. They are typically very large systems that are historically modified without looking at the effect on the whole system. Exchangers with differing design specifications are added, which begin to erode the hydraulic design creating bottlenecks in process units that are difficult to recognize without having a detailed hydraulic model of the entire system. Buried pipe headers with build-up in solids can further reduce the total system capacity. Adding new pumps typically provides very minimal improvement in the system's total capacity. However, this is what is often done without modeling the system in detail. A detailed fluid flow model will allow for the discovery of the root-cause problems of utility systems and provide significant increase in the utility system capacity with a similar impact on process system performance. Fundamental to a cooling water study and any other large utility system study is a comprehensive validation of the "current operations" fluid flow model using field flow and pressure readings (EPCON Software example cooling water model shown in Figure 5.3).

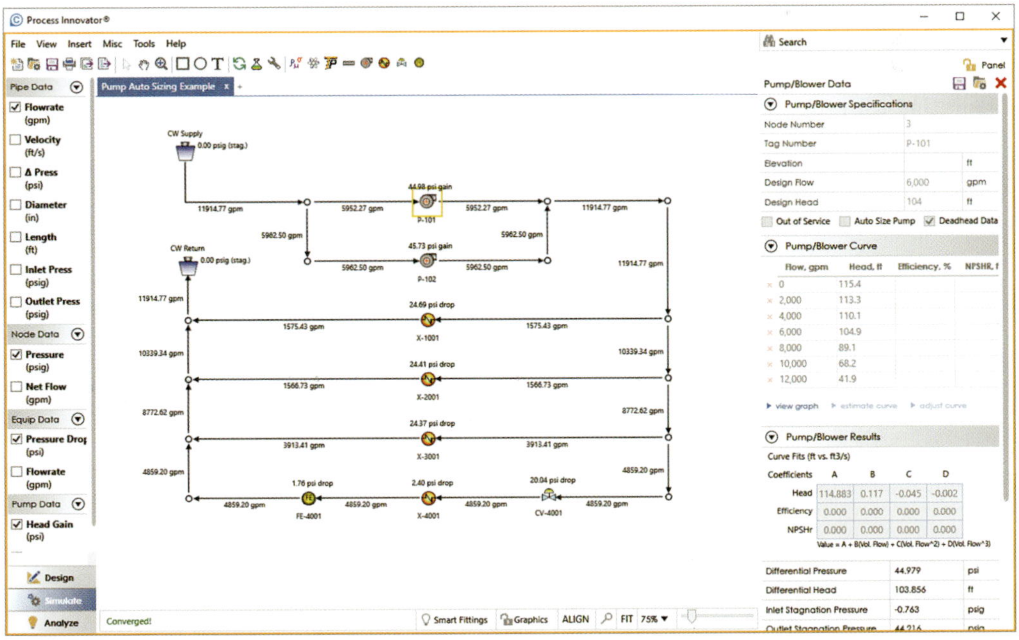

Figure 5.3 Cooling Water System Flow Simulation Model – EPCON Process Innovator.

5.5 Laminar Flow

Laminar flow is not common in the process industries, but there are cases where it does occur. For flow of very viscous fluids, such as polymers, the flow may be laminar. Moreover, there are situations in network flow calculations where we have to deal with low flow rates, even approaching no flow (as in a dead leg). A major difference between turbulent and laminar flow is that in turbulent flow the pressure drop is proportional to the square of the fluid flow rate, whereas in laminar flow the pressure drop is proportional to the first power of the fluid flow rate.

Laminar flow in a tubular geometry is known as "Poiseuille flow" after the nineteenth-century French physicist and physiologist Jean Léonard Marie Poiseuille, who was interested in the flow of blood in capillaries and veins. The standard unit of viscosity, the "poise," is named after him.

We have already seen how elevation changes enter into the energy balance. Here we are simply trying to determine the head loss due to friction in a laminar flow. So, without loss of generality, we will analyze horizontal flow. We model such a flow as a set of cylindrical shells of infinitesimal thickness sliding past each other. Consider one such shell (see Figure 5.4 with flow from left to right).

The shell is acted on by four forces in the x-direction: pressure on the upstream and downstream faces, and shear stresses on the inner and outer cylindrical surfaces. In steady state flow, when there is no acceleration, these forces sum to zero.

The force balance on the shell is:

$$P2\pi r\delta r + \tau 2\pi r\delta x = \left(P + \frac{\partial P}{\partial x}\delta x\right)2\pi r\delta r + \tau 2\pi r\delta x + \frac{\partial}{\partial r}(\tau 2\pi r\delta x)\delta r \tag{5.67}$$

which we simplify to:

$$\left(\frac{\partial P}{\partial x}\right)r + \frac{\partial}{\partial r}(\tau r) = 0 \tag{5.68}$$

Since we are ignoring the effect of gravity on the pressure, we can replace the partial derivatives by ordinary derivatives to find:

$$r\frac{dP}{dx} = -\frac{d}{dr}(\tau r) \tag{5.69}$$

Integrating with respect to r:

$$\frac{r^2}{2}\frac{dP}{dx} + C_1 = -\tau r \tag{5.70}$$

where C_1 is the constant of integration. Solving for τ:

$$\tau = -\frac{r}{2}\frac{dP}{dx} - \frac{C_1}{r} \tag{5.71}$$

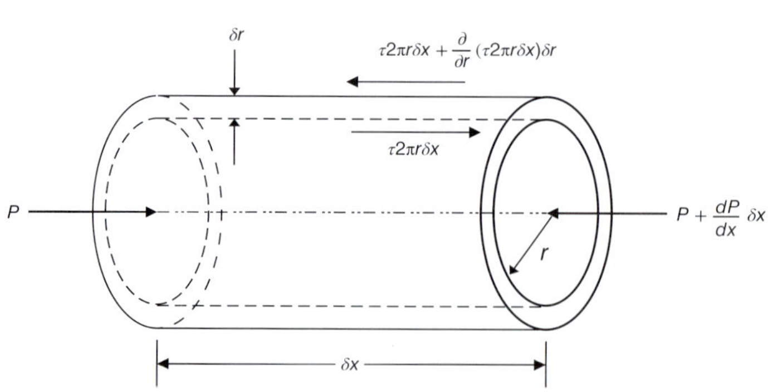

Figure 5.4 Forces on a Laminar Shell.

We cannot allow the shear stress to increase without bound as the centerline is approached. Therefore, we must have $C_1 = 0$, and Eq. (5.71) becomes:

$$\tau = -\frac{r}{2}\frac{dP}{dx} \tag{5.72}$$

The shear stress in the fluid is created by the dynamic viscosity resisting the shear strain (velocity gradient). Mathematically, this is expressed as:

$$\tau = -\mu\frac{du}{dr} \tag{5.73}$$

Inserting Eq. (5.73) into Eq. (5.72) we have:

$$\mu\frac{du}{dr} = \frac{r}{2}\frac{dP}{dx} \tag{5.74}$$

Which we integrate to:

$$\mu u = \frac{r^2}{4}\frac{dP}{dx} + C_2 \tag{5.75}$$

We determine C_2, the constant of integration, by noting that at the wall ($r = \frac{D}{2}$) the fluid velocity goes to zero.

$$C_2 = -\frac{D^2}{16}\frac{dP}{dx} \tag{5.76}$$

Inserting this result into Eq. (5.75) we obtain, for the velocity as a function of radial position:

$$u = \frac{1}{4\mu}\left(r^2 - \frac{D^2}{4}\right)\frac{dP}{dx} \tag{5.77}$$

We have derived the familiar result that the radial velocity profile in laminar pipe flow is a parabola. To obtain the total volumetric flow through the tube we integrate:

$$Q = \int_0^{\frac{D}{2}} 2\pi r u dr = \frac{2\pi}{4\mu}\frac{dP}{dx}\int_0^{\frac{D}{2}} r\left(r^2 - \frac{D^2}{4}\right)dr = -\frac{\pi D^4}{128\mu}\frac{dP}{dx} = -\frac{\pi D^4}{128\mu}\frac{\Delta P}{L} \tag{5.78}$$

The flow is in the positive x-direction, and the negative sign in Eq. (5.78) arises because pressure decreases in the direction of flow, i.e., $\frac{dP}{dx} < 0$. The final equality is justified when the flow is uniform through a pipe of constant diameter.

We can solve Eq. (5.78) for pressure drop expressed in the form of frictional head loss, and we can express the volumetric flow rate in terms of an average velocity.

$$h_f = -\frac{\Delta P}{\rho} = \frac{1}{\rho}\frac{128\mu L}{\pi D^4}\frac{v\pi D^2}{4} = \frac{32\mu L v}{\rho D^2} \tag{5.79}$$

We see from Eq. (5.79) that frictional head loss is linearly proportional to velocity. However, we can write the equation in the following form:

$$h_f = \frac{64\mu g_c}{\rho v D}\frac{L}{D}\frac{v^2}{2g_c} = \frac{64}{N_{Re}}\frac{L}{D}\frac{v^2}{2g_c} \tag{5.80}$$

where $N_{Re} = \frac{\rho v D}{\mu g_c}$ is the Reynolds number previously defined in Eq. (5.62). The units for dynamic viscosity (μ) here are $lb_f\text{-sec/ft}^2$. In Eq. (5.62) the units for μ were centipoise, and we included a numerical factor that accounted for the change of units and the inclusion of the factor g_c.

Eq. (5.80) has the same form as Eq. (5.47) if we define the friction factor to be $f = \frac{64}{N_{Re}}$. What we have just done above is to prove the statement that was made in the text preceding Eq. (5.63) that this is the limiting expression for the Darcy friction factor at low Reynolds numbers.

Using Eq. (5.79) for the frictional head loss, the mechanical energy balance equation, Eq. (5.66), becomes:

$$\frac{P_1}{\rho} + \frac{g}{g_c}z_1 + \frac{v_1^{\ 2}}{2g_c} + h_p = \frac{P_2}{\rho} + \frac{g}{g_c}z_2 + \frac{v_2^{\ 2}}{2g_c} + \frac{32\mu L}{\rho D^2}sv + K\frac{v^2}{2g_c} + h_e \tag{5.81}$$

Note that in Eq. (5.81) we have retained the dependence on the second power of velocity in the K-factor term. It is known that the K-factor method is not accurate in the laminar flow regime. However, there is no obvious way to extend this method as there would be if we were using the "equivalent length" method of handling fittings. In the absence of such an extension, we retain the form above.

5.6 Compressible Pipeline Flow

5.6.1 Introductory Remarks

When the density of a fluid is dependent on its pressure, we say that the fluid is compressible. The difference that this makes in the analysis of fluid flow problems is profound.

Let us return to Eq. (5.49). To represent the velocity in terms of mass flow rate, we recall Eq. (5.32) and write:

$$v = \frac{q}{\rho A} \tag{5.82}$$

Substituting this into Eq. (5.49), with some manipulation:

$$\frac{1}{\rho}dP + \frac{g}{g_c}dz - \frac{q^2}{g_c\rho^3A^2}d\rho + \frac{f}{D}\frac{q^2}{2g_c\rho^2A^2}d\ell = 0 \tag{5.83}$$

For constant density we can integrate this equation. When we posit other relationships between density and pressure, we will not be able to integrate the equation with the elevation term included. This is not much of a problem, as changes in elevation contribute very little to the energy balance in gas flow calculations. So, we will drop the elevation term, and for gas flow calculations our working form of the mechanical energy balance equation will be:

$$\frac{1}{\rho}dP - \frac{q^2}{g_c\rho^3A^2}d\rho + \frac{f}{D}\frac{q^2}{2g_c\rho^2A^2}d\ell = 0 \tag{5.84}$$

To proceed further, we must know how density varies with pressure. In general, we do not know this, but we infer it by making an assumption about how temperature varies with pressure (even though temperature does not appear explicitly in Eq. (5.84)).

5.6.2 Isothermal Flow

From Eq. (5.17) we infer that at constant temperature:

$$\rho = \frac{\rho_1}{P_1}P \tag{5.85}$$

where we use the subscript "1" to refer to conditions at the pipeline entrance. Substituting this into Eq. (5.84) we obtain:

$$\frac{1}{\rho_1}\frac{P_1}{P}dP - \frac{q^2}{g_c\rho_1^{\ 2}A^2}\frac{P_1^{\ 2}}{P^3}dP + \frac{f}{D}\frac{q^2}{2g_c\rho_1^{\ 2}A^2}\frac{P_1^{\ 2}}{P^2}d\ell = 0 \tag{5.86}$$

We multiply the equation through by $\frac{P^2}{P_1^{\,2}}$ to obtain:

$$\frac{P}{\rho_1 P_1}dP - \frac{q^2}{g_c \rho_1{}^2 A^2}\frac{1}{P}dP + \frac{f}{D}\frac{q^2}{2g_c \rho_1{}^2 A^2}d\ell = 0 \tag{5.87}$$

We can integrate this equation to obtain:

$$\frac{1}{2\rho_1 P_1}(P_2{}^2 - P_1{}^2) - \frac{q^2}{g_c \rho_1{}^2 A^2}\ln\frac{P_2}{P_1} + \frac{fL}{D}\frac{q^2}{2g_c \rho_1{}^2 A^2} = 0 \tag{5.88}$$

A little rearrangement puts this equation into a form that will be helpful in the next subsection:

$$\frac{fL}{D}\frac{q^2}{2g_c A^2} = \left(\rho_1 \frac{P_1 + P_2}{2P_1}\right)(P_1 - P_2) + \frac{q^2}{g_c A^2}\ln\frac{P_2}{P_1} \tag{5.89}$$

5.6.3 Bernoulli Approximation

When the pressure drop in a compressible fluid is relatively small, results of reasonable accuracy can be obtained by means of Bernoulli's equation. We rewrite Eq. (5.49) without the elevation term, and with an average, constant, density:

$$\frac{1}{\bar{\rho}}dP + \frac{f}{D}\frac{q^2}{2g_c A^2}\frac{1}{\bar{\rho}^2}d\ell = 0 \tag{5.90}$$

Integrating and rearranging:

$$\frac{fL}{D}\frac{q^2}{2g_c A^2} = \bar{\rho}(P_1 - P_2) \tag{5.91}$$

Now consider the first factor in the first term on the right hand side of Eq. (5.89). If we apply Eq. (5.85) to this factor we have:

$$\rho_1 \frac{P_1 + P_2}{2P_1} = \frac{\rho_1 + \rho_2}{2} = \bar{\rho} \tag{5.92}$$

Comparing Eq. (5.90) to Eq. (5.89) with Eq. (5.92) inserted, we see that by defining the average density in the usual way, Bernoulli's equation replicates the isothermal equation when the acceleration term is small with respect to the frictional term, i.e., when the second term on the right hand side of Eq. (5.89) is small relative to the left hand side. In pipeline flow with moderate pressure drop, say up to 20% of the initial pressure, this is expected to be a reasonable assumption.

5.6.4 Isentropic Flow

For an isentropic process, from Eq. (5.27):

$$\rho = \rho_1 \left(\frac{P}{P_1}\right)^{\frac{1}{\gamma}} \tag{5.93}$$

Substituting this into Eq. (5.84)

$$\frac{1}{\rho_1}\left(\frac{P_1}{P}\right)^{\frac{1}{\gamma}} dP - \frac{q^2}{g_c \rho_1{}^2 A^2}\frac{1}{\gamma}\left(\frac{P_1}{P}\right)^{\frac{2}{\gamma}}\frac{1}{P}dP + \frac{f}{D}\frac{q^2}{2g_c \rho_1{}^2 A^2}\left(\frac{P_1}{P}\right)^{\frac{2}{\gamma}} d\ell = 0 \tag{5.94}$$

We multiply through by $\left(\frac{P}{P_1}\right)^{\frac{2}{\gamma}}$ to obtain:

$$\frac{1}{\rho_1 P_1^{\frac{1}{\gamma}}} P^{\frac{1}{\gamma}} dP - \frac{q^2}{g_c \rho_1{}^2 A^2} \frac{1}{\gamma} \frac{1}{P} dP + \frac{f}{D} \frac{q^2}{2 g_c \rho_1{}^2 A^2} d\ell = 0 \tag{5.95}$$

We can integrate this equation to obtain:

$$\frac{1}{\rho_1 P_1^{\frac{1}{\gamma}}} \frac{P_2^{\frac{\gamma+1}{\gamma}} - P_1^{\frac{\gamma+1}{\gamma}}}{\frac{\gamma+1}{\gamma}} - \frac{q^2}{g_c \rho_1{}^2 A^2} \frac{1}{\gamma} \ln \frac{P_2}{P_1} + \frac{fL}{D} \frac{q^2}{2 g_c \rho_1{}^2 A^2} = 0 \tag{5.96}$$

For the purpose of comparison with Eqs. (5.89) and (5.91) we rearrange this equation to:

$$\frac{fL}{D} \frac{q^2}{2 g_c A^2} = \frac{\gamma}{\gamma + 1} \rho_1 \frac{P_1^{\frac{\gamma+1}{\gamma}} - P_2^{\frac{\gamma+1}{\gamma}}}{P_1^{\frac{1}{\gamma}}} + \frac{q^2}{g_c A^2} \frac{1}{\gamma} \ln \frac{P_2}{P_1} \tag{5.97}$$

5.6.5 Polytropic Flow

Eq. (5.97) is a little unrealistic in the sense that we cannot have an isentropic process and frictional loss at the same time. A "polytropic" process is like an isentropic process, but in place of the exponent γ we use the exponent k, which has no particular thermodynamic significance, but is chosen to fit the data. In general, k will be between 1 and γ.

$$\frac{fL}{D} \frac{q^2}{2 g_c A^2} = \frac{k}{k + 1} \rho_1 \frac{P_1^{\frac{k+1}{k}} - P_2^{\frac{k+1}{k}}}{P_1^{\frac{1}{k}}} + \frac{q^2}{g_c A^2} \frac{1}{k} \ln \frac{P_2}{P_1} \tag{5.98}$$

If we compare Eqs. (5.89), (5.91), (5.97), and (5.98) we discover that Eqs. (5.89), (5.91), and (5.97) are just special cases of Eq. (5.98) with $k = 1$, $k = \infty$ and $k = \gamma$ respectively. Now if we use Eq. (5.93) with $\gamma = k$ in Eq. (5.98) we find all of these equations reducing to a more simple and elegant form:

$$\frac{fL}{D} \frac{q^2}{2 g_c A^2} = \frac{k}{k + 1} (\rho_1 P_1 - \rho_2 P_2) + \frac{q^2}{g_c A^2} \ln \frac{\rho_2}{\rho_1} \tag{5.99}$$

If we consider real gases with density governed by an equation of state rather than by a simplified relationship of the form of Eq. (5.93), it is the author's conjecture that Eq. (5.99) is a better equation to use than Eq. (5.98) or its relatives. In particular, if we want more accuracy than is provided by the use of Eq. (5.91), it is recommended to use Eq. (5.99) with $k = 1$ (isothermal flow), since the use of an equation for isentropic flow ($k = \gamma$) cannot be justified for frictional pipe flow, and we are unlikely to have a reasonable way to estimate a better value of k. For relatively large pressure drop, where an isothermal approximation is inappropriate, the treatment developed in the following subsection is recommended.

5.6.6 Adiabatic Flow

In the adiabatic model we assume that mechanical energy lost to friction stays with the fluid as thermal energy, increasing its temperature. In this model it is total energy that is conserved.

The following development relies heavily on the excellent text by Professor Ascher Shapiro [7].[5] The mathematics is pretty intense. The key finding is that there is a unique relationship

5 For a more recent treatment, see also the book by John D. Anderson, Ref. [8].

between the ratios of entry to exit fluid properties and the entry and exit Mach numbers. Likewise for pipe length. The reader who wants to skip the math and go directly to the results is referred to Eqs. (5.116), (5.120), (5.123), and (5.124).

The key to finding the adiabatic relationship between pressure and density is to recognize that the total energy balance gives a relationship between temperature and fluid velocity, and the continuity equation gives a relationship between velocity and density. We begin by considering the total energy balance, Eq. (5.52). By the assumption of adiabatic flow, the terms on the right hand side are 0. Also, we are neglecting the elevation term, so it drops out. Finally we substitute for dH using Eq. (5.19). The result is:

$$c_p dT + \frac{1}{2g_c} dv^2 = 0 \tag{5.100}$$

Dividing through by $c_p T$ and using Eq. (5.21), (5.61), and (5.58):

$$\frac{dT}{T} + \frac{\gamma - 1}{2} \frac{1}{c^2} dv^2 = \frac{dT}{T} + \frac{\gamma - 1}{2} M^2 \frac{dv^2}{v^2} = 0 \tag{5.101}$$

Taking logarithmic differentials of Eq. (5.17) we have:

$$\frac{dP}{P} - \frac{d\rho}{\rho} = \frac{dT}{T} \tag{5.102}$$

We can eliminate $\frac{dT}{T}$ between Eqs. (5.101) and (5.102):

$$\frac{dP}{P} - \frac{d\rho}{\rho} + \frac{\gamma - 1}{2} M^2 \frac{dv^2}{v^2} = 0 \tag{5.103}$$

Now we note that for constant A, logarithmic differentiation of Eq. (5.32) yields:

$$\frac{d\rho}{\rho} + \frac{1}{2} \frac{dv^2}{v^2} = 0 \tag{5.104}$$

We can substitute for $\frac{dv^2}{v^2}$ in Eq. (5.103) using Eq. (5.104):

$$\frac{dP}{P} = [1 + (\gamma - 1)M^2] \frac{d\rho}{\rho} \tag{5.105}$$

Eq. (5.105) is the relationship between pressure and density that applies to adiabatic flow. However, this relationship involves the Mach number as an additional variable. In order to use it we need to recast the mechanical energy balance, Eq. (5.84), in a form such that all of the fluid variables are expressed as functions of M. We now proceed to do that.

We begin by substituting for q in Eq. (5.84) using Eq. (5.32), then we multiply through by $\frac{\rho}{P}$:

$$\frac{dP}{P} - \frac{1}{P} \frac{\rho v^2}{g_c} \frac{d\rho}{\rho} + \frac{f}{D} \frac{1}{P} \frac{\rho v^2}{2g_c} d\ell = 0 \tag{5.106}$$

Next we note, using Eqs. (5.17), (5.61), and (5.58):

$$\frac{\rho v^2}{g_c} = \frac{M_w P}{RT} \frac{v^2}{g_c} = P\gamma \frac{v^2}{c^2} = P\gamma M^2 \tag{5.107}$$

Substituting Eqs. (5.104) and (5.107) into Eq. (5.106):

$$\frac{dP}{P} + \frac{\gamma M^2}{2} \frac{dv^2}{v^2} + \frac{f}{D} \frac{\gamma M^2}{2} d\ell = 0 \tag{5.108}$$

Our next step is to express $\frac{dv^2}{v^2}$ in the mechanical energy balance equation, Eq. (5.108), in terms of M. Combining Eqs. (5.58) and (5.61) we can form the following logarithmic differential equation:

$$\frac{dM^2}{M^2} = \frac{dv^2}{v^2} - \frac{dT}{T} \tag{5.109}$$

We can substitute for $\frac{dT}{T}$ in Eq. (5.109) using the total energy balance, Eq. (5.101), to obtain:

$$\frac{dv^2}{v^2} = \left[\frac{2}{2+(\gamma-1)M^2}\right]\frac{dM^2}{M^2} \tag{5.110}$$

Then we need to express $\frac{dP}{P}$ in the mechanical energy balance equation, Eq. (5.108) in terms of M. We use Eq. (5.104) to substitute for $\frac{d\rho}{\rho}$ in Eq. (5.105):

$$\frac{dP}{P} = -\left[\frac{1+(\gamma-1)M^2}{2}\right]\frac{dv^2}{v^2} \tag{5.111}$$

Combining Eqs. (5.111) and (5.110):

$$\frac{dP}{P} = -\left[\frac{1+(\gamma-1)M^2}{2+(\gamma-1)M^2}\right]\frac{dM^2}{M^2} \tag{5.112}$$

We are finally ready to cast the mechanical energy balance equation into a form such that all of the fluid variables are expressed in terms of the Mach number, M. We substitute Eqs. (5.110) and (5.112) into Eq. (5.108):

$$-\left[\frac{1+(\gamma-1)M^2}{2+(\gamma-1)M^2}\right]\frac{dM^2}{M^2} + \frac{\gamma M^2}{2}\left[\frac{2}{2+(\gamma-1)M^2}\right]\frac{dM^2}{M^2} + \frac{f}{D}\frac{\gamma M^2}{2}d\ell = 0 \tag{5.113}$$

which we rearrange to:

$$\frac{f}{D}d\ell = \frac{2(1-M^2)}{\gamma M^2[2+(\gamma-1)M^2]}\frac{dM^2}{M^2} \tag{5.114}$$

To integrate Eq. (5.114) we expand the right-hand side in partial fractions:

$$\int_0^L \frac{f}{D}d\ell = \int_{M_1^2}^{M_2^2}\left[-\frac{\gamma+1}{2\gamma M^2} + \frac{1}{\gamma M^4} + \frac{\gamma^2-1}{2\gamma}\frac{1}{2+(\gamma-1)M^2}\right]dM^2 \tag{5.115}$$

Completing the integration, we find:

$$\frac{\bar{f}L}{D} = \frac{1}{\gamma}\left[\frac{1}{M_1^2} - \frac{1}{M_2^2} + \frac{\gamma+1}{2}\ln\frac{M_1^2[2+(\gamma-1)M_2^2]}{M_2^2[2+(\gamma-1)M_1^2]}\right] \tag{5.116}$$

For a given entering Mach number, M_1, Eq. (5.116) gives the corresponding Mach number at any distance, L, down the pipe. The exit Mach number cannot exceed unity, as will be proved in the next subsection.

The value of $\bar{f}$ in Eq. (5.116) is an average value of f over the length of the pipe, defined as:

$$\bar{f} = \int_0^L f d\ell \tag{5.117}$$

Typically, $\bar{f}$ does not vary greatly under the turbulent flow conditions normally encountered, and a reasonable estimate of $\bar{f}$ can be obtained by simply evaluating f at the inlet conditions of the pipe.

To know the pressure at any position in the pipe, we integrate Eq. (5.112), first expanding the right-hand side in partial fractions:

$$\int_{P_1}^{P_2} \frac{dP}{P} = -\int_{M_1^2}^{M_2^2} \left[\frac{1}{2M^2} + \frac{\gamma - 1}{2} \frac{1}{2 + (\gamma - 1)M^2}\right] dM^2 \tag{5.118}$$

Integrating:

$$\ln \frac{P_2}{P_1} = -\left[\frac{1}{2} \ln \frac{M_2^2}{M_1^2} + \frac{1}{2} \ln \frac{2 + (\gamma - 1)M_2^2}{2 + (\gamma - 1)M_1^2}\right] \tag{5.119}$$

which we rearrange to:

$$\frac{P_2}{P_1} = \sqrt{\frac{M_1^2[2 + (\gamma - 1)M_1^2]}{M_2^2[2 + (\gamma - 1)M_2^2]}} \tag{5.120}$$

For any position in the pipe, L, we obtain M_2^2 by solution of the implicit equation Eq. (5.116), and then obtain P_2 from Eq. (5.120). Or, if we know P_2, we can obtain M_2^2 from Eq. (5.120) and obtain the pipe length required to realize the values from Eq. (5.116). Note that in this case there is only one pipe length that will allow a given pressure ratio.

To obtain an equation for temperature at any position in the pipe, we use Eq. (5.110) to substitute for $\frac{dv^2}{v^2}$ in Eq. (5.101) and we obtain:

$$\frac{dT}{T} = -\frac{\gamma - 1}{2 + (\gamma - 1)M^2} dM^2 \tag{5.121}$$

Integrating:

$$\ln \frac{T_2}{T_1} = -\ln \frac{2 + (\gamma - 1)M_2^2}{2 + (\gamma - 1)M_1^2} \tag{5.122}$$

which we rearrange to:

$$\frac{T_2}{T_1} = \frac{2 + (\gamma - 1)M_1^2}{2 + (\gamma - 1)M_2^2} \tag{5.123}$$

Combining Eqs. (5.120) and (5.123) into Eq. (5.17):

$$\frac{\rho_2}{\rho_1} = \frac{V_1}{V_2} = \frac{T_1}{T_2}\frac{P_2}{P_1} = \frac{2 + (\gamma - 1)M_2^2}{2 + (\gamma - 1)M_1^2} \sqrt{\frac{M_1^2[2 + (\gamma - 1)M_1^2]}{M_2^2[2 + (\gamma - 1)M_2^2]}} = \sqrt{\frac{M_1^2[2 + (\gamma - 1)M_2^2]}{M_2^2[2 + (\gamma - 1)M_1^2]}} \tag{5.124}$$

By procedures similar to those followed above it is possible to obtain expressions for other properties such as stagnation pressure and entropy, but we do not need these properties for our present purposes.

5.6.7 Choked Flow

Following Eq. (5.116) above the statement was made that the exit Mach number cannot exceed unity. There are some elegant ways to prove this assertion by applying the Second Law of Thermodynamics. But we can make the same point by showing that as M_2^2 increases, the value of L attains its maximum at M_2^2=1. We begin by rearranging Eq. (5.116) to:

$$L = \frac{D}{f\gamma}\left[\frac{1}{M_1^2} + \frac{\gamma + 1}{2} \ln \frac{M_1^2}{2 + (\gamma - 1)M_1^2} - \frac{1}{M_2^2} + \frac{\gamma + 1}{2} \ln \frac{2 + (\gamma - 1)M_2^2}{M_2^2}\right] \tag{5.125}$$

and taking the derivative with respect to M_2^2:

$$\frac{dL}{dM_2^2} = \frac{D}{\bar{f}\gamma}\left[\frac{1}{M_2^4} - \frac{\gamma+1}{2+(\gamma-1)M_2^2}\frac{1}{M_2^2}\right] = \frac{D}{\bar{f}\gamma}\frac{2}{M_2^2}\left[\frac{1-M_2^2}{M_2^2[2+(\gamma-1)M_2^2]}\right] \tag{5.126}$$

At the maximum value of L, the derivative in Eq. (5.126) is zero, and this clearly happens when the numerator in the bracketed factor is zero, i.e., when $M_2^2 = 1$. That this is a maximum can be seen by noting that $\frac{dL}{dM_2^2} > 0$ for $M_2^2 < 1$, and $\frac{dL}{dM_2^2} < 0$ for $M_2^2 > 1$. If flow in a pipe of constant cross-section begins as subsonic, it will remain subsonic, with Mach number increasing with increasing length, but never exceeding unity. If flow in such a pipe begins as supersonic, it will remain supersonic, with Mach number decreasing with increasing length but never going below unity. The study of supersonic flow is a specialized area, beyond the scope of normal process engineering calculations, so we will not deal with this topic further here.

We can calculate the maximum length of pipe by setting $M_2^2 = 1$ in Eq. (5.125):

$$L_{\max} = \frac{D}{\bar{f}\gamma}\left[\frac{1}{M_1^2} - 1 + \frac{\gamma+1}{2}\ln\frac{(\gamma+1)M_1^2}{2+(\gamma-1)M_1^2}\right] \tag{5.127}$$

If the actual length of a pipe exceeds the value of L_{max}, the fluid will not be able to enter the pipe with a Mach number of M_1. The flow rate will be reduced, and the flow is said to be "choked." The maximum flow that can pass through a pipe of length L can be calculated by setting $L_{max} = L$ in Eq. (5.127) and solving the implicit equation iteratively for M_1^2.

5.6.8 Rationalization with Bernoulli's Equation

When we integrate the mechanical energy balance to obtain Bernoulli's equation, say, in the form of Eq. (5.66), we have the advantage of being able to assign a physical meaning to each term in the resulting equation. When we integrate to obtain the expressions arising out of the various assumptions for the relationship between pressure and density in compressible flow, we have to do so much mathematical manipulation that to some extent we lose the ability to assign physical meaning to the individual terms in the resulting equation. The one term that retains its identity is the frictional loss term. To cast Eq. (5.116) in a form comparable to Eq. (5.66), we multiply through by $\frac{v_1^2}{2g_c} = \frac{q^2}{2g_c\rho_1^2A^2}$. From Eq. (5.107), we know that this is equivalent to multiplying through by $\frac{P_1}{2\rho_1}\gamma M_1^2$, so we can write Eq. (5.116):

$$\frac{1}{\rho_1}\frac{P_1}{2}\left[\frac{M_1^2}{M_2^2} - 1 - \frac{\gamma+1}{2}M_1^2\ln\frac{M_1^2[2+(\gamma-1)M_2^2]}{M_2^2[2+(\gamma-1)M_1^2]}\right] + \frac{\bar{f}L}{D}\frac{q^2}{2g_c\rho_1^2A^2} = 0 \tag{5.128}$$

Now we have all of the equations for compressible flow in a form where we can easily see how to incorporate the additional factors, s and K:

$$\text{Isothermal:}\ \frac{1}{2}\frac{1}{\rho_1}\frac{P_2^2 - P_1^2}{P_1} - \frac{q^2}{g_c\rho_1^2A^2}\ln\frac{P_2}{P_1} + \left(\frac{\bar{f}L}{D}s + K\right)\frac{q^2}{2g_c\rho_1^2A^2} = 0 \tag{5.129}$$

$$\text{Isentropic:}\ \frac{\gamma}{\gamma+1}\frac{1}{\rho_1}\frac{P_2^{\frac{\gamma+1}{\gamma}} - P_1^{\frac{\gamma+1}{\gamma}}}{P_1^{\frac{1}{\gamma}}} - \frac{q^2}{g_c\rho_1^2A^2}\frac{1}{\gamma}\ln\frac{P_2}{P_1} + \left(\frac{\bar{f}L}{D}s + K\right)\frac{q^2}{2g_c\rho_1^2A^2} = 0 \tag{5.130}$$

$$\text{Polytropic:}\ \frac{k}{k+1}\frac{1}{\rho_1}\frac{P_2^{\frac{k+1}{k}} - P_1^{\frac{k+1}{k}}}{P_1^{\frac{1}{k}}} - \frac{q^2}{g_c\rho_1^2A^2}\frac{1}{k}\ln\frac{P_2}{P_1} + \left(\frac{\bar{f}L}{D}s + K\right)\frac{q^2}{2g_c\rho_1^2A^2} = 0 \tag{5.131}$$

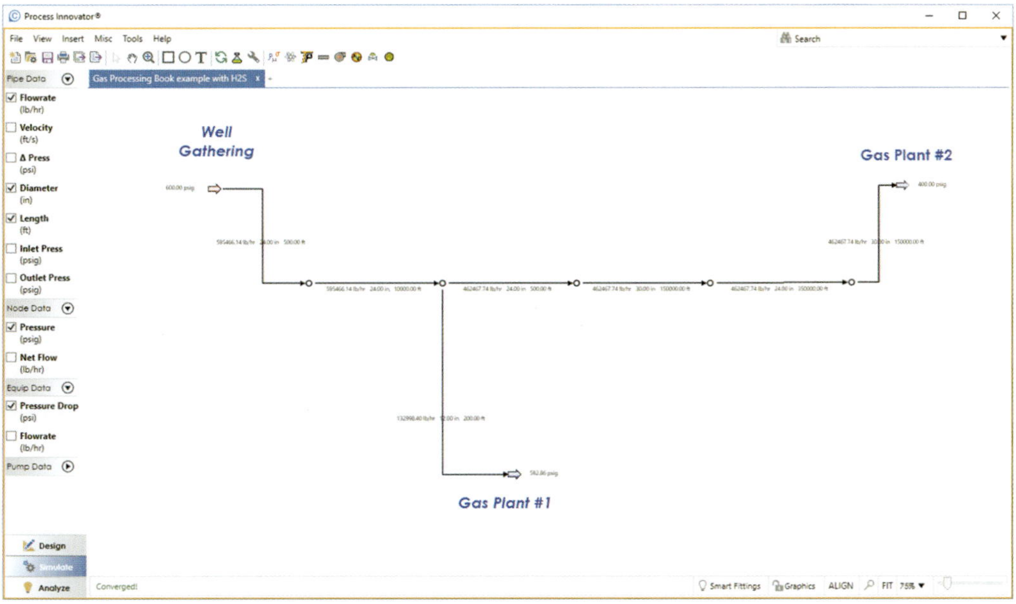

Figure 5.5 Long Distance Gas Pipeline Simulation Model - EPCON Process Innovator.

$$\text{Adiabatic: } \frac{1}{\rho_1}\frac{P_1}{2}\left[\frac{M_1{}^2}{M_2{}^2} - 1 - \frac{\gamma+1}{2}M_1{}^2 \ln \frac{M_1{}^2[2+(\gamma-1)M_2{}^2]}{M_2{}^2[2+(\gamma-1)M_1{}^2]}\right] + \left(\frac{\bar{f}L}{D}s + K\right)\frac{q^2}{2g_c\rho_1{}^2A^2} = 0 \tag{5.132}$$

5.6.9 Example Application

A typical decision in long-distance gas pipeline design is whether to invest in amine treating to remove sulfur at the source of the gas or downstream. The additional mass of the sulfur may significantly limit the capacity of the gas pipeline. The following example gas pipeline model (Figure 5.5) could be used to determine the impact of reduction of sulfur from upstream amine treating. Scenarios can also be run to determine flexibility in supplying multiple users along with variations in production from the well gathering station.

5.7 Comparison with Crane Handbook

Crane Handbook

The major practical field reference for piping system design is Ref. [6], popularly known as the "Crane Handbook." Unfortunately, the Crane Handbook makes a major error in handling Bernoulli's equation. Equation 1–3 of Ref. [6] is:

$$z_1 + \frac{144P_1}{\rho_1} + \frac{v_1{}^2}{2g} = z_2 + \frac{144P_2}{\rho_2} + \frac{v_2{}^2}{2g_c} + h_L \tag{5.133}$$

This is tantalizingly close to our Eq. (5.46), and would seem at first glance to be a logical extension of Eq. (5.44). The difference is that Ref. [6] allows the densities on the two sides of

Eq. (5.133) to be different, whereas Eq. (5.44) was derived under the stipulation that density is constant. This is more than just a minor quibble, as can be readily seen by applying Eq. (5.133) to isothermal flow of a perfect gas. In this case the two pressure terms cancel, and we are left with the clearly erroneous implication that the mechanical energy content of a gas is independent of its pressure. Even in the case of a very slight difference in pressure, where we are inclined to believe that analysis of the flow using Bernoulli's equation is sufficiently accurate for our purposes, the foregoing observation invalidates the use of Eq. (5.133).

This mistake is easy to make, and comes from believing that the left-hand side of Eq. (5.44) represents the total mechanical energy of the system, and that this energy is conserved in flow unless shaft work is withdrawn. In fact, if density changes with flow, pressure-volume work is performed on the environment, and even in the absence of friction or shaft work the quantity $z + \frac{144P}{\rho} + \frac{v^2}{2g}$ is not conserved. The term $\frac{144P}{\rho}$ does not actually represent the mechanical energy content due to pressure for a compressible fluid. That quantity is given by $-\int PdV$. For isothermal compression of a perfect gas, this is $\frac{P}{\rho}\ln P$.

In order to apply Bernoulli's equation to the flow of a gas, we must use the same density on both sides of the equation. The analysis leading to Eq. (5.91) was undertaken to demonstrate that the optimal density to use is simply the arithmetic average of the densities at inlet and outlet conditions ("optimal" in the sense that it comes as close as Bernoulli's equation can come to matching the results of a full integration of Eq. (5.43) for isothermal flow).

The Crane Handbook gives the following guidance with regard to when Bernoulli's equation is adequate for representing the flow of a compressible fluid:

1) If $\Delta P \leq 0.1P_1$, reasonable accuracy is obtained by using Bernoulli's equation, Eq. (5.48), with the velocity in the frictional loss term determined by using the specific volume at either upstream or downstream conditions.
2) If $0.1P_1 < \Delta P \leq 0.4P_1$, reasonable accuracy is obtained by using Bernoulli's equation with the velocity in the frictional loss term based on a specific volume determined by the average of upstream and downstream conditions.
3) If $0.4P_1 < \Delta P$, the energy balance in the form of a differential equation should be integrated with pressure-dependent density.

With regard to case 2, the analysis leading to Eq. (5.91) demonstrates that it is the density rather than the specific volume which should be averaged in order to apply Bernoulli's equation to the flow of a compressible fluid. The two approaches are not equivalent, i.e., in general $\overline{V} \neq \frac{1}{\overline{\rho}}$.

The equations that the Crane Handbook offers to handle case 3 above are all based on the assumption of isothermal flow. This is more appropriate for long-distance pipelines than it is for process piping. Equation 1-6 of Ref. [6] is:

$$w^2 = \frac{144gA^2}{\overline{V}_1\left[\frac{fL}{D} + 2\log\frac{P_1'}{P_2'}\right]}\left[\frac{P_1'^2 - P_2'^2}{P_1'}\right] \tag{5.134}$$

To compare this with the comparable equation derived in this report, we begin with Eq. (5.88):

$$\frac{1}{2\rho_1 P_1}(P_2{}^2 - P_1{}^2) - \frac{q^2}{g_c\rho_1{}^2A^2}\ln\frac{P_2}{P_1} + \frac{fL}{D}\frac{q^2}{2g_c\rho_1{}^2A^2} = 0 \tag{5.135}$$

With some manipulations we can rearrange this to

$$q^2 = \frac{g_cA^2}{\frac{1}{\rho_1}\left[\frac{fL}{D} + 2\ln\frac{P_1}{P_2}\right]}\left[\frac{P_1{}^2 - P_2{}^2}{P_1}\right] \tag{5.136}$$

Eq. (5.136) is identical to Eq. (5.134) when we account for the use of different nomenclature and units of measure.

- Crane uses w to refer to mass flow rate in lb_m/sec where we use q.
- Crane uses g for the gravitational conversion constant where we use g_c.
- Crane uses $\overline{V}$ for specific volume in ft^3/lb_m ;we use $\rho\ (=\frac{1}{\overline{V}})$for density in lb_m/ft^3.
- Crane expresses pressure in psia where we express pressure in psfa.

If we make the simplifying assumption that the acceleration term can be neglected in a long gas pipeline, the second term on the left hand side of Eq. (5.87) drops out to leave:

$$\frac{P}{\rho_1 P_1}dP + \frac{f}{D}\frac{q^2}{2g_c{\rho_1}^2A^2}d\ell = 0 \tag{5.137}$$

which we integrate to:

$$\frac{1}{2\rho_1 P_1}({P_2}^2 - {P_1}^2) + \frac{fL}{D}\frac{q^2}{2g_c{\rho_1}^2A^2} = 0 \tag{5.138}$$

Eq. (5.138) can be rearranged to:

$$q^2 = \frac{g_c DA^2\rho_1}{fL}\left(\frac{{P_1}^2 - {P_2}^2}{P_1}\right) \tag{5.139}$$

Taking account of the differences in nomenclature and units cited above, this is exactly equivalent to the Crane Handbook's Equation 1-7. However, it is this author's contention that the equation:

$$q^2 = \frac{g_c DA^2}{fL}(\rho_1 P_1 - \rho_2 P_2) \tag{5.140}$$

based on Eq. (5.99), will give a better result for real gases.

The volumetric flow rate corresponding to Eq. (5.139) is:

$$\frac{q}{\rho_1} = \sqrt{\frac{g_c DA^2}{fL\rho_1}\left(\frac{{P_1}^2 - {P_2}^2}{P_1}\right)} \tag{5.141}$$

in actual cubic feet per second. Converting this to standard cubic feet per hour:

$$\begin{aligned} SCFH &= 3600\frac{520}{14.7}\frac{P_1}{T}\frac{q}{\rho_1} = 3600\frac{520}{14.7}\frac{\pi}{4}\sqrt{\frac{g_c R}{(5280)(28.966)}}\sqrt{\frac{({P_1}^2 - {P_2}^2)}{f\frac{L}{5280}T\frac{M_w}{28.966}}D^5} \\ &= \frac{114.3}{144}\sqrt{\frac{{P_1}^2 - {P_2}^2}{fL_m TS_g}D^5} \end{aligned} \tag{5.142}$$

where L_m is pipeline length in miles, and S_g is the specific gravity of the gas relative to air. This is exactly equivalent to Equation 1-7a of Ref. [6].

T. R. Weymouth[6] assumed a friction factor of the form:

$$f = \frac{0.032}{\sqrt[3]{D}} \tag{5.143}$$

6 T. R. Weymouth, *Transactions of the American Society of Mechanical Engineers*, Volume 34, 1912, p. 197.

Substituting this into Eq. (5.142):

$$SCFH = \frac{(114.3)}{\sqrt{(0.032)(520)}}\frac{1}{144}D^{\frac{8}{3}}\sqrt{\frac{P_1^{\ 2}-P_2^{\ 2}}{L_m S_g}\frac{520}{T}} = \frac{28.0}{144}D^{2.667}\sqrt{\left[\frac{P_1^{\ 2}-P_2^{\ 2}}{L_m S_g}\right]\frac{520}{T}} \tag{5.144}$$

Eq. (5.144) is the Weymouth formula, given as Equation 1-8 in Ref. [6].

The Panhandle Company[7] proposed the following correlation for the friction factor:

$$f = 0.1225\left(\frac{\mathrm{D}}{(SCFH)S_g}\right)^{0.1461} \tag{5.145}$$

Substituting this into Eq. (5.142):

$$SCFH = 114.3\sqrt{\frac{(SCFH)^{0.1461}}{(0.1225)L_m TS_g^{\ 0.8539}}\frac{P_1^{\ 2}-P_2^{\ 2}}{(144)^2}D^{4.8539}} \tag{5.146}$$

Solving this equation for *SCFH*:

$$SCFH = (515.3)D^{2.6182}\left[\frac{P_1^{\ 2}-P_2^{\ 2}}{(144)^2 S_g^{\ 0.8539}TL_m}\right]^{0.5394} \tag{5.147}$$

The equation given in Ref. [6] (Equation 1-9) assumes that $S_g = 0.6$, but includes no temperature dependence. Setting $S_g = 0.6$ in Eq. (5.147):

$$SCFH = (652.0)D^{2.6182}\left[\frac{(P_1^{\ 2}-P_2^{\ 2})}{(144)^2 TL_m}\right]^{0.5394} \tag{5.148}$$

This equation cannot be compared directly with Equation 1-9 of Ref. [6] because of the lack of a temperature term in the latter. This is as far as we can go with this.

References

1 Streeter VL. Fluid mechanics. 2nd ed. New York: McGraw-Hill; 1958.
2 Brodkey RS. The phenomena of fluid motion. Reading, MA: Addison-Wesley; 1967.
3 Denbigh K. The principles of chemical equilibrium. Cambridge: Cambridge University Press; 1961.
4 API Technical Data Book. Houston: EPCON Software; 2018.
5 Edmister WC, Lee BI. Applied hydrocarbon thermodynamics. Volume 1. 2nd ed. Houston: Gulf Publishing Company; 1984.
6 Flow of fluids through valves, fittings, and pipe. Technical Paper No. 410. Signal Hill, CA: Crane Valves; 1968.
7 Shapiro AH. The dynamics and thermodynamics of compressible fluid flow. New York: Ronald Press; 1953.
8 Anderson JD. Modern compressible flow—with historical perspective. 3rd ed. New York: McGraw-Hill; 2003.

7 B. F. Grizzle, "Simplification of Gas Flow Calculations by Means of New Special Slide Rule", *Petroleum Engineer*, September, 1945.

6

Fluid Flow: Advanced Topics

Paul A. Nelson, Moye Wicks III, Todd J. Willman, and Vinay Gadekar

*EPCON Software™, Houston, USA

CHAPTER MENU

6.1 Introduction

This chapter builds on the material provided in the previous one. Here we present some material that is normally not covered in an academic curriculum, but that is important for understanding fluid flow in industrial applications.

The conventions used with regard to units of measure will be the same as in Chapter 5.

The authors are affiliated with EPCON Software (www.epcon.com), a company that has been providing practical process engineering software and services to the process industries since 1981. The fluid mechanics calculation methods described herein are performed in the EPCON Process Innovator® simulation software used by thousands of engineers worldwide.

6.2 Notation

A = Cross-sectional area, ft^2

b = Parameter in Eqs. (6.53) and (6.55), dimensionless

C_d = Meter discharge coefficient, dimensionless

C_V = Flow coefficient for a control valve, dimensionless

c = Speed of sound, ft/sec

c_p = Heat capacity at constant pressure, $(\text{ft lb}_\text{f})/(\text{lb}_\text{m}\ \text{R})$

c_v = Heat capacity at constant volume, $(\text{ft lb}_\text{f})/(\text{lb}_\text{m}\ \text{R})$

D = Pipe diameter, ft

*www.epcon.com

Natural Gas Processing from Midstream to Downstream, First Edition.
Edited by Nimir O. Elbashir, Mahmoud M. El-Halwagi, Ioannis G. Economou, and Kenneth R. Hall.

$D(i)$ = Set of all pipes in a network for which node i is a destination (see Eq. (6.35))

d = Diameter at the throat of a meter (also called the "bore"), ft

F = Modified Froude number, dimensionless (see Eq. (6.84))

F_γ = Specific heat ratio factor for a control valve (see Eq. (6.80)), $= \frac{\gamma}{1.4}$, dimensionless

f = Darcy friction factor, dimensionless

G = Volumetric fluid flow rate, gallons per minute (gpm)

g = Acceleration of gravity = 32.174 ft/sec^2

g_c = Gravitational conversion constant = 32.174 (lb$_m$ ft)/(sec^2 lb$_f$)

h = Fluid head, (ft lb$_f$)/lb$_m$ (note that this is *not* in units of ft)

= Height of liquid level in a horizontal pipe, ft (Dukler-Taitel analysis)

h_e = Head loss in flow through equipment associated with a pipe, (ft lb$_f$)/lb$_m$

h_f = Frictional head loss, (ft lb$_f$)/lb$_m$

h_p = Head gain from a pump, (ft lb$_f$)/lb$_m$

K = Resistance factor for pipe fitting (K-factor), dimensionless

K_0 = Parameter in Eqs. (6.53) and (6.54), dimensionless

L = Pipe length, ft

ℓ = Element of length in the direction of a streamline, ft

M = Mach number, dimensionless

M_w = Molecular weight, lb$_m$/lbmole

N_n = Number of nodes in a pipeline network simulation

N_p = Number of pipes in a pipeline network simulation

N_{Re} = Reynolds number, dimensionless

P = Pressure, lb$_f$/ft^2 (psfa)

q = Mass flow rate, lb$_m$/sec

q_j = Mass flow rate through pipe j in a pipeline network, lb$_m$/sec (see Eq. (6.35))

$q(i)$ = Net mass flow rate to node i in a pipeline network, lb$_m$/sec (see Eq. (6.35))

q_T = Theoretical mass flow rate (see Eqs. (6.39) and (6.48)), lb$_m$/sec

R = Universal gas constant = 1545.349 (ft lb$_f$)/(lbmole R)

r = Pressure ratio, P_2/P_1, dimensionless

r_c = Critical pressure ratio for choked flow in a meter, dimensionless

$S(i)$ = Set of all pipes in a network for which node i is a source (see Eq. (6.35))

s = Safety factor (or fouling factor), dimensionless

T = Temperature, degrees Rankine (R)

= "Turbulence" parameter in the Dukler-Taitel analysis (see Eq. (6.85)

v = Velocity, ft/sec

X = Parameter in Dukler-Taitel analysis of two-phase flow (see Eq.(6.83))

x = Ratio of pressure drop to initial pressure, $\frac{\Delta P}{P_1}$, dimensionless

= Weight fraction vapor in two-phase flow (Section 6.6), dimensionless

x_T = Value of x for choked flow through a control valve (see Eq. (6.79)), dimensionless

Y = Gas expansion factor, dimensionless

Z = Gas compressibility, dimensionless

z = Elevation, ft

β = Ratio of diameter of a meter at its throat to that of the pipe, dimensionless

γ = Ratio of heat capacities (isentropic exponent) = c_p/c_v, dimensionless

θ = Angle of inclination of pipe from vertical, radians

λ = Parameter in Eqs. (6.53) and (6.56), dimensionless

= Liquid volume fraction in two-phase flow (Section 6.6), dimensionless

μ = Dynamic viscosity, cP (or in English units, lb_f-sec/ft^2)

ρ = Density, lb_m/ft^3

Subscripts:

g Refers to the gas phase in a two-phase mixture

l Refers to the liquid phase in a two-phase mixture

m Refers to the mixture property in a two-phase mixture

s Refers to the superficial velocity of a liquid or gas phase

t Refers to stagnation (total pressure) conditions

6.3 Piping Networks

6.3.1 Network Flow

In Chapter 5, we addressed the flow of fluids through a single pipe. In this section, we will show how the equations for pipeline flow can be linked together to represent a more complex system. We will not address any of the issues associated with calculating the performance of process equipment nor of solving the system of nonlinear equations that is produced. Our attention will be confined to fluid mechanical issues.

One of the major applications of EPCON software is to model flow in systems involving multiple pipes and other pieces of equipment such as are found in the petrochemical process industries, on offshore production platforms, etc. (Ref. [1]).

A network of pipes is defined by a set of nodes and a set of pipes that connect them. At a minimum, each point at which there is a net flow into or out of the system will be a node, and each point at which three or more pipes form a junction will be a node. One would not normally define a point with one pipe in and one pipe out as a node, but there may be good reasons for doing so occasionally:

- When a pipe changes diameter, it is a good idea to define the point of diameter change as a node, so that each pipe connecting two nodes will have a constant diameter.

- When a pipe in compressible flow has a very large pressure drop, it may be a good idea to break it up into several shorter segments, each with a smaller pressure drop.
- When a length of pipe contains more than one piece of equipment requiring a special calculation procedure, e.g., pump, compressor, valve, and/or meter, it is a good idea to break it up into segments such that each segment contains no more than one such item.

For input, each node must be specified with an elevation and with either a pressure or a net flow, i.e., flow either entering (positive) or leaving (negative) the system. A node cannot be specified with both. In practice, the majority of nodes are specified with zero net flow, but there must be at least two exceptions or flow into and out of the system will not be possible. Each pipe is specified by its length, diameter, some parameter which can be used to calculate a friction factor (typically this will be a pipe wall roughness), and the sum of K-factors for any fittings in the pipe.

The solution of a network flow problem will be a mass flow rate through each pipe, and either a pressure or net flow rate for each node (whichever was not specified in the input). If there are N_p pipes and N_n nodes, there will be N_p flow rates and N_n node variables (net flow or pressure) for a total of $N_p + N_n$ problem variables.

The proper approach is to calculate a stagnation pressure for each node. In general, this is not a pressure that can be measured at the node. It is better to think of it as a measure that, when divided by fluid density, gives mechanical energy in terms of (ft lb_f)/lb_m. If each pipe entering or leaving a node is required to do so with the same stagnation pressure, that being the pressure assigned to the node, mechanical energy balance will be achieved at the node.

6.3.2 Stagnation Pressure and Temperature

6.3.2.1 Incompressible

In deriving the momentum balance equation, we made use of Figure 5.2, which depicted flow through a channel of constant cross-sectional area. When we derived Bernoulli's equation, Eq. (5.45), we said that this applies to flow along a streamline. In this case there is no requirement that the flow be of constant cross-sectional area.

Consider the case where the cross-sectional area for flow at point 1 is very large, so much so that the fluid velocity, determined by Eq. (5.33), becomes very small. In the limit as $v_1 \rightarrow 0$, Eq. (5.45) becomes:

$$\frac{P_{t1}}{\rho_{t1}} + \frac{g}{g_c} z_1 = \frac{P_2}{\rho_2} + \frac{g}{g_c} z_2 + \frac{{v_2}^2}{2g_c} \tag{6.1}$$

where we use the notation P_{t1} to denote the pressure at zero flow velocity. This pressure is called the "stagnation pressure," or sometimes the "total pressure." For an incompressible fluid, we can express the stagnation pressure as:

$$P_{t1} = P_1 + \frac{\rho_1 {v_1}^2}{2g_c} \tag{6.2}$$

In the absence of frictional losses, stagnation pressure is affected only by changes in elevation, and temperature is not a function of pressure, so:

$$T_{t1} = T_1 \tag{6.3}$$

Using Eq. (6.2), we could rewrite Eq. (5.45) as

$$\frac{P_{t1}}{\rho_{t1}} + \frac{g}{g_c} z_1 - \frac{P_{t2}}{\rho_{t2}} + \frac{g}{g_c} z_2 \tag{6.4}$$

Stagnation conditions do not require an infinite cross-sectional area. A bluff body inserted into the flow will cause the flow to divide in order to go around it. Somewhere at the front of this body there will be a stagnation point. Flow impinging on the body to the left of this point will continue around the body to the left. Flow impinging to the right will continue around to the right. At the stagnation point itself, the flow is stagnant; the velocity is zero. A tiny hole drilled in the bluff body at exactly this point, connected to a tube leading to a pressure gauge, would constitute a system for directly measuring the stagnation pressure. If this measure were combined with a measurement of the pressure of the flowing fluid, known as the "static pressure" and normally measured using a pressure tap in the pipe wall, we would have a means of measuring the velocity of the fluid. Using Eq. (6.2):

$$v_1 = \sqrt{\frac{2g_c}{\rho_1}(P_{t1} - P_1)} \tag{6.5}$$

This is the principle behind the use of a "Pitot tube" to measure fluid velocity. What is calculated in Eq. (6.5) is the relative velocity between the fluid and the velocity measuring equipment. It makes no difference whether the equipment is stationery and the fluid is moving, or the fluid is stationery and the equipment is moving. It is not very common for Pitot tubes to be used in process equipment, but this is the primary method for measuring the velocity of an airplane relative to the air through which it is flying.

6.3.2.2 Isothermal

When the flow is compressible the calculation of stagnation conditions is somewhat more complicated, and we must impose an assumption about the relationship between pressure and temperature. We begin by assuming the system to be isothermal. Substituting Eq. (5.85) into Eq. (5.43):

$$\frac{P_1}{\rho_1}\frac{dP}{P} + \frac{g}{g_c}dz + \frac{v}{g_c}dv = 0 \tag{6.6}$$

Integrating:

$$\frac{P_1}{\rho_1}\ln\frac{P_2}{P_1} + \frac{g}{g_c}(z_2 - z_1) + \frac{1}{2g_c}(v_2^2 - v_1^2) = 0 \tag{6.7}$$

Rearranging, and noting that $\frac{P_1}{\rho_1} = \frac{P_2}{\rho_2}$:

$$\frac{P_1}{\rho_1}\ln P_1 + \frac{g}{g_c}z_1 + \frac{1}{2g_c}v_1^2 = \frac{P_2}{\rho_2}\ln P_2 + \frac{g}{g_c}z_2 + \frac{1}{2g_c}v_2^2 \tag{6.8}$$

From Eq. (6.8) we can deduce the relationship between static pressure and stagnation pressure:

$$\frac{P_{t1}}{\rho_{t1}}\ln P_{t1} = \frac{P_1}{\rho_1}\ln P_1 + \frac{1}{2g_c}v_1^2 \tag{6.9}$$

Since $\frac{P_{t1}}{\rho_{t1}} = \frac{P_1}{\rho_1}$ we can write Eq. (6.9) as:

$$\ln P_{t1} = \ln P_1 + \frac{1}{P_1}\frac{\rho_1}{2g_c}v_1^2 \tag{6.10}$$

To get a sense of the magnitude of departure from incompressible results, it will be helpful to consider dependence on Mach number alongside dependence on velocity. Eq. (6.10) can be rearranged, and supplemented using Eq. (5.107):

$$P_{t1} = P_1 \exp\left(\frac{1}{P_1}\frac{\rho_1}{2g_c}{v_1}^2\right) = P_1 \exp\left(\frac{\gamma}{2}{M_1}^2\right) \tag{6.11}$$

Since the flow is isothermal, we also note that:

$$T_{t1} = T_1 \tag{6.12}$$

In process operations, the fluid velocity is normally quite low with respect to the speed of sound. If we expand the exponentials in a Taylor series:

$$P_{t1} = P_1 + \frac{\rho_1 {v_1}^2}{2g_c} + \frac{{\rho_1}^2 {v_1}^4}{8P_1 {g_c}^2} + \cdots = P_1\left[1 + \frac{\gamma}{2}{M_1}^2 + \frac{\gamma^2}{8}{M_1}^4 + \cdots\right] \tag{6.13}$$

Comparing Eq. (6.13) with Eq. (6.2), we see that in the limit of low velocity operation compressible isothermal flow behaves like incompressible flow, but as the velocity increases the stagnation pressure for compressible flow increases faster than for incompressible flow.

6.3.2.3 Isentropic

Substituting Eq. (5.93) into Eq. (5.43):

$$\frac{1}{\rho_1}\left(\frac{P_1}{P}\right)^{\frac{1}{\gamma}} dP + \frac{g}{g_c}dz + \frac{v}{g_c}dv = 0 \tag{6.14}$$

Integrating:

$$\frac{P_1^{\frac{1}{\gamma}}}{\rho_1}\frac{\gamma}{\gamma - 1}\left(P_2^{\frac{\gamma-1}{\gamma}} - P_1^{\frac{\gamma-1}{\gamma}}\right) + \frac{g}{g_c}(z_2 - z_1) + \frac{1}{2g_c}({v_2}^2 - {v_1}^2) = 0 \tag{6.15}$$

Rearranging, and noting that $\frac{P_1^{\frac{1}{\gamma}}}{\rho_1} = \frac{P_2^{\frac{1}{\gamma}}}{\rho_2}$:

$$\frac{\gamma}{\gamma - 1}\frac{P_1^{\frac{1}{\gamma}}}{\rho_1}P_1^{\frac{\gamma-1}{\gamma}} + \frac{g}{g_c}z_1 + \frac{1}{2g_c}{v_1}^2 = \frac{\gamma}{\gamma - 1}\frac{P_2^{\frac{1}{\gamma}}}{\rho_2}P_2^{\frac{\gamma-1}{\gamma}} + \frac{g}{g_c}z_2 + \frac{1}{2g_c}{v_2}^2 \tag{6.16}$$

From Eq. (6.16) we can deduce the relationship between static pressure and stagnation pressure:

$$P_{t1}^{\frac{\gamma-1}{\gamma}} = P_1^{\frac{\gamma-1}{\gamma}} + \frac{\gamma - 1}{\gamma}\frac{\rho_1}{P_1^{\frac{1}{\gamma}}}\frac{1}{2g_c}{v_1}^2 = P_1^{\frac{\gamma-1}{\gamma}}\left(1 + \frac{\gamma - 1}{\gamma}\frac{1}{P_1}\frac{\rho_1 {v_1}^2}{2g_c}\right) \tag{6.17}$$

We can solve Eq. (6.17) for P_{t1}, and supplement the solution using Eq. (5.107):

$$P_{t1} = P_1\left(1 + \frac{\gamma - 1}{\gamma}\frac{1}{P_1}\frac{\rho_1 {v_1}^2}{2g_c}\right)^{\frac{\gamma}{\gamma-1}} = P_1\left(1 + \frac{\gamma - 1}{2}{M_1}^2\right)^{\frac{\gamma}{\gamma-1}} \tag{6.18}$$

From Eq. (5.28) we see that:

$$\frac{P}{P_1} = \left(\frac{T}{T_1}\right)^{\frac{\gamma}{\gamma-1}} \tag{6.19}$$

Comparing Eqs. (6.19) and (6.18) we obtain an expression for the stagnation temperature:

$$T_{t1} = T_1\left(1 + \frac{\gamma - 1}{\gamma}\frac{1}{P_1}\frac{\rho_1 {v_1}^2}{2g_c}\right) = T_1\left(1 + \frac{\gamma - 1}{2}{M_1}^2\right) \tag{6.20}$$

At low velocities relative to the speed of sound, we can expand the term in parentheses in Eq. (6.18) as a binomial series:

$$P_{t1} = P_1 + \frac{\rho_1 v_1^2}{2g_c} + \frac{1}{\gamma}\frac{\rho_1^2 v_1^4}{8P_1 g_c^2} + \cdots = P_1\left[1 + \frac{\gamma}{2}M_1^2 + \frac{\gamma}{8}M_1^4 + \cdots\right] \tag{6.21}$$

Comparing Eq. (6.21) with Eq. (6.2), we see that, as for isothermal flow, isentropic flow behaves like incompressible flow in the limit of low velocity operation. For both isothermal and isentropic flow, departure from incompressible behavior enters as the fourth power of the Mach number. The speed of sound in air at normal ambient temperature is in excess of 1000 ft/sec, so at typical process velocities on the order of 100 ft/sec the correction term would show up in the fourth decimal place. Generally, we would feel comfortable ignoring a correction of this magnitude.

As Mach number increases, however, the correction cannot be ignored. For example, as the speed of an airplane increases to where the Mach number is an appreciable fraction of unity, Eq. (6.5) no longer gives the correct velocity. To get that, we solve Eq. (6.18) for the velocity:

$$v_1 = \sqrt{\frac{\gamma}{\gamma-1}\frac{2g_c}{\rho_1}P_1\left[\left(\frac{P_{t1}}{P_1}\right)^{\frac{\gamma-1}{\gamma}} - 1\right]} \tag{6.22}$$

6.3.2.4 Adiabatic

For adiabatic flow, the relationship of stagnation pressure and temperature to static conditions is the same as for isentropic flow. That is, we assume no frictional loss between the stagnation point and the static point. So, the isentropic equations for stagnation pressure and temperature, Eqs. (6.18) and (6.20), apply. However, in adiabatic flow we do have frictional energy loss, so the stagnation pressure decreases along the flow path. To calculate the stagnation pressure along the length of the pipe, we begin with the logarithmic derivative of Eq. (6.18):

$$\frac{dP_t}{P_t} = \frac{dP}{P} + \frac{\gamma M^2}{2+(\gamma-1)M^2}\frac{dM^2}{M^2} \tag{6.23}$$

Substituting Eq. (5.112) into Eq. (6.23):

$$\frac{dP_t}{P_t} = -\frac{1-M^2}{2+(\gamma-1)M^2}\frac{dM^2}{M^2} \tag{6.24}$$

We expand the right-hand side in partial fractions:

$$\frac{dP_t}{P_t} = \left[-\frac{1}{2M^2} + \frac{\gamma+1}{2}\frac{1}{2+(\gamma-1)M^2}\right]dM^2 \tag{6.25}$$

and integrate to obtain:

$$\ln\frac{P_{t2}}{P_{t1}} = -\frac{1}{2}\ln\frac{M_2^2}{M_1^2} + \frac{1}{2}\frac{\gamma+1}{\gamma-1}\ln\frac{2+(\gamma-1)M_2^2}{2+(\gamma-1)M_1^2} \tag{6.26}$$

which we rearrange to:

$$\frac{P_{t2}}{P_{t1}} = \sqrt{\frac{M_1^2}{M_2^2}\left[\frac{2+(\gamma-1)M_2^2}{2+(\gamma-1)M_1^2}\right]^{\frac{\gamma+1}{\gamma-1}}} \tag{6.27}$$

6.3.3 Flow Between Vessels

6.3.3.1 Incompressible

A common occurrence in process piping involves flow from one vessel to another through a pipe. Relative to flow in the pipe, the velocity of flow within the vessels is negligible, and the pressures of the source and destination vessels are the stagnation pressures at the inlet and outlet of the pipe, respectively. For incompressible flow, we can apply Bernoulli's equation, Eq. (5.66), but replace the static pressure terms with stagnation pressures using Eq. (6.2). We obtain:

$$\frac{P_{t1}}{\rho} + \frac{g}{g_c}z_1 + h_p = \frac{P_{t2}}{\rho} + \frac{g}{g_c}z_2 + \left(\frac{fL}{D}s + K\right)\frac{v^2}{2g_c} + h_e \tag{6.28}$$

The elimination of the velocity head terms does not mean, of course, that we can dispense with consideration of velocity altogether, as average velocity enters directly into the calculation of the frictional loss term, and may contribute to the other gain and loss terms as well.

There will be some energy loss associated with the flow contraction entering the pipe from the source vessel. Similarly, there will be some loss associated with the expansion into the receiving vessel. Both of these losses should be incorporated into the calculation by means of contributions to the K-factor for the pipe.

Under some circumstances it may be important to know the actual static pressure within the pipe. For example, if a flowing liquid is near its boiling point, there may be a concern about the possibility of flashing. This is of particular concern at the suction of a centrifugal pump, where it is important to avoid cavitation. Part of a pump specification will be the "net positive suction head" (NPHS) required. This is the difference between the static pressure of the fluid entering the pump and the pressure at which it begins to flash, expressed in terms of feet of fluid head. The designer of a piping system must assure that fluid is supplied to the pump at a pressure that exceeds the NPSH requirement. To calculate the static pressure at the end of a pipe (or at the entrance to a pump), we can use Eq. (6.28) to get the stagnation pressure at the delivery end of the pipe, and then use a rearrangement of Eq. (6.2) to obtain the static pressure:

$$P_2 = P_{t2} - \frac{\rho_2 v_2^{\,2}}{2g_c} \tag{6.29}$$

6.3.3.2 Compressible

For compressible flow in process piping, when the pressure drop is too high to allow the use of Bernoulli's equation, the calculation method of choice assumes adiabatic flow. To start the calculation, we need to know $M_1^{\,2}$. We assume isentropic flow from the vessel to the pipe inlet. The flow rate into the pipe inlet is given by:

$$q = \rho_1 v_1 A = \frac{P_1 M_w}{RT_1} v_1 A = \frac{P_{t1}}{\sqrt{T_{t1}}} A \frac{v_1\sqrt{M_w}}{\sqrt{\gamma g_c R T_1}} \sqrt{\frac{\gamma g_c M_w}{R}} \sqrt{\frac{T_{t1}}{T_1}} \frac{P_1}{P_{t1}} \tag{6.30}$$

Using Eqs. (5.61), (5.58), (6.18), and (6.20):

$$q = \frac{P_{t1}}{\sqrt{T_{t1}}} A \sqrt{\frac{\gamma g_c M_w}{R}} M_1 \left(1 + \frac{\gamma - 1}{2} M_1^{\,2}\right)^{-\frac{\gamma+1}{2(\gamma-1)}} \tag{6.31}$$

which yields the following implicit equation for $M_1^{\,2}$:

$$\frac{q^2 T_{t1}}{A^2 P_{t1}^{\,2}} \frac{R}{\gamma g_c M_w} = M_1^{\,2}\left(1 + \frac{\gamma - 1}{2} M_1^{\,2}\right)^{-\frac{\gamma+1}{\gamma-1}} \tag{6.32}$$

This equation must be solved iteratively for M_1^2. Since it is expected that typical values for M_1^2 are going to be on the order of 10^{-2}, an algorithm that should converge fairly rapidly is to initially assume that $M_1^2 = \frac{q^2 T_{t1}}{A^2 P_{t1}^2} \frac{R}{\gamma g_c M_w}$, and then iteratively recalculate M_1^2 by means of direct substitution into the right-hand side of the equation:

$$M_1^2 = \left(\frac{q}{AP_{t1}}\right)^2 \frac{RT_{t1}}{\gamma g_c M_w}\left(1 + \frac{\gamma - 1}{2}M_1^2\right)^{\frac{\gamma+1}{\gamma-1}} \tag{6.33}$$

With M_1^2 in hand, we use Eq. (5.116) to calculate M_2^2, and Eq. (6.27) to calculate P_{t2}. Note that stagnation temperature does not change in adiabatic flow, since there is no net heat input, so:

$$T_{t2} = T_{t1} \tag{6.34}$$

6.3.4 The System of Equations

To determine $N_p + N_n$ variables, we must have $N_p + N_n$ equations. For each node, *i*, there is an equation expressing material balance. Let $S(i)$ be the set of all pipes for which node *i* is a source, i.e., the set of all pipes carrying fluid away from node *i*. Likewise, let $D(i)$ be the set of all pipes for which node *i* is a destination, i.e., the set of all pipes carrying fluid to node *i*. Let q_j be the flow rate in pipe *j*, and let $q(i)$ be the net flow to node *i*, i.e., the flow to the node from a source external to the pipe network. $q(i)$ is positive for net flow entering the network at the node and negative for net flow leaving the network at the node. Then for each node *i* the mass balance equation is:

$$\sum_{j \subset D(i)} q_j - \sum_{j \subset S(i)} q_j + q(i) = 0 \quad i = 1, \ldots\ldots N_n \tag{6.35}$$

We also have one equation for each of the N_p pipes. For incompressible flow, or for compressible flow with a relatively small pressure drop such that Bernoulli's equation can be used, Eq. (6.28) provides that equation. When the pressure drop is appreciable, such that the adiabatic flow analysis must be applied, the equation is:

$$(P_{t2})_{calculated} = (P_{t2})_{assumed} \tag{6.36}$$

In Eq. (6.36) the right-hand side is the stagnation pressure at the destination node as given by the currently assumed values of pressure and temperature. The left-hand side is the same quantity as calculated using Eqs. (6.33), (5.116), and (6.27).

The end result is a system of $N_p + N_n$ nonlinear equations in $N_p + N_n$ variables. For a network with many pipe legs and nodes, this can be quite a large system of equations, requiring specialized mathematical techniques for its solution. Discussion of such techniques is beyond the scope of this chapter, which focuses on the fluid mechanics.

6.3.5 Example Application

Firewater systems are arguably the most important system at any processing facility to protect the lives of those working there as well as the investments that have been made by the company. These systems typically have buried pipe around key process units to avoid damage from a potential explosion. The looped piping design provides for maximum flexibility to isolate parts of the system that may be compromised by an explosion while providing multiple flow paths for monitors and hydrants and requires a robust pipeline network analysis engine to efficiently handle the required computations. Most industrial fire protection systems are designed

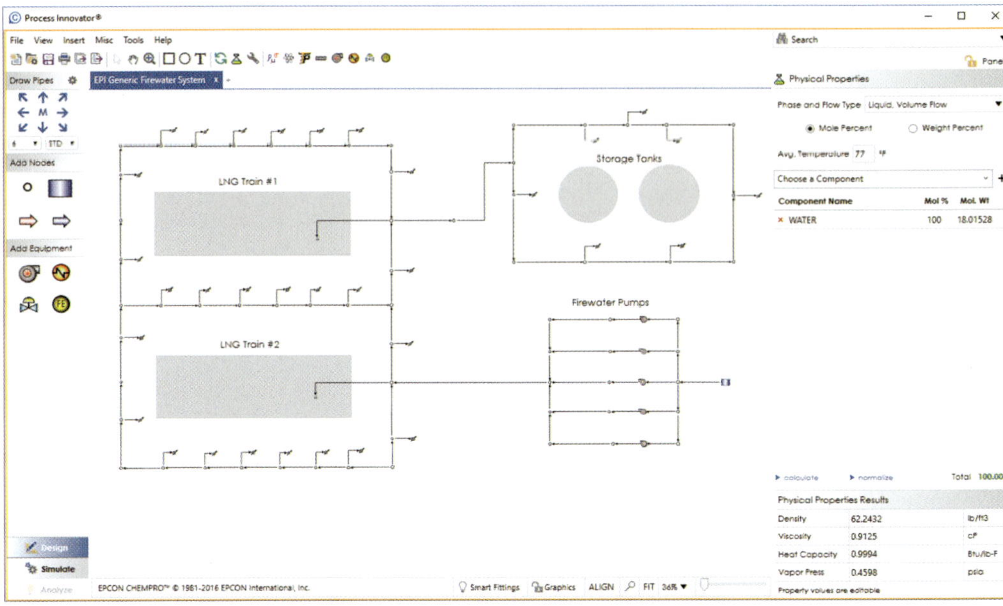

Figure 6.1 Firewater System Flow Simulation – EPCON's CHEMPRO software.

to provide 100 psig pressure at the monitor nozzles. This can be simulated by selecting the desired monitor locations for fighting a fire and specifying the pressure as 100 psig plus the pressure drop through the monitor at its rated flow. With specified pumps in operation, the model will then predict the maximum flow at each nozzle. With accurate field flow and pressure validation, modeling of firewater systems is used to confirm that the fire water system can sufficiently provide the flow and pressure needed for the process facilities they were intended to protect. An example fluid flow simulation model of an LNG facility firewater system is shown in Figure 6.1.

6.4 Meters

6.4.1 Incompressible Flow Through a Meter

There are many different concepts that can be used to design an instrument for measuring the flow rate of fluid in a pipe. But in industrial piping systems the vast majority of flow meters depend on a single principle, and that is the change in pressure that occurs when a change in cross-sectional area causes a change in velocity. We analyze this change using a simplified form of the Bernoulli equation, Eq. (5.66), in which we assume that the flow path is so short that there is no significant head loss due to frictional effects. There is obviously no contribution from the pump head or equipment head terms. And finally, meters are normally mounted in horizontal runs of pipe, and we normally assume that there is no change in elevation between the inlet and the throat (the narrowest point in the flow channel), which are the two points to which we apply the Bernoulli equation (labeled "1" and "2" respectively). Thus simplified, the equation becomes:

$$\frac{P_1}{\rho_1} + \frac{v_1^{\,2}}{2g_c} = \frac{P_2}{\rho_2} + \frac{v_2^{\,2}}{2g_c} \tag{6.37}$$

We express the velocities in terms of the mass flow rate using Eq. (5.32). Since at this point we are considering the flow of an incompressible fluid, $\rho_1 = \rho_2 = \rho$, and we are able to derive the equation

$$\Delta P = P_1 - P_2 = \frac{q^2}{2\rho g_c A_2{}^2}\left[1 - \left(\frac{A_2}{A_1}\right)^2\right] = \frac{1}{2}\frac{q^2}{\rho g_c A_2{}^2}(1 - \beta^4) \tag{6.38}$$

where β is the ratio of meter bore to pipe diameter, i.e., $\beta = \frac{d}{D}$. By rearranging Eq. (6.38), we get an expression for the theoretical rate of flow through a meter, given the pressure drop between inlet and throat:

$$q_T = \frac{1}{\sqrt{1 - \beta^4}} A_2 \sqrt{2\rho g_c \Delta P} \tag{6.39}$$

In actual practice, there are a variety of effects that result in a flow rate less than that implied by Eq. (6.39). These are normally lumped into a single premultiplying coefficient called the discharge coefficient, C_d. Using this factor in Eq. (6.39), we obtain the equation that is actually used to measure flow rate:

$$q = \frac{C_d}{\sqrt{1 - \beta^4}} A_2 \sqrt{2\rho g_c \Delta P} \tag{6.40}$$

We are primarily interested in the way that the meter affects the pressure drop in the pipe, so Eq. (6.38) is of more interest than Eq. (6.40). But Eq. (6.40) is helpful to see how additional factors enter the analysis, and this will be even more useful when we come to the study of compressible flow. Turning Eq. (6.40) back into an equation for pressure drop we have:

$$\Delta P = \frac{1 - \beta^4}{2\rho g_c}\left(\frac{q}{C_d A_2}\right)^2 \tag{6.41}$$

The pressure at the throat of a meter is not the pressure at its outlet. There is a certain amount of pressure recovery, which is different for the different types of meters. The different meter types also have different values for the discharge coefficient. Values are given in Ref. [2]. This is as far as we will take the generic description of flow through a meter. We will examine compressible flow before returning to consideration of the individual characteristics of the three most common flow meter types.

6.4.2 Compressible Flow Through a Meter

For flow of a compressible fluid through a meter, it is customary to assume that the path length is too short for frictional effects to play a significant role, and therefore the flow is treated as isentropic. We therefore substitute for ρ in Eq. (5.43) using Eq. (5.93), and as we did for incompressible flow we assume no change in elevation. Eq. (5.43) becomes:

$$\frac{P_1^{\frac{1}{\gamma}}}{\rho_1}\frac{dP}{P^{\frac{1}{\gamma}}} + \frac{v}{g_c}dv = 0 \tag{6.42}$$

which we integrate between two points to obtain:

$$\frac{P_1^{\frac{1}{\gamma}}}{\rho_1}\frac{\gamma}{\gamma - 1}\left(P_2^{\frac{\gamma-1}{\gamma}} - P_1^{\frac{\gamma-1}{\gamma}}\right) + \frac{v_2{}^2 - v_1{}^2}{2g_c} = 0 \tag{6.43}$$

Using Eq. (5.93) again, Eq. (6.43) is simplified to:

$$\frac{\gamma}{\gamma - 1}\left(\frac{P_2}{\rho_2} - \frac{P_1}{\rho_1}\right) + \frac{v_2{}^2 - v_1{}^2}{2g_c} = 0 \tag{6.44}$$

Now using Eq. (5.33) we can express v_1 in terms of v_2 and write:

$$\frac{\gamma}{\gamma-1}\frac{P_1}{\rho_1}\left(\frac{P_2}{P_1}\frac{\rho_1}{\rho_2}-1\right)+\frac{v_2^2}{2g_c}\left[1-\left(\frac{\rho_2}{\rho_1}\right)^2\left(\frac{A_2}{A_1}\right)^2\right]=0 \tag{6.45}$$

Using Eq. (5.93) we can eliminate the density ratios:

$$\frac{\gamma}{\gamma-1}\frac{P_1}{\rho_1}\left[\left(\frac{P_2}{P_1}\right)^{\frac{\gamma-1}{\gamma}}-1\right]+\frac{v_2^2}{2g_c}\left[1-\left(\frac{P_2}{P_1}\right)^{\frac{2}{\gamma}}\left(\frac{A_2}{A_1}\right)^2\right]=0 \tag{6.46}$$

Eq. (6.46) can be solved for v_2:

$$v_2=\sqrt{\frac{2g_c\frac{\gamma}{\gamma-1}\frac{P_1}{\rho_1}\left[1-\left(\frac{P_2}{P_1}\right)^{\frac{\gamma-1}{\gamma}}\right]}{1-\left(\frac{P_2}{P_1}\right)^{\frac{2}{\gamma}}\left(\frac{A_2}{A_1}\right)^2}} \tag{6.47}$$

With Eq. (5.32) we turn Eq. (6.47) into an equation for the theoretical mass flow rate past point 2:

$$q_T=\rho_2A_2\sqrt{\frac{2g_c\frac{\gamma}{\gamma-1}\frac{P_1}{\rho_1}\left[1-\left(\frac{P_2}{P_1}\right)^{\frac{\gamma-1}{\gamma}}\right]}{1-\left(\frac{P_2}{P_1}\right)^{\frac{2}{\gamma}}\left(\frac{A_2}{A_1}\right)^2}} \tag{6.48}$$

As with incompressible flow, the actual mass flow rate is obtained by multiplying the right-hand side by the discharge coefficient, C_d. We also eliminate ρ_2 using Eq. (5.93):

$$q=C_dA_2\sqrt{\frac{2g_c\frac{\gamma}{\gamma-1}P_1\rho_1\left(\frac{P_2}{P_1}\right)^{\frac{2}{\gamma}}\left[1-\left(\frac{P_2}{P_1}\right)^{\frac{\gamma-1}{\gamma}}\right]}{1-\left(\frac{P_2}{P_1}\right)^{\frac{2}{\gamma}}\left(\frac{A_2}{A_1}\right)^2}} \tag{6.49}$$

We would like to be able to use an equation of the form of Eq. (6.40) to represent the mass flow rate, but with the addition of a factor Y, called the "gas expansion coefficient," which incorporates all of the special effects arising out of the compressibility of the gas phase. We take as point 2 in Eq. (6.49) the throat of the meter, so that $\frac{A_2}{A_1}=\beta^2$. Equating the two expressions for q we obtain:

$$\frac{C_d}{\sqrt{1-\beta^4}}YA_2\sqrt{2\rho_1g_c(P_1-P_2)}=C_dA_2\sqrt{\frac{2g_c\frac{\gamma}{\gamma-1}P_1\rho_1\left(\frac{P_2}{P_1}\right)^{\frac{2}{\gamma}}\left[1-\left(\frac{P_2}{P_1}\right)^{\frac{\gamma-1}{\gamma}}\right]}{\left[1-\left(\frac{P_2}{P_1}\right)^{\frac{2}{\gamma}}\beta^4\right]}} \tag{6.50}$$

which we solve for Y:

$$Y=\sqrt{\frac{\frac{\gamma}{\gamma-1}\left(\frac{P_2}{P_1}\right)^{\frac{2}{\gamma}}\left[1-\left(\frac{P_2}{P_1}\right)^{\frac{\gamma-1}{\gamma}}\right](1-\beta^4)}{\left(1-\frac{P_2}{P_1}\right)\left[1-\left(\frac{P_2}{P_1}\right)^{\frac{2}{\gamma}}\beta^4\right]}} \tag{6.51}$$

The working equation for a meter is thus:

$$q = \frac{C_d}{\sqrt{1-\beta^4}} Y A_2 \sqrt{2\rho g_c (P_1 - P_2)} \tag{6.52}$$

When the fluid is incompressible, $Y = 1$. When the fluid is compressible, Y is given by Eq. (6.51).

6.4.3 Individual Meter Types

6.4.3.1 Orifice Meter

An orifice meter is simply a plate with a hole in it inserted into the pipe between two flanges. It is the simplest and cheapest of the three meter types, but also the least efficient and least accurate. The problem with an orifice meter is that the throat of the meter is not actually the narrowest part of the flow. Due to the inertia of the fluid as it is forced to move from the edge of the pipe inward, its radial motion inward continues for a short distance past the throat. The narrowest part of the flow, and the point with highest velocity and lowest pressure, is called the "vena contracta." The location of the vena contracta depends on several parameters, most notably the ratio of bore to pipe diameter, β. With one pressure tap located 1–2 pipe diameters upstream of the orifice, and the other located at the vena contracta, the discharge coefficient is (Ref [2] § I-5-39):

$$C_d = \sqrt{1-\beta^4}(K_0 + b\lambda) \tag{6.53}$$

where:

$$K_0 = 0.5992 + 0.4252\left(\frac{0.0006}{D^2\beta^2 + 0.01D} + \beta^4 + 1.25\beta^{16}\right) \tag{6.54}$$

$$b = 0.00025 + 0.002325(\beta + 1.75\beta^4 + 10\beta^{12} + 2D\beta^{16}) \tag{6.55}$$

$$\lambda = \frac{1000}{\sqrt{N_{Re}}} \tag{6.56}$$

The Reynolds number in Eq. (6.56) is based on the flow in the full diameter of the pipe. There is some pressure recovery downstream of the vena contracta. The total pressure drop through an orifice meter with vena contracta taps is given approximately by (Ref. [3], Eq. (5-24)):

$$P_1 - P_2 = (1 - \beta^2)\Delta P \tag{6.57}$$

Because the downstream pressure tap for an orifice meter is not at the throat, and the cross-sectional area at the vena contracta is not precisely known, Eq. (6.51) is not the appropriate one to use for the gas expansion factor. Instead, the following empirical equation is recommended (Ref. [2], Eq. (I-5-68)):

$$Y = 1 - (0.410 + 0.350\beta^4)\frac{P_1 - P_2}{P_1\gamma} \tag{6.58}$$

6.4.3.2 Flow Nozzle

A flow nozzle imposes more control over the path of the fluid as it approaches the point of minimum cross-sectional area. A flow nozzle is somewhat more expensive, and somewhat more accurate, than an orifice meter. Flow nozzles have been extensively studied, and their design has been standardized. The discharge coefficient of a flow nozzle is (Ref [2], § I-5-36):

$$C_d = 0.99622 + 0.00059D - (6.36 + 0.13D - 0.24\beta^2)\sqrt{\frac{\beta}{N_{Re}}} \tag{6.59}$$

The total pressure drop through a flow nozzle is approximately (Ref 3, Eq. (5-17)):

$$P_1 - P_2 = \frac{1-\beta^2}{1+\beta^2}\Delta P \tag{6.60}$$

6.4.3.3 Venturi Tube

The Venturi tube is the most accurate and most expensive of the three meter types. It is generally used only in applications where high accuracy is a process requirement. A Venturi tube achieves its accuracy by controlling the rate of contraction and expansion on both sides of the throat. As such, it has the greatest amount of wall in contact with the fluid, and wall friction appears to be the greatest source of loss of efficiency. Several different values for the discharge coefficient are given in Ref. [2], depending on the fabrication method. For a Venturi tube with rough-cast convergent inlet section and diameter in the range 4-32 in, the suggested discharge coefficient is (Ref [2], § I-5-35):

$$C_d = 0.984 \tag{6.61}$$

Higher coefficients are proposed for Venturi tubes with smoother surfaces, but this value is a good one to use because it is conservative in that it will give the highest pressure loss for a given flow rate.

The total pressure drop through a Venturi tube depends on the diameter ratio, β, and on the discharge cone angle. For small cone angles (5 to 7°) it is in the range of 10–15%. For larger cone angles (15°), it is in the range of 10–30%, with larger losses occurring at lower values of β (Ref. [3], p. 5–13). A decent average value for permanent pressure loss would seem to be about 15%.

$$P_1 - P_2 = 0.15\Delta P \tag{6.62}$$

6.4.4 Choked Flow Through a Meter

To simplify our notation, let us rewrite Eq. (6.49) setting $r = \frac{P_2}{P_1}$ and $\beta^2 = \frac{A_2}{A_1}$:

$$q = C_d A_2 \sqrt{\frac{2g_c \frac{\gamma}{\gamma-1} P_1 \rho_1 \left[r^{\frac{2}{\gamma}} - r^{\frac{\gamma+1}{\gamma}} \right]}{1 - r^{\frac{2}{\gamma}} \beta^4}} \tag{6.63}$$

The behavior of q as a function of r is controlled by two terms in the numerator on the right-hand side. When $r = 1$, the flow rate is zero, as would be expected, and as r decreases the flow rate increases. However, as r continues to decrease, the flow rate reaches a maximum and then decreases, returning to zero when $r = 0$. The explanation for this seemingly counterintuitive behavior is that the density at the throat is proportional to the pressure, and as the pressure and density approach zero, the mass flow rate approaches zero, even though the velocity may be very high. We will first show how to calculate the pressure ratio at which the maximum flow rate occurs. At that point the flow is said to be "choked." Further decrease in the outlet pressure does not increase the flow rate. Then we will show that the maximum flow rate corresponds to the speed of sound at the throat. As the pressure ratio continues to decrease beyond the point of maximum flow, the flow rate does not actually decrease because that pressure information cannot be transmitted upstream against the sonic flow. So the shape of a q vs. r curve in this region is horizontal. In a properly designed nozzle, the velocity downstream of the throat can become supersonic, though the mass flow rate stays constant. That is, in fact, how supersonic flow is generated. In a nozzle not designed to generate supersonic flow, the excess energy of throat pressure over outlet pressure dissipates in oblique shock waves.

6.4.4.1 Critical Pressure Ratio

We take the logarithmic derivative of q in Eq. (6.63) with respect to r:

$$\frac{dq^2}{q^2} = \left[\frac{\frac{2}{\gamma}r^{\frac{2-\gamma}{\gamma}} - \frac{\gamma+1}{\gamma}r^{\frac{1}{\gamma}}}{r^{\frac{2}{\gamma}} - r^{\frac{\gamma+1}{\gamma}}} + \frac{\frac{2}{\gamma}r^{\frac{2-\gamma}{\gamma}}\beta^4}{1 - r^{\frac{2}{\gamma}}\beta^4}\right]dr \tag{6.64}$$

At the maximum flow rate, $dq = 0$, and this happens when the term in brackets on the right-hand side is zero. The value of r for which this happens is its critical value, r_c.

$$\frac{\frac{2}{\gamma}r_c^{\frac{2-\gamma}{\gamma}} - \frac{\gamma+1}{\gamma}r_c^{\frac{1}{\gamma}}}{r_c^{\frac{2}{\gamma}} - r_c^{\frac{\gamma+1}{\gamma}}} + \frac{\frac{2}{\gamma}r_c^{\frac{2-\gamma}{\gamma}}\beta^4}{1 - r_c^{\frac{2}{\gamma}}\beta^4} = 0 \tag{6.65}$$

Cross-multiplying to eliminate the denominators:

$$\left(\frac{2}{\gamma}r_c^{\frac{2-\gamma}{\gamma}} - \frac{\gamma+1}{\gamma}r_c^{\frac{1}{\gamma}}\right) - \left(\frac{2}{\gamma}r_c^{\frac{4-\gamma}{\gamma}} - \frac{\gamma+1}{\gamma}r_c^{\frac{3}{\gamma}}\right)\beta^4 + \frac{2}{\gamma}\left(r_c^{\frac{4-\gamma}{\gamma}} - r_c^{\frac{3}{\gamma}}\right)\beta^4 = 0 \tag{6.66}$$

which simplifies to:

$$\frac{2}{\gamma}r_c^{\frac{2-\gamma}{\gamma}} - \frac{\gamma+1}{\gamma}r_c^{\frac{1}{\gamma}} + \frac{\gamma-1}{\gamma}r_c^{\frac{3}{\gamma}}\beta^4 = 0 \tag{6.67}$$

Multiplying through by $\frac{\gamma}{2}r_c^{-\frac{1}{\gamma}}$:

$$r_c^{\frac{1-\gamma}{\gamma}} - \frac{\gamma+1}{2} + \frac{\gamma-1}{2}r_c^{\frac{2}{\gamma}}\beta^4 = 0 \tag{6.68}$$

This is an implicit equation for r_c which must be solved iteratively. One simple way to do this is to rearrange Eq. (6.68) to:

$$r_c = \left[\frac{\gamma+1}{2} - \frac{\gamma-1}{2}r_c^{\frac{2}{\gamma}}\beta^4\right]^{\frac{\gamma}{1-\gamma}} \tag{6.69}$$

We start with a value of r_c arbitrarily set at $r_c = 0.577$, plug this into the right-hand side of Eq. (6.69) to get a new value for r_c, and iterate the process until convergence is achieved. Having calculated this value for flow through a given meter, we know that if $\frac{P_2}{P_1} < r_c$ the flow will be choked.

6.4.4.2 Maximum Flow Rate

To find the limiting velocity through the throat of a meter, it is easier to start from stagnation conditions rather than conditions at the inlet of the meter. This is equivalent to a meter with infinite diameter at the inlet, or, equivalently, $\beta = 0$. With this value of β we can solve Eq. (6.68) explicitly. The pressure ratio that we obtain by so doing is $\frac{P_2}{P_{1t}}$ for critical flow.

$$\left(\frac{P_2}{P_{1t}}\right)_c = \left(\frac{\gamma+1}{2}\right)^{\frac{\gamma}{1-\gamma}} \tag{6.70}$$

Using Eq. (5.100) integrated from stagnation conditions at point 1 to point 2:

$$c_p(T_2 - T_{1t}) + \frac{1}{2g_c}v_2^2 = 0 \tag{6.71}$$

Dividing through by c_pT_2 and using Eq. (5.21) followed by Eqs. (5.61) and (5.58):

$$\frac{T_{1t}}{T_2} = 1 + \frac{\gamma-1}{\gamma}\frac{M_w}{R}\frac{1}{T_2}\frac{1}{2g_c}v_2^2 = 1 + \frac{\gamma-1}{2}\frac{v_2^2}{c_2^2} = 1 + \frac{\gamma-1}{2}M_2^2 \tag{6.72}$$

Using the isentropic relationship in Eq. (5.28) followed by Eq. (6.72):

$$\frac{P_{1t}}{P_2} = \left(\frac{T_{1t}}{T_2}\right)^{\frac{\gamma}{\gamma-1}} = \left(1 + \frac{\gamma - 1}{2} M_2{}^2\right)^{\frac{\gamma}{\gamma-1}} \tag{6.73}$$

Comparing Eqs. (6.70) and (6.73), we see that at the critical pressure ratio:

$$\frac{\gamma + 1}{2} = 1 + \frac{\gamma - 1}{2} M_{2c}{}^2 \tag{6.74}$$

Solving this equation for M_{2c}, we find that $M_{2c} = 1$ at the critical pressure ratio, i.e., the maximum velocity at the throat of the meter is the speed of sound at local conditions.

6.4.5 Example Problem

A new plant air compressor is being installed, and an orifice meter is required to monitor the air flow through the compressor to determine when it is falls below capacity and requires maintenance. The normal flow rate of compressed air is 6690 lbs/hr (3035 kg/hr), and the pipeline is a 6 inch schedule 40. An orifice pressure drop between 70 to 80 inches of water (17.4–19.9 kPa) is desired to provide accurate transmitter readings. Air temperature is 70°F (21.1°C), and upstream pressure is 90 psig (620.53 kPag). Figure 6.2 shows this above example

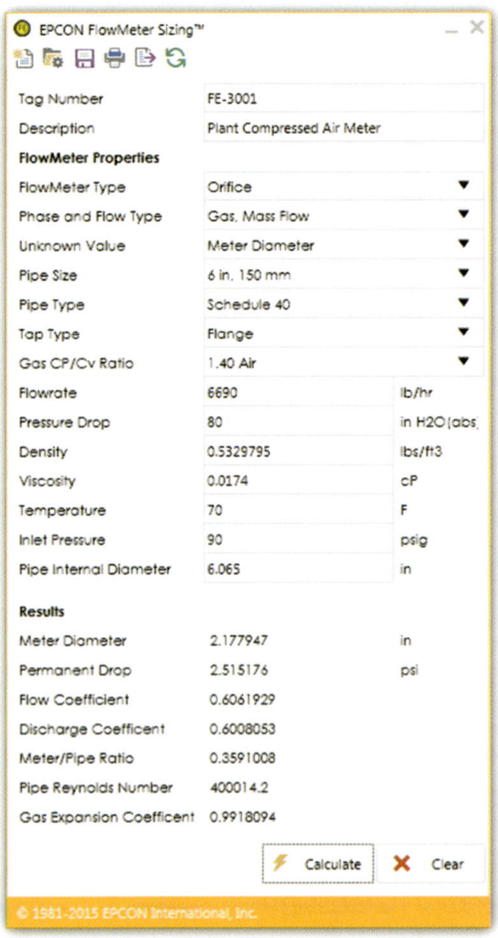

Figure 6.2 EPCON Flowmeter Sizing.

entered into an flowmeter sizing program with the calculated meter bore diameter of 2.18 inches (55.3 mm) and a permanent pressure drop of 2.52 psi (17.37 kPa).

6.5 Control Valves

6.5.1 Incompressible Flow Through a Control Valve

Generally speaking, one cannot know the pressure drop across a control valve without knowing the purpose of the valve. The whole point of a control valve is to restrict the flow as much or as little as required to accomplish a particular task. This might be to maintain a certain pressure upstream or downstream of the valve, or to establish a certain flow rate through the pipe containing the valve. However, we can get some idea of the effect of a control valve from the way it is specified for the purpose of sizing. The process designer typically has some idea of the flow rate that will be required, and the valve is sized to accommodate that flow rate. More specifically, a control valve is specified in terms of its flow coefficient, C_v. By definition, the flow coefficient is the number of gallons per minute of water that will flow through the valve when the pressure drop across the valve is 1 psi. The reference water is at a temperature of 15°C, where its density is 62.4 lb_m/ft^3.

Even though the flow path through a control valve is more complicated than that through a meter, we expect that the relationship between flow rate and pressure drop will be similar to that in Eq. (6.40), except that many of the parameters in that equation will be lumped into the flow coefficient. However, Eq. (6.40) is written in terms of mass flow rate. To be sure that we account for the density properly, we need to convert this to a relationship for volumetric flow rate in units of gpm. Noting that 1 ft^3 = 7.48052 gallons, we find:

$$G = \frac{(7.48052)(60)q}{\rho} = (7.48052)(60)\frac{C_d}{\sqrt{1-\beta^4}}A_2\sqrt{\frac{2g_c\Delta P}{\rho}} = C_v\sqrt{\frac{62.4\Delta P}{144\rho}} \tag{6.75}$$

Now converting Eq. (6.75) back to mass flow units, we have:

$$q = \frac{\sqrt{62.4}}{(7.48052)(60)}C_v\sqrt{\frac{\rho\Delta P}{144}} \tag{6.76}$$

Solving for the pressure drop we have:

$$\Delta P = \frac{144}{62.4\rho}\left[\frac{(7.48052)(60)}{C_v}\right]^2 q^2 \tag{6.77}$$

6.5.2 Compressible Flow Through a Control Valve

Control valves for gas phase applications are specified in exactly the same way as control valves for liquid phase applications, despite the fact that gas flow rates are not normally reported in terms of gpm. However, when the fluid is a gas, other factors also come into play, such as gas expansion and choked flow.

As for gas flow through a meter, gas flow through a control valve requires that we account for gas expansion. So, Eq. (6.76) must be rewritten:

$$q = \frac{\sqrt{62.4}}{(7.48052)(60)}C_vY\sqrt{\frac{\rho_1\Delta P}{144}} \tag{6.78}$$

The path of flow through a control valve is complicated, and it is not possible to derive a theoretical expression for Y as we did for a meter in Eq. (6.51). Instead we use an empirical expression (Ref. [4], Eq. 32):

$$Y = 1 - \frac{x}{3F_\gamma x_T} \tag{6.79}$$

The factor x in Eq. (6.79) is the ratio of pressure drop across the valve to the inlet pressure. The parameter x_T is the value of x for which the flow becomes choked. This is a characteristic parameter of the valve. It is supplied by the valve manufacturer, and is determined by tests using air as the fluid, with $\gamma = 1.4$. When a gas other than air is being used, with a different value of γ, this value of x_T must be corrected using the factor

$$F_\gamma = \frac{\gamma}{1.4} \tag{6.80}$$

The critical value of pressure drop through the valve is:

$$\Delta P = P_1 F_\gamma x_T \tag{6.81}$$

At the critical pressure drop Eq. (6.79) gives $Y = \frac{2}{3}$, and with Eq. (6.78) this gives the maximum mass flow rate through the valve:

$$q_{max} = \frac{\sqrt{62.4}}{(7.48052)(60)} \frac{2}{3} C_v \sqrt{\frac{\rho_1 P_1 F_\gamma x_T}{144}} \tag{6.82}$$

Figure 6.3 EPCON Control Valve Sizing.

6.5.3 Example Problem

A surplus control valve is available for use to replace a malfunctioning one in a new hot water system (water temperature is 185°F). The C_v on the surplus valve is 124, and the normal flow rate is 185 gallons/minute. A maximum of 10 psi of valve pressure drop can be allowed without affecting the process. The inlet pressure is 36.8 psia. Figure 6.3 shows this example entered into a control valve sizing program with the calculated pressure drop of 23.2 psi (160 kPa) and a choked volume flow of 246.2 gal/min (55.9 m^3/h).

6.6 Two-Phase Gas-Liquid Flow

(With contribution by Moye Wicks III)

6.6.1 Introductory Remarks

To this point, we have assumed, without making the assumption explicit, that we are dealing with fluids that are homogeneous (spatially uniform properties) and isotropic (directionally uniform properties), excepting only that we allow the density of a compressible fluid to vary with pressure. When a fluid consists of more than one phase, neither of these assumptions is valid, and the analysis gets much more complex. We could go so far as to say that except in certain specific circumstances, the behavior of multiphase flow defies analysis. and we will have to resort to simplifying assumptions to make any statements about the flow at all.

The presence of two-phase gas-liquid flow can result in reduced throughput and mechanical damage to pipes and equipment. The best practice is to avoid two-phase flow conditions by changing the operating temperature and/or pressure, or by separating the phases and transporting them separately. However, there are cases where these options are not economically feasible so two-phase flow must be dealt with.

The first point to be made is that there are several different patterns in which the gas and liquid phases can arrange themselves, depending on their respective flow rates. These are called "flow regimes." Illustrations of some flow regimes for horizontal flow are given in Figure 6.4. The diagram illustrates how the flow regime changes as the flow rate of the gas phase increases in horizontal flow. If we were to attempt detailed analysis of the physics of flow for any given regime, the results would be specific to that regime, and not of general applicability.

Many methods have been proposed to calculate pressure loss in horizontal flow of gas-liquid mixtures (Ref [5]). These methods span a period of 70-plus years, beginning with the pioneering paper by Lockhart and Martinelli in 1945, and extended by them in 1949 (Ref [6]). They developed a calculation method, a correlation of the two-phase measurements, in terms of the superficial pressure drops that would occur in single-phase flow of the individual phases by themselves. Hundreds of publications followed, by a host of investigators. All the proposed calculation methods are either empirical, based on one or more sets of data obtained under a limited range of conditions (a majority being in small-diameter pipes and using compressed air and liquid water as the fluids), or based on an idealized physical assumption about phase configuration.

A number of investigators have produced maps defining the boundaries of the flow regimes using various functions of gas and liquid flow rates as coordinates. These maps are useful in helping the engineer to picture the way the flow patterns will change as flow rates change. However, corresponding progress in predicting transport phenomena characteristics (pressure gradient, heat flow, and mass transfer) has been disappointing. The phenomena controlling these processes are not easy to quantify. For example, it is known that an interface where waves

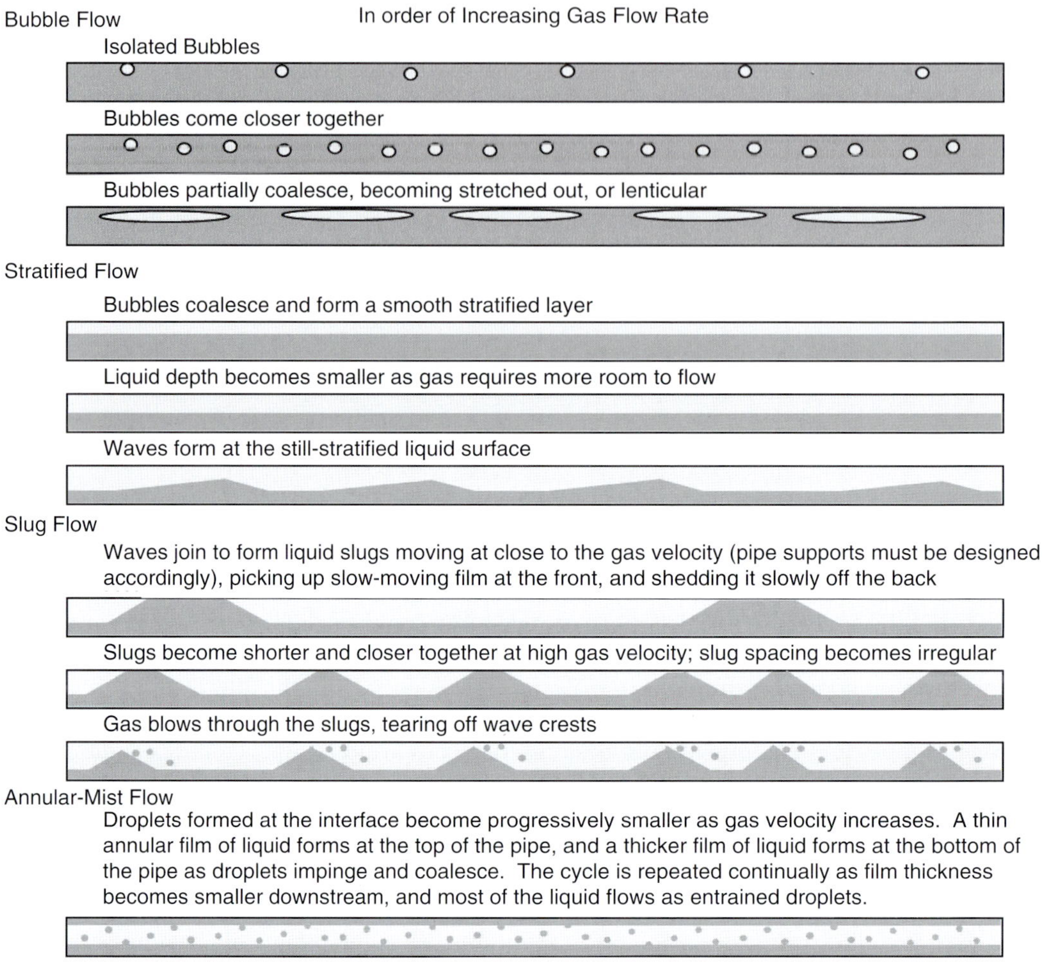

Figure 6.4 Horizontal Gas-Liquid Flow Regimes.

are flowing at the surface as gas flows by, exhibits a roughness effect that adds to gas-phase friction. Wave shape and height distribution, surface spacing, and the three-dimensional effects as a wave laps onto the pipe wall are indeed more complex than our methods can grasp. Methods to attempt simplified models of these phenomena have not really made impressive strides.

6.6.2 The Method of Dukler and Taitel

Beginning in the 1970s the technical community began to realize that progress was evidently not sufficient to improve accuracy of predictions. Consortia of companies and investigators began to search for more realistic models better able to describe the physical phenomena that led to pressure loss. The method employed by Dukler and Taitel (Ref [7]) was based on very simple models of wave formation and growth on a sheared interface. When enough kinetic energy was supplied to the gas flow, waves would form, grow, and seal off the pipe with a large wave that could cause a plug of gas to move the wave rapidly downstream. Other flow types were also modeled, including slug flow, bubble flow, and annular-mist flow. Vertical and inclined flows (+ and –) were also studied, having their own flow regime parameters, since the flow geometry was so different.

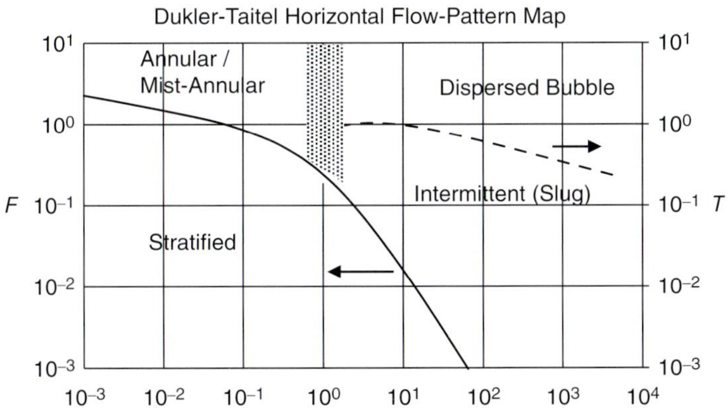

Figure 6.5 Dukler-Taitel Horizontal Flow-Pattern Map.

Dukler and Taitel attempted to define the boundaries between the various flow regimes on the basis of fundamental fluid mechanical principles. Figure 6.5 is such a map based on the equations they developed.

The parameter "X" for the x-axis is essentially the well-known Lockhart-Martinelli parameter, proportional to the ratio of liquid-to-gas flow rates. More exactly, the "X"-parameter can be defined as:

$$X = \left[\left(\frac{dP}{dl} \right)_{ls} \Big/ \left(\frac{dP}{dl} \right)_{gs} \right]^{\frac{1}{2}} \tag{6.83}$$

That is to say, X is the square root of the ratio of pressure drops per unit length for the liquid and gas phases respectively flowing at their superficial velocities. This parameter is dependent on the properties of the liquid and gas. The phase boundaries in Figure 6.5 are based on dimensionless parameters, and are almost independent of material properties, the only exception being a built-in assumption that the friction factor for each phase is proportional to $N_{Re}^{-0.2}$.

Dukler and Taitel started with the condition of stratified flow, and calculated a gas phase velocity such that the reduced pressure of the gas flowing through the reduced area (Bernoulli's principle) over a small wave would be sufficient to lift the weight of the liquid in that wave. To use their findings, one calculates a Froude number modified by a ratio of fluid densities:

$$F = \left(\frac{\rho_g}{\rho_l - \rho_g} \right)^{\frac{1}{2}} \frac{v_{gs}}{\sqrt{Dg}} \tag{6.84}$$

If the "F"-parameter plots below the solid curve in Figure 6.5, the flow is stratified. Dukler and Taitel also calculate a boundary between "smooth stratified" and "wavy stratified" flow, but we are not addressing that distinction here.

If the "F"-parameter plots above the curve, further analysis is required. Dukler and Taitel next address the question of whether a growing wave will fill the cross-sectional area, or be distributed around the inner circumference of the pipe. The answer is assumed to depend on the fraction of the cross-sectional area of the pipe that is filled with liquid, expressed as the ratio of a liquid level to the pipe diameter, h/D. If this ratio is greater than 0.5, it is assumed that enough liquid is present to allow the wave height to build until the pipe is bridged. Since the wave need not be all-liquid, but could contain some entrained bubbles, it is further assumed that a lower value of h/D might allow the pipe to be bridged. Since the degree of frothiness in the flow is unknown, they posit that the transition may be somewhere in the range $0.35 < \frac{h}{D} < 0.5$. As part of their analysis, they developed a unique relationship between X and h/D, and that relationship is used to define the left and right boundaries of the cross-hatched area in Figure 6.5. But it should be noted that this transition is not abrupt, and not very well defined.

For flows that plot above the solid curve, and to the right of the cross-hatched area, the question is whether the gas is dispersed in the liquid phase, or whether gas bubbles separate so that the flow consists of slugs of alternating gas and liquid. The assumption made by Dukler and Taitel is that gas bubbles will disperse in the liquid when turbulent forces in the liquid phase are sufficient to overcome the buoyancy force lifting the bubbles. To assess this, we calculate the parameter "T":

$$T = \left[\frac{\left(\frac{dP}{dl}\right)_{ls}}{(\rho_l - \rho_g)g} \right]^{\frac{1}{2}} \tag{6.85}$$

The dashed curve in Figure 6.5 is plotted with reference to the secondary y-axis on the right side of the plot area. As presented here, this axis is exactly the same as the primary y-axis in Figure 6.5, but it need not be so, and the axes are kept separate to emphasize this point. The "T"-parameter is plotted with reference to the secondary axis, and the flow is classified according to whether it falls above or below the dashed line.

This leads us to an important point about the Dukler-Taitel map. Early investigators produced maps in which a single plot point was supposed to describe the flow regime according to the region in which it fell. The problem is that the boundaries between those regions, being based on limited experimental data rather than fundamental analysis, were subject to question as to their universal applicability. While one can certainly question the Dukler-Taitel model as well, the fundamental analyses on which it is based gives more confidence in its applicability over a wide range of conditions. Moreover, Dukler and Taitel tested their model against such experimental data as came to hand, with generally good agreement. So, the price one pays for a more fundamental analysis is that several plot points, rather than a single one, are required to describe the flow. This seems like a small price to pay.

6.6.3 Pressure Drop in Two-Phase Flow

To assess the validity of models of two-phase flow behavior, it is necessary to have experimental data against which to compare the predictions. However, the acquisition of accurate data is difficult, and there are many pitfalls to be avoided:

- There is always an "entrance effect" associated with the flow at the inlet end of a pipe. To observe the behavior of fully developed flow, measurements must be taken beyond the entrance length. This length is dependent on the pipe diameter and the Reynolds number. However, in single-phase turbulent flow, the dependence on the Reynolds number is weak, and the flow is pretty much fully developed after ten diameters. This is the length required for the boundary layer to reach its steady-state configuration. In two-phase flow, the entry length is much longer because the equilibrium flow pattern is much more complicated than simply stabilizing a single-phase boundary layer. It is beyond the scope of this chapter to recommend an entry length; we simply point out that it is the duty of the experimentalist to determine this length and to assure that data is taken beyond it.
- Surface tension errors can accumulate in the lead lines to a transducer unless kept in a completely purged state. If the investigator controls gas or liquid purge flows to give matching pressure drops in the transducer lines themselves, the effect of the purge on the measured ΔP is nulled out.

- Holdup is notoriously difficult to measure accurately if below 10% fraction liquid.
- The parameters that can influence frictional pressure drop are sometimes elusive. Models developed to account for these parameters are thus also fraught with uncertainty. As models become more complex in their attempt to include flow-regime-specific details, the uncertainty in their predictions of transport variables is compounded.

Rather than wait for some unspecified future date when all of these issues will be resolved, an interim solution is to use the best method available, defined as having the following characteristics:

- It correctly handles the limiting cases where the liquid volume fraction, λ, approaches 0 (all gas flow) or 1 (all liquid flow).
- It scales correctly as pressure increases.
- It reduces to the correct limit as the critical point is approached.
- It does not consume a large amount of empirical data for its establishment.
- It meets limits for compressibility effects.
- It should predict continuous and generally slope-continuous trends of pressure loss as flow rates are systematically changed. Discontinuities are rarely observed in actual flow tests as gas and liquid flow rates vary. This means that flow regime-specific mathematical models must be very accurate or large unrealistic aberrations will enter the predicted results as flow regimes change. This difficulty could magnify in importance if the flow system is coupled with an automatic control system.

The homogeneous flow model meets the criteria noted above. Experience has shown that this model gives reasonable estimates of pressure drop over the range of flow regimes. Further, it requires no experimental measurements of gas-liquid pressure drop to establish empirical constants.

6.6.4 The Homogeneous Flow Model

Homogeneous two-phase flow is subject to the same forces as single-phase compressible fluid flow. So, we can use Eq. (5.83) as our starting point, with properties modified to account for the presence of two phases:

$$\frac{1}{\rho_m}dP + \frac{g}{g_c}dz - \frac{q^2}{g_c{\rho_m}^3A^2}d\rho_m + \frac{f}{D}\frac{q^2}{2g_c{\rho_m}^2A^2}d\ell = 0 \tag{6.86}$$

The mixture density, ρ_m, is the volumetric average of the densities of the liquid and gas phases:

$$\rho_m = \lambda\rho_l + (1-\lambda)\rho_g \tag{6.87}$$

where the liquid volume fraction, λ, can be calculated from the mass flow rates and densities of the gas and liquid phases using the equation:

$$\lambda = \frac{q_l/\rho_l}{q_l/\rho_l + q_g/\rho_g} \tag{6.88}$$

The friction factor, f, in Eq. (6.86) is obtained in the same way as for single-phase flow, by solving the Colebrook-White equation, Eq. (5.63), but the Reynolds number must be calculated from Eq. (5.62) using mixture properties for the density and viscosity. Mixture density is given by volumetric averaging using Eq. (6.87). A variety of relationships have been suggested for the mixture viscosity, but the one that seems most satisfactory is volumetric averaging of this parameter as well:

$$\mu_m = \lambda\mu_l + (1-\lambda)\mu_g \tag{6.89}$$

Multiplying Eq. (6.86) through by mixture density and rearranging, we can obtain:

$$dP = -\frac{f}{D}\frac{q^2}{2g_c\rho_m A^2}d\ell + \frac{q^2}{g_c{\rho_m}^2 A^2}d\rho_m - \frac{g}{g_c}\rho_m dz \tag{6.90}$$

With a little additional rearrangement, and incorporation of the angle of inclination of the pipe, this becomes:

$$dP = -\frac{f}{D}\frac{q^2}{2g_c\rho_m A^2}d\ell - \frac{q^2}{g_c A^2}d\left(\frac{1}{\rho_m}\right) - \frac{g}{g_c}\rho_m \cos\theta\, d\ell \tag{6.91}$$

The first term on the right-hand side represents the pressure loss due to friction; the second term represents the pressure loss due to acceleration, and the third term represents the pressure effect of a change in elevation.

6.6.5 Temperature Effects

In order to proceed further we must know, or postulate, a relationship between pressure and mixture density. We have the same range of considerations as was described in Chapter 5, Section 5.6. However, whereas our objective there was to develop some analytical relationships, here that seems like a hopeless and pointless task. There are several reasons for this:

- The mixture density consists of two contributions, one of which (the gas-phase density) is pressure-dependent, and the other of which is constant, which complicates the analysis.
- The ability to develop analytical solutions in Chapter 5, Section 5.6, was contingent on ignoring the effect of elevation. We could justify this in the case of single-phase gas flow by noting that the gas-phase density is usually sufficiently low that the hydrostatic effect is negligible. As the liquid volume fraction increases, this assumption becomes increasingly untenable. However, in the case of horizontal flow, which is expected to be the majority of cases analyzed, we can still ignore the elevation term.
- There is an additional complication when the liquid flashes as the pressure is reduced. This is not so much a factor with air and water, which seems to be one of the main systems of choice for experimental study of two-phase flow. But if a process stream has two phases, they are typically in a state of thermodynamic equilibrium, which will change as the pressure changes. The latent heat effect associated with a change of phase as the liquid evaporates can be very large. While it is the purpose of this chapter to focus on the fluid mechanics, and not to address process considerations, formulation of Eq. (6.91) in a form that caters to numerical integration makes it easier to modify the analysis to incorporate the effect of change of phase.

Assuming, then, that we can postulate a relationship between pressure and mixture density, we rewrite Eq. (6.91) as:

$$\frac{dP}{d\ell} = -\frac{f}{D}\frac{q^2}{2g_c\rho_m A^2} - \frac{q^2}{g_c A^2}\frac{d}{dP}\left(\frac{1}{\rho_m}\right)\frac{dP}{d\ell} - \frac{g}{g_c}\rho_m \cos\theta \tag{6.92}$$

which we rearrange to:

$$\frac{dP}{d\ell} = -\frac{\frac{f}{D}\frac{q^2}{2g_c\rho_m A^2} + \frac{g}{g_c}\rho_m \cos\theta}{1 + \frac{q^2}{g_c A^2}\frac{d}{dP}\left(\frac{1}{\rho_m}\right)} \tag{6.93}$$

It is actually more convenient to describe the flow in terms of the weight fraction of gas, $x = \frac{q_g}{q_g + q_l}$, rather than in terms of the volume fraction of liquid, λ, since x remains constant in the case where the liquid phase is not evaporating. Noting that $q_l = (1 - x)q$ and $q_g = xq$ in Eq. (6.88) we can easily show that:

$$\lambda = \frac{\frac{1-x}{\rho_l}}{\frac{1-x}{\rho_l} + \frac{x}{\rho_g}} \tag{6.94}$$

and using this result in Eq. (6.87) yields:

$$\frac{1}{\rho_m} = \frac{1-x}{\rho_l} + \frac{x}{\rho_g} \tag{6.95}$$

Since liquid density is, by assumption, independent of pressure, we have, for constant x:

$$\frac{d}{dP}\left(\frac{1}{\rho_m}\right) = x\frac{d}{dP}\left(\frac{1}{\rho_g}\right) \tag{6.96}$$

With this result, Eq. (6.93) becomes:

$$\frac{dP}{d\ell} = -\frac{\frac{f}{D}\frac{q^2}{2g_c\rho_m A^2} + \frac{g}{g_c}\rho_m \cos\theta}{1 + \frac{q^2}{g_c A^2} x \frac{d}{dP}\left(\frac{1}{\rho_g}\right)} \tag{6.97}$$

6.6.6 Comment on the Effect of Change in Elevation

In a truly homogeneous flow, the contribution of elevation change to pressure drop is no different from that in a single-phase fluid. The net effect of elevation change depends only on the difference in elevation between the inlet and outlet of the pipe, without regard for any intervening up-slope and down-slope segments. In general, this is not the case for two-phase flow. Pressure lost in pushing the fluid uphill, is not recovered in downhill flow ($\cos\theta < 0$). Except as noted below, one can anticipate that the downhill flow will be stratified: a continuous gas phase will form at the top of the pipe, and any hydrostatic pressure gain will be associated with the density of the gas phase, and will be negligible. Any pressure gain that might have been achieved due to the presence of the liquid will be dissipated by the liquid trickling down along the bottom of the pipe at constant pressure.

However, experiments have shown that pressure recovery in downhill flow is possible at high liquid/gas volume ratios and at a sufficiently high velocity. Only if λ exceeds 0.75 and mixture velocity exceeds 5 ft/sec should pressure recovery be assumed. Otherwise, assume that the net elevation effect is the sum of all the uphill elevation changes, and take no credit for downhill elevation changes.

6.6.7 Isothermal Flow

The simplest analysis results if we assume that the flow is isothermal. In defense of this assumption, the presence of a liquid phase will have a moderating influence on the temperature change that would occur in its absence. In the absence of a strong reason to believe the assumption to be inadequate, it seems logical to start here. Comparison of isothermal results with results using more complex models (isentropic or adiabatic flow) indicate that the more complex models often yield only a minor change in the calculated results.

Rearranging Eq. (5.29) we have:

$$\frac{1}{\rho_g} = \frac{RTZ}{PM_w} \tag{6.98}$$

If we assume that the flow is isothermal, and that the variation of Z with pressure is so slight that it can be ignored for the purpose of this analysis, then:

$$\frac{d}{dP}\left(\frac{1}{\rho_g}\right) = -\frac{RTZ}{P^2 M_w} = -\frac{1}{P\rho_g} \tag{6.99}$$

Inserting this relationship into Eq. (6.97) we have:

$$\frac{dP}{d\ell} = -\frac{\frac{f}{D}\frac{q^2}{2g_c\rho_m A^2} + \frac{g}{g_c}\rho_m \cos\theta}{1 - \frac{q^2}{g_c A^2}\frac{x}{P\rho_g}} \tag{6.100}$$

and this is the differential equation that we must integrate, starting with inlet pressure at the entrance to the pipe, in order to calculate pressure drop over the length of the pipe.

The relationship between P and ℓ is governed mainly by the denominator on the right-hand side of Eq. (6.100). In order for there to be a stable flow, this expression must be positive, i.e.,

$$\frac{q^2}{g_c A^2}\frac{x}{P\rho_g} < 1 \tag{6.101}$$

As the length increases, pressure decreases, and gas density decreases in proportion to pressure, so that the left-hand side of inequality (6.101) increases proportional to $1/P^2$. If the pipe is long enough, there will come a point at which the inequality becomes an equality. The denominator on the right-hand side of Eq. (6.100) becomes zero, and the pressure change for an incremental change in length is infinite. At this point the flow is choked, meaning that the length at which it occurs, L_{max}, sets an upper limit for the length of pipe through which the specified flow can pass.

In the limit as the flowing fluid is all gas, i.e. $x = 1$, we find, using Eqs. (5.32), (5.17), and (5.61), that:

$$\frac{q^2}{g_c A^2}\frac{x}{P\rho_g} = \frac{\rho_g v^2}{g_c}\frac{1}{P} = \frac{M_w v^2}{RTg_c} = \gamma\frac{v^2}{c^2} \tag{6.102}$$

In this analysis, the maximum Mach number is $M = \sqrt{\frac{1}{\gamma}}$. This is not a correct result, of course, but it is consistent with our assumption of isothermal flow which, as shown in Chapter 5, Section 5.6, is equivalent to considering a fluid with $\gamma = 1$.

It is of more interest to ask what the effect of a liquid phase on the maximum velocity would be. Using Eq. (5.32) we can rearrange inequality (6.101) to:

$$v < \sqrt{\frac{P\rho_g}{x}\frac{g_c}{\rho_m{}^2}} \tag{6.103}$$

As an illustrative calculation, we consider a mixture of air and water discharging to the atmosphere. The results are plotted in Figure 6.6. The maximum velocity drops rapidly as the volume fraction of water is increased from zero, and there is a broad minimum in the range 5% to 95%, with a minimum value of about 66 ft/sec at 50%.

6.6.8 Isentropic Flow

Since we are considering flow with a frictional component of pressure loss, the assumption of isentropic flow is no more realistic here than it was in Chapter 5, Section 5.6. Nevertheless, it at least gives us the opportunity to investigate the extremes of temperature behavior. In the

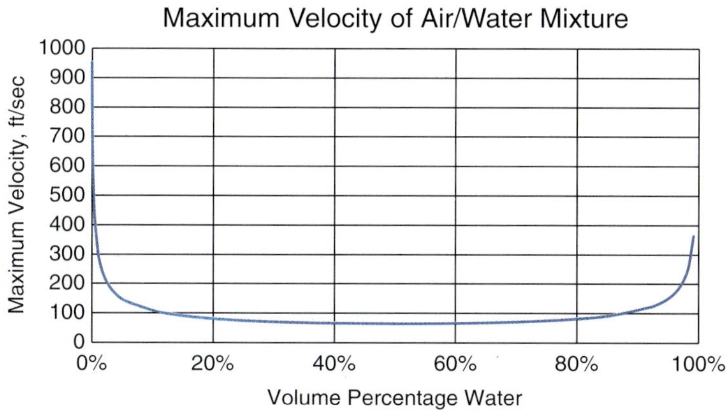

Figure 6.6 Maximum Velocity of Air/Water Mixture.

absence of flashing flow, actual temperature variation will be less than that found using the assumption of isentropic flow.

In isentropic flow, the temperature of a gas drops. In the presence of a liquid phase, the drop will be reduced because the heat capacity of the liquid phase will absorb some of the change. To understand the effect of the liquid, we need to return to the analysis of isentropic flow in Chapter 5, Section 5.2. The heat capacities of the mixture are weighted averages of the heat capacities of the gas and liquid phases, but we note that for the liquid phase we do not press a distinction between c_p and c_v. So, we have:

$$c_{vm} = (1 - x)c_{pl} + xc_{vg} \tag{6.104}$$

$$c_{pm} = (1 - x)c_{pl} + xc_{pg} \tag{6.105}$$

Applying Eq. (5.20) to the gas phase (which requires that we assume, for the purposes of this development, that the gas can be treated as ideal):

$$c_{pm} = (1 - x)c_{pl} + xc_{vg} + x\frac{R}{M_w} = c_{vm} + x\frac{R}{M_w} \tag{6.106}$$

We define an isentropic exponent for the mixture as:

$$\gamma_m = \frac{c_{pm}}{c_{vm}} \tag{6.107}$$

Eliminating c_{pm} between Eqs. (6.106) and (6.107):

$$c_{vm} = \frac{1}{(\gamma_m - 1)}x\frac{R}{M_w} \tag{6.108}$$

We apply Eq. (5.4) to an isentropic expansion of the mixture in the same way that we did in the derivation of Eq. (5.23):

$$c_{vm}dT = -PdV_m = -PxdV_g \tag{6.109}$$

Dividing through by T and using Eqs. (5.17) and (6.108):

$$\frac{dT}{T} = -(\gamma_m - 1)\frac{dV_g}{V_g} \tag{6.110}$$

which we integrate to:

$$\ln T = -(\gamma_m - 1)\ln V_g + C_1 \tag{6.111}$$

and rearrange to:

$$TV_g^{\gamma_m - 1} = C_2 \tag{6.112}$$

Substituting for T in Eq. (6.112) using Eq. (5.17):

$$PV_g^{\gamma_m} = \frac{P}{\rho_g^{\gamma_m}} = C_3 \tag{6.113}$$

from which we derive the result:

$$\frac{d}{dP}\left(\frac{1}{\rho_g}\right) = -\frac{1}{\gamma_m P \rho_g} \tag{6.114}$$

Substituting Eq. (6.114) into Eq. (6.97):

$$\frac{dP}{d\ell} = -\frac{\frac{f}{D}\frac{q^2}{2g_c\rho_m A^2} + \frac{g}{g_c}\rho_m \cos\theta}{1 - \frac{q^2}{g_c A^2}x\frac{1}{\gamma_m P \rho_g}} \tag{6.115}$$

and this is the equation to integrate, in place of Eq. (6.100) if we wish to perform an isentropic analysis.

The limit of choked flow occurs when the denominator in Eq. (6.115) is zero, i.e. when:

$$\frac{q^2}{g_c A^2}x\frac{1}{\gamma_m P \rho_g} = 1 \tag{6.116}$$

Using Eq. (5.32)

$$v_{max} = \sqrt{\frac{P\rho_g}{x}\frac{\gamma_m g_c}{\rho_m^2}} \tag{6.117}$$

In the limit as $x \to 1$, we have $\rho_m \to \rho_g$ and $\gamma_m \to \gamma$, and we find, with the help of Eq. (5.17) and (5.61):

$$v_{max} \to \sqrt{\frac{P\gamma g_c}{\rho_g}} = \sqrt{\frac{\gamma R T g_c}{M_w}} = c \tag{6.118}$$

which is the result that we would expect.

6.6.9 Adiabatic Flow

When two-phase flow is encountered in a process setting, many of the assumptions made up to this point do not hold. The liquid and vapor phase are usually in thermodynamic equilibrium, and as the pressure declines, some of the material in the liquid phase moves into the vapor phase. In addition to the thermal effect resulting from the latent heat of vaporization, other parameters of the flow change as well, namely: ρ_g, x, and ρ_m. It would seem that it is more nearly correct to treat the flow as adiabatic rather than isothermal or isentropic. However, even if we assume the vapor phase to be ideal, the adiabatic analysis which applied to compressible flow in Chapter 5, Section 5.6, won't be possible here. The situation is just too complex. We will have to resort to a numerical analysis in which the behavior of the system is modeled by a thermodynamic computation engine. This takes us into the realm of process simulation, which is beyond the scope of this chapter. But it seems appropriate to provide the general outline of a solution procedure.

The equation that describes the loss of pressure in adiabatic two-phase flow is Eq. (6.93). (Note that we cannot use Eq. (6.97) because we cannot assume that x is constant, and therefore Eq. (6.96) is not valid.) This is a nonlinear first-order ordinary differential equation, which we may write, symbolically, as:

$$\frac{dP}{d\ell} = -f(\underline{\mathbf{C}}, \rho_m) \tag{6.119}$$

where, by $\underline{\mathbf{C}}$, we intend to denote a collection of parameters that are constants for this analysis, i.e., q, D, A, and $\cos\theta$. In the sort of fully developed turbulent flow that we expect to encounter here, the friction factor, f, is generally a rather weak function of flow rate, and we may choose to treat it as constant as well, although it is not necessary to do so.

Without going into detail, we will simply stipulate that the Gibbs Phase Rule dictates that this system will have two degrees of freedom. The reader who is unfamiliar with this rule can easily find it explained in standard thermodynamics texts. For the purposes of our analysis here, enthalpy and pressure provide the two degrees of freedom. If these are specified, all other thermodynamic parameters are determined. If we calculate the enthalpy of the flow at the entrance to the pipe, this becomes another constant (by the assumption of adiabatic flow), and from here on all of the properties of the system are determined by the pressure. We can make the dependence of the right-hand side of Eq. (6.119) on pressure more explicit by writing:

$$\frac{dP}{d\ell} = -f(\underline{\mathbf{C}}, \rho_m(P)) = -f(P) \tag{6.120}$$

A variety of techniques are available for solving an ordinary differential equation of the form of Eq. (6.120). Generally, these require the evaluation of $f(P)$ at a variety of values of P determined by the calculation procedure. The Runge-Kutta-Gill method is popular, and should be satisfactory, although others are available as well.

The calculation by which two parameters are specified and the state of a two-phase system is determined therefrom is known to chemical engineers as a "flash" calculation, and the particular computation required here is known as a "pressure-enthalpy flash." Given the enthalpy, pressure, and chemical composition, the flash calculation will produce the vapor fraction, temperature, and composition of the vapor and liquid phases, from which mixture density can be determined.

Note, however, that evaluation of the right-hand side of Eq. (6.93) also requires the derivative, $\frac{d}{dP}\left(\frac{1}{\rho_m}\right)$, and this is generally not available directly from the flash calculation. We will have to perform a series of flash calculations at different pressures, do a curve fit to approximate the functional form of $\frac{1}{\rho_m(P)}$, and then obtain the derivative by differentiating the fitting function. The same curve fit will also yield the value of $\rho_m(P)$. In this way, we can do a few flash calculations, which are relatively expensive in a computational sense, over the range of pressures likely to be encountered, and then generate the right-hand side of Eq. (6.93) at any pressure needed within that range.

References

1 CHEMPRO reference guide, Houston: EPCON Software; 2018.

2 Bean HS, ed. Fluid meters: their theory and application. Report of ASME Research Committee on Fluid Meters. 6th ed. New York: American Society of Mechanical Engineers; 1971.

3 Perry RH, Green D. Perry's chemical engineering handbook. 6th ed. New York: McGraw-Hill; 1984.
4 ANSI/ISA-75.01.01-2002. Flow equations for sizing control valves. Research Triangle Park, NC: The Instrumentation, Systems and Automation Society; 2002.
5 Dukler AE, Wicks III M, Cleveland RG. Frictional pressure drop in two-phase flow. AICHE J, 10 (1964): 38–43.
6 Lockhart RW, Martinelli RC. Proposed correlation of data for isothermal two-phase, two-component flow in pipes, Chem Engg Prog, 45 (1949): 39–48.
7 Dukler AE, Taitel Y. Flow pattern transition in gas-liquid systems: measurement and modeling. Multiphase Sci Tech, 2, (1986): 1–94
8 Flow of fluids through valves, fittings, and pipe. Technical Paper No. 410. Signal Hill, CA: Crane Valves; 1968.

7

Use of Process Simulators Upstream Through Midstream

Justin C. Slagle

Bryan Research & Engineering, Inc. Texas, USA

CHAPTER MENU

7.1 Introduction

7.1.1 The Origin of Hydrocarbon Process Simulation

The oil and gas industry, born from the demand for energy to fuel continued growth after the Industrial Revolution, has enjoyed immense innovation in the Information Age. In 1846, the lamp oil industry and the demand for heavier lubricating oils led James Young to drill the first oil well in what is now Azerbaijan. Using distillation experimentation, Young was able to create both lamp oil and lubricating oil from the same fluid. Further innovation led to kerosene products followed by gasoline and diesel products after invention of the automobile.

Around the same time, the desire to optimize the efficiency of steam engines created breakthroughs in the area of thermodynamics. These advances migrated to chemical processes and provided mathematical expressions for analyses.

Early refineries emphasized kerosene products, with the remaining hydrocarbons largely considered as waste. After demand for gasoline and diesel products increased dramatically, thermodynamic principles allowed engineers to maximize production of the more valuable products. Gasoline, kerosene, and diesel, being the "middle cuts" of the petroleum, left gas and liquid heavier ends as waste products. Modern refineries now use the lighter ends for chemical feedstock, fuel gas, and liquid petroleum gas (LPG). The heavier ends now provide asphalt, bunker fuel, and lubricating oils. Of course, the most valuable products are still gasoline and diesel. Any way to maximize these two products often is preferred. As such, the introduction of reactors and crackers convert the less valuable products into other products for shipping with gasoline or diesel. Natural gas processing also went through development from simple separation to more complex treating and distillation providing several products.

Naturally, both natural gas processing and refineries have transformed from simple distillation sites with large quantities of waste to highly sophisticated facilities with minimum waste. The Information Age has greatly influenced this transition. Over 100 years of research into

Natural Gas Processing from Midstream to Downstream, First Edition.
Edited by Nimir O. Elbashir, Mahmoud M. El-Halwagi, Ioannis G. Economou, and Kenneth R. Hall.

chemical properties for hydrocarbons and other chemical species present in oil and gas has made thermodynamic calculations more and more accurate. With most species now defined, computational automation can fuel future innovations in the oil and gas industry. For over 40 years, computer software, known as process simulators, have been combining thermodynamic principles, chemical properties, and unit operations to spur innovation.

7.1.2 What Is a Process Simulator?

Process simulation is software capable of conducting unit operation calculations for process engineers across many industries, including oil and gas. While it is one of many tools for solving problems in the oil and gas industry, engineers and other professionals often use process simulation daily to solve numerous design and operation-related challenges.

A process simulator consists of two basic elements: a database of thermodynamic properties called a property package and a suite of unit operations. Every simulation must have a property package chosen in order to complete unit operation calculations.

Process simulation will likely be at the heart of future innovations in the industry. It is important to understand why industry uses simulation and which problems to solve using process simulation. This chapter outlines several uses for process simulation and emphasizes two subcategories of the oil and gas industry: upstream and midstream.

7.2 Upstream

It is best to think of a process simulator as a tool used for solving problems. The first set of problems discussed are in the upstream sector of the oil and gas industry, defined here as the unit operations between the well and the processing facilities. Figure 7.1 shows the typical process flow of fluids in the upstream sector.

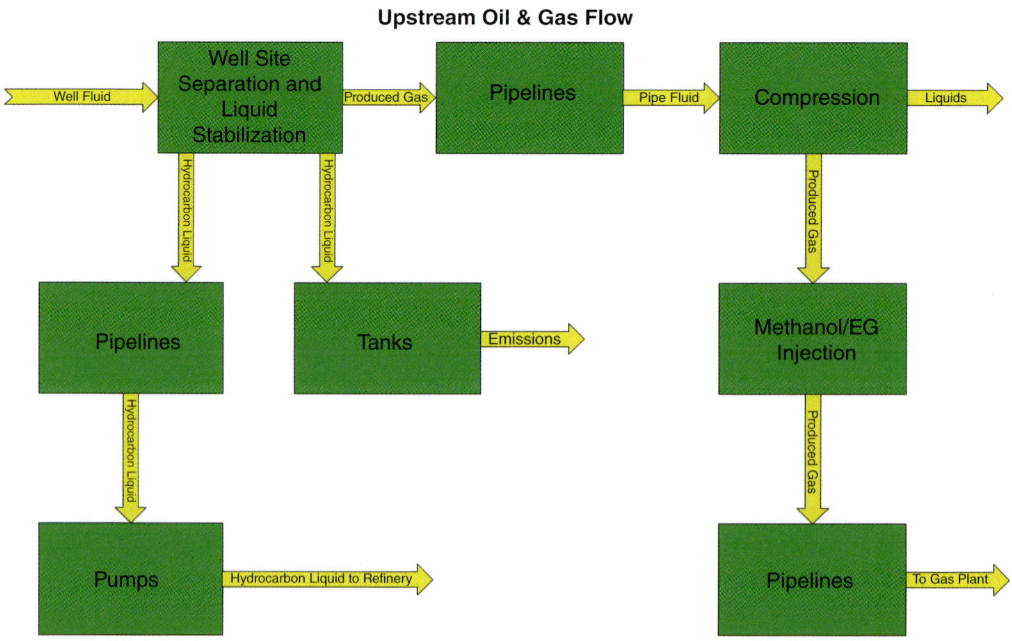

Figure 7.1 Block flow diagram of upstream oil and gas processes.

While this discussion is not exhaustive nor describes every component of an upstream facility in detail, it highlights some of the typical problems of importance that process simulators can solve.

7.2.1 Down Hole PVT

Pressure-volume-temperature (PVT) studies of reservoir fluids are essential for analyzing reserves, potential recoveries, and production quality of the fluids. This information aids in valuing the well's profitability.

While PVT tests provide valuable information for upstream applications, process simulators can offer results for these tests with very little input. Generally, it is necessary to have accurate compositions for the well stream. This may come from down hole samples or from the gas and liquid leaving the separator just downstream of the producing well.

The major advantage of using the simulator is that it introduces no additional experimental error. It minimizes complicated experimental lab tests, which come with large costs and experimental error. Labs are typically well equipped to produce accurate compositions for gases and liquids emanating from the high-pressure separator. The gas and liquid compositions combined with the gas to oil ratio (GOR) and reservoir temperature are often enough information for a simulator to produce PVT charts. Using this information, Figure 7.2 shows the phase loops of

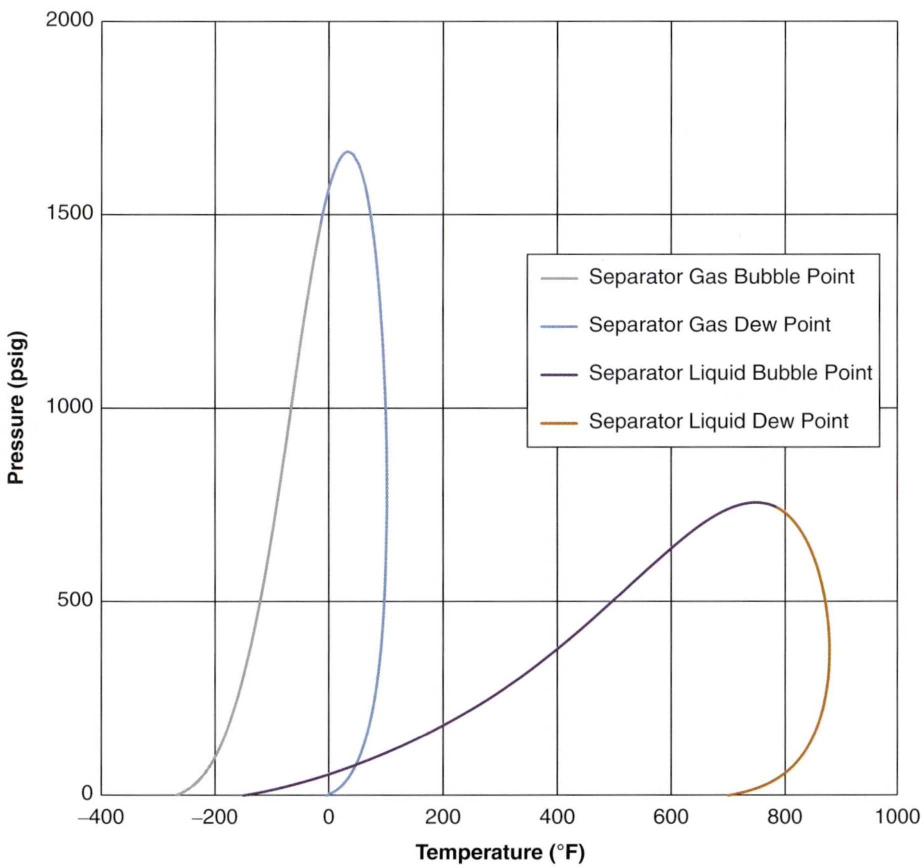

Figure 7.2 Phase envelopes of gas and liquid from inlet separator.

the separator gas and liquid generated using a process simulator, ProMax®, based upon field sample inputs [1].

In this figure, if the intersection of the separator gas dew point and the separator liquid bubble point closely resemble the temperature and pressure of the separator, the lab analysis of the samples are considered accurate.

Although simulators use an equation of state (EOS) model, which has associated error depending upon the system, modern EOS models are actually reasonable in reproducing experimental data. Their errors are local to certain regions of the phase space (usually near the cricondentherm of a phase loop, and exactly at the critical point where it cannot distinguish between liquid and vapor).

The simulator can produce all desired PVT information quickly and accurately. A list of some of the most desirable down hole PVT data that can come from a process simulator are:

- Hoffman table/plot (including all components if desired)
- Separator liquid flash
- Constant composition expansion
- Constant volume depletion
- Differential liberation expansion
- Cumulative recovery
- Separator tests

Even if experimental results were still desirable, it would be prudent to validate them with simulator results given the comparatively low cost of simulation and the reduction of potential errors. Table 7.1 summarizes the information needed to generate a model and the typical problems solved using the model.

7.2.2 Well Site

The first equipment in upstream facilities involves the separation of the gas, oil, and water. The well fluid typically consists of three phases, which requires initial separation. Figure 7.3 presents a well site model developed in ProMax.

The fluid separates into hydrocarbon gas, hydrocarbon liquid, and water. Because the hydrocarbon liquid is typically in a volatile state on site, contained in tanks at ambient conditions, it proceeds through a heater-treater before storage. The purpose of the heater-treater is to recover some of the lighter hydrocarbons that would otherwise flash in the atmospheric tank that could increase volatile organic compound (VOC) emissions significantly and reduce

Table 7.1 Process simulation summary for PVT studies.

Minimum information needed for model
Composition of HC liquid
Composition of produced gas
Gas to oil ratio (GOR)
Temperature of reservoir
Common solutions using studies
Value well

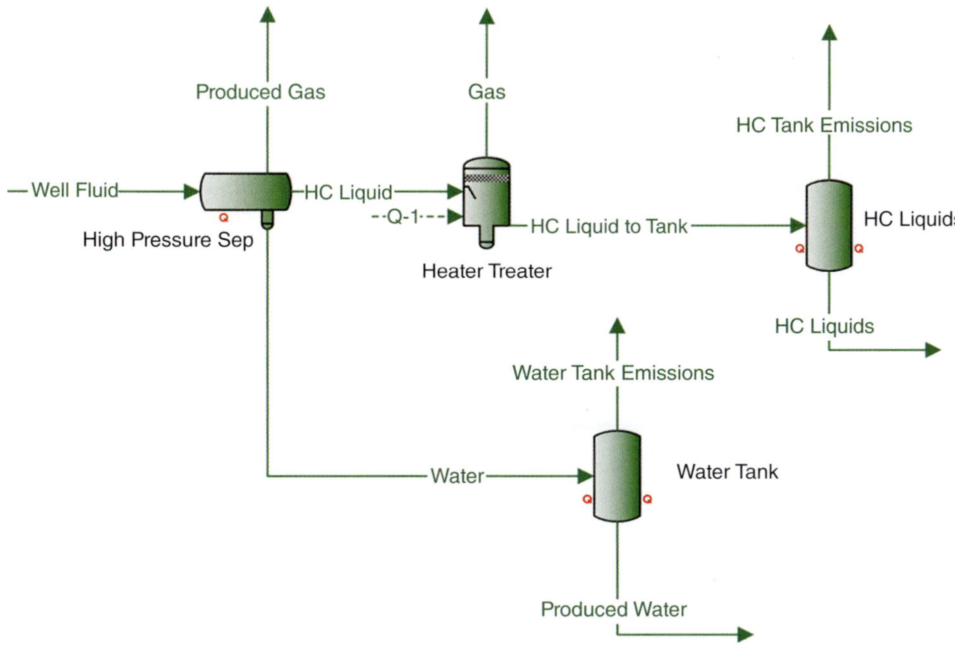

Figure 7.3 Model of simplified well site.

liquid production. Instead, the heater-treater recovers most of the lighter hydrocarbons and re-routes them to join the gas pipeline.

The heater-treater provides some control of the gravity (set according to American Petroleum Institute [API] standard) of the stock tank hydrocarbon liquid and the tank emissions. A process simulator can optimize the temperature and pressure of the heater-treater to meet desired oil specifications and maintain compliance with environmental quality agencies.

The same information needed for the PVT study can provide a well site model. In addition to the high-pressure separator gas and hydrocarbon liquid compositions, only the heater-treater temperature, pressure, and production rates are necessary to create a highly accurate model.

While Figure 7.3 shows a well site model with a single well feeding the tanks, another typical problem faced in the upstream sector concerns allocation of products to the originating well or owners. When several unique well fluids commingle and feed a single facility, it becomes challenging to decide which wells contribute to the final products. If each of the wells has a unique owner, how do you distribute royalties? It is possible to collect samples before commingling, but once comingled, the composition changes, which changes phase equilibrium and the overall liquid production. Imagine four wells commingling in a central processing facility similar to Figure 7.4. The heater-treater provides the desired API gravity and acceptable emissions using the simulator, but then comes the daunting task of allocating the revenue from the hydrocarbon liquid. The simulator can track molecules from each well through each unit operation. Figure 7.4 presents a ProMax model tracking the allocation of hydrocarbon liquids to the various wells of origin. This information allows distribution of royalties and/or designing/sizing production sites.

The summary of information needed to generate a model and the typical problems solved using the model are in Table 7.2.

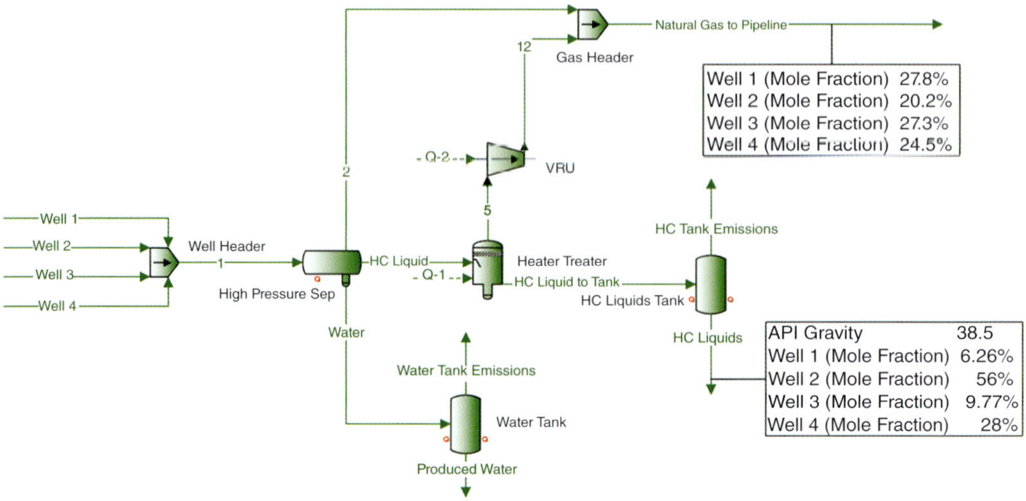

Figure 7.4 Model of well site with comingled well fluids and allocation of products according to source.

Table 7.2 Process simulation summary for well site models.

Minimum information needed for model
Composition of HC liquid
Composition of produced gas
HC liquid production rate
Gas production rate
Water production rate
Common solutions using model
Optimize heater-treater to maximize HC liquid value
Size equipment
Accurately predict and report tank emissions
Allocate royalties to land, mineral, and lease owners

7.2.3 Pipelines

Pipeline modeling plays a large role in upstream and midstream oil and gas operations. Pipeline models routinely provide understanding of phase changes and pressure drops through the pipe. Figure 7.5 is a phase envelope generated using ProMax of a typical natural gas traveling from a well site to a processing facility.

As expected, the gas is at conditions precisely on the dew point curve. A reduction in temperature or an increase in pressure will move this fluid from a gas to a gas/liquid mixture. A process simulator can generate a phase envelope for any fluid mixture in the model. In addition, the pipeline model predicts not only a phase change but also the volume of the liquid that forms.

More specifically, a robust pipeline model should predict phase changes, temperature changes, pressure drops, velocity, density, hydrate formation, and other properties of the fluid.

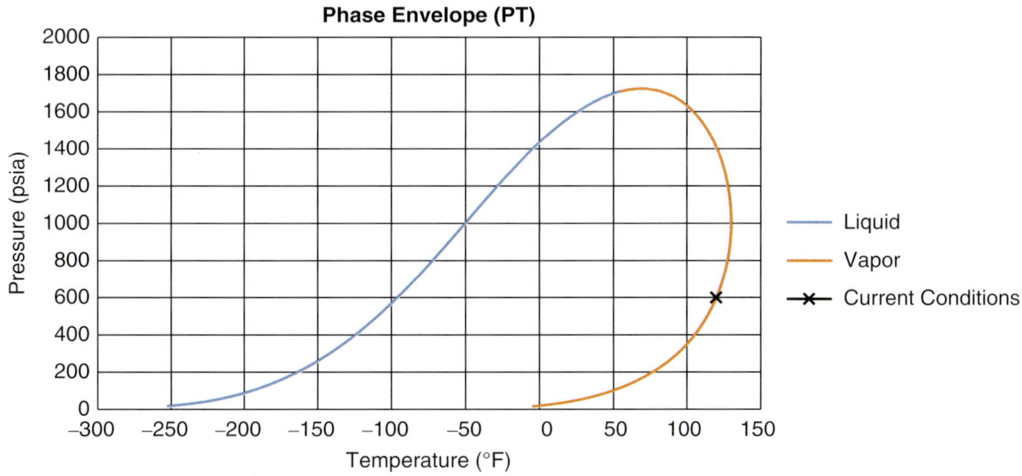

Figure 7.5 Phase envelope of a natural gas.

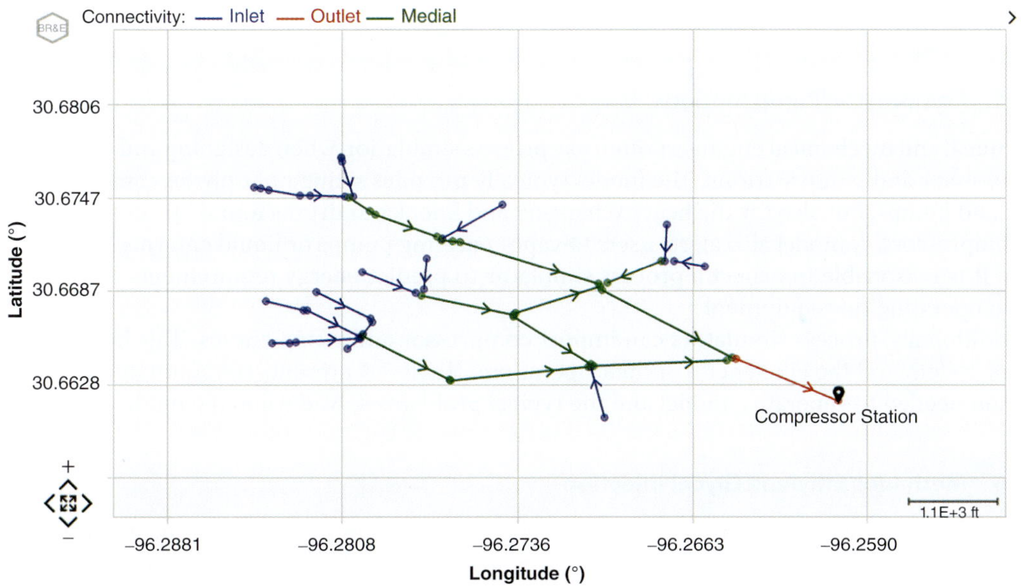

Figure 7.6 Model of a gas gathering pipeline network using GIS data.

It should also incorporate latitude, longitude, elevation, fittings, material of construction, buried pipelines, above-ground pipelines, submerged pipelines, ambient conditions, and various other hydraulic properties of the pipeline. Figure 7.6 illustrates a gas-gathering network model generated in ProMax using Geographic Information System (GIS) data.

This model is particularly useful for sizing pipelines and designing the downstream compressor station. When gas enters a pipeline, it travels the path of least resistance, from high pressure to low pressure. The model assists engineers in managing the pressure profile and in operating the necessary stages of compression downstream. Adding even one more gas inlet

Table 7.3 Process simulation summary for pipelines.

Minimum information needed for model
Pipeline inlet conditions
Pipeline inlet flowrate/desired flowrate
Common solutions using model
Size pipeline
Select material
Calculate throughput
Calculate flow direction, velocity
Calculate where to add booster stations
Calculate phase change

line to this model may significantly affect the liquids produced through the pipeline network and the pressure of the fluid at the compressor station. The model must reflect such changes.

The summary of information needed to generate a model and the typical problems solved using the model appears in Table 7.3.

7.2.4 Compressor/Pump Stations

Chemical and mechanical engineers often use process simulation when designing and operating compressor and pump stations. The model typically provides sizing not only for the compressors and pumps, but also for the heat exchangers and knock out drums found just downstream of compressors. A model also alerts users to vapor entering pumps or liquid entering compressors. It is reasonable to expect a process simulator to predict energy requirements accurately when operating this equipment.

Additionally, process simulators can import compressor and pump curves. This helps engineers understand the efficiencies of their operations. Table 7.4 presents the summary of information needed to generate a model and the typical problems solved using the model.

7.2.5 Methanol/Ethylene Glycol Injection

Various conditions and scenarios may cause hydrate formation to be a concern. When water is present in a gas, it is possible for a non-water molecule to stabilize a water structure around

Table 7.4 Process simulation summary for compressor and pump stations.

Minimum information needed for model
Inlet composition
Inlet conditions
Inlet flowrate/desired flowrate
Common solutions using model
Size compressor/pump
Calculate number of stages
Calculate liquid dropout between stages

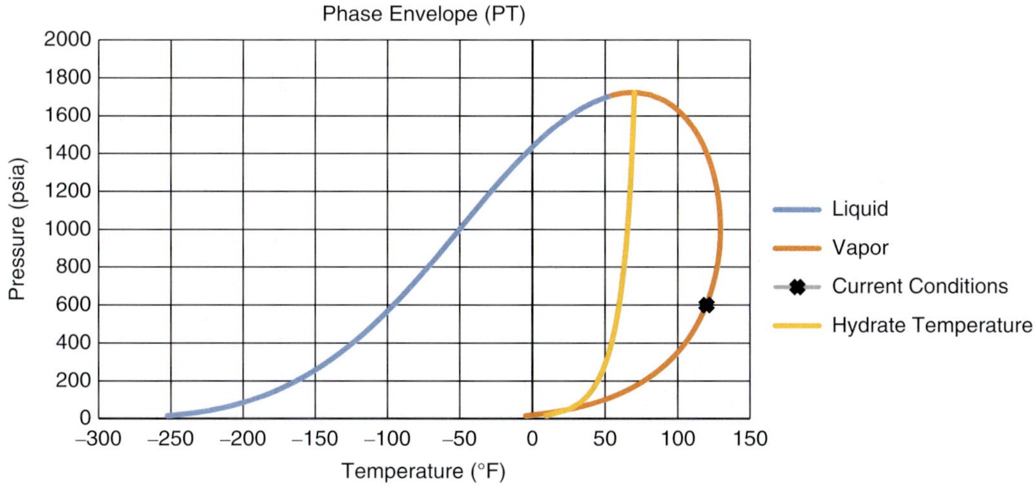

Figure 7.7 Typical phase envelope and hydrate curve for a natural gas.

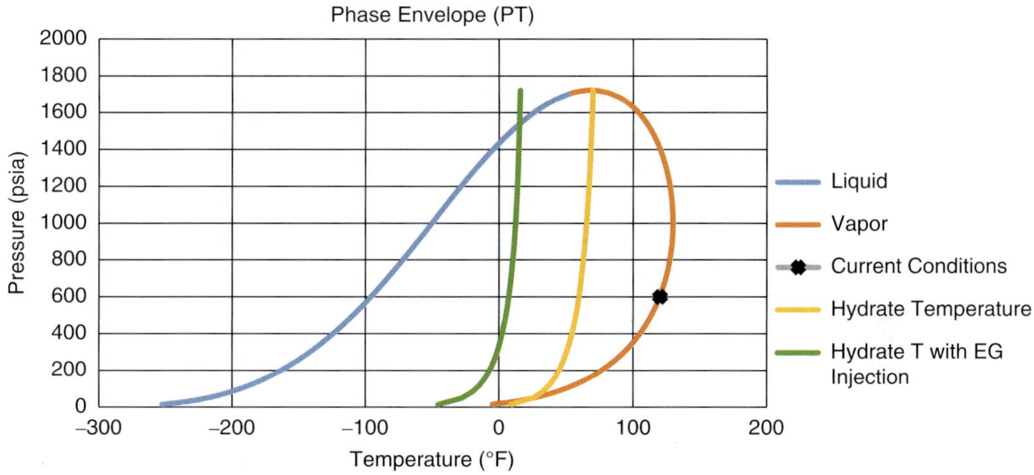

Figure 7.8 Typical phase envelope and hydrate curve shift with EG injection.

itself, causing a solid to form, commonly referred to as a hydrate. A process simulator must be able to predict these conditions accurately. Figure 7.7 illustrates a hydrate curve predicted by ProMax for the phase envelope generated previously.

As the fluid travels through the pipeline, the pressure and temperature change. It is common for a fluid to enter conditions conducive to hydrate formation. Therefore, operators typically inject a hydrate suppressor into the pipeline. Methanol and ethylene glycol (EG) are the two most common chemicals used to avoid hydrate formation. The question is then: how much methanol or EG is necessary? Too little allows hydrates to form; too much can cause downstream products to exceed methanol or EG limits. A process simulator can help determine the methanol and EG flowrates needed to inject enough to suppress hydrates, but not so much that downstream operations or products suffer.

Figure 7.8 again contains the same phase envelope, but this time with hydrate temperatures after EG injection into the pipeline.

Table 7.5 Process simulation summary for methanol/EG injection.

Minimum information needed for model
Fluid composition and conditions
Common solutions using model
Choose hydrate inhibitor
Calculate hydrate inhibitor flowrate
Maintain conditions above hydrate formation temperature

The hydrate curve shifts to the left with methanol or EG injection, as expected. Table 7.5 contains the summary of information needed to generate a model and the typical problems solved using the model.

7.2.6 Tanks

Well site hydrocarbon tanks are common in oil production operations in the United States. Process simulators with thermodynamic phase data for hydrocarbon mixtures and hydrocarbon/water mixtures apply in the world of air emissions. While process simulation has become more robust, the graphic user interface has greatly improved. What was once a tool used only by chemical engineers with savvy computer skills, is now accessible to environmental departments without people having strong chemical engineering or thermodynamic backgrounds.

Air quality calculations previously used assumptions and somewhat arbitrary rules of thumb. Robust thermodynamic calculations are now the norm in calculating tank emissions for atmospheric tanks. This results in well-informed decisions based upon real scientific calculations. Production companies now have a robust tool for reporting and substantiating emissions to regulatory agencies. Air quality agencies now have a tool for monitoring emissions inventories in a robust fashion. The model may also use specific standards such as the U.S. EPA's AP 42 guidance for calculating working and breathing losses, flash losses, and loading losses, as the ProMax model shows in Figure 7.9.

Table 7.6 contains the summary of information needed to generate a model and the typical problems solved using the model.

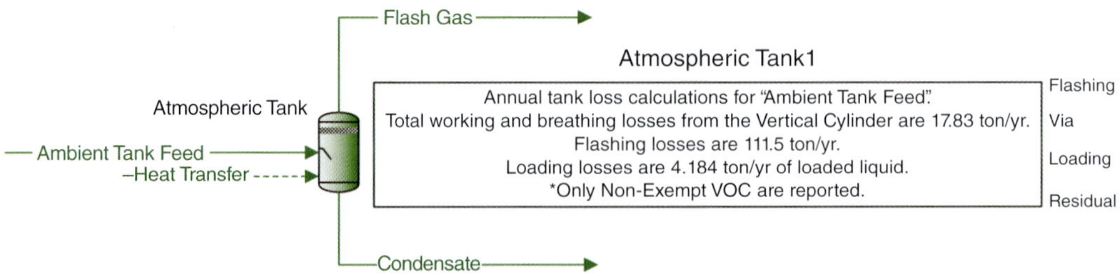

Figure 7.9 Model of an atmospheric tank and the associated working, breathing, and flash losses.

Table 7.6 Process simulation summary for tanks.

Minimum information needed for model
Inlet composition
Inlet conditions
Location of tanks
Tank type/dimensions
Common solutions using model
Size tanks
Calculate emissions accurately

7.3 Midstream

The heart of gas processing occurs in the midstream sector of the oil and gas industry, defined here as the gas processing facility. Figure 7.10 presents a typical gas plant block flow diagram.

When the gas arrives at the plant from the well site, it undergoes initial separation and possibly stabilization of any liquids that formed in the pipeline. Gas treatment then removes impurities such as hydrogen sulfide, carbon dioxide, and water. After treatment, the gas contains several products. Sales gas and natural gas liquids (NGL) are typical. NGL may

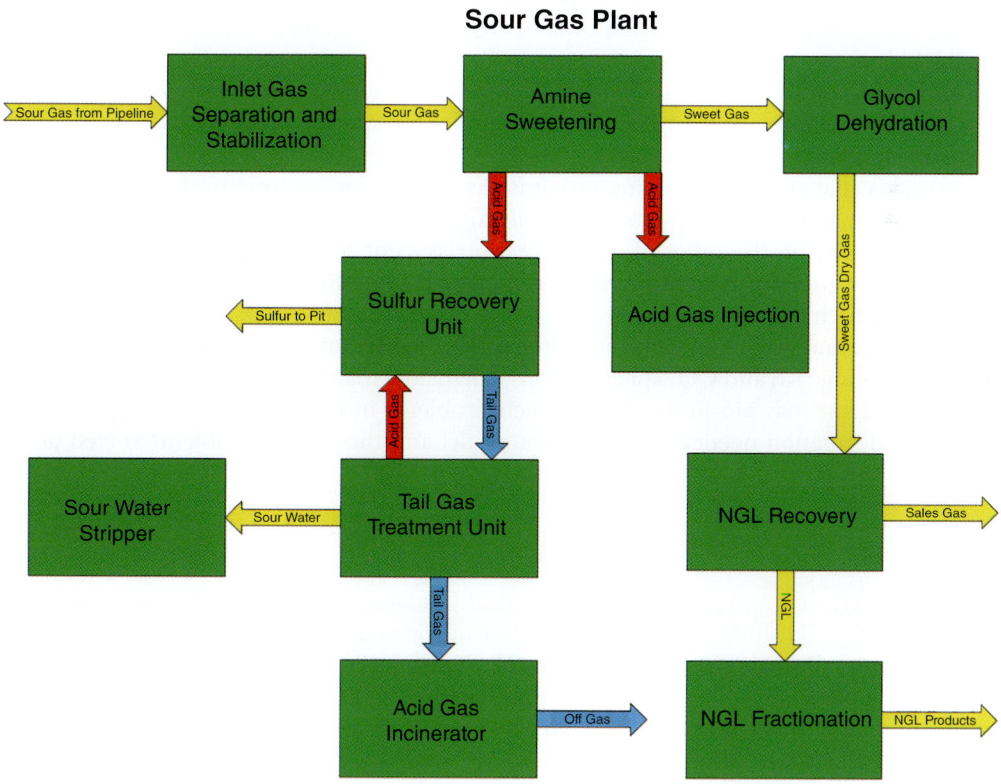

Figure 7.10 Block flow diagram for a sour gas processing facility.

provide additional products on site, but more commonly this occurs in NGL gathering centers, such as the ones located in Hobbes, NM or Mont Belvieu, TX. Process simulation is a normal feature of midstream processing.

Many technologies exist for treating and processing natural gas, a few of which appear in this chapter. However, process simulation is often central for technology selection. A process simulator should model all the processes shown in Figure 7.10, even when evaluating several competing technologies. The model may contain economics, such as product values, energy costs, and capital costs.

7.3.1 Amine Sweetening

If H_2S and/or CO_2 are present in the gas, it is usual to utilize an amine to remove those impurities. Process simulation has become essential in designing, optimizing, and troubleshooting this process, which is a large source of energy consumption in the plant, and product specifications are typically unmet if the amine sweetening unit operates poorly or not at all. Figure 7.11 illustrates a typical amine sweetening unit model built in ProMax.

The model can evaluate numerous configurations, amine types, conditions, concentrations, and energy consumption. The models are valuable for the life of the plant and especially when gas quality or conditions change. Operators also look to engineers to provide guidance for operating conditions. For example, Figure 7.12 shows an operator reference guide generated by engineers using ProMax to provide operating guidance for a specific amine unit operating in the Middle East [2].

Using a process simulator to optimize operations for various inlet gas flowrates allows operators to utilize rigorous calculations in a simple, succinct manner. As the true power of process simulation becomes more widely known, it becomes more common to see these figures hanging in control rooms throughout the industry. While rules of thumb have governed many operations in the past, customized operator reference guides for specific units are likely to become the industry standard in the future.

As conditions change (which is common), it is easy to move away from optimum operation. For an amine unit, that could mean unnecessarily high operating costs on one end of the spectrum or catastrophe on the other. It is simply too risky not to have up-to-date models and custom reference guides for each operating unit. The cost of simulation is low, and the risk of unit failure is high without it.

Additionally, amine sweetening units are often subject to various troubleshooting issues, such as difficulty meeting H_2S and CO_2 specifications, corrosion, foaming, and solvent degradation. A process simulator may aid in diagnosing such problems in the plant. Table 7.7 presents a summary of information needed to generate a model and the typical problems solved using the model.

7.3.2 Sulfur Recovery

After sweetening the natural gas, the acid gas typically passes to a flare or a sulfur recovery unit (SRU) to produce solid sulfur and to reduce sulfur dioxide emissions. Sulfur recovery utilizes a series of reactors. Process simulation is particularly helpful for modeling the many chemical reactions taking place.

Figure 7.13 contains a ProMax model of an acid gas burner with two passes through a waste heat boiler and two Claus sulfur converter beds.

Many different configurations exist for an SRU, which may have different combinations of converter beds, acid gas bypass, oxygen enrichment, low temperature catalysts, direct acid

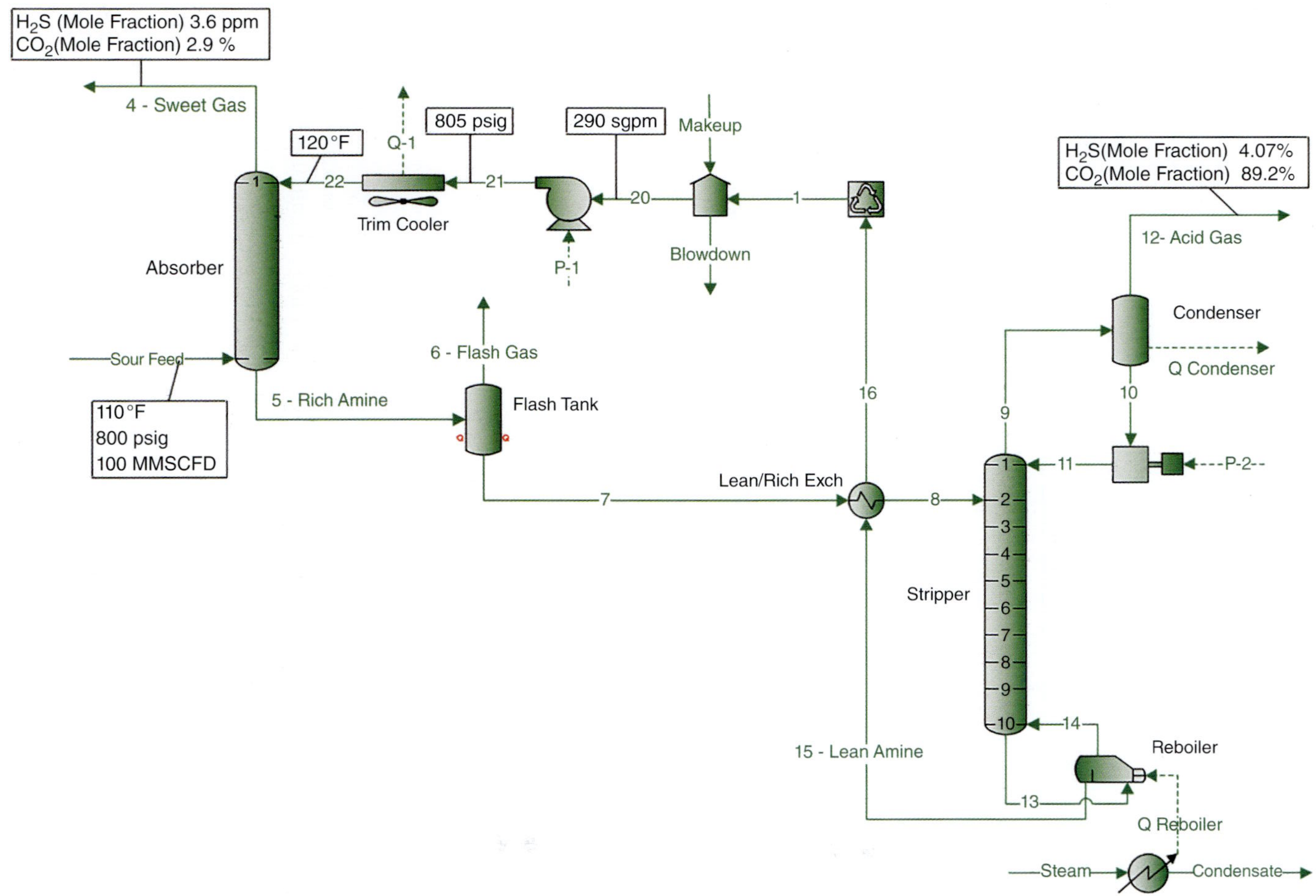

Figure 7.11 Typical amine sweetening unit model.

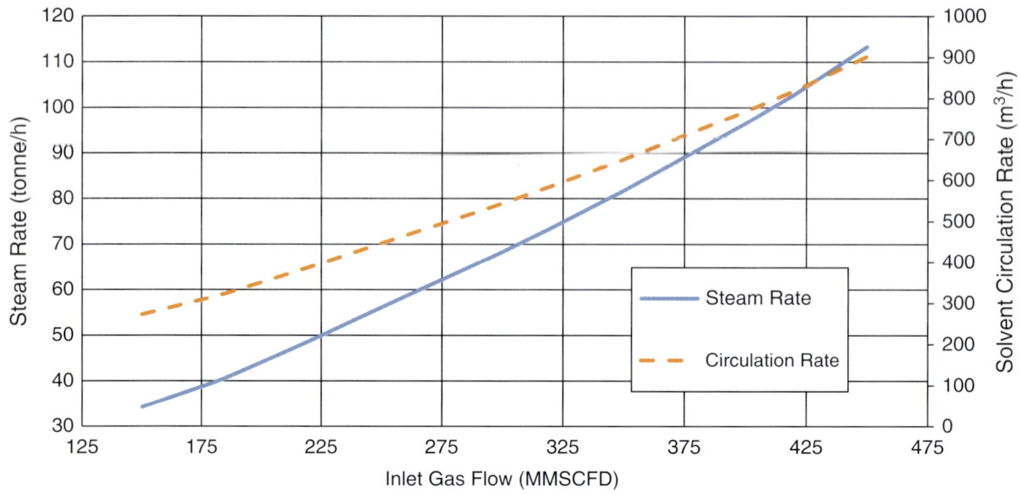

Figure 7.12 Operator reference guide generated from a model to assist operators in selecting steam and solvent circulation rates based on the inlet gas feed rate.

Table 7.7 Process simulation summary for amine sweetening.

Minimum information needed for model
Inlet composition and conditions
Desired sweet gas specifications
Common solutions using model
Choose optimum solvent/solvent mixture
Calculate optimum solvent circulation rate
Calculate optimum reboiler duty
Size/rate all equipment
Troubleshoot process upsets
Diagnose corrosion problems

gas fired re-heaters, etc. A robust process simulator must be fully customizable and flexible to enough to model the above configurations and many more. Table 7.8 contains the summary of information needed to generate a model and the typical problems solved using the model.

7.3.3 Tail Gas Treatment

A tail gas treatment unit (TGTU) is often necessary when very strict environmental regulations are in effect. The tail gas from the SRU typically goes to an incinerator or a TGTU to improve sulfur recovery. As shown in the block flow diagram in Figure 7.10, the tail gas from the SRU passes to the TGTU allowing more H_2S removal, which return to the SRU, leaving a tail gas with significantly less sulfur. The TGTU typically includes a hydrogenation reactor for converting sulfur species back into H_2S and an amine sweetening unit operating at low pressure. Competing technologies exist, most of which are modeled and compared during technology selection.

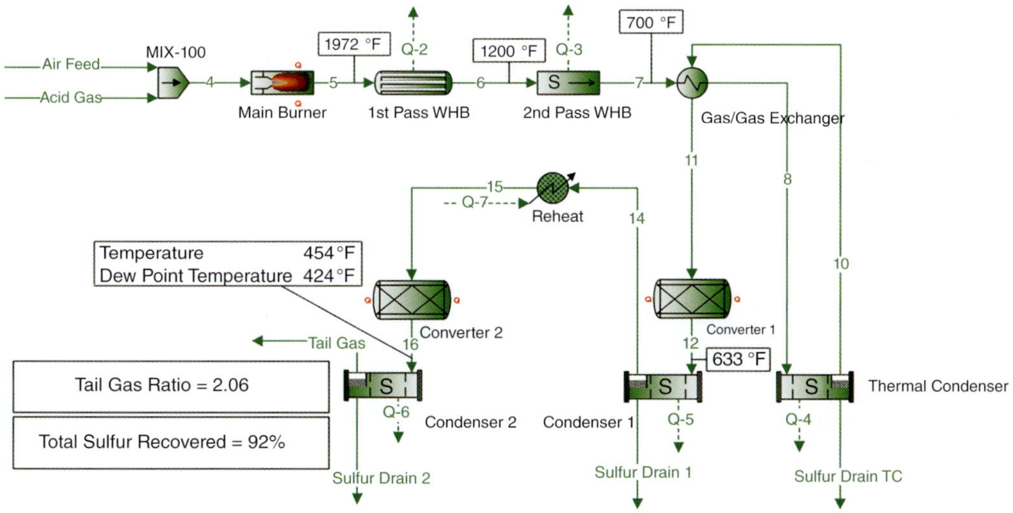

Figure 7.13 Typical sulfur recovery unit model.

Table 7.8 Process simulation summary for sulfur recovery.

Minimum information needed for model
Inlet composition and conditions
Common solutions using model
Calculate optimum air flowrate
Calculate optimum re-heater temperatures
Calculate total sulfur recovery
Choose optimum tail gas handling technology
Troubleshoot process upsets

Modeling the TGTU with a process simulator gives the engineer insight into the operation of the unit. Much like the amine sweetening unit just upstream of the SRU, the model can help engineers and operators optimize solvent selection, concentration, and circulation rate. Process simulation can be particularly helpful when addressing unique challenges associated with TGTUs. For instance, the low pressure and need for selective removal of H_2S while slipping CO_2 into the off-gas often leads to the selection of a MDEA based solvent at relatively high circulation rates. Additives are also common in TGTU solvents to assist with increasing the driving force for H_2S removal. A TGTU amine sweetening unit looks much like the ProMax model shown in Figure 7.11, except it contains fewer absorber stages and a pump in place of the rich flash vessel.

Table 7.9 provides a summary of information needed to generate a model and the typical problems solved using the model.

7.3.4 Sour Water Stripper

Some gas plants use a sour water stripper (SWS) to remove H_2S from various water sources in the plant. If only H_2S and CO_2 are present, a SWS may not be necessary. Typically, it is essential to have an SWS when the water contains both acid and base constituents, such as H_2S and NH_3.

Table 7.9 Process simulation summary for tail gas treatment units.

Minimum information needed for model
Inlet composition and conditions
Desired off-gas and H_2S removal specifications
Common solutions using model
Choice of tail gas treatment technology
Calculate required hydrogen for hydrogenation
Choose optimum solvent and additives
Calculate optimum solvent circulation rate
Calculate optimum reboiler duty
Size/rate all equipment
Troubleshoot process upsets
Diagnose corrosion problems

It is not common for bases like NH_3 to be in gas plants. However, there are cases when an SWS removes H_2S in gas plants, such as in plants with very high throughput and fairly high H_2S content. In these cases, it is relatively easy to move the H_2S out of the water by adding stripping gas. The process simulator may assist in choosing the stripping technology and the stripping gas source, such as fuel gas, steam, or nitrogen.

Figure 7.14 is a ProMax model of an SWS.

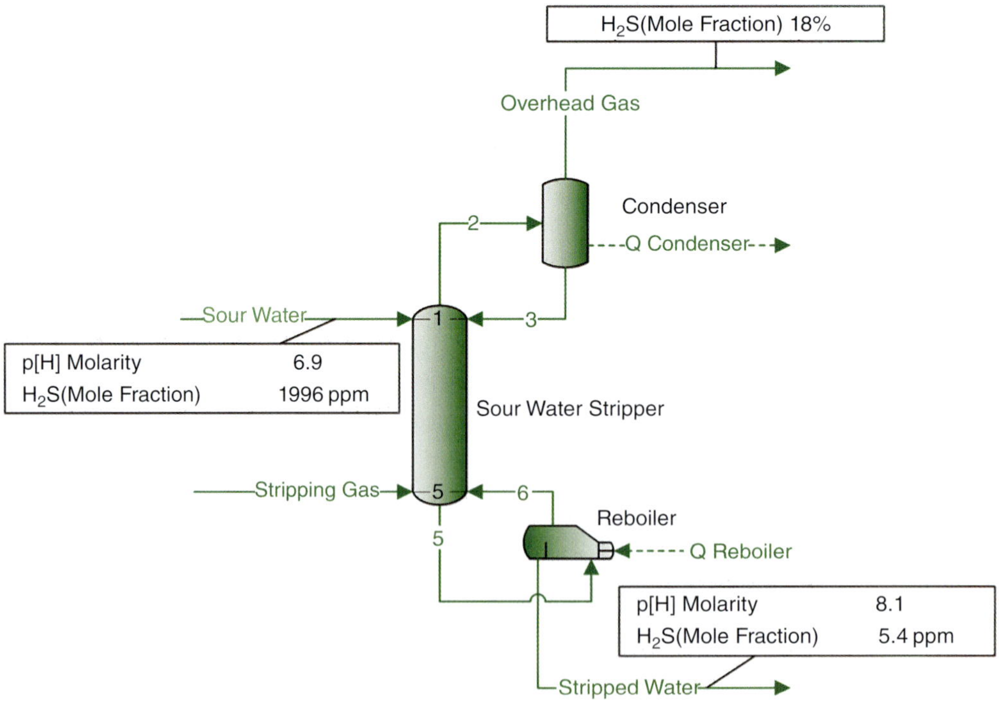

Figure 7.14 Model of a sour water stripper.

Table 7.10 Process simulation summary for sour water stripping.

Minimum information needed for model
Inlet composition and conditions
Desired product specifications
Common solutions using model
Choose sour water stripping technology
Size/rate all equipment
Troubleshoot process upsets
Run scenarios to see necessary changes in operation as feed quality changes

The model is useful for optimizing the equipment for any particular application. For example, the reboiler may not be necessary if enough stripping gas is available. If there is no reboiler and fuel gas or nitrogen is the stripping gas, the condenser may not be necessary. A process simulator can easily evaluate the economics of each case.

The summary of information needed to generate a model and the typical problems solved using the model is in Table 7.10.

7.3.5 Incinerator/Flare

While not discussed in the upstream section, flares are prevalent in upstream and midstream operations. Models of flares and incinerators are often necessary to solve various problems, such as relief valve sizing, ensuring relief network integrity, calculating the amount of air and fuel gas required for combustion, and reducing or reporting air emissions. Figure 7.15 demonstrates an incinerator/flare model in ProMax.

The process simulator helps maximize flare and incinerator operability and flexibility by accurately calculating results with various feed sources. Table 7.11 contains a summary of information needed to generate a model and the typical problems solved using the model.

7.3.6 Glycol Dehydration

The natural gas usually enters the gas plant saturated with water. It is necessary to remove the water to prevent freezing in cold processes used to separate hydrocarbons into different

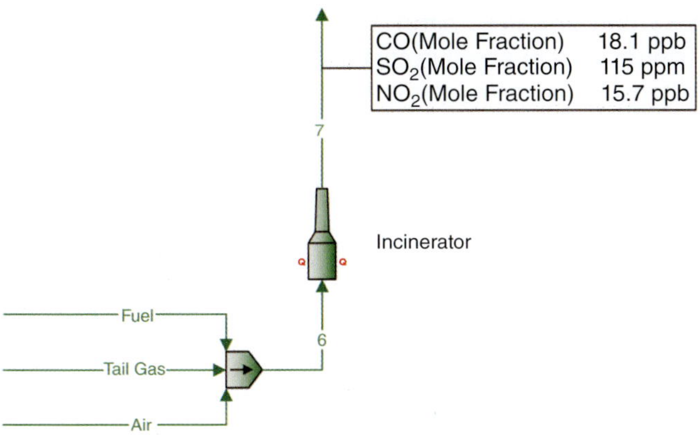

Figure 7.15 Model of an Incinerator.

Table 7.11 Process simulation summary for flares and incinerators.

Flare/incinerator
Minimum information needed for model
Inlet composition and conditions
Common solutions using model
Calculate air flow required for combustion
Calculate fuel gas requirements
Optimize operations to control/report emissions
Relief valve sizing
Relief network sizing and optimization
Size blowers
Troubleshoot process upsets

products. Glycol dehydration and molecular sieves are the two primary methods for water removal. While process simulators do not usually predict molecular sieve results, process engineers use simulation extensively for glycol dehydration. Figure 7.16 is a ProMax model of a typical glycol dehydration unit.

Like amine sweetening, TEG units contain many variables to evaluate, including solvent selection, circulation rate, reboiler duty, operating conditions, and stripping gas. A common problem found in glycol units operating today is overcirculation of the glycol. Not only does this condition lead to additional heavy hydrocarbon absorption and possible emissions, but it also may impact water removal negatively. An operating glycol dehydration unit in West Texas exhibited that effect as confirmed by ProMax in Figure 7.17 [3].

As the glycol circulation rate increases, the water content decreases until reaching a minimum and increasing again. The plant engineer observed the same trend as the model predicted. In this case, the model assisted the engineer in significantly reducing the glycol circulation rate while improving plant results. Table 7.12 contains a summary of information needed to generate a model and the typical problems solved using the model.

7.3.7 NGL Recovery

The natural gas liquids (NGL) recovery unit processes natural gas into a marketable product ready for utilities to consume or distribute with minimal preparation. Therefore, it is essential to process the gas to proper specifications. Process simulation finds extensive use in modeling the many technologies utilized to deliver the desired product.

Figure 7.18 contains a ProMax model of a typical NGL recovery unit.

Market conditions greatly influence the desired natural gas quality, more specifically the ethane content. Liquid ethane or ethane/propane mixture contributes heating value as part of the sales gas product. Because of these market variables, it may be more profitable to operate the NGL recovery unit differently year to year. Moving the ethane from the sales gas to NGL or back again can assist the plant in maintaining the highest profitability even as market demands change. It is essential to maintain a current model that accurately represents current conditions, expected conditions, and "what if?" scenarios that may affect ethane recovery decisions.

Table 7.13 contains a summary of information needed to generate a model and the typical problems solved using the model.

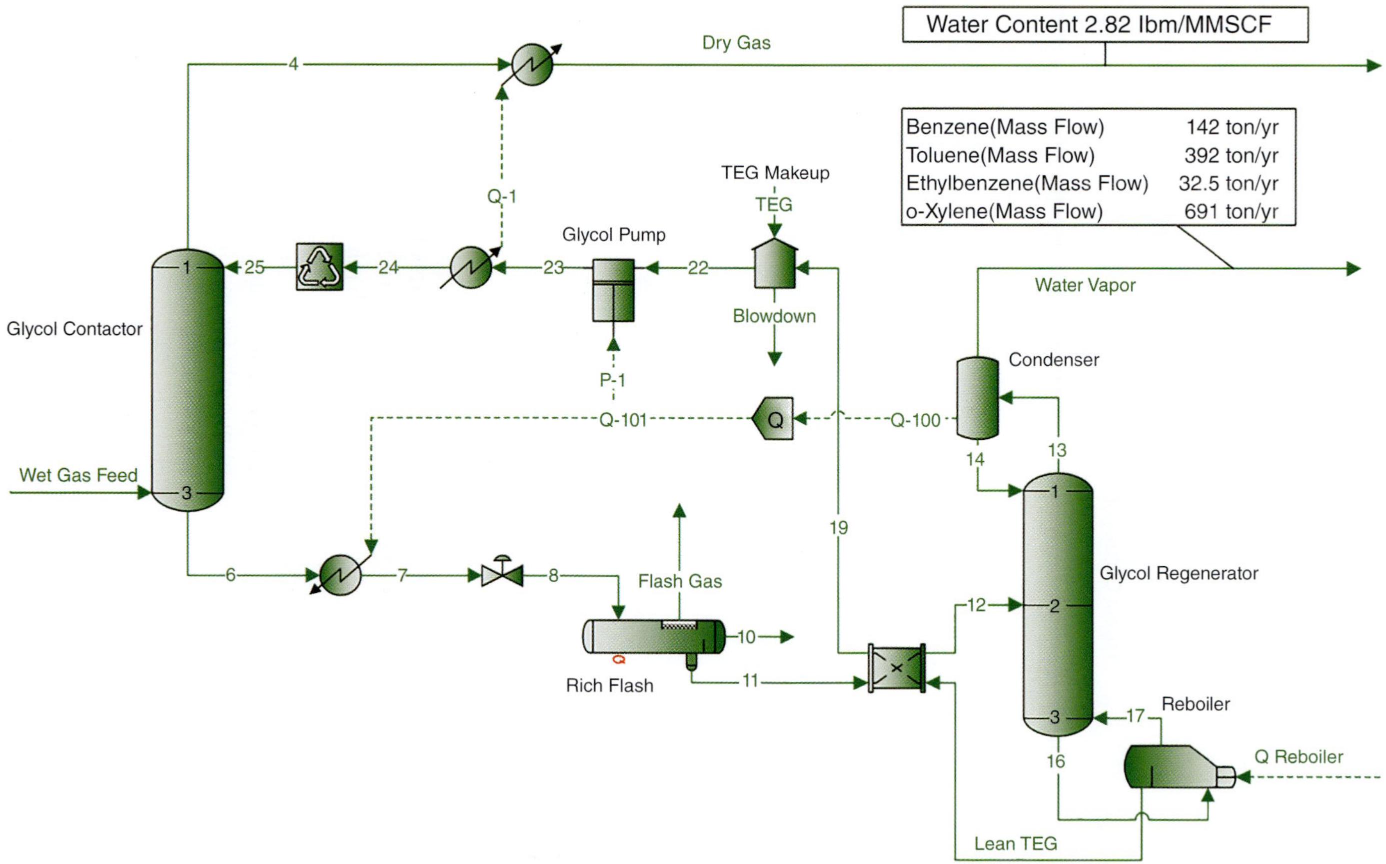

Figure 7.16 Typical model of a TEG dehydration unit.

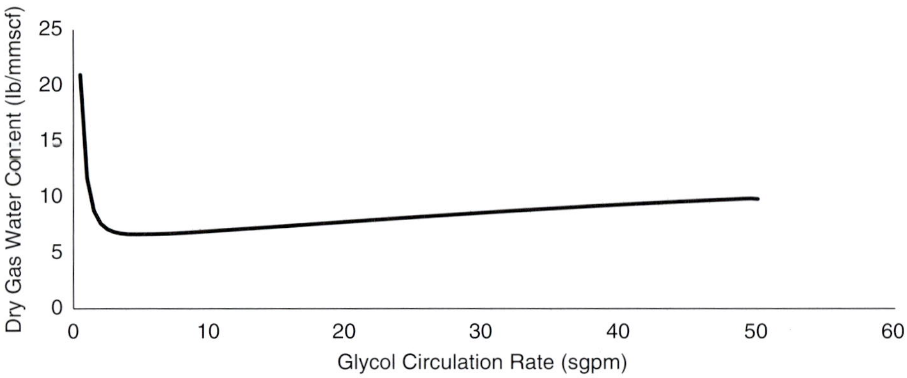

Figure 7.17 Effect of dry gas water content as a function of glycol circulation rate.

Table 7.12 Process simulation summary for glycol dehydration units.

Minimum information needed for model
Inlet composition and conditions
Desired dry gas water content
Common solutions using model
Choose optimum solvent
Calculate optimum solvent circulation rate
Calculate optimum reboiler duty
Size/rate all equipment
Troubleshoot process upsets
Optimize stripping gas
Optimize/report air emissions

7.3.8 NGL Fractionation

The NGL recovered from the natural gas provides more products in the fractionation plant. This plant produces products such as LPG, chemical feedstock, and natural gasoline according to current market demand. It is typically a series of columns with reboilers and condensers with overhead and bottoms products.

Figure 7.19 is a ProMax model of a NGL fractionation train.

Regional demand for specific products naturally control product specifications, but process simulation is essential to operate the plant continuously for maximum profit while meeting consumer needs. Table 7.14 presents a summary of information needed to generate a model and the typical problems solved using the model.

7.4 Going Further

Much like other software seen in everyday life, process simulators have benefited from hardware and operating system advancements over the past 50 years. Not too long ago, Microsoft DOS (MS-DOS) was the standard operating system for process simulators. Having

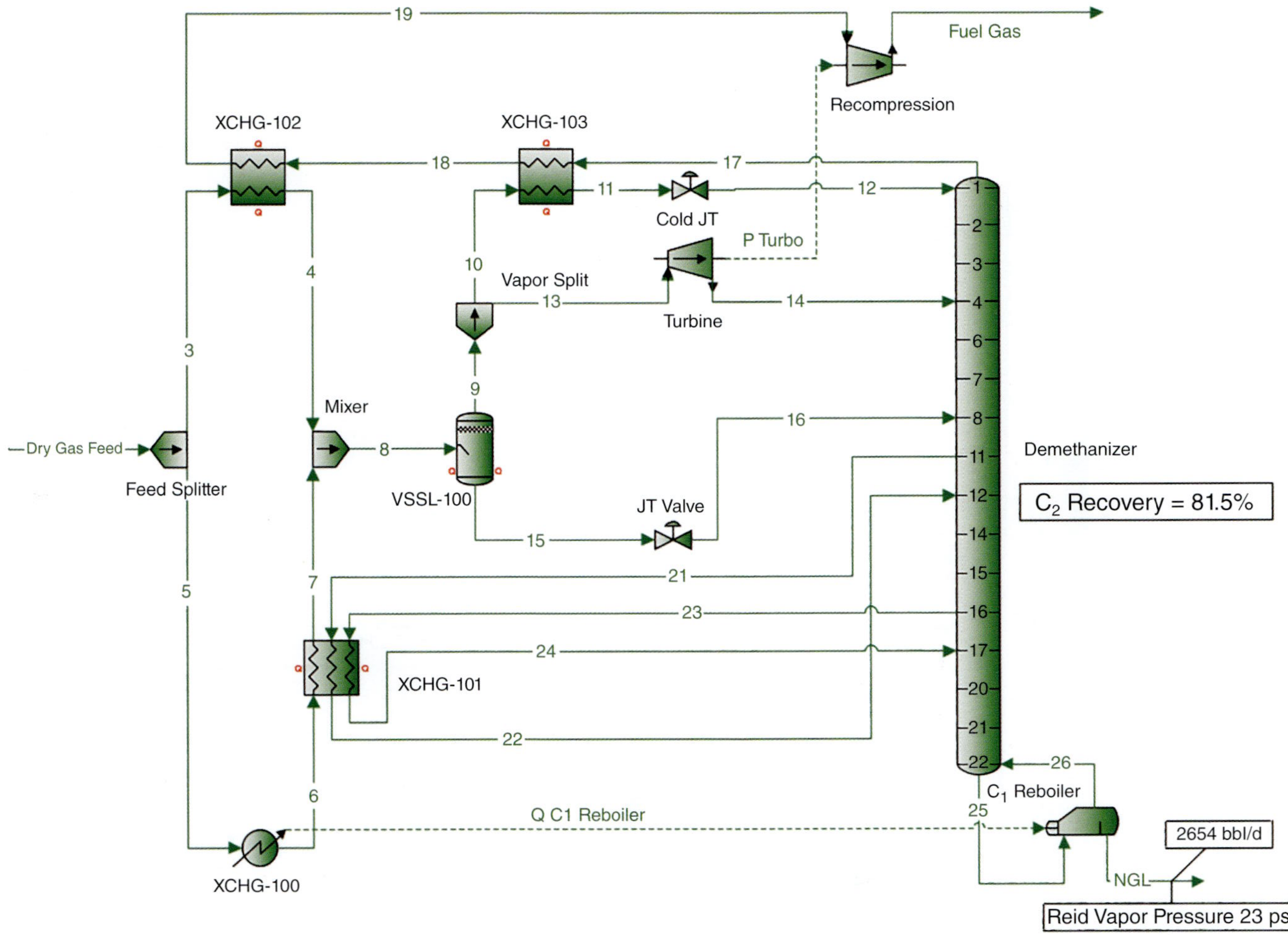

Figure 7.18 Typical model of NGL recovery unit.

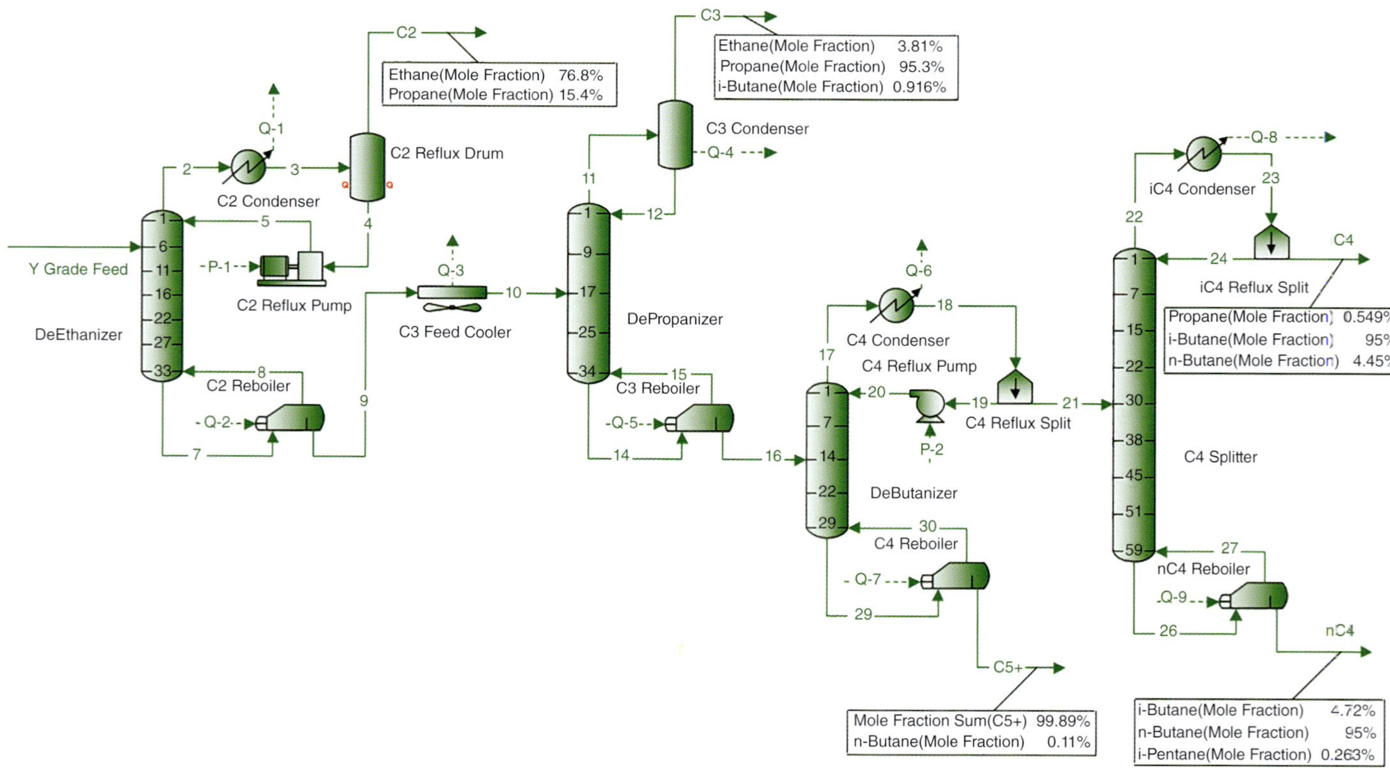

Figure 7.19 Typical model of NGL fractionation train.

Table 7.13 Process simulation summary for NGL recovery units.

Minimum information needed for model
Inlet composition and conditions
Market conditions or desired ethane recovery
Common solutions using model
Choose NGL recovery technology
Calculate when to switch ethane recovery modes
Size/rate all equipment
Troubleshoot process upsets
Run scenarios to improve/optimize turboexpander

Table 7.14 Process simulation summary for NGL fractionation.

Minimum information needed for model
Inlet composition and conditions
Market conditions
Desired product specifications
Common solutions using model
Optimize product composition and flow to maximize profit
Size/rate all equipment
Troubleshoot process upsets
Run scenarios to quickly change operations when market conditions change

rudimentary graphic user interfaces (GUIs), the use of process simulation was primarily in the hands of engineers with notable computer skills. Thus, a relatively small population of engineers used simulation. More recently, the opposite is true. Modern operating systems have extremely simple, yet powerful, GUIs capable of hosting data-intensive software. Almost every household and school in the western world have these systems at the fingertips of rising generations.

With Generation X and Millennials increasingly making up the engineering workforce in the oil and gas Industry, every chemical engineer must have some knowledge and proficiency in process simulation when entering the workforce. With each generation setting a higher computer literacy baseline, more integrated, robust, and powerful models are possible.

Microsoft® Excel®, Visual Basic®, and SQL Server® commonly exist in conjunction with process simulators. In fact, engineers regularly implement all three together with ProMax to automate simulation. A single model can now have thousands of "what if?" scenarios shown in simple, easy to read figures. A model can even use real-time data fed from the processing plant to update automatically every 10 minutes, hour, day, or any time interval needed. Process simulators have empowered engineers to utilize much more data. Engineers are in the process of discovering the power they have at their fingertips. Indeed, it is an exciting time to be a process engineer.

Acknowledgement

Dr. Kenneth R. Hall provided advice, support, and contribution to this chapter.

References

1 ProMax. "Version 5.0," Bryan Research & Engineering, 2018.
2 Alkasem A, Al Zaroun YI, Slagle JC, Berrouk AS, Satyadileep D. Optimization of an existing Habshan II amine sweetening unit. Abu Dhabi International Petroleum Exhibition and Conference, 9-12 November, Abu Dhabi, UAE.
3 Welsh J, Slagle JC, Cheaney S, Shiveler G. Hydrate formation in Chevron Mabee Unit for NGL Recovery and CO2 Purification for EOR. Gas Processors Association, San Antonio; 2017.

8

Optimization of Natural Gas Network Operation under Uncertainty

Emmanuel Ogbe[1,2], Ali Elkamel[1,2], Michael Fowler[1], and Ali Almansoori[2]

[1] *Department of Chemical Engineering, University of Waterloo, Waterloo, Ontario, Canada*
[2] *Khalifa University, The Petroleum Institute, Abu Dhabi, United Arab Emirates*

CHAPTER MENU

With global energy demands on the ascendency, natural gas is being considered as the twenty-first century fuel essential in bridging the gap between traditional fossil-fuel-based energy economy and a completely decarbonized energy economy. This is due to its smaller carbon footprint in terms of CO_2 emission equivalent inventories during production, lower SO_x, NO_x, and particulate emissions, and lower pollution metrics. The optimization of natural gas networks, starting from pipelines connecting sources of natural gas at the natural gas wells, through to preprocessing and processing, to storage and distribution of processed natural gas, at sometimes very remote demand location, is of paramount importance.

In this chapter, we aim at reviewing the modeling and optimization efforts in natural gas network systems, with attention to blending natural gas from various sources under steady-state conditions. The mixing of natural gas from diverse sources, with different quality specifications, leads to a mixed-integer nonlinear programming formulation for the problem, which is solved using a global optimization approach. The strategic and tactical decision issues, related to design and operation of natural gas network systems, are highlighted. The optimization objective is based on minimizing total capital and annualized operating costs. In the optimization model, constraints associated to natural gas flow through pipelines and those related to pressures at compressor stations guarantee the feasibility of the set of operational and design decisions obtained from the optimization search. Further, a two-stage stochastic programming approach to handle uncertainty associated with the design and operation of natural gas

Natural Gas Processing from Midstream to Downstream, First Edition.
Edited by Nimir O. Elbashir, Mahmoud M. El-Halwagi, Ioannis G. Economou, and Kenneth R. Hall.

networks is implemented. A case study of the Ontario natural gas pipeline system in southern Canada, is used to demonstrate the proposed approach. The results from the case study suggest implementing various design configurations and operational conditions for different uncertainty realizations.

8.1 Introduction

Global estimates of greenhouse gas (GHG) emissions were about 10 billion metric tons of carbon equivalent [1]. Renewables such as solar, wind, nuclear, hydro, geothermal, and biomass can all produce zero emissions but at some substantial cost. Natural gas, although a nonrenewable source of fuel, provides a cleaner alternative to fossil-based fuels such as coal and crude oil. Moreover, recent increased domestic production rates in the United States, suggest that the existing natural gas system will continue to provide relatively clean and domestic energy for the near future, especially with increased adoption of energy efficiency measures.

Traditional network flow models, for example the minimum cost network flow problem or the transportation problem, rely on considerations that the flow is uniform or homogenous over the network. In other words, composition of the components in the system do not change. However, in most industrial applications, such as blending of gasoline [2], multicomponent water networks [3, 4], etc., variations of composition of components in the system vary across the network. The variation of natural gas quality plays a significant role in the design and operation of natural gas production systems. This operation can be carried out either under steady state, e.g. [5, 6], or in transient state as in [7], where the authors considered the simulation and analysis of transient gas networks. Modeling uncertainty in process system engineering has received a lot of attention in recent years [8, 9]. More importantly, uncertainty has been considered in natural gas planning models, because of the effect of changing conditions such fluctuating natural gas demands, natural gas quality, product prices, etc. [10]. Different kinds of uncertainty modeling approaches exist in the literature such as stochastic programming [8, 11, 12], which could take the form of the classical two-stage stochastic [13] or multistage approaches [14], probabilistic or chanced constrained programming [15] and robust optimization [16]. In stochastic programming, a set of decisions are made without the event of uncertainty and then a set of recourse decisions are taken after the uncertainty is realized. Chanced constrained programming introduces uncertainty through penalizing violation of some of the constraints in the problem, whereas in robust optimization, a bounded uncertainty set is assumed, and a resulting problem consisting of bounded uncertainty regions is solved to feasibility.

In this chapter, we explore the modeling and optimization efforts in natural gas network systems, with attention to steady state blending of natural gas from various sources, containing different quality specifications. We focus attention on the quality of CO_2 across the nodes and pipelines conveying the natural gas. An optimization model for natural gas production system design and operation in Ontario is developed. The data for the study is obtained largely from literature, and the details of the results are presented.

In Section 8.2, we provide a review of some of the publications in natural gas system design and operation, including steady state gas network design and operation under uncertainty. Section 8.3 discusses the natural gas supply chain, from production through processing to storage and distribution to end users or markets. Section 8.4 presents the generic deterministic natural gas production model, with further discussions on the stochastic model. Section 8.5 discusses the computational study applied to a natural gas network design and operation in the province of Ontario in Canada. Section 8.6 presents the results and discussions while conclusions and recommendations are made in Section 8.7.

8.2 Literature Review

Global energy demand, currently put at around 400 quadrillion BTU of consumption, is expected to increase by 30% in 2040. Growth in oil demand is expected to slow down over the projection period, with an expected peak of 103 million barrels per day (MB/d) by 2040. The demand for coal, another fossil source of energy, is expected to essentially come to a halt because of the strict environmental sanctions, after the rapid expansion in recent years [17]. Natural gas, on the other hand, is expected to fill a total of 180 quadrillion BTU, a rise of 50% or about 30% of the overall total energy demand [18]. Natural gas is expected to satisfy this bulk of demand because it is a less carbon-intense fuel than either crude oil or coal [19]. Other renewables such as solar, wind, geothermal, and nuclear are expected to fill the remainder of the energy demand.

The literature covering natural gas network optimization is vast. These include, among others, operations research and management sciences, chemical engineering, energy economics, and petroleum engineering. These include research on planning the infrastructure for natural gas operation [5], different modes of operation of natural gas systems [5, 7, 21], modeling of quality of natural gas from different sources [10, 22, 23], and consideration of uncertainty [10, 21, 24]. Optimization under uncertainty in natural gas processing is a recent addition to the list because of the more recent developments in the field of stochastic programming [25]. Uncertainty in natural gas design and operation can come from a host of factors including natural gas demand, quality from different sources, natural gas prices, costs of operations, etc. [25]. These factors can be mitigated by making decisions that are inherently robust enough to withstand the effect of uncertainty, using modeling frameworks such two-stage or multistage stochastic programming [14, 25], chance-constrained programming [15], and robust optimization [26, 27]. Ultimately, the aim of optimization under uncertainty or stochastic programming is to make decisions here and now that can withstand the effect of uncertainty in the future.

Several surveys exist on design and operational optimization of gas pipeline transmission systems, the main objective being to minimize the capital and operational costs for the network. The problems of design and operation in natural gas production have been solved as separate problems because of the often large computational burden. The design problem, obtained with fixed operation and the operational operation with fixed design, can be solved by linear programming as in [28]. Problems with combined design and operation decision variables have been solved in the following papers [10, 23]. Another survey in the natural gas literature studies both the steady-state and dynamic (or transient) operation of natural gas infrastructure. For steady state operation, De Wolf and Smeers [5] developed a model to solve the problem of gas distribution at a minimum cost through a pipeline network under nonlinear flow-pressure relations constraints, material balance equations, and pressure bounds. Other steady state models are found in a paper by [6]. An optimal control perspective on the problem can be found in Marqués and Morari [30] and Osiadacz and Bell [31]. A model for planning investment in residential gas networks is presented in Davidson et al. [34]. Abbaspour et al. [35] develop an optimization model for line pack operation of compressor stations and solve it with a sequential unconstrained minimization technique. Kabirian and Hemmati [36] present a nonlinear programming (NLP) model for design of natural gas transmission networks. A nonconvex NLP model for the design of a natural gas distribution network is presented and solved with Floudas's global optimization approach in Wu et al. [37].

Haverly's pioneering work in blending or pooling problems [38, 39] led to model formulations that capture the different components in crude oil blending. Three main formulations of the pooling problem in the literature include (i) p-formulation, (ii) q-formulation, and (iii) p-q formulations. In the p-formulation [38, 39], the variables include flow and quality variables. The q-formulation introduced by Ben Tal et al. [40], replaces the quality variables of the p-formulation variables with flow proportions. This formulation dramatically reduces the number of nonconvex terms present in the model, leading to faster solution times. The p-q formulation was developed in Sherali and coworkers [41] as an extension of the p-formulation, where a set of linear constraints or cuts, otherwise known as RLT cuts (reformulation-linearization technique) are added to the q-formulation. The pq-formulation dominates (i.e., is tighter than) both the q- and p-formulations, and the p-formulation is in turn dominated by the q-formulation [42]. However, the pq-formulation cannot be applied to networks with multilayered of pools. For this type of networks, dry gas pipeline transportation and multiperiod inventory models are applied. Modeling the quality of natural gas across the natural gas network follows the generalized methods for modeling blending problems in the oil network system [10, 23]. The content of natural gas, such as the CO_2 or H_2S content, has to be controlled to meet customer specifications. The resulting problem leads to a nonconvex mixed-integer bilinear program that can be computationally expensive to solve [43]. A detailed review is given by Li et al. [10].

Handling uncertainty in natural gas network problems is another hot topic in the optimization literature [8]. A classical approach to include uncertainty is the two-stage stochastic programming with recourse, which is unlike the deterministic approaches where uncertainty is not considered. A review of deterministic approaches is given by van den Heever [44]. Sahinidis [9] summarized the various approaches for optimization under uncertainty, among which stochastic programming with recourse [8] naturally fits and has been adopted in the integrated design and operation of oil and natural gas production systems. In a typical stochastic programming formulation with recourse, the possible realizations of uncertainty are approximated by a limited number of representative realizations (scenarios), and the decisions are made sequentially over two or more time periods according to the outcome of uncertain variables over these periods. Jonsbraten [45] classified uncertainty in planning problems into two categories: (1) project exogenous uncertainty, where optimization decisions do not affect the resolution of uncertainty or change the scenario tree in the stochastic recourse formulation, for example, the prices of the products, the customer demands; (2) project endogenous uncertainty, where optimization decisions do affect the resolution of uncertainty and changes the scenario tree in the stochastic recourse formulation, for example, capacity and quality of reservoirs. Generally, project endogenous uncertainty is more difficult to handle (for problems with more than two stages), because it requires the incorporation of non-anticipativity constraints to link the decisions (in the second or a later stage) for different scenarios. A typical application of stochastic programming to a natural gas production system design can be found in [23, 46].

8.3 Natural Gas Supply Chains

Natural gas conventionally can be produced as "associated" gas from oil fields or as "non-associated" gas from gas-only fields. If gas is sweet, it requires no further processing. However, if gas is sour, or has a high H_2S and (or) CO_2 content, it needs additional processing to remove them. The processed gas is then sent to regional or national markets relatively close to the fields by short-distance transportation or by long-distance transportation for far-away markets. Transportation as liquefied natural gas (LNG) involves specialized infrastructure to liquefy, transport, and degasify natural gas. Regional networks are linked to local distribution networks that operate on much lower pressure and supply residential and small commercial customers. Industrial consumers (e.g., power plants, petrochemicals plants, fertilizer plants) may be directly supplied by national or regional networks. Natural gas is then sent though treating units or processes where the impurities in the form of CO_2, H_2S, water vapor, and

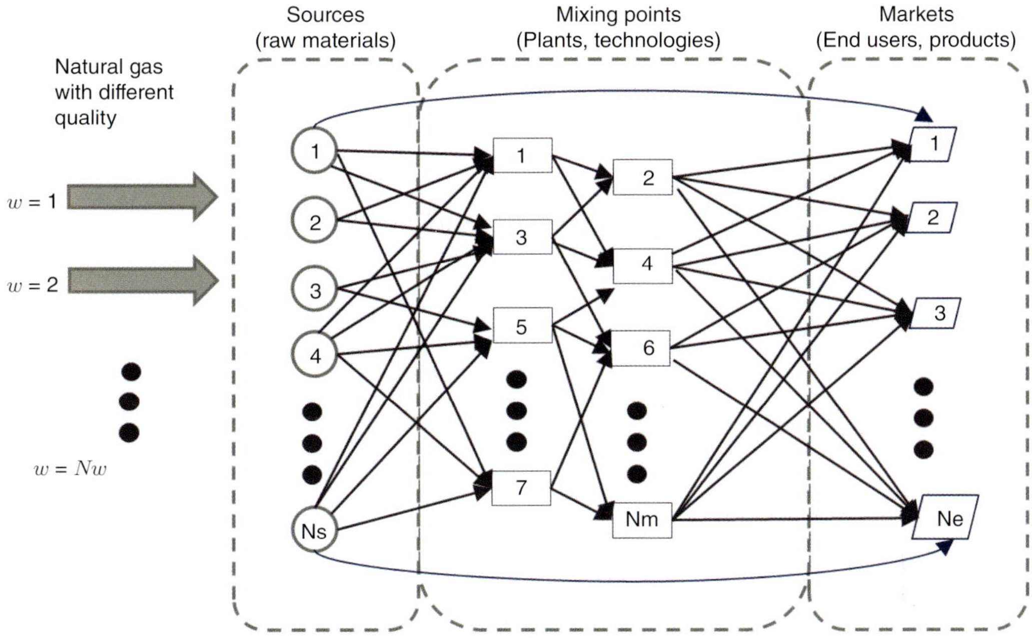

Figure 8.1 A generic natural gas network system.

other trace elements, are separated out of the "clean" methane gas. A recent review of natural gas processing is given in [47].

Figure 8.1 describes conceptually a generic natural gas supply chain. Natural gas, consisting of different components with index $w \in \{1, \ldots, Nw\} = \Omega$, comes in through source nodes $s \in \{1, \ldots, Ns\} = S$, denoting distribution stations. The gas is mixed or split into nodes $m \in \{1, \ldots, Nm\} = M$, containing mixers and splitters, and the gas is then sent to end user or market nodes $e \in \{1, \ldots, Ne\} = E$, containing products. Notice here that Ns, Nm and Ne represent the number of source, mixer and end user nodes respectively.

This generic arrangement shown in Figure 8.1 can be discussed according to the different phases of the natural gas supply chain. The first phase in the supply chain consists of the production and import of natural gas, through which the quantities necessary to meet domestic requirements are obtained [50]. The flow in each pipeline is a decision variable in the model and so is the quality out of each node as shown in Figure 8.2. Natural gas is transported in pipelines by using compressors to create a higher pressure at a pipeline's origin, for easy flow downstream of the pipeline. For a further discussion on transportation in pipelines refer to Li et al. [53]. In the next phase, natural gas containing different components is treated (sweetened) to meet the pipeline specifications and regulatory standards required for transportation [47, 54]. After this step, the sweetened gas is transferred to the dehydration and mercury-removal unit to prevent solidification of water during cooling process and reaction of mercury with aluminum heat exchangers, respectively, and then sent via transport pipelines for storage. The end user locations or markets are facilities for natural gas where the export pipelines converge. Natural gas is delivered at these locations based on specifications of minimum and maximum quality thresholds and energy content.

The natural gas production system in Ontario is operated by Union Gas Limited and Enbridge Gas Distribution Inc. [55]. Union Gas operates the Parkway West Project that involves the installation of a compressor along with 740 km natural gas pipeline and associated facilities in the town of Milton. A second project, known as the Brantford-Kirkwall/Parkway Project is

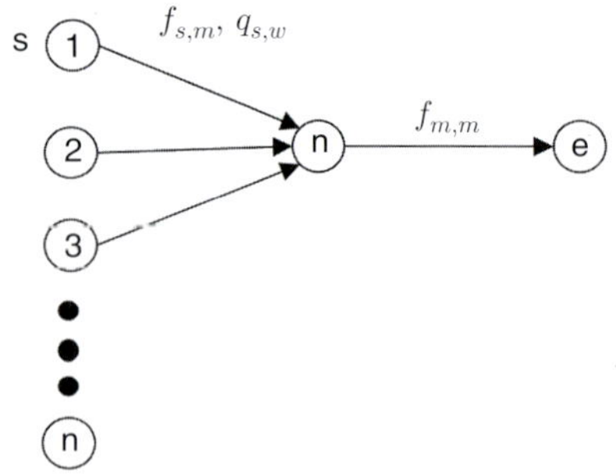

Figure 8.2 Modeling quality in Natural gas production.

also operated by Union Gas. It involves the construction of 13.9 km of a 48″ diameter pipeline with associated facilities between the city of Cambridge and the city of Hamilton, with an installation of a compressor at the Parkway West site. Enbridge's operates the GTA Project, which involves the construction of two segments of natural gas pipeline, segments A and B, and associated facilities, in and around the city of Toronto. Segment A is approximately 27 km long and is in the town of Milton, the city of Mississauga and the city of Toronto. Segment B is approximately 23 km long and is in the city of Vaughan, the city of Markham, the city of Toronto, and the town of Richmond Hill. Figure 8.3 attempts to capture the transportation of natural gas according to Union and Enbridge superstructure map available in the literature for natural gas transportation in Ontario.

8.4 Optimization Model

8.4.1 Mathematical Notation

The natural gas network problem can be represented by a directed graph. Let $D = (N, A)$ be a directed graph, where the node or vertices set $N = S \cup M \cup E$ consists of the disjoint subsets S of sources, M of mixers, and E of end users. The arc or edge set $A \subseteq (S \times M) \cup (M \times E)$ links sources with pools and pools with terminals. Let Ω be a finite set and refer to its elements as w, which denotes index of quality parameters. For each pool $m \in M$, the set of neighbor sources and the set of neighbor end users are denoted by $\Pi_m = \{s \in S : (s, m) \in A\}$ and $\Theta_m = \{e \in E : (m, \text{e}) \in A\}$, respectively. For each source $s \in S$, the set of neighbor pools is denoted by $\Pi_s = \{m \in M : (s, m) \in A\}$, while $\Theta_s = \{m \in M : (m, \text{e}) \in A\}$ denotes the set of neighbor pools of terminal $e \in E$. Finally, for each source $e \in E$, the set of neighbor mixers is denoted by $\Pi_e = \{m \in M : (e, m) \in A\}$, while $\Theta_e = \{s \in S : (s, \text{e}) \in A\}$is the set of neighbor sources connected to the end users.

8.4.2 Considering Gas Quality in Natural Gas Production Operation

Tracking natural gas quality along pipeline segments and nodal elements is therefore an essential aspect of the natural gas production system. However, modeling natural gas quality severely complicates the model because it introduces nonconvexity in the flow. For instance, where two flows meet, the gas quality out of the node to the downstream pipelines is denoted by $q_{i,\,w}$, where $w \in \Omega$, denotes the different components of natural gas. The gas quality depends on flow and quality from all the pipelines going into the node. The flow in each pipeline is a decision variable in the model, and so is the quality out of each node. We assume that the resulting blending

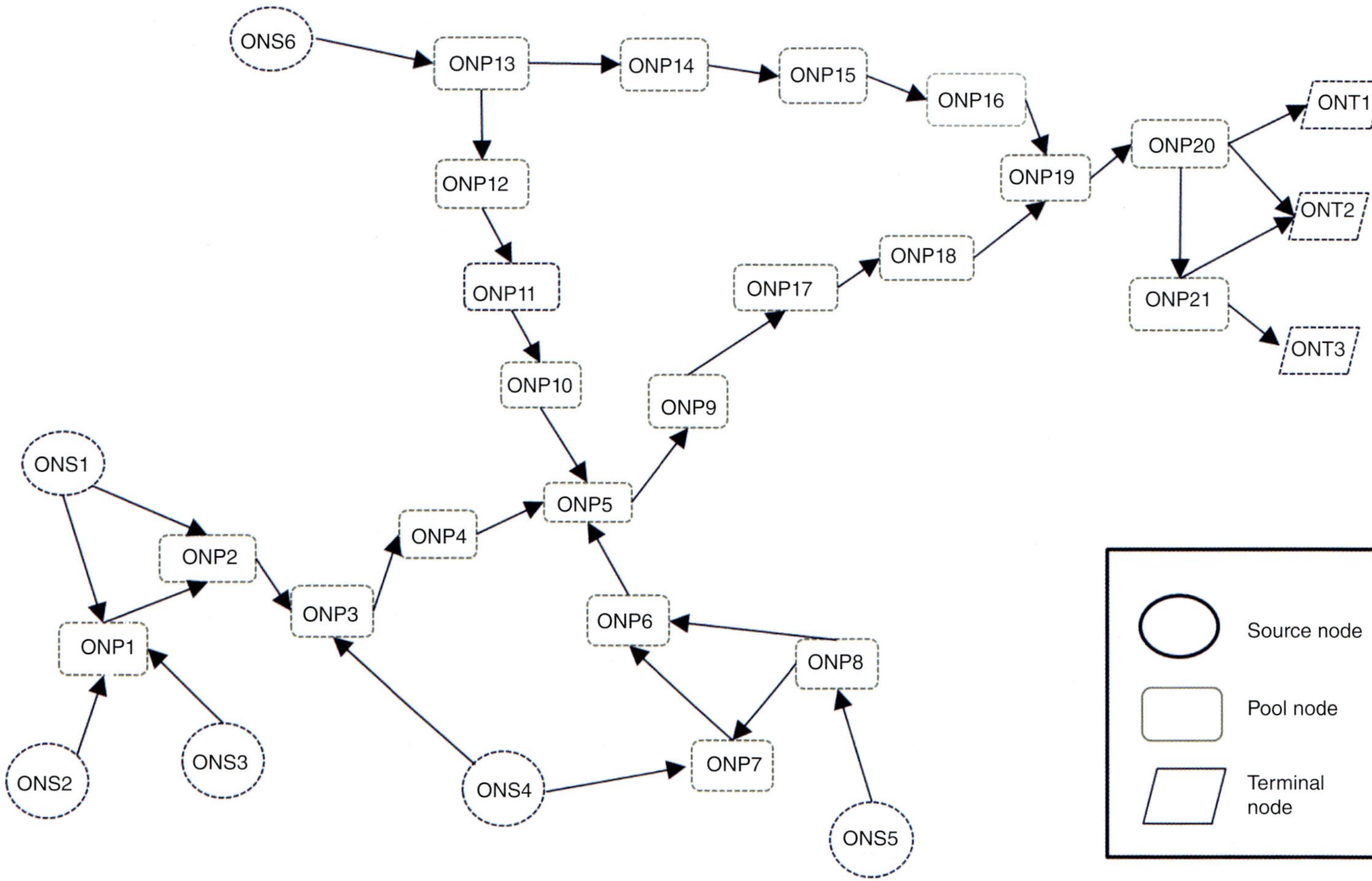

Figure 8.3 Natural gas superstructure in Ontario.

quality is common for all the downstream pipelines being connected to a node, and that it is decided by the convex combination of inflow qualities to the node.

$$q_{s,w} = \frac{\sum_{m \in M} q_{s,m} f_{s,m}}{\sum_{m \in M} f_{s,m}}, \forall s \in S, \forall w \in \Omega \tag{8.1}$$

The bilinear term $q_{s,n} f_{s,n}$, a product of two continuous variables, is responsible for the nonconvexity in the problem.

8.4.3 Model for the Natural Gas Network System

The natural gas network system model is derived based on the pooling problem by Li et al. [10] but with a different objective. The objective of the current problem is to determine the optimal design and operating variable decisions that would minimize the total capital and operating costs, as well as the total power consumption of the system. The design decisions on sources, mixers, end users, and connections between them are denoted by x_s, x_m, x_e, $x_{s,m}$, $x_{s,e}$, $x_{m,e}$, $x_{m,m}$, and can be continuous, integer, or mixed integer. If $x_s \in \{0, 1\}$, then the design decision is to determine whether source s is to be developed, and the related parameter Z_s^{UB} represents the fixed capacity of the source. If y_s is continuous and $x_s \in [0, 1]$, then it is a capacity design decision, specifically it represents the ratio of source s capacity to the maximum allowed capacity of the source (denoted by Z_s^{UB}). Operational decisions are given by $f_{s,m}$, $f_{s,e}$, $f_{m,m}$, $f_{m,e}$ and the maximum allowed capacities allowed for the flowrates are denoted by $F_{s,m}^{\text{UB}}$, $F_{s,e}^{\text{UB}}$, $F_{m,m}^{\text{UB}}$, $F_{m,e}^{\text{UB}}$. A comprehensive list of sets, variables, and parameters is given on Table 8.1.

All design and operational decision variables described in this chapter are nonnegative. The model is divided into three submodels, derived for the sources, mixers, and end users, respectively.

8.4.3.1 Model for the Sources

The following group of constraints represent the submodel for the sources. Eqs. (8.2)–(8.4) limit the flow from the source, with a known upper bound on the flows. Eqs. (8.5)–(8.6) constraints restricting flow from the source. Eqs. (8.6)–(8.7) limit the capacity of a pipeline by the capacity of the source it connects. If $x_s = 0$, then there cannot exist a pipeline connecting it, in other words, the capacity of a pipeline connecting it should be zero. Eq. (8.7) requires that the total capacity of all pipelines connecting to a source should be no less than the capacity of the source.

$$\sum_{m \in \Pi_s} f_{s,m} + \sum_{e \in \Theta_s} f_{s,e} \le x_s Z_s^{\text{UB}}, \tag{8.2}$$

$$f_{s,m} \le x_{s,m} F_{s,m}^{\text{UB}}, \tag{8.3}$$

$$f_{s,e} \le x_{s,e} F_{s,e}^{\text{UB}}, \tag{8.4}$$

$$x_{s,m} F_{s,m}^{\text{UB}} \le x_s Z_s^{\text{UB}}, \tag{8.5}$$

$$x_{s,e} F_{s,e}^{\text{UB}} \le x_s Z_s^{\text{UB}}, \tag{8.6}$$

$$x_s Z_s^{\text{UB}} \le \sum_{m \in \Pi_s} x_{s,m} F_{s,m}^{\text{UB}} + \sum_{e \in \Theta_s} x_{s,e} F_{s,e}^{\text{UB}}, \tag{8.7}$$

$$\forall s \in S, \ \forall m \in \Pi_s, \ \forall e \in \Theta_s$$

Table 8.1 List of symbols.

Symbol	Type	Description
S, M, E, Ω	Sets	Denotes source, mixer, end user and quality sets respectively
Θ_s	Set	Denotes connection between sources
Θ_m	Set	Denotes connection between different mixers
Π_m, Θ_m	Set	The set of neighbor sources and end users for every mixer m
Π_s, Θ_s	Set	The set of neighbor mixers and end users for every source s
Π_e, Θ_e	Set	The set of neighbor sources and mixers for every end user e
$\kappa_{s,m}$	Parameter	Parameter for pressure relation
α_m	Parameter	Parameter for power consumption
$c_i, c_j, c_k, c_{i,j}, c_{i,k}, c_{j,k}, c_{j,j}$	Parameters	Coefficient of investment cost
d_i, d_j, d_k	Parameters	Coefficient of operating cost
Z^{LB}, Z^{UB}	Parameters	Lower and upper bounds on capacity
D^{LB}, D^{UB}	Parameters	Maximum and minimum demands
$U_{s,w}$	Parameter	Quality parameter for natural gas
$W_{s,w}^{UB}, W_{s,w}^{LB}$	Parameters	Upper and lower bounds on the power consumption
$P_{s,w}^{ups,LB}, P_{s,w}^{ups,UB}, P_{s,w}^{downs,LB}, P_{s,w}^{downs,UB} P_{m,w}^{ups,LB}, P_{m,w}^{ups,UB}, P_{m,w}^{downs,LB}, P_{m,w}^{downs,UB} P_{e,w}^{ups,LB}, P_{e,w}^{ups,UB}, P_{e,w}^{downs,LB}, P_{e,w}^{downs,UB}$	Parameters	Upper and lower bounds on the upstream and downstream pressures
$x_s, x_m, x_e, x_{s,m}, x_{s,e}, x_{m,e}, x_{m,m}$	Binary variables	Design variables
$f_{i,j}^{SP}, f_{i,k}^{ST}, f_{j,k}^{PT}, f_{j,j}^{PP}, Q_{i,k}^{ST}, Q_{i,j}^{SP}$	Continuous variables	Operational flow variables
$p_s^{ups}, p_m^{ups}, p_e^{ups}, p_s^{downs}, p_m^{downs}, p_e^{downs}$	Continuous variables	Operational pressure variables
W_s	Continuous variables	Power consumption

8.4.3.2 Model for Mixing Stations

Eqs. (8.8)–(8.10) represents constraints for the material balance. Eq. (8.11) and (8.12) limits the flow to the mixing stations and end users. Eqs. (8.13)–(8.18) are developed to limit the total capacity of a particular mixing station m to be less than flows in all pipelines connecting to that mixing station.

$$f_{m,e,w} = s_{m,e}\left(\sum_{s\in\Pi_m} f_{s,m}U_{s,w} + \sum_{m\in\Theta_m} f_{m,m,w}\right), \tag{8.8}$$

$$f_{m,m,w} = s_{m,m}\left(\sum_{s\in\Pi_m} f_{s,m}U_{s,w} + \sum_{m\in\Theta_m} f_{m,m,w}\right), \tag{8.9}$$

$$\sum_{m\in\Pi_m} s_{m,m} + \sum_{e\in\Theta_m} s_{m,e} = 1, \quad s_{m,m}, s_{m,e} \geq 0, \tag{8.10}$$

$$x_{m,m}F^{\mathrm{LB}}_{m,m} \leq \sum_{w\in\Omega} f_{m,m,w} \leq x_{m,m}F^{\mathrm{UB}}_{m,m}, \tag{8.11}$$

$$x_{m,e}F^{\mathrm{LB}}_{m,e} \leq \sum_{w\in\Omega} f_{m,e,w} \leq x_{m,e}F^{\mathrm{UB}}_{m,e}, \tag{8.12}$$

$$x_m Z^{\mathrm{UB}}_m \geq x_{s,m}F^{\mathrm{UB}}_{s,m}, \tag{8.13}$$

$$x_m Z^{\mathrm{UB}}_m \geq x_{m,m}F^{\mathrm{UB}}_{m,m}, \tag{8.14}$$

$$x_m Z^{\mathrm{UB}}_m \geq x_{m,m}F^{\mathrm{UB}}_{m,m}, \tag{8.15}$$

$$x_m Z^{\mathrm{UB}}_m \geq x_{m,e}F^{\mathrm{UB}}_{m,e}, \tag{8.16}$$

$$x_m Z^{\mathrm{UB}}_m \leq \sum_{m\in\Pi_m} x_{m,m}F^{\mathrm{UB}}_{m,m} + \sum_{s\in\Theta_m} x_{s,m}F^{\mathrm{UB}}_{s,m}, \tag{8.17}$$

$$x_m Z^{\mathrm{UB}}_m \leq \sum_{m\in\Pi_m} x_{m,m}F^{\mathrm{UB}}_{m,m} + \sum_{e\in\Theta_m} x_{m,e}F^{\mathrm{UB}}_{m,e}, \tag{8.18}$$

$$\forall m \in \Pi_m, \;\; \forall e \in \Theta_m, \;\; \forall w \in \Omega.$$

8.4.3.3 Model for End Users

The following group of constraints in Eqs. (19)–(20) limits the flow through the end user locations. Furthermore, Eqs. (21)–(23) are developed to restrain flow to and from the end users based on the flow upper bounds.

$$\sum_{m\in\Pi_e}\sum_{w\in\Omega} f_{m,e,w} + \sum_{s\in\Theta_e} f_{s,e} \leq x_e D^{\mathrm{UB}}_e, \tag{8.19}$$

$$\sum_{m\in\Pi_e} f_{m,e,w} + \sum_{s\in\Theta_e} f_{s,e}U_{s,w} \leq \left(\sum_{m\in\Pi_e}\sum_{w\in\Omega} f_{m,e,w} + \sum_{s\in\Theta_e} f_{s,e}\right) V^{\mathrm{UB}}_{e,w} \tag{8.20}$$

$$x_e D^{\mathrm{UB}}_e \geq x_{s,e}F^{\mathrm{UB}}_{s,e}, \tag{8.21}$$

$$x_e D^{\mathrm{UB}}_e \geq x_{e,k}F^{\mathrm{UB}}_{e,k}, \tag{8.22}$$

$$x_e D^{\mathrm{UB}}_e \leq \sum_{s\in\Pi_e} x_{s,e}F^{\mathrm{UB}}_{s,e} + \sum_{e\in\Theta_e} x_{m,e}F^{\mathrm{UB}}_{m,e}, \tag{8.23}$$

$$\forall e \in E, \;\; \forall w \in \Omega.$$

8.4.3.4 Pressure Model

Let p_s^{ups} and p_s^{downs} represent the upstream and downstream pressure of a pipeline respectively. To consider the pressure model in the natural gas network problem, a pipeline performance and a compression performance set of equations are derived. These equations are based on Selot [23] and are described as follows.

8.4.3.5 Pipeline Performance Model

The flow through the natural gas pipeline is assumed to follow the ideal gas relation. Under this assumption and the fact that that the operation is under steady state, Eq. (8.24) and Eq. (8.25) apply.

$$x_{s,m}[(p_s^{downs})^2 - (p_s^{ups})^2] = \kappa_{s,m}Q_{s,m}, \forall(s,m) \in \Pi_s, \tag{8.24}$$

$$x_{s,m}((p_s^{downs})^2 - (p_s^{ups})^2) = 0, \forall(s,m) \in \Pi_s. \tag{8.25}$$

$\kappa_{s,m}$ in Eq. (24) is the constant for the gas flow equation, and $Q_{s,m}$ is the mass flowrate.

8.4.3.6 Compression Performance model

Eq. (8.26) estimates the power consumption W_s using a gas compressor. The power is the sum total of the power output of all the compressors that are available in the facility. Eq. (8.27) ensures that the upstream pressure is less than the downstream pressure at all times, and Eq. (8.28) imposes bounds the power consumption for a particular compressor

$$W_s - \alpha_m \sum_{s\in S}\sum_{m\in M} f_{s,m,w}\left[\left(\frac{p_m^{downs}}{p_m^{ups}}\right)^v - 1\right] = 0, \forall s \in S, \forall w \in \Omega \tag{8.26}$$

$$p_s^{ups} \le p_s^{downs}, \forall s \in S, \tag{8.27}$$

$$x_s W_s^{LB} \le W_s \le x_s W_s^{UB}, \forall s \in S. \tag{8.28}$$

Other sets of pressure constraints in the natural gas transportation problem include bounds imposed on the inlet and outlet pressures based on the design capacity of the compressors as specified by the manufacturers.

$$x_s P_s^{ups,LB} \le p_s^{ups} \le x_s P_s^{ups,UB}, \forall s \in S, \tag{8.29}$$

$$x_s P_s^{downs,LB} \le p_s^{downs} \le x_s P_s^{downs,UB}, \forall s \in S. \tag{8.30}$$

Eq. (8.29) and Eq. (8.30) impose lower and upper bounds on the upstream and downstream pressures respectively. The parameters $P_s^{ups,LB}$ and $P_s^{ups,UB}$ in Eq. (8.29) and Eq. (8.30) represent the lower and upper bounds of the inlet pressure while $P_s^{downs,LB}$ and $P_s^{downs,UB}$ represent the lower and upper outlet pressure.

The natural gas design and operation model can be stated as:

$$(\text{P})\begin{cases}\text{minimize objective}\\ \text{Flow and pressure model in Eq. 2-34,}\\ x_s, x_m, x_e, x_{s,m}, x_{s,e}, x_{m,m}, x_{m,e} \in \{0,1\},\\ \text{All flow rates and pressures are nonnegative.}\end{cases}$$

Where objective is given by,

$$objective = Cap\cos t + \left(\sum_{t=\{1,\dots T\}} \frac{1}{(1+r)^t}\right)(Op\cos t - Income), \tag{8.31}$$

$$\begin{aligned}Cap\cos t = &\sum_{s\in S} c_s x_s + \sum_{m\in M} c_m y_m + \sum_{e\in E} c_e y_e\\ &+ \sum_{s\in S}\sum_{m\in M} c_{s,m}x_{s,m} + \sum_{s\in S}\sum_{e\in E} c_{s,e}x_{s,e} + \sum_{m\in M}\sum_{e\in E} c_{m,e}x_{m,e}\end{aligned} \tag{8.32}$$

And,

$$Op\cos t = \sum_{s\in S} d_s\left(\sum_{m\in\Theta_s} f_{e,m} + \sum_{e\in\Theta_s} f_{s,e}\right) + \sum_{s\in S} d_s^{elec} W_s \tag{8.33}$$

$$Income = \sum_{e\in E} d_e\left(\sum_{s\in\Pi_e}\sum_{w\in\Omega} f_{m,e,w} + \sum_{s\in\Pi_e} f_{s,e}\right), \tag{8.34}$$

Cap cos *t* is the capital cost of the project; and coefficients c_s, c_m, and c_e are the investment cost of building the natural gas infrastructure; and $c_{s,m}$, $c_{s,e}$, and $c_{m,e}$ are the investment costs of building pipelines connecting the infrastructure units. *Op* cos *t* is the cost of operating the project including a term for the power consumption W_s; d_s is the annual operating cost; and d_s^{elec} is the electricity cost. *Income* is the income generated from the sale of natural gas, and d_e is the natural gas price.

Problem (P) is the deterministic formulation for the natural gas system design and operation problem. Extension to a two-stage stochastic formulation is straightforward according to [8, 10] and the explicit representation is presented in the appendix.

8.5 Computation Study

8.5.1 Implementation

The optimization model was implemented using the General Algebraic Modeling Systems (GAMS) optimization platform, on a computer with a 2.4 GHz processor and 6 GB RAM. The model developed is a mixed-integer nonlinear programming problem and is solved using the global solver called the Algorithm for continuous/integer global optimization of nonlinear equations or ANTIGONE [59], with CONOPT [60] as the nonlinear local sub-solver and IBM's CPLEX as its linear programming sub-solver [61].

8.5.2 Case Study and Description

This problem is the natural gas network system located in Ontario, a province in Canada. Figure 8.3 shows the superstructure of the Ontario Natural Gas Pipeline System (ONGPS) under consideration, where the circles represent source locations, rectangles represent the mixing stations (pools) at which gas flows from different gas fields are mixed and split, and parallelograms represent the domestic gas requirement (product terminals). Symbols are all in dashed lines signifying that the entire superstructure in the problem description needs to be designed. The six potential source locations for the natural gas, denoted by ONS1, ONS2, ONS3, ONS4, ONS5, and ONS6, have different qualities, specified by the gas quality regulations applicable at that node. The 21 potential mixing stations are denoted by ONP1, ONP2, ONP3, …, ONP21, and three terminals location symbolized by ONT1, ONT2, and ONT3. The goal of optimization is to maximize expected net present value (or minimize expected cost) while satisfying specifications for gas qualities at the end user or market locations while considering the influence of uncertainty.

For the stochastic blending problem, we consider one uncertain parameter, i.e., the quality of CO_2 at source ONS1. The distribution of the uncertain quality is assumed to follow independent normal distributions with mean of 0.88%, and standard deviation of 0.51%. In the case study, we consider different realization of uncertain parameter sampled for 3, 5, 7, 9, 11 scenario values, via the same sampling rule described in [10]. The deterministic model contains a total of 92 binary variables that determines where or not sources, mixers, or end users are to be developed, and the 364 continuous variables represent the operational decisions at the different units to be developed.

8.6 Results and Discussion

The optimization results for the Ontario natural gas system design and operation case study are presented in Tables 8.2 and 8.3, and the optimized superstructure configuration is shown in Figure 8.4.

Table 8.2 shows the different pipelines developed and the possible connections across the network, the quality specification for CO_2 in mol % for each pipeline connection and the total gas flowrate (both natural gas and CO_2) in Mmol/day, in the deterministic optimization. As seen from Table 8.1, the quality or concentration of CO_2 starts with some large values of up 0.88 mol % CO_2 at the source nodes, and gradually decreases across the network to about 0.4 mol% at the end user in nodes ONT1, which is of course lower than the 2.8 mol% threshold imposed at the end user node section in the model. Furthermore, we see a maximum flowrate of 750 Mmol/day across the pipelines ONP8–ONP7. This is because of the relatively clean natural gas with less CO_2 content coming from the south of Ontario, or the United States. On the other hand, a minimum flowrate of 10.8 Mmol/day is obtained across the pipelines ONP9–ONP17 and ONP10–ONP5. Here, small-diameter pipelines can be used to achieve this relatively low pipeline flowrate.

The detailed deterministic and stochastic solutions are presented in Table 8.2. In Table 8.2, the result summary including optimization objective values, problem size in terms of the number of binary and continuous variables involved in the model and computational times recorded are shown. As seen from the results, the optimum objective of $7730 million is obtained for the deterministic solution strategy. This value is equivalent to the net present value with a discount rate of 12%. The net present value is obtained over a period, specified as $T = 25$ years in this study. When uncertainty in CO_2 quality at node ONS1 is included in the model, this net present value reduces. Also, note that the net present value reduces steadily as the number of scenarios increases. This behavior is typical of stochastic programming, particularly the two-stage stochastic programming approach, which typically provides a conservative solution in general. The effect of uncertainty affects the pipeline design decisions obtained for the deterministic solutions, which are the first-stage decisions modeled in the two-stage stochastic programming formulation, and require some revision because the solution is inadequate in the face of uncertainty. This adjustment is shown in Figure 8.4 and is discussed later. As can be seen in Table 8.2, the 3-scenario case gives identical optimal objective value to the deterministic case because of the small number of scenarios considered. As the number of scenarios is increased, we obtain a very conservative solution of $7725 million as net present value for the 11-scenario case at the expense of significant computational time. A difference of $5 million, equivalent to the difference between the stochastic and deterministic optimal objective values, roughly represents the price to pay for not considering uncertainty. Furthermore, Table 8.2 presents the computational performance of solving the deterministic equivalent of a two-stage stochastic program. We can see from the table that considering increasing number of scenarios increases the size of the problem, and consequently increases the computation time required to solve the problem. The deterministic solution required 8 seconds of CPU time to solve the problem, while the largest problem with 11 scenarios require 20 minutes to solve the problem. This drastic increase in computational resource typifies the behavior of branch-and-bound based solvers such as ANTIGONE for large-scale MINLP.

Table 8.2 Pipeline Results.

Pipeline Nodes	CO_2 Quality (mol %)	Total gas flowrates (Mmol/day)
ONS1-ONP12	0.72	210.8
ONS5-ONP8	0.41	329.2
ONS6-ONP13	0.68	145.1
ONP5-ONP9	0.42	340
ONP6-ONP5	0.41	329.2
ONP6-ONP7	0.41	10
ONP7-ONP6	0.41	85.12
ONP7-ONP8	0.41	760
ONP8-ONP6	0.49	339.2
ONP8-ONP7	0.41	750
ONP9-ONP17	0.42	340
ONP10-ONP5	0.72	10.79
ONP11-ONP10	0.72	10.79
ONP11-ONP12	0.72	300
ONP11-ONP13	0.72	200
ONP12-ONP11	0.719	510.8
ONP13-ONP12	–	–
ONP13-ONP14	0.7	345.1
ONP14-ONP15	0.7	345.1
ONP15-ONP16	0.7	345.1
ONP16-ONP19	0.7	345.1
ONP17-ONP18	0.4	340
ONP18-ONP19	0.42	324
ONP19-ONP20	0.566	669.2
ONP20-ONP21	0.52	650
ONP20-ONT1	0.56	35.1
ONP21-ONT1	1.18	650

Table 8.3 Computational results.

Scenarios	1	3	5	7	9	11
Optimal Objective ($\$10^6$)	−7730.32	−7730.13	−7667.82	−7728.5	−7725.94	−7724.56
No of binary variables	92	104	114	122	134	144
No of continuous variables	364	1319	2197	2536	3260	4128
Total time (s)	8	52	75	45	82	1234

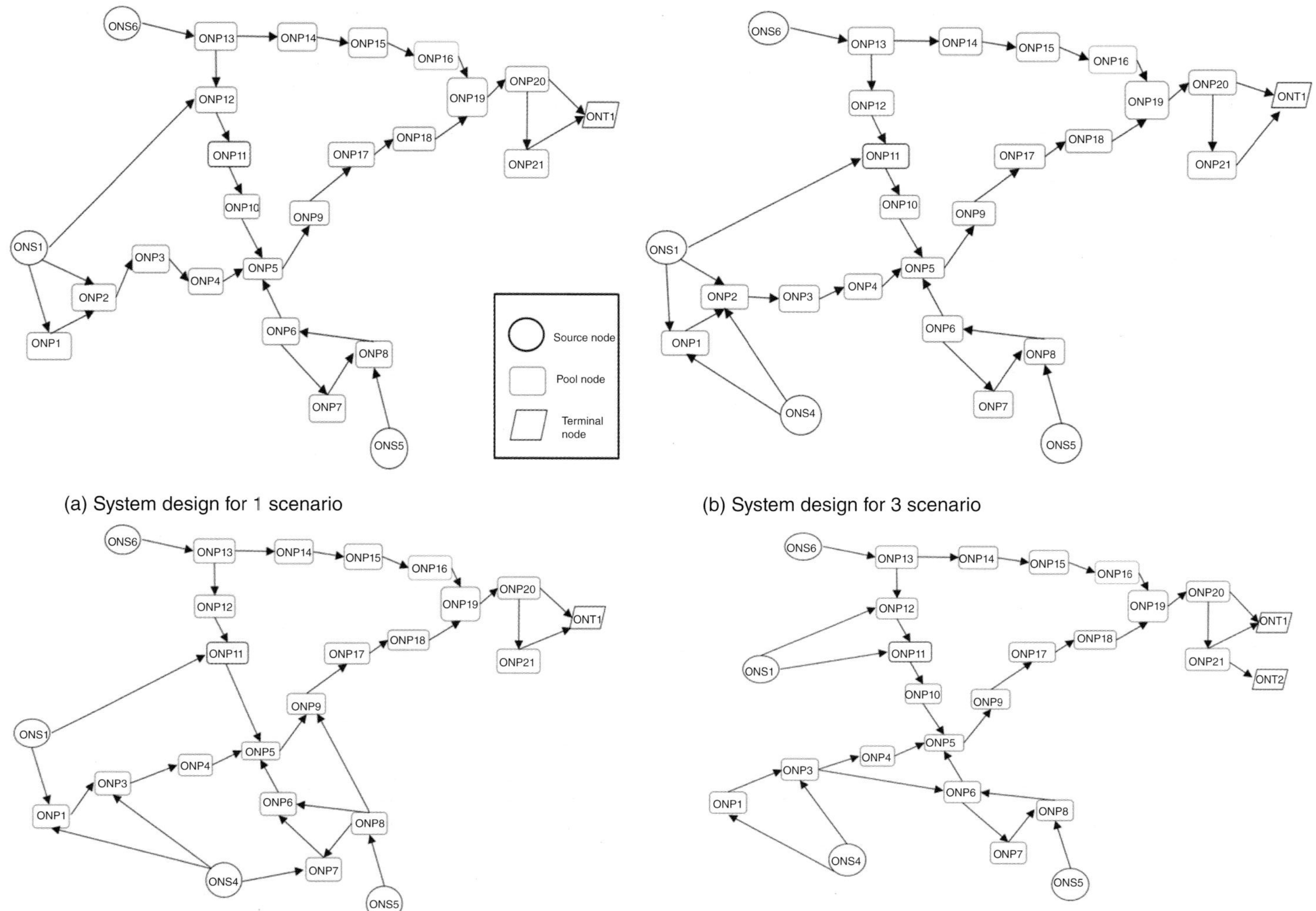

(a) System design for 1 scenario

(b) System design for 3 scenario

(c) System design for 9 scenario

(d) System design for 11 scenario

Figure 8.4 System design for different realization of uncertainty.

Figure 8.4 with subplots (a), (b), (c), and (d) presents the detailed optimized superstructure framework, including nodes and connections between the nodes or pipelines, that provide maximum benefits, for selected scenarios including 1, 3, 9, and 11 scenarios. The choice of design decisions returned by the global optimizer ANTIGONE obviously varies as the number of scenarios are increased. For example, the 1-scenario or deterministic solution in Figure 8.4(a) developed sources ONS1, ONS5, and ONS6; all the mixers ONP1, ONP2, …,ONP21 across the network; and only ONT1 end user. In the 3-scenario case in Figure 8.4(b), only sources ONS2, ONS3 are not developed, while all the other sources, mixers, and ONT1 end user are developed. This is because this design option gives the maximum profitability while maintaining the quality specification at that realization of uncertainty. The 9-scenario case in Figure 8.4(c) is missing the ONS2 and ONS3 source nodes as well as mixing nodes ONP2 and ONP10 and the end user ONT1. In the last case presented in Figure 8.4(d), the 11-scenario case, source node ONS2 and ONS3 are not developed; all mixer nodes and both end users ONT1 and ONT2 are included in the design. The inclusion of end user node ONT2 in the final design suggests that the largest scenario case encourages more end users with variable demands to be developed by the optimization. The unique designs, in terms of source nodes, mixers, and end users, and the possible pipeline connections between them that were developed for the different scenarios, generated different flowrates and pressure conditions for operation of the natural gas system. This is consistent with studies in the literature, that show that the higher the number of scenarios present in the model, the more the system can withstand the impact of uncertainty.

8.7 Conclusions and Recommendations

A review of a natural gas supply chain system is presented in this chapter, with emphasis on the network design and operational under steady state conditions with the consideration of uncertainty. A natural gas network model is developed and applied to the Ontario natural gas network system, located in the southern part of Canada. The global optimization solver, known as the Algorithm for continuous/integer global optimization of nonlinear equations, or ANTIGONE, to solve the mixed-integer nonlinear optimization problem is developed, and an ε-optimal solution to the problem is obtained in finite number of iterations.

The design and operation planning of natural gas is developed, consisting of both deterministic and stochastic planning models. The optimal objective changed from a net present value of \$7730 million for a deterministic formulation to only \$7725 million for a stochastic formulation with 11 scenarios, which translates to a cost of \$5 million if uncertainty is not considered. More so, the computational results vary from about 6 seconds for the deterministic solution to an hour for a stochastic solution. Furthermore, the results show changing design and operational decisions as the number of scenarios increase. The superstructure framework varied as the scenarios change, with the 11-scenario case, which is also the largest scenario case considered, included more than one end user as the optimal superstructure configuration.

This chapter only considers a monolithic approach of solving a large-scale optimization problem under uncertainty. As can be seen from the results on Table 8.3, commercial global optimization approaches such as ANTIGONE suffer from the curse of dimensionality, i.e., as the size of scenarios increases, the computational time required to solve the problem rapidly increases. This is because solvers such as ANTIGONE solve the problem by performing a branch-band-bound search on the variables in the problem, which is known to dramatically become computational slower when the size of the problem is large. This can be a problem for real-time optimization applications in process systems engineering. Hence, application of decomposition algorithms, such as the duality-based decomposition approaches by authors [62–66] to the Ontario natural production system can substantially improve the solution time.

References

1 Boden, T., Marland, G. & Andres, R. J. Global, Regional, and National Fossil-Fuel CO_2 Emissions. (2005). doi:doi 10.3334/CDIAC/00001_V2017

2 Kolodziej, S. P., Grossmann, I. E., Furman, K. C. & Sawaya, N. W. A discretization-based approach for the optimization of the multiperiod blend scheduling problem. Comput. Chem. Eng. 53, 122–142 (2013).

3 Bagajewicz, M., Rodera, H. & Savelski, M. Energy efficient water utilization systems in process plants. Comput. Chem. Eng. 26, 59–79 (2002).

4 Bagajewicz, M. A review of recent design procedures for water networks in refineries and process plants. Comput. Chem. Eng. 24, 2093–2113 (2000).

5 Wolf, D. E. & Smeers, Y. Optimal Dimensioning of Pipe Networks with Application to Gas Transmission Networks. 44, 596–608 (1996).

6 Rios-Mercado, R. Z. & Borraz-Sanchez, C. Optimization problems in natural gas transportation systems: A state-of-the-art review. Appl. Energy 147, 536–555 (2015).

7 Osiadacz, A. Simulation of transient gas flows in networks. Int. J. Numer. Methods Fluids 4, 13–24 (1984).

8 Birge, J. R. & Louveaux, F. Introduction to Stochastic Programming. (Springer, 2010).

9 Sahinidis, N. V. Optimization under uncertainty: state-of-the-art and opportunities. Comput. Chem. Eng. 28, 971–983 (2004).

10 Li, X., Armagan, E., Tomasgard, A. & Barton, P. I. Stochastic pooling problem for natural gas production network design and operation under uncertainty. AIChE J. 57, 2120–2135 (2011).

11 Van Slyke, R. M. & Wets, R. L-Shaped Linear Programs with Applications to Optimal Control and Stochastic Programming. SIAM J. Appl. Math. 17, 638–663 (1969).

12 R. T. Rockafellar, R. J.-B. W. Scenarios and Policy Aggregation in Optimization under Uncertainty. Math. Oper. Res. 16, 119–147 (1991).

13 Birge, J. R. & Louveaux, F. V. A multicut algorithm for two-stage stochastic linear programs. Eur. J. Oper. Res. 34, 384–392 (1988).

14 Birge, J. R. Decomposition and Partitioning Methods for Multistage Stochastic Linear Programs. Oper. Res. 33, 989–1007 (1985).

15 Charnes, A. & Cooper, W. W. Chance-Constrained Programming. Manage. Sci. 6, 73–79 (1959).

16 Ben-Tal, A., Ghaoui, L. El & Nemirovski, A. Robust Optimization. (Princeton University Press, 2009).

17 International Energy Agency. World Energy Outlook 2016. International Energy Agency (2016). doi: http://www.iea.org/publications/freepublications/publication/WEB_WorldEnergyOutlook2015ExecutiveSummaryEnglishFinal.pdf

18 EIA. EIA projects world energy consumption will increase 56% by 2040. (2013).

19 EPA. Emission factors for greenhouse gas inventories. US EPA Cent. Corp. Clim. Leadersh. 1–5 (2014).

20 Khan, M. S., Karimi, I. A. & Wood, D. A. Retrospective and future perspective of natural gas liquefaction and optimization technologies contributing to efficient LNG supply: A review. J. Nat. Gas Sci. Eng. 45, 165–188 (2017).

21 Zavala, V. Stochastic optimal control model for natural gas network operations. 1–22 (2014).

22 Haverly, C. A. Studies of the behaviour of recursion for the pooling problem. ACM SIGMAP Bull. 25, 29–32 (1978).

23 Selot, A., Kuok, L. K., Robinson, M., Mason, T. L. & Barton, P. I. A short-term operational planning model for natural gas production systems. AIChE J. 54, 495–515 (2008).

24 Sahinidis, N. V. Optimization under uncertainty: state-of-the-art and opportunities. Comput. Chem. Eng. 28, 971–983 (2004).

25 Birge, J. R. & Louveaux, F. Introduction to Stochastic Programming. Introduction to Stochastic Programming (Springer, 2011). doi:10.1007/978-1-4614-0237-4

26 Verderame, P. M. & Floudas, C. A. Multisite Planning under Demand and Transportation Time Uncertainty: Robust Optimization and Conditional Value-at-Risk Frameworks. Ind. Eng. Chem. Res. 50, 4959–4982 (2010).

27 Hajimiragha, A. H., Canizares, C. A., Fowler, M. W., Moazeni, S. & Elkamel, A. A robust optimization approach for planning the transition to plug-in hybrid electric vehicles. IEEE Trans. Power Syst. 26, 2264–2274 (2011).

28 Lee, A. S. & Aronofsky, J. S. A Linear Programming Model for Scheduling Crude Oil Production. J. Pet. Technol. 10, 51–54 (1958).

29 Peretti, A. & Toth, P. Optimization of a pipe-line for the natural gas transportation. Eur. J. Oper. Res. 11, 247–254 (1982).

30 Marqués, D. & Morari, M. On-line optimization of gas pipeline networks. Automatica 24, 455–469 (1988).

31 Osiadacz, A. J. & Bell, D. J. A simplified algorithm for optimization of large-sacle gas networks. Optim. Control Appl. Methods 7, 95–104 (1986).

32 Chaczykowski, M. Transient flow in natural gas pipeline - The effect of pipeline thermal model. Appl. Math. Model. 34, 1051–1067 (2010).

33 Zlotnik, A., Chertkov, M. & Backhaus, S. Optimal Control of Transient Flow in Natural Gas Networks. 4563–4570 (2015).

34 Davidson, R. A., Lembo Jr, A. J., Ma, J., Nozick, L. K. & O'Rourke, T. D. Optimization of investments in natural gas distribution networks. J. Energy Eng. 132, 52–60 (2006).

35 Abbaspour, M., Krishnaswami, P. & Chapman, K. S. Transient Optimization in Natural Gas Compressor Stations for Linepack Operation. J. Energy Resour. Technol. 129, 314 (2007).

36 Kabirian, A. & Hemmati, M. R. A strategic planning model for natural gas transmission networks. Energy Policy 35, 5656–5670 (2007).

37 Wu, Y., Lai, K. K. & Liu, Y. Deterministic global optimization approach to steady-state distribution gas pipeline networks. Optim. Eng. 8, 259–275 (2007).

38 Haverly, C. A. Studies of the behavior of recursion for the pooling problem. ACM SIGMAP Bull. 25, 19–28 (1978).

39 Haverly, C. A. Behaviour of recursion model - more studies. ACM SIGMAP Bull. 26, 22–28 (1979).

40 Ben-Tal, A., Eiger, G. & Gershovitz, V. Global minimization by reducing the duality gap. Math. Program. 63, 193–212 (1994).

41 Sherali, Hanif D., Adams, W. P. A Reformulation-Linearization Technique for Solving Discrete and Continuous Nonconvex Problems. (1999).

42 Sherali, H. D. & Alameddine, A. A new reformulation-linearization technique for bilinear programming problems. J. Glob. Optim. 2, 379–410 (1992).

43 Audet, C., Brimberg, J., Hansen, P., Digabel, S. L. & Mladenović, N. Pooling problem: alternative formulations and solution methods. Manage. Sci. 50, 761–776 (2004).

44 van den Heever, S. A. & Grossmann, I. E. A {L}agrangean decomposition heuristic for the design and planning of offshore hydrocarbon field infrastructure with complex economic objectives. Ind. Eng. Chem. Res. 40, 2857–2875 (2001).

45 Jonsbraten, T. W. Optimization Models for Petroleum Field Exploitation. (Stavanger College, 1998).

46 Goel, V., Grossmann, I. E., El-Bakry, A. S. & Mulkay, E. L. A novel branch and bound algorithm for optimal development of gas fields under uncertainty in reserves. Comput. Chem. Eng. 30, 1076–1092 (2006).

47 Mazyan, W., Ahmadi, A., Ahmed, H. & Hoorfar, M. Market and technology assessment of natural gas processing: A review. J. Nat. Gas Sci. Eng. 30, 487–514 (2016).

48 Hamedi, M., Zanjirani Farahani, R., Husseini, M. M. & Esmaeilian, G. R. A distribution planning model for natural gas supply chain: A case study. Energy Policy 37, 799–812 (2009).

49 Selot, A. Short-term Supply Chain Management in Upstream Natural Gas Systems. (Massachusetts Institute of Technology, 2009).

50 The Natural Gas Supply Chains. Available at: http://eng.gruppohera.it/group/business_activities/business_energy/gas/natural_gas_supply_chain/. (Accessed: 28th July 2017)

51 Katz, D. L. & Lee, R. Natural Gas Engineering: Production and Storage. (1999).

52 Ikoku, C. Natural Gas Engineering: A Systems Approach. (1980).

53 Speight, J. G. Natural Gas: A Basic Handbook. (Gulf Publishing Company, 2007).

54 Mokhatab, S. & Poe, W. A. Handbook of Natural Gas. Journal of Chemical Information and Modeling (2012). doi:10.1017/CBO9781107415324.004

55 Limited, U. G. A. S. & Limited, U. G. A. S. Union gas limited. (2014).

56 Gupte, A., Ahmed, S., Dey, S. S. & Cheon, M. S. Relaxations and discretizations for the pooling problem. J. Glob. Optim. 67, 631–669 (2017).

57 Audet, C., Hansen, P., Jaumard, B. & Savard, G. A branch and cut algorithm for nonconvex quadratically constrained quadratic programming. Math. Program. 87, 131–152 (2000).

58 Kolodziej, S. P., Grossmann, I. E., Furman, K. C. & Sawaya, N. W. A Novel Global Optimization Approach to the Multiperiod Blending Problem. Comput. Aided Chem. Eng. 31, (2012).

59 Misener, R. & Floudas, C. ~A. ANTIGONE: Algorithms for coNTinuous / Integer Global Optimization of Nonlinear Equations. J. Glob. Optim. 59, 503–526 (2014).

60 Drud, A. S. {CONOPT} – A Large-Scale {GRG} Code. {ORSA} J. Comput. 6, 207–216 (1994).

61 IBM. {IBM ILOG CPLEX: High-performance mathematical programming engine}.

62 Li, X., Chen, Y. & Barton, P. I. Nonconvex generalized benders decomposition with piecewise convex relaxations for global optimization of integrated process design and operation problems. Ind. Eng. Chem. Res. 51, 7287–7299 (2012).

63 Oliveira, F., Gupta, V., Hamacher, S. & Grossmann, I. E. A Lagrangean decomposition approach for oil supply chain investment planning under uncertainty with risk considerations. Comput. Chem. Eng. 50, 184–195 (2013).

64 Adhya, N., Tawarmalani, M. & Sahinidis, N. V. A Lagrangian Approach to the Pooling Problem. Ind. Eng. Chem. Res. 38, 1956–1972 (1999).

65 Li, X. Parallel nonconvex generalized Benders decomposition for natural gas production network planning under uncertainty. Comput. Chem. Eng. 55, 97–108 (2013).

66 Geoffrion A. Bride, R. M. Lagrangean Relaxation Applied to Capacitated Facility Location Problems. A I I E Trans. 10, 40–47 (1978).

Appendix

The two-stage stochastic formulation for Problem (P) is shown below. The design decisions on sources, mixers, end users, and connections between them, denoted by x_s, x_m, x_e, $x_{s,m}$, $x_{s,e}$, $x_{m,e}$, $x_{m,m}$ are the first-stage decisions. Operational decisions given by $f_{s,m}$, $f_{s,e}$, $f_{m,m}$, $f_{m,e}$, $p^{downs}_{s,w,h}$, $^{ups}_{s,w,h}$

are the second-stage decisions, made after the realization of uncertainty. Each parameter is assumed to follow a normal distribution with a given mean μ and a standard deviation σ and a probability of occurrence Pr_h , where $h \in \{1, \ldots, H\}$ is the realization of uncertainty and H is the total number of scenarios. The uncertainties come about from the parameters in the model, such bounds on the capacity $Z^{\mathrm{UB}}_{s,h}$, $Z^{\mathrm{UB}}_{m,h}$, and $Z^{\mathrm{UB}}_{e,h}$, which are the stochastic upper bounds on the sources, mixers, and end users respectively, the natural gas price $d_{e,h}$, the quality of natural gas $U_{s,w,h}$, etc.

8.A.1 Stochastic Model for the Sources

The following group of constraints represent the submodel for the sources. Eq. (8.A1)–(8.A6) are the group of constraints represent the submodel for the sources, with uncertainty in the upper bounds on the pipeline and nodal capacities.

$$\sum_{m\in\Pi_s} f_{s,m,h} + \sum_{e\in\Theta_s} f_{s,e,h} \leq x_s Z^{\mathrm{UB}}_{s,h}, \tag{8.A1}$$

$$f_{s,m,h} \leq x_{s,m} F^{\mathrm{UB}}_{s,m,h}, \tag{8.A2}$$

$$f_{s,e,h} \leq x_{s,e} F^{\mathrm{UB}}_{s,e,h}, \tag{8.A3}$$

$$x_{s,m} F^{\mathrm{UB}}_{s,m} \leq x_s Z^{\mathrm{UB}}_{s}, \tag{8.A4}$$

$$x_{s,e} F^{\mathrm{UB}}_{s,e} \leq x_s Z^{\mathrm{UB}}_{s}, \tag{8.A5}$$

$$x_s Z^{\mathrm{UB}}_{s} \leq \sum_{m\in\Pi_s} x_{s,m} F^{\mathrm{UB}}_{s,m} + \sum_{e\in\Theta_s} x_{s,e} F^{\mathrm{UB}}_{s,e}, \tag{8.A6}$$

$$\forall s \in S,\ \forall m \in \Pi_s,\ \forall e \in \Theta_s, \forall h \in \{1, \ldots, H\}.$$

8.A.2 Stochastic Model for Mixing Stations

Eqs. (8.A7)–(8.A17) represents constraints for the material balance for the mixing stations with consideration of uncertainty.

$$f_{m,e,w,h} = s_{m,e,h}\left(\sum_{s\in\Pi_m} f_{s,m,h} U_{s,w,h} + \sum_{m\in\Theta_m} f_{m,m,w,h}\right), \tag{8.A7}$$

$$f_{m,m,w,h} = s_{m,m,h}\left(\sum_{s\in\Pi_m} f_{s,m,h} U_{s,w,h} + \sum_{m\in\Theta_m} f_{m,m,w,h}\right), \tag{8.A8}$$

$$\sum_{m\in\Pi_m} s_{m,m,h} + \sum_{e\in\Theta_m} s_{m,e,h} = 1, \quad s_{m,m}, s_{m,e} \geq 0, \tag{8.A9}$$

$$x_{m,m} F^{\mathrm{LB}}_{m,m,h} \leq \sum_{w\in\Omega} f_{m,m,w,h} \leq x_{m,m} F^{\mathrm{UB}}_{m,m,h}, \tag{8.A10}$$

$$x_{m,e} F^{\mathrm{LB}}_{m,e,h} \leq \sum_{w\in\Omega} f_{m,e,w,h} \leq x_{m,e} F^{\mathrm{UB}}_{m,e,h}, \tag{8.A11}$$

$$x_m Z^{\mathrm{UB}}_{m} \geq x_{s,m} F^{\mathrm{UB}}_{s,m}, \tag{8.A12}$$

$$x_m Z_m^{\mathrm{UB}} \geq x_{m,m} F_{m,m}^{\mathrm{UB}}, \tag{8.A13}$$

$$x_m Z_m^{\mathrm{UB}} \geq x_{m,m} F_{m,m}^{\mathrm{UB}}, \tag{8.A14}$$

$$x_m Z_m^{\mathrm{UB}} \geq x_{m,e} F_{m,e}^{\mathrm{UB}}, \tag{8.A15}$$

$$x_m Z_m^{\mathrm{UB}} \leq \sum_{m \in \Pi_m} x_{m,m} F_{m,m}^{\mathrm{UB}} + \sum_{s \in \Theta_m} x_{s,m} F_{s,m}^{\mathrm{UB}}, \tag{8.A16}$$

$$x_m Z_m^{\mathrm{UB}} \leq \sum_{m \in \Pi_m} x_{m,m} F_{m,m}^{\mathrm{UB}} + \sum_{e \in \Theta_m} x_{m,e} F_{m,e}^{\mathrm{UB}}, \tag{8.A17}$$

$$\forall m \in \Pi_m, \quad \forall e \in \Theta_m, \quad \forall w \in \Omega, \forall h \in \{1, \ldots., H\}.$$

8.A.3 Stochastic Model for End Users

The following group of constraints in Eqs. (8.A18)–(8.A22) represents the submodel for the end users.

$$\sum_{m \in \Pi_e} \sum_{w \in \Omega} f_{m,e,w,h} + \sum_{s \in \Theta_e} f_{s,e,h} \leq x_e D_e^{\mathrm{UB}}, \tag{8.A18}$$

$$\sum_{m \in \Pi_e} f_{m,e,w,h} + \sum_{s \in \Theta_e} f_{s,e,h} u_{s,w,h} \leq \left(\sum_{m \in \Pi_e} \sum_{w \in \Omega} f_{m,e,w,h} + \sum_{s \in \Theta_e} f_{s,e,h} \right) V_{e,w,h}^{\mathrm{UB}} \tag{8.A19}$$

$$x_e D_e^{\mathrm{UB}} \geq x_{s,e} F_{s,e}^{\mathrm{UB}}, \tag{8.A20}$$

$$x_e D_e^{\mathrm{UB}} \geq x_{e,k} F_{e,k}^{\mathrm{UB}}, \tag{8.A21}$$

$$x_e D_e^{\mathrm{UB}} \leq \sum_{s \in \Pi_e} x_{s,e} F_{s,e}^{\mathrm{UB}} + \sum_{e \in \Theta_e} x_{m,e} F_{m,e}^{\mathrm{UB}}, \tag{8.A22}$$

$$\forall e \in E, \quad \forall w \in \Omega, \forall h \in \{1, \ldots., H\}.$$

8.A.4 Stochastic Pipeline Performance Model

Eq. (8.A23) and Eq. (8.A24) apply pipeline pressure relations under uncertainty:

$$x_{s,m}[(p_{s,h}^{downs})^2 - (p_{s,h}^{ups})^2] = \kappa_{s,m} Q_{s,m,h}, \forall (s,m) \in \Pi_s, \forall h \in \{1, \ldots., H\} \tag{8.A23}$$

$$x_{s,m}((p_s^{downs})^2 - (p_s^{ups})^2) = 0, \forall (s,m) \in \Pi_s, \forall h \in \{1, \ldots., H\}. \tag{8.A24}$$

8.A.5 Stochastic Compression Performance Model

Eq. (8.A25), Eq. (8.A26), and Eq. (8.A27) are constraints for the power requirement of the compressors, with the necessary bounds:

$$W_{s,h} - \alpha_m \sum_{s \in S} \sum_{m \in M} f_{s,m,w,h} \left[\left(\frac{p_{m,h}^{downs}}{p_{m,h}^{ups}} \right)^{\nu} - 1 \right] = 0, \tag{8.A25}$$

$$p_{s,h}^{ups} \leq p_{s,h}^{downs}, \tag{8.A26}$$

$$x_s W_s^{LB} \leq W_{s,h} \leq x_s W_s^{UB}, \tag{8.A27}$$

$$\forall s \in S, \forall w \in \Omega, \forall h \in \{1, \ldots, H\}$$

Other sets of pressure constraints in the natural gas transportation problem include bounds imposed on the inlet and outlet pressures based on the design capacity of the compressors as specified by the manufacturers.

$$x_s P_s^{ups,LB} \leq p_{s,h}^{ups} \leq x_s P_s^{ups,UB}, \tag{8.A28}$$

$$x_s P_s^{downs,LB} \leq p_{s,w,h}^{downs} \leq x_s P_{s,w}^{downs,UB}, \tag{8.A29}$$

$$\forall s \in S, \forall w \in \Omega, \forall h \in \{1, \ldots, H\}.$$

Eq. (8.A28) and Eq. (8.A29) impose lower and upper bounds on the upstream and downstream pressures respectively.

The stochastic natural gas design and operation model can be stated as Problem (SP) as follows:

$$(\text{SP})\begin{cases}\text{minimize} \quad \text{objective}\\ \text{Flow and pressure model in Eq.8.A1-8.A33,}\\ x_s, x_m, x_e, x_{s,m}, x_{s,e}, x_{m,m}, x_{m,e} \in \{0,1\},\\ \text{All flow rates and pressures are nonnegative.}\end{cases}$$

where objective is given by,

$$objective = Cap\cos t + \left(\sum_{t=\{1,\ldots T\}} \frac{1}{(1+r)^t}\right) \sum_{h\in\{1,\ldots,H\}} \Pr_h(Op\cos t_h - Income_h), \tag{8.A30}$$

$$\begin{aligned} Cap\cos t = &\sum_{s\in S} c_s x_s + \sum_{m\in M} c_m y_m + \sum_{e\in E} c_e y_e \\ &+ \sum_{s\in S}\sum_{m\in M} c_{s,m} x_{s,m} + \sum_{s\in S}\sum_{e\in E} c_{s,e} x_{s,e} + \sum_{m\in M}\sum_{e\in E} c_{m,e} x_{m,e} \end{aligned} \tag{8.A31}$$

And,

$$Op\cos t_h = \sum_{s\in S} d_{s,h}\left(\sum_{m\in\Theta_s} f_{e,m,h} + \sum_{e\in\Theta_s} f_{s,e,h}\right) + \sum_{s\in S} d_{s,h}^{elec} W_{s,h} \tag{8.A32}$$

$$Income_h = \sum_{e\in E} d_{e,h}\left(\sum_{s\in\Pi_e}\sum_{w\in\Omega} f_{m,e,w,h} + \sum_{s\in\Pi_e} f_{s,e,h}\right), \tag{8.A33}$$

Where $Op\cos t_h$ and $Income_h$ are the scenario-dependent operating cost and income, while $d_{s,h}$, $d_{s,h}^{elec}$ and $d_{e,h}$ are the stochastic annual operating cost, electricity cost, and natural gas price respectively.

9

A Multicriteria Optimization Approach to the Synthesis of Shale Gas Monetization Supply Chains

Ahmad Al-Douri[1], *Debalina Sengupta*[2], *and Mahmoud M. El-Halwagi*[1,2]

[1] *Department of Chemical Engineering, Texas A&M University, USA*
[2] *TEES Gas and Fuels Research Center, Texas A&M Engineering Experiment Station, USA*

CHAPTER MENU

9.1 Introduction

With the recent discovery of unconventional shale gas resources, the potential supply of natural gas has increased significantly. Production from shale gas plays has increased from 1.3 trillion cubic feet in 2007 to 15.4 trillion cubic feet in 2015. This accounts for approximately 47% of the total United States natural gas production in 2015 (EIA 2017). There are a multitude of options for the monetization of shale gas including physical and chemical conversion of shale gas for the production of energy and chemicals (Al-Douri, Sengupta, and El-Halwagi 2017; Bamufleh, Noureldin, and El-Halwagi 2016; Marano, Spivey, and Morreale 2015; Siirola, 2014). The production of chemicals is particularly attractive because of the profitability margins. Methane (the primary constituent of shale gas) can be converted to a variety of derivatives either directly or indirectly. Indirect methods of methane utilization require synthesis gas (syngas), which is composed mostly of hydrogen and carbon monoxide. Syngas may be produced by one of the following processes or a combination thereof: steam reforming, partial oxidation, and dry reforming (e.g., Martínez-Gomez et al. 2016; Noureldin, Elbashir, and El-Halwagi 2014, 2015; Martínez et al. 2013). Once syngas is produced, it can be transformed into numerous intermediates and products. Examples of chemicals that can be manufactured from shale gas include methanol (Ehlinger et al. 2013; Julián-Durán et al. 2014), ethylene (Ortiz-Espinoza et al. 2017), propylene (Jasper and El-Halwagi 2015), benzene (Pérez-Uresti et al. 2017), liquid transportation fuels (Bao, El-Halwagi, and Elbashir 2010), and propylene (Agrawal, Sengupta, and El-Halwagi 2018). Similar to natural gas, shale gas can also be monetized via integrated supply chains. An analogous example for natural gas monetization is the supply chain of the Qatar industry, which includes a highly integrated network of intermediate and final products (Alfadala and El-Halwagi 2017). The concepts of industrial symbiosis can be used to guide the emerging infrastructure of shale-gas supply chains (Martinez-Gomez et al. 2017; Hasaneen and El-Halwagi 2017; El-Halwagi 2017a; Noureldin and El-Halwagi 2015).

Natural Gas Processing from Midstream to Downstream, First Edition.
Edited by Nimir O. Elbashir, Mahmoud M. El-Halwagi, Ioannis G. Economou, and Kenneth R. Hall.

The creation of viable supply chains for monetizing shale gas requires a systems perspective that investigates the multitude of conversion technologies and their interconnection. Additionally, in the selection of the recommended supply chains, multiple objectives must be considered including economic, environmental, and safety metrics (El-Halwagi 2017b; Roy et al. 2016; Thiruvenkataswamy et al. 2016). Optimization techniques enable the systematic synthesis, screening, and selection of alternatives while accounting for the multiple design objectives. Recently, some work has been done on multi-objective optimization for process systems. Sengupta and Pike (2012) used material and energy balances, chemical reaction rates, and thermodynamic equilibrium relations to optimize the development of chemical plants that use renewable resources. Santibañez-Aguilar et al. (2014) and El-Halwagi et al. (2013) developed a multi-objective optimization approach for the design of biorefining supply chains. Cafaro and Grossman (2014) addressed the long-term planning of the shale gas supply chain by presenting a MINLP. Gao and You (2012) used a mixed-integer nonlinear programming (MINLP) approach to address the life cycle economic and environmental optimization of a shale gas supply chain network based on the Marcellus shale play. Onel, Niziolek, and Floudas (2015) investigated the co-production of liquid transportation fuels and olefins from a natural gas feedstock. As the review of literature shows, most of the work done hitherto has focused on the optimization of shale gas supply chains for the purposes of electricity production and fuels production. In this work, the focus is on optimization of the production of chemicals from a given resource of shale gas. A case study is solved where the annual production of dry gas in the state of Texas will be used as the resource quantity available to be monetized.

9.2 Methodology

The problem to be addressed in this chapter is described as follows: Given is a certain supply of shale gas with known flows and characteristics. Available for consideration is a set of monetization technologies that can convert shale gas into value-added chemicals. The performance of each technology and high-level economic data are available. Limitations on the market for each chemical are known. It is desired to synthesize a supply chain for monetizing shale gas into chemicals while accounting for the economic, environmental, and safety metrics.

The superstructure representation shown by Figure 9.1 is developed to embed the various supply-chain configurations of interest. The supply from the shale gas play is pretreated and separated into a set of feedstocks = $\{i|i = 1, N_{Feedstocks}\}$. Each candidate conversion technology is given an index j (ranging from 1 to $N_{Conversion}$). The potential value-added chemicals are represented by the set of products = $\{k|k = 1, N_{Products}\}$. An unknown flowrate, $F_{i,j}$, is assigned from each feedstock to each conversion technology. Similarly, an unknown flowrate , $G_{j,j'}$, is assigned from each conversion technology, j, to a different one, j'. Finally, an unknown flowrate, $W_{j,k}$, is assigned from each conversion technology to each candidate product.

The optimization formulation is briefly given by:

$$\text{Maximize Profit} = \text{Product Sales} - \text{Total Annualized Cost of Monetization}$$

The total annualized cost includes the annualized fixed costs of the conversion technologies and the annual operating cost (raw materials, transportation, energy, etc.).

The objective function is subject to the following constraints:

- Shale gas supply (flowrate and composition)
- Performance of each conversion technology (input-output models)
- Material balances for the distribution of feedstocks to conversion technologies

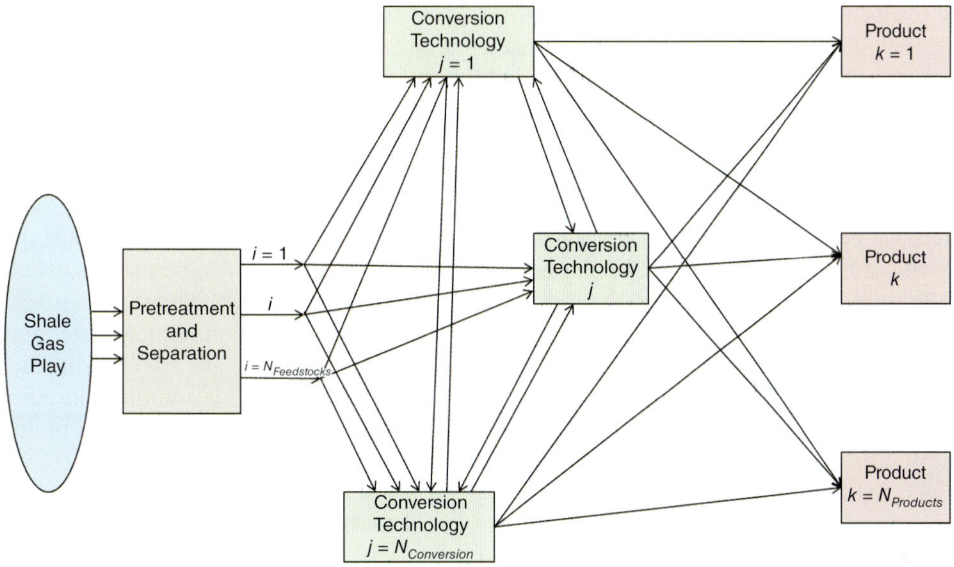

Figure 9.1 Superstructure Representation of the Monetization Network.

- Material balances for the distribution of intermediates among the various conversion technologies
- Materials balances for the distribution of chemicals from conversion plants to final products
- Capacity constraints for each process
- Market constraints for each product
- Environmental limitations

Safety limitations were taken into consideration by calculating the Dow Fire & Explosion Index (DF&EI) for the base case of the plants in the proposed chemical complex. However, these calculations were not included in the objective function. The developed model is a mixed-integer nonlinear programming (MINLP) formulation, which is coded and solved using the software LINGO®.

9.3 Case Study

9.3.1 Problem Statement

The devised approach is illustrated by solving a case study for shale gas monetization in Texas. The annual production of dry gas in the state of Texas (137 MMTA (million metric ton per annum)) was taken as available supply. Various chemicals were considered as potential products including methanol, acetic acid, dimethyl ether (DME), formaldehyde, ethylene, propylene, ammonia, and urea. For some of these chemicals, more than one production process will be considered as will be discussed in further detail later. By representing these production options in a superstructure, the optimal allocation of feedstock (e.g., shale gas) can be achieved to maximize profit, subject to the constraints discussed in the previous section. For the case study, three scenarios will be examined: base case, reduced methanol price, and reduced urea price. The base case scenario will utilize feedstock and product prices from November 2015. The reduced methanol price case is likely to occur when a number of methanol plants come online, while a reduced urea price case is tested to determine the price at which urea will be produced from the proposed chemical complex. A detailed rationale for the testing of the latter two cases will be discussed later.

9.3.2 Environmental and Safety Metrics

Under the environmental criterion, the aim is to find the most environment-friendly processes in the chemical complex. Greenhouse gases (GHG) and criteria air pollutants (CAP) are the primary pollutants of concern in the process industries. Various metrics can be used to assess sustainability and safety individually (e.g., Roy et al. 2016) or augmented with economic criteria (El-Halwagi 2017b; Guillen-Cuevas et al. 2018). The United States Environmental Protection Agency defines GHG as gases that trap heat in the atmosphere, specifically carbon dioxide, methane, nitrous oxide, and fluorinated gases. They also describe CAP as common pollutants that can harm the environment and the population's health. They include six major air pollutants, and they are carbon monoxide (CO), particulate matter (PM), lead, sulfur dioxide (SO_2), and nitrogen oxides. For the purposes of this model, only carbon dioxide emissions for the processes of the chemical complex were taken into account. These CO_2 emissions values were calculated either from available literature or heat of reactions. In either case, it was assumed that natural gas, which emits 117 lbs CO_2/MMBtu, is used as fuel. By determining the amount of energy consumed by a process (in MMBtu/yr), the amount of CO_2 emissions can be calculated using the ratio for natural gas.

Process safety has gained utmost importance in recent years when companies have seen direct relation between brand image and safety culture of a company. In addition to financial losses, any loss in human life is not acceptable, and all precautionary measures should be taken as early as the design and planning stage of a process plant. For this, inherently safer process design has gained importance. In this class of design problems, safety metrics are considered at early stages of design to evaluate the feasibility of a process. The safety metric taken into account in this model is the Dow Fire & Explosion Index (DF&EI). The index is based on the calculation of Material Factor (MF) which is a measure of the intrinsic rate of potential energy release from fire or explosion produced by combustion or other chemical reaction (AIChE 1987). In this metric, a penalty is assigned to General Process Hazards (exothermic chemical reactions, endothermic processes, material handling and transfer, enclosed or indoor process units, access, drainage and spill control) and Special Process Hazards (toxic material, sub-atmospheric pressure, operation in or near flammable range, dust explosion, pressure, low temperature, quantity of flammable/unstable material, corrosion and erosion, leakage—joints and packing, use of fired equipment, hot oil heat exchange system, rotating equipment).

Under "General Process Hazards," a penalty is assessed for "Exothermic Chemical Reactions" and "Material Handling and Transfer." In each process, the reaction that occurs is unique and requires a certain penalty as specified in the Fire & Explosion Index Hazard Classification Guide. This penalty is based on the type of reaction that and the exotherm category it falls under. The categories of exotherms and their corresponding penalties are given in Table 9.1. The "Material Handling and Transfer" takes into account the potential for fire involving the reactor(s) during handling, transfer, and warehousing of materials (AIChE 1987). For the purposes of this study, a penalty of 0.50 will be used for any loading or unloading operation involving Class I flammables or LPG-type materials where transfer lines are connected or disconnected (AIChE 1987).

Under "Special Process Hazards," the penalty for the item labeled "Toxic Material(s)" is determined from the health factor N_h found in Appendix A of the Hazard Classification Guide. N_h

Table 9.1 Exotherm categories and their corresponding penalties.

Exotherm	Penalty
Mild	0.30
Moderate	0.50 except for addition reactions (0.75) and oxidation reactions (1.00)
Critical-to-control	1.00
Particularly sensitive	1.25

Table 9.2 Pressure penalty for flammable and combustible liquids.

Pressure (psig)	Penalty Factor
1,500	0.92
2,000	0.96
2,500	0.98
3,000–10,000	1.00
>10,000	1.50

is the health factor of a material as defined in the National Fire Protection Association 704 standard (AIChE 1987). The health factor used is that of the basic material in the process and the penalty assessed is: $0.20 \times N_h$. Another important item is "Pressure," where a penalty is applied to offset the effect of high pressure on the potential for leaks with higher release rates (AIChE 1987). This item is determined from a chart, a modification of which is shown in Table 9.2, in the Hazard Classification Guide, which gives a penalty based on the operating pressure of the process unit.

The last two items of relevance in the F&EI form in this study were the items "Quantity of Flammable/Unstable Material" and "Corrosion and Erosion." The penalty for the former is determined using Figure 3 on page 24 of the Hazards Classification Guide, while the penalty for the latter is determined by searching for values of corrosion rates of the chemicals being examined in the F&EI study. Penalties for corrosion are assigned on the following basis:

- A penalty of 0.10 is assigned for corrosion rates of less than 0.127 mm/year with a risk of pitting or local erosion.
- A penalty of 0.20 is assigned for corrosion rates over 0.127 mm/year and less than 0.254 mm/year.
- A penalty of 0.50 is assigned for corrosion rates over 0.254 mm/year.

Afterwards, the items under each of the "General Process Hazards" and "Special Process Hazards" sections are summed up, and two values are obtained, F1 and F2. F1 refers to the General Process Hazards Factor, and F2 refers to the Special Process Hazards Factor. Then, they are multiplied to obtain a Process Unit Hazards Factor, labelled F3. Finally, the Dow Fire & Explosion Index is obtained by multiplying the Process Unit Hazards Factor by Material Factor mentioned earlier in this chapter. The degree of hazard of different F&EI values (light, moderate, intermediate, heavy, and severe) is described in Table 6 of the Hazards Classification Guide.

9.3.3 Objectives of the Case Study

The case study in this work has five objectives: (1) examine the results of the optimization program based on market conditions as of November 2015, (2) examine the effects of a reduction in the price of methanol (a major intermediate), (3) determine the impact urea prices have on the monetization pathway chosen, (4) determine carbon dioxide emissions for the base case, and (5) obtain Dow Fire & Explosion Index values to provide a safety metric for the processes in the complex.

Regarding environmental emissions, the case study is limited to carbon dioxide and thus does not consider other emissions that are also causes for concern to industry and regulators. Also, the DFEI results will not be included in the objective function (profit equation), but are intended to provide a starting point for further analysis of the safety aspects of this proposed chemical complex. While the DFEI is useful in identifying process units where hazardous conditions exist, it does not estimate the damage that might result from such an event.

9.4 Case Study Results

9.4.1 Feedstock

Shale gas produced in the state of Texas is used as the feedstock for this chemical complex. According to the Energy Information Administration, Texas shale gas production in 2015 was 4,353 billion cubic feet, or approximately 83.5 MMT. Shale gas has different compositional makeup depending on the formation, as well as the well it is produced from. Bullin and Krouskop (2009) discussed some of the compositional varieties across shale plays in the United States. For the Antrim shale play primarily in Michigan, samples from four wells showed a nitrogen composition ranging between 0.7 and 65 vol%. For the Marcellus play located mainly in Pennsylvania, C_1 composition ranges from 79.4 to 95.5 vol%. Samples from four wells in the Barnett shale play in Texas showed a C_1 composition range between 80.3 and 93.7 vol%. Tests on wells in the New Albany shale play in southern Illinois show it can contain 8–10 vol% of CO_2. For the case study, the average volume percent of methane in the Barnett shale play was used to determine the amount of methane from the available shale gas supply.

9.4.2 Conversion Technologies

The superstructure contains processes for the production of the following chemicals: methanol, acetic acid (2), dimethyl ether (2), formaldehyde, ethylene, propylene, ammonia, and urea. Methanol is a major intermediate and is subsequently used in the production processes of acetic acid, dimethyl ether, formaldehyde, and olefins. While methanol has been produced from syngas previously, it can also be produced from methane directly by a two-step reforming process. For acetic acid, two production processes were included in the superstructure: a two-step process from methanol, and the recently developed integrated three-step SaaBre process by BP. Similarly, two dimethyl ether (DME) production processes were included: dehydration of methanol, and a direct syngas-to-DME process. In addition to acetic acid and dimethyl ether, methanol is also used in the production of formaldehyde, an important precursor in the manufacture of industrial resins. The process can occur by one of two routes, both starting with methanol: oxidation-dehydrogenation, and oxidation with air. In this study, the latter process will be considered in the superstructure. Two processes that have been increasingly used in industry for the production of olefins were included also: methanol-to-olefins (MTO) and methanol-to-propylene (MTP). Another process to produce ethylene, oxidative coupling of methane (OCM), is included in the superstructure. This process is currently being tested at the pilot scale and has not seen commercial-scale production yet. Ammonia production is accomplished by desulfurization of natural gas, converting the gas to hydrogen by steam reforming, then combing hydrogen and nitrogen in the Haber-Bosch process to synthesize ammonia. In this superstructure, the ammonia produced is either sold as is or further processed into urea. As for urea, it is produced in a two-step process: reaction of carbon dioxide and ammonia to produce a carbamate intermediate, and the decomposition of carbamate to urea.

9.4.3 Base Case Product Prices

Various chemicals were considered as potential products. These include methanol, acetic acid, dimethyl ether (DME), formaldehyde, ethylene, propylene, ammonia, and urea. The base-case prices (in $/metric ton) of the chemicals are shown by Table 9.3.

In the superstructure representation of the proposed complex, methanol can be sold to the market as is, or be further processed to produce acetic acid, DME, formaldehyde, or olefins. As stated in the conversion technologies section, acetic acid and DME can be produced directly from the gas feedstock (methane) in processes that do not use methanol. Similarly, ammonia can be sold as is or further processed to make urea, widely used as a fertilizer.

9.4.4 Plant Costs and Capacity Limits

In the objective function, the term "total annualized costs of monetization" includes capital and utility (or energy) costs for the process. Capital costs were considered through the annualized fixed cost (AFC) for the process, as shown in the following equation.

$$AFC = \frac{FCI_0 - FCI_S}{N}$$

where:

FCI_0 = Initial value of depreciable FCI

FCI_S = Salvage or scrap value of FCI at the end of service life (assumed to be zero in this study)

N = Service life of the property (20 years)

Table 9.3 Sales prices of chemicals for the base case.

Chemical	Price ($/MT)
Methanol	366
Acetic Acid	500
DME	946
Formaldehyde	830
Ethylene	1041
Propylene	819
Ammonia	700
Urea	255

Table 9.4 Production capacities for the proposed chemical complex.

Product	Capacity (1000 Metric Tons per year)
Methanol	2,500
Acetic acid	605
Dimethyl ether	187
Formaldehyde	725
Ammonia	2,000
Urea	1,500
Ethylene	2,000
Propylene	1,500

The initial value of depreciable fixed capital investment (FCI) is calculated through the six-tenths factor rule as shown in the equation below. Using a known FCI value and its corresponding production rate, a value of "*a*" for each plant is determined. Then, that value is used along with the production rates from Table 9.4 to calculate the FCI of the plant.

$$FCI = a * (Annual\ Production\ Rate)^{0.6}$$

For each candidate product, an upper limit for the amount produced had to be set. These limits were set to qualify the proposed plants to be close to the largest ten production plants for each product. The values used as capacity limits are shown below in Table 9.4. It should be noted that these values are for the products, not the individual plants. For example, in the case that both processes to produce acetic acid are used, then the acetic acid production limit is 605 KMTA.

9.4.5 Base Case Solution

The results of the chemical production complex for this case can be seen in Figure 9.2. The methanol produced in the complex was distributed between three pathways: the formaldehyde production process, the MTO process, and sold as is. Thus, no methanol was used to produce acetic acid or DME. Instead, acetic acid was produced by the BP SaaBre process, while DME was produced through the syngas route. Ethylene was produced through the oxidative coupling of methane (OCM) process, a new technology for which the first pilot plant started up in April 2015. Propylene, the other major olefin, was produced by the MTO technology. Also, 2.0 MMTA (million metric tons per annum) of ammonia was produced and sold as is, so no urea was manufactured in this case of the complex.

In this case, it is observed that out of 2.50 MMTA of methanol produced, approximately 60,000 MTA of it was sold as is to the market. From the 2.44 MMTA remaining, 1.67 MMTA of methanol is used to produce 0.73 MMTA of propylene (sold for $819/MT) via the MTO process. The remaining 0.773 MMTA of methanol is converted through the formaldehyde production unit to formaldehyde, which is subsequently sold in the market at $830/MT. While 1.16 MMTA of carbon dioxide were consumed, it should be noted that 13.8 MMTA was produced in this case of the production complex. As for water, 2.90 MMTA of it were consumed in the complex compared to 4.06 MMTA of it being produced. Also, there were 0.044 MMTA of hydrogen, an

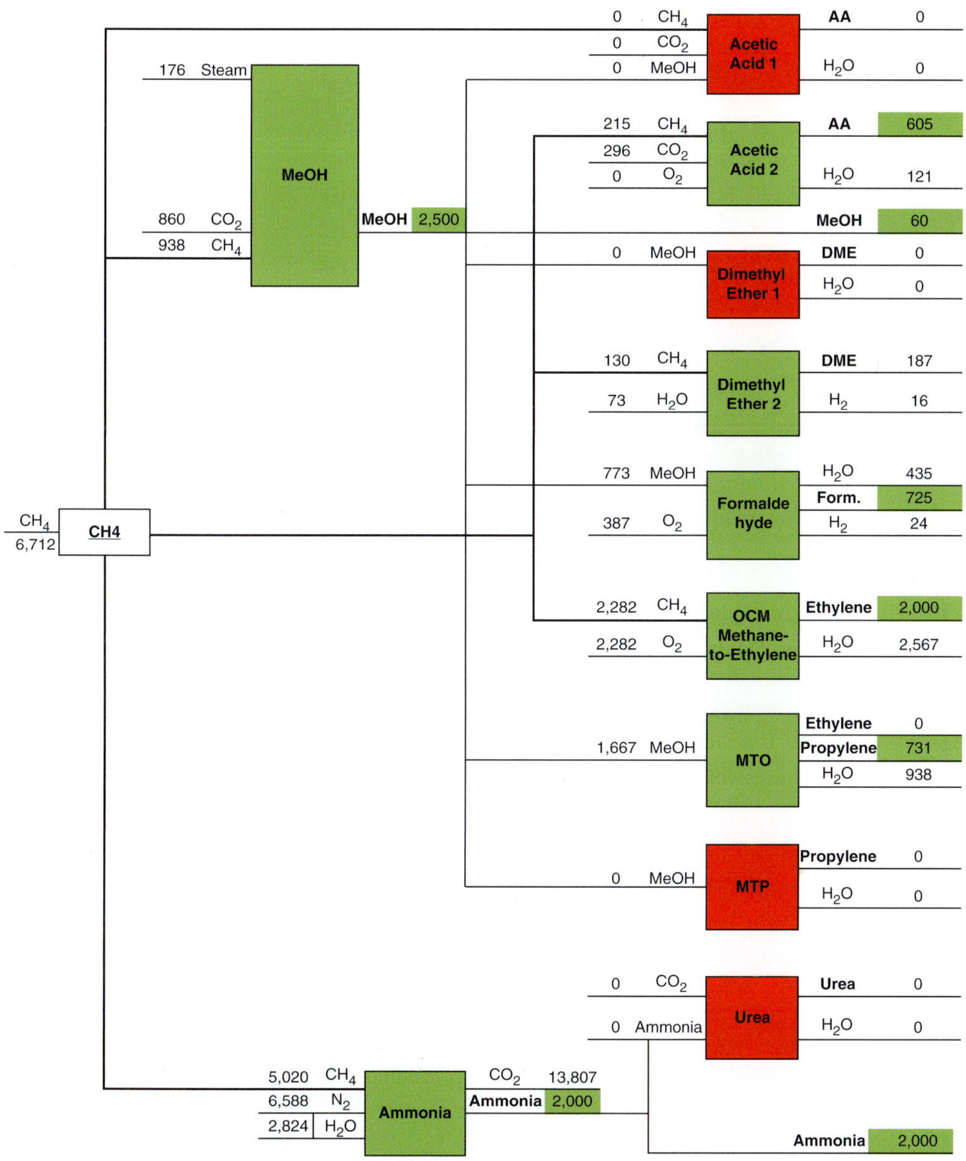

Figure 9.2 Chemical production complex base case results (values are in 1000 MTA).

important chemical building block, produced in the complex. The annual profit obtained in this case was $3,771 million per year. These results are shown in Table 9.5.

9.4.6 Reduced Methanol Price Case Results

After the base case, the objective was to examine the effect of price sensitivity on the type and quantity of chemicals produced. Specifically, the effects of fluctuations in the price of methanol will be examined here. After the middle of 2015, the price of methanol declined again to approximately $161/MT in response to the 60% increase in U.S. production capacity (ICIS January, 15, 2016). Thus, the price of methanol in the program was changed from $366/MT to $161/MT to

Table 9.5 Sales and costs associated with base case.

Category	Value ($ millions)
Income from sales	5,183
Annualized fixed costs	268
Raw material costs	912
Utility costs	232
Gross Profit	3,771

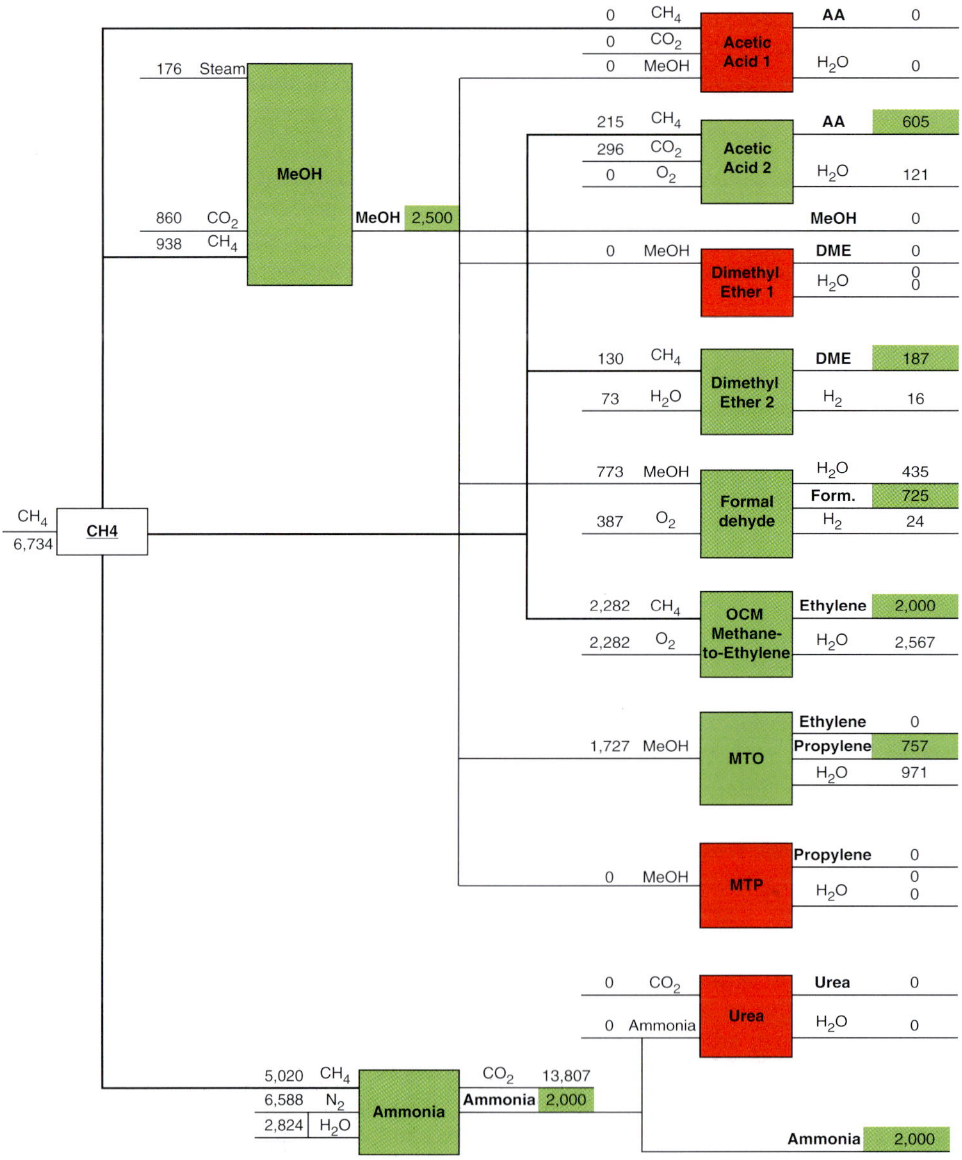

Figure 9.3 Chemical production complex results for reduced methanol price case (values are in 1000 MTA).

Table 9.6 Sales and costs associated with reduced methanol price case.

Category	Value ($ millions)
Income from sales	5,183
Annualized fixed costs	268
Raw material costs	915
Utility costs	234
Gross profit	3,766

examine how the effects of that price change on the chemical production complex. From the results in Figure 9.3, it can be seen that no methanol was sold as-is to the market. This can possibly be explained by the drop in methanol price, thus rendering the product less likely to make a profit than if it was further processed into formaldehyde or olefins. In this case of the complex, of the 2.5 MMTA of methanol that were produced: 0.773 MMTA were used to produce 0.725 MMTA of formaldehyde, and the remaining 1.73 MMTA were used to produce 0.757 MMTA of propylene via the MTO process. By comparing these values to the base case, it can be concluded that the methanol that was previously sold as is (60,000 MTA) was further processed into propylene in the MTO process. As a result, the amount of propylene produced was slightly higher in this case (0.757 MMTA) than in the base case (0.73 MMTA). Similar to the base case, 2.00 MMTA of ammonia, and no urea, were produced and sold as is to the market. As for the amounts of water, carbon dioxide, and hydrogen that were reacted and produced in the complex, the numbers were very similar to the base case, with the exception being a slight increase in the amount of water produced to 4.09 MMTA (up from 4.06 MMTA). This slight increase can be attributed to the increase in the amount of propylene produced in comparison to the base case.

The total profit generated in this case of the production complex was $3,766 million, which is very close to the base case. Table 9.6 shows the breakdown of the sales and costs associated with the profit equation.

9.4.7 Reduced Urea Price Case Results

In both the base and the lower methanol price cases, it was noted that no urea was produced in the production complex. This raises the opportunity to experiment with the LINGO program to determine the conditions under which urea is produced and the changes that will be incurred on the quantity and types of chemicals manufactured in the complex as well as the profit. It should be noted that the methanol price in this case of the program is the same as it was in the base case ($366/MT).

For this case of the production complex, urea was produced when the price reached $500/MT, and the results for the complex can be seen in Figure 9.4. Similar to the base case, 2.00 MMTA of ammonia was produced and was processed in two routes: 1.15 MMTA sold as ammonia, and 0.849 MMTA processed into urea. In comparison to the base case, the production of other chemicals such as acetic acid, DME, and olefins was not affected. The addition of the urea process increased the amount of carbon dioxide consumed in the complex to 2.25 MMTA (up from 1.16 MMTA) and the amount of water produced to 4.51 MMTA (up from 4.06 MMTA).

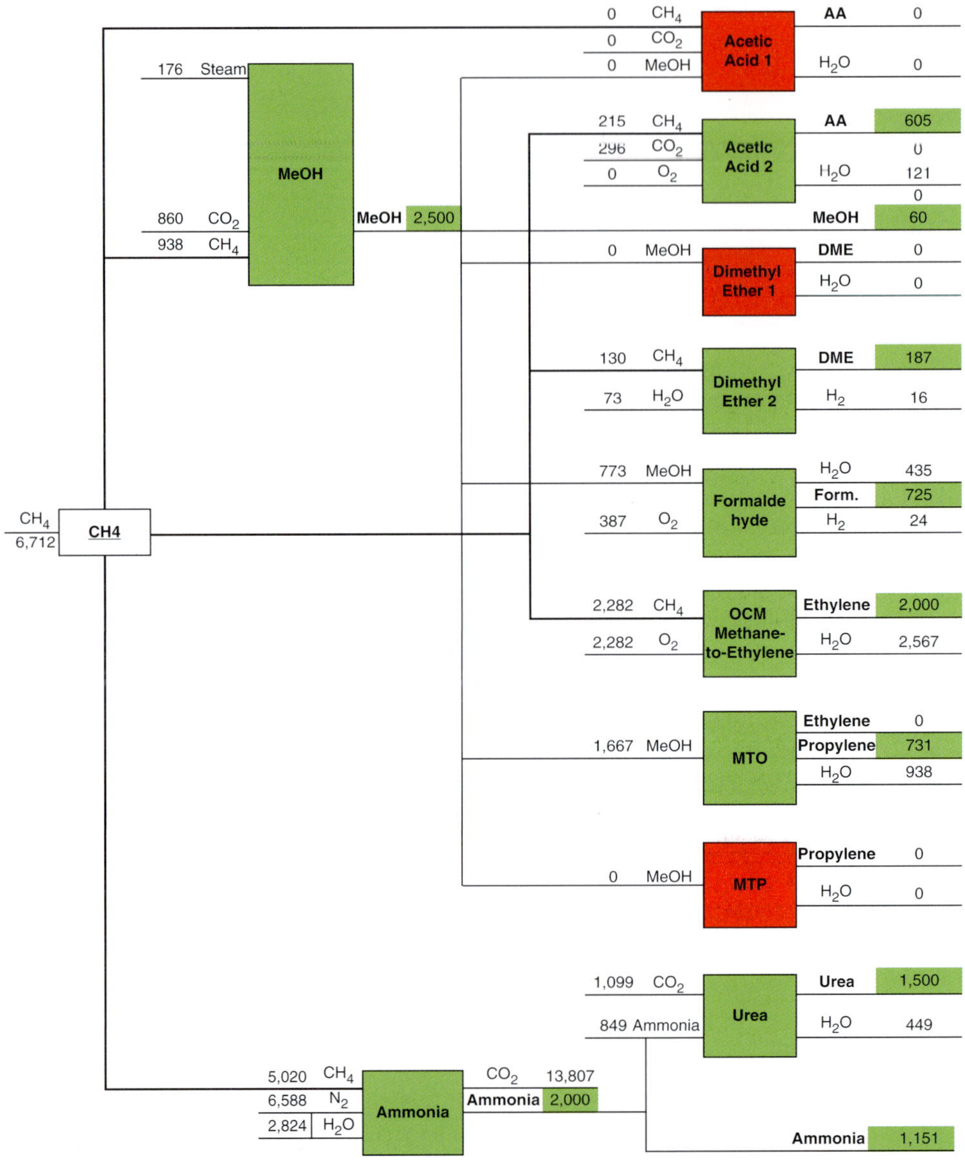

Figure 9.4 Chemical production complex for higher urea price case (values are in 1000 MTA).

For this case, the overall profit as \$3,823 million representing an increase of \$92 million in profits attributable to urea sales. Table 9.7 shows a breakdown of the sales and costs associated with this case of the production complex.

9.4.8 Base Case Environmental Considerations

Aside from economics/profitability, it is desired to see the effect of environmental aspects or emission restrictions on the results of the chemical production complex. This is done by

Table 9.7 Sales and costs associated with urea price sensitivity case.

Category	Value ($ millions)
Income from sales	5,339
Annualized fixed costs	296
Raw material costs	912
Utility costs	268
Gross Profit	3,863

Table 9.8 CO_2 emission values used in the optimization program.

Process	Rate of CO_2 Emissions (MT CO_2/MT product)	Source
Methanol	0.437	Ehlinger 2014
Acetic acid	0.196	ΔH_{rxn}
Syngas-to-dimethyl ether	0.452	Karagoz 2014
Formaldehyde	0.0561	ΔH_{rxn}
Methanol-to-olefins	20.9 (MT of propylene)	Jasper and El-Halwagi 2015

incorporating CO_2 emissions obtained from literature as rates per metric ton of product. In Table 9.8, the CO_2 emission values used in the optimization program are shown. These values were obtained either from published sources or from the heat of reactions of the overall process reaction. In addition to the CO_2 emissions associated with production, there are emissions resulting from the fuel burned for utilities (heating, cooling, and electric power) in the process.

9.4.9 Base Case Safety Considerations

The final set of considerations to be taken into account for this chemical production complex are safety metrics, which will be done through a Dow Fire & Explosion Index analysis of the reactors of each of the processes involved in the base case of the complex. The DFEI results will not be included in the objective function (profit equation) but provide a starting point for further analysis of the safety aspects of this proposed chemical complex. This analysis was done for only the base case of the chemical complex, not for the two alternative cases. While the DFEI is useful in identifying process units where hazardous conditions exist, it does not estimate the damage that might result from such an event. Table 9.4 shows the qualitative hazard levels for various values of the DFEI. Throughout the analysis, the following items in the DFEI form were considered: A–C under General Process Hazards, and A, E, and G under Special Process Hazards. Table 9.9 shows the Dow Fire & Explosion Index values that were obtained using this methodology and their corresponding qualitative hazard levels. From these DFEI values, it is clear that the processes that pose the highest hazard levels are those that produce methanol and ammonia.

Table 9.9 DFEI values and their corresponding qualitative hazard levels.

Process	Dow Fire & Explosion Index	Qualitative hazard level
Methanol	153	Heavy
Acetic acid	110.5	Intermediate
Syngas-to-dimethyl ether	88	Moderate
Formaldehyde	73.6	Moderate
Methanol-to-olefins	98	Intermediate
Ammonia	153	Heavy

9.5 Conclusion

This chapter has provided a systematic approach for the synthesis and screening of shale gas monetization pathways. An optimization formulation has been developed and solved to determine the optimal pathways of a given shale gas resource into a number of candidate products and intermediates. The objective of the optimization formulation is to maximize profit based on capital and operating costs of the candidate processes, feedstock prices, the sales prices of the produced chemicals while accounting for supply, demand, technical, and environmental factors. In addition, safety issues have been taken into account. A case study has been solved to demonstrate the applicability and merits of the proposed approach.

References

Agrawal A, Sengupta D, El-Halwagi MM. 2018. A sustainable process design approach for on-purpose propylene production and intensification. ACS Sustain Chem Engg, 6, 2407–2421.

AIChE (American Institute of Chemical Engineers). 1987. Dow's fire & explosion index hazard classification guide. AIChE.

Al-Douri A, Sengupta D, El-Halwagi MM. 2017. Shale gas monetization—a review of downstream processing to chemicals and fuels. J Nat Gas Sci Engg, 45: 436–455.

Alfadala HE, Halwagi MM. 2017. Qatar's chemical industry: monetizing natural gas. Chem Eng Prog, 113 (2), 38–41.

Bamufleh HS, Noureldin MMB, El-Halwagi MM. 2016. Chapter 6, Sustainable process integration in the petrochemical industries. In: Al-Megren H, Xiao T, editors, Petrochemical catalyst materials, processes, and emerging technologies. Advances in chemical and materials engineering (ACME) Book Series. Hershey, PA: IGI Global, p. 155–168.

Bao B, El-Halwagi MM, Elbashir NO. 2010. Simulation, integration, and economic analysis of gas-to-liquid processes. Fuel Proc Technol, 91(7), 703–713.

Bullin KA, Krouskop PE. 2009. Compositional variety complicates processing plans for US shale gas. Oil Gas J., 107 (10), 50–55.

Cafaro D, Grossmann I. 2014. Strategic planning, design, and development of the shale gas supply network. AIChE J, 60(6): 2122–2142.

Ehlinger VM, Gabriel K J, Noureldin MM, El-Halwagi MM. 2013. Process design and integration of shale gas to methanol. ACS Sustain Chem Engg, 2(1): 30–37.

EIA: Energy Information Administration: https://www.eia.gov/ accessed on July 7, 2017

El-Halwagi AM, Rosas C, Ponce-Ortega JM, Jiménez-Gutiérrez A, Mannan MS, El-Halwagi MM. 2013. Multi-objective optimization of biorefineries with economic and safety objectives. AIChE J, 59 (7): 2427–2434.

El-Halwagi MM. 2017a. A shortcut approach to the multi-scale atomic targeting and design of C-H-O symbiosis networks. Proc Integr Optimiz Sustain, 1 (1): 3–13.

El-Halwagi MM. 2017b. A return on investment metric for incorporating sustainability in process integration and improvement projects. Clean Technol Environ Policy, 19:611–617.

Gao J You, F. 2012. Shale gas supply chain design and operations toward better economic and life cycle environmental performance: MINLP model and global optimization algorithm. ACS Sustain Chem Engg, 3, 1282–1291.

Guillen-Cuevas K, Ortiz-Espinoza AP, Ozinan E, Jiménez-Gutiérrez A, Kazantzis NK, El-Halwagi MM. 2018. Incorporation of safety and sustainability in conceptual design via a return on investment metric. ACS Sustain Chem Engg 6, 1411–1416.

Hasaneen R, El-Halwagi MM. 2017. Integrated process and microeconomic analyses to enable effective environmental policy for shale gas in the United States. Clean Technol Environ Policy, 19(6): 1775–1789.

Jasper S, El-Halwagi MM. 2015. A techno-economic comparison of two methanol-to-propylene processes. Processes, 3, 684–698.

Julián-Durán L, Ortiz-Espinoza AP, El-Halwagi MM, Jiménez-Gutiérrez A. 2014. Techno-economic assessment and environmental impact of shale gas alternatives to methanol. ACS Sustain Chem Engg 2 (10): 2338–2344.

Karagoz S. 2014. Process design, simulation and integration of dimethyl ether (DME) production from shale gas by direct and indirect methods. Master's thesis, Texas A&M University.

Marano J, Spivey JJ, Morreale B. 2015. Chemical synthesis. Chemical Engg Prog, 111(8): 58–62.

Martínez DY, Jiménez-Gutiérrez A, Linke P, Gabriel KJ, Noureldin MM, El Halwagi MM. 2013. Water and energy issues in gas-to-liquid processes: assessment and integration of different gas-reforming alternatives. ACS Sustain Chem Engg, 2(2): 216–225.

Martinez-Gomez J, Sánchez-Ramírez E, Quiroz-Ramírez JJ, Segovia-Hernandez JG, Ponce-Ortega JM, El-Halwagi MM. Involving economic, environmental and safety issues in the optimal purification of biobutanol. Proc Safety Environ Protect, 103B: 365–376.

Martinez-Gomez J, Nápoles-Rivera F, Ponce-Ortega JM, El-Halwagi MM. 2017. Optimization of the production of syngas from shale gas with economic and safety considerations. App Therm Engg, 110: 678–685.

Noureldin MM, Elbashir NO, El-Halwagi MM. 2014. Optimization and selection of reforming approaches for syngas generation from natural/shale gas. Industrial & Engg Chem Res, 53(5): 1841–1855.

Noureldin MMB, El-Halwagi MM. 2015. Synthesis of C-H-O symbiosis networks. AIChE J., 64 (4): 1242–1262.

Onel O, Niziolek A, Floudas C. 2015. Natural gas to liquids transportation fuels and olefins (GTL+C2_C4). 12th International Symposium on Process Systems Engineering and 25th European Symposium on Computer Aided Process Engineering. 31 May–4 June 2015, Copenhagen, Denmark.

Ortiz-Espinoza AP, Jiménez-Gutiérreza A, Nourledin M, El-Halwagi MM., 2017. Design, simulation and techno-economic analysis of two processes for the conversion of shale gas to ethylene. Comp Chem Eng. 107, 237-246 .

Pérez-Uresti SI, Adrián-Mendiola JM, El-Halwagi MM Jiménez-Gutiérrez A. 2017. Techno-economic assessment of benzene production from shale gas. Processes, 5(3): 33–42

Roy N, Eljack F, Jiménez-Gutiérrez A, Zhang B, Thiruvenkataswamy P, El-Halwagi MM, Mannan MS. A review of safety indices for process design. 2016. Curr Op Chem Engg, 14, 42–48.

Santibañez-Aguilar JE, Ponce-Ortega JM, González-Campos JB, Serna-González M, El-Halwagi MM. 2014. Optimal planning and site selection for distributed multiproduct biorefineries involving economic, environmental and social objectives. J Cleaner Product, 65(2): 270–294.

Sengupta D, Pike RW. 2012. Chemicals from biomass: integrating bioprocesses into chemical production complexes for sustainable development. CRC Press, Boca Raton, USA

Siirola J. 2014. The impact of shale gas in the chemical industry. AIChE J, 60(3): 810-819.

Thiruvenkataswamy P, Eljack FT, Roy N, Mannan MS, El-Halwagi MM. 2016. Safety and techno-economic analysis of ethylene technologies. J Loss Prevent Proc Indus, 39, 74–84

Turton R, Bailie R, Whiting W, Shaeiwitz j. 2003. Analysis, synthesis, and design of chemical processes. Upper Saddle River, NJ: Prentice Hall PTR.

U.S. Energy Information Administration (EIA). Annual Energy Outlook 2015. April 14, 2015. Available from: http://www.eia.gov/forecasts/aeo/.

10

Study for the Optimal Operation of Natural Gas Liquid Recovery and Natural Gas Production

Mozammel Mazumder and Qiang Xu

Dan F. Smith Department of Chemical Engineering, Lamar University, USA

CHAPTER MENU

10.1 Introduction

The recent increase in the supply of natural gas is the result of improved recovery methods (fracking) and has been a tectonic development in the energy sector. Natural gas prices have dropped significantly to less than 30% of their previous highs. As reported, cheap natural gas has greatly improved the cost and profitability of the U.S. chemical and petrochemical industry (Luyben et al. 2013). In the meantime, as the once largest energy-consuming country in the world, the United States has been becoming energy independent and may possibly become an important energy resource (especially natural gas) exporter. It is no doubt that natural gas will become a major clean energy resource worldwide in the coming decades. Along with this trend, the technology development for natural-gas processing becomes an increasingly important area in chemical and petrochemical engineering.

The composition of natural gas includes methane, heavier hydrocarbons (e.g., ethane, propane, and butane) and other impurities. Natural gas liquid (NGL) products can be extracted from natural gas through various separation processes to give products such as liquefied petroleum gas (LPG) while meeting the product specifications of sale gases. The NGL fractionation process is the most widely used method for the separation of NGL in natural-gas processing plants, in which a number of distillation columns are used to generate products including ethane, propane, butane, and C5+. However, such distillation processes are energy-intensive and hence improving the energy efficiency of NGL separation is one of the main targets for the reduction of costs (Kim et al. 2016).

NGLs are removed to maintain the product dew points and yield a source of revenue. NGL has significantly greater value than the main product of methane. The lighter NGL fractions, such as ethane, propane, and butane can be used as feedstock to refineries; whereas the heavier

Natural Gas Processing from Midstream to Downstream, First Edition.
Edited by Nimir O. Elbashir, Mahmoud M. El-Halwagi, Ioannis G. Economou, and Kenneth R. Hall.

hydrocarbons can be used as gasoline-blending stock. Meanwhile, the heavier hydrocarbons should also removed for the following reasons:

- Pentane and heavier hydrocarbons with a high freezing point may cause freezing and subsequent plugging of the process equipment during the downstream liquefaction;
- Heavy hydrocarbons (e.g., LPG and C5+) often have higher economic values than the liquefied natural gas (LNG) because they could serve as basic feedstock for petrochemical plants;
- Most LNG contracts specify a range of acceptable heating values, which requires the removal of heavier hydrocarbons. Such that the obtained LNG does not exceed the upper limit on the heating value (Khan et al. 2014a,b).

The increase in the price of energy sources has driven cryogenic natural gas liquid recovery plants to be efficient but more complex as well. In other words, the new generation of NGL plants is created based on decreasing fixed and operating costs for a specific output. To achieve this goal, optimal process design and operation are two important factors. There are various NGL process designs introduced to decrease the operating and capital costs for the high manufacturing efficiency. Decreasing the energy consumption for refrigeration by increasing the performance of the process heat integration is a fundamental procedure for improving the efficiency of such process designs. Heat integration is one of the most widely used methods for improving energy efficiency that has been successfully applied to a wide range of process industries over the last three decades. Heat integration methodology aims to maximize heat recovery in the processes by considering thermodynamic principles (Kim et al. 2016).

As an example of the optimal design for helping NGL recovery, the gas subcooled process (GSP) has been developed to save heat and power energy consumption and overcome problems encountered with the conventional process (Mehrpooya et al. 2009). For another example of the optimal operation for NGL recovery, the temperature in NGL recovery processes may reach to −100°C in the lowest value (this value is related to the demethanizer top gas outlet). In such processes the required refrigeration is gained by propane refrigeration cycles. Several studies have showed that the performance of NGL plants can be improved by adjusting the operating conditions of the process (Mehrpooya et al. 2006; Khan et al. 2014a,b; Chebbi et al. 2010; Kim et al. 2016).

The extraction of ethane and NGL from natural gas is generally based on some of the following alternatives: (i) external refrigeration (ER), (ii) turboexpansion (TE), (iii) Joule-Thompson expansion, and (iv) absorption. In many processing schemes, a combination of these effects is used to improve the energy efficiency or to obtain greater recoveries. The present work is directed to improving the economics and efficiency of the process for the recovery of NGL. A significant cost in NGL recovery processes is related to the refrigeration required to chill the inlet gas (Mehrpooya et al. 2009).

Refrigeration systems for NGL recovery use either propane or butane as refrigerants. In some applications, mixed refrigerants and cascade refrigeration cycles have been used. Refrigeration has also been provided by turboexpansion or work expansion of the compressed natural gas feed with appropriate heat exchange. Traditionally, the gas stream is partially condensed at medium to high pressures with the help of external propane refrigeration, a turboexpander or both. The condensed streams are further processed in a distillation column, e.g., a demethanizer operated at medium to low pressure, to separate the lighter components from the recovered hydrocarbon liquids (Mehrpooya et al. 2009).

Another improvement of the turboexpander-based NGL process is the IPSI enhanced NGL recovery process. This process utilizes a split stream from or near the bottom of the distillation column (demethanizer) as a mixed refrigerant. The mixed refrigerant is totally or partially vaporized, providing refrigeration for inlet gas cooling otherwise normally accomplished using an external refrigeration system. The vapor generated from this "self-refrigeration" cycle is recompressed and recycled back to the bottom of the tower, where it serves as a stripping gas. The innovation not only reduces or eliminates the need for inlet gas cooling via external refrigeration, but also provides the following enhancements to the demethanizer operation (Mehrpooya et al. 2009):

- Lower the temperature profile in the tower, thereby permitting better energy integration for inlet gas cooling via reboilers, resulting in reduced heating and refrigeration requirements;
- Reduce and/or eliminate the need for external reboiler heat, thereby saving fuel plus refrigeration.

The turboexpander is usually employed for the NGL recovery. Turbo-expansion provides the maximum amount of heat removal from a system for a given pressure drop while generating useful work like driving the compressor. The feed gas after precooling is sent to the expander and tower, respectively. The residue recycle (RR) was developed to achieve the NGL recovery higher than 80%, and GSP was developed to overcome the problems encountered with the conventional expander process (Mehrpooya et al. 2014).

The NGL recovery section in a gas processing plant is considered the heart of all processes, where C2+ hydrocarbons should be efficiently stripped from the feed stream. Several process schemes have been developed for high NGL recovery. The main affecting factors for those design options are the inlet condition, product requirement, mechanical constraint, and operational requirements. Among these, the condition for inlet gas has significant effects as it determines not only the types of process configuration to be employed but also the economic performance of the plant (Getu et al. 2015).

There are many extraction processes for natural gas liquids, which include Joule–Thompson (JT) expansion, refrigeration using propane in a chiller, and turboexpansion. Mixed refrigerants can also be used, but the most popular process in the NGL recovery industry is turboexpansion. More often than not, all three processes are used at once. In this study, all three processes are employed simultaneously. The feed is first cooled by the heat exchanger, which takes the heat from the demethanizer different tray, and further cooling of the feed is achieved by heat exchange with the residue gas. The natural gas stream is partially condensed at medium to high pressures with the help of external propane refrigeration. The gas leaving the separator is split, with one fraction subcooled by heat exchange with the overhead stream from the demethanizer, and the other fraction entering the turboexpander. Turboexpansion provides the maximum amount of heat removal from a system for a given pressure drop while generating useful work. The condensed streams are further processed in a demethanizer operated at medium to low pressure, to separate the lighter components from the hydrocarbon liquids. The fraction subcooled by the demethanizer overhead stream is flashed in a valve and fed to the tower as reflux. The maximum ethane and NGL recovery can be carried out by a large number of design variables that include, but are not limited to, demethanizer pressure and split ratio. The present work is directed to improve the economics and efficiency of the proposed process for the NGL recovery and natural gas production.

10.2 Methodology Framework

The framework of the NGL liquefaction process optimization is shown in Figure 10.1. For this new NGL liquefaction process, which simultaneously employs the three process of Joule–Thompson (JT) expansion, refrigeration using propane in a chiller, and turboexpansion. Rigorous modeling and simulation are first conducted for this NGL liquefaction process.

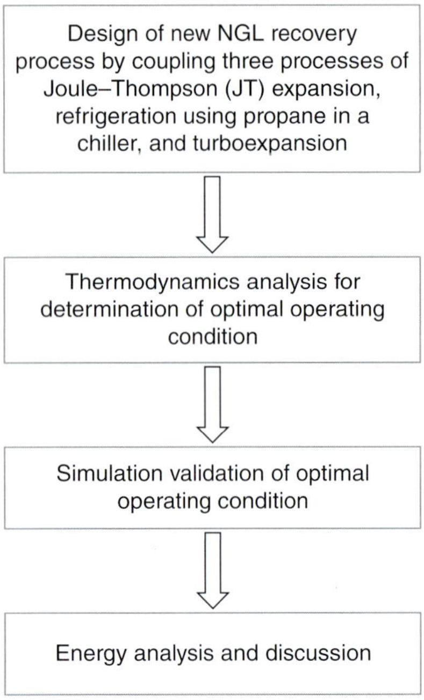

Figure 10.1 Framework of NGL liquefaction optimization.

According to various simulations, the operating states of the refrigerants and natural gas streams are obtained along with the operational condition of the entire NGL liquefaction process. Next, the energy consumption roadmap is disclosed through thermodynamic analysis, where the temperature and specific enthalpy (T-H) diagram is employed. From the thermodynamic studies, the opportunities for energy consumption minimization based on current design and operations could be identified. In this work, the optimal operating condition of the entire NGL liquefaction process has been determined. After that, the optimization results are examined and validated through the rigorous process simulation and thermodynamic analysis again. Finally, the energy analysis and the optimization benefits compared with the base case are also performed.

10.3 New Process Design for NGL Recovery

This study presents the a new NGL recovery process, which is a GSP process coupling Joule–Thompson (JT) expansion, refrigeration using propane in a chiller, and turboexpansion. Figure 10.2 shows a schematic diagram of the GSP process scheme. This is one of the state-of-art improvements made to the ISS (Industrial Single Standard Stage) process scheme. Normally, in an ideal operation, the residue gas stream 450 from the demethanizer column should contain pure methane, and the bottom stream 519 should contain all the remaining C2+ hydrocarbons. The presence of ethane in the residue gas depends on the column top temperature. As the volatility of ethane decreases, and so does the top temperature of the demethanizer column also decreases. Thus, ethane recovery could be increased significantly as a liquid product (Getu et al. 2015).

The feed is firstly cooled by the heat exchanger E-223 and E-224, which respectively take the heat from the demethanizer tray 11 (stream 439) and tray 16 (stream 436). Also, the feed

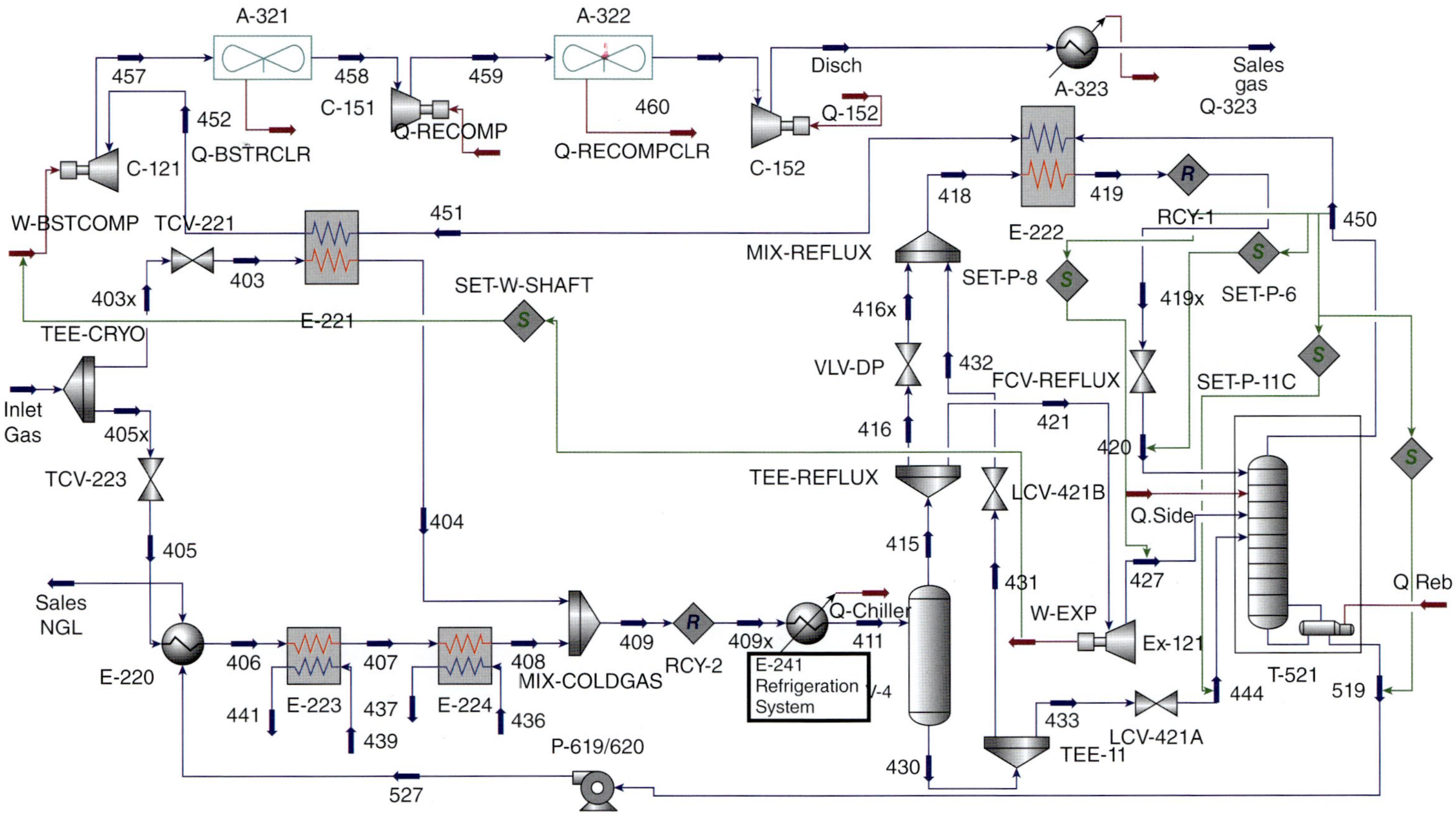

Figure 10.2 A new GSP process for NGL recovery.

is further cooled down by heat exchanger E-221 with the residue gas. The precooled feed is sent to a chiller where propane is used to reduce its temperature to −30°C. This temperature was selected to maximize cooling; taking into account the lowest temperature allowed in the chiller of −40°C required to avoid air leakage into the system and temperature approach in the chiller. The cold feed from the chiller enters a flash separator where the gas is separated from the liquid. A portion of the separated gas is cooled by heat exchange E-222 with a fraction of the overhead stream leaving the demethanizer column. It is then expanded through JT expansion valve (FCV-Reflux) and enters the demethanizer. The other portion of the separated gas is expanded in a turboexpander EX-121 and sent to the demethanizer at a lower stage. A fraction of the liquid leaving the separator is mixed with the separator gas outlet that goes into heat exchange, while the remainder undergoes JT expansion (LCV-421A) to column pressure and enters the demethanizer at a lower stage than the feed stream from the turboexpander. The demethanizer overhead is termed the residue gas. Part of the power required to recompress the residue gas is provided by the turboexpander, but another compressor is needed to bring the residue gas pressure up to 1115 psia.

All operational conditions of equipment are presented in Table 10.1. These data were used to input data into the simulator to perform a simulation as close as possible to the real process. Implementation of exergy analysis in process simulators requires exergy calculation along with the traditional energy and mass balances. To facilitate this step of exergy analysis, an Excel-based exergy calculator has been developed for Aspen Hysys which facilitates the thermo-economic analysis. It calculates exergy of heat, work, and material streams where the mixing exergy is being considered to be a part of physical exergy. The feed composition is a very vital factor that affects NGL recovery. Since the feed normally originates from upstream plants, there is a continuous variation of feed composition which increases the instability of the plant. Feed gas, sales gas, and sales NGL composition are shown in Table 10.2.

Table 10.1 Operating conditions of the NGL process equipment.

Equipment Name	Equipment Index	Operating Pressure (Inlet/Outlet) (psia)	Operating Temperature (Inlet/Outlet) (°C)
Heat exchanger (hot)	E-221	859/859	48.72/−17.55
Heat exchanger (hot)	E-220	862/862	48.72/28.69
Heat exchanger	E-223	858/858	28.69/7.47
Heat exchanger	E-224	856/856	7.47/−23.16
Chiller	E-241	849/849	−20.48/−30
Flash separator	V-421	842/842	−30/−30
Turboexpander	Ex-121	837/230	−30/−79.16
Demethanizer	T-521	225/225	−102.66/−102.66
Heat exchanger	E-222	225/225	−102.66/−31.38
Heat exchanger (cold)	E-221	218/218	−31.36/44.16
Compressor	C-121	208/281	44.22/73.88
Residue compressor	C-151	276/565	51.66/123.89
Residue compressor	C-152	560/1120	48.88/119.44
Cooler	A-323	1,120/1,120	119.44/48.88
NGL pump	P-619/620	228/578	1.66/4.44
Heat exchanger (cold)	E-220	578/573	4.44/29.36

Table 10.2 Operating conditions and composition of feed gas, sales gas, and sales NGL.

Parameter	Feed Gas	Sales Gas	Sales NGL
Flow (MMSCFD)	150	121.5	28.5
Pressure (psia)	864	1,115	573
Temperature (°C)	48.88	48.88	29.36
Nitrogen (mol %)	0.12	0.14	0.00
CO_2 (mol %)	0.01	0.01	0.03
Methane (mol %)	80.13	98.79	0.82
Ethane (mol %)	11.93	1.02	58.31
Propane (mol %)	4.21	0.04	21.93
i-Butane (mol %)	0.96	0.00	5.03
n-Butane (mol %)	1.26	0.00	6.61
i-Pentane (mol %)	0.44	0.00	2.31
n-Pentane (mol %)	0.34	0.00	1.78
n-Hexane (mol %)	0.60	0.00	3.15
H_2O (mol %)	0.01	0.00	0.00

10.3.1 Demethanizer

In all NGL recovery processes, one of the final steps in the plant is the production of the desired liquid product by use of a fractionation column. This column produces the specification product as a bottom product with the overhead stream being recycled to the process or sent out of the plant as residue gas product. Processes employing a demethanizer column separate methane and other more volatile components from ethane and less volatile components in the purified gas stream. An NGL recovery plant design is highly dependent on the operating pressure of the distillation column. At the medium to low pressure, i.e., 400 psia or lower (Mehdi et al. 2012).

The key design variable for distillation columns is the operating pressure of the column, as changing column operating pressure significantly affects different important operating condition such as relative volatilities, vapor loads, and temperature profiles inside the columns. For NGL fractionation it is necessary to simultaneously select both the column pressures and the distillation sequence. To identify optimal operating conditions and column sequences for NGL fractionation, a systematic and integrated design approach is required (Kim et al., 2016).

The study was conducted by first simulating the GSP process. Figure 10.2 demonstrates the process flow sheet for the GSP process. The figure does not depict the refrigeration loop, which is connected to the main process through the chiller. The feed is first cooled by the heat exchanger, which takes the heat from the demethanizer different tray and further cooling of the feed is achieved by heat exchange with the residue gas. During this optimization the number of theoretical stages is considered as seventeen stages, since it has been widely used in practice. The demethanizer pressures considered are 225 psia, which cover the typical range of demethanizer operating pressures, 100–450 psia (Chebbi et al. 2010).

10.3.2 J-T Expansion

One of the key parameters in the recovery of ethane and heavier products is the effect of the extraction on the heat content of the residue gas. The use of the Joule-Thomson (J-T) effect to recover liquids is an attractive alternative in many applications. With the appropriate heat exchange and large pressure differential across the J-T valve, cryogenic temperatures can be achieved resulting in high extraction efficiencies (Mehdi et al. 2012).

10.3.3 Turboexpander

A turboexpander generates the deep, low-temperature refrigeration industrially used for gas separation and liquefaction. It does so by the mechanism of constant entropy expansion, together with the production of power (a byproduct). Unlike J-T expanders, turboexpaners perform work during the process. Whereas J-T expansion is essentially an isenthalpic process, an ideal, thermodynamically reversible turboexpansion is isentropic. Turboexpansion provides the maximum amount of heat removal from a system for a given pressure drop while generating useful work (Mehdi et al. 2012).

10.3.4 Refrigeration

Castillo et al. (Castillo and Dorao 2013) studied suitable choices of refrigerants that are applicable for the precooling cycle by analyzing the effects of various refrigerants (i.e., N_2, CH_4, C_2H_6, and C_3H_8) on the compressor power using the Linde-Hampson process. It was found that compared to other refrigerants, propane has a higher specific refrigerant effect, which makes it the preferred refrigerant to be used in the precooling cycle (Khan et al. 2016).

The propane precooling cycle has been widely used in most LNG plants as the first cooling cycle in the NGL process. Three main LNG processes have been applied in LNG plants: single mixed refrigerant (SMR), propane precooled mixed refrigerant (C3MR), and cascade liquefaction process (Lim et al. 2014). In the last 15 years, the innovations of LNG technologies have drastically progressed whereby new LNG processes have been introduced such as mixed fluid cascade (MFC), Air Products (AP-X™), dual mixed refrigerant (DMR), and parallel mixed refrigerant (PMR) (Castillo et al. 2013). Most of the existing LNG plants have three main cooling cycles, namely the precooling, liquefying, and subcooling cycle. Earlier LNG plants that employed the SMR process did not have the precooling cycle, instead the natural gas was cooled directly to −160°C using a single mixed refrigerant. The precooling cycle is the first cycle in an LNG process that removes the heat from natural gas to a temperature range between −30°C to −55°C depending on the precooling technology applied. As a result of technological advancement, the precooling cycle can now be designed using either pure refrigerant or mixed refrigerant. Castillo et al. (2013) reported that 95% of the current LNG plants employ the precooling cycle; 85% of which are dominated by propane refrigerant compared to mixed refrigerant (Khan et al. 2016).

Ebrahimi, et al. (2015) observed that to have a minimum of work, when the ambient temperature increases, the compositions of both methane and ethane should be decreased, while those of nitrogen, propane, and n-butane should be increased. Single mixed refrigerant (SMR) processes are the simplest MR cycles for liquefaction of natural gas. Compared to mixed refrigerant cycles (MRC) that have been applied for both large and small-scale systems (Cao et al. 2006), SMR systems have only one series of compressors beside the main cryogenic heat exchangers to cover all range of the liquefaction temperatures. Nevertheless, these MR processes are highly energy-intensive, so an optimization procedure is required to minimize energy consumption. In this study we consider 80% propane with 20% butane as the refrigerant (Ebrahimi et al. 2015).

The refrigeration cycle is made up of four basic pieces of equipments: evaporator, compressor, condenser, and expansion valve, which are shown in Figure 10.3. The aim of using a refrigeration cycle is to provide needed cooling load. In other words, evaporator duty and temperature

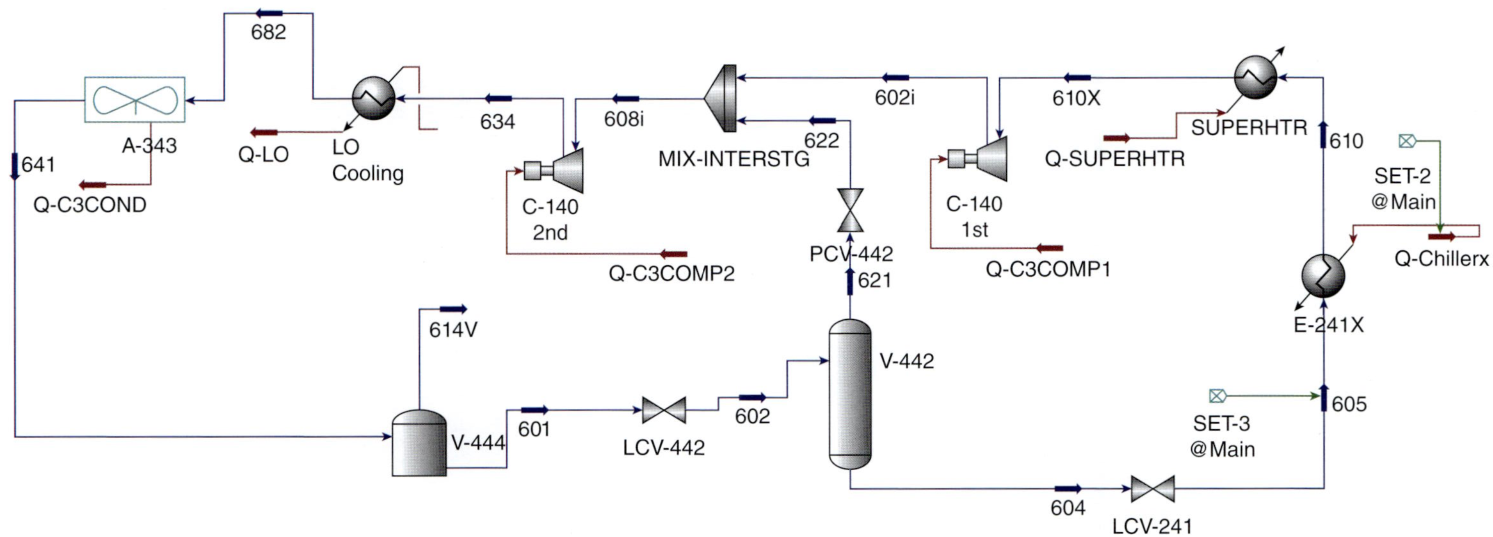

Figure 10.3 Flowsheet of the refrigeration system.

levels determine the ability of the refrigeration cycle. Therefore, in the productive structure the products of chiller (E-241) supposed to be the final products of the cycle (Stream 411) to the environment.

First, refrigerant passes through evaporators (E-241) and evaporates. Next, it flows to the refrigerant compressors (C-140 1st stage and 2nd stage), and its pressure is increased. Later it is led to the condenser (A-343) and condensed, followed by J–T valve (LCV-442) and a sudden decrease of the pressure. Finally, it reaches the evaporators again. The process simulation and exergy analysis of this refrigeration cycle have been completed.

10.3.5 Compression

The first residue compressor C-121 has the compression ratio of 1.353, for which the required duty is 2426hp taken from the Expander EX-121. It reaches a residue gas pressure of 281psia.The second two-stage residue gas compressors (C151 and C152) have the same compression ratio of 2, which suggests a discharge pressure of 1120 psia.

10.4 Thermodynamic Analysis for Propane Refrigeration System

10.4.1 Liquefaction Process Analysis

The natural gas (components in the mole ratio 1% nitrogen, 80.13% methane, 11.93% ethane, 4.21% propane, 0.9% i-butane, 1.26% n-butane, 0.44% i-pentane, and 0.34% n-pentane, 0.6% n-hexane) is precooled by exchanger (E-220, E-223, E-224, and E-221). Meanwhile, the propane refrigerant (in the mole ratio of 80% propane, 20% butane) is used in the propane refrigeration system. The process flowsheet of a typical propane refrigeration system, which contains two refrigeration cycles, is shown in Figure 10.3. One is for the compressor main suction line, and the other is for the compressor inter stage. Propane stream has its own cooling loop. From the accumulator V-444, the liquid propane goes to the Joule-Thompson (JT) valve LCV-442 to reduce the pressure and temperature of the propane stream and accumulate in the flash gas separator V-442. Here the propane vapor and propane liquid are separated. A portion of propane vapor goes to the compressor inter stage to increase the compressor performance and also reduce the energy consumption. Propane liquid from V-442 then again goes to the Joule-Thompson (JT) valve LCV-241 to reduce the pressure and temperature of the propane stream. After the JT valves, the propane streams have a vapor-liquid phase and re-enter chiller E-241 to absorb heat from all the input streams, in the meantime evaporating liquid propane into the saturated vapor stream. After this, all the vapor-phase propane streams enter a two-stage compressor (C-140) main suction to increase their pressure. Both the propane stream's main suction line and inter stage are merged together after compression, and then again go to the compressor second stage for further increase in pressure. The propane stream is then condensed by air condenser fan A-343. Propane is condensed here and changes its phase from gas to liquid. In normal operation, the inlet pressure of C-140 is 15 to 30 psia and inter stage pressure is 50 to70 psia. Compressor discharge pressure is 210 to 250 psia.

10.4.2 Simulation Results and Thermodynamic Analysis

To analyze the performance of propane refrigeration system, a rigorous simulation of the entire refrigeration system was conducted using Aspen HYSYS (version V8.8). The accurate

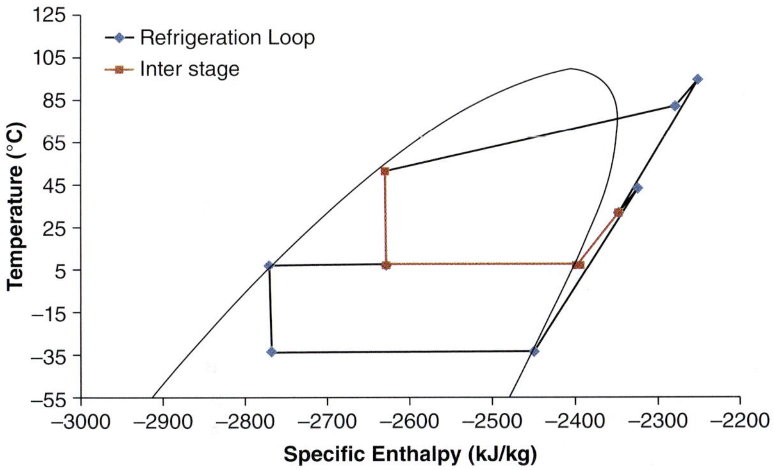

Figure 10.4 T-H diagram of the propane refrigeration cycle.

calculation of the fluid's thermodynamic properties is of indispensable importance for the basic engineering and performance of technical process. For oil, gas, or petrochemical applications, Peng-Robinson is the generally recommended property package. Therefore, in this process model, the built-in Peng-Robinson property method of Aspen Hysys is used to calculate thermodynamic properties of fluids. During the simulation, the ambient temperature is assumed to be maintained at 25°C, such that each stream will be cooled down to 30°C after passing through a water-cooler. All the operating temperatures and pressures of each stream inlets are summarized in Table 10.1.

Based on the simulation results, the operating states of the entire refrigeration system can be mapped into T-H diagrams. For the propane refrigeration cycle, the T-H diagram is given in Figure 10.4. To disclose the phase information, the saturated liquid line and the saturated vapor line of propane are also shown in the figure. Based on the temperature and specific enthalpy information, all the operating states of the propane cycle labeled in Figure 10.4.

From Figure 10.4, the compressor discharge reaches such a high pressure and temperature (94°C) that it has to be liquefied at a high temperature, around 52°C. If compressor discharge is operated at a lower pressure, the condensing temperature will be lower. This means if the last-stage discharge pressure could be reduced to some extent so that the condensing temperature will occur at the temperature of propane, around 45°C, the liquid temperature reduction by water-cooler can be avoided. Thus, the current operation of the propane cycle sets an unnecessarily high compressor final discharge pressure that wastes the compression energy and consumes more water-cooling duty. Therefore, the optimization opportunity is to drive the compressor pressure down to the dew point of propane stream, which is around 45°C.

10.5 Optimization for Natural Gas Liquefaction

10.5.1 Optimization Model Development

The rigorous simulation-based (sequential modular) optimization is conducted in Aspen HYSYS V8.8. To illustrate the optimization model development, the major model equations are presented below.

10.5.1.1 Objective Function

In this work, the objective is to minimize the total shaft work consumption rate, W_{total}. The compressor compression ratio at each stage in both refrigeration cycle and residue gas cycle is considered as the manipulating variable.

$$\min_{r_{j,i}} W_{total} = \sum_{j,i} W_{j,i} \tag{10.1}$$

where $(j, i) \in \{(1,1), (1,2), (2,1), (2,2), (2,3)(2,4)\}$; $r_{j,i}$ and $W_{j,i}$ are the compression ratio and shaft work at the i-th stage compressor of the j-th cycle, respectively. Note that j=1 means the refrigeration cycle while j=2 means the residue gas cycle. During the simulation, the refrigeration compressor adiabatic efficiency is set as 62% based on normal industrial experience, and the polytropic efficiency is assumed to be 65%. The residue gas compressor adiabatic efficiency is set as 75%, and the polytropic efficiency is assumed to be 77%. The compressor work consumption on each stage is equal to the specific enthalpy change of the fluid stream multiplying its mass flowrate. Equation (10.2) calculates the work consumption at each stage of the propane cycle; while Eq. (10.3) calculates the work consumption at each stage of the residue gas cycle.

$$W_{1,i} = \frac{i}{2} f_R \, \Delta H_{1,i}, \quad i = 1, 2, \tag{10.2}$$

$$W_{2,i} = f_{RG} \, \Delta H_{2,i}, \quad i = 1, 2, 3, 4, \tag{10.3}$$

where f_R is the total mass flowrate of refrigerant; f_{RG} is the total mass flowrate of Residue Gas. $\Delta H_{1,i}$ and $\Delta H_{2,i}$ are respectively the specific enthalpy changes at each compression stage in refrigeration and residue gas cycles, respectively. $\Delta H_{1,i}$ or $\Delta H_{2,i}$ are calculated by the outlet enthalpy ($H_{j,i}^{out}$) minus inlet enthalpy ($H_{j,i}^{in}$) as shown in Eq. (10.4). Both $H_{j,i}^{out}$ and $H_{j,i}^{in}$ are the functions ($\delta(\bullet)$) of temperature ($T_{j,i}^{out}, T_{j,i}^{in}$), pressure ($P_{j,i}^{out}, P_{j,i}^{in}$), and compositions ($\mathbf{C}_{j,i}$) as shown in Eqs. (5) and (6), which actually is calculated based on rigorous simulations.

$$\Delta H_{j,i} = H_{j,i}^{out} - H_{j,i}^{in} \tag{10.4}$$

$$H_{j,i}^{out} = \delta(P_{j,i}^{out}, \; T_{j,i}^{out}, \; \mathbf{C}_{j,i}) \tag{10.5}$$

$$H_{j,i}^{in} = \delta(P_{j,i}^{in}, \; T_{j,i}^{in}, \; \mathbf{C}_{j,i}) \tag{10.6}$$

The outlet temperature ($T_{j,i}^{out}$) of each compressor is a function ($\tau(\bullet)$) of the inlet temperature ($T_{j,i}^{in}$), pressure ($P_{j,i}^{out}, P_{j,i}^{in}$), and composition ($\mathbf{C}_{j,i}$) represented by Eq. (7).

$$T_{j,i}^{out} = \tau(P_{j,i}^{out}, \; P_{j,i}^{in}, \; T_{j,i}^{in}, \; \mathbf{C}_{j,i}) \tag{10.7}$$

where $(j, i) \in \{(1,1), (1,2), (2,1), (2,2), (2,3), (2,4)\}$. Note that $\mathbf{C}_{j,i}$ is a vector representing all of the component compositions of the investigated stream.

Since a cooler is employed before compressor C151 and C152, the inlet temperatures of C151 and C152 are both at 48.89°C. Eqs. (10.8) through (10.9) give the related calculations.

$$T_{j,i}^{in} = 48.89, \quad (j, i) \in \{(2,3), (2,4)\}, \tag{10.8}$$

$$T_{j,i}^{in} = T_u, \quad (j, i) \in \{(1,1), (1,2), (2,1), (2,2)\}, \quad u \in \{610x, 608i, 421, 452\} \tag{10.9}$$

10.5.1.2 Pressure Ratio Constraints

Generally, the pressure ratio of a compressor is defined by its outlet pressure divided by its inlet pressure, which can vary in the range from 1.5 to 3.5.

$$r_{j,i} = P_{j,i}^{out}/P_{j,i}^{in} \tag{10.10}$$

$$1.5 \le r_{j,i} \le 3.5 \tag{10.11}$$

where $(j, i) \in \{(1, 1), (1, 2), (2, 1), (2, 2), (2, 3), (2, 4)\}$.

10.5.1.3 Heat Transfer Constraints

To make heat transfer possible in the heat exchanger, the temperature difference between the hot stream and cold stream has to meet some threshold. For the counter-current flow exchanger, the outlet temperature of the hot stream has to be higher than the inlet temperature of the cold stream; similarly, the inlet temperature of the hot stream has to be higher than the outlet temperature of the cold stream. For a cryogenic refrigeration system, 2°C is usually adopted as the minimum temperature difference. Thus, the heat transfer feasible constraints are modeled by Eqs. (10.12) and (10.13).

$$T_{HX,out}^{hot} - T_{HX,in}^{cold} \ge 2 \tag{10.12}$$

$$T_{HX,in}^{hot} - T_{HX,out}^{cold} \ge 2 \tag{10.13}$$

where

$$(T_{HX,out}^{hot},\ T_{HX,in}^{cold}) \in \{(T_{407}, T_{439}), (T_{408}, T_{436}), (T_{419}, T_{450}), (T_{404}, T_{451}), (T_{411}, T_{605})\};$$

$$(T_{HX,in}^{hot},\ T_{HX,out}^{cold}) \in \{(T_{406}, T_{441}), (T_{407}, T_{437}), (T_{418}, T_{451}), (T_{403}, T_{452}), (T_{409x}, T_{610})\}.$$

10.5.1.4 Energy Balance Constraints

For simplicity, heat losses of heat exchangers are ignored. According to the energy balance of all the sub exchangers shown in Figure 10.2, Eqs. (10.14) through (10.34) must be hold.

Equation for E-223:

$$Q_{E-223} = f_{439}(H_{441} - H_{439}) \tag{10.14}$$

$$Q_{E-223} + f_{406}(H_{407} - H_{406}) = 0 \tag{10.15}$$

Equation for E-224:

$$Q_{E-224} = f_{436}(H_{437} - H_{436}) \tag{10.16}$$

$$Q_{E-224} + f_{407}(H_{408} - H_{407}) = 0 \tag{10.17}$$

Equation for E-222:

$$Q_{E-222} = f_{450}(H_{451} - H_{450}) \tag{10.18}$$

$$Q_{E-222} + f_{418}(H_{419} - H_{418}) = 0 \tag{10.19}$$

Equation for E-221:

$$Q_{E-221} = f_{451}(H_{452} - H_{451}) \quad (10.20)$$

$$Q_{E-221} + f_{403}(H_{404} - H_{403}) = 0 \quad (10.21)$$

Equation for E-241:

$$Q_{E-241} = f_{605}(H_{610} - H_{605}) \quad (10.22)$$

$$Q_{E-241} + f_{409x}(H_{411} - H_{409x}) = 0 \quad (10.23)$$

Equation for Expansion Valves:

$$H_{433} = H_{444} \quad (10.24)$$

$$H_{431} = H_{432} \quad (10.25)$$

$$H_{416} = H_{416x} \quad (10.26)$$

$$H_{419x} = H_{420} \quad (10.27)$$

$$H_{421} = H_{427} \quad (10.28)$$

Equation for Mixers:

$$H_{409} = H_{408} + H_{404} \quad (10.29)$$

$$H_{418} = H_{416x} + H_{432} \quad (10.30)$$

$$H_{608i} = H_{602i} + H_{622} \quad (10.31)$$

Equation for TEEs:

$$H_{Inlet\ gas} = H_{403x} + H_{405x} \quad (10.32)$$

$$H_{430} = H_{431} + H_{433} \quad (10.33)$$

$$H_{415} = H_{416} + H_{421} \quad (10.34)$$

where f_{4xx} is the mass flowrate of natural gas at the corresponding stream; f_{6xx} is the mass flowrate of refrigerant at the corresponding stream; Q_{xxx} is the heat duty exchanged in corresponding exchangers. In the above equations, various specific enthalpies are generalized by Eq. (10.35). Again, $\mathbf{C}_u$ is a vector representing all the component compositions of the investigated stream.

$$H_u = \delta(P_u, T_u, \mathbf{C}_u) \quad (10.35)$$

where,

$$u \in \left\{ \begin{array}{l} 441, 409, 407, 406, 437, 436, 408, 451, 450, 419, 418, 452, 404, 403, 610, 605, 411, \\ 409x, 431, 432, 433, 444, 416, 416x, 419x, 420, 409, 408, 404, 608i, 602i, 622, \\ InletGas, 403x, 405x, 430, 415, 421 \end{array} \right\}.$$

10.5.1.5 Other Constraints

To make sure that NGL has to reach −90°C at the outlet of the demethenizer T-521, Eq. (10.37) provides such a process specification. As mentioned before, thermodynamic analysis for energy consumption minimization suggests the last-stage discharge pressure of the propane compressor should be lower than the original value (261 psia); meanwhile, it cannot be lower than the pressure of 156.78 psia, under which the propane has the dew point temperature at around 45°C (see Figure 10.4). Thus, the discharge pressure limits are shown in Eq. (10.38).

$$T_{450} \leq -90 \tag{10.36}$$

$$156.78 \leq P_{1,3}^{out} < 261 \tag{10.37}$$

10.5.2 Optimization Results

In this NGL liquefaction process, a rigorous simulation for base case is first conducted. Then, the energy consumption roadmap is disclosed through thermodynamic analysis, where T-H diagram is employed. Then, a rigorous optimization model is developed and solved. Optimization results are discussed in this section.

10.5.2.1 Optimization Results of Propane Cycle

The propane T-H diagram based on the optimization results is plotted (solid line) together with the base case simulation results (dashed line) in Figure 10.5. The critical operating states labeled in Figure 10.4 are re-remarked in Figure 10.5 based on the optimization results. As shown, the propane condensation occurs at the temperature of 45°C, which is exactly expected from the thermodynamic analysis. Meanwhile, the specific enthalpy of propane after the final stage of refrigeration compression is much lower than the base case, which suggests significant compression work savings. Similarly, the discharge temperature and pressures of the other stages are also reduced. Note that the reductions can still ensure that the propane cycle could provide enough cold utility for the pre-cooling task.

10.5.2.2 Optimization Results of Compressor and Condenser

All the compressors' operational conditions before and after optimization, including discharge pressure, compression ratio, and brake power, are summarized in Table 10.3. The overall

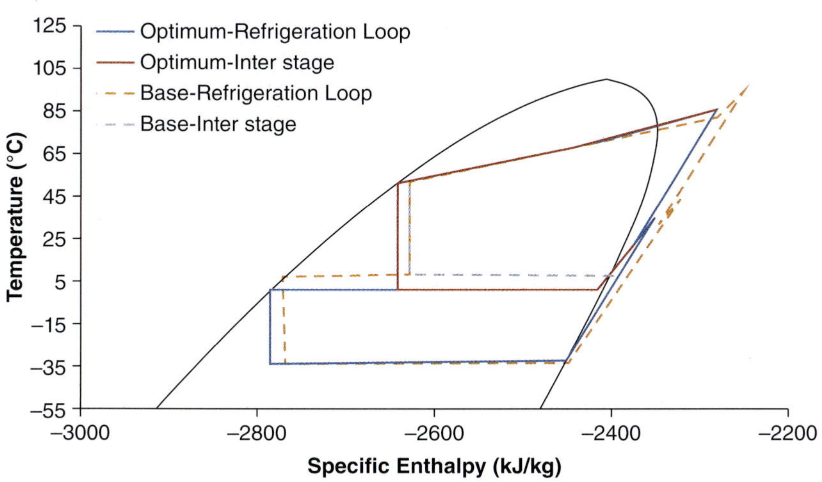

Figure 10.5 T-H diagram of the propane refrigeration cycle for the base and optimization cases.

Table 10.3 Compressor operation comparison between the base and optimization cases.

Compressor Index	Outlet Pressure (Pisa)		Compression Ratio		Shaft Work (KW)	
	Base Case	Optimization Case	Base Case	Optimization Case	Base Case	Optimization Case
C-140	50.79	70.5	3.577	4.65	480.8	619.8
C-141	267	228.6	3.815	3.26	915.2	814.6
Ex-121	230	230	0.2748	0.2748	−1,846	−1,846
C-121	280	281	1.348	1.353	1,809	1,809
C-151	1,120	565	4.09	2.045	9,730	4,508
C-152	1,120	1,120	----	2	----	4,248
Total	----	----	----	----	11,089	10,153

Table 10.4 Comparisons of condenser cooling duties between the base and optimization cases.

Cooler Index	Cooling Duty (KW)	
	Base Case	Optimization Case
A-343	2,826	2,992
A-321	1,687	1,636
A-322	11,430	5,032
A-323	----	5,237
Total	15,943	14,897

optimization result shows that 936 KW of total brake power is saved compare with the base case, accounting for savings of nearly 8.45%. In addition, condenser duties are also compared in Table 10.4. It shows that the condenser duties are also reduced by 1,046 KW, which accounts for about 6.56% condenser duties consumption.

Note that the total consumptions of brake power and condenser duties are reduced almost the same amount. This could be explained by Figure 10.6 for the energy flow overview of the entire LNG liquefaction system. The total need of cooling duty (Q_{NGL}) is actually fixed. The entire process actually is to transfer Q_{NGL} from the natural gas stream to exchanger cooling through the propane refrigeration systems and the turbo expansion system at the expense of compressor brake power, which is represented as (W_{total}). Its calculation is shown in Eq. (10.1). Thus, the overall system energy balance is that the sum of all the condenser cooling duty (Q_{AC}) is equal to the sum of Q_{NGL} and W_{total}, as shown in Eq. (10.38).

$$Q_{AC} = Q_{NGL} + W_{total} \tag{10.38}$$

With the fixed cooling capacity (Q_{NGL}) for the case study, Q_{WC} in Eq. (10.40) will change in the same trend with W_{total}. Thus, when W_{total} is reduced after optimization, Q_{WC} will also reduce in the same magnitude. Therefore, the NGL liquefaction optimization not only minimizes the total shaft work, but also cuts down the condenser heat duty consumption.

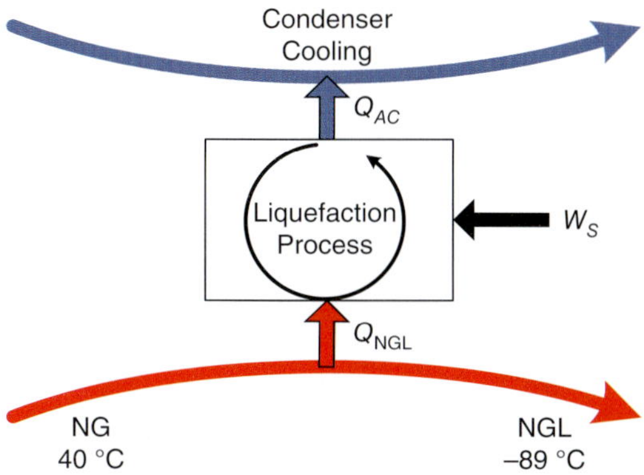

Figure 10.6 Energy flow overview of the NGL liquefaction system.

For further in-depth analysis, a coefficient of performance (COP) analysis is conducted for evaluating the efficiency of the cryogenic system. COP definition is shown in Eq. (10.39).

$$COP = \frac{Q_C}{W_S} \tag{10.39}$$

where Q_C is the total heat removed by refrigerant along the precooling cycle and the turboexpander cooling cycles in the process, which is calculated in Eq. (10.40).

$$Q_C = Q_{E-223} + Q_{E-224} + Q_{E-241} + Q_{E-222} + Q_{E-221} \tag{10.40}$$

where Q_{E-223}, Q_{E-224}, Q_{E-222}, Q_{E-221} and Q_{E-241} are obtained by Eqs. (10.14), (10.16), (10.18), (10.20), and (10.22), respectively. From the calculation, the COP for the base case is 1.39. It increases to 1.53 after being optimized, which also indicates that the liquefaction process efficiency is improved after optimization.

To explain the improvement of this NGL liquefaction process from a thermodynamic point of view, the hot and cold composite curves of the propane refrigeration system and cryogenic turboexpander system are plotted in Figure 10.7. For both the base case and the optimization case, the NG cooling curve (hot composite curve) is the same. The propane refrigeration cycle and cryogenic turboexpander cool down NG in different temperature ranges. The propane refrigeration and cryogenic turboexpander composite curves (cold composite curves) are both under the NG cooling curve. The vertical differences between hot and cold composite curves are actually the temperature drive force for conducting heat transfer. In Figure 10.7, the base case is represented using dashed lines, while the optimization case is represented by the solid line. It can be seen that the optimization case has less driving forces on the cryogenic turboexpander cycle while having larger driving forces on the propane refrigeration cycle. Note that as long as the temperature driving forces are satisfactory (i.e., larger than the threshold value), they should be as small as possible in order to increase the heat transfer efficiency. Thus, the efficiency of cryogenic turboexpander cooling is increased, while the efficiency of propane precooling is actually decreased after optimization. However, because the liquefaction process is mainly controlled by the cryogenic turboexpander cooling cycle, the improvement of the cryogenic turboexpander cooling cycle dominates the final results. Therefore, the performance of the entire NGL liquefaction process is improved.

10.5.2.3 Demethanizer Pressure and Ethane Recovery

Although the ethane recovery process allows for ethane recoveries of 99% or higher values, the process optimization yielded significantly lower recovery values for intermediate and high

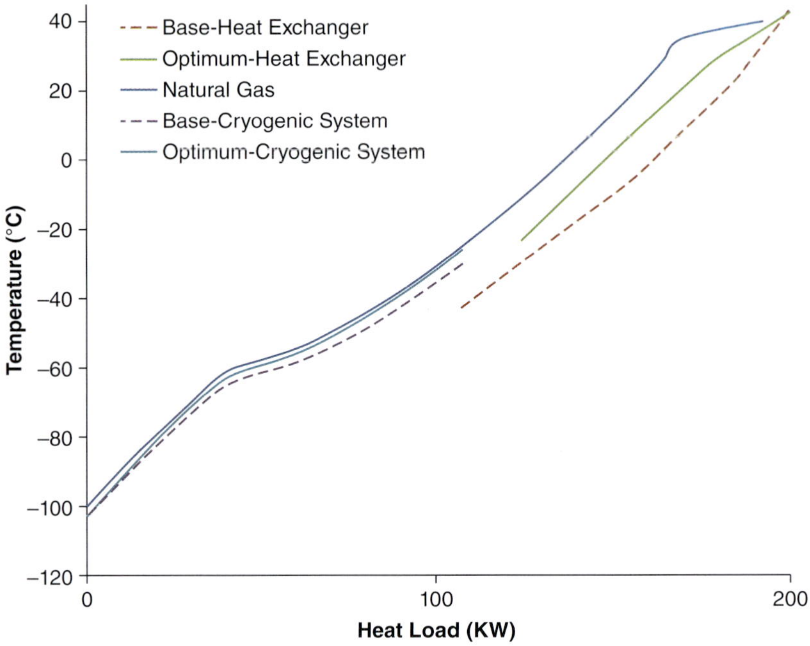

Figure 10.7 Composite curve comparison between the base and optimization cases.

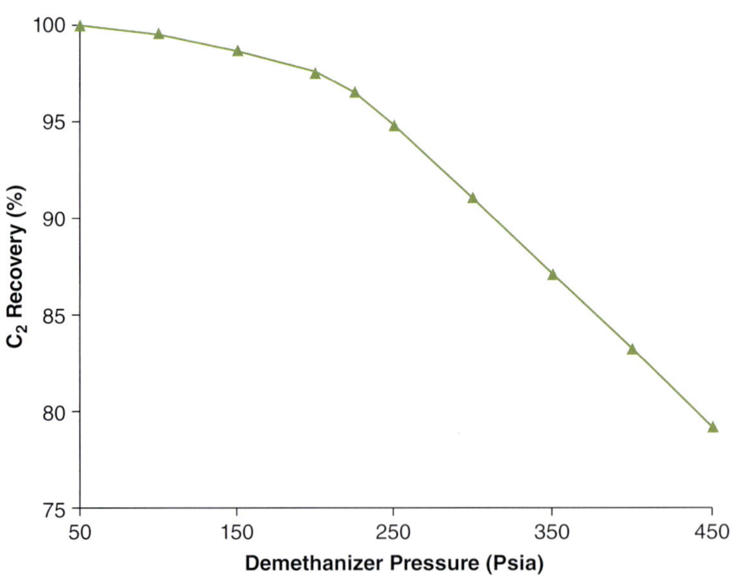

Figure 10.8 Effect of demethanizer pressure on ethane recovery.

demethanizer pressures. Figure 10.8 shows the effect of demethanizer pressure on ethane recovery. Higher ethane recoveries are obtained at low demethanizer pressure. The lowest ethane recovery is at the highest demethanizer pressure (450 psia).

But Figure 10.9 shows the effect of demethanizer pressure on ethane recovery with the energy required. Higher energy required at low demethanizer pressure. Table 10.5 compares

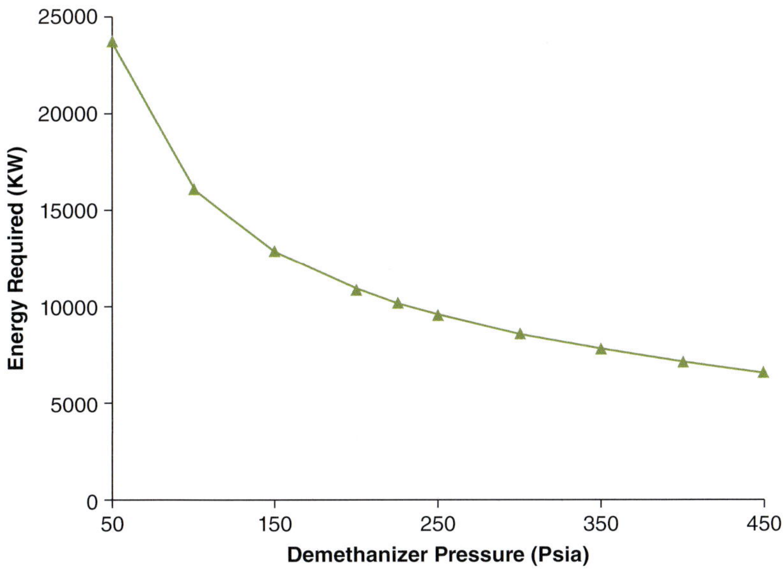

Figure 10.9 Effect of demethanizer pressure with energy required.

Table 10.5 Comparison of demethanizer pressure and NGL recovery rate.

Demethanizer Pressure (psia)	NGL Recovery Rate (%)	Energy Required (KW)
50	100	23,748
100	99.53	16,080
150	98.68	12,855
200	97.53	10,904
225	96.49	10,153
250	94.83	9,568
300	91.05	8,575
350	87.13	7,783
400	83.22	7,129
450	79.20	6,576

the demethanizer pressure and NGL recovery rate with the associated energy required. Ethane recovery is 100% at low demethanizer pressure but required almost double the energy than the optimum condition. Ethane recovery is 96.5% at optimum pressure which we consider 225psia and required less energy.

Table 10.5 Comparison of demethanizer pressure and NGL recovery rate

At the low demethanizer pressure (100 psia), the reflux stream entering the tower is −122.78°C colder than the turboexpander outlet stream entering the tower. For intermediate pressure (225 psia), the temperature difference between the reflux stream and the turboexpander outlet stream entering the tower is 6.11°C. It is speculated that it is the large temperature gap observed at low demethanizer pressure that is responsible for the superior ethane recovery as compression not only enhances pressure but also temperature.

10.6 Conclusion

The feed is first cooled by the heat exchanger, which takes the heat from the demethanizer different tray, and further cooling of the feed is achieved by heat exchange with the residue gas. The natural gas stream is partially condensed at medium to high pressures with the help of external propane refrigeration and a turboexpander. In such processes the required refrigeration is gained by propane refrigeration cycles. The optimization opportunity is to drive the refrigeration compressor pressure down to the dew point of the propane stream, which is around 30°C. Turboexpansion provides the maximum amount of heat removal from a system for a given pressure drop while generating useful work. The first residue compressor C-121 has the compression ratio of 1.353 for which the required duty is 2426hp taken from the expander EX-121. It reaches a residue gas pressure of 281psia. The condensed streams are further processed in a demethanizer operated at medium to low pressure, to separate the lighter components from the hydrocarbon liquids. The process temperature in NGL recovery processes may reach −100°C in the lowest value. For intermediate pressure (225 psia), the temperature difference between the reflux stream and the turboexpander outlet stream entering the tower is 6.11°C. It is the large temperature gap observed at low demethanizer pressure that is responsible for the superior ethane recovery.

For the developed NGL liquefaction process, a rigorous simulation for the base case is first conducted. Then, the energy consumption roadmap is disclosed through thermodynamic analysis, where the T-H diagram is employed. In this way, a rigorous optimization model is developed and solved. Finally, the optimization benefits compared with the given base case are also calculated. The overall optimization result shows that 936 KW of total compressor brake power is saved, compared with the base case, accounting for near 8.45% savings. It shows that the condenser duties are also reduced by 1,046 KW which accounts for about 6.56% condenser duties consumption. The COP for base case is 1.39. It increases to 1.53 after being optimized, which also indicates that the liquefaction process efficiency is improved after optimization.

Acknowledgements

This work was supported in part by a Graduate Student Scholarship, the President Visionary Project, and the Anita Riddle Faculty Fellowship from Lamar University.

Abbreviations

AP-X™	Air Products
C3MR	Propane precooled mixed refrigerant
COP	Coefficient of performance
DMR	Dual mixed refrigerant
GSP	Gas subcooled process
LNG	Liquefied natural gas
LPG	Liquefied petroleum gas

MFC	Mixed fluid cascade
NGL	Natural gas liquid
PMR	Parallel mixed refrigerant
SMR	Single mixed refrigerant

Nomenclature

Sets and indices

i	index of compression stage
j	index of compression cycle (j=1 refrigeration cycle; j=2 residue gas cycle)
k	index of the heat exchanger indicated in Figure 10.2
u	stream status indicated in Figure 10.2

Variables

COP	coefficient of performance for the NGL liquefaction process
f_R	mass flowrate of the refrigerant stream (kg/hr)
f_{RG}	mass flowrate of the residue gas stream (kg/hr)
$H_{j,i}^{in}$, $H_{j,i}^{out}$	inlet and outlet stream–specific enthalpy at the i-th stage compressor of the j-th cycle (kJ/kg)
H_u	stream-specific enthalpy at status u (kJ/kg)
$\Delta H_{j,i}$	specific enthalpy change at the i-th stage compressor of the j-th cycle (kJ/kg)
$P_{j,i}^{in}$, $P_{j,i}^{out}$	inlet and outlet stream pressure for the i-th stage compressor of the j-th cycle (psia)
Q_{E-K}	heat duty of the heat exchanger E-k (KW)
Q_C	total heat removed by refrigerant and turboexpander cooling cycles (KW)
Q_{AC}	total condenser-coolers' cooling duty (KW)
Q_{NGL}	required cooling duty from NGL liquefaction (KW)
$r_{j,i}$	compression ratio for the i-th stage compressor of the j-th cycle
$T_{j,i}^{in}$, $T_{j,i}^{out}$	inlet and outlet stream temperature for the i-th stage compressor of the j-th cycle (°C)
$T_{HX,in}^{cold}$, $T_{HX,out}^{cold}$	inlet and outlet temperature for the cold stream of an exchanger (°C)
$T_{HX,in}^{hot}$, $T_{HX,out}^{hot}$	inlet and outlet temperature for the hot stream of an exchanger (°C)
T_u	temperature at status u (°C)
W_{total}	total shaft work consumption rate (KW)
$W_{j,i}$	shaft work at the i-th stage compressor of the j-th cycle (KW)

Vectors

$\mathbf{C}_{j,i}$	component composition vector of the stream passing through the i-th stage compressor of the j-th cycle

Functions

$\delta(\bullet)$	function for calculating the stream enthalpy rate
$\tau(\bullet)$	function for calculating compressor outlet stream temperature

References

Cao W, Lu X, Lin W, Gu A. 2006. Parameter comparison of two small scale natural gas liquefaction processes in skid mountain package. Appl Therm Eng, 26 (8): 898-904.

Castillo L, Dorao CA. 2013. On the conceptual design of pre-cooling stage of LNG plants using propane or ethane/propane mixture. Energy Convers Manage, 65: 140–146.

Chebbi R, Amoodi NS, Jabbar NMA, Husseini GA, Mazroui KA. 2010. Optimum ethane recovery in conventional turboexpander process. Chem Eng Res Design, 8 (8): 779–787.

Ebrahimi H, Moein P, Sarmad M, Zare M, Pakseresht S, Vakili SZ. 2015. APCI-LNG single mixed refrigerant process for natural gas liquefaction cycle: analysis and optimization. J Nat Gas Sci Egg, 26: 470–479.

Fahmy MFM, Nabih HI, Nigeily M El. 2016. Enhancement of the efficiency of the Open Cycle Phillips Optimized Cascade LNG process. Energ Convers Manage, 112: 308–318.

Getu M, Mahadzir S, Samyudia Y, Khan MS, Bahadori A, Lee M. 2015. Risk-based optimization for representative natural gas liquid (NGL) recovery processes by considering uncertainty from the plant inlet. J Nat Gas Sci Egg, 27 42–54.

He T, Ju Y. 2014. Design and optimization of a novel mixed refrigerant cycle integrated with NGL recovery process for small-scale LNG plant. Ind Engg Chem Res, 53, 5545–5553.

He T, Ju Y. 2016. Dynamic simulation of mixed refrigerant process for small-scale LNG plant in skid mount packages. Energy, 97: 350–358.

Khan MS, Chaniago YD, Getu M, Lee M. 2014a. Energy saving opportunities in integrated NGL/LNG schemes exploiting thermal-coupling common-utilities and process knowledge. Chem Engg Proc, 82: 54–64.

Khan MS, Lee S, Hasan M, Lee M. 2014b. Process knowledge based opportunistic optimization of the N_2-CO_2 expander cycle for the economic development of stranded offshore fields. J Nat Gas Sci Egg, 18: 263–273.

Khan NBN, Barifcani A, Tade M, Pareek V. 2016. A case study: application of energy and exergy analysis for enhancing the process efficiency of a three stage propane pre-cooling cycle of the cascade LNG process. J Nat Gas Sci Egg, 29: 125–133.

Kherbeck L, Chebbi R. 2015. Optimizing ethane recovery in turboexpander processes. J Indus Engg Chem, 21: 292–297.

Kim JK, Yoo H, Binns M, Jang M, Cho, H. 2016. A Design procedure for heat-integrated distillation column sequencing of natural gas liquid fractionation processes. Korean J Chem Eng, 33 (2): 405–415.

Lim W, Lee I, Tak K, Cho JH, Ko D. 2014. Efficient configuration of a natural gas liquefaction process for energy recovery. Ind Eng Chem Res, 53: 1973–1985.

Liu CW, Zhang J, Xu Q, Gossage JL. 2010. Thermodynamic-analysis-based design and operation for boil-off gas flare minimization at LNG receiving terminals. Ind Eng Chem Res, 49: 7412–7420.

Luyben WL. 2013. NGL demethanizer control. Ind Eng Chem Res, 52L 11626–11638.

Mehdi M, Vatani A, Moosavian SMA. 2012. Optimum pressure distribution in design of cryogenic NGL recovery processes. Iran J Chem Chem Engg, 31 (3): 1021–9986.

Mehrpooya M, Gharagheizi F, Vatani A. 2009. Thermoeconomic analysis of a large industrial propane refrigeration cycle used in NGL recovery plant. Int J Energy Res, 33: 960–977.

Mehrpooya M, Gharagheizi F, Vatani A. 2006. An optimization of capital and operating alternatives in a NGL recovery unit. Chem Eng Technol, 29, (12): 1469–1480.

Mehrpooya M, Hossieni M, Vatani A. 2014. Novel LNG-based integrated process configuration alternatives for coproduction of LNG and NGL. Ind Eng Chem Res, 53: 17705–17721.

Mehrpooya M, Vatani A, Mousavian SMA. 2010. Introducing a novel integrated NGL recovery process configuration (with a self-refrigeration system (open–closed cycle)) with minimum energy requirement. Chem Engg Proc, 49: 376–388.

Mortazavi A, Somers C, Alabdulkarem A, Hwang Y, Radermacher R. 2010. Enhancement of APCI cycle efficiency with absorption chillers. Energy, 35: 3877–3882.

Rinc MD, Junca CJ, Duarte CR. 2016. A novel absorption process for small-scale natural gas dew point control and dehydration. J Nat Gas Sci Egg, 29: 264–274.

Shin MW, Shin D, Choi SH, Yoon ES, Han C. 2007. Optimization of the operation of boil-off gas compressors at a liquified natural gas gasification plant Ind Eng Chem Res, 46: 6540–6545.

Shin MW, Shin D, Choi SH, Yoon ES. 2008. Optimal operation of the boil-off gas compression process using a boil-off rate model for LNG storage tanks. Korean J Chem Engg, 25 (1): 7–12.

Tahmasebi S, Abbasabadi AB, Ghasemi N, Javadian H. 2015. Investigation of various feed conditions on NGL recovery plant energy and exergy performance: a case study. J Nat Gas Sci Eggg, 22: 83–89.

Vatania A, Mehrpooyab M, Tirandaz B. 2013. A novel process configuration for co-production of NGL and LNG with low energy requirement. Chem Engg Proc, 63: 16– 24.

Wang M, Khalilpour R, Abbas A. 2014. Thermodynamic and economic optimization of LNG mixed refrigerant processes. Energ Conver Manage, 88: 947–961.

Wang M, Xu Q. 2014. Optimal design and operation for simultaneous shale gas NGL recovery and LNG re-gasification under uncertainties. Chem Engg Sci, 112: 130–142.

Wang M, Zhang J, Xu Q, Li K. 2011. Thermodynamic-analysis-based energy consumption minimization for natural gas liquefaction. Ind Eng Chem Res, 50: 12630–12640.

Wang M, Zhang J, Xu Q. 2012. Optimal design and operation of a C3MR refrigeration system for natural gas liquefaction. Comput Chem Engg, 39: 84–95.

Xiong X, Lin W, Gu A. 2016. Design and optimization of offshore natural gas liquefaction processes adopting PLNG (pressurized liquefied natural gas) technology. J Nat Gas Sci Egg, 30: 379–387.

11

Modeling and Optimization of Natural Gas Processing and Production Networks

Saad A. Al-Sobhi[1,2], Munawar A. Shaik[3,4], Ali Elkamel[1,3], and Fatih S. Erenay[5]

[1] Department of Chemical Engineering, University of Waterloo, Waterloo, ON, Canada
[2] Department of Chemical Engineering, Qatar University, Qatar
[3] Department of Chemical Engineering, Khalifa University, UAE
[4] Department of Chemical Engineering, Indian Institute of Technology (IIT) New Delhi, India
[5] Department of Management Studies, University of Waterloo, Waterloo, ON, Canada

CHAPTER MENU

11.1 Introduction

Natural gas is the cleanest primary fossil fuel. It produces less CO_2, NO_x, SO_x, and particulate emissions when burned to produce energy compared to other fossil fuels like oil and coal (EIA 2017). Specifically, less greenhouse gas (GHG) emissions responsible for global warming are generated. In 2016, natural gas was used and consumed globally to generate electricity by 24% relative to the other energy sources, namely, oil, coal, nuclear energy, hydroelectricity, and renewables. Figure 11.1 shows in percentages the total world consumption of each fuel or energy source of total 13,276.3 million tons oil equivalent in 2016 (BP 2017). Globally, about 85% of energy consumption is supplied and fulfilled by fossil fuels, and this picture will remain true in the foreseeable future. The world total energy consumption will increase from 575 quadrillion Btu in 2015 to 736 quadrillion Btu in 2040. Although this represents a 28% increase in the total energy consumption, 77% of fuel consumption in 2040 will be provided by fossil fuels. Furthermore, natural gas is the fastest-growing fossil fuel with global consumption increasing an average of 1.4% per year during 2015 to 2040. Figure 11.2 provides projection of the total world energy consumption by fuel in quadrillion Btu (EIA 2017).

In this chapter we address (1) development of a framework for natural gas utilization, (2) synthesis of a novel natural gas processing and production network that involves multiple utilization options, (3) simulation and analysis of the network in order to estimate accurately the mass and energy balances, (4) formulation of a mathematical model that addresses the

Natural Gas Processing from Midstream to Downstream, First Edition.
Edited by Nimir O. Elbashir, Mahmoud M. El-Halwagi, Ioannis G. Economou, and Kenneth R. Hall.

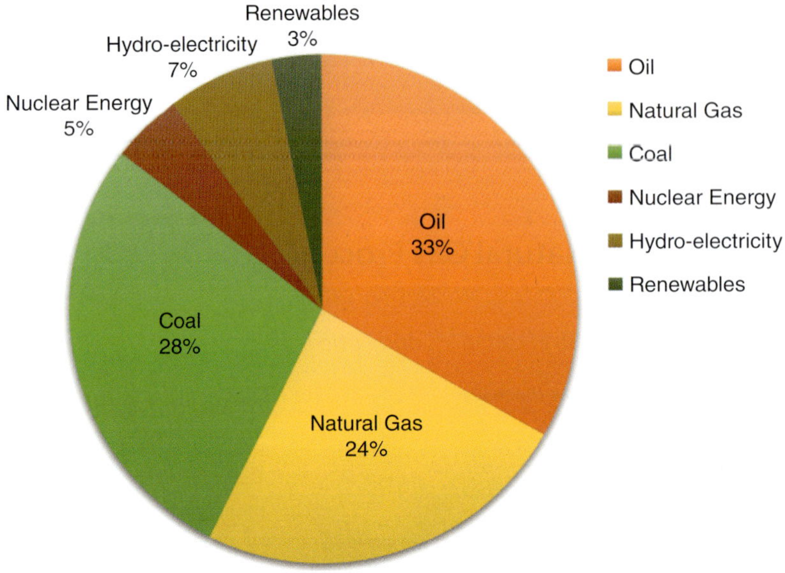

Figure 11.1 World consumption by percentage of each fuel to produce electricity.

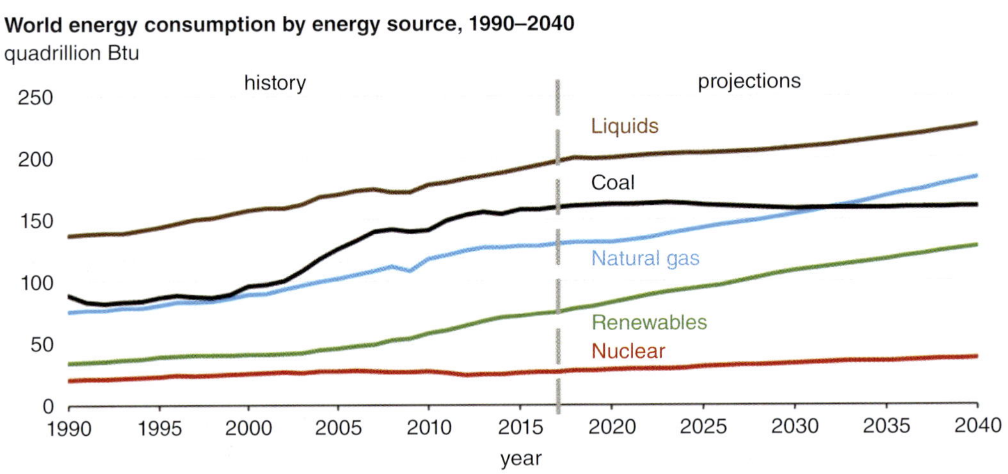

Figure 11.2 Projection of world total energy consumption by fuel (Source: EIA, 2017).

operational strategy to utilize optimal usage of the natural gas resource, (5) design of natural gas processing and production network considering multiple options of technologies, different configurations, and different operating conditions, and (6) assessment and benchmarking of network performance in the context of sustainability.

11.2 Background and Process Description

11.2.1 Natural Gas Supply Chain

Supply chain management is a discipline that integrates the manufacturing plants with their suppliers and customers in an efficient way. The supply chain problems are further divided

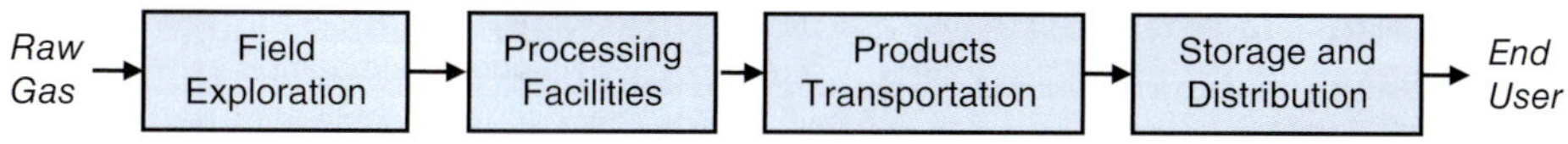

Figure 11.3 Natural gas supply chain.

into supply chain network design, supply chain analysis and policy formulation, and supply chain planning and scheduling categories (Shah 2005). The supply chain network design is a broad problem pertaining to a strategic activity that aims to increase its value by considering typical decisions related to sourcing, allocation, and location. Sourcing decisions are related to finding optimal use of supplies for each facility, allocation decisions are related to finding the optimal product portfolio at each production facility, and location decisions refer to finding optimal location of new facilities. Supply chains or supply networks can be distinguished and separated into production and distribution networks. The production network consists of production plants and storage facilities, while the distribution network consists of storage and distribution facilities responsible for delivering the product to end users. For example, the petroleum industry supply chain consists of exploration, crude procurement and storage logistics, transportation to the refinery, refinery operations, and final distribution of its products. In a similar fashion, the natural gas supply chain starts with field exploration and drilling where the raw material is extracted and processed. Next, it is sent to the processing plant where further processing takes place to make the desired products. Then, the loading and shipping occur to transfer the products to different destinations. Finally, the storage and distribution take place to distribute the product to final consumers. Figure 11.3 shows the natural gas supply chain diagram.

In the literature, models for combined production/distribution—either production or distribution are presented. A decomposition strategy is preferred and justified because when implemented, it eases the complexity involved in solving the combined model. Furthermore, the production cost is dominated in industrial processes (Camm et al. 1997). Thus, the focus of this work is on processing and production network design. Field exploration, transportation and distribution network design are beyond the scope of this chapter.

11.2.2 Natural Gas: Proven Reserves

Natural gas reserves are located and distributed unequally. The top three countries in percentage share are Iran, the Russian Federation, and Qatar, which account for 48% of the total share. Furthermore, the top ten countries own 79% of the total proven reserves (BP 2017). Table 11.1 lists the top ten countries of total proven reserves.

This geographical allocation makes natural gas a stranded energy source in many cases. Stranded energy sources are either far from markets or are located in or near self-sufficient regions where production exceeds the domestic demand. It is worth mentioning that this allocation promotes the advancement in natural gas industry. This advancement is clearly seen either in the way the natural gas is utilized or in the implementation of highly advanced technologies that make the utilization process of natural gas as efficient and cost effective as other fuels.

11.2.3 Natural Gas: Utilization

With high oil prices and environmental considerations, natural gas will play an important role in the energy supply structure. There are a number of ways to monetize natural gas resources. The utilization process depends on the resource location, quantity, quality, and

Table 11.1 World natural gas: proven reserves.

Country	Reserves(tcf)[a]	% share of total
Iran	1183.0	18.0
Russian Federation	1139.6	17.3
Qatar	858.1	13.0
Turkmenistan	617.3	9.4
USA	307.7	4.7
Saudi Arabia	297.6	4.5
United Arab Emirates	215.1	3.3
Venezuela	201.3	3.1
China	189.5	2.9
Nigeria	186.6	2.8
Others	1393	21.1
Total	6588.8	100

a) Trillion cubic feet at the end of 2016.
Source: BP 2017.

so on. Conventionally, pipelines are used to transfer the natural gas to the final consumers. In many cases, this option is not possible or not practical when the resource is stranded and the transportation distance is long. Other promising options include (Thomas 2003) liquefied natural gas (LNG); compressed natural gas (CNG); gas to solid (GTS), i.e., hydrates; gas to wire (GTW), i.e., electricity; gas to liquids (GTL); and gas to chemicals. Moreover, there is a growing recognition of the unconventional gas sources, such as shale gas, deep tight gas, and coal bed methane (CBM), and it is expected that they will contribute significantly in the future (Economides and Wood 2009).

Many researchers have addressed these utilization options, either individually or as a combination. They have assessed the utilization option from design, operation, economic, and environmental impact perspectives. For example, Khalilpour and Karimi (2012) considered LNG, CNG, and GTL as monetization options under the uncertainty of natural gas and oil prices. Wood et al. (2012) presented a review study addressing the available opportunities for the GTL industry. GTL economic viability depends strongly on oil prices because the GTL products such as low-sulfur gasoline and diesel are oil refinery products competitors. Oil prices of at least \$20–30 per barrel will justify the decision of considering GTL process as an option of utilization according to studies done by Al-saadoon (2007). The utilization way varies as being physical compression and mixing processes like the case of LNG, CNG, and GTS or chemical conversion processes like the case of GTL. These utilization options are described as mature, developing, and prospective options. Pipeline and LNG are mature; GTL, CNG, and GTW are developing. Additionally, another highly important factor of consideration in any industry is the capital investment; both fixed and operating costs. For energy-intensive processes, such as refining, chemical, and petrochemical, the operating cost is considered to be significantly important. Many studies have estimated that oil prices should be within \$20-\$60/bbl for other fuel derivatives to be competitive. In particular, GTL industry expansion is predicted if oil prices average above \$20/bbl (Steynberg and Dry 2004). The economic viability analysis of GTL (Al-saadoon 2007) showed \$36/bbl to be the break-even crude oil price for \$2.5/MMBtu natural gas feedstock cost. With oil prices around \$50 per barrel (2016), the decision of considering GTL process is still justified. For the abovementioned reasons, the three most attractive utilization processes, namely LNG, GTL, and methanol are considered in this research to represent the processing and production network.

11.2.3.1 LNG Process

LNG accounts for 47% of global natural gas trade in 2016 (BP 2017). World LNG trade is expected to grow more than 2.5 times from 2015 to 2040. Natural gas is converted physically through compression and liquefaction to LNG. Natural gas is liquefied at −160 °C and 1 atm; this will reduce its volume by 600 times. Figure 11.4 shows a typical block flow diagram of an LNG train. Natural gas feedstock is received at the inlet receiving unit where the initial gas-liquid separation and metering take place.

The hydrocarbon condensate stream is stabilized in the condensate stabilization unit. The acid gas removal unit is designed to reduce the CO_2 and H_2S concentration levels to specification limits. The specifications (Klinkenbijl 2005) are set to be lower than 50 ppmv for CO_2 to prevent freezing in the main cryogenic heat exchanger and below 3.5 ppmv for H_2S to meet sale gas and pipeline specifications. Sulfur is recovered as byproduct. Gas leaving the acid gas removal unit is called sweet gas. This treated gas is passed to the dehydration, mercaptan, and mercury removal unit. The dehydration involves water vapor removal to a very low level for example below 0.5 ppmv. Mercury is removed to a very small level such as 0.01 microgram per normal cubic meter (Nm^3). Heavier hydrocarbons are recovered in the precooling NGL separation unit. They are sent to the fractionation unit where ethane, propane, butane, and plant condensate are recovered. The methane-rich stream leaving the NGL separation unit is liquefied by one of several approved large-scale baseload natural gas liquefaction processes, including the pure refrigerant cascade process, the propane precooled mixed refrigerant process, the propane precooled mixed refrigerant with the back-end nitrogen expander cycle, or other mixed refrigerant processes (Tusiani and Shearer 2007).

11.2.3.2 GTL Process

The gas to liquid process was developed in the early twentieth century when Franz Fischer and Hanz Tropsch developed an integrated synthesized hydrocarbon reaction, now known as the Fischer-Tropsch (FT) reaction. Although coal was used initially in 1923 as the primary feedstock, the modern FT syntheses are still profoundly similar to the original process. For practicality and viability of GTL process from 1950 to 2000 the reader is referred to Dry (2002).

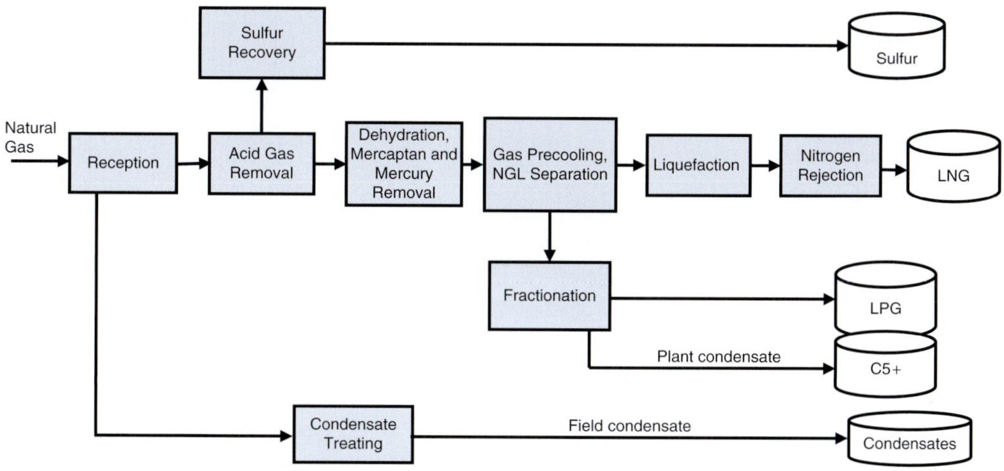

Figure 11.4 A typical LNG train block flow diagram.

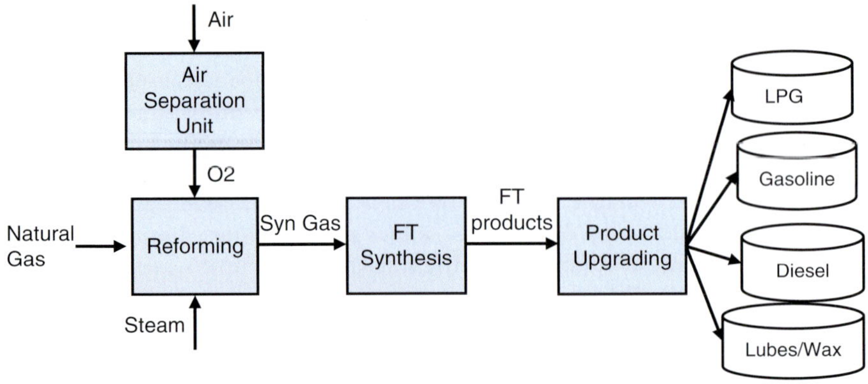

Figure 11.5 A typical GTL train block flow diagram.

Natural gas is converted chemically through the Fischer-Tropsch (FT) reaction into liquid fuels (Steynberg and Dry 2004). Figure 11.5 shows a typical block diagram of a GTL train.

The upstream processing units, such as stabilization, acid gas removal, sulfur recovery, dehydration, and natural gas liquids separation units are similar to those in the LNG train. However, the downstream processing units consist mainly of three basic units. These are (1) the reforming or synthesis gas production unit where the proper syngas ratio (H_2/CO) is produced; the (2) FT synthesis unit where the synthesis fuels (synfuels) are produced; and (3) the product upgrading and separation unit where the hydro-treating/cracking takes place to obtain the final liquid fuels.

11.2.3.3 Methanol Process

Methanol, which is a primary feedstock for the chemical industry, is expected to contribute as a fuel and energy carrier. Methanol when combined with dimethyl ether (DME) creates excellent fuels. Furthermore, methanol and DME can be blended with gasoline/diesel and used in internal combustions engines and in electricity generators (Olah et al. 2006). For the chemical sector, methanol to olefins and methanol to hydrogen are promising applications (Haid and Koss 2001). Natural gas is converted chemically through reactions into methanol. Figure 11.6

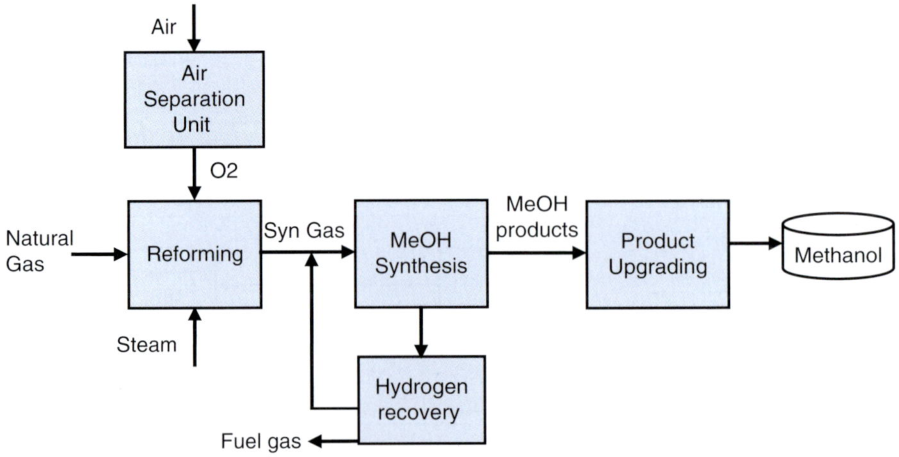

Figure 11.6 A typical methanol train block flow diagram.

shows a typical block flow diagram of a methanol train. The upstream processing units such as stabilization, acid gas removal, sulfur recovery, dehydration, and natural gas liquids separation units are similar to those in the LNG train. However, the downstream processing units consist mainly of three basic units (Olah et al. 2006). These are (1) the reforming or synthesis gas production unit where the proper molar syngas stoichiometric ratio SR= ($H_2-CO_2/CO+CO_2$) is produced, (2) the methanol synthesis unit where the crude methanol is produced, and (3) the product upgrading unit where the separation and distillation take place to obtain the final methanol product.

11.3 Simulation of Natural Gas Processing and Production Network

Process simulation plays an important role in the design and operation of any chemical, petrochemical, and refining process. It helps with the accurate estimation and calculation of mass and energy balances, transport and chemical kinetic equations, and physical properties. It also helps with equipment sizing and costing, profitability analysis, and optimization. Having generated candidate process flowsheets through process synthesis activity, then comes the role of analysis and simulation to assess them. Clearly, this will facilitate the decision to be made by engineers and scientists to choose and select the most practical flowsheet. Process synthesis requires that the input data and process output be provided, and the process flowsheet (structure) is created; whereas, in process analysis, both the input data and the process flowsheet are provided, and process output will be obtained. Different techniques are available to analyze the process including mathematical models, empirical correlations, and computer-aided process simulation tools. Many commercial process simulators (such as CHEMCAD, HYSYS, ASPEN Plus, and PRO/II) can be used to estimate the material and energy balances.

A number of scientific publications have addressed the importance and the implementation of simulation, either in the design/operation of the energy system or for economic assessment, as will be seen. Different simulation packages such as CHEMCAD, ASPEN Plus, HYSYS, and PRO/II have been shown to be useful in many applications and applicable in many industries. These include the oil, gas, petrochemical, paper and pulp, and power industries. Here, some relevant work that utilized the simulation will be presented. Some were focused on a single natural gas conversion process such as GTL, LNG, gas to power, or gas to methanol. As mentioned earlier, the simulation package can be used to design a process, improve the design, or improve the operation through validation. For example, Zheng and Furimsky (2003) developed a detailed flow sheet of the combined cycle cogeneration plant fueled by natural gas. They used ASPEN Plus to simulate the cogeneration plant and compared their results with a commercial plant generating about 43.6 MW of electricity by gas turbine and 28.6 MW of electricity by steam turbine. The key data generated using the ASPEN models were in good agreement with the operating data. Another simulation study was carried out by Kim et al. (2009) on natural gas conversion to FT synthetic fuel. The ASPEN HYSYS software simulation package was used to find optimum reaction conditions for maximum production of synthetic fuel. Optimum reaction temperature in FT synthesis unit was found to be 255 °C.

Bao et al. (2010) developed a simulated case study of a GTL plant using natural gas to produce 118,000 bbl/d of products. The ASPEN Plus software simulation package was used to simulate key processing units in GTL plants that will help in assess the economics of the process. Their study highlighted that, depending on the price of natural gas, the return on investment (ROI) ranges from 7.4% to 19.4% for the cost of natural gas being \$8 and 5/1000 SCF of natural gas, respectively. Also, a break-even point analysis considering current market conditions, shows that the production capacity should be at least 57,000 bbl/d to be profitable.

Ehlinger et al. (2014) presented a simulated work of shale gas-to-methanol production using ASPEN Plus along with published data for techno-economic analysis of shale gas-to-methanol production. They found that a desirable 31% ROI is achieved for a methanol selling price of \$2/gal and shale gas price of \$3.5/kSCF. Fissore and Sokeipirim (2011) presented a study addressing the simulation and energy consumption of a cryogenic plant using ASPEN Plus for recovering C3+ compounds from natural gas. The optimized plant allows for 25% heat reduction and up to 60% refrigeration reduction compared to the base case. Al-Sobhi et al. (2009) presented simulation and heat integration for a synthesized LNG plant using ASPEN Plus to model flow, temperatures, and heat duties for various hot and cold streams. Then, heat integration was performed using pinch analysis and reported reduction of hot and cold utilities by 15% and 29%, respectively.

Hao et al. (2008) used ASPEN plus to simulate an integrated GTL process with iron and cobalt-based catalysts for which the optimal flowsheet structures were selected using thermal, carbon efficiency, and product distributions. Sudiro and Bertucco (2009) developed a hybrid process producing synthetic gasoline and diesel from coal and natural gas. Three processes were modeled and simulated using ASPEN Plus: namely, gas to liquids (GTL), coal to liquids (CTL) and the hybrid process that processes both natural gas and coal as feedstock. It was shown that the hybrid process produces 44.4% as fuel produced per unit of feedstock compared to 66.7% and 32.5% from GTL and CTL process, respectively. Also, environmental assessment in terms of CO2 emission was estimated for the three processes. It was shown that 2.45 kg CO2/kg liquid fuel from hybrid process, 0.63 kg CO2/kg liquid fuel for GTL, and 4.66 kg CO2/kg liquid fuel for CTL were emitted.

Although the products yield values of selected processes can be obtained from previous works, the variations in operating conditions will result in different yield values. Thus, we need to simulate our own flowsheet to address the variation in operating conditions and their impact on the profitability of the processing and production network. Furthermore, in order to estimate the capital and operating cost using ICARUS, we need to converge the flowsheet, first using the ASPEN Plus simulation package.

11.3.1 Problem Statement

Given natural gas feedstock with its composition, flow rate, and operating conditions as shown in Table 11.2, it is desirable to simulate the key processing units of the processing and production network using a simulation package. The main key processing units of LNG, GTL, and methanol will be considered for simulation and evaluation. The rigorous steady state simulation of the processing network is carried out using the ASPEN Plus V7.3 version. A novel natural gas processing and production network has been synthesized. The network starts with stabilization, unit (A) and ends with methanol upgrade, unit (M) as shown in Figure 11.7.

11.3.2 Steady State Process Simulation of Natural Gas Processing and Production Network

11.3.2.1 LNG Process Simulation

The LNG process is represented by the processing units (A-F) and units (H,I) as shown in Figure 11.7. Natural gas feedstock with flow rate, operating conditions, and mole composition, as shown in Table 11.2, is received at the inlet receiving unit. The steady state simulation of LNG

Table 11.2 Natural gas feed operating conditions and composition.

Flow rate	1,500 MMSCFD 74,700 kmol/h 1.5×10^6 kg/h
Temperature	20 °C
Pressure	70 bar
Component	mol%
H_2S	1
CO_2	2
N_2	4
H_2O	0.05
CH_4	83
C_2H_6	5
C_3H8	1.8
i-C_4H_{10}	0.4
n-C_4H_{10}	0.7
i-C_5H_{12}	0.3
n-C_5H_{12}	0.3
n-C_6H_{14}	1.45
Total	100

Source: Al-Sobhi et al., 2009.

process was carried out using ASPEN Plus. In general, a cubic equation of state, for example, the Peng Robinson or Soave-Redlich-Kwong equation (SRK) is appropriate and recommended for gas processing by ASPEN Plus. The Peng Robinson thermodynamic property method is considered and selected as the global thermodynamic property method. However, NRTL was selected as the thermodynamic property method for the amine system. The primary gas/condensate separation takes place in stabilization unit (A) and condensate C_{5+} product is stabilized.

The primary purpose of the stabilization unit is to recover the intermediate and heavy C_{5+} components to generate greater revenues. Figure 11.8 shows the ASPEN Plus flowsheet representation of a typical stabilization unit (A). The residual sour gas is then fed to acid gas removal unit (B) where H_2S and CO_2 are removed using the mixed amine solution. Figure 11.9 shows the ASPEN Plus flowsheet representation of a typical acid gas removal unit. The composition of mixed amine solution is 15 wt% DEA, 30 wt% MDEA, 45 wt% water. The CO_2 and H_2S mole fraction in the sweet gas stream are 7.9 E-5, and 1.8 E-25. The elemental sulfur is recovered from the H_2S-rich stream by the conventional straight or split flow Claus process in unit C shown in Figures 11.10 and 11.11. The sweet gas is fed to dehydration unit (D). Figure 11.12 shows an ASPEN Plus flowsheet representation of a typical dehydration unit. TEG glycol is used to remove the water content to 9.0 E-05 of water mole fraction.

The sweet dry gas stream is fed to NGL unit (E) where the heavy hydrocarbons and LPG are separated from the gas, which is mainly methane. The heavy hydrocarbons are sent to fractionation unit (F) to be separated into ethane, propane, butane, and pentane products as shown in the ASPEN Plus flowsheets of Figures 11.13 and 11.14 for conventional and nonconventional cases, respectively. The methane gas is liquefied to −160°C using a mixed refrigerant (MR).

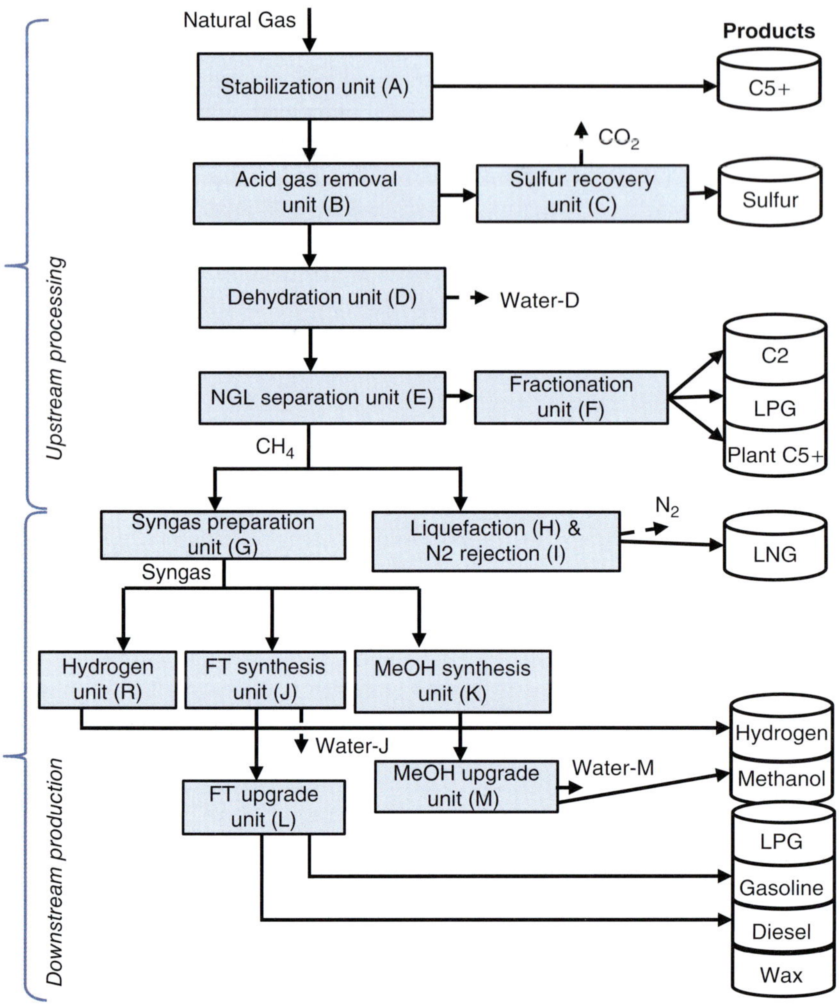

Figure 11.7 Key processing units of natural gas processing and production network.

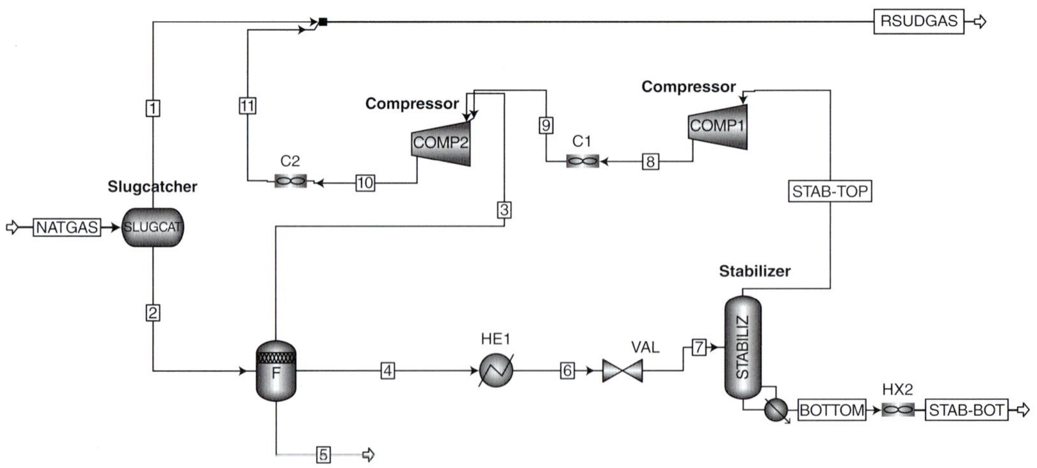

Figure 11.8 ASPEN Plus flowsheet for stabilization unit.

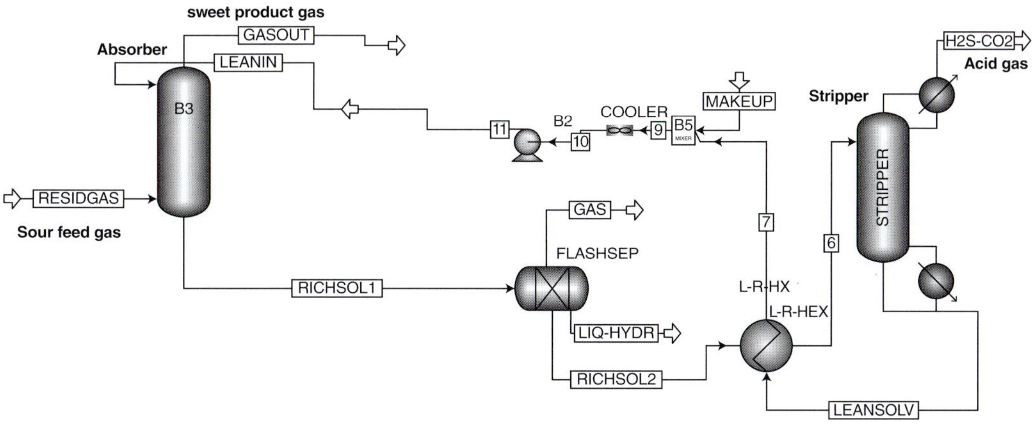

Figure 11.9 ASPEN Plus flowsheet for acid gas removal unit.

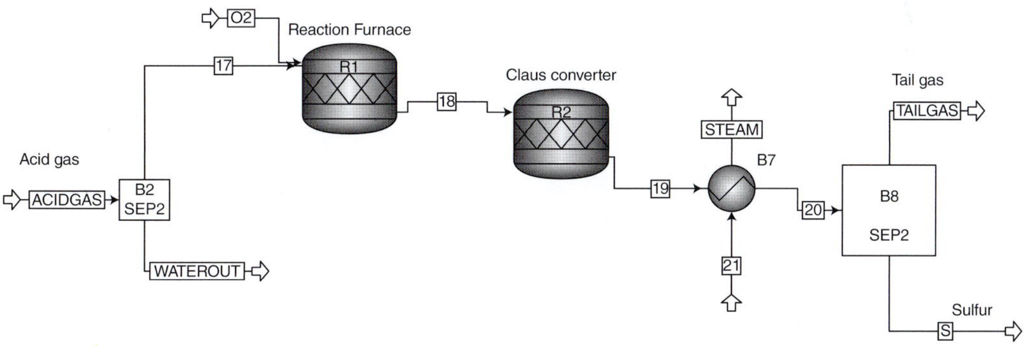

Figure 11.10 ASPEN Plus flow sheet for sulfur recovery unit (Claus straight).

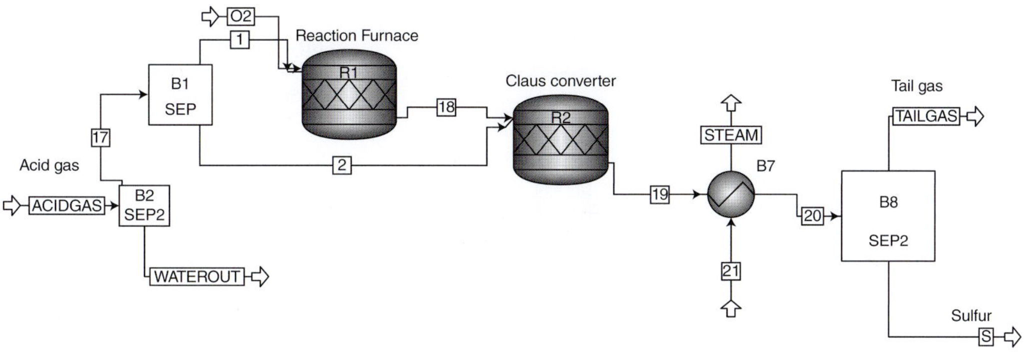

Figure 11.11 ASPEN Plus flowsheet for sulfur recovery unit (Claus split).

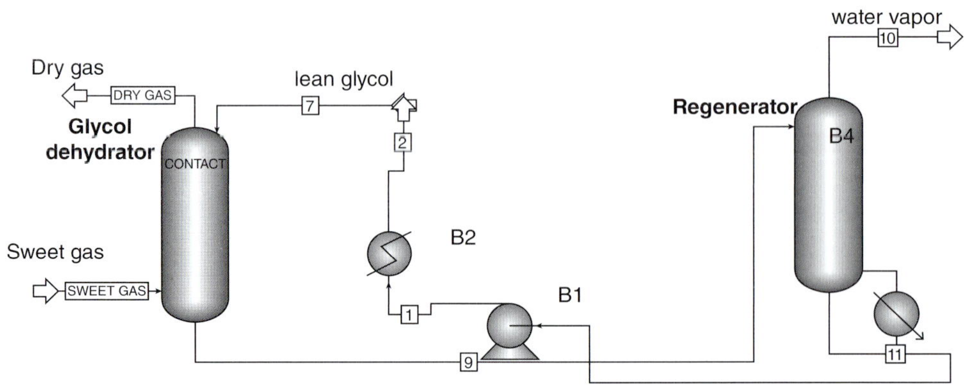

Figure 11.12 ASPEN Plus flowsheet for dehydration unit.

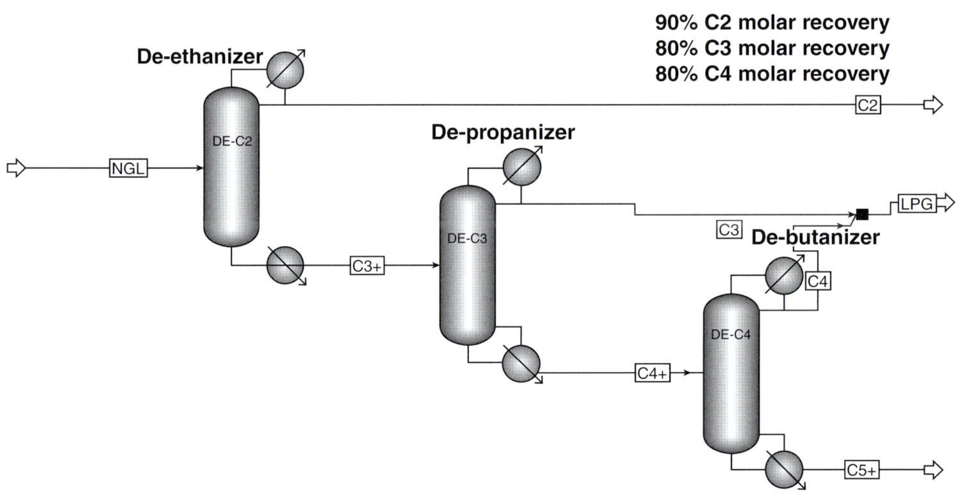

Figure 11.13 ASPEN Plus flow sheet for fractionation unit (conventional).

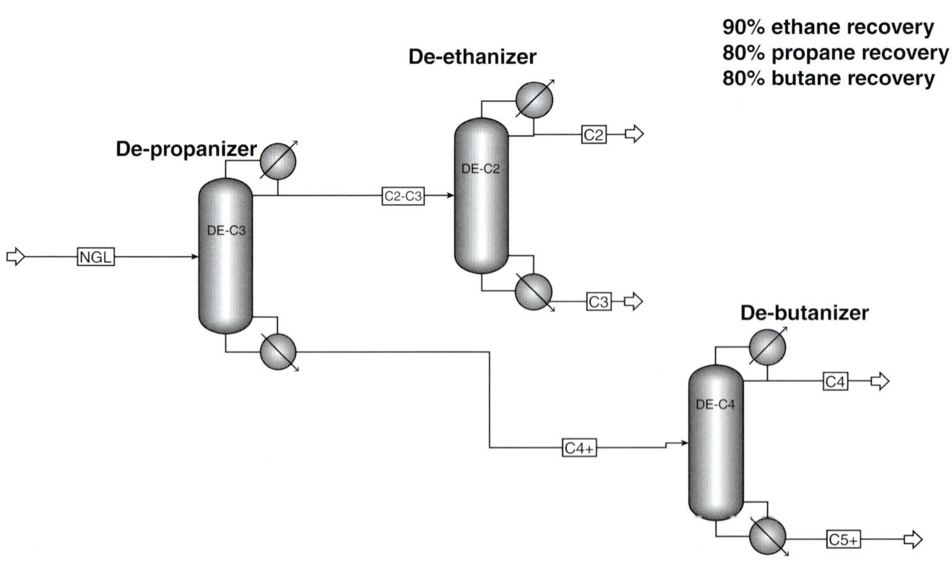

Figure 11.14 ASPEN Plus flowsheet for fractionation unit (nonconventional).

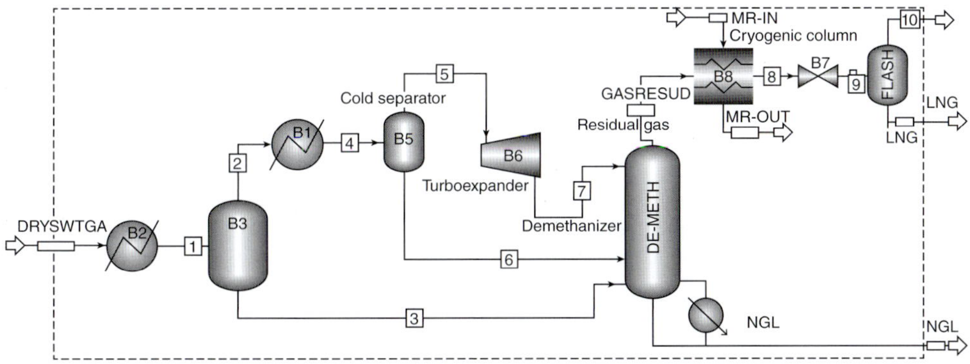

Figure 11.15 ASPEN Plus flowsheet of the NGL and liquefaction unit.

The optimized MR composition is obtained from Alabdulkarem et al. (2011). Figure 11.15 shows the ASPEN Plus flowsheet representation of NGL recovery and liquefaction unit. A detailed discussion about cryogenic processes can be found in Venkatarathnam (2008). Then, the nitrogen is rejected to obtain the desired heating value of LNG product.

11.3.2.2 GTL Process Simulation

The GTL process is represented by units (A-F) and units (G, J and L) as shown in Figure 11.7. The treated methane gas from unit (E) is preheated, mixed with steam and pure oxygen, and fed to the auto-thermal reactor (ATR). The ATR converts the natural gas, which is mainly methane, steam, and pure oxygen into a syngas (H_2/CO). The ATR reaction scheme is complex, but the overall reaction is represented by (Steynberg and Dry 2004):

$$CH_4 + 3/2\ O_2 \rightarrow CO + 2\ H_2O - \Delta H^\circ{}_{298} = +519\ KJ/mole \tag{11.1}$$

$$CH_4 + H_2O \leftrightarrow CO + 3\ H_2 - \Delta H^\circ{}_{298} = -206\ KJ/mole \tag{11.2}$$

$$CO + H_2O \leftrightarrow CO_2 + H_2 - \Delta H^\circ{}_{298} = +41\ KJ/mole \tag{11.3}$$

The ATR is modeled as an equilibrium reactor in ASPEN Plus as shown in Figure 11.16. For a given natural gas flowrate, the steam to CH_4 is set to be 0.6 as the operating ratio. This very low ratio around 0.6, rather than the previously used high ratio of 1.5-2.0, becomes the state-of-the-art syngas ratio for FT application in modern plants in Europe and Middle East (Steynberg and Dry 2004).

Now, the O_2 flowrate sensitivity analysis was performed by varying the O_2 flowrate to obtain the desired H_2/CO ratio. The syngas H_2/CO ratio of 2 was achieved by using the ATR with 70,000–72,500 kmol/h as O_2 flowrate. The high syngas temperature is utilized by producing high pressure steam. The syngas is fed to slurry phase FT reactor which is modeled as a yield reactor. The FT reactor operating conditions are 240°C and 20 bar. The FT synthesis

$$nCO + 2nH_2 \rightarrow \{-CH_2-\}n + nH_2O \tag{11.4}$$

This study is considered as a low- temperature slurry-phase Fischer-Tropsch reactor with cobalt catalyst. The product distribution follows the chain growth probability function known as Anderson-Schulz-Flory (ASF) distribution (Steynberg and Dry 2004).

$$W_n/n = (1-\alpha)^2\alpha^{n-1} \tag{11.5}$$

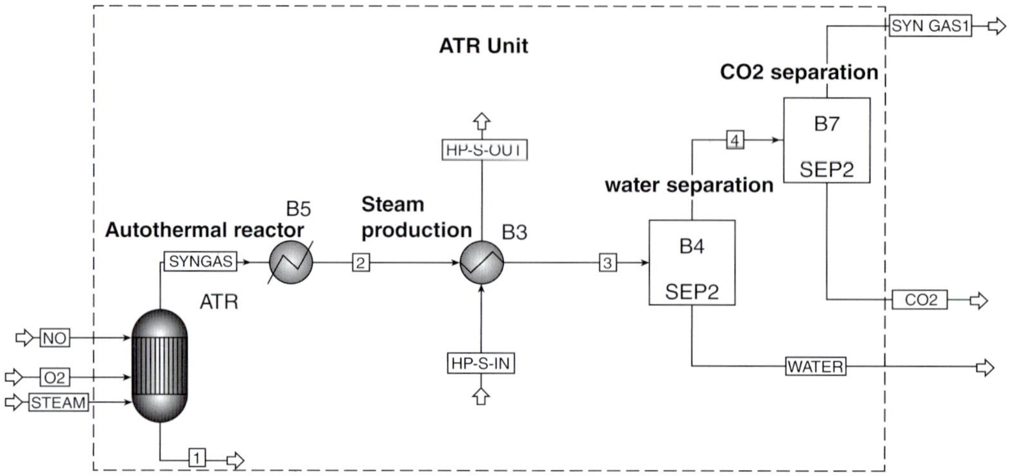

Figure 11.16 ASPEN Plus flowsheet for ATR unit.

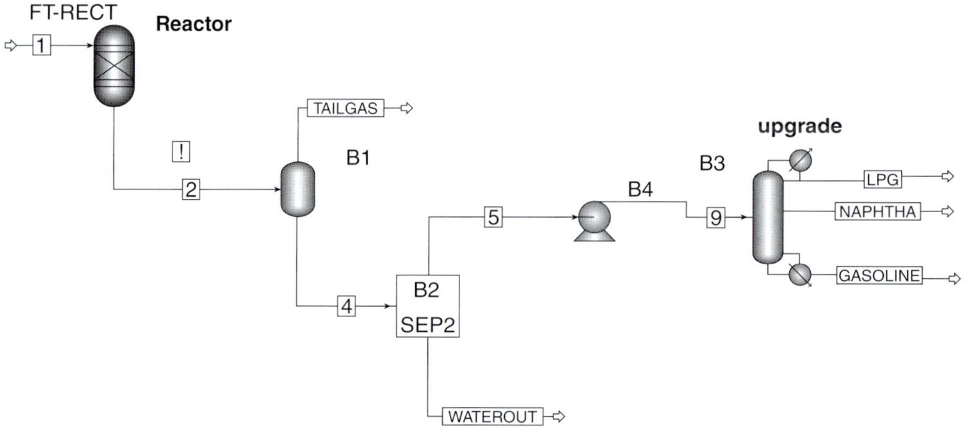

Figure 11.17 ASPEN Plus flowsheet of FT synthesis and upgrading unit.

where W_n is the mass fraction of the hydrocarbon molecular with carbon number n and α is the chain growth probability. An alpha value of 0.9 is selected to find the mass fraction of the hydrocarbons. The FT reactor is assumed to produce only paraffin (no olefin). The hydro-treating/cracking of the waxes takes place to obtain the final desired products normally LPG, synthetic gasoline and diesel. Figure 11.17 shows ASPEN Plus flowsheet of FT synthesis and upgrading unit.

11.3.2.3 Methanol Process Simulation

The methanol process is represented by units (A-F) and units (G, K, and M) as shown in Figure 11.7. The treated methane gas from unit (E) is preheated, mixed with steam and pure oxygen, and fed to an auto-thermal reactor (ATR) similar to the one used in the GTL process. The ATR converts the natural gas, which is mainly methane, steam, and pure oxygen into the proper syngas stoichiometric ratio SR= (H_2−CO_2/CO+CO_2). Now, one of the advantages of

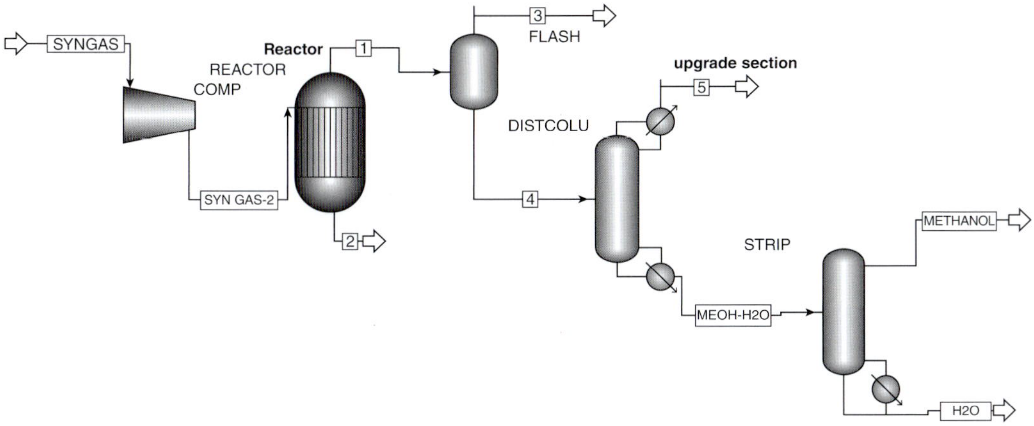

Figure 11.18 ASPEN Plus flowsheet of methanol synthesis and upgrading unit.

using the rigorous simulation is highlighted by using the sensitivity analysis modelling tool in ASPEN plus. For the O_2 flowrate, sensitivity analysis was performed by varying the O_2 flowrate to obtain the desired H_2/CO and SR ratio of 2 for GTL and methanol applications. The syngas (H_2/CO) and SR ratio of 2 is achieved by using the ATR with 72,500 and 27,500 kmol/h as O_2 flowrate for GTL and methanol applications, respectively. The syngas with SR of 2 is fed to methanol synthesis reactor which is modeled as the equilibrium reactor. The methanol synthesis typically consists of mainly three reactions, where the two reactions in equation (11.6) and (11.7) are exothermic, with heat of reaction equal to −21.7 kcal/mol and −9.8 kcal /mol, respectively:

$$CO + 2H_2 \leftrightarrow CH_3OH - \Delta H^\circ_{298} = -21.7 \text{ kcal/mole} \tag{11.6}$$

$$CO_2 + 3H_2 \leftrightarrow CH_3OH + H_2O - \Delta H^\circ_{298} = -9.8 \text{ kcal/mole} \tag{11.7}$$

$$CO + H_2O \leftrightarrow CO_2 + H_2 - \Delta H^\circ_{298} = 11.9 \text{ kcal/mole} \tag{11.8}$$

The synthesis reactions are exothermic, hence, the heat released should be either recovered for power generation or absorbed by cooling water to obtain an isothermal operation. Equation (11.8) describes the endothermic reverse water gas shift reaction (RWGSR) that also occurs during methanol synthesis, producing CO, which can further react with hydrogen to produce methanol. The methanol reactor operating pressures range from 50 to 100 atm and temperatures of 200–300°C, which is the basis for most methanol production processes.

Crude methanol leaving the reactor contains water and some impurities depending on the feed gas composition, reaction conditions, and type and lifetime of the catalyst, such as dissolved gases (methane, CO, CO_2), higher alcohols (ethanol, propanol, butanol) and long-chain hydrocarbons. Methanol is available in three grades of purity: (1) fuel grade, (2) "A" grade, used as a solvent, (3) "AA" grade or chemical grade with highest purity with 99.85% methanol content (Olah et al. 2006). The distillation systems using one or more distillation columns will be used to purify the methanol product. Figure 11.18 shows the ASPEN Plus flowsheet of methanol synthesis and upgrading unit. The products' mass flowrates and yields are shown in Table 11.3 (Al-Sobhi and Elkamel, 2015). ICARUS, cost estimator and analyzer, has been used to evaluate and estimate the cost of the processes. Both the capital and operating cost of key processing units were estimated as shown in Table 11.4.

Table 11.3 Yields and flowrates of products from LNG, GTL, and methanol processes.

Product (kg/h)	NG1	NG2	NG3	Yield[a]	Min. demand	Max. demand	LP model output
Sulfur1	22,900	-----	-----	0.015	20,000	30,000	20,000
LNG1	1,044,157	-----	-----	0.690	1,000,000	1,200,000	1,090,000
Ethane1	62,800	-----	-----	0.040	60,000	70,000	60,000
LPG1	75,200	-----	-----	0.050	70,000	80,000	80,000
Plant Condesate1	12,700	-----	-----	0.010	10,000	15,000	10,000
Field Condensate1	64,900	-----	-----	0.043	60,000	70,000	60,000
Losses[b]1	232,343	-----	-----	0.152	230,000	235,000	2,30,000
Sulfur2	-----	22,900	-----	0.015	20,000	30,000	21,150
Ethane2	-----	62,800	-----	0.040	60,000	70,000	70,000
LPG2	-----	114,000	-----	0.075	110,000	120,000	120,000
Gasoline2	-----	227,911	-----	0.150	220,000	230,000	230,000
Diesel2	-----	174,730	-----	0.110	170,000	180,000	180,000
Wax2	-----	99,900	-----	0.070	95,000	100,000	100,000
Plant Condesate2	-----	12,700	-----	0.010	10,000	15,000	15,000
Field Condensate2	-----	64,900	-----	0.043	60,000	70,000	70,000
Losses[b]2	-----	735,159	-----	0.480	733,000	735,200	733,000
Sulfur3	-----	-----	22,900	0.015	20,000	30,000	30,000
Ethane3	-----	-----	62,800	0.040	60,000	70,000	70,000
LPG3	-----	-----	75,200	0.050	70,000	80,000	80,000
Methanol	-----	-----	870,000	0.574	850,000	900,000	900,000
Field Condensate3	-----	-----	64,900	0.043	60,000	70,000	70,000
Losses[b]3	-----	-----	419,200	0.270	400,000	410,000	400,000
Available NG supply(kg/h)	1,515,000	1,515,000	1,515,000	-----	-----	-----	-----

a) Yield defined as product flowrate divided by natural gas feedstock flowrate.
b) Losses represent other byproducts such as CO_2, N_2, water, etc.

11.4 LP Model for Natural Gas Processing and Production Network

A strategic planning linear programming (LP) model for maximizing the profit of the natural gas processing and production network, given by Al-Sobhi and Elkamel (2015), is presented here. Biegler and Grossmann (2004) highlighted that the design and synthesis of processes have been dominated by NLP and mixed integer linear programming (MINLP) models, although simpler targeting models give rise to LP and MILP problems. Modeling the natural gas supply chain is a challenging task. Therefore, researchers try to model the main two components: production and distribution networks separately. The production network includes both upstream and downstream processing. This technique simplifies the formulation, convergence, and increases the level of accuracy significantly (Zhang and Zhu 2000).

The following questions are addressed through modeling the processing and production network shown in Figure 11.19:

- What is the optimal natural gas flow rate to each process?
- What is the optimal production level of each product? Which process can increase the economic portfolio of processing network? How to plan under the price fluctuations?

Table 11.4 Capital cost, operating cost, and net CO_2 emission of major processing units.

Key unit	Capital cost ($MM)	Operating cost ($MM/yr)	Utility cost ($MM/yr)	Net CO_2 equiv[a)] (kg/h)
Stabilization	12.1	2550	2360	−2.43E+00
Acid gas removal	34.5	380	350	1.27E+04
Sulfur recovery	3.29	0.96	0.0355	0
Dehydration	27.6	12.1	9.87	−9.823E-09
NGL recovery	15.4	2310	2140	1.52E-03
Fractionation	8.05	1770	1640	−2.56E-11
Liquefaction	7.26	1.01	0.0355	−1.31E-08
Syngas production	32	24.6	21.4	−2.33E+07
FT syn. and upgrade	15.1	125	115	2.31E+05
Meth. syn. and upgrade	43.9	259	237	8.32E+05
Total cost LNG	108	7030	6500	1.27E+04
Total cost GTL	148	7170	6630	−3.55E+07
Total cost methanol	177	7310	6750	−3.49E+07

a) Net CO_2 equiv value is obtained from Aspen plus and defined as output (products) equivalent minus input equivalent.

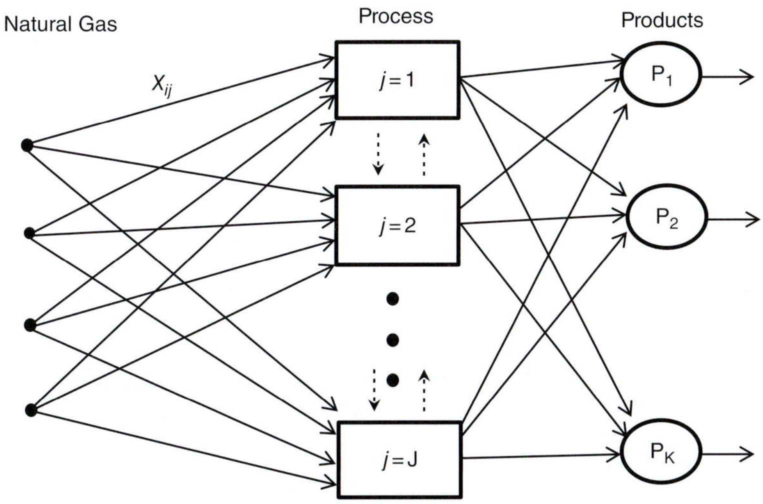

Figure 11.19 A schematic superstructure of natural gas network.

- Which process is more sustainable; for example, producing less CO_2 emissions? How to make the processing network more sustainable?
- The effect of gas composition on the overall structure and optimal planning strategy?
- The effect of demand on available capacity and expansion strategy in the processing network?

A natural gas processing and production network consisting of LNG, GTL, and methanol facilities is considered. The network involves the processing of natural gas feedstock to produce a different set of main products such as LNG, gasoline, diesel, methanol, and by-products such as, sulfur, LPG, and condensate. The network is envisioned as a set of interconnected nodes representing the main processing units of three facilities. These nodes are connected by arcs that represent the material flows within each facility. The utilization options are specified, and the problem boundaries are defined, for example, a specific flow rate, composition, and operating conditions. It is worth mentioning that considering a specific number of utilization processes will just represent the base case. Furthermore, any promising processes can be included in the future planning to address process flexibility. For example, hydrogen as a main product from natural gas, methanol to olefin (MTO) or methanol to gasoline (MTG) production process from methanol (Haid and Koss 2001; Wood et al. 2012) or even hydrogen from methanol. The available technologies as shown in Table 11.5 are then screened.

After consulting some published works and books (Mokhatab and Poe 2012; Olah et al. 2006; Robert and John 1998; Steynberg and Dry 2004; Tusiani and Shearer 2007) in natural gas processing industry, the most applicable one is selected for steady-state simulation. In other words, we fix the topology of the superstructure for steady-state simulation. The use of any process simulators (such as CHEMCAD, Aspen Plus, HYSYS, and PRO/II) will be beneficial at this step. The steady-state simulation of the selected process flowsheet is carried out using ASPEN Plus

Table 11.5 Different possible technologies for industrial processing units.

Major processing unit	Possible processes/technologies	Selected Process/ technology
Condensate stabilization	a) Flash vaporization b) Stabilization by fractionation	Stabilization by fractionation
Acid gas removal	a) Indirect conversion b) Direct conversion (dry bed or liquid phase) c) Separation technologies (membrane or cryogenic fractionation)	Direct conversion (liquid phase)
Sulfur recovery	a) Gas phase-Claus type b) Liquid-phase	Gas phase Claus type
Dehydration	a) Liquid desiccant (Glycol) b) Solid desiccant c) Cooling the gas	Glycol
NGL recovery (extraction)	a) Refrigeration process b) Lean oil absorption c) Solid bed adsorption d) Membrane separation e) Twister supersonic	Refrigeration process
NGL fractionation	a) Direct sequence b) Indirect sequence	Direct sequence
Liquefaction	a) Pure-refrigerant cascade b) Propane-precooled mixed-refrigerant c) Propane-precooled mixed-refrigerant, with back-end nitrogen expander-based d) Nitrogen expander-based	Propane-precooled mixed-refrigerant
Reforming (syngas production)	a) Steam reforming b) Adiabatic oxidative reforming c) Auto-thermal reforming	Auto-thermal reforming
FT synthesis	a) Low temperature FT b) High temperature FT	Low temperature FT
Methanol synthesis	a) Quench b) Steam raising c) Gas cooled (tubular)	Quench
Product upgrading	a) Direct sequence distillation b) Indirect sequence distillation	Direct sequence distillation

V7.3 (2011). Essentially, the material and energy balances are calculated at this step for each process. Products yield is obtained to be used in the mathematical programming step. After the steady-state simulation is converged, equipment sizing is estimated. Then, Aspen process Economic Analyzer is used to estimate both the fixed capital and operating costs of each key unit of selected processes to be used in the mathematical model. The yields obtained from the simulation are used as parameters in the optimization step. The modeling software LINGO version 14.0 (LINGO 2013) is used to run the model and obtain the optimal results.

The general solution strategy for simulation-optimization is depicted in Figure 11.20 for using ASPEN Plus simulation platform and mathematical programming approaches sequentially to derive an optimal solution.

Prior to the selection of optimal technology, the set of conditions for each key processing unit should be specified such as flow rate, composition, pressure, and temperature of feedstock, specification levels for product purity and quality, capital and operation costs for each process. We start with a specific natural gas flowrate, composition, and operating conditions (model boundary) for the simulation. The yields for different technologies and operating modes are determined after running ASPEN simulation for each unit, along with estimation of annual capital and operating costs using ICARUS software. Then we employ this information as input to the formulation and solution of optimization model. The output from the solution is compared to the set of conditions specified for running the ASPEN simulation to begin with. The simulation is carried out with updated operating conditions and revised yield data is obtained for input to optimization and the procedure is continued until convergence.

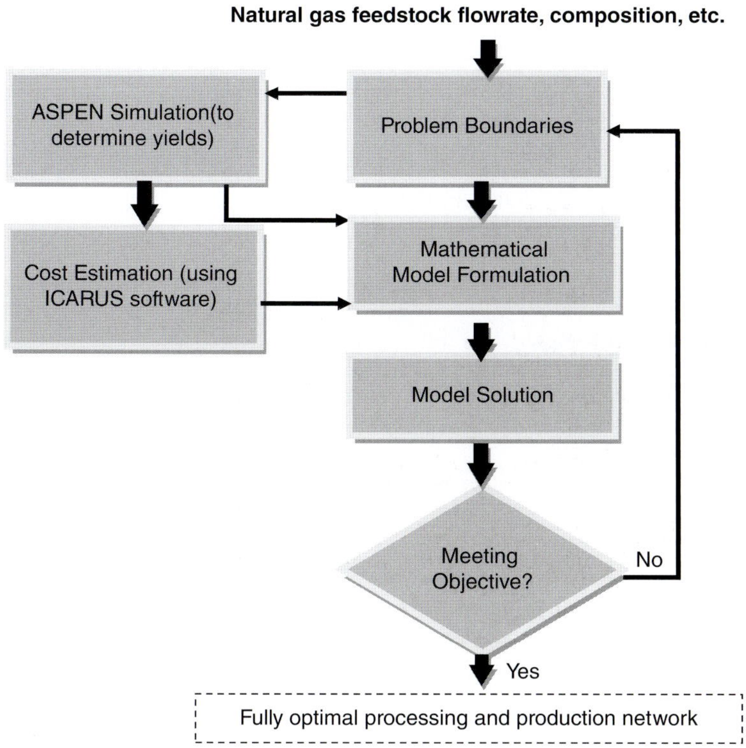

Figure 11.20 The sequential use of simulation and optimization methods for natural gas network design and operation.

11.4.1 LP Model Formulation

The gas processing and production network consists of a set of plants or nodes $j \in J$ that are of main types LNG, GTL, and methanol (Al-Sobhi and Elkamel, 2015). Each node consists of a set of main processing units that are connected in a specified way and this gives the identity to the node in terms of consumption of raw materials, production of final products, utility requirements, and environmental impact as shown in Figure 11.19. In order to propose an appropriate mathematical model, we first define the following sets, indices, parameters, and variables. $i \in I$ is a set of nodes feeding to the processing network consisting of different processes $j \in J$ (such as LNG, GTL, and methanol), producing different set of products $k \in K$, by using different utilities $u \in U$.

Variables:
X_{ij} = mass flowrate of natural gas i feeding process j
Y_{jk} = mass flowrate of product k from process j
$X_i = \sum_{j \in J} X_{ij}$ = total mass flowrate of natural gas fed to processing node i
$X_j = \sum_{i \in I} X_{ij}$ = total mass flowrate entering process/node j
E_{uj} = utility requirement per unit feed of unit j for utility u
Objective function is set to maximize profit:

$$\sum_{j \in J} \sum_{k \in K} C_{jk} Y_{jk} - \sum_{i \in I} \sum_{j \in J} C_{ij} X_{ij} - \sum_{j \in J} f(X_j) \cdot X_j - \sum_{j \in J} \sum_{u \in U} C_u \cdot g(E_{uj}) \quad (11.9)$$

The objective function is calculated as the sum of product values minus the cost of raw material, the operating cost, and the cost of utility consumption. The functions $f(X_j)$ and $g(E_{uj})$ are in general nonlinear, where $f(X_j)$ is indicating the variable of operating cost with amount produced by unit j and $g(E_{uj})$ is indicating the utility consumption as a function of utility requirement per unit j.

The main constraint covering the processing network is the material balance constraint and will be represented by the yield obtained from the steady-state simulation.

i) *Supply constraint*
Total natural gas allowable usage from the field per day

$$\sum_{j \in J} X_{ij} \le X_i^U \quad (11.10)$$

ii) *Process capacity constraint*
Lower and upper capacity for each process

$$X_j^L \le X_j \le X_j^U \quad (11.11)$$

iii) *Demand constraint*

$$Y_k^L \le Y_k \le Y_k^U \quad (11.12)$$

iv) *Material balance: fixed plant yield*

$$Y_k = \sum_{j \in J} Y_{jk} \quad (11.13)$$

$$Y_{jk} = \sum_{j \in J} \sum_{i \in I} Y_{ij} X_{ij} \quad (11.14)$$

v) *Utility consumption constraint*

$$E_{uj} \le E_u^U \quad (11.15)$$

vi) *Nonnegativity constraint*
All the variables are nonnegative.

11.4.2 Illustrative Case Study for LP Model

An illustrative case study is presented to show the applicability of the framework proposed earlier. The steady-state simulation of the processing and production network was carried out using Aspen Plus as discussed in Section 11.3. A typical natural gas composition (mol%) as shown in Table 11.2 with specific flowrate and operating conditions is used in our analysis. Also, the product yields and flowrates shown in Table 11.3 are used for the implementation of LP model. The model has been solved for different scenarios.

11.4.2.1 Scenario 1: Network Optimization (Base Case)

After carrying out the steady-state simulation of LNG, GTL, and methanol processes, the products mass flowrate in kilograms per hour, yield, and demand range of LNG, GTL, and methanol processes products are tabulated for the base case conditions in Table 11.3. The economic data is shown in Table 11.6. The LP model has 25 variables and 81 constraints. It has been solved in LINGO 14.0 version. The optimal values of product flowrates are tabulated in the last column of Table 11.3. The optimal hourly profit was found to be \$219,038. The optimal natural gas feedstock flowrate to LNG, GTL, and methanol facilities was 1,515,000 kg/h.

11.4.2.2 Scenario 2: Natural Gas Feedstock Flowrate Increment

In Scenario 2, we consider increase in natural gas feedstock flowrate by running the simulation at higher flowrate from 1500 to 1800 MMSCFD fed to the stabilization unit. With the new higher flow rate and updated operating and utility cost, the optimal hourly profit was found to be \$155,946. The lower value of profit is justified by the higher operating and utilities costs and fixed demand of products similar to base case scenario. If we lower the availability of natural gas feedstock from 5,400,000 to 5,200,000 and 5,100,000 kg/day, we find that the hourly net profit is \$41,000 and \$41,500, respectively.

11.4.2.3 Scenario 3: Natural Gas Feedstock and Product Price Increments

In Scenario 3, we consider the increase in both the natural gas feedstock and product prices. The natural gas price fluctuated between \$2 and \$9 per MMBtu for the period 1994 to 2014. Thus, we need to consider this by solving the model for 50% and 100% incremental from base case prices. The optimal hourly profit was found to be \$1,339,721 and \$2,479, 711 for 50% and 100% incremental, respectively, as shown in Table 11.7. The optimal natural gas feedstock flowrate to

Table 11.6 Economic data.

Natural gas feed stock	\$4.4 per MMBtu
Sulfur	\$200 per ton
Ethane	\$10 per MMBtu
LNG	\$7 per thousand cubic feet
LPG	\$2.5 per gallon
Plant condensate	\$15 per MMBtu
Field condensate	\$12 per MMBtu
Gasoline	\$2.8 per gallon
Diesel	\$3 per gallon
Wax	\$2 per gallon
Methanol	\$500 per ton

Table 11.7 Comparison of results for the three scenarios.

	Scenario 1	Scenario 2	Scenario 3	
			50%	100%
Sulfur	71,150	60,000	60,000	60,000
LNG	1,090,000	1,245,000	1,090,000	1,090,000
Ethane	200,000	180,000	180,000	180,000
LPG	280,000	420,000	280,000	280,000
Plant condensate	25,000	22,000	25,000	25,000
Field condensate	181,334	175,000	181,334	181,334
Losses**	230,000	300,000	230,000	230,000
Gasoline	230,000	220,000	230,000	230,000
Diesel	180,000	222,900	180,900	180,900
Wax	100,000	90,000	100,000	100,000
Losses**	733,000	950,000	733,000	733,000
Methanol	900,000	1,002,500	900,000	900,000
Losses**	400,000	500,000	400,000	400,000
Profit ($/h)	219,037	155,946	1,339,721	2,479,711

*Yield defined as product flow rate divided by natural gas feedstock flow rate.
**Losses represent other byproducts such as CO_2, N_2, water, etc.

LNG, GTL, and methanol facilities was 1,550,000, 1,523,666, and 1,526,334 kg/h respectively, for both 50% and 100%.

We observe from Table 11.7 that we can maximize the network profit by processing more natural gas feedstock and when the product selling prices increased by 50% or 100% from base case prices. Also, the model solves for lower values for all loss streams because they do not contribute to the processing network profit. The loss stream from the GTL process is the highest where CO_2, and water are produced in large amounts as waste products. Thus, further consideration of capturing the CO_2 and utilizing it within the network and incorporating waste water management will improve the performance significantly.

11.5 MILP Model for Design and Synthesis of Natural Gas Upstream Processing Network

Although modeling, simulation, and optimization design studies for other fuel types and products have been addressed previously, the design and operation of enterprise-wide natural gas upstream processing network has not been addressed in the literature adequately. For example, Liu et al. (2009) presented a MINLP model for the optimal design of poly-generation energy systems of a coal-based plant producing electricity and methanol. A suitable superstructure was proposed by partitioning a general poly-generation energy system into four major blocks, where for each block alternative technologies and types of equipment were considered. Schulz et al. (2005) addressed the supply chain optimization of a petrochemical complex consisting of NGL, ethylene, chlorine, VCM, PVC, polyethylene, ammonia, and urea plants. They formulated two multiperiod finite-horizon MINLP models coordinating production, product delivery, and inventory management with different levels of rigorousness. The objective function was set to maximize the total profit; the difference between sales revenue and the total operating cost plus any penalty for not meeting demands and inventory targets. Quaglia et al. (2012) presented a MINLP formulation to simultaneously solve both the business and engineering aspects of soybean resources allocation problems. Floudas et al. (2012) and Wood et al. (2012) presented review of alternate processes for conversion of natural gas to multiple liquid fuels (GTL).

In this section, we present the comprehensive design and operational mixed integer linear programming model of Al-Sobhi et al. (2018) for superstructure optimization to optimally select the most cost-effective pathway in natural gas upstream processing networks (Figure 11.7). The key processing units considered include stabilization, acid gas removal, dehydration, sulfur recovery, NGL recovery, and NGL fractionation. Many process configurations are available for selection and the choice of technologies can be vast. There is no single technology or process configuration that is superior in all aspects. Thus, there is a need of a mathematical model that considers different flowsheet configurations and operating mode options and selects optimally among them. In the LP model presented in Section 11.4, the network topology was fixed. In other words, a prespecified technology and operating mode were utilized for the major processing units. The earlier work is extended by considering a superstructure of the natural gas upstream processing network with various design alternatives for simulation and optimization as shown in Figure 11.21.

The natural gas stream (F_{natgas}) enters the processing network at a stabilization unit (A), producing a C_{5+} product whose quality depends on the operating mode of selected technology. Then, the resulting residual gas (F_{resgas}) enters the acid gas removal unit (B), for H_2S and

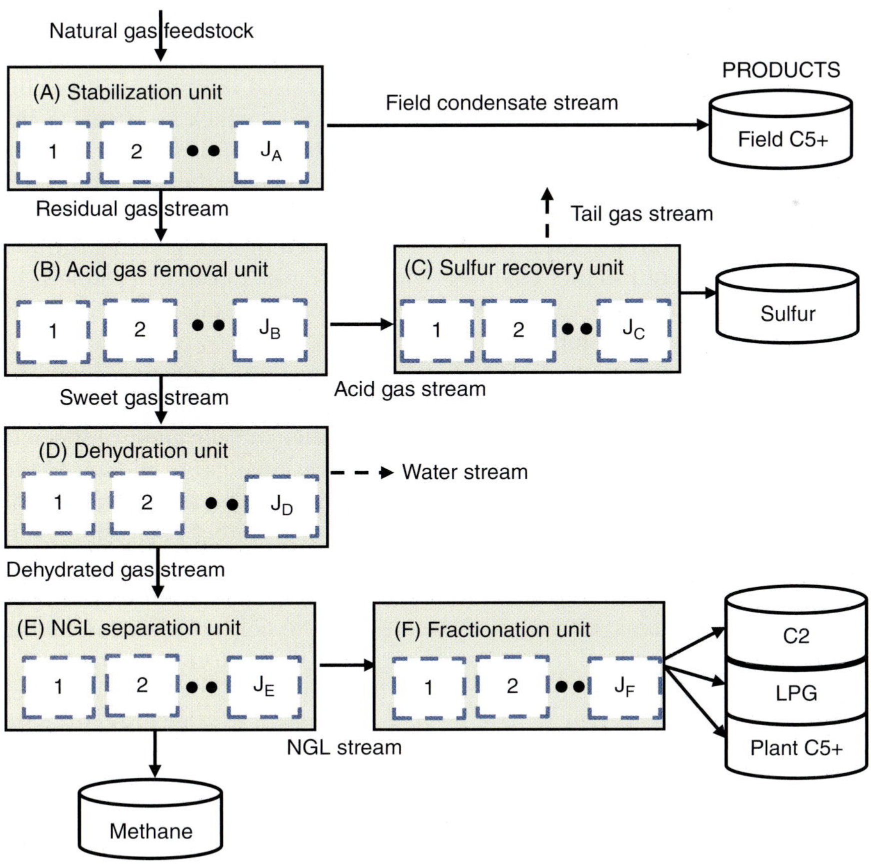

Figure 11.21 Superstructure representation for natural gas upstream processing network.

CO_2 capturing. The resulting acid gas stream ($F_{acidgas}$) enters the sulfur recovery unit (C) to produce sulfur and tail gas. The resulting sweet-gas stream ($F_{sweetgas}$) from unit (B) with the required H_2S and CO_2 concentration level enters the dehydration unit (D) for water removal to reduce pipeline corrosion and prevent line blockage caused by hydrate formation. The dehydrated stream (F_{dehgas}) enters the NGL recovery unit (E), to recover NGL products from the stream. The resulting NGL stream (F_{NGL}) is finally fractionated into ethane, LPG, and plant C_{5+} products in unit (F).

11.5.1 Process Descriptions of Major Processing Units

The main processes, available technologies, other design alternatives, and possible operating modes for each major processing unit of the natural gas upstream processing network of Figure 11.21 are described.

11.5.1.1 Stabilization Unit (A)

The primary purpose of the stabilization unit is to recover the intermediate and heavy C_{5+} by-products early from the natural gas feedstock. The contribution of revenue from these liquid by-products to the profit can be maximized by adjusting the fraction of by-product recovery while meeting the specifications of the gas feedstock (e.g. residual gas) fed to the next major processing units. The stabilized liquid is characterized by its vapor pressure and hydrogen sulfide content. The stabilization process can be performed in industry through either flash vaporization or fractionation. In flash vaporization, the feed is flashed through two or three flash tanks. The separation of vapor and condensate phases occurs due to equilibrium principles. However, flash vaporization is an old technology and is not currently used in modern gas plants (Mokhatab and Poe 2012), compared to stabilization by fractionation, which is a modern and widely accepted technology in the natural gas industry. The stabilization is typically carried out in an absorber with a reboiler and internal trays. Moreover, a refluxed distillation tower is used for better separation. The C_{5+} condensate product is sold based on a specified Reid vapor pressure (RVP) defined by the customer. The RVP is controlled by manipulating the bottom reboiler temperature. Different product qualities (with varying RVP) can be produced by changing the operating conditions (particularly the column pressure) of stabilization column. A detailed discussion can be found in Campbell (2000) and GPSA (2004).

11.5.1.2 Acid Gas Removal Unit (B)

The primary purpose of the acid gas removal unit is to reduce the concentration of the acid gases, carbon dioxide (CO_2) and hydrogen sulfide (H_2S), in the residual gas from Unit (A) to very low levels. Natural gas with H_2S or other sulfur compounds is called sour gas, whereas gas with only CO_2 is called sweet gas (Mokhatab and Poe 2012; Klinkenbijl et al. 2005). Furthermore, H_2S and CO_2 are referred to as acid gas components because they form a weak acid when dissolved in water. The acid gas removal is conducted to meet specifications, and prevent corrosion and plugging problems. Currently, there are four possible contexts for natural gas purification through acid gas removal (Kidnay and Parrish 2006): (1) CO_2 removal from a gas that contains no H_2S, (2i) H_2S removal from a gas that contains no CO_2, (3) simultaneous removal of both CO_2 and H_2S, and (4) selective removal of H_2S from a gas that contains both CO_2 and H_2S.

Since the acid gas concentration and the final product specification varies based on region and consumers, many different purification processes are available for selection. Liquid-phase absorption processes are the most common sweetening technologies and classified into chemical, physical, and hybrid solvent categories (Mokhatab and Poe 2012). In the chemical solvent process, alkanolamines or alkaline salt of various weak acids are used for absorption of acid gases. Physical solvent processes use an organic solvent without any chemical reaction, i.e. acid gas removal depends on physical absorption. In hybrid solvent processes, a mixture of chemical and physical solvents is used. In all liquid absorption processes, solvent circulation rate (affecting equipment size and operating costs) and energy consumed in solvent regeneration are major cost factors (Mokhatab and Poe 2012). In this study, we focus on chemical solvent method based on diethanolamine (DEA) and methyldiethanolamine (MDEA) amines, in different operating modes based on concentrations ranging from approximately 10 to 65 wt% amines. Additional option based on two absorbers in sequence (e.g. first MDEA and then DEA) is also considered. A detailed discussion may be found in Campbell (2000), GPSA (2004), and Kohl and Nielson (1997).

11.5.1.3 Sulfur Recovery Unit (C)

Currently, two options exist for handling gases with significant H_2S component: (1) disposal by underground injection and (2) conversion of H_2S to sulfur. The main purpose of the sulfur recovery unit is converting the H_2S component in the acid gas to elemental sulfur which is commonly used for sulfuric acid production. This conversion is done industrially using Claus process or one of its modifications. The two modified Claus processes are straight-through and split-flow. The straight-through process is preferred when the feedstock has a high H_2S concentration, e.g. up to 55 mol% H_2S. The split-flow configuration can process feedstock that contains lower concentrations, e.g. 5–30 mol% H_2S. The straight-through process provides the highest sulfur-recovery efficiency (GPSA 2004). All Claus units involve a two-step process of initial combustion in a furnace followed by processing of combustion products through a series of catalytic converters, to produce elemental sulfur. A tail gas cleanup unit is used to eliminate the last sulfur compounds to meet environmental regulations. The most commonly used Claus processes are Shell Claus Offgas Treating, SUPERCLAUS, and cold-bed adsorption (GPSA 2004). In this study, we focus on straight-through and split-flow Claus processes with operating modes based on different temperatures.

11.5.1.4 Dehydration Unit (D)

The water removal processes considered in the industry are mainly based on absorption and adsorption. The water level in sweet gas can be reduced to 10 ppmv range via physical absorption where gas is contacted with a liquid to absorb water vapor (Kidnay and Parrish 2006). In practice, different glycols such as ethylene-, diethylene-, triethylene-, tetraethylene-, and propylene-glycols are used as absorbents. In this study, we focus on the use of triethylene glycol being most common in industry (GPSA 2004).

11.5.1.5 NGL Separation Unit (E)

The primary purpose of NGL separation unit is to separate the feedstock gas from liquids to reach a specific heating value because pipeline-quality gas specification requires a high heating value of around 950–1150 Btu/scf (35,400–42,800 kJ/S m^3) as well as limiting the sulfur and water content (Klinkenbijl et al. 2005). The process elements for hydrocarbon recovery vary based on the desired products, gas volume being processed, and inlet gas composition/pressure. Broadly, the commercially available technology options for NGL separation are (1) refrigeration processes, (2) lean oil absorption, (3) solid bed adsorption, (4) membrane separation, and (5) twister supersonic separation. A detailed discussion can be found in Campbell (2000) and GPSA (2004). In this study, we focus on use of refrigeration process, being most common in industry, with different operating modes based on Demethanizer pressures in expander plant.

11.5.1.6 Fractionation Unit (F)

The primary purpose of this unit is to fractionate the NGL stream into its various components (ethane, propane, isobutene, normal butane, and gasoline), which have a higher market value as pure products (Mokhatab and Poe 2012). The bottom liquid from Unit (F) is fractionated by heating and passing it through a series of distillation columns. There are many design options related to the sequence of distillation columns for product recovery. The number of possible sequences can be calculated (Seider et al. 2004) based on the number of components (K), as $[2(K-1)!/K!(K-1)!]$. In this study, we focus on two process schemes: direct (conventional) and indirect (nonconventional) configurations with different product recoveries. In the direct configuration, de-ethanizer, de-propanizer, and de-butanizer are placed in sequence to separate NGL into different products, whereas in the nonconventional configuration, the de-propanizer, de-ethanizer, and de-butanizer are placed in sequence.

The different process alternatives considered for each processing unit in the overall superstructure of a natural gas upstream processing network are summarized in Table 11.8 (Al-Sobhi et al. 2018).

11.5.2 Problem Statement and Solution Strategy

Consider a superstructure with different design alternatives and operating modes for each key processing unit, such as shown in Table 11.8. It is desired to determine the optimal natural gas processing network configuration that maximizes production (e.g. product yields), minimizes the capital investment and operating costs, meets the product specifications (e.g. RVP value) along with environmental constraints (e.g. CO_2 and H_2S concentration levels), using the

Table 11.8 Different possible technologies and operating modes considered in the superstructure.

Major processing unit (i)	Possible processes/ technologies	Process/technology considered here (J_i)	Operating modes (M_i)
Condensate stabilization (A)	1) Flash vaporization 2) Stabilization by fractionation	• Stabilization by fractionation	• Operating pressure
Acid gas removal (B)	1) Indirect conversion 2) Direct conversion (i.e. dry-bed or Liquid phase) 3) Separation technologies, (i.e. membrane or cryogenic fractionation)	• Direct conversion (Liquid phase)	• Solvent type • Solvent concentration • Absorber arrangements, etc.
Sulfur recovery (C)	1) Gas phase Claus type 2) Liquid phase	• Gas phase Claus	• Claus straight at different temperatures • Claus split at different temperatures
Dehydration (D)	1) Liquid desiccant (Glycol) 2) Solid desiccant 3) Cooling the gas	• Glycol	• Glycol concentration and circulation rate
NGL recovery (E)	1) Refrigeration process 2) Lean oil absorption 3) Solid-bed adsorption 4) Membrane separation 5) Twister supersonic	• Refrigeration process	• Demethanizer pressure in expander plant
NGL fractionation (F)	1) Direct sequence 2) Indirect sequence	• Direct sequence • Indirect sequence	• Molar recovery

simulation-optimization framework shown in Figure 11.20. The yields for different technologies and operating modes are determined after running ASPEN simulation for each unit, along with estimation of annual capital and operating costs using ICARUS software. Then we employ this information as input to the formulation and solution of MILP model.

11.5.3 MILP Model Formulation

The MILP model of Al-Sobhi et al. (2018) includes maximization of a profit function subject to the model constraints for balancing material flow between the major processing units, satisfying end-product demands as well as complying with CO_2 and H_2S-quality specifications and operational restrictions. The following constraints in the model mathematically describe the operational and material flow in each block of the superstructure in terms of overall mass balance, yield, quality, use of different technologies, and capacity constraints. The details of ASPEN simulation flowsheets for each unit under different design alternatives are provided in Section 11.3.

$X^i_{j,m}$ is a binary decision variable specifying the design alternative (e.g. technology) and operating mode selection for unit $i \in I$, where J_i and M_i refer to sets of design alternatives and operating modes, respectively.

$$X^i_{j,m} = \begin{cases} 1, & \text{if technology } j \text{ with operating mode } m \text{ is selected for unit } i \\ 0, & \text{otherwise} \end{cases}$$

$$\forall i \in \{I : A, B, \dots, F\}, j \in \{1, 2, \dots, |J_i|\}, m \in \{1, 2, \dots, |M_i|\}.$$

Index s or s' is defined for all streams; $s \in S^{\text{in}}_i = \{\text{natgas, resgas, acidgas, sweetgas, dehgas, NGL}\}$ refers to input streams entering unit $i \in I^{\text{in}}_s$; $s \in S^{\text{out}}_i$ refers to output streams leaving unit $i \in I^{\text{out}}_s$; $s \in S^{\text{int}} = \{\text{resgas, acidgas, sweetgas, dehgas, NGL}\}$ refers to connecting/intermediate streams between units, specifying plant topology and unit sequence/connectivity. We define $F^{\text{in}}_{s,j,m}$ as the flow rate of incoming stream $s \in S^{\text{in}}_i$ with selected technology j in operating mode m. F^{out}_s is the flow rate of outgoing stream $s \in S^{\text{out}}_i$ for each unit i.

Linearization is required to avoid the typical bilinear terms appearing in mass balances. Since the number of output streams is generally more than the number of input streams for any unit, we consider model formulation using linearization of bilinear terms based on input streams for each unit. In general, we can consider linearization based on either input or output streams for each unit of other networks, whichever are less in number. For instance, for Unit A, rather than defining three bilinear terms $F_{\text{natgas}_{j,m}}$, $F_{\text{resgas}_{j,m}}$, and $F_{\text{field}C_{5+j,m}}$ we define only one term based on input stream (natural gas), $F_{\text{natgas}_{j,m}}$. The output streams are simply calculated from input streams using yield values. Similarly for unit B, the linearization is done based on input stream, i.e. residual gas, and so on.

The following MILP model (Al-Sobhi et al. 2018) optimizes the network design and operation to maximize the annual profit in Eq. (11.16).

$$\max \sum_{s\in P} \text{sp}_\text{p} * F^{\text{out}}_s - \text{np} * \sum_{m\in M_i} \sum_{j\in J_i} F^{\text{in}}_{\text{natgas},j,m} - \sum_{i\in I} \sum_{m\in M_i} \sum_{j\in J_i} X^i_{jm} * (\text{ACC}_{ijm} + \text{AOC}_{ijm}) \tag{11.16}$$

The total annual profit is represented as the difference between the overall sales revenue and total annual cost that includes annualized capital cost, variable annual operating costs, fixed annual operating costs of major units, and the annual cost of natural gas usage. $P \equiv \{\text{field}C_{5+}, S, C_1, C_2, \text{LPG}, \text{plant}C_{5+}\}$ denotes the set of commercial end-products, sp_p is selling price of each product. np is natural gas feedstock price, ACC_{ijm} is amortized capital cost of unit i for technology j in operating mode m, AOC_{ijm} is annual operating cost of unit i for technology j in operating mode m. It is assumed that capital costs are amortized over the project lifetime of 20 years with 10% compound interest rate.

The selection of technology and mode for each unit is given in Eq. (11.17). The activation of flowrates based on input streams is given in Eq. (11.18) for selection of technology j and mode m. The same equation can be used for capacity constraints in each unit with updated bounds. The calculation of output flowrates based on yields is an equality constraint given in Eq. (11.19). It can be eliminated along with the outflow variable, F_s^{out}, by substituting it in the objective function and relevant constraints.

$$\sum_{m \in M_i} \sum_{j \in J_i} X_{j,m}^i = 1 \quad \forall i \in I \tag{11.17}$$

$$F_{j,m}^{i,\text{LB}} * X_{j,m}^i \leq F_{s,j,m}^{\text{in}} \leq X_{j,m}^i * F_{j,m}^{i,\text{UB}} \quad \forall i \in I, j \in J_i, m \in M_i, s \in S_i^{\text{in}} \tag{11.18}$$

$$F_s^{\text{out}} = \sum_{i \in I_s^{\text{out}}} \sum_{m \in M_i} \sum_{j \in J_i} Y_{sjm}^i * \sum_{s' \in S_i^{\text{in}}} F_{s',j,m}^{\text{in}} \quad \forall s \in S^{\text{out}} \tag{11.19}$$

In the above model, so far there is no connection across different units. Therefore, the following plant topology-based constraint is added, which ensures adherence to the sequence of units as per the specified plant topology.

$$\sum_{i \in I_s^{\text{out}}} F_s^{\text{out}} = \sum_{i \in I_s^{\text{in}}} \sum_{m \in M_i} \sum_{j \in J_i} F_{s,j,m}^{in} \quad \forall s \in S^{\text{int}} \tag{11.20}$$

Equation (11.20) states that the total outgoing flow rate from the units, which produce each connecting/intermediate stream, should be equal to the total incoming flow rate entering the next unit in sequence, which consumes this stream. The above plant topology constraint replaces the conventional material balances across units used in literature. The demand constraint is given in Eq. (11.21) for all products.

$$F_s^{\text{out}} \geq D_s \quad \forall s \in P \tag{11.21}$$

The mole fraction of component k (CO_2 and H_2S) in sweet gas from unit B for technology j in mode m should be within the acceptable limit as given in Eq. (11.22), where the parameter, $MF_{B,j,m}^{k,\text{sweetgas}}$, is obtained from ASPEN simulations.

$$MF_{B,j,m}^{k,\text{sweetgas}} \leq X_{j,m}^B * q_{\text{sweetgas}}^k \quad \forall k, j, m \tag{11.22}$$

11.5.4 Illustrative Case Study

Consider the natural gas upstream processing network shown in Figure 11.21 and Table 11.8 for which ASPEN steady-state simulations of each major processing unit are presented in Section 11.3. Typical natural gas composition (mol%) used in this analysis is shown in Table 11.2. The flow rate of natural gas was bounded in the range 1.50×10^6 to 1.55×10^6 kg/h. The sale prices of end-products and cost of raw-materials are shown in Table 11.6. The plant-related cost data (i.e. capital and operating costs) are estimated based on ICARUS software. Different product qualities with varying RVP can be produced by changing the stabilization column's operating conditions (particularly the column pressure) as reported in Campbell (2000).

For this case study, three operating modes were considered for stabilization for 16 RVP field condensate at column pressures of 180, 200, and 220 psia. Three configurations were considered for acid gas removal unit (B): (1) using MDEA only, (2) using a mixed amine solution of MDEA and DEA, (3) using two absorbers in sequence, first MDEA and then DEA. Two sulfur recovery technologies were considered for unit C: Claus straight vs. split flow with different operating conditions (at 5 bar and temperatures of 200, 300, and 400 °C). For units D and E one operating mode was considered. For fractionation unit (F) two process schemes are considered: the direct (conventional) and indirect (nonconventional) configurations. In both sequences, at least 90% ethane mole recovery, 80% propane mole recovery, and 80% butane mole recovery are achieved. After running the steady-state simulation using ASPEN Plus V7.3 for each key processing unit under various configurations and operating modes, the necessary material balances are obtained. The total capital, annuity of capital, and total operating cost of each processing unit are shown in Table 11.9.

Based on the different yields obtained from ASPEN Plus V7.3 for different products and various capital and operating costs for each option, the MILP model was applied on the processing network. The MILP model has 17 continuous variables, 14 binary variables, and 38 constraints. The MILP model was solved in LINGO 14.0 via the branch-and-bound method. The optimal annual profit was found to be MM$3207 and the optimal flowrate values are 1,550,000, 68,815.35, 22,281.26, 1,122,050.702, 61,945.583, 72,682.809, and 15,092.626 kg/h for natural gas feedstock, field condensate, sulfur, methane, ethane, LPG, and plant condensate

Table 11.9 Total capital, annuity of capital, and total operating cost of each processing unit.

Processing unit	Operating conditions	Total capital cost (MM $)	Annuity of capital, ACC_{ijm} (MM $/yr)	Annual operating cost, AOC_{ijm} (MM $/yr)
Condensate stabilization (A)				
	12.4 bar, 151 °C	12.2	1.48	2810
	13.8 bar, 157 °C	11.4	1.39	5.32
	15.2 bar, 163 °C	11.9	1.45	5.44
Acid gas removal (B)				
1) MDEA 35 wt%	55.2 bar, 35 °C	34.2	4.17	45.5
2) MDEA30 wt% + DEA15 wt%	55.2 bar, 35 °C	34.5	4.20	380
3) MDEA first + DEA second	55.2 bar, 21.1 °C	80.4	9.78	1380
Sulfur recovery (C)				
1) Claus straight	5 bar, 200 °C	3.06	0.372	0.959
	5 bar, 300 °C	3.07	0.373	0.959
	5 bar, 400 °C	3.29	0.401	0.960
2) Claus split	5 bar, 200 °C	3.27	0.306	0.959
	5 bar, 300 °C	3.29	0.307	0.959
	5 bar, 400 °C	3.46	0.421	0.960
Dehydration (D)	50 bar, 21.1 °C	27.6	3.36	12.1
NGL separation (E)	14 bar, −62.2 °C	16.8	2.04	2480
Fractionation (F)				
1) Conventional (direct)	20 bar, 143 °C	8.05	0.979	1770
2) Nonconventional (indirect)	20 bar, −9.4 °C	12.8	1.55	2.46×10^5

products, respectively. Furthermore, MILP results showed that the optimal operating mode for the stabilization unit (Unit A) is 13.8 bar and 157 °C for producing 16 RVP condensate. Also, operating acid gas removal unit with (MDEA 30 wt% + DEA15 wt%) at 55 bar and 35 °C, operating the sulfur recovery unit at 5 bar and 200 °C with Claus straight configuration, operating the dehydration unit at 50 bar and 21.1 °C with TEG, operating the NGL recovery unit at 14 bar and −62.2 °C, and operating the fractionation unit at 20 bar and 143 °C with direct sequence configuration were the optimal configurations for units (B) to (F) according to MILP model solution.

11.6 MILP Model for Design and Synthesis of Natural Gas Production Network

The production system we consider in this section includes three main conversion options: LNG, GTL, and methanol systems. The process descriptions for these three systems and their flowsheet simulations are presented in sections 11.2 and 11.3. Figure 11.22 shows the superstructure of a natural gas production network with multiple alternatives for each processing block. The main processing units are syngas preparation (G), liquefaction (H), N2 rejection (I), hydrogen production (R), Fischer-Tropsch synthesis (K), methanol synthesis (L), FT upgrade (M), and methanol upgrade units (N). The different possible alternatives for unit designs include different LNG liquefaction cycles, syngas production technologies, different types of catalysts and reactors, etc. with wide range of operational conditions as shown in Figure 11.22.

Different technologies are available to produce syngas from natural gas (Aasberg-Petersen et al. 2001; Luyben 2014; Rostrup-Nielsen 2000; Wilhelm et al. 2001). These include catalytic steam methane reforming (SMR), two-step reforming, partial oxidation (POX), auto-thermal reforming (ATR), combined reforming (CR), ceramic membrane reforming (CMR), and dry reforming (DR). The combined reforming consists of a combination of steam methane reforming and auto-thermal reforming. Julia et al. (2014) assessed syngas technologies for methanol production from shale gas. Four reforming technologies were considered in their work, namely, partial oxidation (POX), steam methane reforming (SMR), auto-thermal reforming (ATR), and combined reforming (CR). The results showed that the use of POX or ATR provides the best alternative with the high profitability potential for methanol production. However, from an environmental aspect, CR turned out to be the option with the lowest carbon footprint. Noureldin et al. (2014) addressed the optimization and selection of natural or shale gas reforming. They considered SMR, POX, DR, and CR. They found that CR (including tri-reforming) improved process aspects, such as energy usage, safety, and flexibility.

Many liquefaction technologies exist; they mainly differ in the types of different refrigeration cycles used such as the APCI propane pre-cooled mixed refrigerant (PPMR) process, Phillips Optimized Cascade LNG Process (OCLP), and Shell Dual Mixed Refrigerant (DMR) process (Tusiani and Shearer 2007). Mokhatab and Economides (2006) have presented an overview of LNG processes and provided a comprehensive analysis of methods for liquefaction of natural gas in an onshore LNG plant.

Nitrogen separation or rejection is required under three scenarios: (1) nitrogen-rich gas processing to produce pipeline-quality gas, (2) removal of nitrogen for enhanced oil recovery (EOR) purposes, and (3) removal of helium from nitrogen for helium-recovery purposes (Kidnay and Parrish, 2006). The methods available for nitrogen removal are (1) cryogenic distillation, (2) adsorption, and (3) membrane separation, with cryogenic distillation being the most common. For low concentrations (below 20% N_2), a single column can be used. For high concentrations, a dual column is preferred (GPSA 2004). Kuo et al. (2012) summarized the selection criteria of

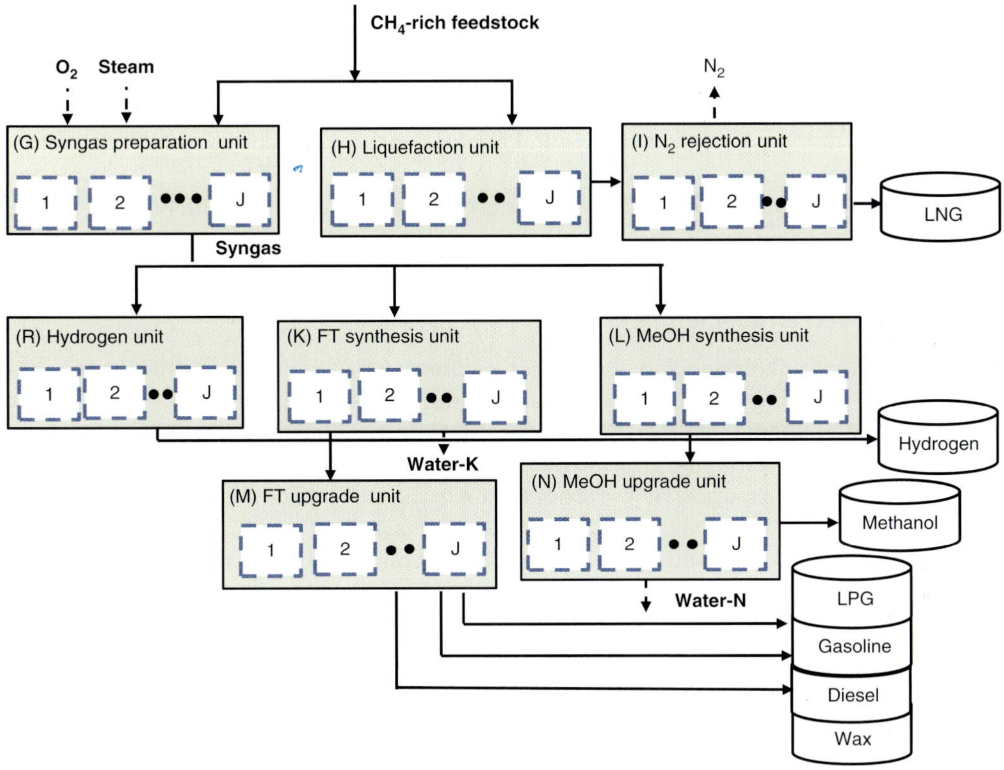

Figure 11.22 Superstructure representation for natural gas downstream production network.

an optimum NRU for all currently available technologies including both those that have been commercialized and those in the development stage.

Typically, hydrogen is produced in three main steps: (1) Syngas preparation, using steam reforming of natural gas, accounting for more than 50 % world hydrogen production (Mueller-Langer et al. 2007); (2) water gas-shift reaction, where CO reacts with steam, producing product gas with hydrogen, CO_2, and some impurities (unconverted CH_4 and CO); and (3) separation, where CO_2 is removed using alkanolamines via chemical absorption producing hydrogen-rich gas that is purified by pressure swing adsorption (PSA). There are two major categories of natural gas–based FT processes: (1) the high-temperature Fischer-Tropsch (HTFT) and (2) the low-temperature Fischer-Tropsch (LTFT) (Dry 2002). In HTFT, the typical operating conditions range from 300°C to 350°C with a pressure of approximately 2.5 MPa (Steynberg et al. 1999). Typical operating conditions for LTFT range around 200–240°C at approximate pressures of 2.0–2.5 MPa (Espinoza et al. 1999; Jager and Espinoza 1995).

There are three major categories of methanol synthesis reactors: (1) quench reactors, (2) adiabatic reactors in series, and (3) boiling water reactors (BWR). A quench reactor consists of up to five adiabatic catalyst beds installed in series in one pressure shell, and has small production capacity. The adiabatic reactor system normally consists of 2 to 4 fixed-bed reactors in series with intermittent cooling. The BWR is in principle a shell and tube heat exchanger with catalyst on the tube side, and circulating boiling water on the shell side (Olah et al. 2006).

The purpose of the methanol upgrading unit is purification of the raw methanol to produce a methanol product with a specific purity. The crude methanol from the synthesis unit contains water and other byproducts. The byproducts include DME, higher alcohols, other oxygenates, and minor amounts of acids and aldehydes. Different designs of distillation column systems are available to be implemented depending on the desired product grade. Usually, two or three distillation columns are used to achieve the commonly accepted specification, grade AA.

The objective is to determine the optimal natural gas production network configuration that maximizes production (e.g., product yields), and minimizes the capital investment and operating costs. The environmental impact represented in the analysis by CO_2 or CO_{2eq} emission is also obtained using the ASPEN Plus simulator for each flowsheet configuration under different operating modes.

11.6.1 MILP Model Formulation

The mathematical formulation includes an objective function that maximizes profit and several constraints to satisfy energy demand, CO_2 emission limits, operational restrictions, and other limitations. In the mathematical model, the operations and associated material flow in each block of the natural gas production superstructure are represented by constraints based on factors such as overall mass-balance, yields, product quality requirements, available technologies, demand, and capacities.

The selection of technology and mode for each unit remains same, as given in Eq. (11.17), using the binary variable $X^i_{j,m}$ defined over set $i \in \{I : G, H, R, K, L, M, N\}$.

The methane-rich stream, F_{CH4}, is sent to both syngas preparation and liquefaction units simultaneously. The exact percentage of methane fed to each unit is determined by optimization; however, lower bound values are set for both units to ensure that they are both operational. The overall mass-balance of methane rich stream can be expressed as:

$$F_{CH4} = F^G_{CH4} + F^H_{CH4} \tag{11.23}$$

Although other technologies are available to produce syngas from natural gas, we will consider only ATR and SMR as competing technologies for the reforming unit due to their applicability in large-scale production. The rich methane stream directed to the syngas unit (F^G_{CH4}) is fed along a flow of steam with rate F^G_{steam} in case of SMR, and with a flow of oxygen with the rate F^G_{O2} in the case of ATR to produce the required syngas ratio shown in Eq. (11.24). We will write two different types of balances for each technology such as j = ATR, SMR. Output of each option is H_2, CO, and CO_2 as shown in Eq. (11.24) with different flowrate, and accordingly we get different syngas (H_2/CO) ratios. The desirable composition of the syngas for the low-temperature FT corresponds to a ratio H_2/CO of two.

$$F^G_{CH4_j} + F^G_{steam_j} + F^G_{O2_j} = F^G_{H2_j} + F^G_{CO_j} + F^G_{CO2_j} + F^G_{water_j} \tag{11.24}$$

whereas, the syngas flowrate is defined in Eq. (11.25).

$$F^G_{syngas_j} = F^G_{H2_j} + F^G_{CO_j} + F^G_{CO2_j} \tag{11.25}$$

For j=ATR, the operating steam-to-CH_4 ratio is set to 0.6 (Steynberg and Dry 2004) as shown in eq. (11.26).

$$F^G_{steam_j} = 0.6.F^G_{CH4_j} \quad j = \text{ATR} \tag{11.26}$$

$$F^G_{O2_j} = f(F^G_{CH4_j}) \quad j = \text{ATR} \tag{11.27}$$

Equation 11.27 states that the required oxygen flowrate is a function of the methane flowrate. In order to produce the required syngas ratio, we need to generate O_2 and CH_4 flowrate data from the simulation and get different syngas ratio values by changing the oxygen flowrate for

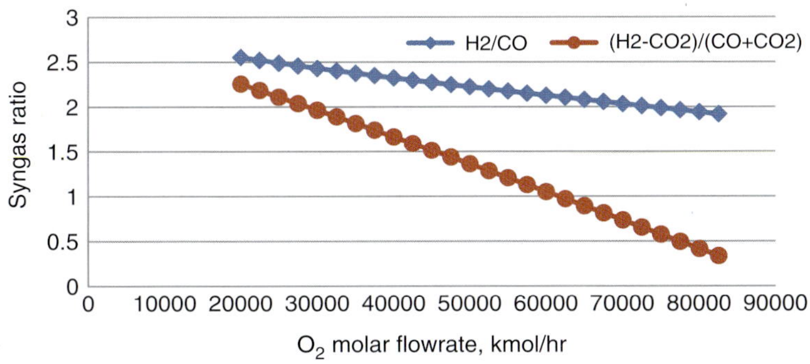

Figure 11.23 Sensitivity analysis for oxygen flowrate.

a given methane flowrate. Then, by plotting syngas ratio vs. O_2 flowrate, we can find the right value of O_2 that corresponds to a syngas ratio of two. Also, the simulator sensitivity analysis modeling option can be used for this purpose as shown in Figure 11.23.

For j=SMR, the overall material balance is given in equation (11.28).

$$F^G_{CH4_j} + F^G_{steam_j} = F^G_{H2_j} + F^G_{CO_j} + F^G_{CO2_j} \quad j = \text{SMR} \tag{11.28}$$

$$F^G_{steam_j} = f(F^G_{CH4_j}) \qquad j = \text{SMR} \tag{11.29}$$

$$F^G_{syngas} = \sum_{m \in M_G} \sum_{j \in J_G} F^G_{syngas_j} \tag{11.30}$$

Furthermore, the produced syngas is distributed among the candidate receiving units: namely, hydrogen unit (R), FT synthesis unit (K), and methanol synthesis unit (L) as shown in Eq. (11.31). Again the exact amount of F^G_{syngas} fed to these three units is an optimization variable. A lower bound value is set for the three units to ensure that all units receive the syngas flowrate.

$$F^G_{syngas_j} = F^R_{syngas_j} + F^K_{syngas_j} + F^L_{syngas_j} \tag{11.31}$$

$$F^R_{syngas_j} = F^R_{H2} + F^R_{CO} + F^R_{CO2} \tag{11.32}$$

Because we need to select just one hydrogen production technology from the different available technologies, we have the following constraint in Eq. (11.33),

$$F^R_{syngas_j} \le X^R_{jm}.F^{R,U}_{syngas} \qquad \forall j \in J_R, m \in M_R \tag{11.33}$$

Similarly, we have the following constraints shown in Eq. (11.34) for FT synthesis, and Eq. (11.35) for methanol synthesis.

$$F^K_{syngas_j} \le X^K_{jm}.F^{K,U}_{syngas} \qquad \forall j \in J_K, m \in M_K \tag{11.34}$$

$$F^L_{syngas_j} \le X^L_{jm}.F^{L,U}_{syngas} \qquad \forall j \in J_L, m \in M_L \tag{11.35}$$

Hydrogen from unit R is shown in Eq. (11.36), which states that the produced hydrogen flowrate is a function of syngas flowrate and selected syngas technology.

$$F_{H2} = f(F^R_{syngas_j}) \tag{11.36}$$

The methane-rich stream fed to the liquefaction unit, F^H_{CH4} is compressed and cooled down to −160 C. Now, this stream is in liquid form but with nitrogen content, $F^I_{CH4(l)\_N2}$, and is sent to the N_2 rejection unit to reject N_2 stream and get LNG stream, F_{LNG} as shown in Eqs. (11.37) and (11.38).

$$F^{H}_{CH4} = F_{N2} + F_{LNG} \tag{11.37}$$

$$F_{LNG_j} \leq X^{H}_{jm}.F^{H,U}_{LNG} \qquad \forall j \in J_H, m \in M_H \tag{11.38}$$

Equations (11.39) to (11.45) give the overall material balances around units K, L, M and N.

$$F^{K}_{H2} + F^{K}_{CO} + F^{K}_{CO2} = F^{K}_{syncrude} + F^{K}_{water} \tag{11.39}$$

$$F^{L}_{H2} + F^{L}_{CO} + F^{L}_{CO2} = F^{L}_{methanolcrude} + F^{L}_{water} \tag{11.40}$$

$$F^{N}_{methanol} = F^{L}_{methanol\_crude} - F^{N}_{water} \tag{11.41}$$

$$F^{M}_{LPG} = sf_1.F^{K}_{syncrude} \tag{11.42}$$

$$F^{M}_{Gasoline} = sf_2.F^{K}_{syncrude} \tag{11.43}$$

$$F^{M}_{Diesel} = sf_3.F^{K}_{syncrude} \tag{11.44}$$

$$F^{M}_{Wax} = sf_4.F^{K}_{syncrude} \tag{11.45}$$

where sf_1, sf_2, sf_3 and sf_4 are prespecified selectivity factors for LPG, gasoline, diesel, and wax, respectively.

Consumption of the methane stream through the network should be within the specified lower and upper bounds as shown in Eq. (11.46).

$$F^{LB}_{CH4} \leq F_{CH4} \leq F^{UB}_{CH4} \tag{11.46}$$

The annual demand constraints for all main products, $P = \{\text{LNG}, H_2, \text{methanol, LPG, gasoline, diesel, wax}\}$, are given in Eq. (11.47).

$$F_p \geq D_p \quad p \in P \tag{11.47}$$

Capacity constraints of main processing units are given in Eqs. (11.48–11.54).

$$F^{G}_{CH4} + F^{G}_{steam} + F^{G}_{O2} \leq F_{CH4j} \leq Q_{Gj}.X^{G}_{jm} \qquad \forall j \in J_G, m \in M_G \tag{11.48}$$

$$F^{H}_{CH4} + F_{N2} \leq Q_{Hj}.X^{H}_{jm} \qquad \forall j \in J_H, m \in M_H \tag{11.49}$$

$$F^{R}_{H2} + F^{R}_{CO} \leq Q_{Rj}.X^{R}_{jm} \qquad \forall j \in J_R, m \in M_R \tag{11.50}$$

$$F^{K}_{H2} + F^{K}_{CO} \leq Q_{Kj}.X^{K}_{jm} \qquad \forall i \in J_K, m \in M_K \tag{11.51}$$

$$F^{L}_{H2} + F^{L}_{CO} \leq Q_{Lj}.X^{L}_{jm} \qquad \forall i \in J_L, m \in M_L \tag{11.52}$$

$$F^{M}_{syncrude} + F^{M}_{water} \leq Q_{Gj}.X^{M}_{jm} \qquad \forall i \in J_M, m \in M_M \tag{11.53}$$

$$F^{N}_{methanolcrude} + F^{N}_{water} \leq Q_{Gj}.X^{N}_{jm} \qquad \forall i \in J_N, m \in M_N \tag{11.54}$$

where Q_{ij}, denotes the upper capacity limit for units, $i = \{G,H,I,R,K,L,M, \text{and } N]$ for technology j. All continuous variables are nonnegative.

The objective of the optimization model is to maximize the annual profit of production network. The total production cost is represented by an annualized capital cost, variable annual operating costs, and fixed annual operating cost of major units plus the annual cost of methane stream feedstock. It is assumed that capital costs are amortized over the lifetime of the project life of 20 years with 10% as a compound interest rate. Annual revenue is calculated as the sum of all product sales minus processing cost.

$$max \sum_{p} sp_p.F_p - np.F_{CH4} - \sum_{i}\sum_{m}\sum_{j} ACC_j.X^i_{jm} - \sum_{i}\sum_{m}\sum_{j} AOC_j.X^i_{jm} \tag{11.55}$$

where, sp_p is the selling price of each network products, np is methane feedstock price, ACC_j is amortized capital cost for technology j, AOC_j is annual operating cost for technology j.

11.6.2 Case Study

An illustrative case study is presented to show the applicability of the overall framework and the formulated model presented in the previous sections. Different rigorous simulations of the production network pathway were carried out using Aspen Plus to obtain surrogate models or appropriate yield equations for the production flowrate. The main key processing units such as syngas preparing unit (G), liquefaction unit (H), N2 rejection unit (I), hydrogen unit (R), FT synthesis unit (K), methanol synthesis unit (L), FT upgrade unit (M), and MeOH upgrade unit (N) are considered. The methane stream comes from NGL recovery unit with a specific flowrate and operating conditions is used in our analysis. The cost data used in the case study are given in earlier Table 11.6 based on ICARUS software. The demand, capacity, and yield values from ASPEN simulation were reported earlier in Table 11.3.

We present below the results of some of the simulation scenarios. Different percentage values for methane utilization considering different levels of utilization such as LNG, GTL, and methanol are considered. For example, first we assume that LNG is the most promising option and 100%, 70%, 50%, and 30% of methane stream is utilized to produce just LNG. Table 11.10 shows the total capital cost, total operating cost, total utilities cost, yield values, and objective function values for each planning mode considering LNG production.

Second, we assume that methanol is the most promising option and 100%, 70%, 50%, and 30% of methane stream is utilized to produce just methanol. Table 11.11 shows the total capital cost, total operating cost, yield values, and objective function values for each planning mode

Table 11.10 LNG different planning mode results.

LNG	100%	70%	50%	30%
Total capital cost, $MM	18.45	14.30	11.60	8.90
Amortized capital cost, $MM/yr	2.24	1.74	1.42	1.08
Total operating cost, $MM/yr	248.10	174	124	74.5
Total utilities cost, $MM/yr	229.60	161	115	18.9
Desired rate of return, %/yr	10	10	10	10
Lifetime of the project, yr	20	20	20	20
LNG mass flow rate, kg/h	1,128,350	789,845	564,175	338,505
LNG yield	0.88	0.62	0.44	0.26
Objective function, $MM	196	134	92.60	51.10

Table 11.11 Methanol different planning mode results.

Methanol	100%	70%	50%	30%
Total capital cost, $MM	44.8	33.7	25	19.3
Amortized capital cost, $MM/yr	5.45	4.10	3.04	2.35
Total operating cost, $MM/yr	71.5	58.5	50.4	42.3
Total utilities cost, $MM/yr	12.0	50.6	43.4	36.2
Desired rate of return, %/yr	10	10	10	10
lifetime of the project, yr	20	20	20	20
Methanol mass flowrate, kg/h	688,053	481,615	344,011	206,401
Water mass flowrate, kg/h	480	336	240	144
Methanol yield	0.67	0.47	0.33	0.20
Objective function, $MM	1100	845	568	291

Table 11.12 GTL (LTFT) different planning mode results.

GTL LTFT	100%	70%	50%	30%
Total Capital Cost, $MM	86.4	14.1	44.0	30.4
Amortized capital cost, $MM/yr	10.5	7.8	5.35	3.7
Total operating cost, $MM/yr	31.6	22.9	16.2	10.6
Total utilities cost, $MM/yr	24.3	17.0	11.7	1.96
Desired rate of return, %/yr	10	10	10	10
Lifetime of the project, yr	20	20	20	20
LPG mass flowrate, kg/h	12,850	4,471	2,725	2,395
Gasoline mass flowrate, kg/h	83,664	57,696	41,179	27,773
Diesel mass flowrate, kg/h	162,909	114,612	76,502	40,226
Wax mass flowrate, kg/h	610,756	443,272	310,548	188,175
Water mass flowrate, kg/h	65,799	34,693	32,462	19,477
LPG yield	0.012	0.004	0.003	0.002
Gasoline yield	0.081	0.056	0.040	0.027
Diesel yield	0.158	0.111	0.074	0.040
Wax yield	0.592	0.430	0.301	0.182
Objective function, $MM	1780	1400	935	513

considering methanol production. For 100% utilization of natural gas to produce methanol the total capital cost estimate using ICARUS software is found to be around $45 MM for production of 688,053 kg/h of methanol (about 16,500 ton/day), which is very low compared to literature. The investment cost for a 5000 ton/day methanol plant in North America was reported to be about $1.3 billion (HCP 2013). Therefore, a scale-up factor of about 70 is suggested to match reported costs in the literature.

Third, we assume that GTL is the most promising option and 100%, 70%, 50%, and 30% of methane stream is utilized to produce just FT products. Now, for FT process we have two distinct operating modes, LTFT and HTFT. Tables 11.12 and 11.13 show the total capital cost, total

Table 11.13 GTL (HTFT) different planning mode results.

GTL HTFT	100%	70%	50%	30%
Total capital cost, \$MM	90.5	57.7	47.2	34.4
Amortized capital cost, \$MM/yr	11.0	7.02	5.74	4.18
Total operating cost, \$MM/yr	4560	1510	1150	923
Total utilities cost, \$MM/yr	4170	1400	1060	852
Desired rate of return, %/yr	10	10	10	10
lifetime of the project, yr	20	20	20	20
LPG mass flowrate, kg/h	67,898	308,035	202,597	205,452
Gasoline mass flowrate, kg/h	308,035	223,772	136,950	137,400
Diesel mass flowrate, kg/h	202,597	158,931	103,560	94,112
Wax mass flowrate, kg/h	205,452	77,533	50,213	47,776
Water mass flowrate, kg/h	124,745	87,321	62,504	61,502
LPG yield	0.066	0.299	0.196	0.199
Gasoline yield	0.2987	0.217	0.133	0.133
Diesel yield	0.196	0.154	0.100	0.091
Wax yield	0.199	0.075	0.049	0.046
Objective function, \$MM	282	1960	1410	1480

operating cost, total utilities cost, yield values, and objective function values for each planning mode considering LTFT and HTFT production, respectively.

It can be observed from the tabulated results, that 1.96E+09, 1.10E+09, 1.78E+09, 1.86E+09 are the objective function values (profit) for the 100% utilization mode of LNG, methanol, LTFT, and HTFT, respectively. However, many other combinations are possible that can potentially lead to a higher profit, such as: (70%LNG, 30% HTFT) with a profit of \$2.82 E+09, (70% HTFT, 30% LNG) with a profit of \$2.47E+09, and (50% LNG, 50%HTFT) with a profit of \$2.33 E+09. Since some combinations of utilization modes result in better objective function values, we will use the formulated MILP model to find the optimal combination.

11.6.2.1 Economic Planning Using Formulated MILP Model

In this section, we show the benefits of the developed model in helping us select and find the optimal production pathway. Based on the different yields obtained for different products, and different capital and operating costs for each utilization option, the formulated model was applied on the production network. The formulated MILP has 24 variables with 4 integer variables and 39 constraints. It has been solved in LINGO 14.0 version with branch and bound solver type. It was found that \$ 4.20 E+09 is the optimal annual profit as defined by the objective function. The optimal solution selects (16% methanol, 84% HTFT) combination mode with 196,079 kg/h and 100,3921 kg/h flow rates. The optimal results such as product flowrate values are 130,817, 66,094, 299,854, 200,000, 200,784 kg/h for methanol, LPG, gasoline, desiel, and wax, respectively.

11.6.2.2 Sustainable Planning Using Formulated Model

In this section, we consider environmental impact as an equally important issue while designing the production network in addition to the economic objective. The environmental impact

Table 11.14 CO_2 equivalent values in tons/yr for different utilization options.

Utilization option/ percentage	100%	70%	50%	30%
LNG	0.01471095	−7.23E+07	−1.21E+08	−1.69E+08
Methanol	−1.14E+04	−7.95E+03	−5.68E+03	−3.41E+03
LTFT	3.94E+05	1.57E+05	1.96E+05	1.17E+05
HTFT	1.94E+06	1.36E+06	9.72E+05	9.95E+05

represented by CO_2 or CO_{2eq} is incorporated in the analysis. The CO_{2eq} values for each utilization mode are obtained from the ASPEN Plus simulator. ASPEN Plus reports greenhouse gas (GHG) emissions in terms of CO_2 equivalents of global warming potential (GWP) for the streams based on data from three popular standards, the IPCC's 2nd (SAR), the 4th (AR4) Assessment reports, and the U.S. EPA's proposed rules from 2009. The greenhouse gases are carbon dioxide, methane, nitrous oxide, hydrofluorocarbons, perfluorocarbons, sulfur hexafluoride, and nitrogen triflouride. In our analysis we consider the standard carbon cost as $40 per ton of CO_2 equivalent emitted for the base case (BP 2017). Then, different carbon prices such as $20 (low), and $80 (high) are considered to address possible foreseeable scenarios. Table 11.14 shows the CO_2 equivalent values in tons/yr for different utilization options considering SAR standard.

Now, the process with positive values will be discredited for carbon equivalent cost, and a negative cost will be shown in their profit equation as they are emitting GHG according to their corresponding ASPEN Plus flowsheet. Whereas, the process with negative values will be credited for carbon equivalent cost, and a positive cost will be shown in their profit equation as their output product streams are emitting less CO_2 equivalent than their input streams according to their corresponding ASPEN Plus flowsheet. After incorporating the carbon equivalent cost value of $40 per ton emitted, it was found that $4.165E+09 is the optimal annual profit as defined by the objective function. The optimal solution still selects a methanol and HTFT combination with 196079 kg/h and 1003921 kg/h as a utilized methane stream. This is (16% methanol, 84% HTFT) as a combination mode. Furthermore, we found that $4.183E+09 and $4.127E+09 are the optimal annual profit for $20 and $80 per ton emitted, respectively.

11.7 Sustainability Assessment of Natural Gas Network

Recently, sustainability has emerged as a key issue in process system engineering (PSE). The widely known definition is, "*Sustainable development is development which meets the needs of the present population without compromising the ability of future generations to meet their own needs*" (UNGA 1987). Sustainability requires a holistic view of processes by considering interactions among industrial processes, and human and ecological systems. Batterham (2006) highlighted that five levels are required for connecting global and individual activities: global objectives (Level 1), industry strategy (Level 2), enterprise targets (Level 3), specific projects (Level 4), individual actions/measured outcomes (Level 5). Sustainable chemical engineering must address greenhouse gas (GHG) reduction (level 1) rather than just modeling system (level 4) and process measures and analysis (level 5).

There are three main aspects of sustainability metrics: ecological, economical, and sociological metrics. Two classes of *metrics* or *indicators* are used to indicate the state and performance

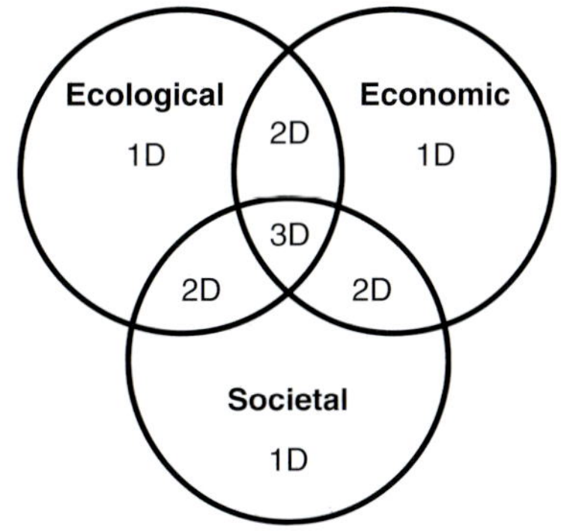

Figure 11.24 A schematic representation of the three dimensions of sustainability.

of the system. State metrics of a system are known as content indicators, and those that measure the behavior of a system are known as performance indicators (Sikdar 2003). The metrics measuring only one aspect of the system, are referred as one-dimensional (1-D). The metrics measuring interactions of any two aspects, such as eco-efficiency metrics, socio-ecological metrics, and socio-economic metrics, are referred as 2-D metrics as shown in Figure 11.24. The 3-D metrics obtained from the intersection of all three aspects, are called true sustainability metrics.

Six basic indicators of sustainability (Sikdar 2003) are:

1) *Material intensity,* expressed as pounds of material wasted (not converted to desirable product) per unit output. This metric is calculated by subtracting the mass of product and saleable co-products from the mass of raw materials input to the process.
2) *Energy intensity,* expressed as Btu per unit output. It is a measure of the net fuel-energy consumed to provide the heat and power requirements for the process.
3) *Water consumption,* expressed as gallons of fresh water, excluding rainwater, consumed per unit output.
4) *Toxic emissions,* expressed as pounds of toxic material emitted by the process per unit output.
5) *Pollutant emissions,* expressed as pounds of pollutants emitted by the process per unit output.
6) *Greenhouse gas emissions,* expressed as pounds of carbon dioxide equivalents emitted per unit output.

The framework proposed by Azapagic and Perdan (2000) contributed to standardization of the indicators of sustainable development for industry. The different applications of sustainability metrics (Azapagic and Perdan 2000; Schwarz et al. 2002) include comparison of similar products made by different companies, evaluation of different processes, benchmarking performance, tracking progress towards sustainable development, stacking along the supply chain, calculation metrics for facilities, and calculating metrics with other tools.

11.7.1 Case Study 1

In this section we assess the sustainability metrics for a natural gas network comprised of LNG, GTL, and methanol facilities, based on the indicators of resource, energy, and water consumption, and pollution emissions. A lower metric indicates effective process: if the numerator is smaller, then the impact of the process is less; if denominator is larger, then the output of the process is more. Table 11.15 gives the metrics values for LNG, GTL, and methanol production processes.

Table 11.15 Metrics for comparing LNG, GTL, and methanol production processes.

Product (Production Process)	Unit	LNG Per pound of product	GTL Per pound of product	Methanol Per pound of product
Material	lb	0.228	0.909	0.814
Energy	MMBtu	0.058	0.0933	0.111
Water	gal	0.0	0.0	0.0
CO_{2eq}	lb	0.0	36.798	25.524

For material intensity metrics, the simulated values for products and losses for the base case shown in Table 11.3 are considered. The LNG process turns out to be better in terms of material consumption with 0.228856 lb per pound of product index, implying that less wasted materials are produced from the LNG process. GTL and methanol processes have higher material intensity indices of 1.163492 and 0.814444, due to wastage of water produced and CO_2 emission. For the energy-intensity metric, the input energy for key processing units obtained from the ASPEN Plus simulations are considered. It was found to be 0.05783, 0.0933, 0.11053 MMBtu per pound of product for LNG, GTL, and methanol processes, respectively. All the processes have zero water consumption index, as fresh water is not used as input, although it is produced and used in other utility applications. Gabriel et al. (2014) provide details for targeting of water and energy issues in gas-to-liquid processes. Next, CO_{2eq} is evaluated for all processes, which is obtained as lb produced from Aspen Plus simulations. The CO_{2eq} is found to be 0, 36.7978, 25.524 lb per pound of product for LNG, GTL, and methanol processes, respectively.

11.7.2 Case Study 2

Next, we consider acid gas removal unit (B) for sustainability assessment, benchmarking process performance and evaluation of different processes. We consider the three different operating modes for evaluation as shown in Figure 11.25. In the first configuration, MDEA is used only to absorb CO_2 and H_2S to the desired levels; in the second configuration, a mixed amine solution of MDEA and DEA is used; and in the third configuration, two absorbers are used, with MDEA in the first and DEA in the second.

Material intensity index, energy intensity index, water consumption, and pollution emission metrics are used to benchmark and evaluate the three schemes, as shown in Table 11.16. Although scheme 1 is superior in terms of sustainability metrics, since it does not meet the H_2S specification level, scheme 2 is the best configuration, which is in agreement with the mathematical model results.

11.7.3 Case Study 3

Similarly, here we consider the two distinct classifications of FT reactor, low- and high-temperature Fisher-Tropsch (LTFT and HTFT), to evaluate their relative performance in terms of impact per unit output. Table 11.17 shows the sustainability metrics of each process.

For material intensity metrics, LTFT has 0.175 lb per pound of products, whereas HTFT has 0.304 lb per pound of products. For energy-intensity metrics, the input energy for both LTFT and HTFT reactors are obtained from the ASPEN Plus simulation. It was found to be 0.0003 and 6.0838E-05 MMBtu per pound of product for LTFT and HTFT, respectively. Now,

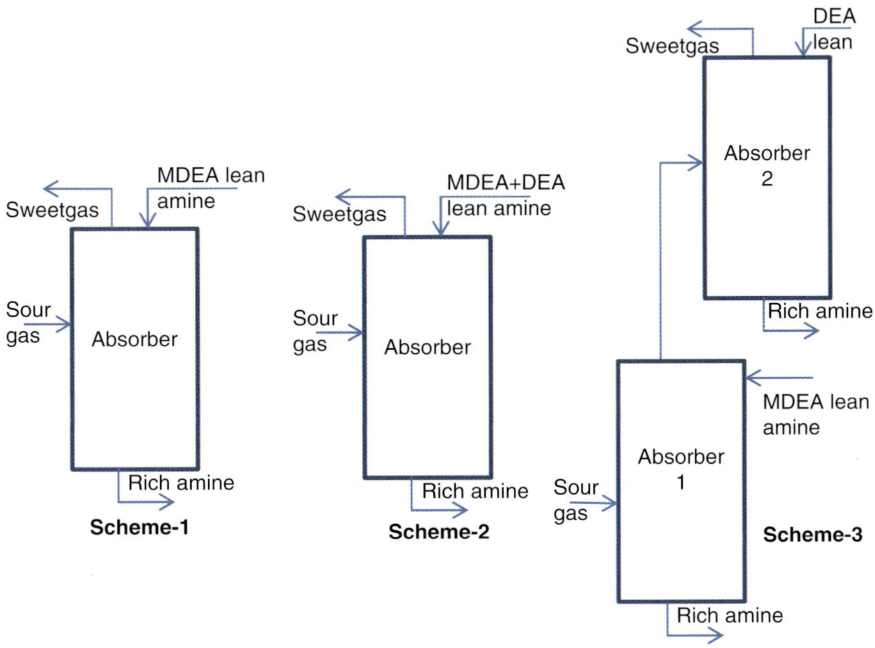

Figure 11.25 Three operating schemes for acid gas removal unit (B).

Table 11.16 Metrics for comparing different schemes for acid gas removal unit.

Amine system	Unit	Scheme 1 Per pound of sweetgas	Scheme 2 Per pound of sweetgas	Scheme 3 Per pound of sweetgas
Material	lb	0.038	0.110	0.305
Energy	MMBtu	1.23E-04	0.001	0.005
Water	lb	1.383	0.875	2.512
$\mathbf{CO_{2eq}}$	lb	−1.31E-03	9.68E-03	0.028

Table 11.17 Metrics for comparing LTFT and HTFT production processes.

Product (Production Process)	Unit	LTFT Per pound of product	HTFT Per pound of product
Material	lb	0.175	0.304
Energy	MMBtu	0.0003	6.0838E-05
Water	gal	0.0	0.0
$\mathbf{CO_{2eq}}$	lb	0.047	0.257

HTFT has a lower energy index. Both have a zero water consumption index, since water is not used/consumed in production. LTFT has a lower CO_{2eq} index of 0.047. This example presents a different case where not all metrics considered are lower for one process, and other metrics are needed to get fair evaluation.

11.8 Conclusion

Natural gas with its abundance and relatively cheap prices has a clear advantage and is expected to maintain 23% contribution of global energy demand. A framework for the design, synthesis, analysis, and planning of a natural gas processing and production network has been presented. Simulation and optimization techniques were applied sequentially to the network of gas processing and production plants, consisting of LNG, GTL, and methanol facilities. Furthermore, the developed framework has a general sense that can be applied to other processing and production networks such as paper and pulp production, oil refineries, or petrochemical industries. ASPEN Plus is used to rigorously simulate and evaluate the initial design of key processing units of selected production facilities: namely, LNG, GTL, and methanol. The steady state simulation helped to determine the mass and energy balances, operating conditions, and equipment specifications. The flowsheet simulations provided important inputs to the decision-making process, for example, determining product yields, natural gas feedstock requirements, capital and operating cost estimates (using Aspen's Economic Analyzer) linked to plant capacities, environmental impacts in terms of quantified emissions, and so on. An operational LP model was shown to be helpful to optimize the processing and production network and to address variations and changes. It also evaluates the processing and production network over a wide range of forecast economic changes such as gas market prices, product prices, and so on, with the objective function set to maximize profitability. The initial network is expanded further into a superstructure that consists of multiple technologies and operating conditions. MILP models are presented to select optimally the sustainable processing and production routes among different available utilization pathways, multiple technologies, and operating conditions to maximize annual profit. Trade-offs between material consumption, energy usage, and environmental impact are assessed by the sustainability metrics. Sustainability metrics are helpful for performance benchmarking of production network; they highlight the opportunities for process improvements.

References

Aasberg-Petersen K, Bak Hansen JH, Christensen TS, et al. 2001. Technologies for large-scale gas conversion. *Appl Catal A: Gen, 221*, (1–2): 379–387.

Alabdulkarem A, Mortazavi A, Hwang Y, Radermacher R, Rodgers P. 2011. Optimization of propane pre-cooled mixed refrigerant LNG plant. *Appl Therm Engg*, 31 (6): 1091–1098.

Al-Saadoon, FT 2005. *Economics of GTL Plants*. Presented at SPE, 94380, Dallas.

Al-Sobhi SA, Alfadala H, El-Halwagi MM. 2009. Simulation and energy integration of a liquefied natural gas (LNG) plant. In: Alfadala H, Reklaitis G, El-Halwagi MM, editors, *Advances in gas processing. Proceedings of the 1st International Gas Processing Symposium*, p. 131–135, Elsevier.

Al-Sobhi SA, Elkamel A. 2015. Simulation and optimization of natural gas processing and production network consisting of LNG, GTL, and methanol facilities. *Journal of Natural Gas Science and Engineering*, 23, 500–508.

Al-Sobhi SA, Elkamel A., Erenay FS, Shaik MA. 2018. Simulation-Optimization Framework for Synthesis and Design of Natural Gas Downstream Utilization Networks. *Energies*, 11 (2): 362–381.

Al-Sobhi SA, Shaik MA, Elkamel A, Erenay FS. 2018. Integrating simulation in optimal synthesis and design of natural gas upstream processing networks. *Indus Engg Chem Res*, 57: 5792–5804.

ASPEN Plus V7.3. 2011, *Aspen Technology Inc*. MA, USA.

Azapagic A, Perdan S. 2000. Indicators of sustainable development for industry: a general framework. *Proc Safe Environ Protect*, 78 (4): 243–261.

Bao B, El-Halwagi MM, Elbashir NO. 2010. Simulation, integration, and economic analysis of gas-to-liquid processes. *Fuel Process Technol*, 91, (7): 703–713.

Batterham RJ. 2006. Sustainability: the next chapter. *Chem Engg Science*, 61, (13): 4188–4193.

Biegler LT, Grossmann IE 2004. Retrospective on optimization. *Comput Chem Engg*, 28 (8): 1169–1192.

BP 2017, *British Petroleum statistical review of world energy, June* 2017.

Camm JD, Chorman TE, Dill FA, Evans JR, Sweeney DJ, Wegryn GW. 1997. Blending OR/MS, judgment, and GIS: restructuring P&G's supply chain. *Interfaces*, 27, (1): 128 –142.

Campbell J. 2000, *Gas conditioning and processing*. 8th ed. Norman, OK: Campbell Petroleum Series.

Dry ME. 2002. The Fischer-Tropsch process: 1950–2000. *Catal Today*, 71, (3–4): 227–241.

Economides MJ, Wood Ad. 2009. The state of natural gas. *J Nat Gas Sci Engg* 1, (1–2): 1–13.

Ehlinger VM, Gabriel KJ, Noureldin MMB, El-Halwagi MM. 2014, Process design and integration of shale gas to methanol. *ACS Sustain Chem Engg*, 2, (1): 30–37.

EIA. 2017. *International Energy Outlook*. Washington, DC.

Espinoza RL, Steynberg AP, Jager B, Vosloo AC. 1999. Low temperature Fischer-Tropsch synthesis from a Sasol perspective. *Appl Catal A: Gen*, 186, (1–2): 13–26.

ExxonMobil. 2017. *Outlook for Energy: a view to 2040.*

Fissore D, Sokeipirim D. 2011. Simulation and energy consumption analysis of a propane plus recovery plant from natural gas. *Fuel Proc Technol*, 92 (3): 656–662.

Floudas CA, Elia JA, Baliban RC. 2012. Hybrid and single feedstock energy processes for liquid transportation fuels: a critical review. *Comput Chem Engg*, 41: 24–51.

Gabriel KJ, Linke P, Jiménez-Gutiérrez A, Martínez DY, Noureldin M, El-Halwagi MM. 2014. Targeting of the water-energy nexus in gas-to-liquid processes: a comparison of syngas technologies. *Indus Engg Chem Res*, 53, (17): 7087–7102.

GPSA. 2004. *Engg Data Book 12th ed.* Tulsa, OK: GPSA.

Haid J, Koss U. 2001. Lurgi's mega-methanol technology opens the door for a new era in down-stream applications. *Studies Surface Sci Catal*, 399–404.

Hao X, Djatmiko ME, Xu Y, Wang Y, Chang J, Li Y. 2008. Simulation analysis of a gas-to-liquid process using aspen plus. *Chem Engg Technol*, 31 (2): 188–196.

HCP. 2013, *Hydrocarbon Processing*. Available from: http://www.hydrocarbonprocessing.com/news/2013/03/ louisiana-group-plans-to-build-largest-methanol-plant-in-north-america.

Jager B, Espinoza R. 1995. Advances in low-temperature Fischer-Tropsch synthesis. *Catal Today*, 23, (1): 17–28.

Julia LM, Ortiz-Espinoza AP, El-Halwagi MM, Jime A. 2014. Techno-economic assessment and environmental impact of shale gas alternatives to methanol. *ACS Sustain Chem Engg*, 2, (10): 2338–2344.

Khalilpour R, Karimi IA. 2012. Evaluation of utilization alternatives for stranded natural gas. *Energy*, 40 (1): 317–328.

Kidnay AJ, Parrish WR. 2006, *Fundamentals of natural gas processing*. Boca Raton, FL: Taylor & Francis Group.

Kim YH, Jun KW, Joo H, Han C, Song IK. 2009. A simulation study on gas-to-liquid (natural gas to Fischer-Tropsch synthetic fuel) process optimization. *Chem Engg J*, 155 (1–2): 427–432.

Klinkenbijl AJ, Grootjans H, Rajani J. 2005. *Best practice for deep treating sour natural gases (to LNG and GTL)*, Paper presented at GasTech. March 15.

Kohl A, Nielsen R, 1997, *Gas Purification. 5th ed.* Houston: Gulf Publishing Company.

LINGO. 2013, *LINDO System Inc.*, Chicago, IL.

Liu P, Pistikopoulos EN, Li Z. 2009. A mixed-integer optimization approach for polygeneration energy systems design. *Comput Chem Engg*, 33 (3): 759–768.

Luyben WL. 2014. Design and control of the dry methane reforming process. *Indus Engg Chem Res*, 53, (37): 14423–14439.

Mokhatab S, Economides MJ. 2006. Onshore LNG production process selection. *Presented at SPE Annual Technical Conference and Exhibition*, San Antonio, TX, 1, 1–11.

Mokhatab S, Poe WA. 2012, *Handbook of natural gas transmission and processing*. Waltham, MA: Gulf Professional Publishing.

Mueller-Langer F, Tzimas E, Kalschmitt M, Peteves S. 2007. Techno-economic assessment of hydrogen production process for the hydrogen economy for the short and medium term'. *Intl J Hydrogen Energy*, 32: 3797–3810.

Noureldin MMB, Elbashir NO, El-Halwagi MM. 2014. Optimization and selection of reforming approaches for syngas generation from natural/shale gas. *Indus Engg Chem Res*, 53 (5): 1841–1855.

Olah GA, Goeppert A, Prakash GKS. 2006. *Beyond oil and gas: the methanol economy*, Weinheim, Germany: WILEY-VCH Verlag GmbH & Co. KGaA.

Quaglia A, Sarup B, Sin G, Gani R. 2012. Integrated business and engineering framework for synthesis and design of enterprise-wide processing networks. *Comput Chem Engg*, 38: 213–223.

Robert M, John M. 1998. Gas conditioning and processing. 4, *Gas treating and liquid sweeting*. Campbell Petroleum Series. Norman, OK.

Rostrup-Nielsen JR. 2000. New aspects of syngas production and use. *Catal Today*, 63 (2–4): 159–164.

Schulz EP, Diaz MS, Bandoni JA. 2005. Supply chain optimization of large-scale continuous processes. *Comput Chem Engg*, 29 (6): 1305–1316.

Schwarz J, Beloff B, Beaver E. 2002. Use sustainability metrics to guide decision-making. *Chem Engg Progress*, 98 (7): 58–63.

Seider WD, Seader JD, Lewin DR. 2004. *Product & process design principles synthesis, analysis, and evaluation*. hoboken, NJ: John Wiley & Sons Inc.

Shah N. 2005. Process industry supply chains: advances and challenges. *Comput Chem Engg*, 29: 1225–1235.

Sikdar SK 2003. Sustainable development and sustainability metrics. *AIChE J*, 49 (8): 1928–1932.

Steynberg AP, Espinoza RL, Jager B, Vosloo AC. 1999. High temperature Fischer-Tropsch synthesis in commercial practice. *Appl Catal A: Gen*, 186, (1–2): 41–54.

Steynberg A, Dry M. 2004. *Fischer-Tropsch technology*. Elsevier B.V.

Sudiro M, Bertucco A. 2009. Production of synthetic gasoline and diesel fuel by alternative processes using natural gas and coal: process simulation and optimization. *Energy*, 34, (12): 2206–2214.

Thomas S 2003. Review of ways to transport natural gas energy from countries which do not need the gas for domestic use. *Energy*, 28, (14): 1461–1477.

Tusiani M, Shearer G. 2007. *LNG, a nontechnical guide*. Tulsa, OK: PennWell Corporation.

UNGA. 1987. Report of the World Commission on Environment and Development: our common future. *United Nations Gen Assembly, Development and International Co-operation: Environment*, Norway.

Venkatarathnam G. 2008, *Cryogenic mixed refrigerant processes.* International Cryogenics Monograph Series, Springer Science & Business Media.

Wilhelm DJ, Simbeck DR, Karp AD, Dickenson RL. 2001. Syngas production for gas-to-liquids applications: technologies, issues and outlook. *Fuel Proc Technol*, 71, (1–3): 139–148.

Wood DA, Nwaoha C, Towler BF. 2012. Gas-to-liquids (GTL): a review of an industry offering several routes for monetizing natural gas. *J Nat Gas Sci Engg*, 9: 196–208.

Zhang N, Zhu XX. 2000. A novel modelling and decomposition strategy for overall refinery optimization. *Comput Chem Engg*, 24: 1543– 1548.

Zheng L, Furimsky E. 2003. ASPEN simulation of cogeneration plants. *Energy Conver Manage*, 44, (11): 1845–1851.

12

Process Safety in Natural Gas Industries

Monir Ahammad and M. Sam Mannan

Mary Kay O'Connor Process Safety Center, Artie McFerrin Department of Chemical Engineering, Texas A&M University, USA

CHAPTER MENU

12.1 Introduction

Natural gas streams may contain H_2S and organic volatiles (i.e., mercaptans) that have toxic health effects. Refrigerants (e.g., ammonia and propane are often used for lowering process temperature in LNG facilities) can also have both toxic and flammable effects when released to the environment. The American Industrial Hygiene Association (AIHA) has issued Emergency Response and Guidelines (ERPGs) for air contaminants such as H_2S, NH_3 and methyl-mercaptans. An exposure of H_2S at a concentration level of 100ppm up to an hour is considered as life threatening for human beings. Process Safety is the science, technology, and management system to prevent, and mitigate unwanted events such as exposure of H_2S.

The application of process safety principles begins with the identification of hazards. A hazard can be a material, or an activity, or a procedure that can cause harm to human, or environment, or incur economical loss. For example, a large release of liquefied natural gas (LNG) may create an asphyxiating environment. LNG being stored at −162°C, its vapor is heavier than air, and therefore an unignited vapor cloud resulting from a loss-of-containment event may engulf a large area, causing an asphyxiating environment. An ignition of such flammable vapor cloud may lead to a flash fire returning to the source of the leak. Depending on the congestion or confinement level in the facility, a delayed ignition can cause deflagration to mild vapor cloud explosion. This scenario would be completely different if the released flammable

Natural Gas Processing from Midstream to Downstream, First Edition.
Edited by Nimir O. Elbashir, Mahmoud M. El-Halwagi, Ioannis G. Economou, and Kenneth R. Hall.

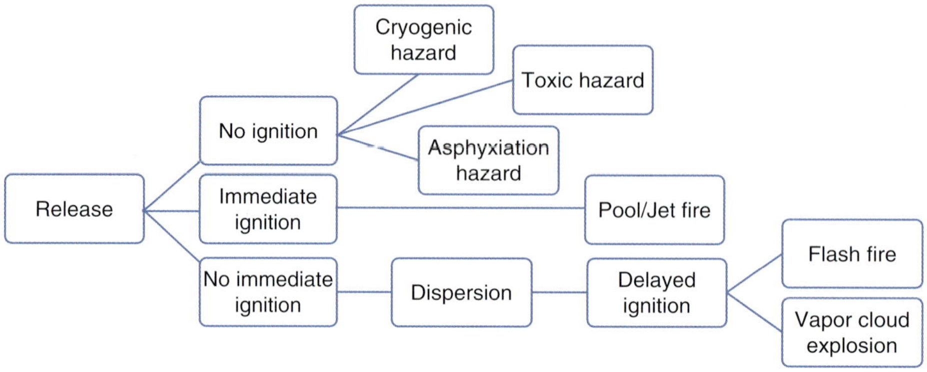

Figure 12.1 Hazards related to natural gas processing.

material is propane or other refrigerants with detonating characteristics. A propane vapor cloud explosion may result in detonation and can severely damage the facility. Figure 12.1 depicts different hazards associated with natural gas industries. An immediate ignition of pressurized flammable gas/liquid leak would result in a jet fire. Flashfire occurs during a delayed ignition of flammable vapor mixed with air. Fire burning a liquid pool, whether spreading or fixed diameter, are referred as pool fire.

12.2 Incident History

Process safety incidents are prevalent in all types of industries including natural gas. Hirschberg et al. (2004) analyzed ENSAD (Energy-related Severe Accident Database), established by the Paul Scherrer Institute (PSI), for a comparative perspective on severe incidents in the energy sector. The study separates LPG from natural gas. Figure 12.2(a) shows a comparison of an aggregated normalized number of affected people per gigawatt (GW) of energy production among different industries. There is no significant difference between the personal injuries among the countries worldwide. The study shows that among the energy sectors, LPG has the largest damage indicator and nuclear power is the lowest. Though the natural gas industry lies in a moderate region based on the statistics, in the LNG liquefaction process, propane and butane are separated from the natural gas stream, an accidental leak from those streams has the damage potential as high as LPG production. Figure 12.2(b) shows the societal risk curve for different energy sectors in the OECD countries. Among the fossil energy chains, the natural gas industry has the lowest frequency of severe incidents involving fatalities, whereas LPG has the highest frequency of fatality. Despite favorable statistics, incidents in natural gas industries keep happening. Some well-publicized disasters related to natural gas processing and transportation are discussed below.

12.2.1 Cleveland, Ohio, 1944

On October 20, 1944, LNG storage tank number four collapsed catastrophically due to the brittle fracture of the tank wall. The released content flowed into the streets and sewers due to the absence of secondary containment (e.g., dike). The resulting vapor cloud ignited, causing explosions and fires. Jets of fire erupted from the manholes of sewers, and the manhole covers were thrown miles away. The fire caused significant structural damage to a nearby spherical

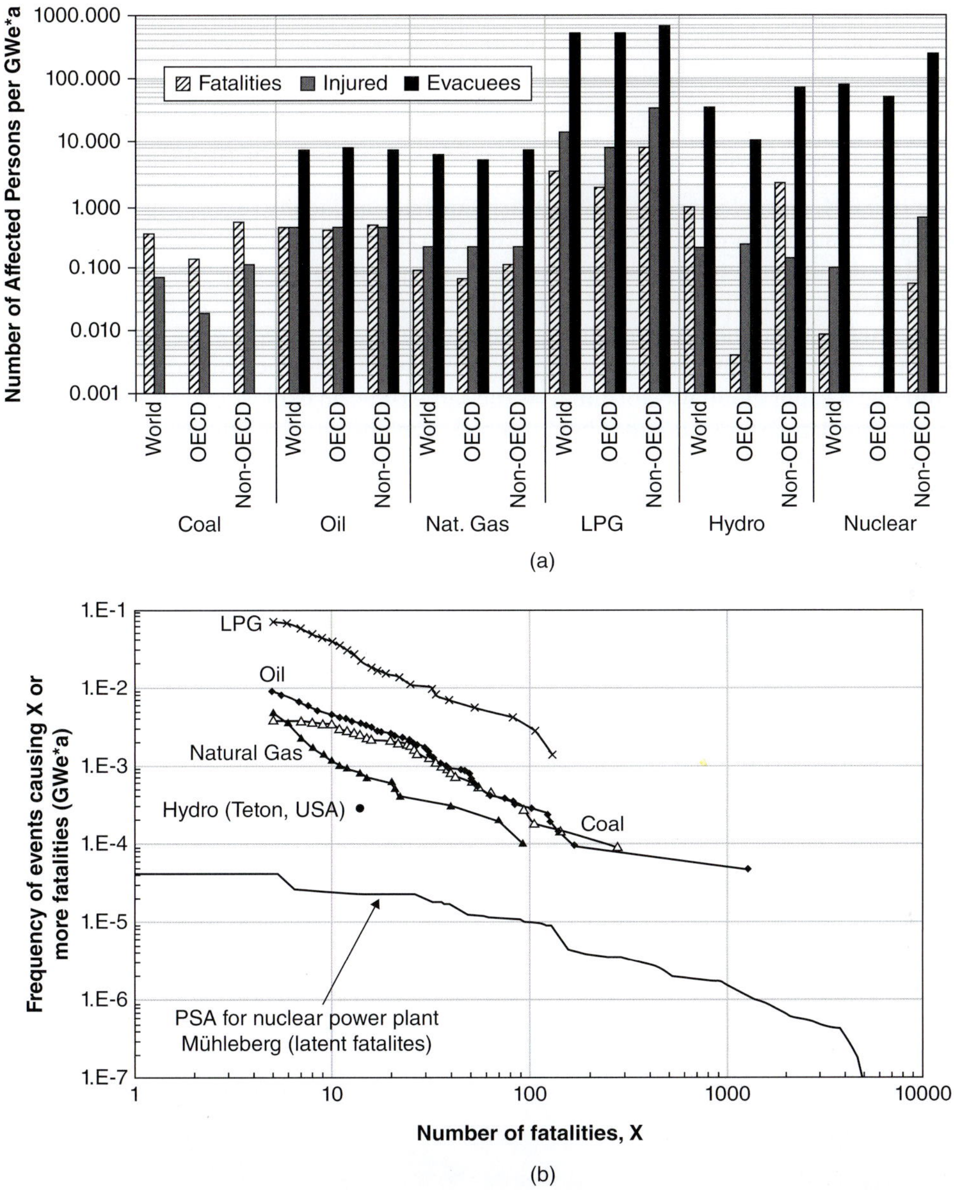

Figure 12.2 (a) Comparison of energy-related damage rates occurred worldwide in the period of 1969-1996; (b) Frequency – Consequence (F-N) curve for severe incidents in various energy chains in OECD countries during 1969-1996. Reprinted from Journal of Hazardous Materials, Vol 111, Stefan Hirschberg, Peter Burgherr, Gerard Spiekerman, and Roberto Dones, Severe accidents in the energy sector : comparative perspective, pp. 57-65, Copyright (2004), with permission from Elsevier.

LNG storage tank because it had no insulation against fire. Once the fire was extinguished, people started returning to their homes and businesses thinking that the matter has been taken care of. A second release occurred from the damaged tank of an additional 15,000 kg of LNG. The explosions and fire resulting from the ignition of the evaporating LNG continued to occur engulfing homes and traveling through sewers. The incident killed at least 200 people and caused extensive property damage. Improper choice of tank material (3.5% Ni- Steel) led to a low-temperature embrittlement, which was exacerbated by vibrations from the nearby railroad.

12.2.2 Skikda, Algeria, 2004

On January 19, 2004, an explosion at the Sonatrach liquefaction facility resulted in 27 fatalities, 72 injuries, and destruction of 3 LNG manufacturing trains, a marine berth, 76% down production, and $900 million in total losses. At 6:39 PM, an operator noticed a pressure rise in the steam boiler 40 unit, opening safety relief valves. Despite reducing the pressure of the fuel gas flow to the burner to a minimum setting, the operator was not able to reduce the pressure. After one minute, the operator from the nearby liquefaction train 30 informed the train 40 operator via intercom that a vapor cloud was forming. At the same time, an explosion was heard following another immediate explosion and a huge fireball. The ensuing fire and explosions totally destroyed liquefaction trains 40, 30, and 20.

The incident occurred due to the leakage of refrigerant (cold hydrocarbon) near the boiler of liquefaction train 40. The uncontrolled gas entrance in the boiler formed an explosive mixture, which eventually exploded, destroying the boiler and igniting the vapor cloud formed outside. Investigations found that the facility siting and layout optimization, control of ignition sources, and appropriate placement of gas detectors could have reduced the damage (Dweck et al. 2004).

12.2.3 San Bruno, California, 2010

On September 10, 2010, the natural gas transmission pipeline operated by Pacific Gas and Electric Company (PG&E) ruptured. The subsequent underground explosion caused 8 fatalities, multiple injuries, destruction of 38 homes, and damage to additional 70 houses (Richards 2013). The ruptured pipeline had a 30-inch diameter at an operating pressure of 400 psig. The U.S. National Transportation Safety Board (NTSB) reported that a big piece of steel pipe was blown about hundred feet away producing a 72-foot crater. It is also reported that the mechanical integrity of a piece of pipe section was lost due to improper welding. The released gas from the leak accumulated in the underground airspace around the pipeline and formed a flammable gas-air mixture, which ultimately found the ignition source (Peekema 2013).

Figure 12.3 Destruction after San Bruno incident, 2010 (Source: Wikipedia).

Figure 12.4 Havoc caused by the Kaohsiung gas explosion, Taiwan, 2014. Reprinted with permission from Process Safety Progress, Vol 35, Issue 3, Horng-Jang Liaw, Lessons in process safety management learned in the Kaohsiung gas explosion accident in Taiwan: 228-232, Copyright (2016) John Wiley & Sons.

12.2.4 Kaohsiung, Taiwan, 2014

On July 31, 2014, a series of underground gas explosions due to the leak of propylene killed 32 persons and injured another 321. The blast destroyed 6 km of roads, overturned and trapped cars and fire trucks. The pipeline operated by China General Terminal and Distribution (CGTD) was delivering propylene to the LYC Chemical Corporation. Investigations found that following an observation of abnormal pressure in a 4-inch pipeline, the gas transmission was shut down after 2 hours. Meanwhile, an estimated amount of 20 tons of propylene was released. The leak seeped into the sewer system and found an ignition source. The leak was the result of corrosion in the presence of moisture (Liaw 2016).

12.3 Process Safety Methods

The principles of process safety ensure the design intent of equipment, unit, plant, procedure, and management systems are satisfied during all the stages of the project. Loss of containment of hazardous materials, unplanned energy release, and unplanned human activity may lead to unwanted consequences. Therefore, process safety methods seek to identify associated hazards and risks in order to develop appropriate measures for implementing, measuring, evaluating, and monitoring safety. Figure 12.5 graphically categorizes existing process safety methods according to their purposes. The methods are classified into five groups, namely methods to (1) identify process hazards, (2) implement safety, (3) measure process safety, (4) evaluate process safety, and (5) monitor process safety performance. As shown in the figure, the knowledge

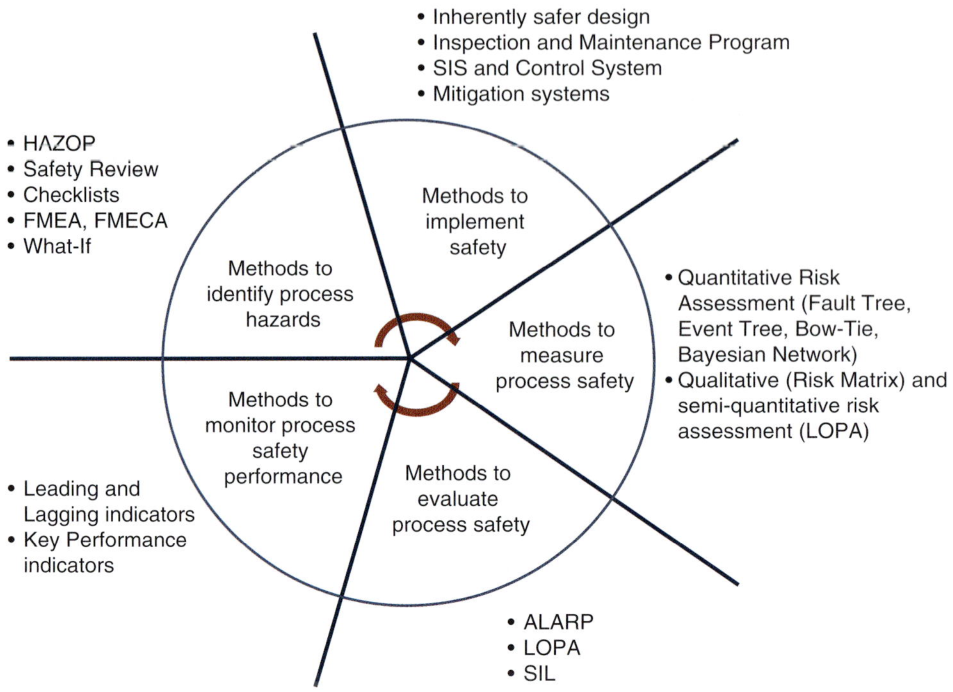

Figure 12.5 Process safety methods and their purpose.

of process design, management, and systems such as monitoring, inspection, control, mitigations, safety barriers, and operating procedures are needed for the effective use of process safety methods.

The methods to identify process hazards are intended to link relationships between the hazards, causes, and the potential consequences. Hazard and Operability (HAZOP) Study, Safety Review, Checklists, Failure Mode and Effect Analysis (FEMA), Failure Mode, Effect and Criticality Analysis (FMECA) and What-If are used frequently. HAZOP, a structured method that is most used, starts with identifying a node; the keywords are used to understand the potential consequences of deviation from the design intent for that node. However, it is not suitable for identification of mechanical integrity failures and external events (Khan and Hashemi 2017). The safety review is a less rigorous study than HAZOP and intends to find initiating events that can cause incidents. It also reviews previous incidents and recommends safety measures (Crowl and Louvar 2011). Often checklists are used to identify process hazards due to ease of implementation. The effectiveness of checklist methods depends on the quality of checklists and personal experiences. The What-If method asks "what-if" questions of potential deviations from design intents for the cause and consequence relationships. As the name suggests, FMEA and FMECA are very structured methods, applied to critical and complex equipment at the component level, to identify the cause and effect of component failure.

After identification of the causes and consequences of potential incidents, safety measures also known as control measures or barriers need to be implemented to prevent the incident from occurring. The barrier strategy can be classified into three categories. Inherently Safer Design is the first category, which is highly effective and uses the principle of elimination, substitution, minimization, and moderation of hazards. Engineering barriers, also recategorized as passive and active barriers, are the second choice in implementing safety barriers. Passive barriers do not require any human interventions, and therefore the effectiveness of preventing an incident is much higher, e.g., dikes, firewalls, gravity drains, flame arrestors. Control systems, shut-down systems, and relief devices are considered active barriers. Finally, administrative barriers such as rules, standard operating procedures, and training are considered in implementing safety.

The risk is used as a measure of safety, which is defined as the combination of the probability of a certain scenario combined with a certain consequence. Existing risk assessment methods can be classified as: a) quantitative—e.g., Event Trees, Fault Trees, Bow-tie, Bayesian Network; b) semiquantitative—e.g., Layer of Protection Analysis (LOPA); and c) qualitative—e.g., Risk Matrix. In quantitative risk assessment (QRA), both the probability of the incident and the associated consequences are estimated quantitatively. In semiquantitative risk assessment, usually the probability estimation is quantitative, and the consequence is determined qualitatively. Both the probability and the consequences are estimated qualitatively for a qualitative risk assessment, and acceptability of risk is determined by comparing to a risk matrix. The assessed risk is evaluated to choose whether it is acceptable for the plant personnel, public, or environmental damage. There are no universal criteria for evaluating risk. Existing practice includes a demonstration that the adopted barriers reduce risk to a level called As-Low-As-Reasonably-Practicable (ALARP). In LOPA, the criteria of acceptable risk are defined based on the company's tolerable risk criteria. Safety Integrity Level (SIL) target assessment further studies the risk reduction credited for the implementations of safety instrumented functions (SIF).

The health of safety systems is monitored using leading and lagging indicators. Use of leading indicators provides a proactive measure in preventing incidents or deteriorating conditions by detecting early signs. On the other side of the process safety spectrum, lagging indicators measure performance based on fatality, injury, first aid, and the number of releases and therefore are not effective in prevention and mitigation of incidents. Figure 12.6 illustrates lagging and leading indicators on the same scale. Most of the downstream oil and gas industries have adopted API RP 754 (API 2010) for safety indicators, but recent advancements propose risk-based performance indicators (Khan et al. 2009).

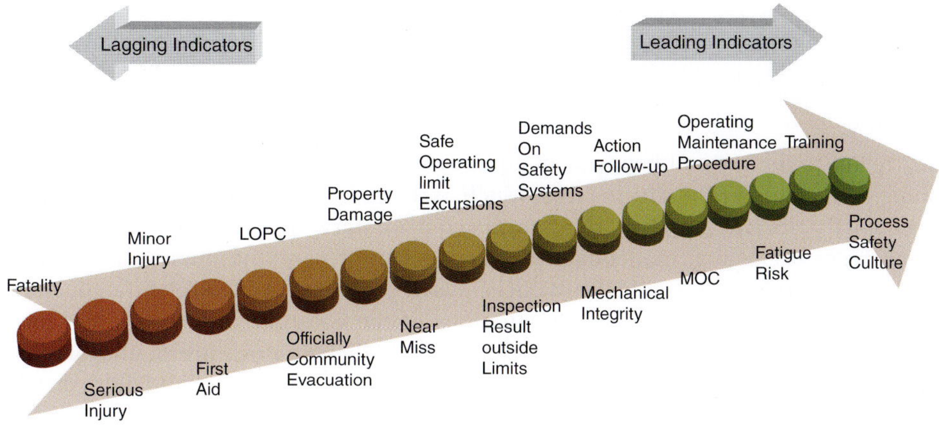

Figure 12.6 Leading and lagging indicators to measure safety performance. Reprinted with permission from Process Safety Progress, Vol 32, Issue 4, Wang., M., Mentzer, R.A., Gao, X., Richardson, J., & Mannan, M.S., Normalization of the process safety lagging metrics: 337-345, Copyright (2013) John Wiley & Sons.

12.4 Equipment and Plant Reliability

Among many reasons, one major reason for incident occurrence is a failure of equipment. Equipment can fail instantly or over a long period of time due to its use. When a piece of equipment fails to perform its intended design purpose, it leads to unwanted consequences. The functionality of a system (e.g., level control loop) depends on the performance of the individual components (e.g., sensors, valves, logic controller, and wiring). Failure characteristics of these components determine the functionality of the overall system.

The reliability of a plant is determined based on probabilistic analysis of the failure of systems to minimize the cost of system downtime, spares, repair equipment, personnel, and warranty claims. It is an important factor in designing engineering systems to perform satisfactorily throughout their life cycle. The high reliability of a plant often also indicates higher safety factors. It ensures that individual pieces of equipment are reliable enough to ensure a safer processing plant. Thus, probabilistic assessments of the individual elements are performed to establish plant reliability. Reliability analysis is also the basis of other safety studies such risk assessment, failure mode analysis, and ensuring the Safety Integrity Level (SIL) of Independent Protection Layers (IPL). If the existing safety barriers do not mitigate the risk to a tolerable limit, a higher reliable system is often recommended during the hazard analysis stage. Such systems need verifications to ensure the validity of risk reduction measures and therefore reliability analysis is conducted.

The definition of reliability implies that if after t period of operation, n_f number of pieces of equipment out of n failed, then reliability of the equipment is given by,

$$R(t) = 1 - \frac{n_f(t)}{n}$$

The failure rate is a function of the quantity of the equipment, also called "failure density function," is expressed as,

$$f(t) = \frac{1}{n} \cdot \frac{dn_f(t)}{dt} = -\frac{dR(t)}{dt}$$

After a time t, of testing period, we know how many pieces of the equipment survived. Therefore, the conditional instantaneous failure rate, commonly known as hazard rate, becomes a very useful parameter, and it is defined as,

$$h(t) = \frac{f(t)}{R(t)}$$

Figure 12.7 refers to the general behavior of equipment. The failure rate is high during the commissioning of the equipment, due to factors such as defective equipment, incorrect installations, substandard parts and materials, poor quality control, start-up human error, inadequate debugging, incorrect packaging, inadequate process, and poor handling methods. It is also known as the burn-in region in the bathtub curve. During the useful life region, a constant failure rate can be caused by the random fluctuations of the equipment operations. It may result from undetected defects, abuse, low safety factors, higher random stress, unavoidable conditions, and human error. If equipment consists of many pieces with different failure distributions, the overall effect may also result in a constant failure rate. This is the longest period of the equipment life, for which $h(t) = \lambda$, gives –

$$R(t) = \exp(-\lambda t)$$

$$f(t) = \lambda \exp(-\lambda t)$$

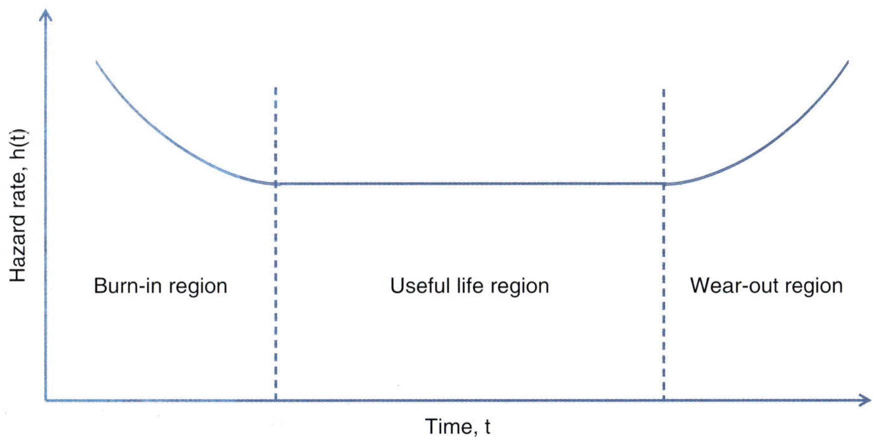

Figure 12.7 Bathtub curve for hazard rate.

The equipment wears out at the end of its life increasing its failure rate. Contributing reasons are poor maintenance, friction, aging, corrosion and creep, wrong overhaul, and short design-in life.

The mean life of equipment, also known as mean time to failure (MTTF) or mean time between failure (MTBF), can be calculated as:

$$m = \int_0^\infty t f(t)dt = \int_0^\infty t\lambda \exp(-\lambda t)dt = \frac{1}{\lambda}$$

For a system consisting of n components in series, the reliability of the whole system can be expressed in terms of individual components reliability as follows:

$$R = \prod_{i=1}^{n} R_i$$

Thus, the failure rate of the system is

$$\lambda = \sum_{i=1}^{n} \lambda_i$$

A parallel system fails only if all components fail simultaneously. Thus, the reliability and MTTF of such systems are:

$$R = 1 - \prod_{i=1}^{n}(1 - R_i)$$

$$\frac{1}{\lambda} = \sum_{i}^{n} \frac{1}{\lambda_i} - \left(\frac{1}{\lambda_1 + \lambda_2} + \ldots + \frac{1}{\lambda_{n-1} + \lambda_n}\right) + \ldots (-1)^{n-1}\frac{1}{\sum_{i=1}^{n} \lambda_i}$$

The reliability of a bridge network of components as shown in Figure 12.8 is expressed as follows:

$$R_b = 2R_1R_2R_3R_4R_5 + R_2R_3R_4 + R_1R_3R_5 + R_1R_4 + R_2R_5 - R_2R_3R_4R_5 - R_1R_2R_3R_4 - R_5R_1R_2R_3 - R_1R_3R_4R_5 - R_1R_2R_4R_5$$

A system where n components are in parallel operation but r components are functional, binomial distributions can be applied to determine the failure rate of such system. In a system

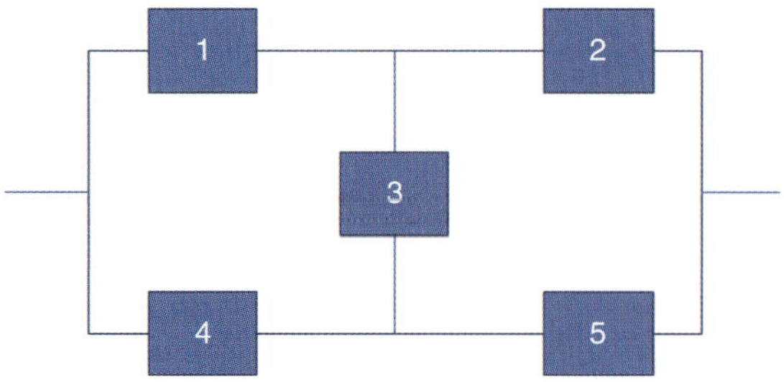

Figure 12.8 Bridge network of components.

with redundant components, where one component is operating and others are standby, it can be assumed that the standby component has zero failure rates. A Poisson distribution can be used for estimating the reliability of such a system.

There are many reliability evaluation techniques available. The most commonly used methods are:

- Network Reduction Method
- Fault Tree Analysis (FTA)
- Markov Method
- Decomposition Method
- Failure Mode and Effects Analysis (FMEA)

The network reduction method simply evaluates the system reliability, as it is composed of independent series and parallel subsystems. A bridge configuration can be converted to equivalent series and parallel system using delta-Y conversion.

Fault tree analysis (FTA) is very popular for reliability analysis and identifying the hazardous pathways leading to incidents. Modern FTA can be very large and complex, and thus computer software is often used. It is basically based on "AND" and "OR" logic operations. For example, in Figure 12.9, a heat exchanger overpressurization can occur due to the failure of alarm and failure to initiate emergency shutdown system. Each of these items can further be analyzed for the causes. Thus, there are four ways in which the over pressurization can occur. Each individual path is called cut-sets. The probability of the top event is the summations of all cut-sets. The minimum number of ways the event can occur is called "minimal cut-sets."

The Markov method assumes that the transition from a system state (e.g., operational) to another state (e.g., failed equipment) after a finite time interval Δt occurs at λ transition rate.

$$P(t + \Delta t) = P(t) + \lambda \Delta t\, P(t)$$

This gives,

$$MTTF = \frac{1}{\lambda}$$

The decomposition method simplifies a system into subsystems and applies conditional probability theory. The system probability can be measured by combining the subsystem probabilities. If A is a key element of a system S, which can have both a good state A, and a failed state, A', then the probability of the system failure can be determined by-

$$P(S) = P(A)\, P(S/A) + P(A')P(S/A')$$

FMEA is a step-by-step method to analyze all possible failure modes and their effect on the functionality of the system. Details on FMEA/FMECA can be found in (Dyadem Press., 2003).

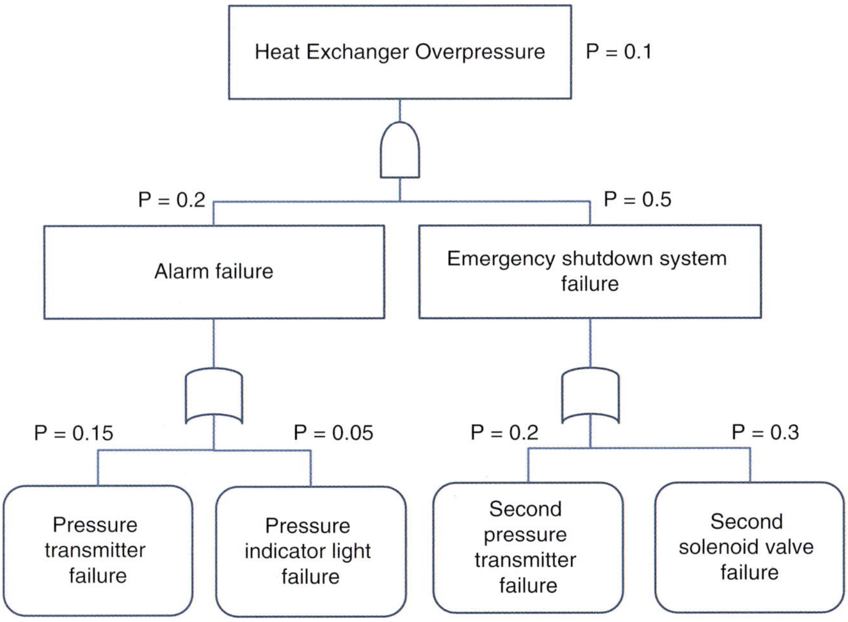

Figure 12.9 Example Fault Tree.

12.5 Facility Siting and Layout Optimization

The primary consideration of facility siting is safety. Other factors that impact the siting of the natural gas processing plant and transportation system are sources of raw material, distance to market, access (e.g., raw materials, products, maintenance, operators, and emergency response), availability of land, sources of utility and effluent disposal, integrating with other plants, permits and regulatory policies, and investment incentives. Topography also becomes a crucial factor in the facility siting of the natural gas processing plant. As an example, during a loss of containment of LNG storage tank or a leak in a liquefaction process, heavy vapor (flammable or toxic) may flow towards a downhill, populated area and may cause catastrophic consequences such as Bhopal, 1984.

Separation distances between the hazards and potentially affected areas, within the plant boundary as well as to the public, can reduce the risk of the operation. This measure does not replace the importance of best practices in designing and operating a plant; however, it becomes a very effective measure as it is a passive protection layer in mitigating the consequences of major incidents. It is also an inherently safer strategy to separate the sources of potential fire, explosion, or toxic release from nearby populated areas. For example, a plant in a sparsely populated area will cause many fewer causalities compared to a plant in an urban area, in case of a major incident. Impacts on the surrounding area can be obtained from studies such as consequence analysis or hazard assessment. Facility siting improves safety in ways such as a) segregation of risks; b) limitation of exposure; c) mitigation of incident impacts; d) efficient and safe construction, operation, maintenance; e) better control room design; f) emergency control such as firefighting, access for emergency response; and g) security. Despite many advantages of large separation distances, plant economics determine the siting and optimum layout of the plant. Also, there are some disadvantages associated with large separation distances because additional space increases the costs in terms of land, pipework, and operating cost. Safety issues may also be impacted in terms of increased corrosion and maintenance requirements and greater opportunities for failure.

There are various ways of generating plant layout. A structured approach follows four stages. In the first stage, a 3D model is prepared to analyze the space and equipment placement based on operations access, maintenance access, and piping connections. In the second stage, the concept of "flow," meaning the progression of process materials or utilities towards the final products or completion of the loop is applied to minimize the transfer of materials for economic reasons and minimizing potential release locations. In the third stage, the relationships between the items that share common factors, in the viewpoint of discipline, are identified. Example of relationships can be (1) process, (2) operational, (3) mechanical, (4) electrical, (5) structural, and (6) safety. In this stage, separation distances between equipment due to hazard consideration, fire and explosions, population density, and weather conditions, are applied. CCPS guidelines (CCPS 2003) provide a more detailed account of such relationships. In the fourth stage, the identified relationships are prioritized into groups to arrange items on the plot plan. A typical group should not include more than seven items. For example, a distillation column group should consist of the column itself and associated heat exchangers. The synthesized layout is then analyzed for

1) Classification, rating, and ranking
2) Critical examination
3) Hazard assessment
4) Consequence modeling
5) Economic optimization

The plant layout is classified into areas such as fire hazard area, storage, firefighting facilities, access, and electrical area classification. In most cases, the quantity of material in storage, e.g., LNG, is much greater than the amount in the process. Storage classification is needed to meet safety requirements such as diking and venting. Storage sites are segregated from the process, in an open location to allow dispersion, diked to contain 110% of the total stored volume and designed to withstand heavy loads. Flammable liquid storage area classifications include: (1) liquid at atmospheric conditions, (2) pressurized storage, (3) refrigerated storage, and (4) gas under pressure. The electrical area classification is aimed to isolate ignition sources from flammable or combustible leaks and identifying potential safeguards to prevent fire. The International Electro-Technical Commission (IEC) classifies process hazardous area into four categories: (1) zone 0: presence of flammable atmosphere for long periods; (2) zone 1: flammable atmosphere of short periods in normal operation; (3) zone 2: flammable atmosphere is not likely in normal operations but may occur for a short time; and (4) non-hazardous area. In the USA, hazardous areas are classified as: (1) Class I: combustible material is gas or vapor; (2) Class II: combustible material is dust; and (3) Class III: combustible material is fiber in suspended form. Each class is subdivided into two divisions for distinctions between the frequencies of the substance presence in the atmosphere.

In the critical examination of a plant layout, the following questions are asked: Where is the plant equipment placed? Why is it placed there? Where else could it go? The alternatives are considered and critically examined for the best possible plant layout. Hazard assessment and consequence analysis are conducted to determine the risk in terms of the whole site, the effect on the occupied buildings due to leak, fire, vapor cloud explosion, and escalation of other scenarios. For more detail of the hazard assessment and consequence analysis please refer to Mannan (2012). Economic analysis of the alternative layouts is performed for decision

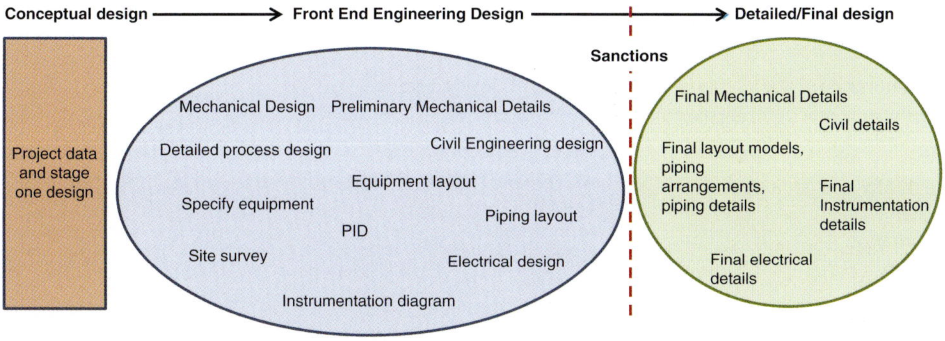

Figure 12.10 Stages of facility siting and plant layout.

making and determining the factors in terms of the cost of the plant layout, such as foundations, structures, piping and pipe tracks, pumps, and power consumptions.

The facility siting and plant layout activities can be broadly categorized as site layout, plot layout, and equipment layout, which undergo three stages: stage one layout, stage two layout, and final layout. In the stage one layout, significant consideration is given to items that may threaten project viability. Plots are generally considered as 100m X 200 m separated by 15m wide roads. Once the stage one preliminary plot layout is completed, it is then subjected to hazard assessments of smaller and frequent leaks with possible sources of ignition. For stage one site layout, separation distances, access requirements, traffic characteristics, material and utilities flow to the site are considered. Hazard assessment of site layout is performed to consider the escalation of incidents and vulnerability of service buildings both inside and outside of the boundary.

Figure 12.10 depicts the details of stage two and the final stage of plot and site layout in contrast with other process design activities. In stage two site layouts, reworking is done on stage one layout for the specific site, hazard assessment, economic optimization, and critical analysis. Legal requirements such as soil and drainage, meteorological conditions, environmental aspects, insurance, and emergency services are incorporated in this stage. International and national standards, company standards, contractor standards are considered in evaluating the process engineering design data: e.g., flow sheets, equipment drawings, pipeline lists. Table 12.1 summarizes the details of site layout and equipment layout considerations.

Many factors determine the layout of equipment, e.g., usage of common facilities. Stacks, for disposal of gaseous and liquid effluents, might be common to different furnaces and fired heaters. The transfer line minimization, avoidance of trenches or pits that may hold flammable liquid under the furnace area, proper ventilation, minimization of peepholes and observation doors are also very important factors. Connecting pipe between heat exchangers is kept at a minimum with the provisions of pipe lengths to allow stress relief and access for maintenance. Pumps handling hot liquids are separated from pumps handling flammable and volatile liquids or from compressors handling flammable gases. Open or closed design of pumps and compressor houses also plays a significant role in ensuring appropriate ventilation. Corrosive materials cause a significant number of incidents among various industries.

Almost all the world's natural gas is transferred through networks of pipelines; thus, siting of transmission pipelines by itself requires a detailed study. The general considerations of piping layout within a processing plant should be minimum for both economic and safety reasons. Cost of emergency isolation valves, manual valves, relief valves and bursting disks, liquid drains, and associated instruments may increase with the increase of pipe runs.

Table 12.1 Site layout and plot layout considerations.

Site Layout Features	Plot Layout Considerations
Constraints 1) Topography and geology 2) Weather 3) Environment 4) Transport 5) Services 6) Legal constraints Standards 1) Separation Distances 2) Building lines 3) Building construction, finish 4) Road dimensions 5) Service corridors 6) Pipe bridges Site services Use of buildings Location of buildings Limitation of exposure Segregation Fire containment Effluents Transport Emergencies Security	1) Process considerations 2) Economic considerations 3) Construction 4) Operations 5) Maintenance 6) Hazards 7) Fire fighting 8) Escape

12.5.1 Separation Distances

The primary constraint of plant layout is the separation distances. Approaches in considering appropriate separation distances can be classified into three categories: a) use of industry standards, b) use of ranking methods, and c) estimation of separation distances based on engineering calculations for the case. More description of using codes and standards are given by CCPS (2003) and Mannan (2012). Lewis (1980) used the Mond Index to determine the separation distances based on two values of overall risk ratings (ORR), R_1 and R_2. The minimum separation distances recommended in the industry standards are not a good practice for new projects. Thus, he used the index-based methodology to minimize the risk to personnel, minimize escalation, and ensure adequate access for emergency response, allowing flexibility in combining units of similar hazards potential. Table 12.2 indicates the separation distances based on using the Mond index between process units and storage units.

Recent advancements in modeling the consequences of hazardous releases allow more accurate estimation of separation distances. Such approaches are often referred as hazard models. Area contours based on threshold limiting value (TLV) for toxic or asphyxiating gas release, thermal radiation from fires, and overpressures from explosions are used to determine the separation distances. Special consideration or modeling needs to be done for liquefied flammable gas such as LPG. Consisting primarily of propane, the flammable vapor from LPG releases can cause vapor cloud detonations.

12.5.2 Advances in Facility Siting and Layout Optimizations

A significant amount of research has been conducted in the last decade to address the current shortcomings in safety and improvement of the facility siting and layout optimization

Table 12.2 Minimum separation distances between a storage unit and a processing unit using the Mond Index, synthesized from Mannan (2012).

Overall risk rating R_2 of storage unit	Overall risk rating R_2 of process unit						
	mild	low	medium	high	very high	extreme	very extreme
Mild	3	7	10	13	18	23	38
Low	6	9	12	17	23	30	50
Moderate	9	12	17	21	31	44	66
High	12	17	21	28	43	56	84
Very high	17	23	31	43	56	72	110
Extreme	23	30	44	56	72	97	145
Very extreme	38	50	66	84	110	145	197

techniques. A study (Mannan, West, and Berwanger 2007) of several recent incidents and historical data (which includes: 1) the Texaco Refinery, Milford Haven Wales, 1994; 2) the Bellingham pipeline rupture, 1999; 3) the BP Texas City Refinery, 2005; 4) Terra industries, IA, 1994; 5) Tosco Martinez Refinery, CA, 1999; 6) Shell chemical, TX, 1997; and 7. Surpass Chemical, NY, 1997) draws attention to the fact that the probability and severity of catastrophic incidents can be greatly impacted by facility siting issues. Despite the fact that many of the studied industries are refineries and downstream chemical companies, the findings are prevalent in all industries including natural gas midstream to downstream operations. The weakness of regulations, standards, and management systems pertaining to facility siting were evaluated. For an effective facility siting management system, the authors recommended an easy-to-understand method, consideration of protection from vapor cloud explosions, shelter-in-place for toxic releases, and removal of temporary buildings before start-up.

Vázquez-Román et al. (2010) have presented a stochastic approach in determining the optimal facility layout by considering the effect of toxic release. The objective of their mathematical model is to minimize the cost of piping, land, and risk as expressed below.

$$\text{Minimize } (C_{piping} + C_{land} + C_{risk})$$

where,

$$C_{piping} = \sum_{(i,j)\in M_{ij}} C_p d_{i,j}$$

$$C_{land} = c_l A_x A_y$$

$$C_{risk} = C_{pp} t_l \sum_{s} \sum_{r_i(i,r)} f_{i,r} p_s Pr_{i,r,s}$$

C_p	Cost of a pipe, \$/m
$d_{i,j} = (x_i - x_j)^2 + (y_i - y_j)^2$	Center-to-center distance between two units, m
C_l	Land cost per square meter, \$/m^2
$A_x, A_y = \max\left(\frac{x_s + Lx_s}{2}\right)$	Length of subunits in the x, y-direction, m
$f_{i,r}$	Frequency of the type of release, r, in unit i
P_s	Expected population in the facility
C_{pp}	Compensation cost per fatality
t_l	Expected lifespan of the plant

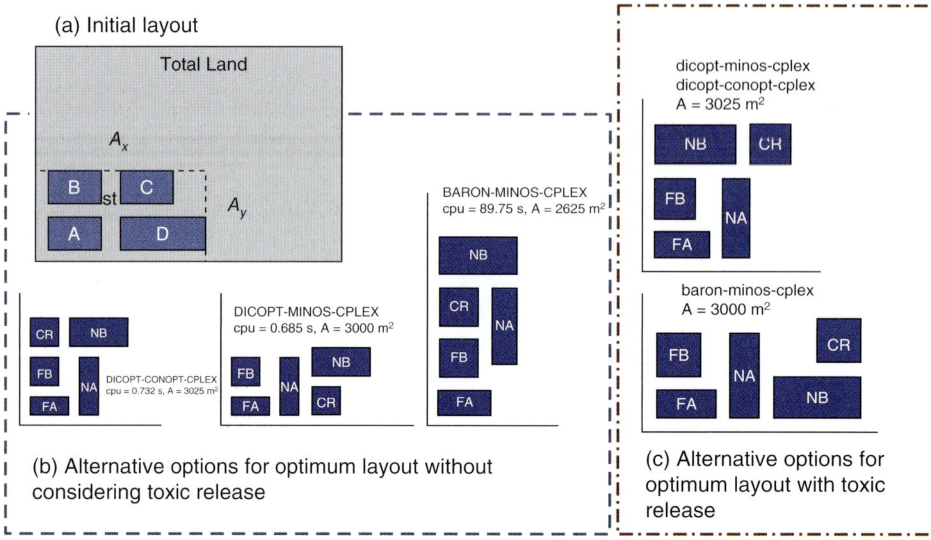

Figure 12.11 Optimum layouts for piping, land and toxic release risk minimization. Reprinted from Richart Vazquez-Roman, Jin-Han Lee, Seungho Jung, M. Sam Mannan, Optimum facility layout under toxic release in process facilities: A stochastic approach, Computers & Chemical Engineering, Vol 34, pp. 122–133. Copyright (2010), with permission from Elsevier.

Constraints such as land, nonoverlapping of two units, minimum separation distances, and toxic release are used to simulate real facility siting issues. Weather conditions significantly affect the consequences of a toxic release; therefore, to evaluate the risk of such release, a stochastic model of weather conditions using wind speed, wind direction, and atmospheric stability, in addition to the geometry of containment vessel hole are considered. Actual meteorological data for the location of Corpus Christi, TX, from the National Oceanic and Atmospheric Administration, were used for determining the probability of meteorological calculations. Figure 12.11 summarizes the results of possible alternative optimum layouts. From Figure 12.11(b) and Figure 12.11(c) alternatives, the spacing of the plant area increases due to the consideration of toxic release scenarios. However, without such considerations, the risk of the plant could be much higher than the tolerable criteria.

Díaz-Ovalle, Vázquez-Román, and Mannan (2010) modified this problem based on the worst-case toxic release scenario. They assumed that a set of facilities is already installed on the site and a new set of facilities need to be accommodated. The worst-case scenario was defined as F type Pasquill-Gifford wind stability, and wind speed of 6 m/s and 1.5 m/s.

Jung et al. (2011) further extended this study to systemically integrate fire and explosion scenarios in safety and economic analysis of plant layout optimization. The mathematical model is formulated by considering a grid-based approach for single unit placement under the constraints of the risk score, cost, and separation distances. The impacts of blast overpressure from hazardous events such as boiling liquid expanding vapor explosion (BLEVE) and vapor cloud explosion (VCE) were simulated. Monte Carlo simulation is considered for the realistic representation of the weather conditions, which is used for determining the flammable gas dispersion zone. For a case of heptane and hexane distillation separation, the mathematical model becomes a mixed integer linear programming problem, which is solved using CPLEX. Jung also pointed out the limitations of the current advanced techniques and proposed a framework for solving such problem (2016). Figure 12.12 depicts the proposed framework. It recommends conducting

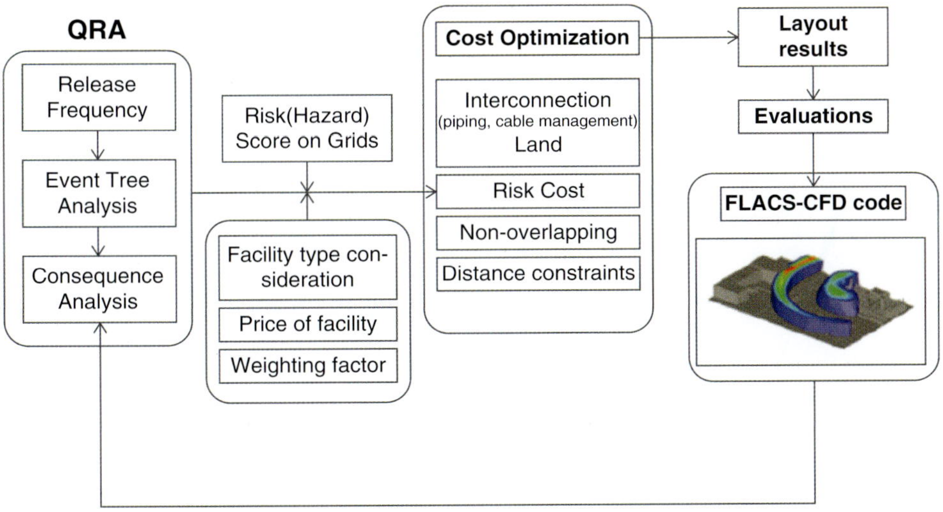

Figure 12.12 Flowchart for the proposed methodology. Kor J Chem Engg, Facility siting and plant layout optimization for chemical process safety, Vol 33, 2016: 1-7, Seungho Jung, Copyright (2016) Korean Institute of Chemical Engineers, Seoul, Korea, With permission of Springer.

QRA as a first step to determine the risk level, and then optimizing the program to select unit locations. In the final step, computational fluid dynamics modeling of hazardous releases were conducted for dispersion of materials and overpressurization due to the explosion. Similar studies have been conducted by Martinez-Gomez et al. (2014), where they showed the effectiveness of solving a Multi-Objective Mixed Integer Non-Linear Programming (MO-MINLP) problem in mitigating the risk via layout optimization.

Domino effects often result in much more catastrophic subsequent damage, compared to the initial event. However, it is very difficult to determine the factors and relationships between initial events with another. So et al. (2011) introduced the concept of quantitative estimation of a domino effect in facility siting and formulated a framework for improved incident prevention based on proper design (López-Molina et al., 2013) and also studied the principles to reduce the risk of a domino effect in plant layout.

Han et al. (2013) proposed a risk index approach to minimize risk to humans in optimizing process layout. A modified definition of individual risk is proposed. For a location "i," person "k," the modified risk is defined as

$$IR_{i,k} = \theta_k \, P_{loc,i,k} \, FoF_i$$

where, θ_k is the fraction of time and $P_{loc,\,i,\,k}$ is the probability of the presence of that person in the ith location. Being a person in that place doesn't mean the person will die. Therefore, it is multiplied with the frequency of fatality, FoF_i, which is defined as;

$$FoF_i = \sum_j f_{eo,j} \, P_{fat,i,j} \, P_{wea,i,j} P_{dir,i,j}$$

Here, $f_{eo,j}$ indicates incident frequency, $P_{fat,\,i,\,j}$ fatality rate, $P_{wea,\,i,\,j}$ weather condition, and $P_{dir,\,i,\,j}$ the directional effect. The authors optimized the siting of a dimethyl ether (DME) filling station, which consisted of four storage tanks, pump, compressor, dispersion and control room. Figure 12.13 presents the key findings of this study. Figure 12.13(a) indicates the individual risks against distance from equipment, whereas Figure 12.13(b) shows the risk map for a certain

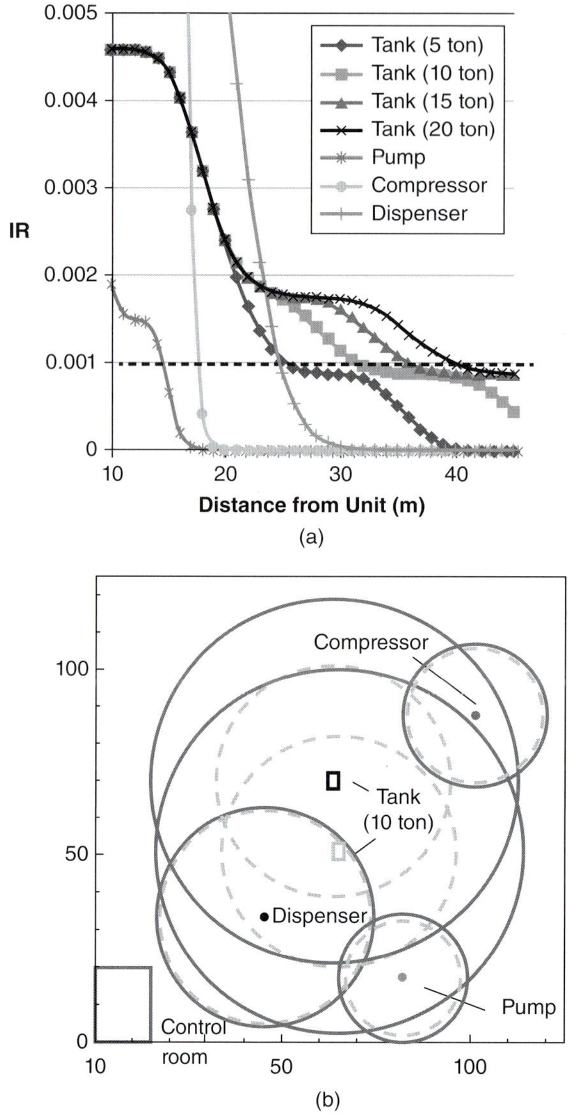

Figure 12.13 (a) Individual risks against distances from each equipment (b) Risk contours for storage tanks. Reprinted with permission from Industrial & Engineering Chemistry Research, Kyusang Han, Young Hun Kim, Namjin Jang, Hosoo Kim, Dongil Shin, En Sup Yoon, Risk index approach for optimal layout of chemical processes minimizing risk to humans, Vol 52 (22): 7274–7281. Copyright (2013) American Chemical Society.

storage tank. Optimum locations for hazardous units are usually in places where the risk level is lowest.

Facility siting study is a complex mathematical problem, and as such oftentimes it is very difficult to communicate with upper management and convince them to pursue advanced techniques. James (2017) presented a best practice on how the resulting information of facility siting and layout optimization can be communicated. The most notable recommendations are using a visual aid to summarize issues and present solutions, historical facility siting fines, maintenance issues, and recommendations.

12.5.3 Lessons Learned from Past Incidents

During the Flixborough disaster (June 1, 1974), 18 of the 28 fatalities occurred in the control room. An estimated overpressure of 0.7 bar (10 psi), resulting from an explosion at 100 m distance, knocked out 1½ story roof of a 2½ storied control room. The building was constructed using a reinforced concrete frame, a brick panel with considerable size window. The control room was located in a building alongside a production block, amenities building, laboratory, and manager's office. Cable duct and electrical switchgear ran through above the control room. The occupant of the control room died because of the collapse of the roof. If a detailed facility siting and layout optimization study had been performed and appropriate recommendations implemented, the number of casualties could have been minimized during the Flixborough incident.

Another classic, but more recent incident, where injuries and the loss of human lives could have been minimized, by conducting a proper facility siting study, is the BP Texas City refinery incident on March 23, 2005. This incident caused 15 fatalities, 170 injuries, and major economic losses. The impact of the incident would have been much lower if proper spacing among units were allowed. The U.S. Chemical Safety and Hazard Investigation Board (CSB) has investigated the incident and recommended that facility siting should be addressed in the Process Hazard Analysis under the Process Safety Management program. The key learnings of this incident are facility siting issues regarding temporary buildings for reducing the risk of personnel during unit malfunctions and blast proofing of occupied buildings in hazard zones.

12.6 Relief System Design

Relief systems provide a means of controlled release of the flammable, toxic, or reactive materials in gas, liquid, or both phases, to a safe location during equipment failures, or operator errors, which leads to overpressurization of the tanks. It provides protection against damage of the overpressurizing equipment, minimizes the loss of materials, prevents damage to adjacent facility or units, reduces insurance premiums, and facilitates compliance with regulations. Recommended codes and standards such as API RP 520, 521, DIERS (Little, Sundmacher, and Kienle 1992; API 1997, 2000), are widely practiced.

The selection of relief devices is specific to its applications. The general classification falls into four broad categories: spring-operated valves (conventional and balanced-bellows), rupture disks, buckling-pin valves, and pilot-operated reliefs. Spring-operated valves used in the liquid service are called relief valves, which begin to open at a set pressure and fully open at 25% overpressure. The safety valve is for gas service. The spring-operated valve used for both liquid and gas service is called a safety relief valve. Rupture discs burst at the set pressure and remain open. Depending on the service, rupture discs need to be corrosion resistant; care is needed to avoid plugging, and discs are often installed in series with spring-loaded relief valves. Buckling pin devices are similar to rupture discs but provide precise response to the set pressure. In a pilot-operated relief, the main valve is controlled by a smaller pilot valve. When the pilot valve reaches the set pressure, it opens the main valve. This kind of relief device is preferred for the large relieving area.

Overpressurization of equipment or tanks can occur under various operating conditions, requiring various sizes of relief area. However, it is not practical to have multiple relief devices for multiple scenarios. Therefore, the worst-case event is considered for relief sizing. Considerations of all relief scenarios need to be evaluated to ensure the adequacy of the relief system. Oversized relief devices cause problems known as "chatter" due to frequent opening and closing of the valves. This can significantly reduce the life of the valve. API RP 520 provides guidelines on the selection and installations of relieving devices. After an opening of a relief valve, the hazardous materials must be contained to prevent further unwanted events such as fire and explosion. As a result, the design of overflow header, the blow-down tank, the subsequent condenser, scrubber, and flare stack should be able to handle the released mass.

API RP 520 recommends the following formula for a conventional spring-operated relief in liquid service.

$$A = \left[\frac{in^2(psi)^{0.5}}{38.0\ gpm}\right] \frac{Q_v}{C_o K_v K_p K_b} \sqrt{\frac{(\rho/\rho_{ref})}{1.25\ P_s - P_b}}$$

For a rupture disc in liquid service, the relief area is calculated by using

$$A = \left[\frac{in^2(psi)^{0.5}}{38.0\ gpm}\right] \frac{Q_v}{C_o} \sqrt{\frac{(\rho/\rho_{ref})}{\Delta P}}$$

For conventional spring-loaded reliefs in vapor or gas service, it is assumed the flow is choked flow, and therefore the relief area is calculated as follows.

$$A = \frac{Q_m}{C_o \chi K_b P} \sqrt{\frac{Tz}{\mathrm{M}}}$$

where,

$$\chi = 519.5 \sqrt{\gamma \left(\frac{2}{\gamma + 1}\right)^{\frac{\gamma+1}{\gamma-1}}}$$

For a gas or vapor service, the above equation simplifies to

$$A = \frac{Q_m}{\chi P} \sqrt{\frac{Tz}{\mathrm{M}}}$$

12.7 Toxic and Heavy Gas Dispersion

The most infamous example of toxic release is the Bhopal (1984) disaster, which killed more than 3,000 people and led to the ultimate sale and demise of Union Carbide. The 30-ton leak of the hazardous intermediate chemical known as methyl isocyanate (MIC) caused the havoc. Fortunately, less hazardous materials are involved in natural gas processing. Nevertheless, there is a potential of toxic releases such as H_2S and toxic refrigerants (e.g., NH_3). In 1950, in Poza Rica, Mexico, due to the malfunctioning of a sulfur-recovery unit that was processing 60 MMSCF natural gas per day, a 10 MMSCF per day gas stream was released to the environment which contained 16% H_2S for 20 minutes (McCabe and Clayton 1952). The exposure caused 22 fatalities and sent 320 people to hospitals. These incidents emphasize the importance of consequence analysis, such as toxic and heavy gas dispersion studies, for further improvement in planning and designing facilities.

Dispersion models play a crucial role in the consequence analysis. It is an integral part in developing an emergency response plan for the surrounding community, modification in the design to eliminate hazards at the source, adding appropriate engineering barriers, reducing the inventories of hazardous materials, and detecting and monitoring for leaks to take actions. A wide variety of parameters affect the airborne transport of the released material in the downwind directions, e.g., wind speed, atmospheric stability, terrain types, the height of released location, and momentum and buoyancy of the release. High wind speed carries the substance downwind at a faster rate and enhances dilution. Atmospheric stability refers to the vertical mixing of the air due to the temperature gradient. In a stable condition, the sun cannot heat the ground as fast as it cools. Ground conditions affect the mixing and turbulence, and the wind profile with the height. The mixing is higher in areas with trees and buildings, whereas open areas such as lakes decrease it. Higher release height allows more dispersion and therefore reduces the ground level concentration. The buoyancy and momentum of the release also affect the effective release height and consequently affect the ground level concentration.

Air mixed gas becomes neutrally buoyant and disperses in a streamline. A plume model is used to describe the steady-state concentration from a continuous release, whereas a puff model is used for the instantaneous release of material. A simplified mathematical model of neutrally buoyant dispersion of gas also known as the Gaussian model is given as follows (Crowl and Louvar 2011).

$$\frac{\partial \overline{C}}{\partial t} + \overline{u}_j \frac{\partial \overline{C}}{\partial x_j} = \frac{\partial}{\partial x_j}\left(K_j \frac{\partial \overline{C}}{\partial x_j}\right)$$

For a steady-state release, the mass flux at any point is equal to the release rate (Q_m). Thus, different scenarios such as steady-state point release, puff release, non–steady state point release, wind and no-wind conditions, and release location at a height and on the ground can be studied by solely modifying the above equations. The Pasquill-Gifford model for weather conditions is the most popularly used model among the community. The accuracy and usefulness of this dispersion model greatly depends on the input (e.g., Q_m) also known as source-term (Ahammad, Liu, et al. 2016; Ahammad, Olewski, et al. 2016).

The Gaussian model is only valid for neutrally buoyant substances and limited to a dispersion distance of 0.1–10 km. The local concentration may exceed the estimated time average concentration. The effect of the structures on the ground is ignored even though it can significantly change the concentration estimates. Commercially available software (e.g., PHAST, CANARY) somewhat improves these limitations by incorporating semi-integral models for source-terms, ground structures and are validated against the large-scale experimental releases (Pandya, Gabas, and Marsden 2012).

A dense gas (heavier than air) has higher damage potential than the neutrally buoyant one because its cloud slumps towards to ground and can blanket occupied areas. Britter and McQuaid (1988) developed a model by performing a dimensional analysis and experimental data for dense gas dispersion. The model works best for ground-level releases and assumes no flashing or aerosol formation. The model is validated using the experimental data for flat terrain and rural area. As the gas moves in a downwind direction, the concentration of the released material decreases, resulting in a transition to a neutrally buoyant dispersion. To determine the transition location, the following criteria can be used (Crowl and Louvar 2011).

$$\left(\frac{C_x}{C_o}\right)^{1/6}\left(\frac{g_o q_o}{u^3 D_c}\right)^{1/3} \geq 0.15$$

The limitations of integral and semi-integral models are overcome by using CFD models. More accurate estimations are achievable using such tools. The effects of buildings or equipment in the dispersion path, simulation for the exact weather conditions of the site, incorporation of terrain properties such as hills and valleys are regularly considered in the modern dispersion calculations. CFD software such as FLACS is validated for dense gas (e.g., LNG) to conduct more accurate dispersion calculations (Hansen et al. 2010). (Cormier et al. 2009; Qi et al. 2010) have shown that ANSYS CFX can be used to faithfully simulate the LNG vapor cloud dispersion experimental study performed at the Mary Kay O'Connor Process Safety Center.

12.8 Fire and Explosion

The most frequent event of a serious nature in the process plant is fire, which occurs almost on a daily basis. It is usually thought to have less potential to cause damage compared to explosions or toxic release. However, under certain circumstances, a fire can lead to more catastrophic consequences, such as boiling liquid expanding vapor explosions (BLEVE) or secondary vapor cloud explosions. Failures of the critical structure or pipelines due to fire exposure could cause secondary releases and domino effects. In general, fire is categorized as jet fire, flash fire, pool fire, and fireball. In a natural gas processing plant, the expected type of fire would be NFPA class B (fire from flammable gas and combustible liquid) and class C (fire involving energized electrical equipment). Fire can be the cause and a consequence of explosion in such plant. Therefore, another convenient fire classification might be based on the type of fire, i.e., Type 1 (fires with no explosion), Type 2 (fires resulting from explosions), and Type 3 (fires resulting in explosions). Buildings, critical structures, and pipelines may need to be rated for intense radiative heat.

No fire can occur without the simultaneous existence of fuel gas, oxygen (i.e., air) and heat (i.e., ignition source). Flammability of a gas determines the concentration range within which a gas-air mixture is combustible. Natural gas, which is predominantly methane, has a lower flammability limit (LFL) of 5% and upper flammability limit (UFL) of 15%. For mixtures of hydrocarbon stream, the flammability can be estimated using Le Chatelier equation (Chatelier 1891).

$$LFL_{mix} = \frac{1}{\sum_{i=1}^{n} \frac{y_i}{LFL_i}}$$

$$UFL_{mix} = \frac{1}{\sum_{i=1}^{n} \frac{y_i}{UFL_i}}$$

The temperature and pressure dependence of the flammability can be estimated using the following empirical correlations (Zabetakis, Lambiris, and Scott 1959; Zabetakis 1965).

$$LFL_T = LFL_{25°C} - \frac{0.75}{\Delta H_c}(T - 25)$$

$$UFL_T = UFL_{25°C} + \frac{0.75}{\Delta H_c}(T - 25)$$

$$UFL_P = UFL + 20.6\,(\log P + 1)$$

The term "auto-Ignition temperature" (AIT) refers to a minimum temperature above which flammable gas-air mixture combustion occurs spontaneously. Releases from the process streams where the temperature is above AIT will catch fire spontaneously. Table 12.3 refers to the AIT of paraffinic compounds.

Figure 12.14 depicts the relationships between different flammability properties, e.g., LFL, UFL, AIT, flash point.

There are numerous ignition sources in a natural gas processing plant, and it is practically impossible to eliminate all of them. Many of the fires can be attributed to well-known ignition sources, such as electrical, smoking, friction, hot surface, hot works, mechanical sparks, and static charges.

U.S. regulations require that the radiant heat release from the fire should not be higher than 5kW/m^2 at the plant boundary. Fire models are the essential component in studying the geometry and heat release rates. An excellent review of the LNG pool fire models can be found in the white paper published by the Mary Kay O'Connor Process Safety Center (MKOPSC 2008), which is based on a workshop held for industry experts. The basic fire models are empirical or

Table 12.3 Flammability limits and AIT of natural gas compounds (Mannan, 2012).

	LFL (Vol % in air)	UFL (Vol % in air)	AIT (°C)		LFL (Vol % in air)	UFL (Vol % in air)	AIT (°C)
Methane	5.0	15.0	537	n-Heptane	1.0	7.0	223
Ethane	3.0	12.5	515	n-Octane	0.8	6.5	220
Propane	2.1	9.5	466	n-Nonane	0.7	5.6	206
n-Butane	1.8	8.5	405	n-Decane	0.8	5.4	208
n-Pentane	1.4	7.8	258	H_2S	4.3	45.0	260
n-Hexane	1.2	7.5	223	NH_3	16.0	25.0	651

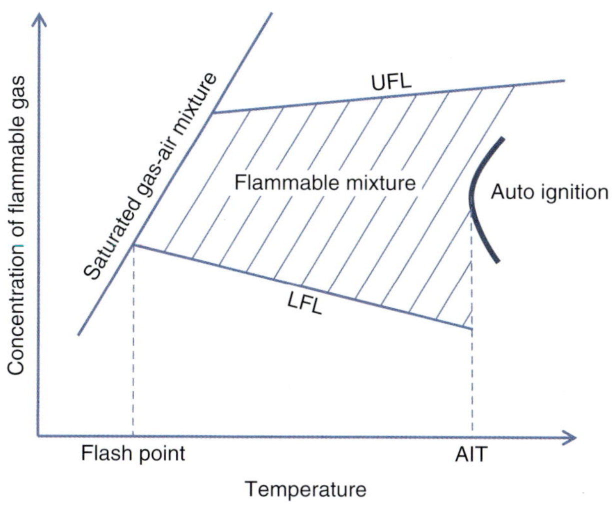

Figure 12.14 Flammability relationships.

semi-empirical; for example, the following pool fire model (Thomas 1963) correlates the flame length to diameter ratio, L/D, with the mass burning rate m.

$$\frac{L}{D} = 42\left[\frac{m}{\rho_a\sqrt{gD}}\right]^{0.6}$$

The actual fire scenario is quite different as many parameters interact with each other, e.g., wind speed, evaporation rate from the liquid pool, the shape of the pool, plume length, and tilt. Therefore, such models require experimental validations. Some notable large-scale LNG pool fire experiments are China Lake experiments, Maplin Sand, Montoir, and Lake Charles.

The major identified research needs related to LNG pool fire modeling are large-scale experiments (>35m), measurement data for large fires, and overcoming the limitations of the geometric structure model such as cylindrical fire (MKOPSC 2008). Well-validated CFD models of pool fire are much more useful than the estimates of semi-empirical models, and these models can overcome such limitations. Figure 12.15 depicts a case simulation using ANSYS CFX, where the temperature profile of the large storage tanks (e.g., T2) is studied for a domino effect.

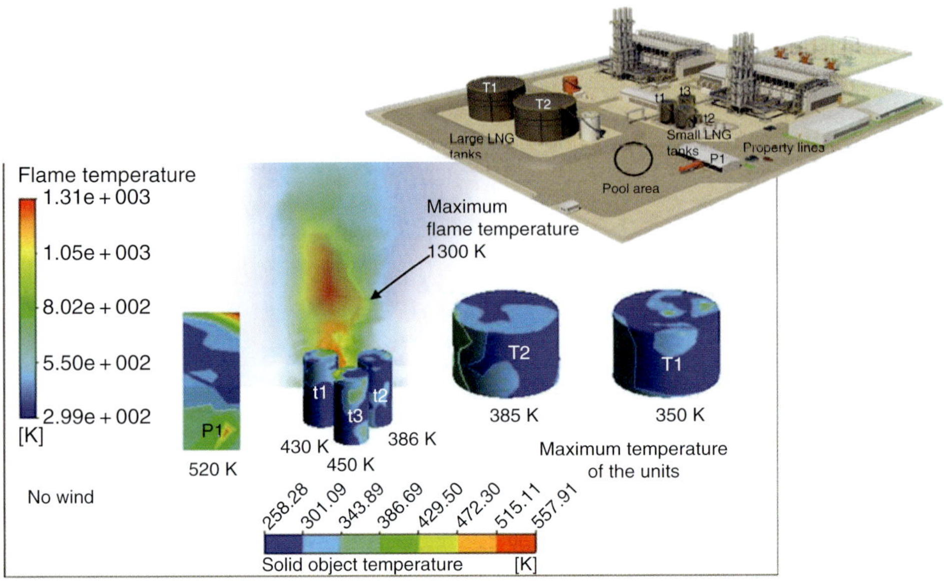

Figure 12.15 Maximum temperature distribution of flame and solid units for quiescent conditions. Reprinted from Muhammad Masum Jujuly, Aziz Rahman, Salim Ahmed, Faisal Khan, LNG pool fire simulation for domino effect analysis, Reliability Engineering & System Safety, Vol 143, pp. 19-29. Copyright (2015), with permission from Elsevier.

Generally, leaks from a pressurized vessel would result in a jet fire, but they can also occur from a wide variety of situations. A low-velocity jet flame is usually attached to the release point, but high-velocity flame can be detached. If the velocity increases, it may become unstable and lift off, or get extinguished. The impingement distance of a jet flame can be more than 50 m. API correlates the impingement length L (m) with the mass flowrate $\dot{m}$ (kg/s) as follows.

$$L \approx 6\sqrt{\dot{m}} \quad 0 < \dot{m} < 50$$

Computer code SAFETI models jet fire based on API RP 521 (Cook, Bahrami, and Whitehouse 1990). The relationships are given below.

$$L = 0.00326[\dot{m}(-\Delta H_c)]^{0.478}$$

$$R_S = 0.29s\left[\log\left(\frac{L}{s}\right)\right]^{0.5}$$

The effects of wind on flame bending are also considered in the geometry of the flame. For a wind velocity of u_w, jet velocity of u_s, the rate of vertical height z in respect to the horizontal distance x can be expressed as:

$$\frac{dz}{dx} = \frac{3.2\,\pi R_s u_s}{u_w}\left(\frac{1}{s} - \frac{1}{L}\right)$$

Bagster and Schubach (1996) proposed the following correlations for methane jet fire dimensions.

$$\frac{H}{d} = 23\,Fr^{0.2}$$

$$\frac{D}{d} = 3.55\,Fr^{0.197}$$

External heating of a tank could lead to a BLEVE and fireball. TNO suggests the following model estimate the dimension and heat flux received in a certain distance from a fireball (TNO 1997).

$$r = 3.24\, m^{0.325}$$

$$t = 0.852\, m^{0.26}$$

$$h = 2r$$

$$q = SEP \times F_{view} \times \tau_a$$

An explosion results from the rapid release of energy at a very high rate to cause local energy accumulation, which is dissipated via pressure wave, projectiles, thermal radiation, and acoustic energy. A shock wave or pressure wave faster than the reaction front propagates through the air. In a deflagration explosion, the reaction front moves at a speed less than sonic, and the shock wave moves at sonic velocity. In the case of a detonation, both the reaction front and the shock wave move at a speed higher than sonic.

The damage associated with the explosion depends on the maximum blast overpressure and its duration. Generally, an overpressure of 0.3 psig is considered to be a "safe distance." Overpressures of 1.0 psig can cause partial demolition to a house. At 3 psig, heavy machines in the industrial building suffer minor damage, and steel frame buildings distort and can be pulled away from the foundation. At 10 psig, total destruction of buildings and heavy machine damage is probable.

The most commonly used explosion models are TNT and TNO multi-energy. In the TNT model, the energy released from the explosion is converted to equivalent TNT mass, and then empirical scale law is used to determine the blast overpressure at certain distances. The energy released in a VCE is due to the burning of dispersed gas over a large volume; therefore, the TNT model overpredicts the result for VCE in the near distance and underpredicts at large distances. Confinement and congestion in the facility increase the severity of the explosion. In the TNO method, the degree of confinement/congestion and semiempirical correlations are used to determine blast overpressure.

Tang and Baker (1999) performed a series of CFD simulations to study VCE for detailed real scenarios without simplifications. Material reactivity, flame expansion, and obstacle density are used to characterize flame speed. The blast pressure decay curves were generated from the CFD simulations. The validations are carried out for vapor cloud detonation, supersonic deflagration, and subsonic deflagration. A more detailed description is given by Tang and Baker (1999).

12.9 Effective Mitigation System

An effective mitigation system can significantly reduce the consequences of the loss of primary containment event. Such systems should be based on the principle of detection–decision–action. The system detects the loss of primary containment, gas release, and presence of fire to determine appropriate measures that need to be taken. The system then actuates the mitigation action, such as a water curtain to disperse flammable gas or absorb toxic gas or foam to reduce the excessive pool vaporization thus reducing the heat flux. The mitigation system can be automatic, but human action often plays a crucial role in the detection and decision-making steps. Therefore, the reliability of mitigation system becomes dependent on the reliability of human actions.

The effectiveness of the detection process depends on the placement of sensors near the leak locations. Unfortunately, detector placement has traditionally been done by following rules of thumb rather than analyzing the propagation of the event, e.g., the dispersion behavior of gas. Benavides-Serrano, Mannan, and Laird (2015) presented a novel approach regarding the placement of gas detectors and compared its efficacy with random approach (RA), volumetric approach (VA), the minimization of the distance between the detectors and the leak sources approach (MSDP), the greedy scenario coverage approach (GC), maximum covering location problem (MCLP) approach, and a novel stochastic programming formulation considering unavailability and voting effects (SP-UV). In the random approach (RA), N number of detectors is randomly placed in L number of possible locations. In the volumetric approach (VA), the goal is to detect a vapor cloud, which is assumed to grow spherically before getting larger than a predetermined diameter. Therefore, the number of detectors N is calculated from the facility volume, and the placement is carried out by either following a regular or a staggered pattern. In the minimum source distance problem (MSDP), the detectors are placed near the potential leak location, a method which conjectures that the effectiveness of the mitigation system will depend on the earlier detection. Placement locations are determined based on the

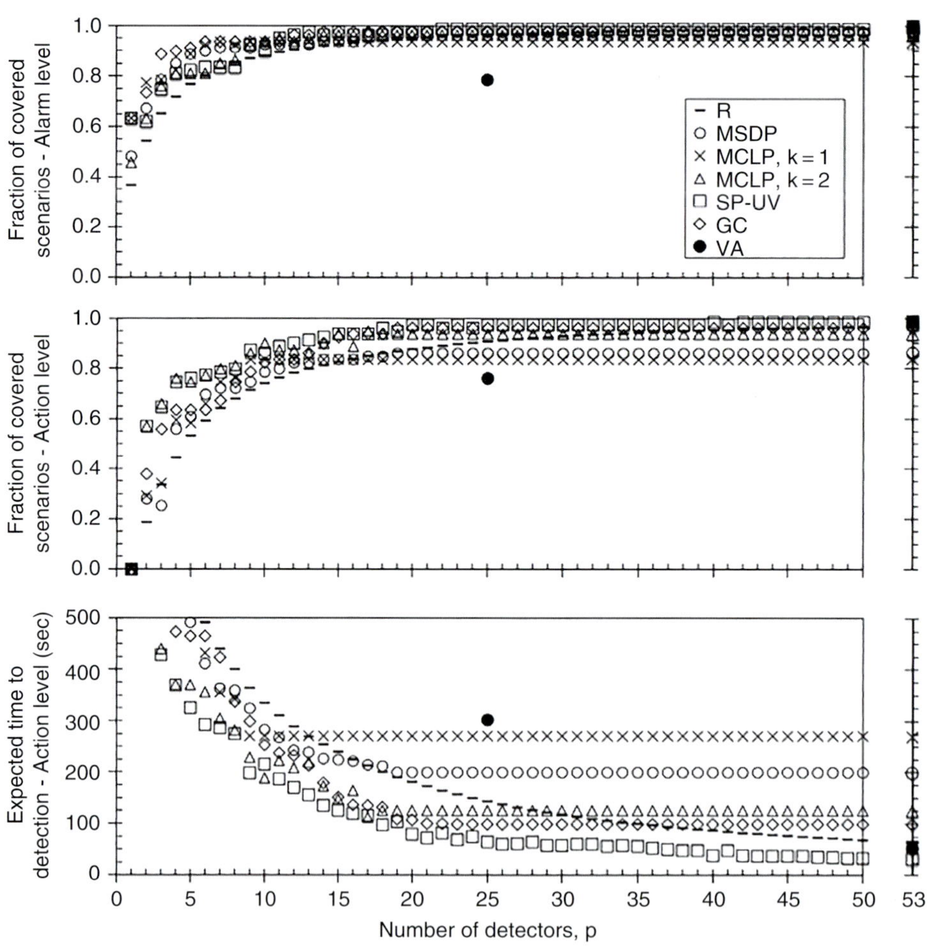

Figure 12.16 Effectiveness of gas detector placement methods under a randomly selected subset of scenarios. Reprinted from Vol 35, A.J. Benavides-Serrano, M.S. Mannan, C.D. Laird, A quantitative assessment on the placement practices of gas detectors in the process industries, Journal of Loss Prevention in the Process Industries, pp. 339–351. Copyright (2015), with permission from Elsevier.

rule of thumb industry practice. A greedy coverage algorithm (GC) and maximum coverage location problem (MCLP) are based on the principle of maximum release scenario coverage. In the GC approach, a simple rule is followed where at each stage of placement the candidate detector location should cover the largest number of uncovered leak scenarios. Once all scenarios are covered, redundant sensors are placed until each scenario is covered *k* times. In the MCLP approach, a weighted sum of the covered scenarios is maximized by selecting most of the candidate locations and therefore avoiding suboptimal greedy placement approach (Church and Revelle 1974). A stochastic programming formulation originally developed by Legg et al. (2012) is modified for SP-UV approach. The assumptions of detector failures such as false negative or false positive alarms, and detector redundancy for emergency work are incorporated in the formulation of SP-UV approach by Benavides-Serrano, Mannan and, Laird (2015). Figure 12.16 depicts the effectiveness of the gas detection for different placement approaches. For a subset of randomly selected scenario, the SP-UV formulations give the quickest response with the fewest number of sensors.

The water curtain has been recognized as one of the most effective and economical systems for dispersing dense gas, absorbing toxic components, and preventing flammable gas from reaching ignition sources. It has been proven to enhance the dispersion and reduce the safety distance for LNG vapors. Rana and Mannan (2010) performed an experimental study of the physical mechanisms of water curtains dispersing LNG vapor cloud. A flat fan and full conical upward water sprays were used for these tests. The results indicate that the water curtain reduced LNG vapor concentration and pushed the vapor in the upward direction as shown in Figure 12.17.

Kim et al. (2013) applied CFD using the Eulerian-Lagrangian spray model to evaluate global key parameters, e.g., droplet characteristics, installation distances, nozzle configurations, and air entrainment rates for designing effective forced mitigation system. Figure 12.18 shows how the LNG vapor concentration decreases with the water spray installation distance.

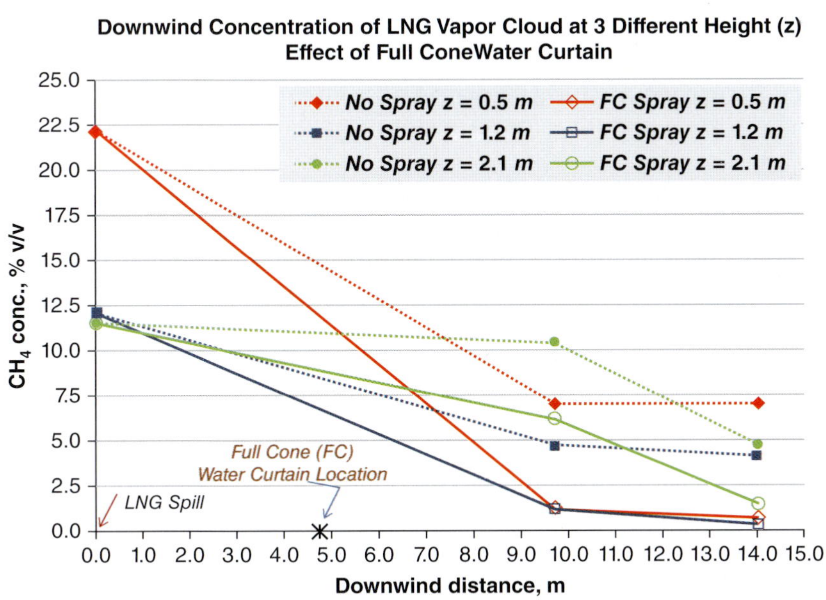

Figure 12.17 Downwind concentration at three different heights with and without full-cone water curtain. Reprinted from Morshed A. Rana, M. Sam Mannan, Forced dispersion of LNG vapor with water curtain, Journal of Loss Prevention in the Process Industries, Vol 23, pp. 768-772. Copyright (2010), with permission from Elsevier.

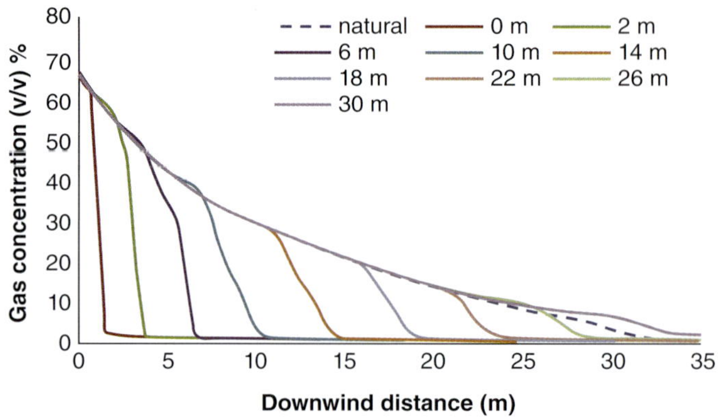

Figure 12.18 LNG vapor concentration in downwind distances at ground level for different water spray installation distances. Reprinted from Vol 26, Byung Kyu Kim, Dedy Ng, Ray A. Mentzer, M. Sam Mannan, Key parametric analysis on designing an effective forced mitigation system for LNG spill emergency, Journal of Loss Prevention in the Process Industries, pp. 1670-1678. Copyright (2013), with permission from Elsevier.

Yun, Ng, and Mannan (2011) studied high expansion foam as a mitigating measure for a boiling and evaporating LNG pool and subsequent fires. They studied the temperature changes in the pool and the fire and the profile of the radiant heat fluxes for determining the effectiveness of foam application. The experiments were carried out at the Brayton Fire Training Field, College Station Texas. LNG is poured into a 78.5 m^3 concrete pool, which was instrumented with 166 thermocouples, and radiometers. A portable weather station was used to record experimental conditions. Figure 12.19 shows one of the results of this experimental series. After foam application, the flame height was significantly reduced from 17 m to 6 m and the radiative heat flux at 15 m downwind distance was reduced from 14k W/m^2 to 1 kW/m^2.

12.10 Regulatory Program and Management Systems for Process Safety and Risks

Historically, the approach towards process safety has been reactive rather than proactive. Well-publicized catastrophic events have shaped the process safety regulations and technology developments (Mannan et al. 2016). Figure 12.20 pictorially shows the evolution of process safety technology, management systems, standards, and regulations in the context of well-publicized process safety incidents.

The European Union adopted the Seveso Directives on June 24, 1982, after a serious incident in Seveso, Italy. The amendments to the Seveso Directives resulted as a regulatory response to incidents at Basel, Switzerland (1986), and Toulouse, France (2001), to formulate Seveso II (1996) and Seveso III (2012).

Following the 1974, Flixborough incident, the UK government responded by promulgating and implementing Control of Industrial Major Accident Hazards (CIMAH) regulations. The key objective of CIMAH and SEVESO I was to address the direct environmental impact of major incidents. The Control of Major Accident Hazards (COMAH) regulation came into effect from 1999 and was later amended in 2005. This regulation is also known as the Safety Case. Australia also adopted regulations similar to the Safety Case of the UK.

In the United States, process safety regulations became a statutory requirement in 1990 under the Clean Air Act Amendments, which directed the Occupational Safety and Health Administration (OSHA) and the U.S. Environmental Protection Agency (EPA) to establish

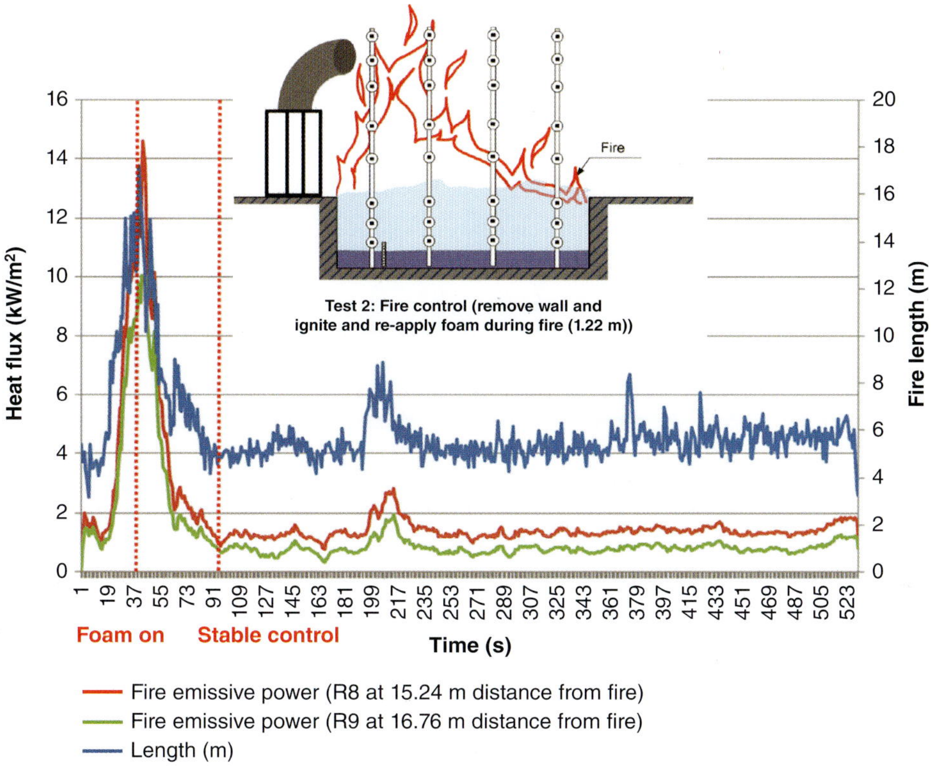

Figure 12.19 Fire length and heat flux for the experimental study of fire radiation mitigation by using high expansion foam. Reprinted with permission from Industrial & Engineering Chemistry Research, Vol 50 (4), Yun, G., Ng, D. and Mannan, M.S., Key findings of liquefied natural gas pool fire outdoor tests with expansion foam application: 2359-2372, Copyright (2011) American Chemical Society.

regulations, to protect workplace employees, the public, and environment, respectively. As a result, in 1992, OSHA promulgated the Process Safety Management (PSM) regulations for facilities handling flammable and toxic chemicals over specified limits. It also recommended adherence to recognized and generally accepted good engineering practice (RAGAGEP). Industry associations such as the American Petroleum Institute (API) published standards such as API 750 and API 754 as proactive measures to improve safety. While OSHA regulations were focused on plant personnel and equipment safety, the EPA fulfilled its mandate by promulgating the Risk Management Program (RMP) in 1996 to address public safety and environmental damage due to loss-of-containment incidents. Both the PSM and RMP regulations are very similar regarding the Prevention Program and Emergency Planning requirements. The RMP regulation additionally requires an offsite consequence analysis, which includes worst-case and alternative-scenario analysis and incident history. In addition, the EPA RMP regulation requires the development of a risk management plan (RMPlan), which must be submitted to the EPA. Table 12.4 compares the various elements of the PSM and RMP regulations. Other U.S. federal regulations that are relevant to Chemical Process Safety are described by Crowl and Louvar (2011). In Canada, the Process Safety Management Division (PSMD) of the Canadian Society for Chemical Engineering (CSChE) published PSM standards in 2012 in accordance with the USA PSM regulations.

In the United States, natural gas and LNG facilities are regulated by the federal government under regulations promulgated by the Pipeline and Hazardous Materials Administration

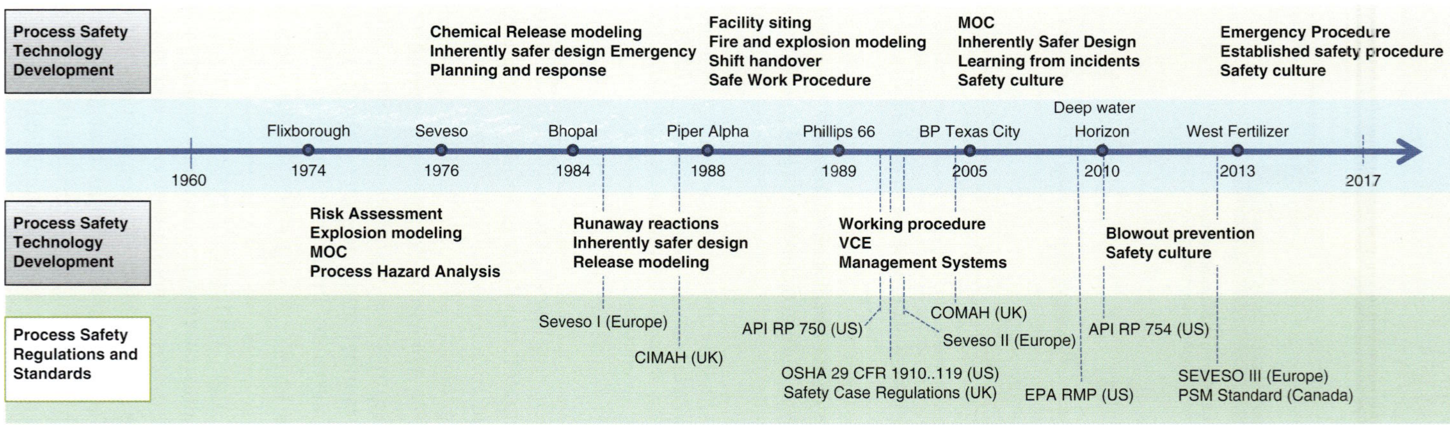

Figure 12.20 Major incidents and evolution of process safety technology, management system and regulations.

Table 12.4 Elements of US PSM and RMP regulations.

OSHA PSM Elements	EPA RMP Elements
Employee participation	–
Process safety information	Process safety information
Process hazard analysis	Hazard evaluation
Operating procedure	Standard operating procedure
Training	Training
Contractors	–
Pre-startup safety review (PSSR)	Pre-Startup Safety Review (PSSR)
Mechanical integrity	Maintenance
Hot Work Permit	–
Management of change (MOC)	Management of change (MOC)
Incident investigation	Accident investigation
Emergency response and planning	Emergency response
Compliance audits	Safety audits
Trade secrets	–
–	Offsite consequence analysis - Worst-case scenario - Alternative release scenario - 5-yr incident history

(PHMSA) of the U.S. Department of Transportation (DOT). Current DOT regulations (49 CFR, part 193) on LNG are outlined in NFPA 59A, "Standard for production, storage, and handling of liquefied natural gas." The original version (2001 edition) of this regulation is prescriptive and inflexible. The later edition (2009) offers risk-based evaluation requirements for land-based facilities. Raj and Lemoff (2009) studied the regulation and proposed alternative methods for performing better LNG facility siting risk assessment. They also compared the practices in other countries, particularly in Europe, and provided recommendations.

12.11 Concluding Remarks

Being a relatively new area of science and engineering, the scope of future research and development in process safety is very wide. However, there are several outstanding challenges that need to be addressed to advance process safety performance (Mannan et al. 2016). Some of the key issues are as follows:

- Historical failure data are not good enough to portray the current picture of risk in the plant. Real-time management data, or operational data, or data-independent mechanistic approaches based on first principles are needed throughout the industry.
- Despite proper facility siting and plant layout optimization, a severe consequence of rare events can occur, due to the migration of population centers towards the plant. Thus, land-use planning based on engineering calculations should be incorporated in the permitting of industry, housing, business centers, education centers, and hospitals.
- Establishment of a chemical incident surveillance system for process safety incidents is an essential necessity for appropriate tracking of incidents for trending analysis, identification

of improvement needs, and other actions. There are very limited data sources to distinguish the efforts and returns in mitigating risks. The surveillance system can also be used to create a knowledge base for improving infrastructure, planning, enhancing the response capability, and incorporating lessons learned into the system.

- Regulations are the minimum requirement, whereas individual companies should strive for the best-in-class safety culture (Mannan, Mentzer, and Zhang 2013; Mannan et al. 2016). However, the regulations should be based on risk-based studies and sound science. Once adopted, regulations must be enforced strictly to ensure compliance. A cost-effective certified third-party enforcement program can be adapted to reduce the burden of the regulators.
- Building process safety competency for the leadership, plant management, engineers, operators, and contractors is the key to success in preventing and mitigating process safety incidents. Academic research and dissemination of knowledge to students and practitioners alike will also go a long way in paving the pathway for preventing future incidents (Olewski et al. 2016). Risk communications to the surrounding community to generate awareness and preparation for response to potential events is another essential need for twenty-first-century society.

Nomenclature

Symbols

A_x, A_y	length of subunits in the x, y-direction, m
$\overline{C}$	mean concentration of the substance, kg/m^3
C_l	land cost per square meter, \$/m^2
C_o	concentration at the source, kg/m^3
C_o	discharge coefficient, dimensionless
C_p	cost of pipe, \$/m
C_{pp}	compensation cost per fatality, \$
C_x	concentration at x distance downwind, kg/m^3
D_c	characteristics dimension of the release, m
F_{view}	geometric view factor, radian
K_b	backpressure correction factor, dimensionless
K_j	eddy diffusion co-efficient, dimensionless
K_p	overpressure correction factor, dimensionless
K_v	viscosity correction factor, dimensionless
P_b	backpressure, psia
$P_{dir,\,i,\,j}$	factor for directional effect
$P_{fat,\,i,\,j}$	factor for fatality rate
P_s	expected population in the facility
P_s	set pressure, psia
$P_{wea,\,i,\,j}$	factor for weather condition
Q_m	mass flowrate, discharge rate, kg/s
Q_v	volumetric flowrate, m^3/s
R_S	radius of flame, m
$d_{i,\,j}$	center-to-center distance between two units, m
$f_{eo,\,j}$	incident frequency, Hz
$f_{i,\,r}$	frequency of the type of release, r, in unit i
$\dot{m}$	mass flowrate, kg/s
t_l	expected lifespan of the plant, yr

u_s	jet velocity, m/s
u_w	wind velocity, m/s
x_j	distance in j^{th} co-ordinate, m
y_i	flammable composition, volume % in air
ρ_a	density of air, kg /m^3
τ_a	atmospheric transmissivity, dimensionless
Fr	Froude number, dimensionless
h	hazard rate, Hz
M	molecular weight, gm
s	distance along the centerline of fire, m
u_i	velocity in i^{th} direction, m/s
ΔH_c	heat of combustion, J/kg
A	pressure relief area, in^2
D	diameter of pool fire, m
H, D	height and diameter of jet fire, m
L	pool fire flame height, m
P	pressure, Pa
P	probability of failure of equipment or component, dimensionless
R	reliability of an equipment, dimensionless
S, A, A'	a system, its good state and failed state
SEP	surface emissive factor, kW/m^2
T	temperature, K
d	hole diameter, m
f	failure density function, Hz
m	flammable mass of fireball, kg
m	Mean Time To Failure (MTTF), or Mean Time Between Failure (MTBF), s
n	total numbers of equipment
r	total numbers of functional equipment
r, h	radius and liftoff height of fireball, m
t	duration of fireball, m
t	time period, s
u	wind speed, m/s
x	horizontal distance, m
z	compressibility factor, dimensionless
z	vertical height, m
γ	heat capacity ratio
λ	a constant failure rate, Hz
ρ/ρ_{ref}	specific gravity, dimensionless

Subscripts

$_f$	identify as failed
$_{i,j}$	index variable
$_l$	lifetime
$_{mix}$	mixture
$_{T,P}$	value at Temperature T or Pressure P

Constants

g 9.8 m/s^2

References

Ahammad M, Liu Y, Olewski T, Vechot L, MS. 2016. Application of computational fluid dynamics (CFD) in simulating film boiling of cryogens. Indus Engg Chem Res, 55: 7548–7557. doi: 10.1021/acs.iecr.6b01013.

Ahammad M, Olewski T, Véchot LN, Mannan MS. 2016. A CFD based model to predict film boiling heat transfer of cryogenic liquids. J Loss Preven Proc Indus, 44: 247–254. doi: 10.1016/j.jlp.2016.09.017.

API. 1997. RP-521, Guide for pressure-relieving and depressuring systems. American Petroleum Institute, Washington DC.

API. 2000. RP-520, Sizing, Selection and Installation of Pressure-Relieving System in Refineries. American Petroleum Institute.

API. 2010. Process safety performance indicators for the refining and petrochemical industries. Washington DC.

Bagster DF, Schubach SA. 1996. The prediction of jet-fire dimensions, J Loss Preven Proc Indus, 9(3): 241–245.

Benavides-Serrano AJ, Mannan MS, Laird CD. 2015. A quantitative assessment on the placement practices of gas detectors in the process industries. J Loss Preven Proc Indus, 35: 339–351.

Britter R, McQuaid J. 1988. Workbook on the dispersion of dense gases. Health and Safety Executive. Available from: http://publications.eng.cam.ac.uk/332178/.

CCPS. 2003. Guidelines for facility siting and layout. New York: Center for Chemical Process Safety of the AIChE (CCPS/AIChE).

Le Chatelier H. 1891. Estimation of firedamp by flammability limits. Annals o Mines, 19 (8), 388–395.

Church R, Revelle C. 1974. The maximal covering location problem. Papers of the regional Science Association, 32. Available from: http://www.geog.ucsb.edu/~forest/G294download/MAX_COVER_RLC_CSR.pdf.

Cook J, Bahrami Z, Whitehouse RJ. 1990. A comprehensive program for calculation of flame radiation levels. J Loss Preven Proc Indus, 3(1): 150–155.

Cormier BR, Qi R, Yun G, Zhang Y, Mannan MS. 2009. Application of computational fluid dynamics for LNG vapor dispersion modeling: a study of key parameters. J Loss Preven Proc Indus, 22(3): 332–352.

Crowl D, Louvar JF. 2011. Chemical process safety: fundamentals with applications. 3rd ed. Boston: Pearson Education.

Díaz-Ovalle C, Vázquez-Román R, Mannan MS. 2010. An approach to solve the facility layout problem based on the worst-case scenario. J Loss Preven Proc Indus, 23(3): 385–392.

Dweck J, Boutillon S, Asbill S, Brennan LLP. 2004. Deadly LNG incident holds key lessons for developers, regulators. Pipe Gas J, (May).

Dyadem Press. 2003.Guidelines for failure mode and effects analysis for automotive, aerospace and general manufacturing industries. Dyadem Press. Available from: https://www.crcpress.com/Guidelines-for-Failure-Mode-and-Effects-Analysis-FMEA-for-Automotive/Press/p/book/9780849319082.

Han K, Kim YH, Jang N, Kim H, Shin D, Yoon ES. 2013. Risk index approach for the optimal layout of chemical processes minimizing risk to humans. Indus Engg Chem Res, 52(22): 7274–7281.

Hansen OR, Gavelli F, Ichard M, Davis SG. 2010. Validation of FLACS against experimental data sets from the model evaluation database for LNG vapor dispersion. J Loss Prevent Proc: 857–877.

Hirschberg S, Burgherr P, Spiekerman G, Dones R. 2004. Severe accidents in the energy sector: comparative perspective. J Haz Mat, 111(1–3): 57–65.

James M. 2017. The eye opener: Preparing for and performing facility siting to best communicate with upper management. Proc Safety Prog, 36 (2): 158–164.

Jung S. 2016. Facility siting and plant layout optimization for chemical process safety. Kor J Chem Engg, 33(1): 1–7.

Jung S, Ng D, Diaz-Ovalle C, Vazquez-Roman R, Mannan M S. 2011. New approach to optimizing the facility siting and layout for fire and explosion scenarios. Indus Engg Chem Res, 50(7): 3928–3937.

Khan F, Abunada H, John D, Benmosbah T. 2009. Development of risk-based process safety indicators. Proc Safe Prog, 29(2): 133–143.

Khan F, Hashemi SJ. 2017. Chapter One - Introduction. In: Methods in chemical process safety. Elsevier. p 1–36.

Kim BK, Ng D, Mentzer RA, Mannan MS. 2013. Key parametric analysis on designing an effective forced mitigation system for LNG spill emergency. J Loss Preven Proc Indus, 26(6): 1670–1678.

Legg SW, Benavides-Serrano AJ, Siirola JD, Watson JP, Davis SG, Bratteteig A, Laird CD. 2012. A stochastic programming approach for gas detector placement using CFD-based dispersion simulations. Comput Chem Engg, 47: 194–201.

Lewis D. 1980. The Mond fire, explosion and toxicity index applied to plant layout and spacing. Loss Prevent, 13: 20–26.

Liaw H-J. 2016. Lessons in process safety management learned in the Kaohsiung gas explosion accident in Taiwan. Proc Safe Prog, 35(3): 228–232.

Little A, Sundmacher K, Kienle A. 1992. Emergency relief system design using diers technology: the design institute for emergency relief systems (DIERS) Project Manual.

López-Molina A, Vázquez-Román R, Mannan MS, Félix-Flores MG. 2013. An approach for domino effect reduction based on optimal layouts. J Loss Preven Proc Indus, 26(5): 887–894.

Mannan MS, Mentzer RA, Zhang J. 2013. Framework for creating a best-in-class safety culture. J Loss Preven Proc Indus, 26(6): 1423–1432.

Mannan MS, Reyes-Valdez O, Jain P, Tamim N, Ahammad M. 2016. The evolution of process safety: current status and future directions. Ann Rev Chem Biomolec Engg, 7(1), 000–000.

Mannan MS, West HH, Berwanger PC. 2007. Lessons learned from recent incidents: Facility siting, atmospheric venting, and operator information systems. J Loss Preven Proc Indus, 20(4–6): 644–650.

Mannan MS. 2012. Lees' loss prevention in the process industries. 4th edn. Edited by S. Mannan. Butterworth-Heinemann.

Martinez-Gomez J, Nápoles-Rivera F, Ponce-Ortega JM, Serna-González M, El-Halwagi MM. 2014. Siting optimization of facility and unit relocation with the simultaneous consideration of economic and safety issues. Indus Engg Chem Res, 53 (10): 3950–3958.

Masum Jujuly M, Rahman A, Ahmed S, Khan F. 2015. LNG pool fire simulation for domino effect analysis. Reliab Engg Sys Safe, 143: 19–29.

McCabe LC, Clayton GD. 1952. Air pollution by hydrogen sulfide in Poza Rica, Mexico: an evaluation of the incident of Nov. 24, 1950. Arch Indus Hyg Occup Med, 6(3): 199–213.

MKOPSC. 2008. LNG pool fire modeling. College Station. Available from: http://psc.tamu.edu/files/links/lng-white-paper-on-pool-fire-modeling/LNG_White_Paper.pdf

Olewski T, Ahammad M, Quraishy S, Gan N, Vechot LN. 2016. Building process safety culture at Texas A&M University at Qatar: a case study on experimental research. J Loss Preven Proc Indus, 44: 642–652.

Pandya N, Gabas N, Marsden E. 2012. Sensitivity analysis of Phast's atmospheric dispersion model for three toxic materials (nitric oxide, ammonia, chlorine). J Loss Preven Proc Indus, 25(1): 20–32.

Peekema RM. 2013. Causes of natural gas pipeline explosive ruptures. J Pipe Sys Engg Prac, 4(1): 74–80.

Qi R, Ng D, Cormier BR, Mannan MS. 2010. Numerical simulations of LNG vapor dispersion in Brayton Fire Training Field tests with ANSYS CFX. J Haz Mat, 183(1–3): 51–61.

Raj PK, Lemoff T. 2009. Risk analysis based LNG facility siting standard in NFPA 59A. J Loss Preven Proc Indus, 22(6): 820–829.

Rana MA, Mannan MS. 2010. Forced dispersion of LNG vapor with water curtain. J Loss Preven Proc Indus, 23(6): 768–772.

Richards F. 2013. Failure analysis of a natural gas pipeline rupture. J Fail Anal Prevent, 13(6): 653–657.

So W, Kim Y-H, Lee CJ, Shin D, Yoon ES. 2011. Optimal layout of additional facilities for minimization of domino effects based on worst-case scenarios. Kor J Chem Engg, 28(3): 656–666.

Tang MJ, Baker QA. 1999. A new set of blast curves from vapor cloud explosion. Proc Safe Prog, 18(4): 235.

Thomas PH. 1963. The size of flames from natural fires. Symposium (International) on Combustion, 9(1): 844–859.

TNO. 1997. Methods for the calculation of physical effects. 3rd ed. Edited by V. den Bosch and R. A. P. Weterings. Hague: Gevaarlijke Stoffen. Available from: http://content.publicatiereeksgevaarlijkestoffen.nl/documents/PGS2/PGS2-1997-v0.1-physical-effects.pdf.

Vázquez-Román R, Lee J-H, Jung S, Mannan MS. 2010. Optimal facility layout under toxic release in process facilities: a stochastic approach. Comput Chem Engg, 34(1): 122–133.

Wang M. 2012. Normalization of process safety metrics. Texas A&M University. MS Thesis.

Yun G, Ng D, Mannan, MS. 2011. Key findings of liquefied natural gas pool fire outdoor tests with expansion foam application, Indus Engg Chem Res, 50(4): 2359–2372.

Zabetakis MG. 1965. Fire and explosion hazards at temperature and pressure extremes. In: AICHE - ICHEME symposium Series -2: 99–104.

Zabetakis MG, Lambiris S, Scott GS. 1959. Flame temperatures of limit mixtures. in 7th Symposium on Combustion. London: Butterworths.

13

Thermodynamic Modeling of Relevance to Natural Gas Processing

*Georgios M. Kontogeorgis and Eirini Karakatsani**

Center for Energy Resources Engineering (CERE), Department of Chemical and Biochemical Engineering, Technical University of Denmark, Denmark
**Current address: Haldor Topsoe A/S, Nymøllevej 55, Kgs. Lyngby, Denmark*

CHAPTER MENU

13.1 Introduction to the Problem

Even though present in small quantities only, water can cause various problems during the production, transportation, and distribution of natural gas. For example the presence of water is one of the three conditions that promote hydrate formation together with high pressures and low temperatures. Natural gas hydrates are solid, nonstoichiometric compounds that consist of water molecules forming "cages" containing guest molecules. They may be stable above the normal melting temperature of water by as much as 50K at high pressures [1]. Water condensation and ice formation are also common problems observed in natural gas production systems [2, 3]. The precipitation of all possible condensed and solid phases containing water may have many negative effects, such as increased corrosion, liquids in dry processes, two-phase flow problems, safety hazards and flow assurance issues, slugging of the flow lines, valves and instrumentation, resulting in reduced capacity and shutdowns, and reduction of the oil recovery efficiency because of reduction of the reservoir permeability.

For all these reasons it is important to control the amount and composition of the aqueous phase in the pipeline systems as accurately as possible. Often the moisture specification is expressed as a water dew point and a very useful list of dew point temperatures is given in [3] and is being repeated here:

- **Water dew point temperature**: the highest temperature, at a specified pressure, where water spontaneously can condense from the natural gas.
- **Aqueous dew point temperature**: the highest temperature, at a specified pressure, where a solution of water and trace chemicals (e.g., monoethylene glycol (MEG) and triethylene glycol (TEG)) can spontaneously condense from the natural gas.

Natural Gas Processing from Midstream to Downstream, First Edition.
Edited by Nimir O. Elbashir, Mahmoud M. El-Halwagi, Ioannis G. Economou, and Kenneth R. Hall.

- **Hydrate point temperature**: the highest temperature, at a specified pressure, where natural gas hydrate can spontaneously form in a natural gas mixture.
- **Maximum water precipitation temperature**: the highest temperature, at a specified pressure, where water can spontaneously precipitate in any form (liquid water, aqueous solution, ice, or hydrate) in a gas mixture.

In a schematic way the different dew point temperatures can be seen in Figure 13.1.

The water found in pipelines can come from free water produced from the reservoir or from water condensed by cooling the hydrocarbon fluid in various gas processing operations. While the first can be determined only from experimental data from the well, the second can be calculated [2]. The need to use calculation methods for water content determination is reinforced by the fact that the analysis of water in natural gas remains one of the most demanding trace gas analyses [3].

It is thus not surprising that many methods have been developed for the estimation of water content and water dew point of natural gas. Among them, many empirical models (e.g., Bukacek's method [4]) and charts (e.g., the diagram of McKetta and Wehe [5]) are widely used by the industry due to their simplicity and low numerical requirements [2]. However these methods are, in general, not accurate enough, because they do not take into account the gas compositional dependency of water content, they are not applicable at high pressures, and they cannot predict hydrate precipitation, ice formation, or water and trace chemicals simultaneous condensation (they are only developed to predict the water dew point, i.e., the highest temperature, at a specified pressure, where water spontaneously can condense to liquid from the natural gas) [2].

Even the GERG-water correlation [6], which has been developed with the specific objective to allow water content and dew point estimations with a good accuracy, suffers from the latter disadvantage.

On the other hand, association thermodynamic models have been shown to give more accurate predictions for a large number of gas compositions, and pressure and temperature conditions. Such models take into account explicitly molecular chemical association and solvation phenomena in organic fluids, such as the hydrogen bonding among water molecules and alcohols/glycols.

In this chapter we present selected results using such an association model, the CPA EoS (Cubic-Plus-Association Equation of State [7]) in combination with the van der Waals-Platteeuw (vdW-P) model for hydrates [8], for the calculation of the water content of several

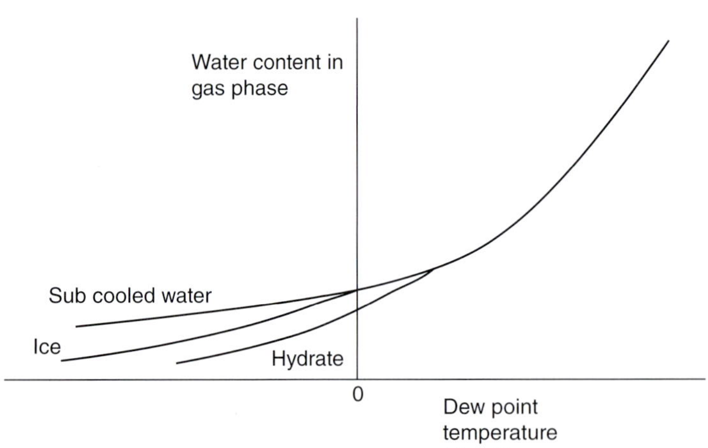

Figure 13.1 Different water dew point temperatures in a natural gas mixture.

binary and multicomponent mixtures present in natural gas streams. The specific model has been previously tested over a limited database for predicting the water content of nitrogen, methane, and a few natural gas mixtures with excellent results [9].

We will present the CPA model validation against extensive experimental data, both water content and hydrate phase equilibrium data, with and without chemicals. For water content / dew point calculations without chemicals, the model results were compared against results obtained using the GERG-water calculation method. In case of chemicals like methanol and glycols (used as gas hydrate inhibitors or in gas dehydration), the GERG-water method cannot yet be applied, as it was designed to be applied only to systems containing 12 specific natural gas components and water. In some cases, results obtained with other models found in the literature or obtained through a limited time use of commercial thermodynamic modeling software packages were available and are also presented or commented in this chapter. The temperature range of interest was (−80 to 100)°C, the pressure range of interest was up to 250bar and the compositions were typical NG compositions found in Northern Europe.

13.2 The Models

13.2.1 GERG-Water

The GERG-water EoS [6] is an ISO-standard model developed by GERG (the European Gas Research Group) to accurately calculate water content and water dew point for natural gas mixtures. It is based on the Peng – Robinson equation of state, which in terms of pressure is given by the following equation:

$$P = \frac{RT}{V_m - b} - \frac{\alpha(T)}{V_m^2 + 2bV_m - b^2} \tag{13.1}$$

The energy parameter is given by the following equations:

$$\begin{aligned} \alpha(T) &= 0.45724\frac{R^2T_c^2}{P_c}[f(T_r)]^2 \\ f(T_r) &= 1 + \kappa(1 - \sqrt{T_r}) \\ \kappa &= 0.37464 + 1.54226\omega - 0.26992\omega^2 \end{aligned} \tag{13.2}$$

The co-volume parameter is calculated using the following equation:

$$b = 0.0778\frac{RT_C}{P_C} \tag{13.3}$$

When extending the model to mixtures, the classical one-fluid mixing rules are used for the energy and co-volume parameters:

$$\alpha(T) = \sum_i \sum_j x_i x_j \alpha_{ij} \tag{13.4}$$

$$b = \sum_i \sum_j x_i x_j b_{ij} \tag{13.5}$$

where the classical combining rules are used:

$$\alpha_{ij} = \sqrt{\alpha_i \alpha_j}(1 - k_{ij}) \tag{13.6}$$

$$b_{ij} = \frac{b_i + b_j}{2} \tag{13.7}$$

Contrary to the original PR EoS, in order to ensure an accurate calculation of water vapor pressure above ice and liquid, the following energy term is used for water:

$$\sqrt{\alpha(T_r)} = 1 + A_1(1 - \sqrt{T_r}) + A_2(1 - \sqrt{T_r})^2 + A_3(1 - \sqrt{T_r})^4 \tag{13.8}$$

The energy term of equation (13.8) is divided in two parts. In the temperature range of 223.15 – 273.16K the energy term is fitted to vapor pressure data above ice (with parameters A1 = 0.106025, A2 = 2.683845 and A3 = −4.75638), while in the temperature range of 273.16–313.15K vapor pressure data over liquid water were used (with parameters A1 = 0.905436, A2 = −0.213781 and A3 = 0.26005). Finally, in the cases of water–methane, water–ethane and water–carbon dioxide systems a temperature-dependent binary interaction parameter is used, which is of the form:

$$k_{ij}(T) = k_{ij,o} + k_{ij,1}\left(\frac{T}{273.15} - 1\right) \tag{13.9}$$

The developed relationship is validated for temperatures ranging from −15°C to +5°C and pressures ranging from 5bar(a) to 100 bar(a). The applicability range can be extended to temperature of −50°C to +40°C and pressure from 1 to 300bar(a) with unknown uncertainties. The components accepted as input parameters in the correlation are the following ones: water, nitrogen (≤55.0%), carbon dioxide (≤30.0%), methane (≥40.0%), ethane (≤20.0%), propane (≤4.5%), 2-methyl propane (≤1.5%), *n*-butane (≤1.5%), 2,2-dimethyl propane (≤1.5%), 2-methyl butane (≤1.5%), n-pentane (≤1.5%), hexane (≤1.5%), and argon. All required coefficients for the calculation of the binary interaction parameter from equation (13.9) as well as the critical component properties that should be used are given by GERG [6]. In this work, calculations will be often shown both with an in-house developed GERG-water version and a commercial version called GasCalc developed by E.ON Ruhrgas AG. The GasCalc software allows the user to perform calculations with the GERG-water method over an extended range of conditions (beyond the limited pressure, temperature, and composition conditions GERG-water was originally checked through the choice "no check"). The GasCalc software gives the chance to the user to perform both water content calculations given the dewpoint temperature and pressure, and dewpoint temperature calculations given the water concentration. The user can choose among three application ranges: a limited one, an extended one, and a "no check" range.

13.2.2 CPA

The CPA EoS can be expressed for mixtures in terms of pressure, P, as [7, 10]:

$$P = \frac{RT}{V_m - b} - \frac{\alpha(T)}{V_m(V_m + b)} - \frac{1}{2}\frac{RT}{V_m}\left(1 + \rho\frac{\partial \ln g}{\partial \rho}\right)\sum_i x_i \sum_{A_i}(1 - X_{A_i}) \tag{13.10}$$

where R is the gas constant, T is the temperature, V_m is the molar volume, α is the energy parameter and b the co-volume parameter, ρ is the molar density of the mixture, g is the radial distribution function, X_{A_i} is the fraction of A-sites on molecule i that do not form bonds with other active sites, and x_i is the mole fraction of component i. The letters i and j are used to index the molecules while A and B indicate the bonding sites on a given molecule.

The key element of the association term is X_{A_i}, which represents the mole fraction of site A on molecule i not bonded at other sites, while x_i is the mole fraction of component i. X_{A_i} is related to the association strength $\Delta^{A_iB_j}$ between two sites belonging to two different molecules, e.g., site A on molecule i and site B on molecule j, determined from:

$$X_{A_i} = \frac{1}{1 + \rho\sum_j x_j \sum_{B_j} X_{B_j}\Delta^{A_iB_j}} \tag{13.11}$$

where the association strength $\Delta^{A_iB_j}$ in CPA is expressed as:

$$\Delta^{A_iB_j} = g(\rho)\left[\exp\left(\frac{\varepsilon^{A_iB_j}}{RT}\right) - 1\right] b_{ij}\beta^{A_iB_j} \tag{13.12}$$

with the radial distribution function $g(\rho) = \frac{1}{1-1.9n}$ and $n = \frac{1}{4}b\rho$ while $b_{ij} = \frac{b_i+b_j}{2}$

Finally, the energy parameter of CPA is given by a Soave-type temperature dependency, while b is temperature independent:

$$\alpha(T) = a_o(1 + c_1(1 - \sqrt{T_r}))^2 \tag{13.13}$$

where $T_r = \frac{T}{T_c}$ and T_c is the critical temperature.

In the expression for the association strength $\Delta^{A_iB_j}$, the parameters $\varepsilon^{A_iB_j}$ and $\beta^{A_iB_j}$ are called the association energy and the association volume, respectively. These two parameters are only used for associating components, and together with the three additional parameters of the SRK term (α_o, b, c_1), they are the five pure compound parameters of the model. They are usually obtained by fitting vapor pressure and liquid density data. For inert components, e.g., hydrocarbons, only the three parameters of the SRK term are required, which can be obtained either from vapor pressures and liquid densities or calculated in the conventional manner (critical data, acentric factor).

When the CPA EoS is used for mixtures, the conventional mixing rules are employed in the physical term (SRK) for the energy and co-volume parameters. The geometric mean rule is used for the energy parameter α_{ij}. The interaction parameter k_{ij} is usually fitted to experimental data:

$$\alpha(T) = \sum_i \sum_j x_i x_j \alpha_{ij}, \text{where}\, \alpha_{ij} = \sqrt{\alpha_i \alpha_j}(1 - k_{ij}) \tag{13.14}$$

$$b = \sum_i \sum_j x_i x_j b_{ij}, \text{where}\, b_{ij} = \frac{b_i + b_j}{2} \tag{13.15}$$

In case of cross-associating systems, e.g., water-alcohol or water-glycol systems, combining rules for the association energy and volume parameters are needed in order to calculate the value of the association strength in equation (13.12).

From previous investigations, two choices have been identified as very successful in various cases, the so-called CR-1 rule:

$$\varepsilon^{A_iB_j} = \frac{\varepsilon^{A_iB_i} + \varepsilon^{A_jB_j}}{2}$$

$$\beta^{A_iB_j} = \sqrt{\beta^{A_iB_i}\beta^{A_jB_j}} \tag{13.16}$$

or alternatively, the so-called Elliott combining rule (ECR):

$$\Delta^{A_iB_j} = \sqrt{\Delta^{A_iB_i}\Delta^{A_jB_j}} \tag{13.17}$$

The CR-1 and ECR rules are functionally similar; the only difference is the function of b in the expression for the cross-association volume (which is important for size-asymmetric systems, e.g., water with heavy alcohols or glycols).

In cases where VIE experimental data are available or water content / dew point data of a vapor phase in equilibrium with an unknown condensed phase, which could be ice, the water fugacity in the ice phase needs to be calculated:

$$\ln f_w^{ICE} = \ln f_{w,P\_ref}^{ICE} + \frac{1}{RT}\int_{P\_ref}^{P} V_W^{ICE}\, dP \tag{13.18}$$

where f^{ICE}_{w,P_ref} is the fugacity of water in the ice phase at the reference pressure P_ref (which is the atmospheric pressure) while V^{ICE}_{W} in the Poynting term correction, is the molar volume of ice, which is obtained by the correlation suggested by Avlonitis [11]. The fugacity of water in the ice phase f^{ICE}_{w,P_ref} can be computed via such a difference term involving the heat of fusion ΔH^{fus} and the difference of heat capacity ΔC_p^{fus} between solid and liquid [12], similar to the way the fugacity of the solid phase is calculated in the case of SLE:

$$\ln(f^{ICE}_{w,P\_ref}) = \ln(f^{L_0}_{w}) + \frac{\Delta H^{fus}}{RT_m}\left(1 - \frac{T_m}{T}\right) + \frac{\Delta C_p^{fus}}{R}\left(\frac{T_m}{T} - 1 - \ln\left(\frac{T_m}{T}\right)\right) \tag{13.19}$$

The value of the melting temperature for water used in equation (13.19) is $T_m = 273.15\text{K}$, the value of the heat of fusion at the melting temperature is $\Delta H^{fus} = 6010$ J. mol^{-1} while the difference in the heat capacity between the liquid and solid phase is $\Delta C_p^{fus} = 37.29$ J/mol.K.

A major difference between the two models, GERG-water and CPA, is that the CPA modeling approach enables differentiation between the various heavy phases (i.e., liquid, ice, or hydrate phase) that could equilibrate with the gas phase. More specifically, CPA's dew points for all possible equilibria (VLE, VHE, VIE) are being calculated, and the equilibrium with the highest equilibrium temperature is taken as the stable equilibrium.

13.2.3 Van der Waals-Platteeuw Hydrate Model

In order to obtain the fugacity of water in the hydrate phase when using CPA EoS, an appropriate hydrate model is needed. In this work, the well-established van der Waals-Platteeuw (vdW-P) hydrate model has been used [8], which provides the following equation for the chemical potential of the hydrate μ^H_W:

$$\mu^H_W = \mu^{EH}_W + RT\sum_i \nu_i \ln\left(1 - \sum_{\text{guests } m} \Theta_{mi}\right) \tag{13.20}$$

where R is the universal gas constant, ν_i is the number of type i cavities per water molecule (which are $\nu_1 = 1/23$ and $\nu_2 = 3/23$ for structure I hydrate and $\nu_1 = 2/17$ and $\nu_2 = 1/17$ for type II hydrates) and the summation is over all cavity types (both 1 and 2). The occupancy of cavity m by a component i, Θ_{mi}, is calculated as following:

$$\Theta_{mi} = \frac{C_{mi} f_m}{1 + \sum_{\text{guests } k} C_{ki} f_k} \tag{13.21}$$

where f_k is the fugacity of a component k in the equilibrium vapor phase obtained from an equation of state (CPA in this work), the summation is over all hydrate-forming components, and C_{mi} is the Langmuir adsorption constant for guest m in i type cage (in other words, the hydrate formation is considered to be similar to Langmuir adsorption). There are two main approaches for the computation of the key quantities of the theory, which are the Langmuir constants. The simplified approach, suggested by Parrish and Prausnitz (PP) [13], is one of them and enables the use of the empirical equation:

$$C_{mi}(T) = \frac{A_{mi}}{T}\exp\left(\frac{B_{mi}}{T}\right) \tag{13.22}$$

where A_{mi} and B_{mi} are fitted parameters (two parameters per guest molecule and structure).

The second, more rigorous approach is to introduce a model potential experienced by the guest molecule in the cage, based on water-guest interactions. The Kihara potential is commonly used [14]. However, if the square-well (SW) potential is used, the Langmuir constant is given by:

$$C_{mi}(T) = \frac{4\pi}{kT} V_{mi} \exp\left(\frac{\varepsilon_{mi}}{kT}\right) \tag{13.23}$$

where V_{mi} is the free volume of guest molecule i inside m type cage and ε_{mi} is the depth of the SW potential, and the resemblance between equation (13.23) and the "empirical" equation (13.22) is obvious. In addition, the "van't Hoff-type temperature dependence" of equation (13.22) can be expected based on quite general thermodynamic considerations according to Bazant and Trout [15]. Papadimitriou et al. [16] recently obtained Langmuir-type mathematical expressions for the cavity occupancies of the methane hydrate of the following type:

$$\Theta_i(P) = \frac{C_i(T)\cdot P}{1 + C_i(T)\cdot P} \tag{13.24}$$

They obtained optimum values of A_{mi} and B_{mi} by matching them to the cavities occupancies resulting from Grand Canonical Monte Carlo simulations [16].

In this work the computationally cheaper and still acceptable from a theoretical point of view approach of PP is used (eq. 2.22).

13.2.4 Model's Pure Component Parameters and Comments on Database

The GERG-water pure component parameters used in this work are taken from [6]. The CPA pure component parameters were also taken from literature [17]. The literature CPA parameters are based on the Soave expression (eq. (13.13)). We have also used alternative to Soave equations for the temperature-dependency of the energy parameter, e.g., Mathias-Copeman [18] and Twu [19] (recommended by some authors, see Refs. [20] and [21]) without any significant change in the results, thus the Soave equation is used in all results shown here.

The database used for the regression of the Langmuir constant parameters in the PP equation 2.22 is given by Karakatsani and Kontogeorgis [22] for methane, ethane, propane, and CO_2. All PP parameters used for CPA are summarized in Table 13.1.

Table 13.1 Optimized values of A_{mi} and B_{mi} parameters for calculating the Langmuir constant parameters for eq.(13.22) using the CPA EoS coupled with VdW-P theory (all parameters are from Ref. [22]).

Component	Average error in T (%)	Small cavity		Large cavity	
		A*10^3 (K/bar)	B (K)	A*10^3(K/bar)	B (K)
methane (str.I)	0.0694	9.99997	1969.77	316.434	2048.66
ethane (str.I)	0.0420	0	0	57.0184	3040.91
ethane (str.II)	0.0858	0	0	9.54095	4000
propane (str.II)	0.3098	0	0	79.9249	3852.69
CO_2 (str.I)	0.3347	0.108734	3400	43.1146	2804.85
CO_2 (str.II)	0.5540	0.1	3119.67	63.11	3500
H_2S (str.I)	0.2046	0.395412	4329.13	99.999	3071.25
H2S (str.II)	0.2597	17.56605	3222.94	3.51321	3222.94

The new parameters were tested under different conditions, including low temperatures, VHE, LHE and structural transitions conditions, with clear improvement compared to literature parameters. These comparisons are shown in Ref. [22].

Experimental data are not always accurate and there are serious deficiencies. For example, in case of the CO_2-H_2O system, at least three cases of inconsistencies among experimental data have been recently reported, concerning the water content of pure CO_2 in equilibrium with hydrate [23], the pressure dependence of the water content in liquid CO_2 [24], and the finding of inconsistent data series of CO_2 solubilities using thermodynamic consistency tests [25]. For such cases, thermodynamic models can be used as an additional test in favor of the one or the other data set.

13.3 Systems Studied and Selected Results: Part 1. No Chemicals

We have considered a significant amount of experimental data of relevance to water content / dew point of natural gas, thus providing valuable information on the phase behavior of NG systems at a useful range of compositions, temperatures, and pressures, as defined previously (−80 to 100°C, up to 100bar, typical NG compositions found in northern Europe). The systems studied have been categorized in a systematic way, starting from binary and passing to ternary and then to multicomponent systems, including systems with chemicals (see Section 13.4).

13.3.1 Binary Systems of NG Components with Water

The accurate description of the phase equilibria of water/hydrocarbon mixtures is the first step towards the phase equillibria modeling of aqueous mixtures found in the oil and gas industry, and that is why many equations of state (EoS), pure compound parameter sets, and binary interaction parameters (BIP) have been suggested over the years. In the case of CPA, some relevant works include references [26–29]. In this work, it was decided to compare the performance of the different parameters sets / BIP values against the available experimental data (VHE, VIE, and VLE) for the binary systems including water with methane, ethane, propane, *n*-butane, *n*-pentane-water, and nitrogen. These are some of the main NG components (at least when looking at the NG compositions found in Northern Europe). The water content of methane and nitrogen using CPA has been previously studied as well by Folas et al. [9].

In parallel, the internal consistency of the data themselves was checked and a clear picture concerning the temperature and pressure conditions covered by the available experimental data was formed, as shown schematically in Figure 13.2.

In some cases, comparisons to models' results obtained by using commercial software packages (HydraFLASH [30], MULTIFLASH [31], GasCalc [32]) were also possible. The use of those software packages allows comparisons to other models, e.g., VPT, SRK, and PR, but also to different versions of the models of interest (CPA and GERG-water). As mentioned previously, the use of GasCalc software is of special value as it allows GERG-water calculations beyond the applicability range of the model.

As it becomes obvious from Figure 13.2, methane-water is the most extensively studied system, probably because methane is the main component of NG (24 data sources found). Most experimental data refer to VLE (12 data sources), fewer to VHE (9 data sources), and very few to unknown condensed phases (liquid, hydrate, or ice) (3 data sources). Our main conclusion is that CPA performs overall better than GERG-water in terms of percent average absolute deviation (%AAD) in water vapor content (7.9 (CPA with a constant BIP) compared to 11.7 (GERG-water with a T-dependent BIP)). The use of a T-dependent versus a T-independent

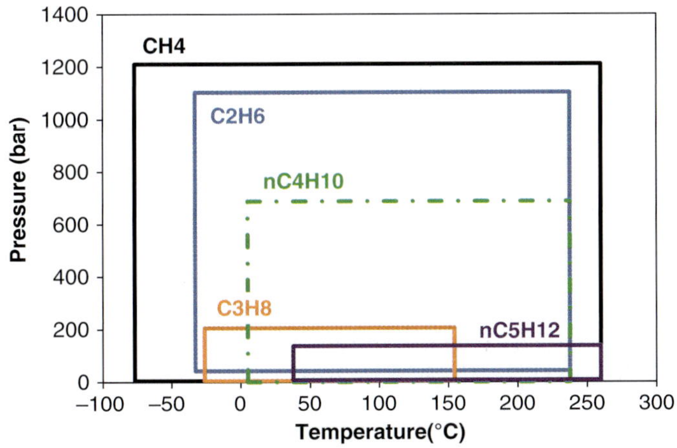

Figure 13.2 Temperature and pressure ranges covered by available VLE, VIE and VHE data for the binary systems CH4-H2O (black frame), C2H6-H2O (blue frame), C3H8-H2O (orange frame) and nC4H10-H2O (green frame).

binary interaction parameter does play a role and is recommendable when looking at the low temperature VHE calculation results.

When looking at VLE modeling results, we can distinguish the following cases/conditions:

- **Low pressures, large T-range**: In case of experimental data which extend over a large T-range but are limited to low pressures, the T-dependent BIP suggested by Yan et al., which had also been derived using an extended T-range database (274-633K) [27], leads to the most accurate results.
- **Low pressures, limited T-range at low temperatures**: In that case the T-dependent BIP of Tsivintzelis et al. [29] work best, which had also been fitted to a limited T-range database.
- **High pressures, limited T-range at high temperatures**: It was found that either a constant BIP or a zero BIP works best at HTHP conditions.

When looking at VHE/VIE results, the use of a T-dependent BIP leads to better results, especially when looking at high pressures (an overall improvement of about 3% in AAD is obtained when using an appropriate T-dependent BIP instead of a constant one.

Concerning the internal consistency test, three data sources were found to disagree with other data sources and/or with model results in the order of magnitude.

In the case of ethane-water, 12 data sources were located for VLE and VHE and 3 sources for LHE data. One data source with undefined condensed phase was located. GERG-water performs slightly better than CPA with a T-independent BIP in the case of VLE calculations (9.5% vs. 11.5%), while CPA with a T-dependent BIP performs slightly better in case of VHE calculations (17.0% vs. 17.8%) and clearly better in case of LHE calculations using a T-dependent BIP (almost twice more accurate results). If the GasCalc calculations are taken into account, then GERG-water proves to be orders-of-magnitude off. This problem has to be examined by the software developers.

In the case of propane-water, three data sources were located with VLE data, two data sources with LHE data, and one with LLE. CPA performs clearly better in case of LHE (18.2% vs. 25.0%) and slightly better in case of VLE calculations (6.3% vs. 7.9%), with a T-dependent BIP (although phase equilibria calculations in the absence of a vapor phase (as; e.g., LHE) are beyond the scope of the project, they were sometimes included in order to get a better picture of the models' performance).

In the case of n-butane-water, only five data sources were located, and the data of two of them were considered to be questionable. CPA performs clearly better than GERG-water with a T-dependent BIP (6.3% vs. 12.3%), but this is a conclusion based on only three experimental data points that are considered reliable and fall within the temperature applicability range of GERG-water.

In the case of n-pentane-water, one data source was located, and a T-independent BIP was estimated for CPA EoS.

Many results for this type of systems (three-phase equilibrium pressures for diverse gas-water at low temperatures, water content of hydrocarbons and CO_2 as function of temperature as well as water solubility in gases in equilibrium with hydrates and hydrate-vapor equilibrium curves of various gases) have been presented by Karakatsani and Kontogeorgis [22], while some additional characteristic results are shown in Figures 13.3–13.5.

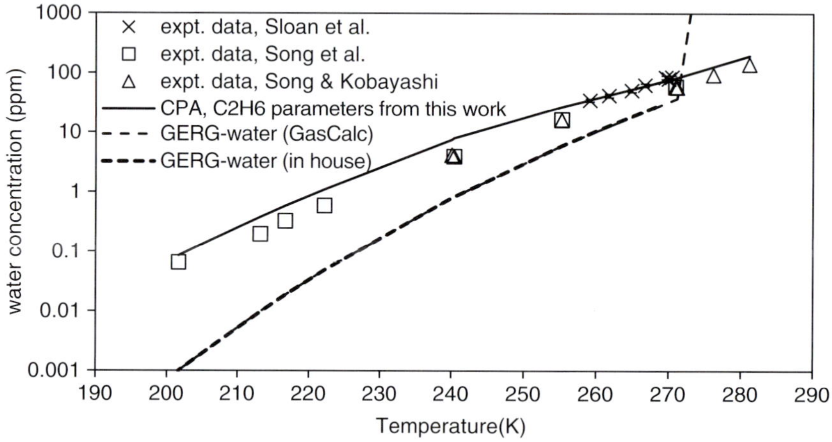

Figure 13.3 Water content of liquid C_2H_6 in equilibrium with hydrate at 3.45MPa. Points are experimental data, the solid curve is CPA predictions using the new C_2H_6 structure I parameters The thick dotted line is GERG-water calculations with GERG-water (in house) and the thin dotted line is GERG-water calculations but produced using the GasCalc software. Experimental data are from Sloan et al. (*FPE* **1986**, *29*, 233), Song et al. (*FPE* **2004**, *224*, 271) and Song and Kobayashi (*FPE* **1994**, *95*, 281).

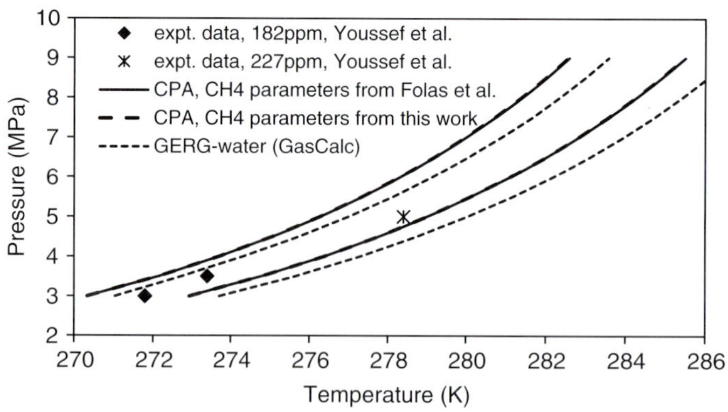

Figure 13.4 Hydrate-vapor equilibrium curves of CH_4 calculated using CPA and GERG-water (GasCalc) at 2 different water vapor concentrations: 182 ppm and 227 ppm. Experimental data are from Youssef et al. (*Ind. Eng. Chem. Res.* **2009**, *48(8)*, 4045).

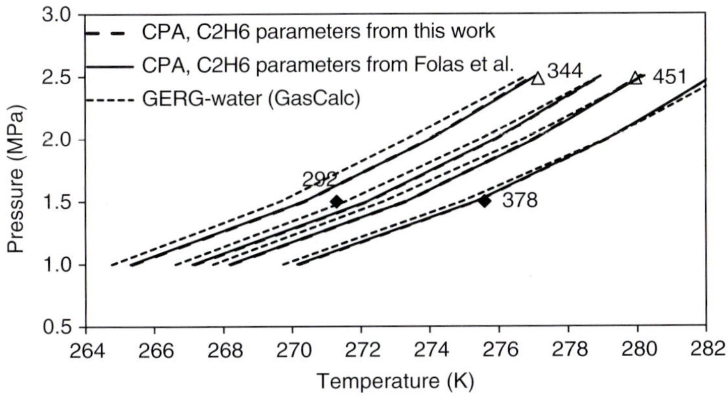

Figure 13.5 Hydrate – vapor equilibrium curve of C_2H_6 calculated using the CPA EoS at 4 different water vapor concentrations: 292 ppm and 378ppm (*Ind. Eng. Chem. Res.* **2009**, *48(8)*, 4045)., 344ppm and 451ppm (*Fluid Phase Equilib.* **1994**, *95*, 281).

In Figure 13.3 are shown results with the new ethane parameters for the system ethane-water at 3.45MPa against three sources of experimental data. CPA performs much better than either version of GERG. It is interesting to note that the CPA model predictions agree more with the older data of Sloan et al. than with the more recent ones of Song et al. (see references in Figure 13.3 caption).

Concerning HVE, two typical results are shown in Figures 13.4 and 13.5. We used the data of Youssef et al. (ref. in Figure 13.4 caption), who determined the hydrate dissociation temperature in the absence of any aqueous phase for CH_4, C_2H_6, and CO_2 hydrates at different pressures and for different water amounts. The results using the literature and the new PP parameters in case of CH_4 and C_2H_6 are similar, almost identical in case of C_2H_6. The agreement with GERG-water is better in this case (results provided by GasCalc when activating the "no check" applicability range option), still CPA performs best. For CO_2 the performance of CPA is even better.

Finally, it is interesting to compare GERG-water and CPA for the calculation of the light alkane solubilities in water in equilibrium with hydrate or vapor. One typical result is shown in ref. [22] for propane and similar results have been obtained for other light gases. While CPA performs very well with deviations around 10–30% (an impressive result for these very low solubilities), GERG-water dramatically fails especially for VHE calculations.

13.3.2 Ternary Systems of NG Components with Water

The next step is the study of ternary systems based on the knowledge gained for binary subsystems containing water. In some cases, comparisons to other models' results obtained by using commercial software packages (HYDRAFLASH [30], MULTIFLASH [31]) were also possible. The effect of mixture composition was also examined. The following systems were considered:

- methane–ethane–water
- methane–propane–water
- methane–carbon dioxide–water
- ethane–carbon dioxide–water
- ethane–propane–water
- methane-*n*–butane–water
- methane-*n*–pentane–water
- methane–nitrogen–water

For C_1-C_2-H_2O (four data sources) and C_1-C_3-H_2O (two data sources), CPA was found to perform better than GERG-water (8.3(26.9)% with T-dependent BIP values vs. 10.4(48.4)% in case of C_1-C_2-H_2O(C_1-C_3-H_2O)). For C_1-CO_2-H_2O, both low temperature–low pressure and high temperature–high pressure data were found, and both models fail to capture the water content quantitatively at the very extreme conditions. However, CPA is qualitatively closer to the experimental data, especially when solvation between CO2 and H2O is taken into account at intermediate temperature and pressure conditions. In the case of C_2-C_3-H_2O, GERG-water describes better the data of the one data source located and CPA the data of the other, so no clear conclusion can be made. It should be remembered, though, that the data found for this system are just LHE and LLE data. In the case of C_1-nC_4-H_2O, one data series was located, and CPA could describe it more accurately than GERG-water (8.4% in water content vs. 39.8% at 310.93K), the latter showing the largest deviations at the largest nC_4 contents. In the case of the C_1-nC_5-water system, only one source with questionable data was located and at conditions outside the applicability range of GERG-water. CPA performed satisfactorily (4.1%). Finally, for C1-N2-water, CPA performs very well with deviations around 4.5% for two systems (with 80–90% C1, T-range: 310–344 K), while GERG-water has error around 21%.

Karakatsani and Kontogeorgis [22] presented satisfactory results for methane-nitrogen and methane-ethane (hydrate formation) and methane-propane (water content in this gas mixture). In all cases the results are as good as or better than GERG-water. They also presented water content of CO_2 in equilibrium with hydrates at 137.9 bar and various temperatures. This system was evaluated with five models; SRK, PR and PVT (calculated made with the commercial software HydraFlash) and CPA with commercial software (HydraFlash, MULTIFLASH) and this work. It was shown that only when solvation between CO_2 and water is accounted for are satisfactory results obtained using CPA. This will be further verified by the results shown here for some typical ternary systems in Figures 13.6–13.9 and Table 13.2.

Indeed the solvation effect and its importance in CO2-water containing mixtures can be discussed also with reference to these figures and tables. The solvation effect between CO_2 and H_2O has been considered in the framework of CPA in line with the findings of Tsivintzelis et al. [33] who clearly showed that very good correlation is obtained for binary mixtures of CO_2 and water or alcohols when the solvation between CO_2 and the polar compound is explicitly accounted for, in line with experimental evidence for such cross-associating interactions. The solvation scheme used to produce the results shown here is the so-called case B1 by Tsivintzelis et al. [33], meaning that the acid gas is modeled as a non-self-associating fluid with cross-association interactions with water and other polar compounds, since it has two proton acceptor sites able to cross-associate.

It can be seen that considering solvation explicitly leads to more accurate water content calculations with the exception of very high temperatures, where we know that hydrogen bonds are weaker [34], so it may be assumed that the solvation should not be considered explicitly at such conditions. Another interesting observation is that when CO_2 concentration increases, considering the solvation explicitly has a more pronounced effect. Looking at the GERG-water model, we observe some unrealistically high deviations and strange behavior in the GasCalc version model, but not in the in-house version of the model.

Finally, concerning the data of Velasco et al. (used in Figure 13.8; ref. provided in figure caption) (i.e., dew point temperature series with a known reference water content), our investigations showed that including the solvation effect for the correlation of such data does not have a great effect on the CPA results, but more investigations should be done before confirming that conclusion. We believe this is because such measurements have a large degree

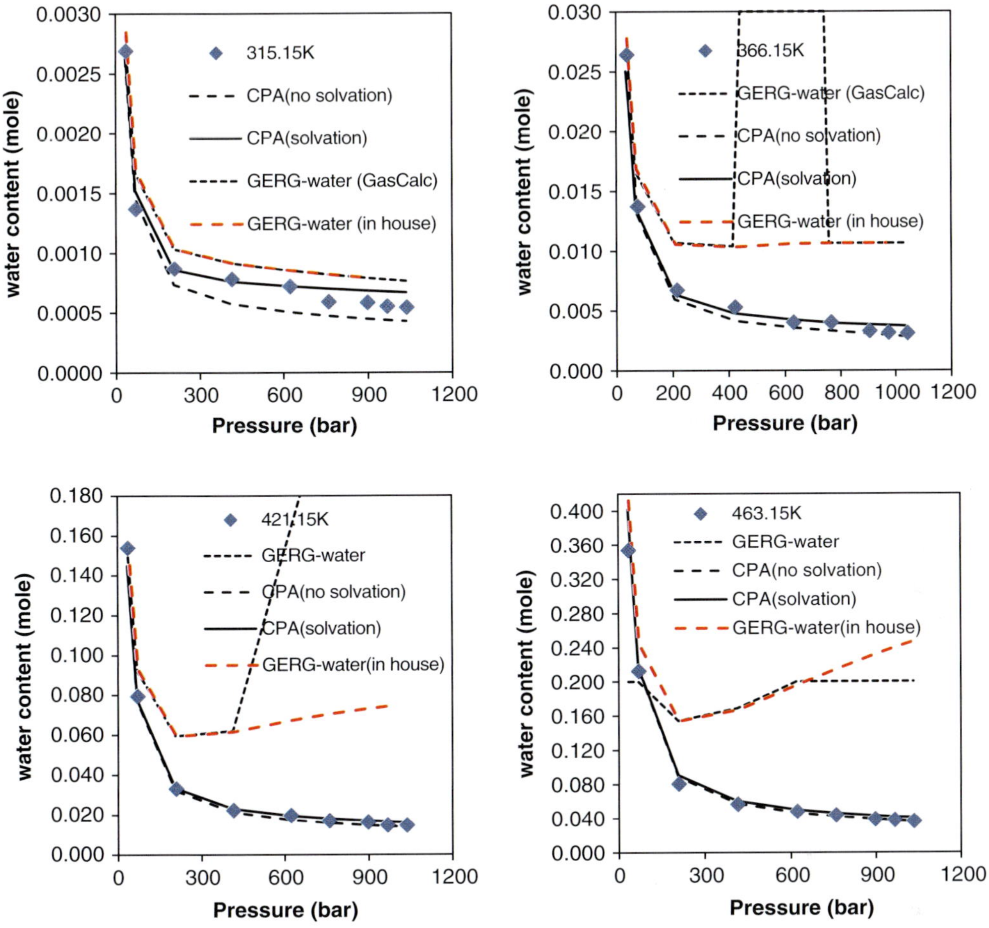

Figure 13.6 Water content of the gas phase of a C1-CO2-H2O system (90% C1, 10% CO2 on a dry basis) as a function of pressure (bar). The experimental data are indicated as points and the calculations as lines.

of uncertainty since very small traces of water are being measured and theoretically kept constant, and also that following the observation of Løkken [35] in the T-range between 0°C and −20°C it is difficult to predict whether the moisture will condense as dew or precipitate as ice, so what the authors call "dew points" could very well be "ice points" at this temperature range. Indeed by comparing VLE and VIE temperatures at <0 °C we found that in many cases ice is the stable phase (the phase with the highest equilibrium temperature). Finally, it is may be worth mentioning that the fact that solvation does not improve the CPA results at these low T & P conditions is in agreement with the carbonic acid formation theory at low pressure of Rushing et al. [36] according to which "at low pressures, water and CO_2 form carbonic acid in the liquid water phase, thus lowering the liquid water's vapor pressure and suppressing the water vaporization process. The increased solubility of water vapor in gases containing CO_2 at higher pressures and temperatures may be due to stronger molecular interactions and forces such as hydrogen bonding. However, it appears as if the enhanced water vapor solubility due to higher temperatures partially balances or offsets the effects of carbonic acid and its effects on the water vaporization process."

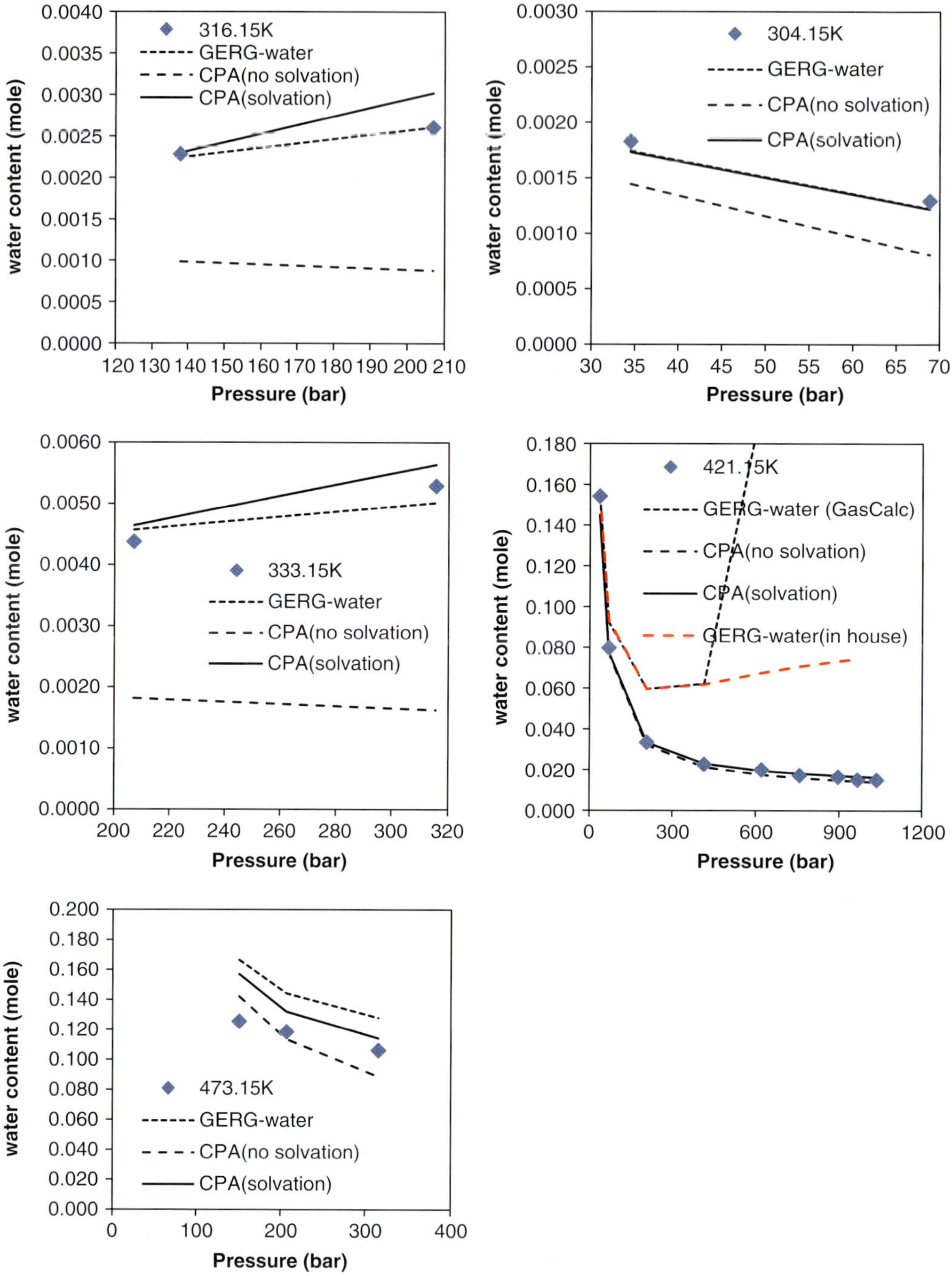

Figure 13.7 Water content of the gas phase of a C1-CO2 -H2O system (30% C1, 70% CO2 on a dry basis) as a function of pressure (bar). The experimental data are indicated as points and the calculations as lines.

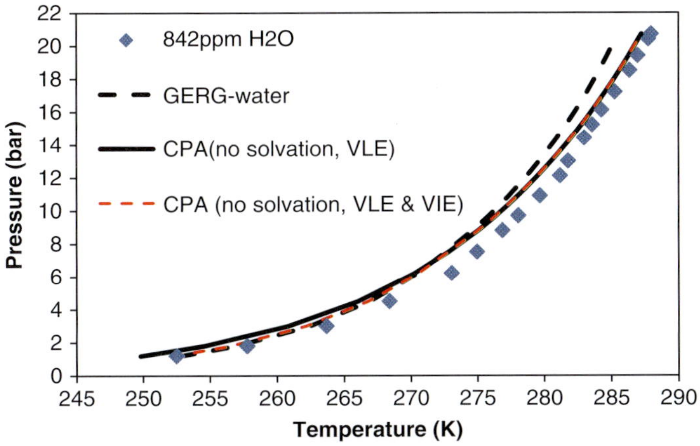

Figure 13.8 Comparison between experimental dew point temperatures for a C1-CO2-H2O system (30% C1-70% CO2 on a dry basis) with 842ppm H2O in the vapour phase. Data are from Velasco et al., Energy & Fuels, 2004, 18, 396.

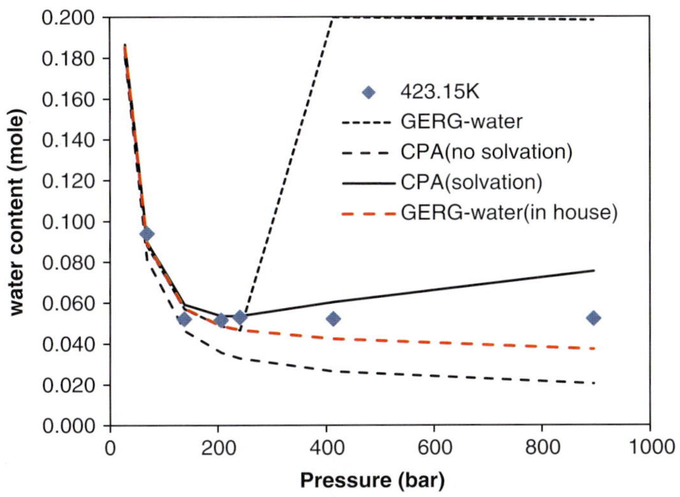

Figure 13.9 Water content of the gas phase of a C2-CO2-H2O system (5% C2-95% CO2 on a dry basis) as a function of pressure (bar) at 423.15K. The experimental data are indicated as points and the calculations as lines (black full lines for CPA with solvation, black dotted lines for CPA without solvation, red dotted lines for the in-house GERG-water version and dotted lines for the GERG water (GasCalc version).

13.3.3 Systems with ≥ 4 NG Components and Water

Experimental water vapour content data were located for the following systems:

- methane–ethane–propane–water
- methane– ethane–*n*-butane–water
- methane–ethane–propane–carbon dioxide–water
- methane–ethane–propane–nitrogen–water
- methane–ethane–propane–*n*-butane–water
- methane–ethane–propane–*i*-butane–*n*-butane–water
- methane–ethane–propane–nitrogen–carbon dioxide–water

Table 13.2 Average absolute deviations (%) in water vapor content using different models for the system ethane-carbon dioxide-water (C_2-CO_2-H_2O). 1.1, 1.2 and 1.3 are from GPA-RR 200 (Yarrison et al., 2008) for 5% C2-95% CO2 at three T (316, 364, 423 K) and diverse pressures and Ref. [2] is from Adissaminto and Sloan, J. Chem. Eng. Data, 37, 1992, 343–349 for six different C2-CO_2 mixtures ranging from 4% C2 to 80% C2 and T: 273–287 K.

	%AAD of water content (error ±5%)			
Ref.	**CPA, no solvation**	**CPA, solvation**	**GERG-water**	**Temperature(K)**
1.1	51.61	52.20	**19.82**	316.15
1.2	36.47	**21.43**	211.33/4.48**	364.15
1.3	31.15	**13.81**	112.85/6.57**	423.15
	39.75	**29.14**	114.67	All temperatures
2		0.24K	Average absolute error in T (VLH data)	

** This number has been calculated excluding the GERG-water results for which the GERG-water model (GasCalc version) returned unrealistically high water content values.

- methane–ethane–propane–*n*-butane–*n*-pentane–nitrogen–water
- methane–ethane–propane–*n*-butane–nitrogen–carbon dioxide–water
- methane–ethane–propane–*i*-butane–*n*-pentane–i-pentane–water
- methane–ethane–propane–*i*-butane–*n*-butane–carbon dioxide–nitrogen–water
- methane–ethane–propane–*i*-butane–*n*-butane–*n*-pentane–carbon dioxide–nitrogen–water
- methane–ethane–propane–*i*-butane–*n*-butane–neopentane–*i*-pentane–nitrogen–carbon dioxide–hexane-water

It should be mentioned that a few cryogenic data for the quaternary system C_1-C_2-C_3-H_2O were found in GPA RR187 [37] and given the low temperatures, the Mathias-Copeman parameters for water were also used to study their effect on the CPA calculations. Data for the same system were also located at HTHP conditions in a study on the catalytic effect of nonhydrocarbon contaminants (CO_2 and N_2) on equilibrium water vapor content [36].

Data for the quaternary system H_2O-C_1-C_2-nC_4 were located from three different sources. In both cases CPA performed better than GERG-water. CPA performs surprisingly better than GERG-water (23.2%) when using zero BIPs in case of the C_1-C_2-C_3-nC_4-H_2O system (one data source found) (9.5%AAD) but not when using *T*-dependent parameters (31.5%AAD). All deviations are in water content. For the same five-component system with iC4 instead of nC4 (C_1-C_2-C_3-iC_4-nC4-H_2O) two data sources were found, and CPA has an error in water content around 5–10% with GERG-water 12–14%.

The data for systems C_1-C_2-C_3-N_2-CO_2-water (one data source), C_1-C_2-C_3-nC_4-nC_5-N_2-water (two data sources), C_1-C_2-C_3-iC_4-nC_4-CO_2-N_2-water (one data source), C_1-C_2-C_3-nC_4-CO_2-N_2-water (two data sources), C_1-C_2-C_3-iC_4-nC5-iC5-water (one data source), and C_1-C_2-C_3-iC_4-nC_4-neoC_5-iC_5-N_2-CO_2-C_6+ (seven different compositions, one data source) were also located, and comparisons between the two models were performed. CPA overall performs similarly to GERG-water in some cases for the water vapor content or better for the systems C_1-C_2-C_3-nC_4-nC_5-N_2-water (6% vs. 16% for GERG-water), C_1-C_2-C_3-iC_4-nC_4-N_2-CO2-water (10% vs. 21% for GERG-water), system C_1-C_2-C_3-N_2-CO_2-water (13% vs. 25% for GERG-water), see also Ref. [22], where CPA performs much better at low temperature–high pressure conditions. Considering the solvation between CO_2 and water improves the results with CPA.

Of particular interest are the seven different systems, actually compositions, of the real multi-component NG mixture: C1-C2-C3-iC4-nC4-neoC5-iC5-N2-CO2-C6+ -H2O (including one mixture composition with 0% CO2 and 0% C6+). Detailed results and comparisons between CPA, GERG-water, other equations of state, and results from diverse commercial software are shown in Reference [22]. These experimental data are from Oellrich and Althaus, GERG-water Correlation (GERG Technical Monograph TM 14), VDI Verlag GmbH-Düsseldorf 2001, covering a T-range: 248–293 K and pressures up to 100 bar. In almost all cases, it has been found that CPA performs better than GERG-water, which underlines the robustness of the approach, irrespective of the method chosen for Langmuir constants' estimation and the nonuse of BIPs. The only case where GERG-water performs better than CPA is for mixture no.7 of Oellrich and Althaus. Mixture no.7 is the carbon dioxide–richest mixture examined (25% in mole). This result illustrates the importance of solvation in modeling the phase behavior of mixtures containing water and CO_2. The average deviation in water vapor content for these seven systems with CPA is 6.3 % (4.7% if no.7 is excluded) vs. 6.4% with GERG-water (6.2% if no.7 is excluded).

Some typical examples are shown in Figures 13.10–13.13 and Table 13.3. Figures 13.10 and 13.11 present results for the C1-C2-C3-H_2O system. The data shown in Figure 13.10 are cryogenic. The use of the MC expression for the alpha function of water instead of the Soave expression improves a bit the water-content correlation accuracy but when looking at all calculated compositions' accuracy, the effect seems to be not so important. GERG-water performs quite a bit worse at these very low temperatures. For Figure 13.11, we have HPHT conditions, so the GERG-water calculation method is expected to fail, from the point of view that it is very far from its applicability range. This is the case especially for the GasCalc version, which returns some erroneous results at the higher pressures.

For the system shown in Table 13.3 and Figure 13.11, the experimental data have an uncertainty of 12% (Ref. [1]) and 4% (Ref. [2]), so all models (except for the GERG-water results) are within the range of the reported uncertainty. The results of CPA and GERG-water can be seen schematically in Figure 13.11.

For the C1-C2-C3-CO2-water system, shown in Figure 13.12, there is some uncertainty with the experimental data, only available in tabulated form. Again GERG-water fails to describe these HPHT data, even when the GasCalc version with the "no check" option is being used. On the other hand, CPA does "see" an increase in the water vapor content at higher concentrations

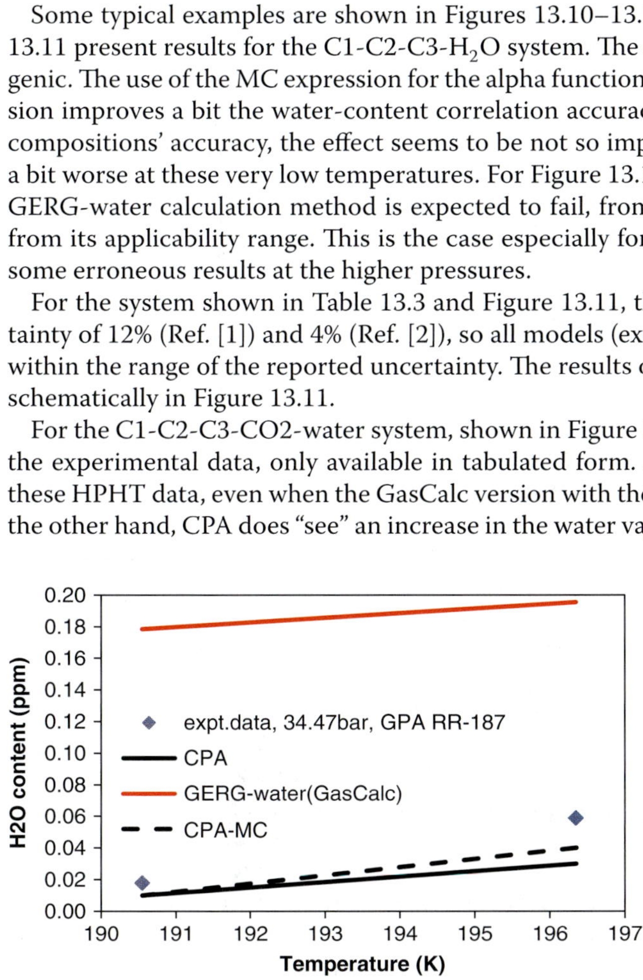

Figure 13.10 Water content of the gas phase of a C1-C2-C3-H2O system (85% C1, 5% C2, 10% C3 on a dry basis) as a function of temperature (K). The experimental data are indicated as points and the calculations as lines (black full lines for CPA with Soave expression, black dotted lines for CPA with MC expression, red full lines for the GERG-water version of GasCalc commercial software).

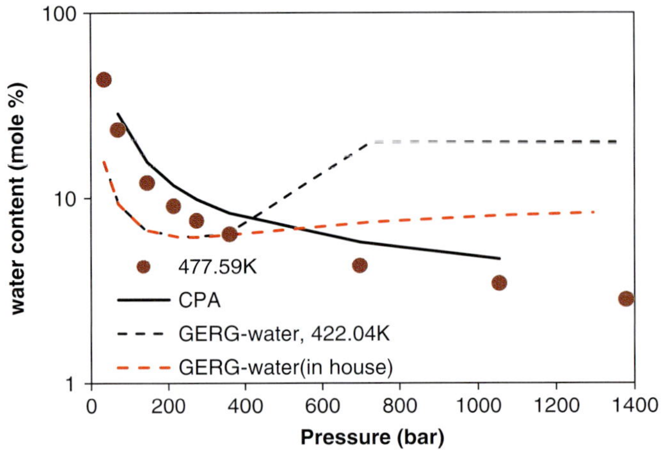

Figure 13.11 Water content of the gas phase of a C1-C2-C3-H2O system (96% C1, 3% C2, 1% C3 on a dry basis) as a function of pressure (bar). The experimental data are indicated as points and the calculations as lines (black full lines for CPA, black dotted lines for GERG-water (GasCalc version), red dotted lines for the GERG-water (in house version)).

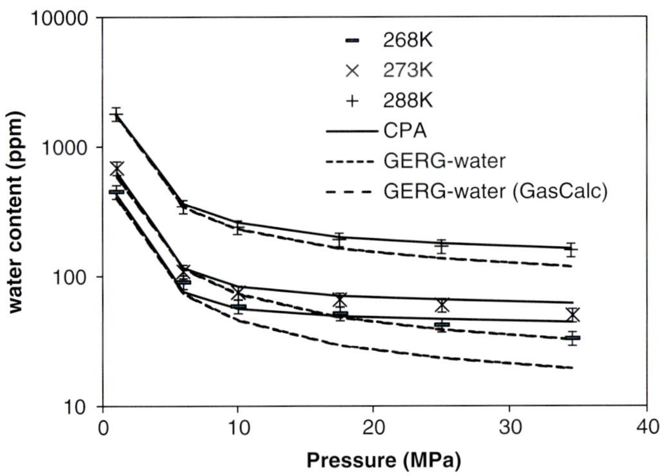

Figure 13.12 Water content of the gas phase of a C1-C2-nC4-H2O system (94% C1, 4% C2, 2% nC4 on a dry basis) as a function of temperature (K) and pressure (MPa). The experimental data are indicated as points and the calculations as lines (full lines for CPA, dotted lines for GERG-water, dashed lines for the GERG-water version of GasCalc commercial software).

of CO_2 and higher temperatures. The solvation does not seem to play a big role in this case, probably due to the high temperatures. Some differences between the calculations with and without solvation are seen only for the highest CO_2 concentration mixture (20% CO_2 in mole).

The authors [36] relate the effect of CO2 to pressure: according to their statements, "at low pressures water and CO_2 form carbonic acid in the liquid water phase, thus lowering the liquid water's vapor pressure and suppressing the water vaporization process. The increased solubility of water vapor in gases containing CO_2 at higher pressure and temperatures may be due to stronger molecular interactions and forces such as hydrogen bonding. However, it appears as if the enhanced water vapor solubility due to higher temperatures partially balances or offsets the effects of carbonic acid and its effects on the water vaporization process."

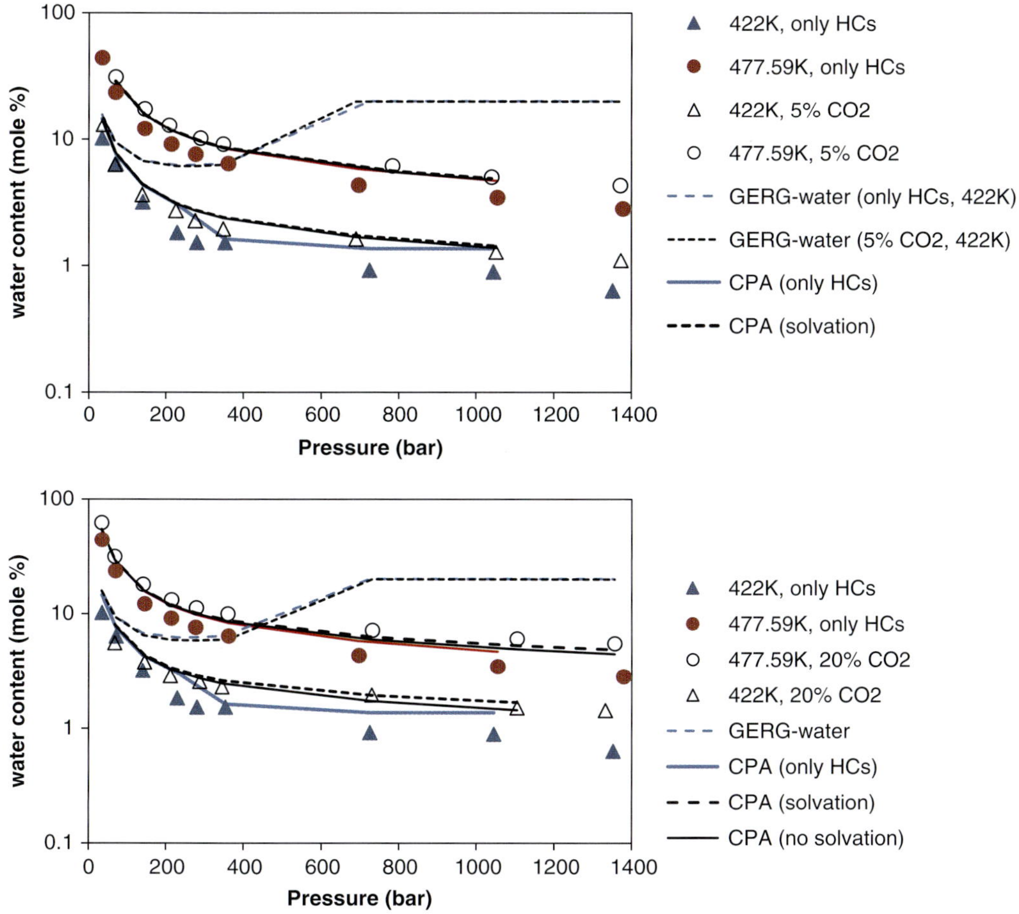

Figure 13.13 Water content of the gas phase of two C1-C2-C3-H2O systems (91.2% C1, 2.85% C2, 0.95% C3, 5% CO2 (top) and 76.8% C1, 2.4% C2, 0.8% C3, 20% CO2 (below)) as a function of pressure (bar). The experimental data are indicated as points and the calculations as lines.

Table 13.3 Average absolute deviations (%) in water vapour content using different BIPs for the system methane–ethane–*n*-butane–water (C1-C2-nC4-H2O), and different models.

	%AAD of water content							
Ref	CPA (kij = 0)	CPA (k_{ij} = f(T)(WY))	GERG-water	Commercial software/other calculations				
1				SRK(HYDRA)	PR(HYDRA)	VPT(HYDRA)	CPA(HYDRA)	CPA(HYDRA)
	10.00	**8.40**	16.70	11.05	10.97	8.46	6.12	
2				SRK(HYDRA)	PR(HYDRA)	VPT(HYDRA)	CPA(HYDRA)	VPT (article)
	1.06	1.12	2.12	1.59	1.47	1.35	1.36	1.4
3	**1.53**	1.55	14.43[a)]	Semi-empirical approach[b)c)]			VPT[b)]	
				1.94			1.93	
4		2.39K	1.81K	Average absolute error in DP T				

a) GasCalc version of GERG-water was used.
b) Calculated based on the results presented in Ref. [3].
c) Developed by Mohammadi et al. (Ind. Eng. Chem. Res. 2004, 43, 7137).

A somehow different explanation but leading to the same qualitative behavior is given by Clark et al. [38]: The ability of the pure compound (CO2) to hold water in the vapor phase decreases as the pressure increases up to about 60bar. At higher pressures the water-holding capacity of the gases increases, corresponding to a higher water absorption capacity in the liquid phase or dense phase compared to the vapor phase. In both cases, increasing the temperature allows more water to be absorbed in the gas phase. Small amount of methane substantially reduce the water absorption ability of both components. If it is assumed that solubility of water in the gas mixtures mimics the trend of the individual components, then a minimum water-holding capacity exists at some pressure.

Rather similar conclusions were obtained for the system C1-C2-C3-N2-water, with CPA performing again much better than GERG-water. However, in this case CPA seems unable to distinguish between mixtures with and without N_2. It must be said that the differences in the water content values are indeed small. And there are further complicating factors. The results themselves are somehow confusing since, despite what one would expect, the water content of gas mixture with 5mol % at 422K is slightly greater than that for gas mixture with no nitrogen. We have to remember here that temperature and N_2 have opposite effects on water content and that temperature balances the suppression caused by higher nitrogen contents. From a theoretical point of view, Rushing et al. [36] state that the presence of N_2 in the gas phase suppresses the water vaporization process and reduces the quantity of water vapor dissolved in the gas. Further increases in N_2 content reduce the water vapor content below that for the same gas with no N_2. However, the enhanced water vapor solubility at higher T counters the suppressive effects of N_2 in the gas phase.

13.4 Systems Studied and Selected Results: Part 2. With Chemicals

13.4.1 Systems of NG Components with Water and Alcohols

Vapour water content data were located in the open literature for the following systems:

- Methane–MeOH–water
- Methane–EtOH–water (only hydrate formation conditions available)
- Ethane–MeOH–water
- Propane–MeOH–water
- CO2–MeOH–water
- Methane–propane–MeOH–water (also hydrate data from GPA RR74)
- Methane–n-hexane–MeOH–water
- Methane–n-heptane–MeOH–water
- Methane–CO2–MeOH–water (GPA RR149)
- Propane–CO2–MeOH–water (only hydrate formation conditions available)
- Methane–propane–n-heptane–MeOH–water (+GPA RR149)
- Methane–ethane–n-butane–MeOH–water
- Methane–ethane–propane–i-butane–n-butane–MeOH–water,
- Methane–ethane–propane–n-butane–n-pentane–carbon dioxide–nitrogen–MeOH–water
- Methane–ethane–propane–i-butane–n-butane–i-pentane–n-pentane–n-hexane–carbon dioxide–MeOH–water (only hydrate formation conditions available)
- Methane–ethane–propane–i-butane–n-butane–i-pentane–n-pentane–nitrogen–carbon dioxide–MeOH-water (only hydrate dissociation conditions available).

Concerning water vapour content results, satisfactory accuracy could be obtained with CPA in most cases using a T-independent BIP for the water–MeOH binary system. The water vapor content calculation accuracy itself was found to be almost insensitive to the choice of BIPs. A strange result was produced when calculating the water content of C_1-C_2-nC_4-MeOH-H2O system at the highest concentration of 49% wt MeOH, where CPA fails to capture the decrease of water content values with increasing pressure, and the same was seen when using the CPA version of the MULTIFLASH software [31].

In the case of hydrate phase boundary calculations, T-independent BIPs also provided satisfactory results. BIPs optimized from H2O-MeOH/EtOH SLE data were found to be necessary for such calculations, especially at high inhibitor concentrations. At low inhibitor concentrations, the sensitivity of the calculations to the MeOH/EtOH-H_2O system's BIP is rather small.

The satisfactory performance of T-independent BIPs in case of natural gas mixtures with alcohols is one more proof of the suitability of CPA EoS for such calculations, especially when considering the different binary parameters needed during the study of the phase behavior of hydrocarbon/water/alcohol systems with other EoS, such as PR [39].

Some representative results are shown in Figures 13.14–13.24.

We offer here a more detailed discussion of these results. The system C1-MeOH-water is rather extensively studied, and many data exist, not all necessarily very accurate. For example, although the CPA model describes very well the data of Mazloum et al., as shown in Figure 13.14 (ref. in the figure caption), significant deviations are observed when it is being applied to the older Russian experimental data mentioned in Figure 13.15. This issue has been analyzed, and various factors have been studied. As shown in Figure 13.15, although methanol concentration in the gas phase and methane concentration in the polar phase are fairly well predicted at all conditions, water vapor concentration is seriously underpredicted at the lowest temperature and highest MeOH concentration in the aqueous phase (in between these extreme conditions results are also of an intermediate accuracy). There are indications based on discussions with industry that the water content in the gas phase is wrong, and specifically it is one order of magnitude too high. Indeed, we also observe the overestimation of the water content by one order of magnitude when comparing the lowT – high MeOH content data to the experimental data.

Figures 13.16–13.18 are concerned with another important issue, which kij (BIP) should be used for methanol-water at these low temperatures, e.g., hydrate dissociation conditions, the one obtained from VLE, or the one regressed from SLE data. In Figure 13.16 the suitability of

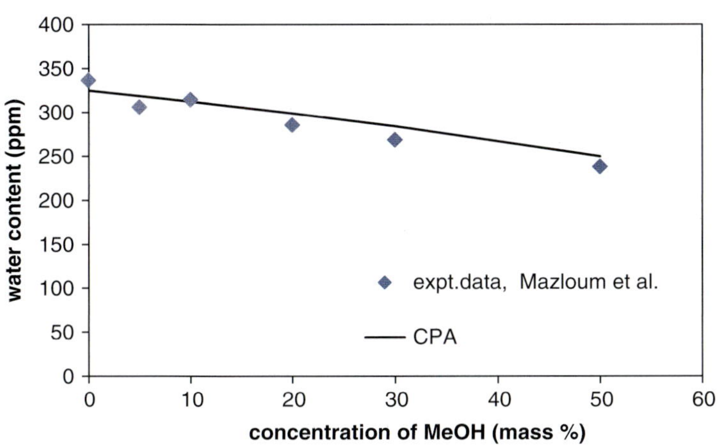

Figure 13.14 Experimental and predicted water content (ppm mole) of the water-methane system at different methanol concentrations at 293.15K and 103.4bar. Experimental data are from Mazloum, S.; Chapoy, A.; Yang, J.; Tohidi, B. Online monitoring of hydrate safety margin. Proceedings of the 7th International Conference on Gas Hydrates (ICGH 2011), Edinburgh, UK, 2011.

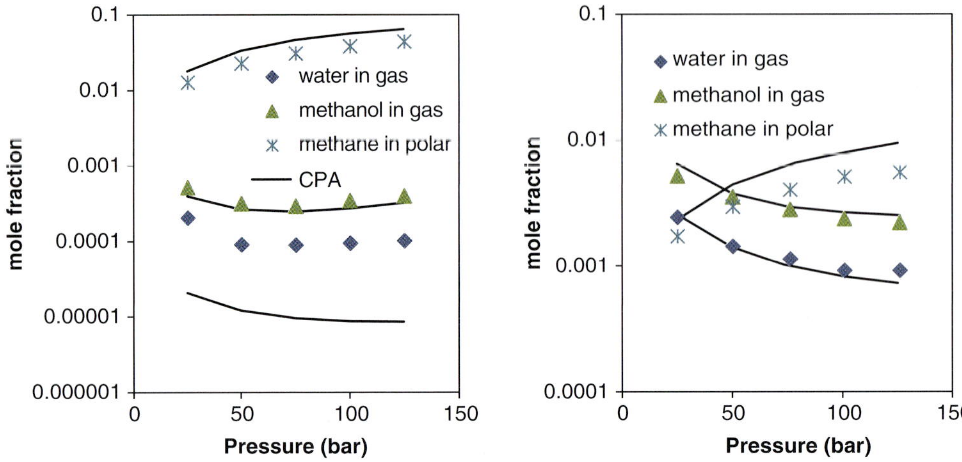

Figure 13.15 Prediction with CPA of methanol and water concentration in the gas phase and methane concentration in the liquid polar phase for the methanol-water-methane system at 251.95K and 84.63% mass MeOH in the aqueous phase (left) and at 313.15K and 39.06% mass MeOH in the aqueous phase (right). Experimental data are points from Sinyavskaya et al., Gazovaya Prom. V2 26-27 (1985) and Sinyavskaya et al., Gazovaya Prom. V1 39-40 (1984).

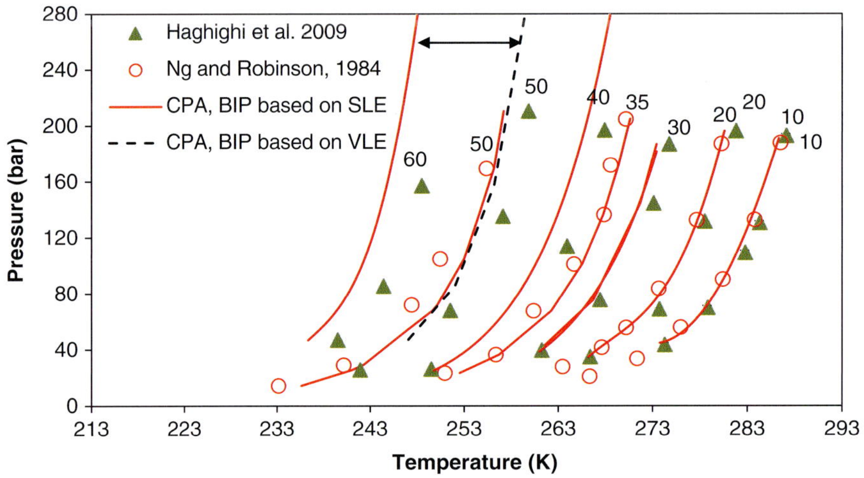

Figure 13.16 Experimental and predicted methane hydrate dissociation conditions in the presence of methanol aqueous solutions. Experimental data from Ng and Robinson (GPA RR-74 and FPE 21(1985), 145) and Haghighi et al. (FPE 278(2009), 109).

the BIP based on SLE data is obvious, as well as the fact that our CPA calculations agree better with the experimental data of Ng and Robinson (GPA RR-74 and FPE 21(1985), 145) than with the experimental data of Haghighi et al. (FPE 278(2009), 109) In Figure 13.17, it can be seen that the BIP fitted to VLE data performs better when it comes to water content correlations, while the BIP fitted to SLE data performs slightly better for the methanol gas content and the methane solubility calculations. For the C1-EtOH-H2O system in Figure 13.18, the use of a BIP fitted to SLE data has a small impact on the accuracy, especially at the highest inhibitor concentrations. Here the BIP used was the one fitted to MeOH–H2O SLE data instead of EtOH-H2O, as there has not been estimated a BIP for EtOH-H2O based on SLE data to our knowledge.

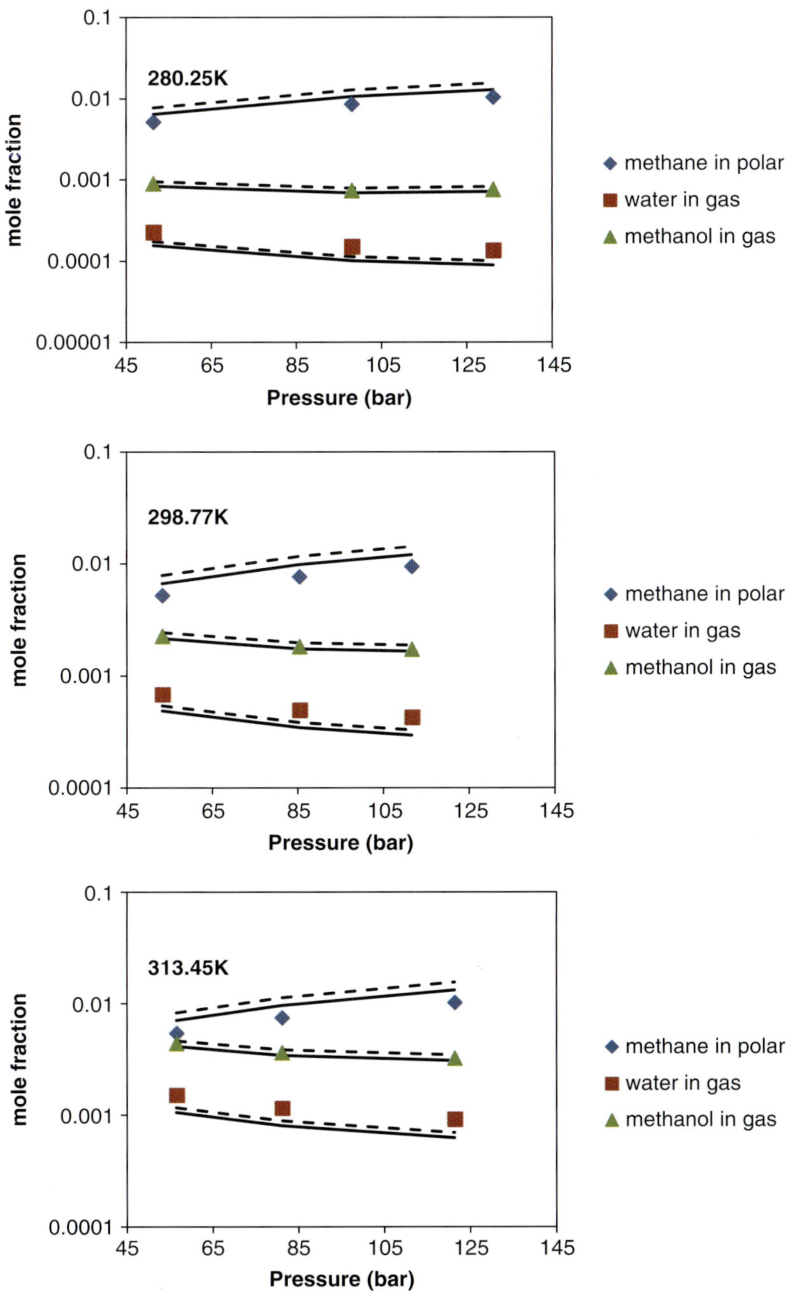

Figure 13.17 Prediction of methanol solubility in the vapor phase, methane solubility in the liquid phase and gas phase water content with CPA for the methanol-methane-water system at 280.25K (top), 298.77K (middle) and 313.45K (bottom). Points are experimental data from M. Frost (2014, PhD Thesis, Technical University of Denmark), solid lines are correlations with water-methanol BIP equal to -0.153 (fitted to SLE data) and dotted lines are correlations with water-methanol BIP equal to -0.09 (fitted to VLE data).

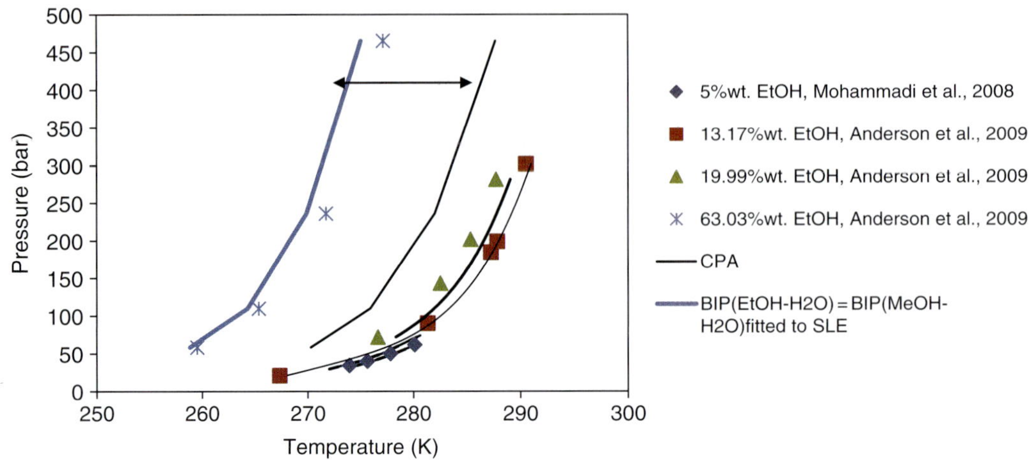

Figure 13.18 Hydrate formation conditions for the system C1-EtOH-H2O. Points are experimental data (Mohammadi et al., J. Chem. Eng. Data 2008, 53, 73), Anderson et al., J. Phys. Chem. C 2009, 113, 12602) and lines are CPA correlations.

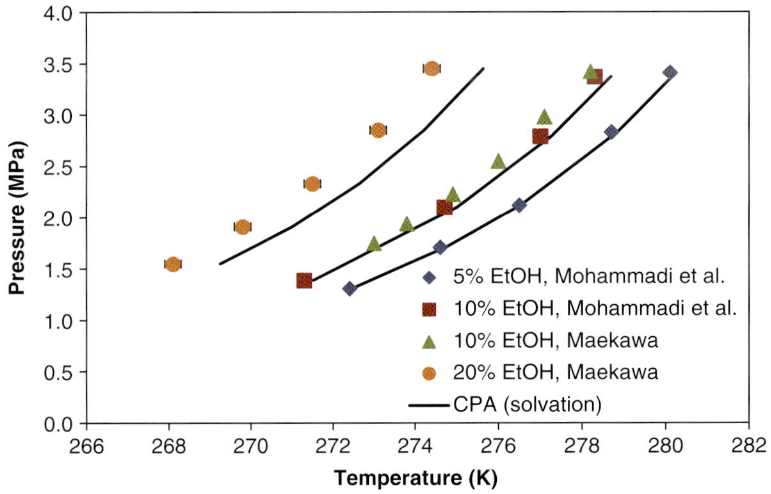

Figure 13.19 Prediction of CO2 hydrate formation in the presence of ethanol as inhibitor with CPA, taking into account the solvation between CO2-H2O and CO2-ethanol. Points are experimental data (Mohammadi et al., J. Chem. Eng. Data (2008), 53, 73 and Maekawa, J. Chem. Eng. Data (2010) 55(3), 1280) and lines CPA predictions.

There are data for CO2 hydrate formation with methanol and ethanol as inhibitors and Figure 13.19 shows one example. Best results are obtained when solvation between CO2–water is accounted for. No solvation is needed between CO2 and either alcohols. There are significant differences between literature data for CO2 hydrates (especially with methanol as inhibitor), but this will not be discussed further here. Another example is shown in Figure 13.20 for a multicomponent system. Taking into account the solvation between CO2 and H2O seems to improve the results, especially at the highest MeOH concentration.

For the system methane–propane–methanol–water (C1-C3-MeOH-H2O), there are VLLE data from Ng & Chen, GPA RR-149, 1995 in the T-range 284–311 K and pressures up to 207 bar. The average absolute deviations (%) in water vapor content is about 35–40% and does not

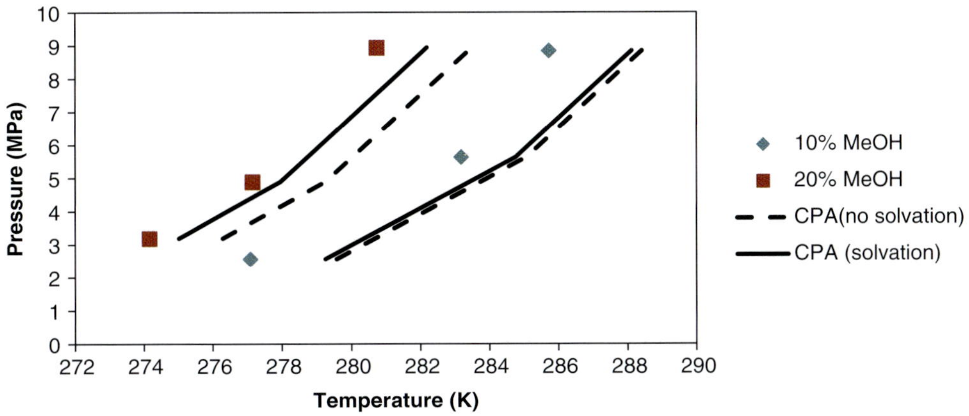

Figure 13.20 Experimental hydrate equilibrium conditions for the ternary C1-C3-CO2 system (78% C1, 20% CO2, 2% C3) inhibited by methanol.

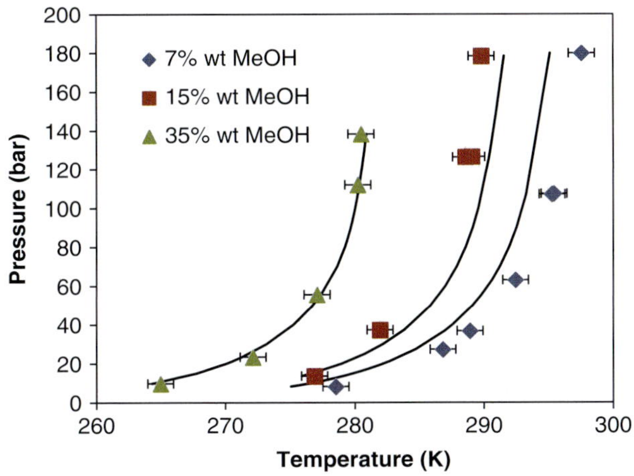

Figure 13.21 Hydrate decomposition conditions for a C1-C3 system (11.87% C1-88.13% C3) in the presence of methanol aqueous solutions. Points are experimental data (Song and Kobayashi (FPE 47(1989) 295) and lines are CPA calculation results obtained using T-independent BIPs.

depend much on whether we use T-dependent or T-independent BIPs. The absolute accuracy of MeOH and H2O concentration values in vapor or nonaqueous liquid phases or hydrate former in the aqueous phase is ±20ppm (repeatability 10%). Given the low water content values especially at 50% wt MeOH, the experimental uncertainty is quite high (of the order of 9.2%).

For the specific systems hydrate decomposition conditions have also been published by Song and Kobayashi (FPE 47(1989) 295) for a mixture containing 11.87mol % propane in methane gas in the presence of 7, 15, and 35% wt. MeOH solutions. The use of a T-independent BIP for MeOH–H2O works well in that case too, as can be seen in Figure 13.21.

The methane-n-hexane (or n-heptane)-methanol-water systems have also been measured, but only the VLLE data of Ng and Chen (GPA RR-117) are reliable (with n-heptane). Very good agreement is obtained for the water content as shown in Figure 13.22.

All multicomponent and NG systems mentioned at the start of the section were considered. The results are satisfactory overall. Only two characteristic results will be shown here, first for the system methane–propane–*n*-heptane–methanol–water (C1-C3-*n*C7-MeOH-H2O);

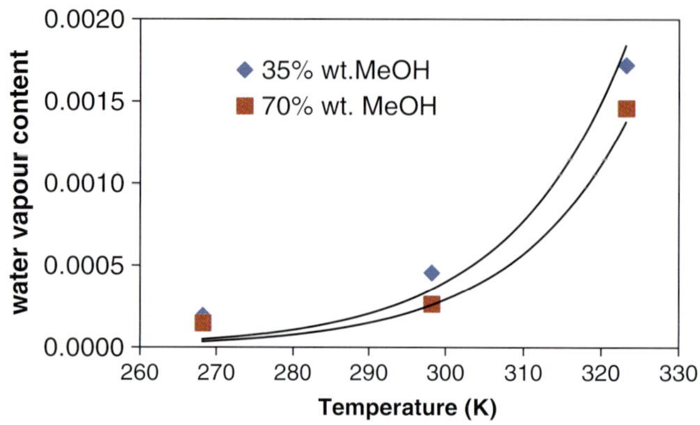

Figure 13.22 Water vapour content against temperature for the C1-nC7 system inhibited by MeOH. Points are experimental data (Ng et al., GPA RR-117, 1988) and lines are CPA results.

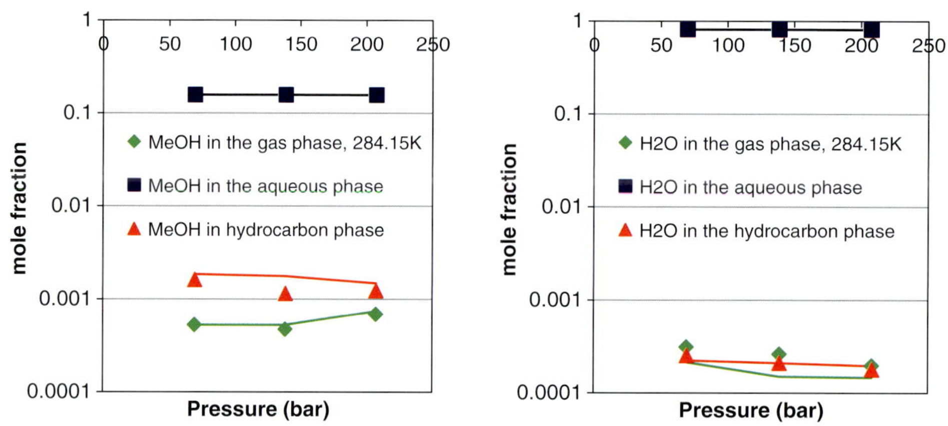

Figure 13.23 VLLE predictions for the system of methanol-water-methane-propane-heptane at 284.15K. Methanol (left) and water (right) mole fractions in the three phases. Data are from Ng and Chen, GPA RR-149, 1995.

see Figure 13.23. The results have been found satisfactory, especially taking into consideration that the concentration of the major components is known to within ± 0.003 mole fraction; the minor components are known to within $\pm 10\%$ of the reported value; methanol and water concentrations in the vapour are estimated to be within ± 20ppm. The methanol concentration in the hydrocarbon liquid and vapor phase is estimated to be known to within ± 2ppm.

Results are satisfactory for the other multicomponent systems and when significant CO2 content is included, we recommend explicitly accounting for solvation between CO2-water as this improves the results in almost all cases.

Finally, for the two NG systems shown in Figure 13.24, very good predictive results are obtained using a constant BIP for MeOH-H2O fitted to SLE data. This is the right choice, which was to be expected, especially for the system where the concentration of MeOH is 60% wt. The results without solvation between CO2 and H2O are very satisfactory (as CO2 content is rather low, around 2–3%).

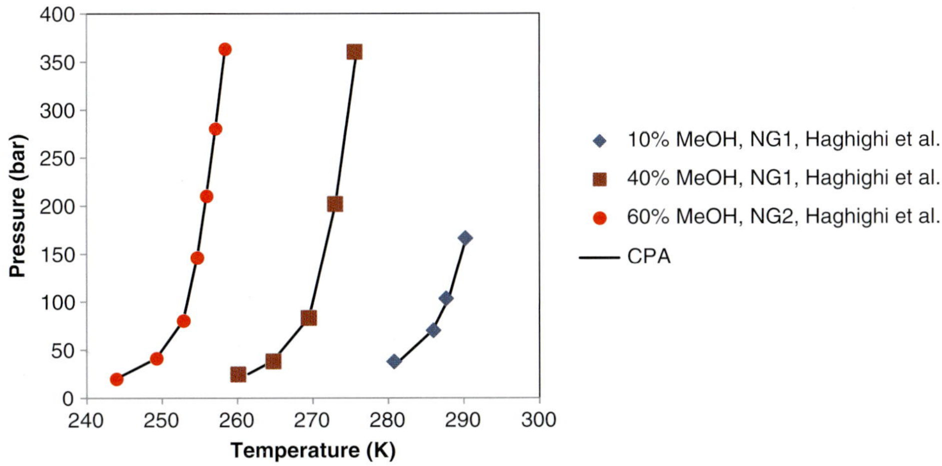

Figure 13.24 Hydrate dissociation conditions for two natural gases in the presence of methanol aqueous solutions. Points are experimental data (Haghighi et al., FPE 278 (2009), 109) and lines are CPA correlations.

13.4.2 Systems of NG Components with Water and Glycols

Vapor water content data were located in the open literature for the following systems of NG components with water and glycols:

- Methane–MEG–water
- Methane–DEG–water
- Propane–MEG–water (only hydrate formation data)
- Ethane–MEG–water (only hydrate formation data)
- CO2–MEG–water
- CO2–DEG–water
- CO2–TEG–water
- Methane–ethane–MEG–water (only hydrate dissociation data)
- Methane–ethane–propane–MEG–water (only hydrate dissociation data)
- Methane–propane–MEG–water
- Methane–propane–TEG–water (only hydrate formation data)
- Methane–*n*-octane–MEG–water
- Methane–ethane–*n*-butane–MEG–water
- Methane–ethane–propane–*i*-butane–*n*-butane–MEG–water
- Methane–ethane–propane–*n*-butane–*n*-pentane–nitrogen–MEG–water
- Methane–ethane–propane–*n*-butane–*n*-pentane–nitrogen–carbon dioxide–MEG–water (hydrate dissociation conditions)
- Methane–ethane–propane–*i*-butane–*n*-butane–*i*-pentane–*n*-pentane–hexane–heptane–nitrogen–carbon dioxide–MEG–water (hydrate dissociation conditions)

Concerning vapor and liquid compositons' prediction, satisfactory accuracy could be obtained with CPA in most cases using a T-dependent BIP for the water–MEG binary system (~10–20%AAD in water content for most systems and acceptable deviations for the other components' concentrations; it should be reminded that 1K deviation corresponds to around 5–8% change in water content above 0°C and to about 10–11% below [9]). However, the water vapor content calculation accuracy itself is almost insensitive to the use of T-dependent vs. the use of T-independent BIPs (a difference of ~1% seen when using the one or the other). An exception was the C1-nC8-MEG-H2O water system for which VLLE data exist. The available water content data for this system do not follow the linear trends observed for many data versus the inverse temperature and/or pressure [40]. The hydrocarbon–water partition coefficient of MEG also does not increase with lowering the pressure as expected so there are strong indications that the data should be seen as "questionable."

In the case of hydrate phase boundary calculations, results produced by different thermodynamic models tend to be more similar to each other at low inhibitor concentrations, while the scatter is more pronounced at high concentrations. In mixtures without inhibitors, the scatter is greater at high pressures and low temperatures (HPLT) where hydrates usually form.

Most results with MEG, corresponding to 14 out of 18 systems with glycols (and all multicomponent ones), have been recently presented and discussed by Tzirakis et al. [41] and will not be further presented. Only a short discussion is offered here. Overall the performance of CPA is excellent, both for water content and hydrate formation/dissociation curves. Equally good performance is obtained for the systems containing and those non-containing acid gases like CO2, and it is not observed any particular change for the many multicomponent MEG systems (with mixtures containing diverse hydrocarbons and acid gases). Solvation between CO2 and water was again considered to be important, and accounting for it improves the results. As mentioned above, the C1-C8-MEG-water data are possibly not reliable. In some systems we see some increasing deviations for the hydrate systems at high concentration values of MEG used as inhibitor (esp. above 40% or 50%). As mentioned, for further discussion see Ref. [41].

Some representative results for DEG and TEG-containing systems are shown in Figures 13.25–13.32. Please notice that all these systems include either DEG or TEG (in mixtures with water, hydrocarbons, and sometimes acid gases) and no mixed glycol systems data have been found.

In Figure 13.25 we see comparisons of CPA and SAFT-VR, another association-based model, with calculations from Dufal et al. [42], for methane hydrate equilibria in the presence of TEG aqueous solutions. It seems that there is a small disagreement between the experimental data available for 20% wt TEG by Ross and Toczylkin and Mohammadi and Richon (refs. in the figure caption). CPA is more in favor of the data of Ross and Toczylkin, while SAFT-VR curve lies closer to the data of Mohammadi and Richon.

The agreement for the vapor water content for C1-DEG-water shown in Figure 13.26 for two temperatures is very good with an average deviation 18% over the whole pressure range. The performance does not seem to depend much on the T, P, or DEG concentration. On the other hand, concerning the prediction of hydrate dissociation conditions of methane in the presence of DEG, these are not predicted as accurately as the corresponding MEG-methane-water conditions, especially at high DEG concentrations. The same is observed however when using the CPA version of the MULTIFLASH commercial software. Some of the literature experimental data do not agree well with each other for this system.

One data source provides equilibrium conditions of ethane hydrates in the presence of DEG (Mohmoodaghdam and Bishnoi, J Chem Eng Data (2002), 47(2), 278) and in the presence of TEG (Ross and Toczylkin, J Chem Eng Data (1992), 37, 488). CPA results are presented in Figure 13.27. The average absolute error in equilibrium temperature is found to be 0.35K in case of DEG and 0.65 in case of TEG. These results are as accurate as in the case of ethane hydrates in aqueous solution of MEG.

Figures 13.28–13.31 show water content and hydrate results for DEG-CO2-water and TEG-CO2-water.

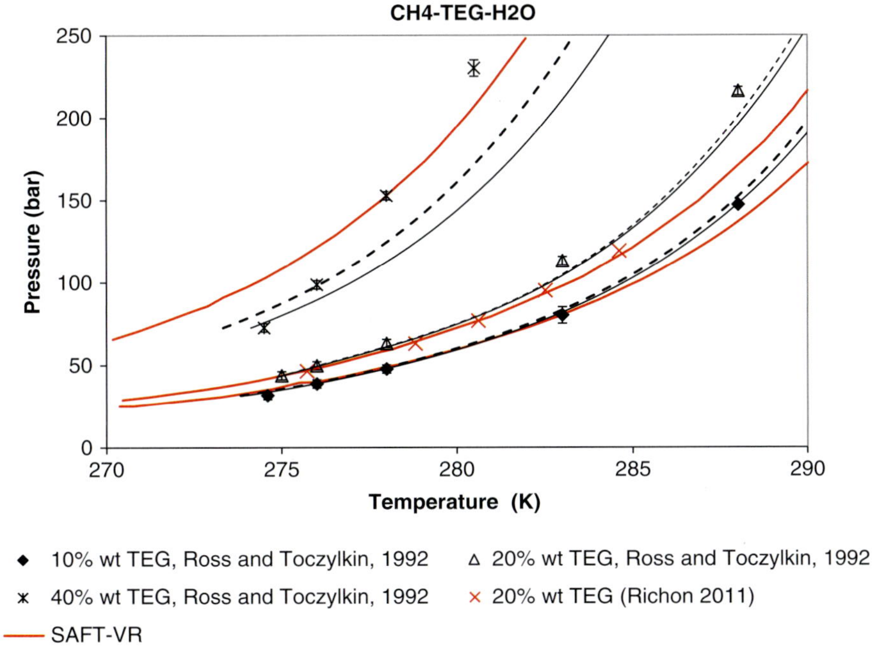

Figure 13.25 Hydrate dissociation curves for methane in the presence of tri-ethylene glycol. Symbols represent experimental data (Ross & Toczylkin, J. Chem. Eng. Data 37, 1992, 488; Mohammadi and Richon, J. Chem. Eng. Data 2011, 56, 4544); curves represent predictions using **SAFT-VR** method (red lines) and CPA with a T-dependent (black full lines) and a T-independent (black dotted lines) BIP between ethylene glycol and water.

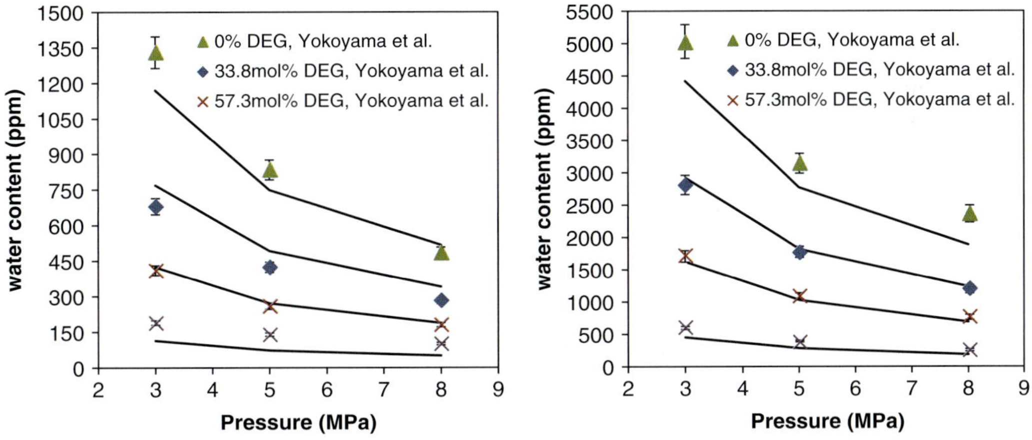

Figure 13.26 Experimental data and model correlations for the vapour water content of the system C1-DEG-H2O at 298.15K (left) and 323.15K (right).

For DEG-CO2-water, the improvement of the results for the water content calculations is obvious when solvation is taken into account, 9% vs. 22% without solvation (the opposite is the case for CO_2 solubility prediction; 26% with solvation vs. 18% without solvation. For the specific system also hydrate equilibrium data exist (Maekawa, J Chem Eng Data 2010, 55, 1280) and have been tested using CPA (no solvation taken into account). There is good agreement, see Figure 13.29, except at the highest glycol concentration.

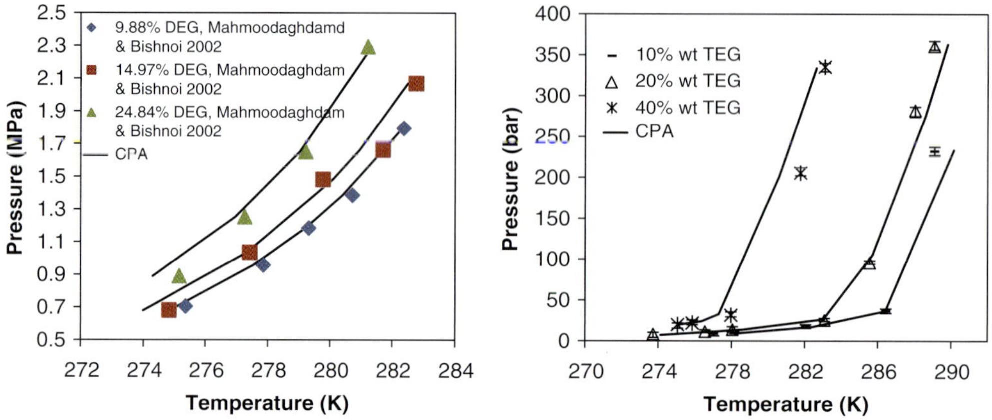

Figure 13.27 Equilibrium conditions of ethane hydrates in aqueous solutions of DEG (left) and TEG (right) (experimental data are points and CPA results are lines).

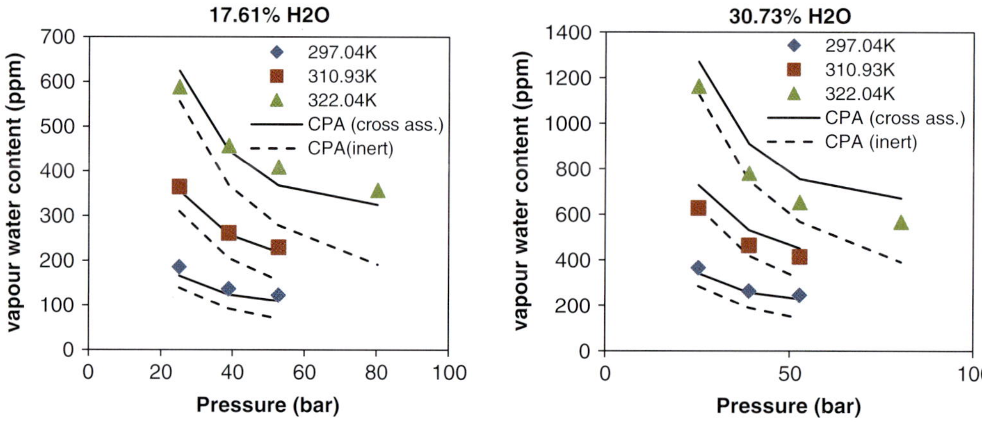

Figure 13.28 Water content in the vapour phase of system DEG-CO2-H2O at 3 different temperatures (297.04K, 310.93K, 322.04K) and 2 different water concentrations (17.61 and 30.73 mole % H2O in solution). Points are experimental data from Takahasi et al., J. Chem. Eng. Data, 1984, 29(1), 23, full lines CPA correlations when solvation taken into account and dotted lines CPA correlations without solvation.

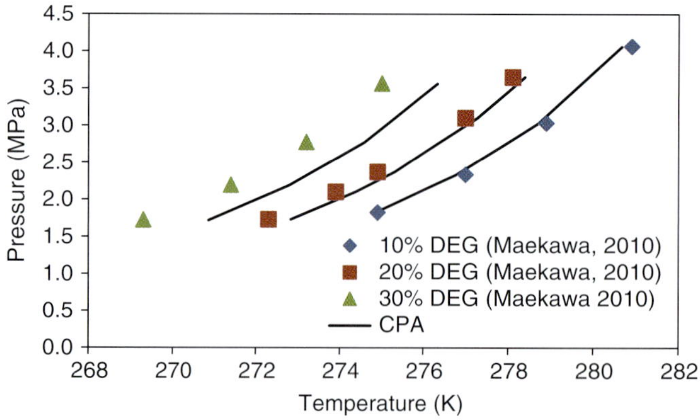

Figure 13.29 Incipient hydrate forming conditions for carbon dioxide in the presence of DEG. Points are experimental data and lines are CPA results.

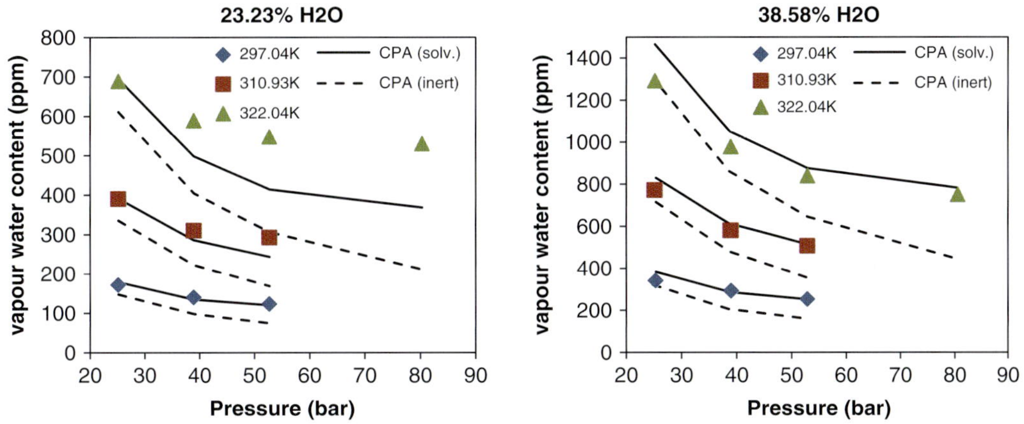

Figure 13.30 Water content in the vapour phase of system **TEG-CO2-H2O** at 3 different temperatures (297.04K, 310.93K, 322.04K) and 2 different water concentrations (23.23 and 38.58 mole % H2O in solution). Points are experimental data from Takahasi et al., J. Chem. Eng. Data, 1984, 29(1), 23, full lines CPA correlations when solvation taken into account and dotted lines CPA correlations without solvation.

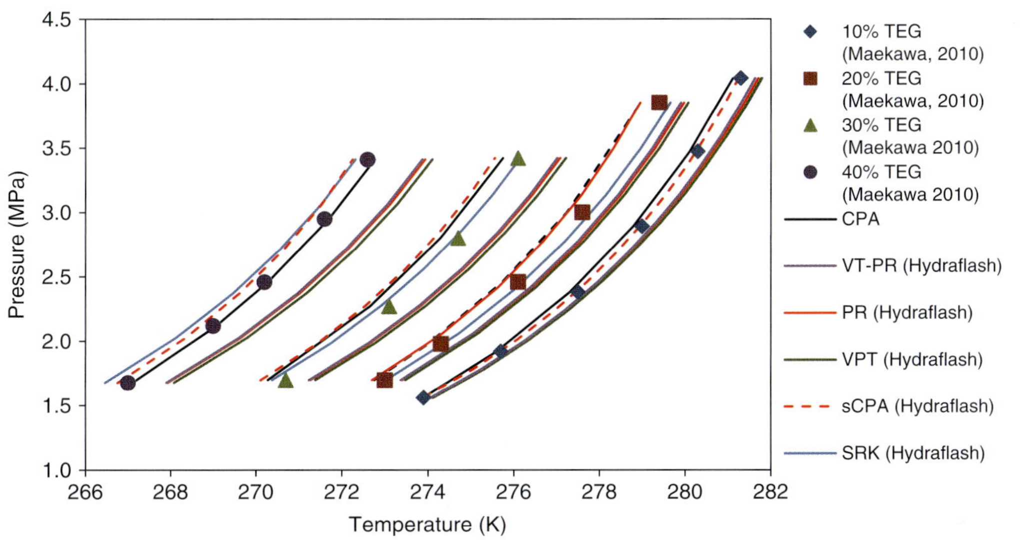

Figure 13.31 Incipient hydrate forming conditions for carbon dioxide in the presence of TEG. Points are experimental data (Maekawa, J. Chem. Eng. Data 2010, 55, 1280) and lines are modeling results computed with CPA and models included in HYDRAFLASH commercial software (PR and a volume translated PR, SRK, VPT and simiplied CPA (sCPA)).

For TEG-CO2-water, the data from the two sources yield different conclusions on whether solvation is needed or not. Still, for water content, for both series of data it is recommended to use solvation, while mixed results are obtained for the CO2 solubility. These findings are more or less in line with the carbonic acid formation theory at low pressure of Rushing et al. [36] according to which at low pressure carbonic acid is formed and the vapour water content does not increase because of the CO2 presence. Another relevant observation is that the prediction of CO2 solubility deteriorates to a bigger degree compared to the system CO2-DEG-H2O.

For the specific system also hydrate equilibrium data have been measured by Maekawa (J Chem Eng Data 2010, 55, 1280). Calculations have been performed using also the

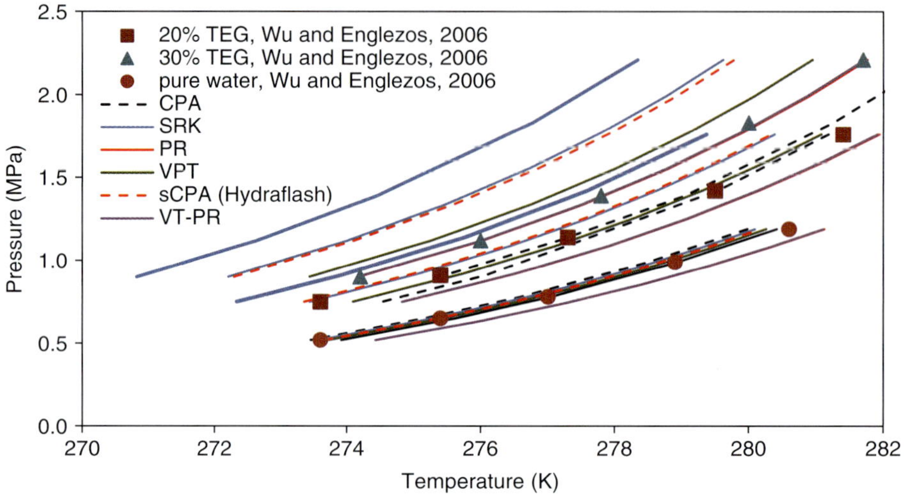

Figure 13.32 Hydrate formation predictions in a C1-C3 gas system (90% C1-10% C3) inhibited by TEG (also data and predictions without inhibitors are shown). Points are experimental data from Wu and Englezos, J. Chem. Eng. Data 2006, 51, 1811 and lines are different models' correlations.

HYDRAFLASH software. All results can be seen in Figure 13.31. An important conclusion here is that different thermodynamic models' results scatter more at higher inhibitor concentrations, so the existence of more experimental data covering high concentrations would significantly help the validation of the different models.

Finally, it should be mentioned that hydrate formation data exist for the C1-C3-TEG-H2O system (gas composition, 90% C1-10% C3) (Wu and Englezos, J Chem Eng Data 2006, 51, 181). Results with CPA and models included in two commercial software packages, HYDRAFLASH and MULTIFLASH can be seen in Figure 13.32. Once again, it is seen that results produced by different models deviate more at higher inhibitor concentrations. Another interesting observation is that the same model, CPA, seems to perform very differently, depending on the software package in which it is included! This is in line with the conclusion of Herri et al. (FPE 301(2011), 171) according to which "it is difficult to compare the models dealing with hydrate equilibria because too many points need to be clarified: what are the data on which the models were regressed? What is the method retained in the integration of the cell potential? Do the models have the same reference parameters?" and so on [36a]. Therefore, it would be very useful for model comparison purposes if commercial software packages were accompanied by a list of the modifications made to the "original" model versions.

13.5 Conclusions and Future Perspectives

The CPA model coupled with the vdW-P theory was initially applied to various types of phase equilibria (VHE, LHE, HIV, HLV) of binary systems containing hydrate-forming NG components and water in order to evaluate the accuracy of the solid parameters previously suggested and to improve them if possible. Solid parameters for hydrate former CO_2 are also presented.

The best parameters determined were then used for NG systems calculations and results were compared against the GERG-water model, when the latter was within its applicability range (calculations outside the applicability range have also been performed in some cases using the GasCalc commercial software [32] and our "in-house" model).

The effect of the BIP values used in the framework of CPA has been studied in order to evaluate to what extent the use of different BIPs influences the accuracy of dew point / water content calculations. As a general conclusion, one may say that T-dependent BIPs work better at low temperature regimes where hydrates also form, while for VLE / high temperature calculations satisfactory accuracy can be obtained with T-independent BIPs.

Concerning the comparison between different models, it was found that in most cases CPA performed better than GERG-water. CPA performed worse than GERG-water in the case of a CO_2–rich natural gas mixture, which underlines the need of taking the solvation effect between CO_2 and H_2O explicitly into account. Among other models used in a few cases, the VPT EoS combined with non-density-dependent mixing rules showed a satisfactory performance, but one has to remember that it uses three BIP values per binary system. The use of different CPA versions or CPA coupled with the Kihara potential instead of the PP approach does not seem to make a very big difference to the model accuracy. In general results produced by different thermodynamic models tend to scatter more at high inhibitor concentrations and/or low temperatures and/or high pressures.

The water content calculations of NG systems with chemicals were found to be almost insensitive to the use of T-dependent BIPs vs. the use of T-independent BIPs, although a T-dependent BIP for H_2O-MEG leads to overall more accurate phase composition results when it comes to NG systems inhibited by MEG, and a T-independent BIP for H_2O–MeOH leads to better results when it comes to NG systems inhibited by methanol.

Concerning NG with CO_2, the solvation effect in CPA has proven to improve significantly the accuracy of our calculations. At low P (<20bar) - low T (<Tc = 304.21K) and very high P (>200bar)-very high T (>373.15K) conditions, the solvation effect between CO2–H2O does not play a significant role when calculating vapor H2O content. In between, solvation should be taken into account according to our calculations.

When solvation is taken into account, the B1 approach suggested by Tsivintzelis et al. [43], where CO2 is modeled as a non–self-associating compound with two proton acceptor sites provides results of acceptable accuracy, besides being physically realistic.

In the T-range between 0°C and –20°C it is difficult to predict whether the moisture will condense as dew or precipitate as ice, so it is always advisable to use CPA and perform both VLE and VIE to define the DP T.

Concerning the GERG-water calculation method, an obvious necessary future step is the extension of its applicability range, both in terms of temperature and pressure conditions, but also in terms of components (as, e.g., alcohols, glycols, and odorants). Only then would a fair and complete comparison to advanced EoS like CPA be possible.

In the future, it would be also of interest to study closer the effect of BIPs of NG subsystems with nitrogen, for which a gap in CPA BIPs was found. The phase behavior prediction of systems containing alkanes and nitrogen is a challenging task. Except for the binary system N2+C1 which exhibits a type I phase behavior in the classification scheme of van Konynenburg and Scott, all the other alkanes mixed with N2 exhibit a type III phase behavior. It is worth mentioning that no CPA BIPs for N2-alkanes or N2–CO2 are found in the literature, although Li and Yan analyzed the impacts of binary interaction parameter for CO2-mixtures and found it to have a clear effect on the calculating accuracy of an EoS in the property calculations of CO2 mixtures (Applied Energy 86(2009), 826). According to the authors, one problem is that the determination of BIPs requires a large amount of experimental data, which are not always available. An inappropriate BIP may cause poor calculating accuracy of an Equation of State. As a conclusion it may be necessary to carry out more experiments to close the experimental gaps in order to improve the CPA EoS accuracy. One should add the fact that many of the available experimental data are quite old (e.g., for CH4-N2 there are two references dating from 1953, one from 1974, one from 1975, and one from 1993, for C3-N2 the references are from 1977, 1966, 1974 and 1999) and for nC4-N2 from 1961 to 1990, so maybe new experimental data could help tuning the current thermodynamic models.

Finally, some room for improvement is believed to exist within the framework of the vdW-P theory used to describe hydrate phases. Besides the nonvalidity of the assumptions of the theory for all cases (e.g., the assumption of single occupancy of hydrate cavities does not hold for nitrogen, the assumption of nondistortion of the cavities by the encaged molecules does not hold for large guest molecules like carbon dioxide, the assumption of lack of interactions between encaged molecules has been questioned based on Monte-Carlo simulation results by Kvamme and Lund, and of course the spherical symmetry assumption only works well for hydrates formed by small, monatomic gases like argon), the "traditional" choice of the Kihara potential could also be re-considered, especially given the recent results of Paricaud [44] or Anderson et al. [45] who solved for the potential directly for hydrates for which the Langmuir constants are computed, either from experimental data or from ab initio data.

Nomenclature

List of abbreviations

%AAD:	average absolute deviation $= \frac{1}{NP}\sum_{i=1}^{NP}\left\lvert\frac{x_i^{\exp}-x_i^{cal}}{x_i^{\exp}}\right\rvert \times 100$
BIP:	binary interaction parameter
CO2:	carbon dioxide
COS:	carbonyl sulfide
CPA:	Cubic-Plus-Association equation of state
CR-1:	combining rule 1
DEG:	di-ethylene glycol
DMS:	dimethyl sulfide
ECR:	Elliott combining rule
EoS:	equation of state
GERG:	the European Gas Research Group
GERG-water:	the calculation method based on Peng-Robinson equation of state developed by GERG
H2S:	hydrogen sulfide
HLV:	hydrate-liquid-vapor equilibrium
HTHP:	high temperature-high pressure
HYDRA:	abbreviation for HYDRAFLASH commercial software (HydraFLASH Model V2.2. Hydrafact Ltd; Edinburgh, UK, 2012. www.hydrafact.com).
LHE:	liquid-hydrate equilibrium
MC:	Mathias-Copeman alpha function
mCR-1:	a modified CR-1 combining rule
MEG:	(mono)ethylene glycol
MULTI:	abbreviation for MULTIFLASH commercial software (MULTIFLASH Model V.4.1, Infochem Computer Services Ltd; London, UK, 2012, www.infochemuk.com).
MeOH:	methanol
NG:	natural gas
PP:	simplified Parish and Prausnitz approach
PR:	Peng-Robinson equation of state

SAFT-VR: the statistical associating fluid theory for fluids interacting through potentials of variable range
SLE: solid-liquid equilibrium
SRK: Soave-Redlich-Kwong equation of state
SW: square-well potential
TEG: tri-ethylene glycol
THT: tetrahydrothiophene
vdW-P: van der Waals-Platteew theory
VHE: vapor-hydrate equilibrium
VIE: vapor-ice equilibrium
VLE: vapor-liquid equilibrium
VLLE: vapor-liquid-liquid equilibrium
VPT: the Valderrama modification of the Patel and Teja equation of state

List of symbols

A_1, A_2, A_3: parameters in the energy term of the GERG-water calculation method
A_{mi}, B_{mi}: fitted parameters in the PP approach
α_o: parameter in the energy term of CPA
b: co-volume parameter
c_1: parameter in the energy term (α)
$C_{m,i}$: Langmuir constant for component i in cavity m
f: fugacity
g: radial distribution function
k_{ij}: binary interaction parameter
P: pressure
P_c: critical pressure
P_ref: reference pressure
R: gas constant
T: temperature
T_c: critical temperature
T_m: melting temperature of water
T_r: reduced temperature (T/T_c)
V_m: molar volume
V_W^{ICE}: molar volume of ice
X_{Ai}: the fraction of A-sites of molecule i that are not bonded with other active sites
x(i): hydrocarbon phase mole fraction of component i
xaq(i): aqueous phase mole fraction of component i
y(i): vapour phase mole fraction of component i
ΔC_p^{fus}: difference of heat capacity between solid and liquid
ΔH^{fus}: heat of fusion

Greek letters

β^{AiBj}: association volume
Δ^{AiBj}: association strength between site A on molecule i and site B on molecule j
ε^{AiBj}: association volume
ε_{mi}: the depth of the SW potential of guest molecule i inside m type cage

Θ_{mi}: the occupancy of cavity m by a component i
v_i: is the number of type i cavities per water molecule
μ_W^H: chemical potential of water in the hydrate phase
ρ: density
ω: acentric factor

Subscript and superscript

i : component i
j : component j
cal.: calculated
EH: empty hydrate
exp.: experimental
H: hydrate
w: water

Acknowledgment

The project was supported by GERG (the European Gas Research Group) in co-operation with Statoil, GASSCO, Snam Rete Gas, Open Grid Europe and DGC.

References

1 Hsieh M-K, Ting W-Y, Chen Y-P, Chen P-C, Lin S-T, Chen L J. Explicit pressure dependence of the Langmuir adsorption constant in the van der Waals-Platteeuw model for the equilibrium conditions of clathrate hydrates. Fluid Phase Equilibr. 2012, 325, 80–89.
2 Sloan Jr, ED. Hydrate engineering monograph. Vol. 21; Henry L. Doherty Series SPE: Richardson, Texas, 2000.
3 Løkken TV, Bersås, Christensen KO, Nygaard CF, Solbraa E. Water content of high pressure natural gas: data, prediction and experience from field. IGRC 2008 report. StatoilHydro.
4 Bukacek RF. Equilibrium moisture content of natural gases, Instit Gas Technol, Res Bulletin 8, 1955.
5 McKetta JJ, Wehe AH. Pet Refiner, 37 (8) (1958): 15.
6 ISO 18453, Natural Gas-Correlation between water content and water dew point (2002) and GERG-Water Correlation. GERG Technical Monograph TM 14 (2000).
7 Kontogeorgis GM, Voutsas E, Yakoumis I, Tassios DP. An equation of state for associating fluids. Ind Eng Chem Res, 1996, 35 (11), 4310–4318.
8 van der Waals JH, Platteeuw JC. Clathrate solutions. Adv Chem Phys, 2, 1959, 1–57.
9 Folas G K, Froyna EW, Lovland J, Kontogeorgis GM, Solbraa E. Data and prediction of water content of high pressure nitrogen, methane and natural gas. Fluid Phase Equil. 2007, 252, 162–174.
10 Kontogeorgis GM, Michelsen ML, Folas GK, Derawi S, von Solms N, Stenby EH. Ten years with the CPA (Cubic-Plus-Association) equation of state. Part 1. Pure compounds and self-associating systems. Ind Eng Chem Res. 2006, 45, 4855–4668.
11 Avlonitis D. The determination of Kihara potential parameters from gas hydrate data. Chem Eng Sc 1994, 49, 1161–1173.

12 Prausnitz JM, Lichtenthaler RN, De Azevedo EG. Molecular thermodynamics and fluid phase equilibria. 3rd ed. Englewood Cliffs, NJ: Prentice-Hall; 1999.

13 Parrish WR, Prausnitz JM. Dissociation pressures of gas hydrates formed by gas mixtures. Ind Eng Chem Process Des Develop 11 (1) (1972): 26–35.

14 Sloan Jr, ED, Koh CA. Clathrate hydrates of natural gases. 3rd ed. CRC Press/Taylor & Francis Group, 2008.

15 Bazant MZ, Trout BL. A method to extract potentials from the temperature dependence of Langmuir constants for clathrate-hydrates. Physica A, 300, (2001): 139–173.

16 Papadimitriou NI, Tsimpanogiannis IN, Stubos AK. Monte Carlo simulations of methane hydrates. Proc. of 7th Int. Conf. on Gas Hydrate, Edinburgh, Scotland, 2011.

17 Kontogeorgis GM, Folas K. Thermodynamic models for industrial applications: from classical and advanced mixing rules to association theories. Chichester, UK: Wiley; 2010.

18 Mathias PM, Copeman TW. Extension of the Peng-Robinson equation of state to complex mixtures: evaluation of the various forms of the local composition concept. Fluid Phase Equilibr, 13, (1983): 91–108.

19 Twu CH, Bluck D, Cunningham JR, Coon JE. A cubic equation of state with a new alpha function and a new mixing rule. Fluid Phase Equilibr, 69 (1991): 33–50.

20 Nasrifar Kh, Bolland O, Moshfeghian M. Predicting natural gas dew points from 15 equations of state. Energ Fuels, 2005, 19(2), 561–572.

21 Chapoy A, Tohidi B. Hydrates in high MEG concentration systems. Proceedings of the 3rd International Gas Processing Symposium, March 5–7 2012, Qatar.

22 Karakatsani EK, Kontogeorgis GM. Thermodynamic modeling of natural gas systems containing water. Ind Eng Chem Res, 52 (2013): 3499–3513.

23 Do Seo M, Kang JW, Lee CS. Water solubility measurements of the CO2-rich liquid phase in equilibrium with gas hydrates using and indirect method. J Chem Eng Data, 56 (2011): 2626–2629.

24 Haghighi H, Chapoy A, Burgass R, Tohidi B. On the phase behavior of the carbon dioxide-water systems at low temperatures. Proceedings of the 7th International Conference on Gas Hydrate 2011.

25 Eslamimanesh A, Mohammadi AH, Richon D. Thermodynamic consistency test for experimental solubility data in carbon dioxide/methane + water system inside and outside gas hydrate formation region. J Chem Eng Data, 56 (2011): 1573–1586.

26 Voutsas EC, Boulougouris GC, Economou IG, Tassios DP. Water/hydrocarbon phase equilibria using the thermodynamic perturbation theory. Ind Eng Chem Res, 39 (2000): 797–804.

27 Yan W, Kontogeorgis GM, Stenby EH. Application of the CPA equation of state to reservoir fluids in presence of water and polar chemicals. Fluid Phase Equil, 276 (2009): 75–85.

28 Tabasinejad F, Moore RG, Mehta, Mehta SA, Van Fraassen KC, Barzin Y. Water solubility in supercritical methane, nitrogen, and carbon dioxide: measurement and modeling from 422 to 483K and pressure from 3.6 to 134 MPa. Ind Eng Chem Res, 50 (2011): 4029–4041.

29 Tsivintzelis I, Maribo-Mogensen B. The cubic-plus-association EoS: parameters for pure compounds and interaction parameters. Technical Report, CERE, DTU, September 2012 and Water-light hydrocarbons (internal CHIGP report).

30 HydraFLASH Model V2.2. Hydrafact Ltd; Edinburgh, UK, 2012. Available from: www.hydrafact.com.

31 MULTIFLASH Model V.4.1, Infochem Computer Services Ltd; London, UK, 2012, www.infochemuk.com.

32 GasCalc Model V.2.3, E.ON Ruhrgas AG; Essen, Germany, 2012. Available from: www.eon-ruhrgas.com.

33 Tsivintzelis I, Kontogeorgis GM, Michelsen ML, Stenby EH. Modeling phase equilibria for acid gas mixtures using the CPA equation of state. II. Binary mixtures with CO2. Fluid Phase Equilibria, 306 (2011): 38–56.

34 Jorgensen WL, Matura JD. 1985. Temperature and size dependence for Monte Carlo simulations of TIP4P water. Mol. Phys. 1985, 56(6), 1381–1382.

35 Løkken TV. 2012. Comparison of hygrometers for monitoring of water vapour in natural gas. J Nat Gas Sci and Eng. 6 (24), 24–36.

36 Rushing JA, Newsham, KE, Van Fraassen KC, Mehta, SA, Moore GR. The catalytic effects of nonhydrocarbon contaminants on equilibrium water vapor content for a dry gas at HP/HT reservoir conditions. SPE 114517 (2008). CIPC/SPE Gas Technology Symposium 2008 Joint Conference, 16–19 June, Calgary, Alberta.

36a Herri J-M, Bouchemoua A, Kwaterski M, et al, Gas hydrate equilibria for CO2–N2 and CO2–CH4 gas mixtures—Experimental studies and thermodynamic modelling. Fluid Phase Equilibria, 301 (2): p. 171–190.

37 Song KY, Yarrison M, Kobayashi R, Chapman WG. Low temperature V-L-E data, CO_2 and light hydrocarbon systems. GPA RR-187. GPA Midstream Association; 2005.

38 Clark MA, Svercek WY, Monnery WD, et al. Designing and optimized injection strategy for acid fgas disposal without dehydration. 77th GPA Annual Convention, 1998.

39 Bang V, Pope GA, Sharma, MM. Phase behavior study of hydrocarbon/water/methanol mixtures at reservoir conditions. SPE 102100, 2006.

40 Mohammadi AH, Richon D. Use of artificial neural networks for estimating water content of natural gases. Ind Eng Chem Res, 46 (4) (2007): 1431–1438.

41 Tzirakis F, Karakatsani EK, Kontogeorgis GM. Evaluation of the cubic-plus-association equation of state for ternary, quaternary, and multicomponent systems in the presence of monoethylene glycol. Ind Eng Chem Res, 55 (43) (2016): 11371–11382.

42 Dufal S, Galindo A, Jackson G, Haslam AJ. Modelling the effect of methanol, glycol inhibitors and electrolytes on the equilibrium stability of hydrates with the SAFT-VR approach. Mol Phys 110 (11–12), 1223–1240.

43 Tsivintzelis I, Kontogeorgis G, Michelsen ML, Stenby EH. Modeling phase equilibria for acid gas mixtures using the CPA equation of state. I. Mixtures with H2S. AI Ch E J, 56 (11) (2010): 2965.

44 Paricaud P. Modeling the dissociation conditions of salt hydrates and gas semiclathrate hydrates: application to lithium bromide, hydrogen iodide, and tetra-n-butylammonium bromide + carbon dioxide systems. J Phys Chem B, 115 (2) (2011): 288–299.

45 Anderson BJ, Bazant MZ, Tester JW, Trout BL. Application of the cell potential method to predict phase equilibria for multicomponent gas hydrate systems. J Phys Chem B, 109 (16) (2005): 8153–8163.

14

Light Alkane Aromatization: Efficient use of Natural Gas

Swarom R. Kanitkar and James J. Spivey

Cain Department of Chemical Engineering, Louisiana State University, Baton Rouge, USA

CHAPTER MENU

14.1 Introduction

Energy and environment have been two major focuses of research in the twenty-first century that are closely related. As population grows and cities expand, the demand for energy inevitably increases, as does the incentive for clean energy. Despite significant research efforts for biofuel development, and for alternative energy sources (wind, solar, hydrothermal, etc.), fossil fuels still dominate world energy demands. Coal and oil are widely used, but natural gas is generally considered to be cleaner and more efficient in many applications. The shale gas revolution has presented a unique opportunity to address these challenges by expanding the use of natural gas. The immense, previously unattainable, reserves of natural gas have already had major impacts on the energy and chemical industries and have helped countries like the United States, reduce CO_2 emissions [1].

14.1.1 Shale Gas Revolution

Advances in horizontal drilling and hydrofracking has led to an increase in shale gas production. In the United States alone, it rose from 1% to 20% from 2000 to 2010 causing a decrease in the prices of natural gas. It is this dramatic effect that is often termed the "Shale gas revolution" [1d].

This has significantly lowered natural gas prices, and projections show that they will stay at current low levels through 2023 [2]. Recently, a 5% drop in prices of natural gas was reported just for the change from 2015 to 2016, whereas European and Asian markets saw a 20–25% drop in natural gas prices to compensate for the supply of natural gas [3]. The shale gas revolution has also provided an alternative to coal-fired power plants, thereby helping reduce CO_2 emissions [4]. This makes natural gas an economic alternative feedstock over conventional fuels. This has resulted in providing a significant incentive for new research in the area of natural gas conversion: in particular, methane and light alkanes.

Natural Gas Processing from Midstream to Downstream, First Edition.
Edited by Nimir O. Elbashir, Mahmoud M. El-Halwagi, Ioannis G. Economou, and Kenneth R. Hall.

14.1.2 Composition of Natural Gas

Natural gas composition can vary significantly based on location of the reserve and also depending on whether it is associated or non-associated with oil. In either case, primary components of natural gas include C_1, C_2, C_3, CO_2, N_2, H_2S, and trace levels of higher hydrocarbons. Table 14.1 shows composition of natural gas from some of the fields around the world. Regardless of the variety in composition of various wells, CH_4 remains the principal component, ~49–84%.

Methane conversion to higher-value products is one of the grand challenges for the scientific community in the area of catalysis and electrocatalysis. Despite the strong incentives, and as indicated by recent reviews [5], direct non-oxidative conversion of methane is yet to be widely commercialized.

Secondary important components of natural gas are ethane and propane that typically constitute 2–5% (combined) of natural gas. Although in some wells, up to 8–10% of ethane along with 1–3% propane can also be seen. N-butane on the other hand, can be observed at around ~0.1% of natural gas in composition. These light alkanes are relatively easier to activate than methane and are principal feedstocks for the production of olefins and plastics, through dehydrogenation or oxidative coupling processes. Ethane is the principal feedstock for the production of ethylene ($C_{2=}$), which is subsequently converted to variety of valuable chemicals such as polyethylene, ethylene oxide, and ethylene dichloride. The production of ethylene from ethane is carried out commercially via steam cracking, which is a well-established process.

Production of ethane has increased significantly over the past few years as a result of shale gas revolution. This has caused costs of ethane to drop to a low level, and this downward trend in price is expected to continue over the next few years (Figure 14.1). This makes ethane an attractive feedstock for aromatization into benzene, toluene. and higher-value hydrocarbons, along with pure H_2. Therefore, the chemical industry recognizes that an economically desirable goal will be to replace naphtha with light alkanes as feedstock, which would also reduce the environmental footprint of aromatics synthesis.

The aromatization of methane is one key potential process of interest to the chemical industry:

$$6\,CH_4 \rightarrow C_6H_6 + 9\,H_2(\Delta G_{298\,K} = +102.4\,\text{kcal/mol}, \Delta H_{298\,K} = +116.3\,\text{kcal/mol}) \quad (14.1)$$

Current research has demonstrated only net yields of 5–10% of benzene, and rapid deactivation of coke has limited commercialization of aromatization [5]. Hydrogen is a valuable

Table 14.1 Composition of natural gas from various wells around the world [6].

Composition (mol %) ↓	Field Name				
	Xifeng, China	Tengiz, Kazakhstan	Ekofisk, Norway	Salt Creek, USA	Boscan, Venezuela
N_2	2	2	0	1	2
CO_2	2	4	3	1	6
C_1	49	60	79	81	84
C_2	13	10	10	9	4
C_3	19	3	4	4	2
C_{4+}	15	6	3	3	1
H_2S	0	15	1	0	1

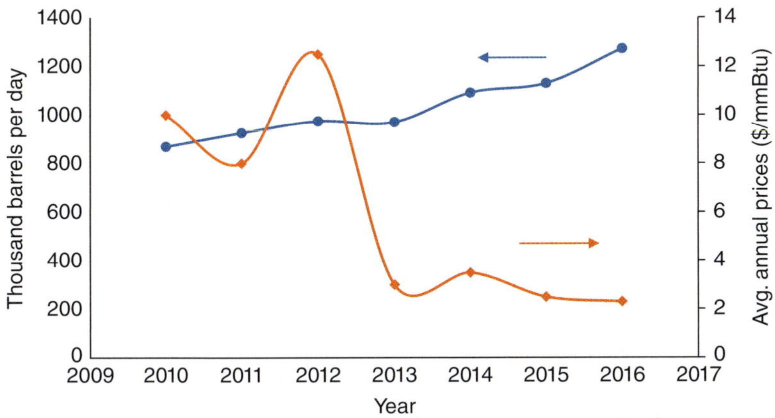

Figure 14.1 Ethane production and prices over last few years [7].

byproduct of the aromatization process, and the appeal of a reaction that produces both benzene and hydrogen suggests that industrial laboratories are developing a viable process.

In addition to aromatization of methane, aromatization of light alkanes is a potential process. In general, acid catalysts have been effective in the aromatization of light alkanes, in particular the pentasil types: ZSM-5. Various types of catalysts based on bi-functionality in which metals like Mo, Ga, Pt, Re, Zn doped on the ZSM-5 are studied. Despite being widely investigated, ethane aromatization is not yet commercialized due to poor catalyst stability, high temperature requirements, and coking of the catalysts. Although several reviews [8] exist on light alkane conversion to aromatics, no recent review show progress in the last decade. The chapter here reviews recent progress in the area of acid catalyzed light alkane aromatization (in particular, ethane). This focuses on the effect of acidity, pore structure, and metal functions on the aromatization of ethane to aromatics.

14.2 Aromatization of Light Alkanes

14.2.1 Thermodynamics and Short History

Alkanes are a class of hydrocarbons having only sigma bonds. Although singly bonded, these are more difficult to activate than the stronger double-bonded olefins, thus requiring more severe conditions, e.g., high temperatures. Alkanes with a high number of carbon atoms are easier to activate than their lower counterparts [9] because the C-C bond is weaker than the C-H bond, and a surrounding carbon influences the C-H bond dissociation energy. Thus, the more the number of "C" atoms in an alkane, the easier it is to break the C-H bond. Often, this breaking of the first C-H bond is the activation step for alkane activation. This exemplifies the reactivity for the first four alkanes (light alkanes):

$$CH_4 < C_2H_6 < C_3H_8 < C_4H_{10}$$

This is clearly reflected by the thermodynamics. Figure 14.2 shows aspects of thermodynamics for light alkane conversion into aromatics. For simplicity, reactions and products shown in eq. (14.2) were considered for the calculations:

$$\text{Alkane } (C_1 - C_4) \rightarrow \text{Benzene, Toluene, Ethylbenzene, Xylene (BTEX)} + H_2 \uparrow \qquad (14.2)$$

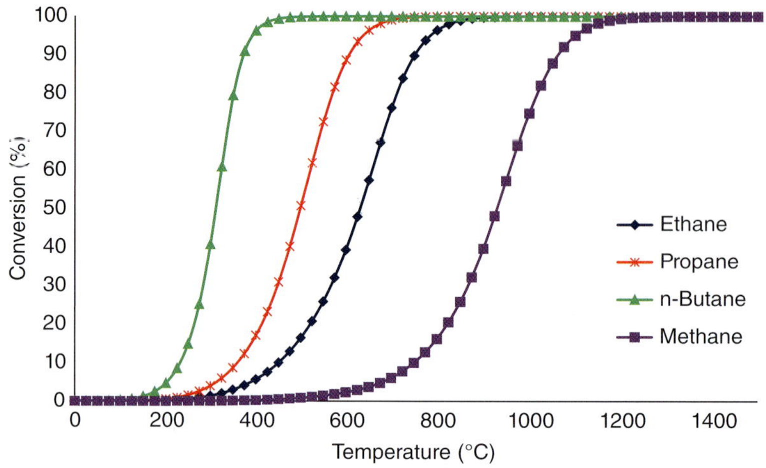

Figure 14.2 Thermodynamics for light alkane aromatization (Plotted using HSC Chemistry 8).

As seen from Figure 14.2, the reactivities for all the four basic alkanes differ quite considerably. Appreciable levels of conversion (20–30%) could thermodynamically be possible only at ~700 °C (for methane), ~550 °C (for ethane), ~450 °C (for propane), and ~250 °C (for n-butane). Because high activation temperatures are required for methane and ethane, efforts are focused on aromatization of C_3 and C_4, as will be discussed in section 14.2.2.

$$21\ C_2H_6 \rightarrow 2C_6H_6 + 2C_7H_8 + 2C_8H_{10} + 39H_2$$
$$(\Delta G_{298K} = +336.6\ \text{kcal/mol}, \Delta H_{298K} = +458.7\ \text{kcal/mol}) \quad (14.3)$$

For the reaction depicted in equation (14.3), a plot of ΔG vs. temperature is shown in Figure 14.3. High levels of conversion for the forward reaction are possible for $\Delta G \leq 0$ after which reaction becomes spontaneous. By observing Figure 14.3, for ethane aromatization from equation (14.3), this is spontaneous only at temperatures ≥ 700 °C.

One way to overcome thermodynamic limitations is to add oxidant along with ethane on the catalyst. This reduces the energy barrier, making it relatively easier to activate ethane. Common oxidants involve CO_2, NO_x, and O_2. However, the presence of an additional reactant changes the reaction pathway considerably, and dehydrogenation reactions (oxidative dehydrogenation) can be favored over aromatization reactions [10].

Aromatization of olefins was first reported around 1946 by Russian workers [8b], and it was not until 1970s when aromatization of light alkanes (C_3-C_5) was first reported by Csicsery [9, 11]. He described aromatization on Pt/Al_2O_3 and transition metal oxides (MoO_3, Cr_2O_3, V_2O_5, WO_3, MnO) doped on Al_2O_3 at 550–600 °C. Primary products observed were xylenes and toluene, but the product distribution was found to be influenced by the strength of the acid sites over the catalyst. Coking was evident, and the rate of coke formation was influenced by temperature, residence time, and catalyst on-stream time. Although transition metal oxides were found to be better additives to the alumina compared to Pt due to high stability during regeneration cycles, the high rate of coking prevented further application of this chemistry.

The first successful aromatization of < C_3 paraffin (ethane) was done by Bragin et al. [12] over Pt/Al_2O_3 in a pulse reactor in 1974. After this, a number of papers have emerged (Table 14.2). It provides a qualitative analysis of the work done by various groups and on various catalysts. Data recorded were obtained from aromatization reactions at different space velocities so the quantitative comparison is not direct.

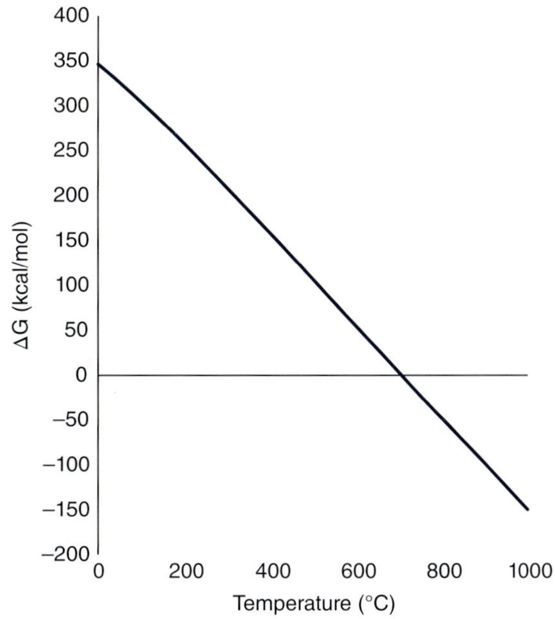

Figure 14.3 Gibbs free energy plot for ethane aromatization (plotted using HSC Chemistry 8).

Table 14.2 Summary of selected ethane conversion attempts.

Catalyst	Temperature (°C)	Ethane Conversion (%)	Aromatics Selectivity (%)	Reference
Pt/H-gallosilicate	500	14.0	41.5	Inui et al. [13]
Ga-pentasil	600	46.0	65.0	Minachev et al. [8c]
H-ZSM-5	600	2.0	–	Ono et al. [8c]
Zn-ZSM-5	600	52.0	31.0	Ono et al. [8c]
Zn-Pd-ZSM5	600	48.4	28.9	Ono et al. [8c]
Re/HZSM5	550	32.0	65.0	Krogh et al. [14]
Ga/HZSM5	600	29.0	17.0	Lapidus et al. [15]
Pt-Ga/HZSM5	600	48.0	30.0	Lapidus et al. [16]
Zn-ZSM-5	650	55.3	27.9	Vosmerikova et al. [17]
Mo-ZSM-5	650	92.8	30.7	Wong et al. [18]

Bifunctional catalysts have gained attention in which the noble metal activates alkane and converts it to olefins, which are then converted into aromatics over the Brønsted acid sites from the support. Activity is due to factors that involve both physical parameters (temperature, pressure, space velocity, or even the pretreatment conditions) and chemical parameters (type and amount of metal exchanged, type of support, concentration of acid sites, Si/Al ratio). Some of the key chemical parameters will be discussed here.

14.2.2 Existing Technologies

Aromatization processes began to appear in commercial practice around the 1970s after the discovery of ZSM-5 zeolite. This zeolite found widespread applications, ranging from FCC additive for propylene production to octane number improvement [19]. Many notable aromatization processes were developed, including those of M-2 forming developed by Mobil, Aroforming by IFP, and Salutec, Z-forming by Mitsubishi and Chiyoda, Alpha forming by Toyo and Sanyo, Cyclar process by BP-UOP, GTA technology by Sinopec Luoyang, Aromax by Chevron Phillips. Most of them have similar feedstocks (light naphthas, lower olefins, LPGs, or light alkanes), similar aromatic yields albeit with different unit operations or reactor setups. Catalysts used are primarily ZSM-5 or modified ZSM-5.

Mobil developed an M-2 forming process in 1986 that involved aromatization of olefins as well as C_5, C_6 alkanes over a HZSM5 catalyst [20]. However the feedstock can be as different as LPG or light olefinic cuts, or paraffinic or unsaturated gasolines [21]. One limitation with the M-2 process is that the side reaction of dehydrogenation shifts the equilibrium giving less aromatics and less hydrogen yield, making it less economical [22].

The Aroforming process is similar to Cyclar, and was developed around 1980s as a joint venture between France's IFP and Salutec from Australia. Similar to M-2 forming, the feedstock for this process also involved LPGs and light naphthas that were converted into BTX and hydrogen [21]. This process uses a shape selective zeolite catalyst doped with a metal oxide but has a product distribution similar to that of the Cyclar Process [23].

Mitsubishi and Chiyoda had developed a "Z forming" process around 1983 that converted light naphtha or LPG into aromatic hydrocarbons over an acidity-adjusted catalyst. The process was used by Nippon Oil till 2008 but was abandoned due to low yield [24]. This process is based on a bifunctional catalyst consisting of a metal (most likely zinc) modified zeolite (ZSM-5). The reforming reactors are of an adiabatic, fixed-bed type. Operating conditions are 500–600 °C at < 100 psia. BTX yields of up to 55–58 wt% are observed [25].

The Chevron Phillips' Aromax process that converts light naphtha into aromatic hydrocarbons was developed around the 1980s. The first commercial demonstration of the Aromax technology was in 1992. Feedstock for this process primarily consists of C_6–C_8 hydrocarbons with high selectivity to benzene, toluene, and hydrogen. The process also involves a highly efficient sulfur control system to eliminate catalyst poisoning. Currently four plants around the world (U.S., Spain, Japan, and Saudi Arabia) make use of this technology [26]. A process flow diagram for the Aromax process is shown in Figure 14.4.

Sinopec Luoyang's GTA technology makes use of Olefinic C_4 streams to convert into aromatics at around 500 °C over a bifunctional metal modified ZSM5 catalyst. This process adopts a

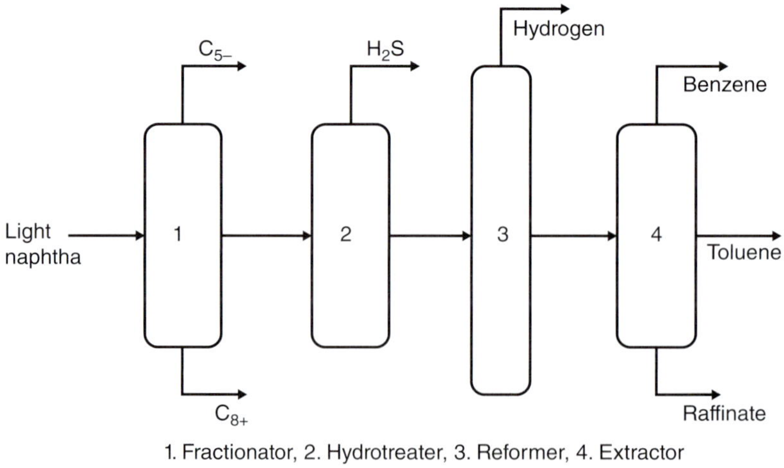

Figure 14.4 Chevron Philips' Aromax Technology Process [26].

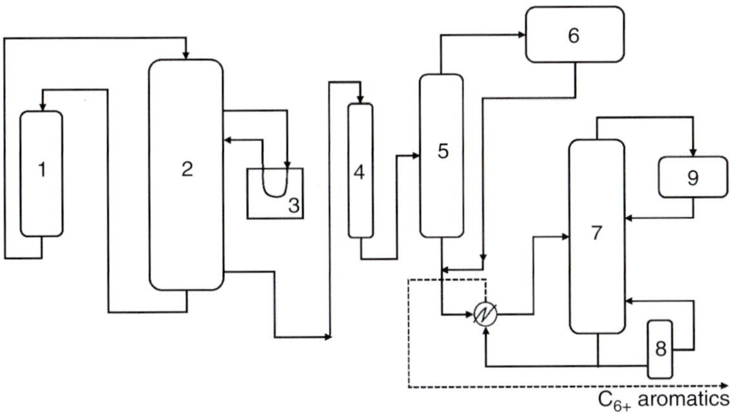

Figure 14.5 Schematic for CYCLAR process [27].

fixed-bed reaction-regeneration system, and it is feasible with short flows. This process is highly selective to aromatics with 55–60 wt% yield, and it does not generate non-aromatic components, eliminating the need for an aromatic extraction unit [25].

BP-UOP's successful "CYCLAR" process, which involves aromatization of propane and butanes, was first demonstrated at BP's refinery at Grange-Mouth, UK, in 1991. This commercial set-up with a capacity of 1000 barrels per day was operated for almost eight years. A second commercial unit was set up in the Middle East around 1999 with a capacity of 45,000 barrels per day and was in operation as of 2013 [27]. The Cyclar process has aromatic (BTX) yields of around 58–60% on a metal-doped (mostly Ga) H-ZSM5 catalyst that is easily coked and requires continuous regeneration. This regeneration is carried out using UOP's trademarked CCR (continuous catalyst regeneration) technology. A schematic for a typical CYCLAR process unit is shown in Figure 14.5.

14.2.3 Role of Metals (Ga, Pt, Mo, Zn, Re)

It is well understood that the MFI zeolite by itself is not active for the aromatization reactions. In order to enhance its activity, doping of metals into the zeolite framework is necessary. These metal dopants lead to distinct changes in the properties of the undoped zeolite and often form a bifunctional catalyst. It is commonly believed that ethane is first activated on the metal species and is then dehydrogenated into ethylene. This ethylene forms olefinic intermediates and undergoes various oligomerization reactions, and cyclization reactions take place on the Brønsted acid site from the zeolite itself. The bifunctional mechanism is shown in Figure 14.6.

Since the first step of ethane activation is crucial, it often determines the rate of formation of olefinic species and subsequently the rate of aromatization. Thus, it becomes important that the metal be well dispersed in the zeolite structure [28]. Different metals have different activation energies for activation of ethane, and introduction of these metal ions in the zeolite structure changes the concentration of both types of acid sites: Lewis and Brønsted.

Various noble and base metals have been tested so far including Cr, Fe, Mn, Pd, Pt, Re, Ru, W, Zn, Mo, Ga. However, the most active and very widely studied metal dopants are Pt, Ga, Zn, and Mo. One relatively new metal dopant is Re. A small metal dopant (as a promoter) is added to an already metal-doped zeolite, which enhances the activity of that bifunctional catalyst. Some examples include addition of Pt into Ga/ZSM-5 [16, 29], addition of W into Mo/ZSM-5 [18] etc.

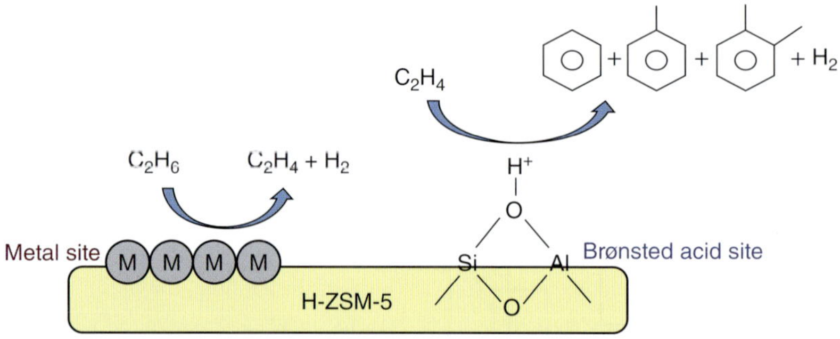

Figure 14.6 Bifunctional mechanism over metal doped HZSM5 catalysts.

14.2.3.1 Mo/ZSM-5

Molybdenum is an active metal dopant for the dehydroaromatization of light alkanes. Mo-based zeolite catalysts are active for methane dehydroaromatization at around 700–800 °C [30]. It is hypothesized that methane is activated into CH_x that subsequently forms a dimer on the Mo_2C species and the dimerized form is oligomerized, then cyclized on the Brønsted acid sites of the zeolite [5a]. H-ZSM-5 and H-MCM-22 have been the two prominent zeolitic/molecular sieve supports for incorporating Mo, demonstrating high activity in the aromatization of methane/ethane.

Molybdenum can typically be doped on the zeolite using incipient wetness impregnation (IWI), often with the ammonium salt of Mo [30b, 31]. Other procedures involved could be solid state ion exchange [32] that involves physical mixing of MoO_3 and zeolite and heating the mixture. However, for aromatization reactions, at least on methane, IWI-prepared Mo catalyst showed greater selectivity to aromatics than the ion exchanged catalysts [32b]. The low selectivity of ion-exchanged catalysts was attributed to the formation of large MoO_x clusters that were unable to diffuse/migrate inside the ZSM5 pores.

There have been few reports of ethane or propane aromatization on Mo-based catalysts. Wong et al. [18] studied Mo/HZSM5 and W/HZSM5 for ethane aromatization that showed very high levels of ethane conversion (93%) and (25%) respectively. However, selectivity for aromatics differed significantly with Mo-doped catalyst, which favored cracking to CH_4 more than aromatization. This was attributed to the strong activity of Mo in both the dehydrogenation reaction (first step in ethane activation) and strong activity in the reaction between H_2 from first step combining with ethane to get cracked into methane. A sequence of reactions [18] could be depicted as shown in eqns. (14.4) and (14.5).

$$C_2H_6 \rightarrow C_2H_4 + H_2 \quad (\Delta G_{298K} = +23.9\ \text{kcal/mol}, \Delta H_{298K} = +32.5\ \text{kcal/mol}) \tag{14.4}$$

$$H_2 + C_2H_6 \rightarrow 2\,CH_4 \quad (\Delta G_{298K} = -16.5\ \text{kcal/mol}, \Delta H_{298K} = -15.6\ \text{kcal/mol}) \tag{14.5}$$

Samanta et al. [29a] reported reduction in number of strong Brønsted acid sites when Mo was doped into H-ZSM-5 despite the increase in the overall acidity that most likely occurs due to the exchange of zeolitic proton with the metal itself. However, generated extra acidity was attributed to the MoO_3 Lewis acid sites. Around 40% conversion of ethane with 35% selectivity to aromatics was reported at 650 °C over these catalysts. Compared to Ga- or GaPt-doped catalysts, Mo-doped catalysts had lower activity. Mo-doped catalysts favored ethylene more than Ga- or GaPt-doped catalysts. This likely would have been due to the generation of extra Lewis acidity, which would have favored dehydrogenation over aromatization. Since dehydrogenation is thought to be the initial activation step, reduction in strong Brønsted acid sites would have caused less aromatization of the olefinic species generated from dehydrogenation.

14.2.3.2 Pt/H-ZSM-5

Platinum can be incorporated onto the ZSM5 support typically by IWI of H_2PtCl_6 (chloroplatinic acid) salt [29a] or by ion exchange [33] of NH_4-ZSM5 using ammonium salts of Pt like $Pt(NH_3)_4Cl_2$. Another method that could be used for depositing Pt into the pores of zeolite is the vapor phase impregnation using platinum acetyl acetonate, $Pt(acac)_2$ as a precursor [34].

Platinum (Pt)-based catalysts have been effective in dehydrogenation of propane [35] and also for dehydrogenolysis activity [36] most likely due to H_2 abstraction. Despite this, Pt by itself shows very low but stable activity (10% conversion) with very high (97%) selectivity to ethylene and almost negligible amounts of aromatics [29a]. Lapidus et al. [29b], however, found around 27% ethane conversion with aromatic selectivity of about 51.5% on Pt-loaded ZSM-5 catalyst. These differences could be due to different Pt loading (0.3 wt%) [29b] in one case whereas (0.5 wt%) [29a] in other. Also, the parent H-ZSM5 zeolite material had different SiO_2/Al_2O_3 ratios, $SiO_2/Al_2O_3 = 30{:}1$ [29b] and $SiO_2/Al_2O_3 = 50{:}1$ [29a], which would have more of an influence than the Pt loading itself.

Despite a lack of strong activity in ethane aromatization by itself, Pt can be a very effective promoter [29a] and can form bimetallic nanoclusters as depicted in Figure 14.7 along with Ga, thereby enhancing the activity of the overall catalyst multiple times with lower deactivation rates.

14.2.3.3 Ga/H-ZSM-5

Gallium can be incorporated into the zeolite support by numerous methods including IWI of $[Ga(NO_3)_3.xH_2O]$ salt [29], ion exchange with Gallium salts [37], vapor phase deposition using volatile GaX_3 (X = halide) precursors [38], physical mixing of Ga_2O_3 with ZSM-5 [39], or directly adding gallium salt during the synthesis of ZSM-5 to form gallosilicate or galloaluminosilicates.

Gallium has been an effective and a well-known dopant in the zeolites for the aromatization of propane [8a]. Gallium is favored over platinum because of greater aromatization activity. Ono et al. [8c, 40] reported ethane aromatization over Ga/ZSM-5 based catalysts. After this, numerous reports [15, 16, 29, 36, 41] have been reported. Gallium activates ethane by abstracting hydrogen from ethane thereby forming ethylene. The mechanism for hydrogen removal from ethane is presented by Keipert et al. [36] as shown in Figure 14.8. However, like Mo, Ga also has a tendency to exchange with the protons on the zeolite surface thereby causing reduction in the strong acid sites on the surface of the zeolite [29a].

There have been various reports of the nature of active Ga species for ethane aromatization. Some reports [41] speculate it to be multinuclear gallium clusters rather than mononuclear ones. On the other hand, Yakerson et al. [42], based on the extensive IR study and electron

Figure 14.7 Possible representation of Ga-Pt nano-clusters.

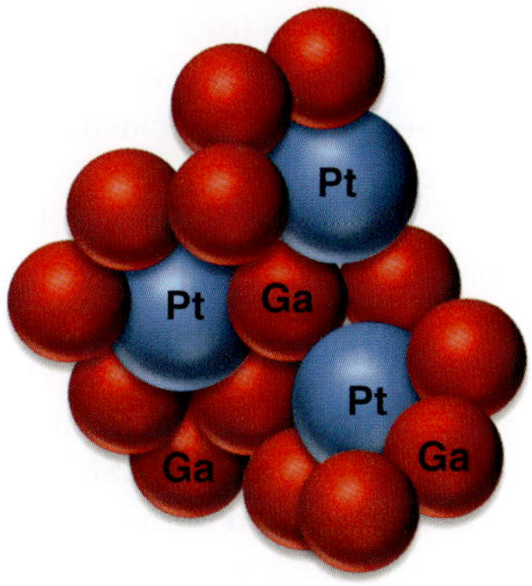

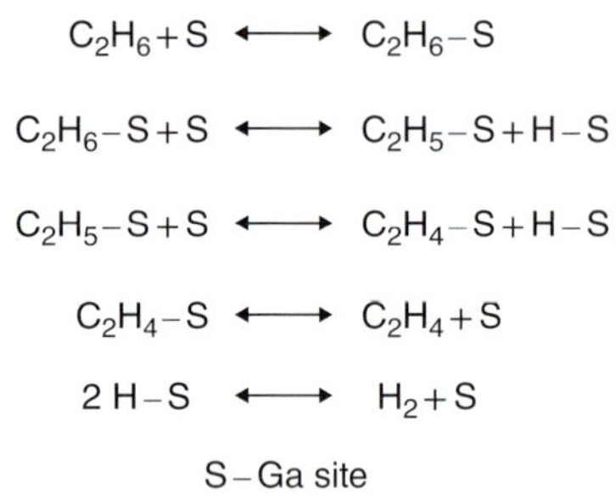

Figure 14.8 Elementary steps in ethane dehydrogenation over Ga based catalysts [36].

microscopy, pointed out that Ga_2O_3 on the external surface is the active species for this reaction. It is widely accepted that Ga_2O_3 favors dehydrogenation through H_2 recombinative desorption. However, Dooley et al. [43] correctly pointed out that transformations in the catalyst can occur while reduction under H_2 or during pretreatment with lower hydrocarbons or even under reaction conditions. Sometimes, these treatments or transformations can result in the formation of Ga^+ that can replace H^+ while migrating into the zeolite channels. In a more recent study [44], ethane dehydrogenation was studied over Ga/ZSM5 specifically to probe the role of various Ga species including: Ga_2O_3, GaO^+, Ga^+, and $GaH_2{}^+$. A noticeable drop in the acidity was observed when Ga/ZSM-5 was reduced under H_2; this was attributed to the formation of reduced Ga species (Ga^+ and $GaH_2{}^+$). These reduced species can undergo exchange with the protons and replace the acid sites, which was confirmed using NH_3-TPD and H_2-TPR. Further, these reduced species showed greater dehydrogenation activity than the oxide species Ga_2O_3/GaO^+ for ethane aromatization. Authors [44] ranked the dehydrogenation activity for different species as follows:

$$GaO^+ \sim Ga_2O_3 < Ga^+ < GaH_2{}^+$$

Because dehydrogenation is the initiation step in the aromatization of light alkanes, it becomes important to reduce Ga species (Ga^+, $GaH_2{}^+$). However, these reduced species can cause reduction of strong acid sites, and this may hamper the oligomerization step that forms aromatics. Therefore, a balance between the number of reduced Ga species and Brønsted acid sites might be necessary for an optimal activity of Ga/ZSM-5 catalyst. For example, Lapidus et al. [29b] found the optimal loading of Ga to be around 2.0% for the pentasil zeolites.

14.2.3.4 Re/H-ZSM-5

Rhenium can typically be introduced into the zeolite (H-ZSM-5) using IWI of ammonium perrheneate salt $[(NH_4)_2ReO_4.4H_2O]$. Like most other metallic species, Re in metallic state is often the active specie for alkane activation [45]. Encapsulated Re clusters have also been reported to be active for CH_4 dehydroaromatization and in fact, can be 30% more active than the MoC_x catalyst [46]. Also, these clusters have been shown to be active for propane activation irrespective of Re content.

Two studies have reported doping of rhenium into the ZSM5 zeolite [14, 28a] for ethane aromatization. Both studies doped Re onto ZSM-5 with a SiO_2/Al_2O_3 ratio of 30:1, although the loading of Re was different. Krogh et al. [14] compared Re/ZSM-5 to Zn/ZSM-5 and found that although Re exhibited a lower activity initially, it deactivated less rapidly than the Zn/ZSM-5 and thus showed higher and stable ethane conversion values over long runs. Conversion of Re/ZSM-5 against Zn/ZSM-5 could be observed in Figure 14.9. The difference was attributed to the sublimation of the Zn species that did not occur in the case of Re/ZSM-5, which is more stable at those temperatures.

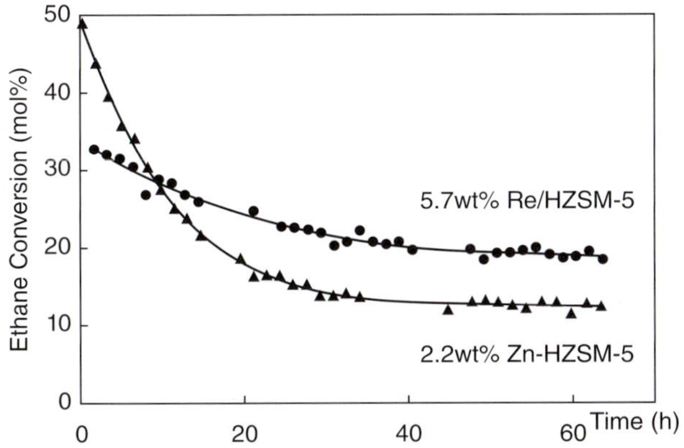

Figure 14.9 Ethane conversion on Re/ZSM5 and Zn/ZSM5. (1 atm, 650 °C, 400 mg catalyst, WHSV = 1.81 h^{-1}). Reprinted from "Re/HZSM-5: a new catalyst for ethane aromatization with improved stability", Catalysis Communications, 4(12), 627–630, 2003, with permission from Elsevier.

Solymosi et al. [28a] reported that Re promotes the activation of the C-H bond, which resulted in greater ethane conversion and higher aromatics selectivity. However, higher loadings of Re resulted in catalysts with lower activity. It could have been that higher loading led to poor dispersion of Re metal, making the catalyst less active.

14.2.3.5 Zn/H-ZSM-5

One of the most widely studied dopants in the zeolite framework for aromatization of ethane is zinc. Zinc can be introduced onto/into the ZSM5 support using various techniques [8b]: IWI using precursors $Zn(NO_3)_2.6H_2O$ or $ZnCl_2$, ZnO, ZnS, etc., ion exchange with a corresponding salt, or mechanical mixing of Zn salt with the zeolite. Heemsoth et al. [47] found that the catalyst prepared by a solid state reaction of Zn dust and HZSM5 gave a similar activity for ethane aromatization as of the one where Zn was impregnated onto HZSM5. Zn salt can also be added to the gel during the synthesis of the zeolite to incorporate Zn into the zeolite framework [48]. Newer techniques of deposition have also been tested such as CVD using dimethyl zinc, $Zn(CH_3)_2$, which allows for almost stoichiometric substitution of Brønsted acid sites (BAS) with Zn^{2+} ions [49]. This leads to the formation of isolated Zn^{2+} species in contrast to heterogeneous distribution of extra-framework Zn^{2+} species obtained during ion exchange (IE) or incipient wetness impregnation (IWI). However, the catalysts prepared by CVD technique were less active in propane aromatization than the catalysts prepared with IE or IWI due to the lesser abundance of active Zn^{2+} and multinuclear zinc oxide species [49].

Zn^{2+} species have proven to be an effective dehydrogenation site for light alkanes including methane. Apart from Zn^{2+}, other divalent Zn species that can activate light alkanes include $[Zn\text{-}O\text{-}Zn]^{2+}$ or small clusters of ZnO [49, 50]. Arzumanov et al. [48] found a similar mechanism of ethane activation using ^{13}CMAS NMR in which dissociative adsorption of ethane occurs on ZnO species to form Zn-ethyl species. A similar result was obtained by Gabrienko et al. [51] and Xu et al. [52], in which the catalyst prepared by partial substitution of Brønsted acid sites with Zn^{2+}, could activate methane by forming Zn-methyl species even at room temperature. Some of the proposed Zn sites as suggested by Mehdad and Lobo [50] are depicted in Figure 14.10.

The amount of zinc on the zeolite support thus has substantial influence on the activity of the catalyst. This is due to generation of Lewis acid sites when Zn is introduced, and at the same time reduction in Brønsted acid sites of the zeolite. Lapidus et al. [53] studied the effect of Zn content

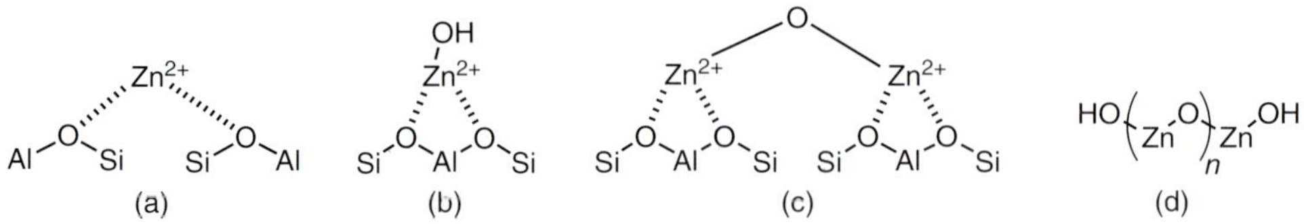

Figure 14.10 Various Zn^{2+} sites in ZSM5 and Silicalite. (a) Zn(II) stabilized by two framework Al atoms, (b) $Zn(OH)^{+}$ stabilized by one framework Al, (c) oxygen-bridged Zn dimer, (d) $(ZnO)_n$ cluster. Reproduced from Ethane and ethylene aromatization on zinc-containing zeolites, Catalysis Science and Technology, 7(16), 3562–3572, 2017, with permission of The Royal Society of Chemistry.

Table 14.3 Effect of Zn content on the ethane conversion and selectivity to aromatics [53] (T = 600 °C, Volume hourly space velocity = 450 h^{-1}) (also conversions and yields of aromatics are per pass of ethane).

Zn content (wt %)	C_2H_6 conversion (%)	Aromatics Yield (%)	Products composition (wt %)				
			C_6	C_7	C_8	C_9	C_{10-12}
0	8.4	2.3	37.2	39.0	11.5	1.4	10.9
0.75	30.7	15.5	33.2	27.0	6.0	1.9	31.9
1.5	41.9	20.1	29.4	27.0	6.4	1.1	41.8
3.0	49.5	21.9	25.3	25.2	6.6	1.1	41.8
5.0	51.5	23.7	26.7	24.2	6.0	1.3	41.8
7.5	50.9	21.4	25.6	25.0	5.9	0.8	42.7

Reproduced from Zinc-containing zeolite catalysts for ethane aromatization prepared by solid-state modification, Russian Chemical Bulletin, International Edition, 52(5), 1094–1099, 2003, with permission of Springer.

into the HZSM5 zeolite and found the optimal loading of the Zn to be in between 3–5 wt%. The effect of Zn content on the ethane conversion and aromatics selectivity as found by the same authors is shown in Table 14.3.

Similar results were observed by Mehdad and Lobo [50] who found out the optimal concentration to be around 5 wt%. Above 5 wt% loading of Zn, blockage of pores due to ZnO deposition was observed. They found that at low Zn content, ethylene was favored instead of aromatics, which was attributed to the presence of a high density of Brønsted acid sites. At high Zn content, the concentration of Brønsted acid sites had decreased, and the presence of generated Lewis acid sites would have favored aromatization instead of ethylene. However, higher zinc content also favored CH_4 formation. This is in contrast with Anunziata et al. [48] observations with the ion exchanged Zn-ZSM-11, where increase in the Zn content decreased CH_4 formation. This might have been due to the difference in the support materials: ZSM-5 versus ZSM-11. Although, they are quite similar, the channel system differs in that the ZSM5 has cross-linked pores as compared to the later in which the pores are straight linked as could be seen in Figure 14.11. Also, the shape selectivity and acidity on these two supports are different, leading to differences in activity.

Although Zn-doped ZSM5 catalyst is an active catalyst, one of the main challenges with its use is the volatilization of Zn, which occurs under reducing reaction conditions, most likely due to the low melting point of 420 °C (high vapor pressure). For stable and long runs of ethane aromatization over Zn-ZSM-5, a constant source of Zn is required to replenish its loss.

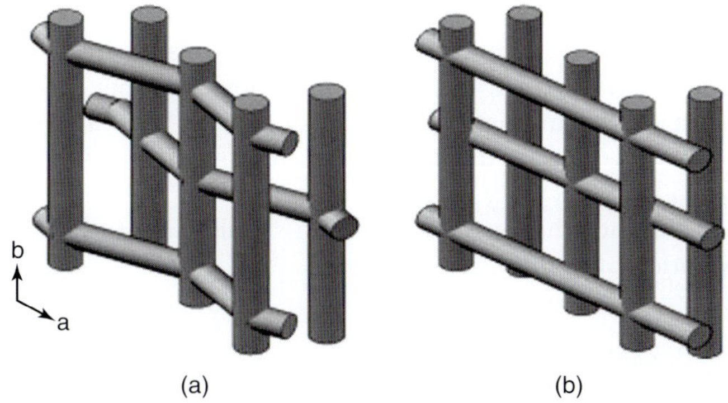

Figure 14.11 Pore channel system for ZSM-5 (a) and ZSM-11 (b) materials. Reprinted from "Hierarchical ZSM-11 with intergrowth structures: Synthesis, characterization and catalytic properties", Journal of Energy Chemistry, 22(5),761–768, 2013, with permission from Elsevier.

Table 14.4 Effect of promoters on concentration of acid sites [48] (pyridine-based FTIR, after desorption at 400 °C for 4 hr), site concentration is in absorbance units.

	Zn-ZSM-11 (Si/Al = 17)	Zn-Pb-ZSM-11 (Si/Al = 17)	Zn-Mo-ZSM-11 (Si/Al = 17)	Zn-Fe-ZSM-11 (Si/Al = 17)
Brønsted sites	0.009	0.004	0.009	0.020
Lewis sites	0.030	0.033	0.044	0.029

14.2.3.6 Promoters

Pb, Mo, and Fe have been used as promoters in Zn-doped ZSM-11 catalysts [48]. It could be done through IE. Most of these promoters generate Lewis acid sites that increase the ethane conversion as well as the selectivity to aromatics. The effect of promoters on concentration of Lewis and Brønsted acid sites as observed by Anunziata et al. [48] is shown in Table 14.4. It could be seen that except for Fe, promoter doping increased the concentration of Lewis acid sites.

W was also used as a promoter for Mo-based ZSM5 catalysts [18], which improved the overall stability of the catalyst and increased selectivity to ethane aromatization, and was attributed to the formation Mo-W oxide, either bimetallic or closely interacted. Authors further concluded that W might be providing an ensemble effect, thus reducing the rate of formation of coke and preventing the sintering of Mo active species.

Pt has also been used as a very effective promoter for Ga-based ZSM-5 catalysts to enhance its activity in the aromatization of lower alkanes. Also, introduction of Pt into Ga-based zeolites considerably reduces the formation of byproducts, methane and ethylene [15]. It is believed that the dehydrogenation step is accelerated in the presence of Pt in Ga-doped zeolites [41] through reduction in the activation energy for dehydrogenation and reduction in the hydrogen desorption energy as well [16]. Further it is speculated that Pt forms bimetallic alloy particles with the Ga and yields active sites in the zeolite framework [16, 41]. Although the presence of Pt does not help prevent coke formation, it partially removes coke that occurs due to hydrogenolysis through hydrogen spillover [29a]. This results in less refractive coke as compared to Ga-based ZSM-5 without Pt.

Pd has also been tested as a promoter for Ga-based ZSM-5 catalysts by Zaikovskii et al. [54]. However, Pd did not promote the aromatizing activity of Ga-ZSM-5 significantly because of poor dispersion of Pd over the catalyst. In contrast, the same study showed higher activity for Pt-promoted Ga-ZSM-5 catalyst due to uniform dispersion of Pt over the catalyst. Similar results were obtained for Pt- and Pd-promoted Zn-ZSM-5 catalysts [55] in which a Pt-promoted Zn-ZSM-5 catalyst showed higher activity in ethane aromatization as compared to a Pd-promoted Zn-ZSM-5 catalyst. Pd, though, can have effect on stabilization of Zn by alloying with it to suppress the volatilization of Zn [56].

14.2.4 Effect of Pore Structure (ZSM-5, ZSM-8, ZSM-11, ZSM-12)

Pore structure can be important for the aromatization reaction as it involves complex mechanisms; thus, the nature of the support can have considerable effects on reaction rates of different stages of the process. This results in considerably different activities for ethane aromatization over these catalysts.

MFI or pentasil-type zeolite ZSM-5 is arguably the most active and the most extensively studied material for creating bifunctional catalysts. A pentasil unit typically consists of 8 five-membered rings. The ZSM-5 has interconnected 10-membered straight and sinusoidal pores that result in three-dimensional pore networks. It is a medium-pore zeolite with pore dimensions of around 0.53 nm × 0.56 nm (straight pores) and 0.51 nm × 0.56 nm (sinusoidal pores) [8a]. One advantage with the medium-pore zeolites such as ZSM-5 is that it can selectively perform conversions of light hydrocarbons into monocyclic aromatics (formation of higher hydrocarbons is restricted) while minimizing coke formation. This is not possible with the small-pore zeolites (requires severe conditions) or large-pore zeolites (heavy aromatics production, faster deactivation) [57]. Some of the zeolite structures that have been studied in ethane aromatization are depicted in Figure 14.12.

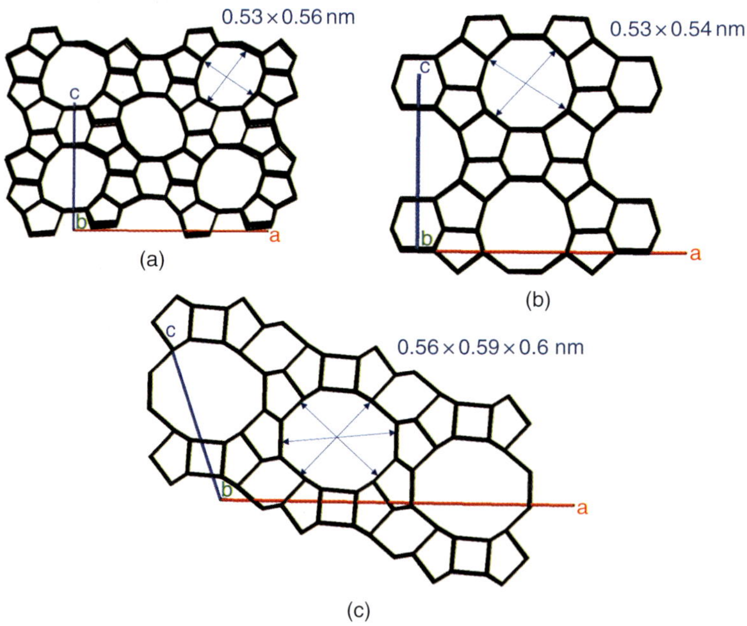

Figure 14.12 Various zeolite structures [58] (a) ZSM-5, (b) ZSM-11, (c) ZSM-12.

Pore structure can be a more important factor as compared to the acidity of the catalysts as observed by Vosmerikova et al. [17], who studied ethane aromatization on Zn-modified zeolites of the type (ZSM-5, ZSM-8, ZSM-11, ZSM-12). Zn-modified ZSM-12 had the highest concentration of acid sites, both strong and weak, but still Zn-modified ZSM-5 catalyst showed the highest levels of conversion and selectivity to aromatics. Catalysts with the ZSM-8 and ZSM-11 pore structure showed lower ethane conversion as well as lower selectivity to aromatics. ZSM-8 based catalyst differs from ZSM-5 only in terms of available surface area (lower than for ZSM-5) and pore volume (lower than for ZSM-5) but these two factors seem to have a significant effect on the catalyst [17]. ZSM-8-based catalyst showed lower ethane conversions and low aromatic yields. Similar results in terms of ethane conversion and aromatics selectivity were obtained on the ZSM-11–based catalyst. However, the ZSM-11–based catalyst showed high selectivity towards naphthalene, which might have been due to slightly larger pore volume that allows condensation of carbenium ion–type intermediates and olefinic species that eventually lead to higher hydrocarbons [59].

Despite stronger and higher distribution of acid sites, ZSM-12–based catalysts had the lowest aromatization activity as compared to dehydrogenation (~80% selectivity). Authors [17] observed C_2-C_4 alkenes on the ZSM-12–based catalyst. Despite low ethane conversions ZSM-12 formed large amounts of coke, possibly due to polymerization of intermediate compounds and slightly larger pore channels. It is possible that high acidity might have favored side reactions such as cracking leading principally to ethylene.

Zn-doped SSZ-13 or Chabazite-type zeolite, having very small pores, has also been tried for ethane aromatization [50] but had little apparent activity. This inactivity was not clearly attributed but was thought to be due to geometric differences. In the same study, authors also studied Zn doped on Beta (BEA) zeolite, but this catalyst was 2.5 times less active than the Zn-ZSM-5 for the aromatization of ethane. The exact values of conversion and selectivity to aromatics were not reported. Structures for Chabazite (SSZ-13) and Beta-type zeolites are shown in Figure 14.13.

14.2.5 Effect of Acidity (Si/Al Ratio etc.)

Mehdad and Lobo [50] pointed out the importance of balance between concentration of Lewis acid sites (LAS) and Brønsted acid sites (BAS). Metal doping introduces or increases Lewis

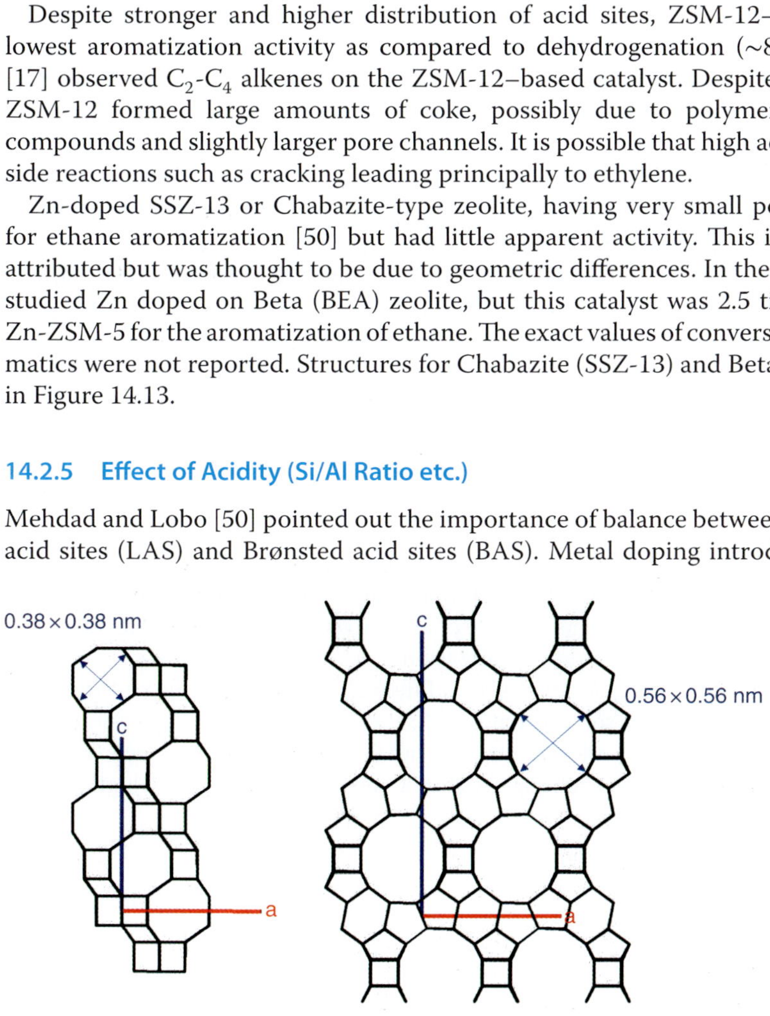

Figure 14.13 Structures of zeolites [58]: (a) CHA (SSZ-13) and (b) BEA (Beta).

Table 14.5 Ethane aromatization on Pt-Ga pentasils having different SiO_2/Al_2O_3 molar ratios [29b] (T = 600 °C, volume hourly space velocity = 450 h^{-1}).

SiO_2/Al_2O_3 mol.	Metal, wt%		X_{Ethane} (%)	Y_{Ar} (%)	S_{Ar} (%)	$S_{methane}$ (%)	$S_{ethylene}$ (%)	$S_{C3\text{-}C4}$ (%)
	Ga	Pt						
30	0.5	0.3	47.1	30.3	64.3	19.5	7.0	–
56	0.5	0.3	39.9	24.0	60.2	23.3	6.5	–
90	0.5	0.3	27.2	14.8	54.5	11.1	19.5	4.1
30	2.0	0.3	47.6	30.1	63.2	20.1	7.8	–
56	2.0	0.3	24.7	12.7	51.4	7.7	21.5	8.9
90	2.0	0.3	17.0	4.2	24.8	3.5	53.9	7.6

acid sites depending on the type of dopant (e.g., Zn). These are useful in activation of ethane through dehydrogenation, whereas Brønsted acid sites are useful for aromatization of olefinic intermediates. BAS can also play a role as an ion exchange site for metal dopants.

Based on their studies over ZSM-11 supports and different metal dopants, Anunziata et al. [48] pointed out that high LAS produced high concentrations of hydrogen acceptor species thus favoring dehydrogenation and ultimately ethane conversion and aromatic hydrocarbon production. On the contrary, BAS promoted the cracking activity and C_1 production.

Si/Al ratio also influences the conversion of ethane and selectivity to aromatic hydrocarbons [28a]. As the Si/Al ratio increases, conversion drops down as well as the selectivity to aromatics. However, dehydrogenation route is favored resulting in more ethylene yield. This effect was observed in the case Ga-Pt systems [29b]. Aluminum content in the catalyst is directly related to the strong Brønsted acid sites that are in turn necessary for the oligomerization or cyclization of the olefinic species, yielding aromatics. Lapidus et al. [29b] recommended optimal Si/Al ratio to be ≤ 30 for Ga-Pt–based system based on the ethane conversion data as shown in Table 14.5.

In another study, Lapidus et al. [53] studied 5% Zn loaded onto ZSM-5 with different Si/Al ratio. High conversions of ethane and highest yield of aromatics were obtained on the catalyst with Si/Al ratio between 30 and 100. Mehdad and Lobo [50] on the other hand, studied 5% Zn on ZSM-5 catalyst with Si/Al ratio below 30. They found that as the ratio increased from 11.5 to 25, the rate of ethane conversion as well as the rate of aromatization decreased. However, the rate of deactivation also decreased (i.e., less aromatization).

Acidity or the Si/Al ratio, in some cases, can also influence the bifunctionality of the catalyst. Ausavasukhi and Sooknoi [44] observed that Si/Al ratio influenced the dispersion of Ga species in the zeolite structure, thereby affecting the overall dehydrogenation activity. This was due to less availability of acid sites for exchange with Ga species in the case of higher Si/Al ratio zeolites.

14.3 Future Perspective

The main challenges associated with the conversion of natural gas are rapid catalyst deactivation through coking, loss of acid sites, and also in some catalysts loss of active dopant metal, such as Zn, through volatilization. All these factors demand constant regeneration of the catalysts as well as constant source of replenishment of the active dopant metal. These processes can be expensive but are often employed on industrial scales for other light alkanes (e.g., the propane-based Cyclar process). Approaches to reduce the formation of coke include the addition of a promoter that lowers the rate of formation of coke or by addition of an oxidant (CO, CO_2, O_2, NO_x). Other options to improve the process include shifting of the thermodynamic equilibrium with the use of a permselective membrane that can selectively remove produced H_2 to drive the reaction forward. H_2–absorbing materials such as Zr_2Fe can also be used, as pointed out in the earlier reviews [8b].

Palladium membrane has been tested for propane aromatization by some research groups [60], and it was found that the activity increased several times by extracting hydrogen from the reaction products. Other membranes reported in the literature include ceramic materials ($La_{5.5}W_{0.6}Mo_{0.4}O_{11.25\text{-}\delta}$) [61]. However, in these hydrogen-selective membranes, along with an increase in the rate of aromatization, the rate of coke formation also increases [62], causing a rapid deactivation of the catalyst. Introduction of an oxidant like oxygen has been shown to inhibit coke production without affecting the selectivity to benzene. Thus, the membranes that can be employed should be multifunctional, i.e., they should selectively be able to remove H_2 from the reaction and at the same time, provide O_2 to remove coke formation from the system. These membranes can be multifunctional or two separate membranes for O_2 and H_2 can be used (from inside and outside of the reaction zone). A typical schematic for a multifunctional membrane (separate membranes for O_2 and H_2) system is depicted in Figure 14.14.

Very recently, a similar multifunctional membrane reactor model was proposed and optimized by Fouty et al. [63]. A variety of factors determine the overall performance of this multifunctional membrane reactor system. For high performance of this system the membrane should have a high permeation rate (0.01 mol/s·m^2·atm$^{1/4}$) and high selectivity to H_2 (> 10^5). Other factors that will also affect the performance included selectivity to O_2 (slow permeation), reaction temperatures, feed conditions, reactor dimensions, and membrane thicknesses.

Practical application of a multifunctional $BaZrO_3$ membrane in the methane aromatization reaction was recently shown by Morejudo et al. [62]. This membrane was electrochemically able to remove H_2 while, at the same time, provide O_2 to remove carbon deposits. This resulted in high aromatics yield as well as high stability of the catalysts. Membranes like these could also be employed for the conversion of ethane or even for the conversion of a mixed feed: methane and ethane or directly natural gas. Ethane being easier to activate, a high-yield, low-temperature process might be feasible.

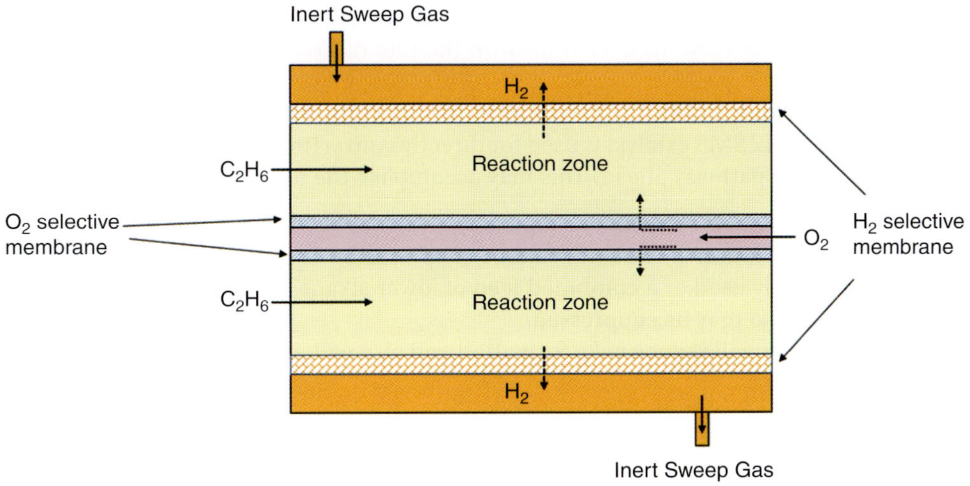

Figure 14.14 Schematic for a multifunctional membrane reactor.

Addition of oxidants such as CO_2 has led to increase in the aromatization activity in terms of both conversion as well as the aromatics yield [64], while at the same time suppressing ethylene yield. In the same study, authors also observed that addition of steam enhanced the dehydrogenation activity of Ga-promoted catalysts. This positive effect was attributed to the additional reaction of steam and CO_2 with the coke on the catalyst surface. CO_2 reacts with desorbed H_2 to undergo reverse water gas shift reaction (eq. 14.6):

$$CO_2 + H_2 \leftrightarrow CO + H_2O \;\; (\Delta G_{298K} = +6.8\ \text{kcal/mol}, \Delta H_{298K} = +9.8\ \text{kcal/mol}) \tag{14.6}$$

as CO and H_2O were observed in the reaction products along with aromatics. The other reason is the Boudouard reaction (eq. 14.7), which removes deposited coke.

$$CO_2 + C \leftrightarrow 2\,CO \; (\Delta G_{298K} = +28.7\ \text{kcal/mol}, \Delta H_{298K} = +41.2\ \text{kcal/mol}) \tag{14.7}$$

Although some activity is observed due to CO_2 addition, it could also be due to steam generated from the reverse water gas shift reaction as pointed out by Nakagawa et al. [64]. On the contrary, Nishi et al. [65] had proposed that the improvement in the activity of the catalyst under the presence of CO_2 was due to a kinetic effect instead of a thermodynamic effect. Also, the authors thought that the increase in activity was due to an increase in acid sites in the presence of CO_2 that can interact with zeolite lattice and form Brønsted acid sites.

As pointed out in an earlier review [8b], addition of an oxidant leads to additional side reactions such as formation of CO and H_2 like a syngas mixture along with methane and other cracking products. Controlling these reactions then becomes important. The ideal catalyst should have high activity in the aromatization of ethane with low selectivity for side reactions. This will ensure high yield to desired aromatics under the presence of an oxidant. Steam, when added as a co-reactant, enhances the conversion of ethane in the dehydrogenation reaction as observed by Nakagawa et al. [64], but at the same time, selectivity to CH_4 also had increased.

Zn-doped ZSM-5 catalysts have shown high activity in ethane aromatization. As discussed before, the primary issue with these catalysts is the volatilization of Zn because of high vapor pressure of Zn at reaction temperatures. There have been numerous attempts to minimize the Zn volatilization. Earlier attempts focused on reducing Zn vapor pressure by alloying it with expensive metals such as Pd [56], Au [56], Ag [56] or some attempts focused on Ga [66]. These attempts resulted in reduction of the volatilization rate of Zn, but improvements were not significant. In another subsequent patent, the authors [67] proclaimed a novel method of addition of either nonmetallic oxides such as CO_2, CO, NO_x, steam, etc. or of nonmetallic sulfur compounds such as H_2S. These seem to have improved the rate of loss of Zn considerably, if not completely.

H_2S or some of the oxidants like CO_2 are already found as constituents of natural gas. Therefore, if an alloyed Zn on HZSM5 catalyst is used for directly converting natural gas to aromatics, this may provide a viable pathway ahead. This may accomplish the following:

1) Promoted or alloyed Zn on HZSM5 has high activity in the aromatization reaction, and at the same time, volatilization of the Zn would be suppressed.
2) If natural gas is directly used or a combined feed of lower alkanes along with CO_2 and H_2S, volatilization of Zn also may be suppressed.
3) Presence of an oxidant will reduce coke formation and also will enhance the overall activity of the catalyst.

Other approaches for minimizing zinc volatilization could include a regeneration unit wherein Zn or ZnO could be resupplied to the spent catalyst through solid state ion exchange [8b]. Catalysts prepared by these methods are effective, and this type of synthesis, could be done in situ by adding the appropriate amounts of Zn or ZnO along with the heat treatment.

Although the scientific advancements in terms of membrane technologies, other separation technologies, and even catalyst development have been impressive, significant progress is needed before the ethane aromatization process is put into commercial practice. Ethane is an important commodity for the production of ethylene and other valuable chemicals. The range of products that can be produced from ethane can further be exploited if aromatization of ethane becomes commercially feasible. This would allow an efficient use of natural gas, much of which is flared for lack of a practical, competitive process. A strong research emphasis from the scientific community should be given to the ethane aromatization. Methane aromatization is the next grand challenge, but ethane should be given an equal attention because it is much easier to activate than methane.

References

1 (a) Jenner S, Lamadrid AJ. Shale gas vs. coal: policy implications from environmental impact comparisons of shale gas, conventional gas, and coal on air, water, and land in the United States. Energy Policy, 53 (0) (2013): 442–453; (b) Wang Q, Chen X, Jha AN, Rogers H. Natural gas from shale formation—the evolution, evidences and challenges of shale gas revolution in United States. Renew Sustain Energ Revs, 30 (0) (2014): 1–28; (c) Boersma T, Johnson C. The shale gas revolution: U.S. and EU policy and research agendas. Rev Policy Res, 29 (4), (2012): 570–576; (d) Aruga K., The U.S. shale gas revolution and its effect on international gas markets. J Unconven Oil Gas Res, 14 (2016): 1–5.
2 Boersma T, Johnson C. The Shale gas revolution: U.S. and EU Policy and Research Agendas. Rev Policy Res, 29 (2012): 570–576.
3 BP Natural gas prices. Available from: https://www.bp.com/en/global/corporate/energy-economics/statistical-review-of-world-energy/natural-gas/natural-gas-prices.html.
4 Middleton RS, Gupta R, Hyman JD, Viswanathan HS. The shale gas revolution: barriers, sustainability, and emerging opportunities. Appl Energ, 199, (2017): 88–95.
5 (a) Spivey JJ, Hutchings G., Catalytic aromatization of methane. Chem Soc'y Revs, 43 (3) (2014): 792–803; (b) Taifan W, Baltrusaitis J. CH4 conversion to value added products: potential, limitations and extensions of a single step heterogeneous catalysis. Appl Catal B: Environ, 198 (2016): 525–547; (c) Karakaya C, Kee RJ. Progress in the direct catalytic conversion of methane to fuels and chemicals. Prog Energ Combust Sci, 55 (2016): 60–97; (d) Choudhary TV, Aksoylu E, Goodman DW. Nonoxidative activation of methane. Catal Revs—Sci Engg, 45 (1) (2003): 151–203.
6 http://journals.plos.org/plosone/article/asset?unique&id=info:doi/10.1371/journal.pone.0144141.s001.
7 EIA. Available from: https://www.eia.gov/energyexplained/index.cfm?page=hgls_prices#tab1.
8 (a) Bhan A, Nicholas Delgass W. Propane aromatization over HZSM-5 and Ga/HZSM-5 Catalysts. Catal Revs, 50 (1) (2008): 19–151; (b) Hagen A, Roessner F. Ethane to aromatic hydrocarbons: past present, future. Catal Revs, 42 (4) (2000): 403–437; (c) Ono Y. Transformation of lower alkanes into aromatic hydrocarbons over ZSM-5 zeolites. Catal Revs, 34 (3) (1992): 179–226; (d) Bañares MA., Supported metal oxide and other catalysts for ethane conversion: a review. Catal Today, 51 (2) (1999): 319–348.
9 Csicsery SM., Dehydrocyclodimerization: III. Dehydrocyclodimerization of butanes over transition metal oxide catalysts. J Catal, 17 (3), (1970): 315–322.
10 Cheng Y, Gong H, Miao C, Hua W, Yue Y, Gao Z. Ga2O3/HSSZ-13 for dehydrogenation of ethane: Effect of pore geometry of support. Catal J, 71 (Supplement C) (2015): 42–45.

11 (a) Csicsery S. M. Dehydrocyclodimerization: II. Dehydrocyclodimerization of propane and pentane over supported platinum catalyst. J Catal, 17 (2) (1970): 216–218; (b) Csicsery SM. Dehydrocyclodimerization: I. Dehydrocyclodimerization of butanes over supported platinum catalysts. J Catal, 17 (2) (1970): 207–215; (c) Csicsery SM. Dehydrocyclodimerization: V. The mechanism of the reaction. J Catal, 18 (1) (1970): 30–32; (d) Csicsery SM. Dehydrocyclodimerization: IV. The reactions of butenes. J Catal, 17 (3) (1970): 323–330.

12 Bragin OV, Vasina TV, Preobrazhenskii AV. Catalytic aromatization of ethylene and ethane. Bull Acad Scis USSR, Division Chem Sci, 33 (1) (1984): 56–63.

13 Inui, TM., Akihiko. In Selective aromatization of light hydrocarbons on polyfunctional metallosilicate catalysts, ACS Fall National Meeting, New York, NY, American Chemical Society: New York; 1991.

14 Krogh A, Hagen A, Hansen TW, Hviid Christensen C, Schmidt I. Re/HZSM-5: a new catalyst for ethane aromatization with improved stability. Catal. Commun., 4 (12) (2003): 627–630.

15 Lapidus AL, Lapidus AL, Mikhailov MN, Mikhailov MN, Dergachev AA, Dergachev AA, Zhidomirov GM, Zhidomirov GM, Mishin IV, Mishin IV. The nature of active sites in Pt promoted GaZSM-5 catalysts. React. Kinet. Catal. Lett., 87 (2) (2006): 249–254.

16 Lapidus AL, Mikhailov MN, Dergachev AA, Mishin IV. Structure of active sites of Ga-Pt zeolite catalysts of alkane aromatization. Dokl Phys Chem, 408 (2) (2006): 175–177.

17 Vosmerikova L, Barbashin Y, Vosmerikov A. Catalytic aromatization of ethane on zinc-modified zeolites of various framework types. Petrol Chem, 54 (6) (2014): 420–425.

18 Wong S-T, Xu Y, Wang L, Liu S, Li G, Xie M, Guo X. Methane and ethane activation without adding oxygen: promotional effect of W in Mo-W/HZSM-5. Catal Letters, 38 (1) (1996): 39–43.

19 Degnan TF, Chitnis GK, Schipper PH. History of ZSM-5 fluid catalytic cracking additive development at Mobil. Microporous Mesoporous Materials, 35 (Supplement C) (2000): 245–252.

20 Chen NY, Yan TY. M2 forming—a process for aromatization of light hydrocarbons. Indus Engg Chem Proc Design Develop, 25 (1) (1986): 151–155.

21 Marcilly C. Acido-Basic Catalysis. Volume 2. Application to Refining and Petrochemistry. Editions Technip, 2013.

22 Olah GA, Molnár Á. Hydrocarbons from Petroleum and natural gas. In: Hydrocarbon Chemistry. Hoboken, NJ: John Wiley & Sons, Inc.; 2003; p. 30–84.

23 Yuzo F, Ken T, Jun W, Yutaka U, Hiroshi T. Thermal activation of alkane C–H Bonds by palladium catalysts. carbonylation of alkanes with carbon monoxide. Chem Letters, 18 (9) (1989): 1687–1688.

24 Comyns AE., Z. In: Encyclopedic Dictionary of named processes in chemical technology. 4th ed. CRC Press; 2014. p. 381–384.

25 Syed N. Process economics program—aromatics from light hydrocarbons IHS: 2014.

26 Chevron Phillips. Process. Available from: http://www.cpchem.com/en-us/rnt/licensing/aromaxtech/pages/process.aspx.

27 Foley T. CYCLAR™ process produces high-quality aromatic products. Available from: https://www.uop.com/cyclar-process-produces-high-quality-aromatic-products/.

28 (a) Solymosi F, Tolmacsov P. Conversion of ethane into benzene on Re/ZSM-5. Catal Lett, 93 (1–2) (2004): 7–11; (b) Chetina OV, Vasina TV, Lunin VV., Aromatization of ethane over Pt,Ga/HZSM-5 catalyst and the effect of intermetallic hydrogen acceptor on the reaction. Appl Catal A: General, 131 (1) (1995): 7–14.

29 (a) Samanta A, Bai X, Robinson B, Chen H, Hu J., Conversion of light alkane to value-added chemicals over ZSM-5/metal promoted catalysts. Indus Engg Chem Research 2017; (b) Lapidus AL, Dergachev AA, Kostina VA, Silakova AA., Ethane aromatization on Ga-Pt pentasil zeolites. Neftekhimiya, 48 (2) (2008): 83–86.

30 (a) Wu Y, Emdadi L, Wang Z, Fan W, Liu D., Textural and catalytic properties of Mo loaded hierarchical meso-/microporous lamellar MFI and MWW zeolites for direct methane conversion. Appl Catal A: General, 470 (2014): 344–354; (b) Matus EV, Sukhova OB, Ismagilov IZ, Tsikoza LT, Ismagilov ZR. Peculiarities of dehydroaromatization of CH4-C2H6 and CH4 over Mo/ZSM-5 catalysts. React Kinet Catal Lett, 98 (1) (2009): 59–67.

31 (a) Solymosi F, Szo"ke, A., Conversion of ethane into benzene on Mo2C/ZSM-5 catalyst. Appl Catal A: General, 166 (1) (1998): 225–235; (b) Osawa T, Nakano I, Takayasu O. Dehydrogenation of methane over Mo/ZSM-5: effects of additives in the methane stream. Catal Lett, 86 (1–3) (2003): 57–62.

32 (a) Ding W, Li S, D Meitzner G, Iglesia E. Methane conversion to aromatics on Mo/H-ZSM5: structure of molybdenum species in working catalysts. J Physical Chem B, 105 (2) (2001): 506–513; (b) Petkovic LM, Ginosar DM. Comparison of two preparation methods on catalytic activity and selectivity of Ru-Mo/HZSM5 for methane dehydroaromatization. J Fuels, 7 (2014).

33 Folefoc GN, Dwyer J. Dispersion of platinum in Pt/ZSM-5 zeolites. J Catal, 136 (1) (1992): 43–49.

34 Weitkamp J, Puppe L. Catal and zeolites : fundamentals and applications. New York : Springer; 1999.

35 (a) Yu C, Ge Q, Xu H, Li W. Propane dehydrogenation to propylene over Pt-based catalysts. Catal Letters, 112 (3) (2006): 197–201; (b) Liu G, Zeng L, Zhao Z-J, Tian H, Wu T, Gong J. Platinum-modified ZnO/Al2O3 for propane dehydrogenation: minimized platinum usage and improved catalytic stability. ACS Catal, 6 (4) (2016): 2158–2162.

36 Keipert OP, Wolf D, Schulz P, Baerns M. Kinetics of ethane aromatization over a gallium-doped H-ZSM-5 catalyst. Appl Catal A: General, 131 (2) (1995): 347–365.

37 Li Z, Lepore AW, Salazar MF, Foo GS, Davison BH, Wu Z, Narula CK. Selective conversion of bio-derived ethanol to renewable BTX over Ga-ZSM-5. Green Chem, 19 (18) (2017): 4344–4352.

38 (a) Kwak BS, Sachtler WMH. Characterization and testing of Ga/HZSM-5 prepared by sublimation of GaCl3 into HZSM-5. J Catal, 141 (2) (1993): 729–732; (b) Kwak BS, Sachtler WMH, Haag WO. Catalytic conversion of propane to aromatics: effects of adding Ga and/or Pt to HZSM-5. J Catal, 149 (2) (1994): 465–473.

39 (a) Price GL, Kanazirev V. Ga2O3/HZSM-5 propane aromatization catalysts: formation of active centers via solid-state reaction. J Catal, 126 (1) (1990): 267–278; (b) Gnep NS, Doyemet JY, Seco AM, Ribeiro FR, Guisnet M. Conversion of light alkanes to aromatic hydrocarbons: II. Role of gallium species in propane transformation on GaZSM5 catalysts. Appl Catal, 43 (1) (1988): 155–166; (c) Gnep NS, Doyemet JY, Guisnet M. Conversion of light alkanes into aromatic hydrocarbons. 3. Aromatization of propane and propene on mixtures of HZSM5 and of Ga2O3. In: Karge HG, Weitkamp, J., eds. Studies in surface science and catal Vol. 46. Elsevier; 1989, p. 153–162.

40 Ono Y, Nakatani H, Kitagawa H, Suzuki E. The role of metal cations in the transformation of lower alkanes into aromatic hydrocarbons. In: Inui, T., ed., Studies in surface science and catal. Volume 44, Elsevier; 1989, p. 279–290.

41 Mikhailov MN, Mishin IV, Kustov LM, Lapidus AL. Structure and reactivity of Pt/GaZSM-5 aromatization catalyst. Microporous Mesoporous Mat–er. 104 (1–3), (2007): 145–150.

42 Yakerson VI, Vasina TV, Lafer LI, Sytnyk VP, Dykh GL, Mokhov AV, Bragin OV, Minachev KM. The properties of zinc and gallium containing pentasils—the catalysts for the aromatization of lower alkanes. Catal Letters, 3 (4) (1989): 339–345.

43 Dooley KM, Price GL, Kanazirev VI, Hart VI. Gallium-loaded zeolites for light paraffin aromatization: evidence for exchanged gallium cation active centers. Catal Today, 31 (3) (1996): 305–315.

44 Ausavasukhi A, Sooknoi T. Tunable activity of [Ga]HZSM-5 with H2 treatment: ethane dehydrogenation. Catal Communs, 45 (Supplement C) (2014): 63–68.

45 Wang L, Ohnishi R, Ichikawa M. Novel rhenium-based catalysts for dehydrocondensation of methane with CO/CO2 towards ethylene and benzene. Catal Letters, 62 (1) (1999): 29–33.

46 Lacheen HS, Cordeiro PJ, Iglesia E. Isolation of rhenium and ReOx Species within ZSM5 channels and their catalytic function in the activation of alkanes and alkanols. Chem – A European Journal, 13 (11) (2007): 3048–3057.

47 Heemsoth J, Tegeler E, Roessner F, Hagen A. Generation of active sites for ethane aromatization in ZSM-5 zeolites by a solid-state reaction of zinc metal with Brønsted acid sites of the zeolite. Microporous Mesoporous Materials, 46 (2) (2001): 185–190.

48 Anunziata OA, Eimer GA, Pierella LB. Catalytic activity of ZSM-11 zeolites modified with metal cations for the ethane conversion. Catal Letters, 75 (1) (2001): 93–97.

49 Almutairi SMT, Mezari B, Magusin PCMM, Pidko EA, Hensen EJM. Structure and reactivity of Zn-modified ZSM-5 zeolites: the importance of clustered cationic Zn complexes. ACS Catal, 2 (1) (2012): 71–83.

50 Mehdad A, Lobo R. F. Ethane and ethylene aromatization on zinc-containing zeolites. Catal Sci & Technol, 7 (16) (2017): 3562–3572.

51 (a) Gabrienko AA, Arzumanov SS, Toktarev AV, et al. Different efficiency of Zn2+ and ZnO species for methane activation on Zn-modified zeolite. ACS Catal, 7 (3) (2017): 1818–1830; (b) Gabrienko AA, Arzumanov SS, Luzgin MV, Stepanov AG, Parmon VN. Methane activation on Zn2+-exchanged ZSM-5 zeolites: the effect of molecular oxygen addition. J Physical Chem C, 119 (44) (2015): 24910–24918.

52 Xu J, Zheng A, Wang X, Qi G, Su J, Du J, Gan Z, Wu J, Wang W, Deng F. Room temperature activation of methane over Zn modified H-ZSM-5 zeolites: insight from solid-state NMR and theoretical calculations. Chem Sci, 3 (10) (2012): 2932–2940.

53 Lapidus AL, Dergachev AA, Kostina VA, Mishin IV. Zinc-containing zeolite catalysts for ethane aromatization prepared by solid-state modification. Russ Chem Bull., 52 (5) (2003): 1094–1099.

54 Zaikovskii VI, Vosmerikova LN, Vosmerikov AV. Ethane aromatization on galloaluminosilicate modified with platinum and palladium. Kinet Catal, 53 (6) (2012): 731–736.

55 Mroczek U, Reschetilowski W, Pietzsch K, Steinberg KH. Aromatization of ethane on Pt and Pd supported Zn- and Mn-ZSM-5 zeolites. React Kineti Catal Letters, 43 (2) (1991): 539–544.

56 Haag WO, Huang TJ. Conversion of olefinic naphtha. Google Patents: 1978. Available from: https://patents.google.com/patent/CA1105494A/en.

57 Bartholomew CH, Farrauto RJ. Fundamentals of industrial catalytic processes. [electronic resource]. 2nd ed. Hoboken, N.J. : Wiley; 2006.

58 Baerlocher CM., L.B. Database of zeolite structures. Available from: http://www.iza-structure.org/databases/.

59 Derouane EG, Dejaifve P, Gabelica Z, Vedrine JC. Molecular shape selectivity of ZSM-5, modified ZSM-5 and ZSM-11 type zeolites. Faraday Discussions of the Chemical Society, 72 (0) (1981): 331–344.

60 Natesakhawat S, Means NC, Howard BH, Smith M, Abdelsayed V, Baltrus JP, Cheng Y, Lekse JW, Link D, Morreale BD. Improved benzene production from methane dehydroaromatization over Mo/HZSM-5 catalysts via hydrogen-permselective palladium membrane reactors. Catal Sci Technol, 5 (11) (2015): 5023–5036.

61 Xue J, Chen Y, Wei Y, Feldhoff A, Wang H, Caro J. Gas to liquids: natural gas conversion to aromatic fuels and chemicals in a hydrogen-permeable ceramic hollow fiber membrane reactor. ACS Catal, 6 (4) (2016): 2448–2451.

62 Morejudo, SH, Zanón, R, Escolástico, S., et al. Direct conversion of methane to aromatics in a catalytic co-ionic membrane reactor. Sci, 353 (6299) (2016): 563–566.

63 Fouty N, Carrasco J, Lima F. Modeling and Design Optimization of Multifunctional Membrane Reactors for Direct Methane Aromatization. Membranes, 7 (3) (2017): 48.

64 Nakagawa K, Kajita C, Ide Y, et al. Promoting effect of carbon dioxide on the dehydrogenation and aromatization of ethane over gallium-loaded catalysts. Catal Letters, 64 (2) (2000): 215–221.

65 Nishi K, Masato E., Atsushi S, Tadashi H, Yuichi M. Effect of carbon dioxide on aromatization of ethane over metal-loaded HZSM-5 catalysts. Sekiyu Gakkaishi, 39 (4) (1996): 260–266.

66 Chu YF, Chester AW. Process for converting propane to aromatics over zinc-gallium zeolite. Google Patents: 1984. Available from: https://www.google.com/patents/US4490569.

67 McCullen SB, Rodewald PG. Stabilization of zinc on catalysts. Google Patents: 1989. Available from: https:/www.google.com/patents/US4849568.

15

Techno-Economic Analysis of Monetizing Shale Gas to Butadiene

Ecem Özinan and Mahmoud M. El-Halwagi

Artie McFerrin Department of Chemical Engineering, Texas A&M University, USA

CHAPTER MENU

15.1 Introduction

Butadiene or 1,3 butadiene (C_4H_6) is a colorless diolefin, which is gaseous at 298 K. The global production capacity of butadiene is about 13 million tons per year. It is a major building block in the petrochemical industry that is used in manufacturing important products such as styrene-butadiene rubber (SBR), polybutadiene, styrene-butadiene latex, acrylonitrile-butadiene-styrene resins, adiponitrile, and chloroprene. There are three common routes to produce 1,3 butadiene. These are [1]:

- Steam cracking or naphtha cracking
- Catalytic dehydrogenation of n-butane and n-butene (also known as the Houdry process)
- Oxidative dehydrogenation of n-butene

Among these production routes, steam cracking is the most conventional production route where naphtha, ethane, propane, or butane is cracked to produce ethylene as the main product and 1,3 butadiene as a side product of this process. Naphtha or ethane is fed to a pyrolysis furnace where cracking takes place at a relatively high temperature (around 1,100 K) [2]. The cracking products such as hydrogen, ethylene, propylene, butadiene, and heavier hydrocarbons are quenched. The cooled gases are compressed, and the heavier components are first separated. Next, a separation train is used to obtain relatively pure streams of ethylene, propylene and 1,3 butadiene. The most important part of the steam cracking process is the pyrolysis/cracking. The type of the pyrolysis furnace depends on the feedstock of the process unit. For example, if the process unit uses naphtha as the feed, then the cracker is designed to produce heavier components and is classified as a heavy cracker. On the other hand, if the process unit utilizes ethane as the feed, the cracker is designed to crack lighter components and is classified as a light cracker [3, 4]. The distribution of the products can be controlled through design and operating

Natural Gas Processing from Midstream to Downstream, First Edition.
Edited by Nimir O. Elbashir, Mahmoud M. El-Halwagi, Ioannis G. Economou, and Kenneth R. Hall.

conditions (most notably temperature). Market conditions and the price gap between ethylene and butadiene are the key factors to determine the rate of butadiene production.

The recent discoveries of substantial reserves of shale gas have resulted in a major impact on butadiene production. Naphtha has been gradually replaced with ethane as a feedstock to the crackers [3]. This shift has led to the increase in the ethylene fraction in the cracking products. The shortfall in butadiene production has led to a global shortage and price increase. Furthermore, the abundant supply of shale gas has offered a competitive feedstock with clear economic and environmental advantages [5]. Therefore, the objective of this paper is to assess the production of butadiene from shale gas. A high-level techno-economic analysis is presented with comparison of two scenarios to illustrate the potential for using shale gas as a feedstock for the production of butadiene.

15.2 Process Description

The proposed process is based on the integration of two processing concepts that are related to two technologies: UOP's methane conversion to butadiene [6] and the Synfuels' cracking of methane to ethylene [7]. There are three primary reactions in the process:

Methane cracking to acetylene:

$$2CH_4 \rightarrow C_2H_2 + 3H_2$$

Hydrogenation of acetylene to ethylene:

$$C_2H_2 + H_2 \rightarrow C_2H_4$$

Dimerization of ethylene to butadiene:

$$2C_2H_4 \rightarrow C_4H_6 + H_2$$

In the first reaction, pipeline-quality shale gas (mostly methane) and oxygen are fed to a supersonic reactor whereby combustion first takes place leading to a high rise in temperature (to about 2500 K). At this high temperature, methane is cracked to produce a mixture with acetylene being the main product. Other cracking products include H_2, CO, and CO_2, and a variety of hydrocarbons. The hot gases leaving the cracker are quenched, and the cooled stream is pressurized to 12 atm before entering an amine sweetening part to remove CO_2. The separate syngas (H_2 and CO) is recycled back to the cracker to provide some of the heat of combustion. Next, acetylene is reacted with hydrogen (separated from the cracker products) to produce ethylene, which is separated to a 99.99 mol% purity. Figure 15.1 is a block flow diagram showing the key steps for converting shale gas to ethylene.

Next, ethylene is dimerized to butadiene and hydrogen. The product distribution is highly impacted by temperature. Table 15.1 shows the impact of temperature on the product distribution based on ASPEN Plus simulation using RGIBBS (equilibrium-based) reactor model.

As can be seen from Table 15.1, butadiene yield is enhanced with increasing temperature. Nonetheless, higher temperature entails higher cost. To assess the economic trade-offs, two scenarios are considered for the dimerization temperature: 1,273 K (Scenario I) and 2,773 K (Scenario II). In both scenarios, atmospheric pressure is used. To reduce the risk hazards, the hot reactor effluent is cooled down to 361 K using a water quenching tower. The cooled stream is then compressed to 25 atm and then cooled down to 298 K. The compressed and cooled stream is fed to a flash column that removes water as the bottom product. The top product is further separated in a second flash column where hydrogen is separated from the rest of the mixture (mostly ethylene and butadiene). Finally, a distillation unit is used to separate ethylene from butadiene and to obtain butadiene as a bottom product of the distillation column with a 99.5 wt.% purity. Figure 15.2 is a schematic representation of the ethylene-to-butadiene portion of the process.

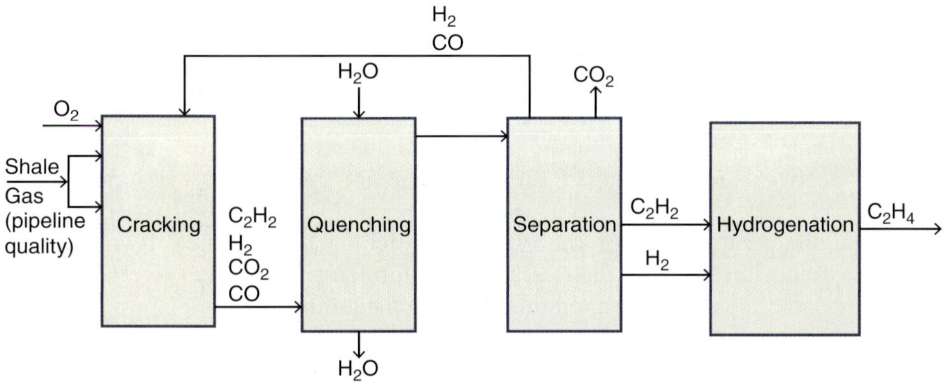

Figure 15.1 A Block Flow Diagram for the Shale Gas-to-Ethylene Portion of the Process.

Table 15.1 Effect of dimerization temperature on product distribution.

Temperature (K)	Ethylene (mol%)	Butadiene (mol%)	H_2 (mol%)
3,273	37.2872	31.3564	31.3564
2,273	43.4089	28.2956	28.2956
1,773	48.6405	25.6798	25.6798
1,573	51.4675	24.2662	24.2662
1,273	56.8765	21.5618	21.5618

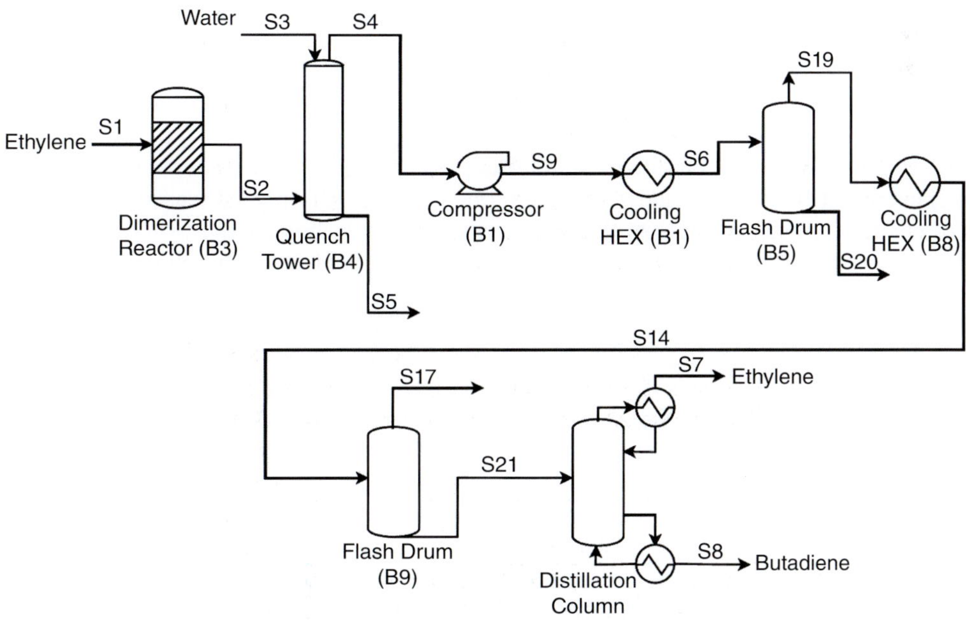

Figure 15.2 Flowsheet for the Ethylene-to-Butadiene Portion of the Process.

15.3 Techno-Economic Analysis

The process was simulated using ASPEN Plus® for a base case processing 10 MM standard cubic feet (SCF) per hour. The following assumptions were used in the simulation:

- Pipeline-quality shale gas was approximated with a methane stream.
- Conversion of acetylene to ethylene in the hydrogenation reactor was assumed to be complete.
- For the ethylene dimerization reactor, the RGIBBS model reactor was used. It is based on identifying the equilibrium product distribution by minimizing Gibbs free energy. Because of lack of kinetic data, the reaction was assumed to reach equilibrium.
- For the butadiene-ethylene distillation column, the RADFRAC model column was used with 20 stages, a kettle reboiler, and a reflux ration of 0.9. The use of RADFRAC is a reasonable choice for distilling similar mixtures as reported by various studies (e.g., [3, 4]).
- The number of operating hours per year was taken as 8,160 (corresponding to 93% on-stream efficiency).

Tables 15.2 and 15.3 summarize the stream data and key results from the simulation of the two scenarios.

Next, an economic analysis was carried out to assess the profitability of the process. Table 15.4 summarizes the economic data used for the raw materials, products, and utilities in the base-case assessment. A 10-year linear depreciation scheme was used for annualizing the fixed cost and a tax rate, of 30% was used in evaluating the net profit.

Table 15.5 shows the main results of the economic evaluation for the two scenarios of the dimerization temperatures (1,273 and 2,773 K). For safety reasons, the dimerization temperature was not allowed to exceed 2,773 K. The higher dimerization temperature leads to higher sales of butadiene but incurs additional fixed and operating costs due primarily to the increased size of heat exchangers and the additional utility cost. The benefits from the higher yield at the higher temperature outweigh the additional costs as evident by the higher return on investment (ROI) for Scenario II compared to Scenario I. In many cases, a minimum (threshold) ROI is required before a process is recommended. For a process of this type, the minimum acceptable ROI is typically around 10% [8]. Other factors such as safety and sustainability may be used to override this minimum [9].

An important factor in the butadiene market is price volatility. Market reports show that between January 2016 and February 2017, butadiene prices sharply increased from $750 to $2,000 per ton [10]. Such large volatilities can significantly impact the process viability. Figure 15.3 shows the ROI for both scenarios as a function of the butadiene selling price.

15.4 Conclusions

This work has assessed the techno-economic viability of manufacturing butadiene from shale gas. A process flowsheet was synthesized to convert shale gas to ethylene, which is subsequently dimerized to butadiene. Computer-aided process simulation was used to design the process and to obtain the data necessary for economic evaluation. Capital and operating costs were evaluated for two scenarios involving two dimerization temperatures (1,273 and 2,773 K). The second scenario was shown to be more profitable because of the enhanced yield of butadiene, which results in higher sales that offset the increase in capital and operating costs. A sensitivity analysis was carried out to assess the impact of price volatility of butadiene on the economic feasibility of the process.

Table 15.2 Simulation results and stream data – Scenario I.

	Units	S1	S14	S17	S19	S2	S20	S21	S3	S4	S5	S6	S7	S8	S9
From			B8	B9	B5	B3	B5	B9		B4	B4	B2	B6	B6	Bl
To		B3	B9		B8	B4		B6	B4	Bl		B5			B2
Substream: MIXED															
Phase:		Vapor	Mixed	Vapor	Vapor	Liquid	Liquid	Liquid	Liquid	Vapor	Liquid	Mixed	Vapor	Liquid	Vapor
Component Mole Flow															
ETHYLENE	KMOL/HR	1683.315	970.4187	62.3229	970.4187	970.4369	0.0170738	908.0958	0	970.4357	0.0011349	970.4357	908.0893	0.0065796	970.4357
BUTADIEN	KMOL/HR	0	356.4386	0.188337	356.4386	356.4391	0.0003345	356.2502	0	356.4389	0.0001528	356.4389	18.47516	337.7749	356.4389
H2	KMOL/HR	0	356.4379	343.8721	356.4379	356.4391	0.0011189	12.5658	0	356.439	5.26E-05	356.439	12.5658	4.87E-12	356.439
WATER	KMOL/HR	0	0.5285489	7.67E-05	0.5285489	0	2731.761	0.5284723	3366.63	2732.289	634.3408	2732.289	0.0002556	0.5282164	2732.289
Mole Flow	KMOL/HR	1683.315	1683.824	406.3834	1683.824	1683.315	2731.779	1277.44	3366.63	4415.603	634.3422	4415.603	939.1305	338.3097	4415.603
Mass Flow	KG/HR	47223.32	47232.3	2451.786	47232.3	5	49213.93	44780.51	60650.78	96446.23	11427.87	96446.23	26500.01	18280.5	96446.23
Volume Flow	L/MIN	2970530	5717.567	4469.252	19803.41	3920.958	805.9251	1248.314	1016.992	2199630	205.0902	24131.26	10013.57	664.9128	177401
Temperature	C	1000	−73.15	−73.15	0	1000	0	−73.15	25	88.80075	90.11547	25	−8.701455	118.9733	462.9901
Pressure	BAR	1	25	25	25	1	25	25	1	1	1	25	25	25	25
Vapor Fraction		1	0.2413456	1	1	0	0	0	0	1	0	0.3817999	1	0	1
Liquid Fraction		0	0.7586544	0	0	1	1	1	1	0	1	0.6182001	0	1	0

(Continued)

Table 15.2 (Continued)

Solid Fraction		0	0	0	0	0	0	0	0	0	0	0	0	0	0
Molar Enthalpy	CAL/ MOL	31231.68	8836.551	1203.258	12093.95	32618.03	−69202.68	11264.88	−68725.85	−30274.28	−67454	−37765.3	11887.44	24174.09	−25951.49
Mass Enthalpy	CAL/ GM	1113.28	315.0216	199.44	431.1472	1162.697	−3841.319	321.3498	−3814.865	−1386.049	−3744.261	−1729.011	421.2776	447.3799	−1188.138
Enthalpy Flow	MMBTU/ HR	208.6257	59.04542	1.940445	80.81117	217.8862	−750.1969	57.10497	−918.1683	−530.4819	−169.8001	−661.7434	44.30177	32.45423	−454.7356
Molar Entropy	CAL/ MOL–K	12.47247	−33.31037	−10.46099	−20.52812	1.336322	−41.83414	−40.57929	−40.11986	−8.058812	−36.26272	−32.18164	−21.29392	−41.96282	−6.304887
Mass Entropy	CAL/ GM–K	0.4445918	−1.187509	−1.733909	−0.731824	0.0476343	−2.32214	−1.157593	−2.22699	−0.368957	−2.012884	−1.473374	−0.7546326	−0.7765887	−0.283657
Molar Density	MOL/ CC	9.44E-06	0.0049083	0.0015155	0.0014171	0.0071552	0.0564936	0.0170555	0.055173	3.35E-05	0.0515498	0.0030497	0.0015631	0.0084801	0.0004148
Mass Density	GM/CC	0.000264955	0.1376818	0.0091432	0.0397509	0.2007303	1.017752	0.5978797	0.993957	0.0007308	0.9286863	0.0666122	0.0441068	0.458218	0.009061
Average Molecular	Weight	28.05376	28.05062	6.033183	28.05062	28.05376	18.01534	35.05488	18.01528	21.84214	18.01531	21.84214	28.2176	54.03481	21.84214

Table 15.3 Simulation results and stream data – Scenario II.

	Units	S1	S14	S17	S19	S2	S20	S21	S3	S4	S5	S6	S7	S8	S9
From			B8	B9	B5	B3	B5	B9		B4	B4	B2	B6	B6	Bl
To		B3	B9		B8	B4		B6	B4	Bl		B5			B2
Substream: MIXED	D														
Phase:		Vapor	Mixed	Vapor	Vapor	Liquid	Liquid	Liquid	Liquid	Vapor	Liquid	Mixed	Vapor	Liquid	Vapor
Component Mol Flow															
ETHYLENE	KMOL/HR	1683.315	675.417	0.2916998	675.417	675.5221	0.1045989	675.1253	0	675.5216	0.0005637	675.5216	675.1225	0.0028205	675.5216
BUTADIEN	KMOL/HR	0	503.8885	1.14E-05	503.8885	503.8964	0.0077504	503.8885	0	503.8963	0.0001678	503.8963	0.8661751	503.0223	503.8963
H2	KMOL/HR	0	503.8836	497.2255	503.8836	503.8964	0.0127994	6.658062	0	503.8964	5.37E-05	503.8964	6.658062	3.43E-12	503.8964
WATER	KMOL/HR	0	2.606007	9.87E-10	2.606007	0	9039.706	2.606007	10000	9042.312	957.6884	9042.312	8.96E-05	2.605918	9042.312
Mole Flow	KMOL/HR	1683.315	1685.795	497.5172	1685.795	1683.315	9039.831	1188.278	10000	10725.63	957.6891	10725.63	682.6468	505.6311	10725.63
Mass Flow	KG/HR	47223.32	47266.86	1010.531	47266.86	47223.32	162856	46256.33	180153	210123	17253.05	210123	19000	27256.33	210123
Volume Flow	L/MIN	2970530	4371.539	3297.281	23425.05	4024.866	2730.818	1074.258	3020.801	5457110	312.2215	26155.87	6546.976	990.487	486966
Temperature	C	1000	−153.15	−153.15	25	2500	25	−153.15	25	96.88343	97.45347	25	−20.28678	119.0322	557.1701
Pressure	BAR	1	25	25	25	1	25	25	1	1	1	25	25	25	25
Vapor Fraction		1	0.2951232	1	1	0	0	0	0	1	0	0.1571745	1	0	1
Liquid Fraction		0	0.7048768	0	0	1	1	1	1	0	1	0.8428255	0	1	0
Solid Fraction		0	0	0	0	0	0	0	0	0	0	0	0	0	0
Molar Enthalpy	CAL/MOL	31231.68	7560.754	−1208.643	12439.92	76509.9	−68714.87	11232.39	−68725.85	−46058.56	−67309.31	−55959.4	11539.2	23863.33	−41440.3
Mass Enthalpy	CAL/GM	1113.28	269.6579	−595.0541	443.6758	2727.26	−3814.229	288.5487	−3814.865	−2351.036	−3736.232	−2856.42	414.5894	442.6878	−2115.299
Enthalpy Flow	CAL/SEC	14603600	3540520	−167030	5825320	35775100	172550000	3707560	190910000	137220000	−17906000	166720000	2188110	3351680	123460000
Molar Entropy	CAL/MOL-K	12.47247	−41.21995	−12.60525	−19.68727	22.02398	−40.12712	−53.20057	−40.11986	−8.168241	−35.86812	−36.9145	−21.75846	−41.87009	−6.49392
Mass Entropy	CAL/GM-K	0.4445918	−1.470129	−6.205975	−0.702156	0.7850635	−2.227378	−1.366668	−2.22699	−0.4169437	−1.990982	−1.884282	−0.7817551	−0.7767305	−0.3314789
Molar Density	MOL/CC	9.44E-06	0.0064272	0.0025148	0.0011994	0.0069705	0.0551717	0.0184356	0.055173	3.28E-05	0.0511223	0.0068344	0.0017378	0.0085081	0.0003671
Mass Density	GM/CC	0.000265	0.1802068	0.0051079	0.0336298	0.1955482	0.9939405	0.717648	0.993957	0.0006417	0.9209833	0.1338916	0.0483683	0.4586351	0.0071916
Average Molecular	Weight	28.05376	28.03832	2.031147	28.03832	28.05376	18.0154	38.9272	18.01528	19.59075	18.01529	19.59075	27.83284	53.90556	19.59075

Table 15.4 Cost data used in the base-case assessment.

Item	Price	Unit
Shale gas	2.51	$/1000 SCF
Oxygen	200	$/metric ton
Hydrogen	1,500	$/metric ton
Butadiene	2,000	$/metric ton
Low-pressure steam	4	$/metric ton
Cooling utility	2	$/MM 10^9 J removed
Electric energy	0.05	$/kWh

Table 15.5 Summary of the economic evaluation results.

Item	Results for Scenario I (dimerization temperature = 1,273 K)	Results for Scenario II (dimerization temperature = 2,773 K)
Raw material cost (MM$/yr)	262.26	262.26
Utility cost (MM$/yr)	321.91	347.21
Annual operating cost (MM$/yr)	584.17	609.47
Fixed capital investment (MM$)	441.46	599.54
Working capital investment (MM$)	77.91	105.80
Total capital investment (MM$)	519.37	705.34
Annual sales of butadiene (MM$/yr)	300.00	444.06
Annual sales of hydrogen (MM$/yr)	296.16	298.06
Annual sales of products (MM$/yr)	596.16	742.12
Annual after-tax profit (MM$)	21.64	110.84
ROI (yr^{-1}%)	4.17	15.71

Figure 15.3 Effect of Price Volatility of Butadiene on ROI for the two Dimertization Temperature Scenarios.

References

1 Al-Douri A, Sengupta D, El-Halwagi MM. Shale gas monetization: a review of downstream processing to chemicals and fuels. J Nat Gas Sci Engg, 45 (2017): 436–455.

2 Gary JH, Handwerk GE, Kaiser MJ. Petroleum refining: technology and economics. 5th ed. Boca Raton, FL: CRC Press; 2007.

3 Thiruvenkataswamy P, Eljack FT, Roy N, Mannan MS, El-Halwagi MM. Safety and techno-economic analysis of ethylene technologies. J Loss Prevent Proc Indus, 39 (2016): 74–84.

4 Jasper S, El-Halwagi MM. A techno-economic comparison of two methanol-to-propylene processes. Processes, 3 (2015): 684–698.

5 Hasaneen R, El-Halwagi MM. Integrated process and microeconomic analyses to enable effective environmental policy for shale gas in the United States. Clean Technol Environ Policy, DOI: 10.1007/s10098-017-1366-5 (in press, 2017).

6 Bricker J, Chen JQ, Coughlin PK, Majumder D. Production of butadiene from a methane conversion process. U.S. Patent Application 13/915, 151 (2013).

7 Hall KR. A new gas to liquids (GTL) or gas to ethylene (GTE) technology. Catal Today, 106 (1) (2005): 243–246.

8 El-Halwagi MM. Sustainable design through process integration: fundamentals and applications to industrial pollution prevention, resource conservation, and profitability enhancement. 2nd ed. Amsterdam: Elsevier/IChemE; 2017.

9 El-Halwagi MM. A Return On Investment Metric For Incorporating Sustainability In Process Integration And Improvement Projects. Clean Technol Environ al Policy, 19 (2017): 611–617.

10 Dang T, Yan H, Wang J, Ong C, Weddle N, Hurley M, Love D. Butadiene value chain. ICIS. Available from: www.icis.com.

16

Fractionation of the Gas-to-Liquid Diesel Fuels for Production of On-Specification Diesel and Value-Added Chemicals

Mostafa Shahin[1], *Shaik Afzal*[2], *and Nimir O. Elbashir*[2,3,4]

[1] *ORYX GTL Company, Ras Laffan, Qatar*
[2] *Chemical Engineering Program, Texas A&M University at Qatar, Qatar*
[3] *Petroleum Engineering Program, Texas A&M University at Qatar, Qatar*
[4] *TEES Gas and Fuels Research Center, Texas A&M University, USA*

CHAPTER MENU

16.1 Introduction

The successful implementation of the gas-to-liquid (GTL) fuels in the existing infrastructure has to go through various steps. Since crude-oil-derived fuels have been around for much longer, all of its products are well studied in terms of their specifications and safe usage. However, GTL-derived fuels are relatively newer and have not yet been studied as comprehensively as crude-oil-based products. Additionally, GTL-based fuels offer some challenges that should be overcome before they can be used in the same supply chain for the same transportation vehicles.

The heart of GTL technology is the Fischer-Tropsch (FT) process. This process converts synthesis gas or syngas (a mixture of hydrogen and carbon monoxide) over a catalyst to produce pure synthetic crude or syncrude. Synthetic crude is a mixture of mainly paraffinic and other hydrocarbons. The synthetic crude produced from FT can be further processed in hydrocrackers to produce liquefied petroleum gas (LPG), naphtha, jet fuel, diesel fuel, and lubricant oils. The synthetic crude reacts with hydrogen in the hydrocracker units and is then separated into different fractions in a distillation unit to produce the required hydrocarbon products' slates. Typically, the carbon number distribution for FT products ranges from C_1 to C_{60} (boiling range $-164\,°C$ to $620\,°C$) based on the selectivity of the FT catalyst and reaction media. Since GTL products are formed from the FT process, which produces mostly straight chain hydrocarbons, it lacks some of the components that contribute to important fuel specifications.

This chapter illustrates an approach to studying GTL fractions through experimental and modeling studies with the objective of bringing GTL fuel into the existing transportation sector. In this study, various fractions of GTL diesel have been studied for compliance with

Natural Gas Processing from Midstream to Downstream, First Edition.
Edited by Nimir O. Elbashir, Mahmoud M. El-Halwagi, Ioannis G. Economou, and Kenneth R. Hall.

the American Standard for Testing Material (ASTM) Diesel specifications so it can be used seamlessly in the diesel supply chain. Based on U.S. Energy Information Administration estimates for the year 2040, diesel will continue to contribute a sizeable 33% to the transportation sector [1]. This will correspond to approximately 36 million b/d by 2040 [2] indicating that diesel/gasoil will continue to play an important part in the transportation sector.

Conventional diesel is a complex combination of hydrocarbons produced by the distillation of crude oil and conversion process units such as hydrocracker and coker units in typical crude oil refineries. Conventional diesel carbon number distribution ranges from C_9 to C_{20} (boiling points are in the range of approximately 163–357 °C). For European markets, carbon numbers and final boiling-points values could reach up to C_{28} and 390 °C, respectively. Conventional diesel contains normal and branched-chain alkanes (paraffins), cycloalkanes (naphthenes), aromatics, and mixed aromatic cycloalkanes with normal alkanes being the most predominant constituent [3].

On the other hand, GTL diesel consists of a much simpler hydrocarbon combination, which mainly consists of normal and iso-paraffinic compounds. GTL diesel has a lesser density than conventional diesel and contains negligible amounts of aromatics, polycyclic aromatics, olefins, sulfur, nitrogen, and metals. The fuel is colorless as well as almost odorless and has a higher Cetane number than conventional diesel [4]. Table 16.1 shows the typical composition of GTL diesel and ultra low sulfur diesel (ULSD).

Conventional diesel fuels are roughly categorized into the following grades based on composition and application:

- Grade No. 1-D S15/S500/S5000: A special-purpose, light middle distillate fuel for use in diesel engine with 15/500/5000 ppm sulfur (maximum), respectively.
- Grade No. 2-D S15/500/5000: A general-purpose, middle distillate fuel for use in diesel engine with 15/500/5000 ppm sulfur (maximum), respectively.

Table 16.1 GTL diesel typical composition compared to ULSD [5].

General carbon type	Chemical structure	GTL diesel carbon content (Mole% C)	ULSD carbon content (Mole% C)
Aromatic carbon		0.5	12.9
Aliphatic carbon		99.5	87.1
CH carbon		0.8	7.0
CH_2 carbon		78.5	61.1
CH_3 carbon		20.1	18.9
Aromatic carbon breakdown	C_nH_{2n-6}	0.5	17.0
Peripheral unsubstituted aromatic carbon		0.3	
Heteroaromatic carbon		0.2	
Total (*n* + *iso*) paraffinic carbon	C_nH_{2n+2}	93.0	43.7
Total cycloparaffinic carbon	C_nH_{2n}	6.7	39.3
Cycloparaffinic CH		0.2	9.1
Cycloparaffinic CH_2		6.1	26.9
Cycloparaffinic CH_3		0.4	3.3

Table 16.2 ASTM D 975 requirements for diesel fuels [6].

Property	Test method	S15, S500, S5000 No. 1-D	S15, S500, S5000 No. 2-D	S15, S500, S5000 No. 4-D
Flash point, °C (°F), min	D 93	38 (100)	52 (126)	55 (131)
Water and sediment, % volume, max	D2709 D1796	0.05	0.05	 0.5
Distillation temperature, °C (°F), 90% volume recovered: Min/ max	D 86		Min–282 (540) Max–338(640)	
Kinematic viscosity, mm^2/sec at 40 °C (104 °F):	D 445			
Min		1.3	1.9	5.5
Max		2.4	4.1	24
Ash, % mass, max	D 482	0.01	0.01	0.1
Sulfur, ppm (μg/g), max	D 5453	15.0	15.0	–
% Mass, max	D 2622	0.05	0.05	–
% Mass, max	D 129	0.50	0.50	2.0
Copper strip corrosion rating, max. after 3 hours at 50 °C (122 °F)	D 130	No. 3	No. 3	–
Cetane number, min	D 613	40	40	30
One of the following must be met:				
(1) Cetane Index, min	D 4737-10	40	40	–
(2) Aromaticity, % volume, max	D 1319	35	35	–
Cloud point, °C (°F), max or	D 2500	Varies	Varies	–
LTFT/CFPP, °C (°F), max	D 4539/ D 6371			
Ramsbottom carbon residue, max (% mass on 10% distillation residue)	D 524	0.15	0.35	–
Lubricity, 60 °C, WSD, microns, max	D 6079	520	520	–

- Grade No. 4-D: A heavy distillate fuel, or a blend of distillate and residual oil, for use in low- and medium-speed diesel engines in applications involving predominantly constant speed and load.

ASTM D975 identifies the conventional diesel properties with respect to each grade as shown in Table 16.2.

Compared to conventional diesel, there are no ASTM specifications for GTL diesel; however, manufacturing and consuming companies for GTL diesel have their own specifications based on their targeted distribution markets, and a typical specification data sheet is shown in Table 16.3.

The next section describes the experimental study performed to better understand the properties of different fractions of GTL diesel and their compliance with important specifications. The modeling work that follows will help in supporting the industry in developing newer superior generations of GTL diesel fuels, in terms of specifications compared to those currently in the market. Building data banks for GTL diesel properties at different boiling

Table 16.3 Typical GTL diesel specifications [7].

Property	Unit	Min	Max	Typical (estimated)	Test Method
Specific Gravity @ 15 °C		0.75	0.81	0.78	ASTM D1298/ ASTM D4052
Color			1.5	1	ASTM D1500
Appearance			Clear & Bright	Clear & Bright	Visual
Flash point	°C	60		70	ASTM D93
Water & sediment	Vol %		0.05		ASTM D2709
Sediment by extraction	Mass %		0.01		ASTM D473
Total acid number	mg KOH/g		0.1		ASTM D974
Strong acid number	mg KOH/g		0		ASTM D974
Cetane number		70		74	ASTM D613
Distillation					ASTM D86
10% volume recovered	°C	Report			ASTM D86
50% volume recovered	°C		300		ASTM D86
90% volume recovered	°C	282	355		ASTM D86
95% volume recovered	°C	Report			ASTM D86
Final boiling point	°C		380		ASTM D86
Residue	Vol %		2.0		ASTM D86
Cloud point	°C		+5	2	ASTM D2500/ ASTM D5771
Pour point	°C		+1	-2	ASTM D97/ ASTM D5950
Carbon residue on 10% bottoms	Mass %		0.10		ASTM D4530
Sulphur content	Mass ppm		8	0	ASTM D4294/ ASTM D5453
Ash	Mass %		0.01	0	ASTM D482
Copper strip corrosion resistance (3h at 50 °C)			1	1	ASTM D130
Kinematic viscosity at 50 °C	cSt	1.9	4.0		ASTM D445
Oxidation stability	mg/100ml		1.0		ASTM D2274
Polycyclic aromatics	Vol%		0.1	0	ASTM D5186/ IP 391
Total aromatics	Vol%		5	0	ASTM D5186/ IP 391

ranges and carbon distributions will boost advanced and future research work in the field of GTL technology in general, and GTL fuels specifically.

16.2 Experimental Study to Measure Properties of GTL Diesel for Different Specifications

This section describes the experimental study undertaken to investigate the properties of different fractions of GTL diesel. The first aim of this work was to compare the properties of GTL

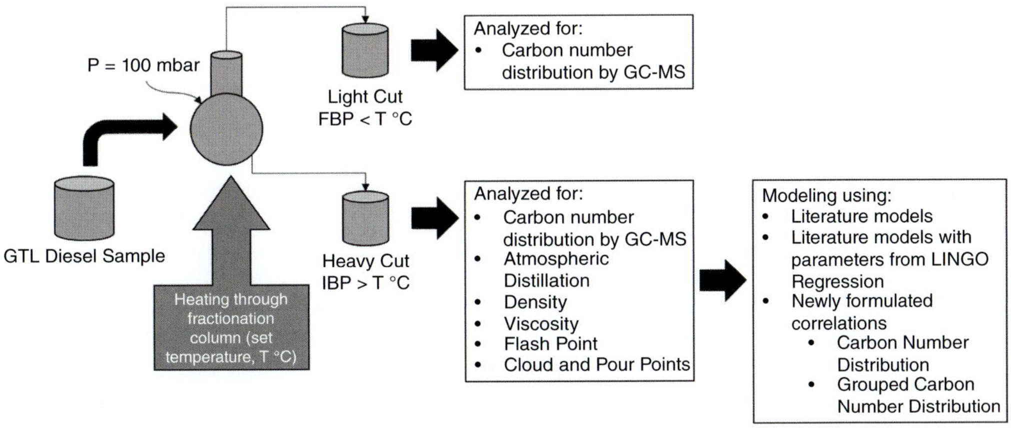

Figure 16.1 Schematic of Experimental and Modeling Study.

Table 16.4 GTL diesel IBP cuts.

Cut No.	1	2	3	4	5	6	7	8	9	10
IBP, °C	>130	>140	>150	>160	>170	>180	>190	>200	>210	>215

diesel cuts at different boiling ranges against their respective market specifications. The second aim was to develop mathematical correlations that will help in predicting the GTL diesel properties at various boiling ranges and carbon number contents. The modeling study is described in detail in section 16.4. Based on the results of the work, a recommendation is proposed to modify an existing GTL plant to produce more value-added products.

The schematic of this work is illustrated briefly in Figure 16.1 and shows the flow of GTL diesel samples after fractionation to experimental and modeling analyses.

The following steps summarize the proposed methodology used to perform this study:

- The first step involved the preparation of different samples of GTL diesel fuel at different boiling ranges (these will be referred to as cuts), by using a fractionation column module in Texas A&M University at Qatar's Fuel Characterization Lab (FCL). A typical GTL diesel fuel was used for the study and was fractionated at 100 mbar pressure into two sets of cuts: light cuts and heavy cuts based on different Initial Boiling Points (IBP), as shown in Table 16.4. The light cut is the distillate obtained after fractionation as shown in Figure 16.1, and the heavy cut refers to the residual fraction left behind in the fractionation column.

The greater than sign (>) is used to indicate that the actual IBP obtained from distillation at atmospheric conditions will be more than the value given here since vacuum was used in this fractionation to speed up the process. More details on this are mentioned in section 16.2.1.

- GTL diesel heavy cuts were analyzed for several physical properties of critical importance for the fuel's certification by the automotive industry. The analyzed properties included carbon number distribution, atmospheric distillation, density, viscosity, flash point, cetane index, pour point, and cloud point. Table 16.5 lists the standard testing methods that were used for this purpose. The results obtained for GTL diesel heavy cuts were compared with the market specification for conventional diesel engines to identify cuts that met the requirements.

Table 16.5 Standard testing methods.

Analysis	ASTM method/other	Test title	Test device
Fractionation	N/A	Fractionating of Petroleum Products	Spinning Band Distillation System, B/R Instrument Corporation
Distillation	D86	Distillation of Petroleum Products at Atmospheric Pressure	Petrotest ADU4+
Carbon distribution	D3238–95	Calculation of Carbon Distribution and Structural Group Analysis of Petroleum Oils by the n-d-M Method	Agilent 5975 Series MSD
Density	D4052	Density, Relative Density, and API Gravity of Liquids by Digital Density Meter	Anton Paar DMA 4100
Viscosity	D7042	Standard Test Method for Dynamic Viscosity and Density of Liquids by Stabinger Viscometer. (and the Calculation of Kinematic Viscosity)	Anton Paar SVM 3000
Flash point	D93 Procedure A	Flash Point by Pensky-Martens Closed Cup Tester	Automated Pensky-Martens Flash Point, Seta PM-93
Cetane index	D4737	Calculated Cetane Index by Four Variable Equation	Calculation
Cloud Point / Pour Point	D5773	Cloud Point	Phase Technology 70Xi series LAB analyzer

- For the light cuts, only the carbon number distribution was investigated, which is essential if the light cuts are routed to the production of petrochemicals.
- In the modeling part of the study described in section 16.4, conventional diesel models were tested for GTL diesel. For models which did not predict GTL diesel properties satisfactorily, new mathematical correlations were developed. Regression analysis was performed wherever applicable by a modeling software package, LINGO by LINDO Systems Inc.

The following sub-sections briefly describe the experimental setup used for the measurement of various GTL diesel specifications and the results are discussed in section 16.3.

16.2.1 Distillation

The first part of the experimental work was to prepare ten samples of GTL diesel at different initial boiling points (IBP) to represent various fractions that can be obtained from the particular sample of GTL diesel. This task was performed by utilizing a fractionation column that operated under regulated vacuum pressure. The fractionation column was a typical laboratory fractionation unit with a temperature controller for heating the pot, reflux system, vacuum pump, condenser, and receivers' flasks. Ten aliquots of 500 ml were taken from a typical GTL diesel tank. Each aliquot was split in the fractionation column, and the programmed temperature, T was set in the pot. According to ASTM petroleum-analysis distillation procedures, initial boiling point (IBP) is the recorded temperature when the first drop of distilled vapor is liquefied and falls from the end of the condenser [8].

Ten different programs were defined for the fractionator controller to have ten different GTL samples at different initial boiling points. All runs were under vacuum atmosphere of 100 mbar to accelerate the fractionation process. The heavy GTL diesel cut remaining in the pot flask was stored in a clean 500 ml glass bottle and labeled with "IBP > T °C". The light GTL diesel cut collected in the top receivers was stored in separate clean glass bottles and labeled with "FBP < T °C". All labeled samples were kept in a refrigerator to ensure the safety of the samples and to avoid any vaporization of light components from the samples at elevated temperatures. This completes the separation of the ten samples into the light and heavy components, ready for further analysis using other tests.

16.2.2 Atmospheric Distillation Analysis

After generating the ten different GTL diesel heavy cuts samples at different IBPs, the whole boiling range of each cut was tested with the approved test method of ASTM D86. The atmospheric distillation test[8] was performed using Petrotest ADU4+ automatic distillation unit.

16.2.3 Carbon Distribution

An Agilent gas chromatograph (GC) with mass selective detector was used to determine the carbon distribution and content for both light and heavy GTL diesel samples. The Agilent GC/MS system fitted with a mass selective detector performed quantitative and qualitative analyses of petroleum products. The column involved has the following specifications: HP5 column, MS grade, 30 m long, 250 μm diameter, and 0.25 μm thickness. The area normalization method for quantification of results was used. The NIST built-in library and MS built-in software were used for verification.

16.2.4 Density Analysis

Density was the first property of interest in comparing the ten different GTL diesel heavy cut samples. Normally, GTL diesel density is lower than that of conventional diesel. The experimental procedure involves injecting the sample into the testing tube that has a U-shaped passage. This U-shaped tube oscillates at a specific frequency to excite the electrons. A digital analyzer receives the signal carried by excited electrons and translates it into density [9]. All heavy GTL diesel samples from the fractionation were tested for density using Anton Paar DMA 4100 according to ASTM D4052.

16.2.5 Viscosity Analysis

The second property of interest was viscosity. Viscosity is the resistance of a fluid to flow and has the units of mPa.S for dynamic viscosity and mm^2/s (cSt) for kinematic viscosity. It is an important parameter for any fuel, especially in the design of any internal combustion engine, where the abrasion of the internal parts of the engines will be dramatically affected by the viscosity of the fuel used. All GTL diesel heavy cut samples generated from the fractionation column were analyzed using Anton Paar SVM 3000 by test method of D7042 [10].

16.2.6 Flash Point Analysis

The third characteristic specification of GTL diesel studied was the flash point. The flash point of a volatile material is the lowest temperature at which vapors of a fluid will ignite. It is an important parameter for any fuel for the design of internal combustion engines, where the flash point is needed to determine the design of the ignitor and its ignition temperature. Another important aspect for the flash point measurement is the storage safety temperature. In hot countries, diesel fuel flash point temperature should be high enough to avoid any fuel flashing and vaporization leading to hazardous scenarios. All the GTL diesel heavy cut samples generated from the fractionation column were analyzed using Automatic Pensky-Martens Closed Cup Flash Point Tester Model 35000-0, by test method ASTM D93 Procedure A [11]

16.2.7 Cloud and Pour Points Analysis

The fourth and the fifth specifications studied were the cloud and pour points. Cloud point is the temperature below which waxy material in diesel starts to give a cloudy appearance, while the pour point is the temperature at which the diesel becomes semi-solid and loses its fluidity. Both cloud and pour points are important properties in the design of diesel engines especially in cold countries where the diesel fuel can plug the injection nozzles if it has high cloud and pour points. All GTL diesel heavy cuts samples were analyzed following the test method of ASTM D5773 using the apparatus Phase Technology 70Xi series.

16.3 Experimental Study Results and Discussion

The trends of the GTL diesel properties indicate that they are dependent on both, the carbon number distribution and boiling range. The results of each analysis are described in separate sections:

16.3.1 GTL Diesel Fractionation

As indicated earlier in section 16.2, the ten samples were fractionated into two parts: light and heavy cuts. Table 16.6 and Figure 16.2 show that the percentage of light cuts volume increased by increasing the initial boiling point of the cut. It was noticed that the actual IBP of the ten samples when analyzed with ASTM D86 distillation was different from that entered in the fractionation column program. This can be attributed to different fractionating conditions, mainly applied pressure (i.e., 100 mbar vacuum vs. atmospheric pressure).

16.3.2 Atmospheric Distillation

The second test performed on the GTL diesel heavy cuts was the atmospheric distillation using ASTM D86 method. In distillation terminology, T_{10} refers to the temperature at which 10% of

Table 16.6 Light and heavy cuts percentage volume at different IBP's and 100 mbar.

Cut #	1	2	3	4	5	6	7	8	9	10
Test Set Temp., °C	130	140	150	160	170	180	190	200	210	215
Actual IBP at 1 atm	164	170	179	190	198	210	218	227	237	240
% Light Cut	0.8%	2.2%	5.2%	8.8%	12.0%	16.0%	19.6%	23.6%	26.8%	28.6%
% Heavy Cut	99.2%	97.8%	94.8%	91.2%	88.0%	84.0%	80.4%	76.4%	73.2%	71.4%

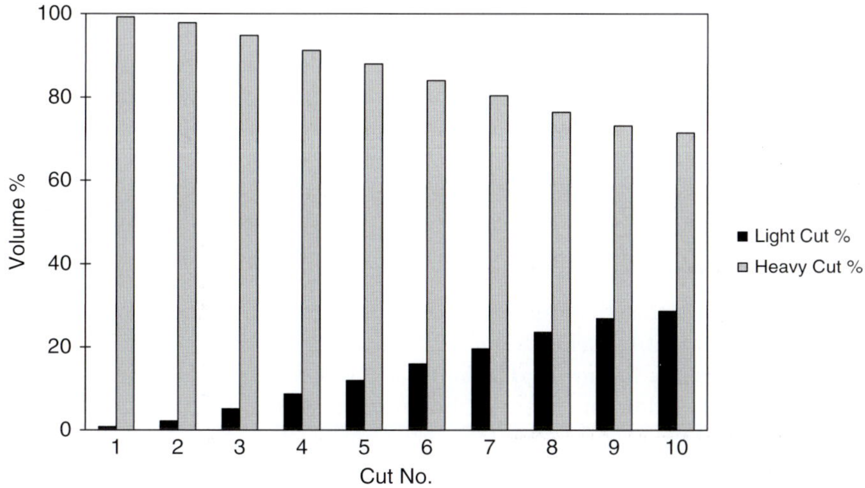

Figure 16.2 Volume % of Light and Heavy GTL diesel for different cuts.

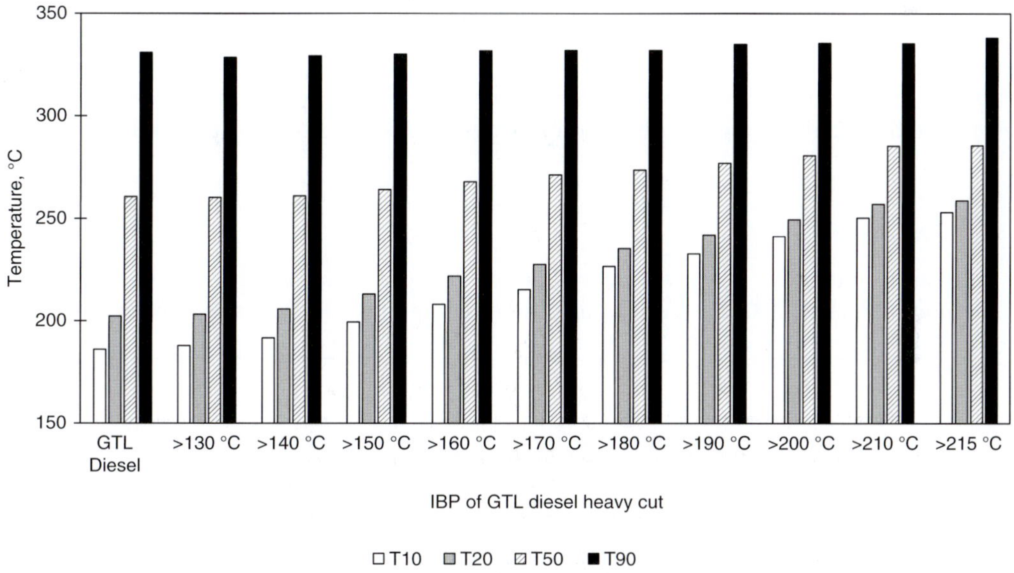

Figure 16.3 Change of T_{10}, T_{20}, T_{50} and T_{90} distillation data with IBP of the cut.

the given mixture vaporizes. Figure 16.3 shows that the distillation data of T_{10} and T_{20} values increased appreciably by increasing the IBP while the T_{50} values slightly increased. The T_{90} values were almost constant for all the cuts. The increase is due to separation of the light ends by the fractionation process, whereas the heavier hydrocarbon content remained in the heavy cuts.

The values of T_{10}, T_{20}, T_{50} and T_{90} are used in many mathematical correlations to calculate different physical properties; e.g., in the calculation of cloud points, equation (16.1) is used:

$$CP = a + bT_{50} + cT_{90} + dT_{90-20} + eT_{50} * T_{90} + fSG + gSG * T_{90} \tag{16.1}$$

Where CP is the cloud point, a, b, c, d, e, f, and g are regression coefficients, and SG is the specific gravity.

16.3.3 Carbon Distribution for GTL Diesel Heavy Cuts

The heavy cuts were analyzed using GC-MS for the carbon number distribution. The carbon number distribution data is represented using carbon number grouping as shown in Figure 16.4. The classification below is also related to specific fuel cuts which are better known to industry:

- (C_7+C_8+C_9+C_{10}+C_{11}) as Light Kerosene
- (C_{12}+C_{13}+C_{14}+C_{15}) as Heavy Kerosene
- (C_{16}+C_{17}+C_{18}+C_{19}) as Light Diesel
- (C_{20}+ C_{21}+C_{22}+C_{23}) as Heavy Diesel

The trend of Figure 16.4 shows that the Light Kerosene wt% reduced as the IBP increased while the Heavy Kerosene and Light Diesel wt% increased; in all cases, Heavy Diesel wt% was found to be less affected. This is due to the separation process of light cuts where light carbon numbers were removed, and the middle and heavy carbon numbers remained in the heavy cuts. The GTL diesel sample refers to the original sample without any fractionation.

16.3.4 Carbon Distribution for GTL Diesel Light Cuts

GTL diesel lights cuts were analyzed for carbon distribution by GC-MS. It was found that the light carbon numbers (C_7–C_9) reduced, while the heavier carbon numbers (C_{10}–C_{13}) increased by increasing the IBP. Figure 16.5 represents the overall carbon number distribution for the light cuts. This data can be used to investigate the feasibility of the light GTL diesel cuts for feedstock to the linear alkyl benzene (LAB) industry. GTL diesel Light cuts may become attractive because their carbon number distribution between C_9–C_{14} is the most suitable feedstock for LAB industry [12].

16.3.5 Density Analysis

The third test conducted was the density analysis using Anton Paar DMA 4100 based on test method of ASTM D4052. Figure 16.6 represents the measured density of each cut while

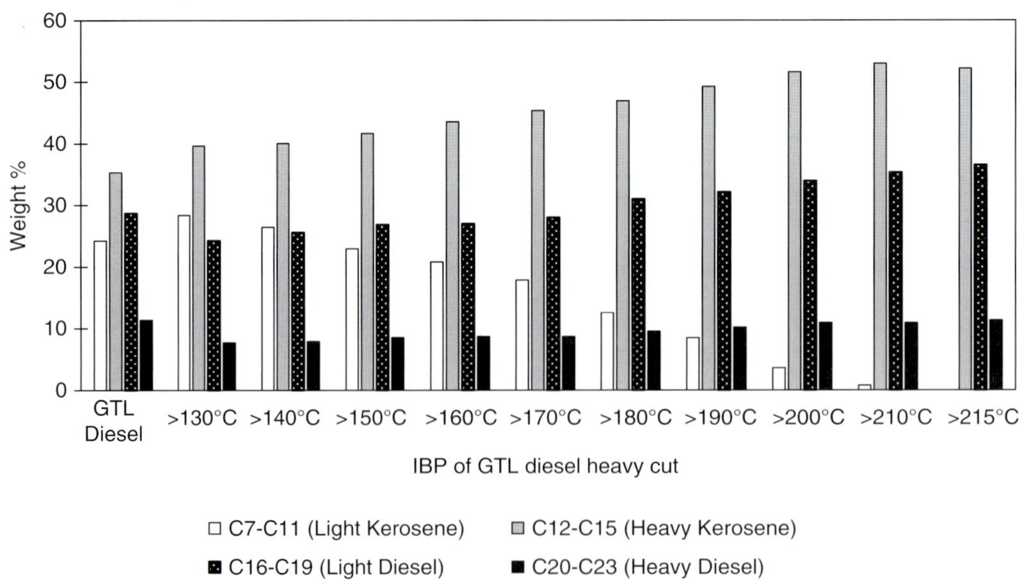

Figure 16.4 Effect of IBP on GTL diesel carbon number grouping for heavy cuts.

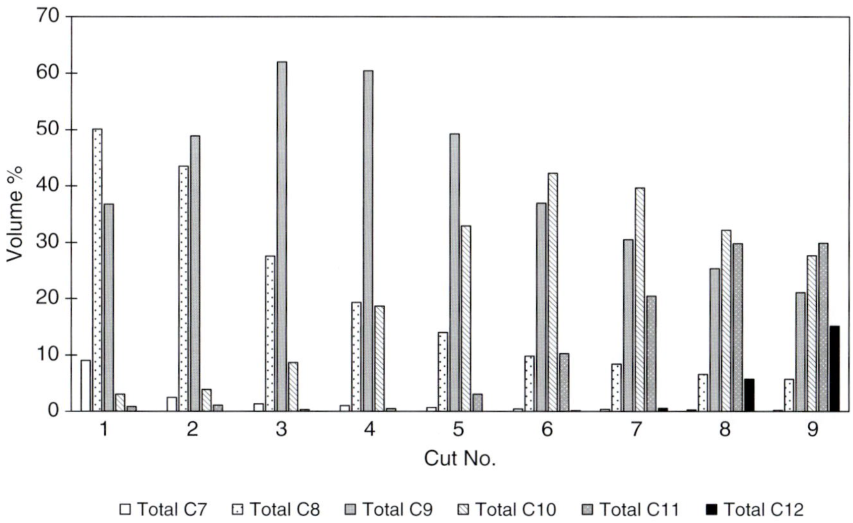

Figure 16.5 Carbon Number Distribution for GTL diesel Light cuts.

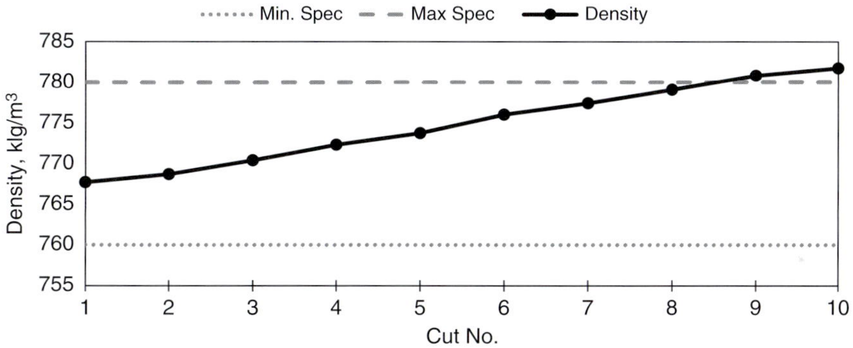

Figure 16.6 Density of GTL diesel heavy cuts compared with typical GTL market specification (min. 760 kg/m^3 and max. 780 kg/m^3).

comparing the minimum and maximum limits as per typical market specifications of min. 760 kg/m^3 to the max. 780 kg/m^3. The recorded readings for the density of GTL heavy cuts were bounded by a lower value of 767.7 kg/m^3 and an upper value of 781.7 kg/m^3.

The density profile from Figure 16.6 is quite representative since it shows that the higher the IBP of the cut, the higher the density, which is as expected [6]. All the density values fell between the limits of market specifications except the cuts numbered 9 and 10 (IBP>210 °C and >215 °C) respectively, where their densities exceeded the maximum limit of 780 kg/m^3.

Furthermore, as shown in Figure 16.7, the density change is almost linear with the change of the average boiling range. Figure 16.7 also includes a data point of the original GTL diesel sample without any fractionation. The average boiling range is the average of the distillation temperatures of IBP, T_{10}, T_{20}... to FBP.

16.3.6 Viscosity Analysis

The fourth test conducted for the GTL diesel heavy cuts was the viscosity analysis using Anton Paar SVM 3000 based on ASTM D-7042. As shown in Figure 16.8, the kinematic viscosity of all

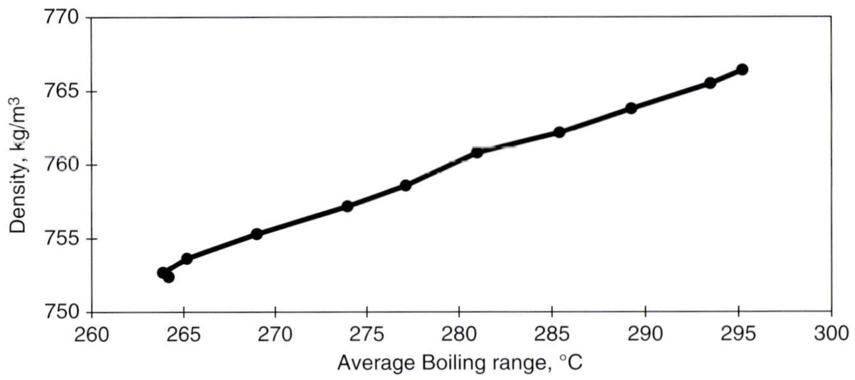

Figure 16.7 Change of GTL diesel heavy cuts with the change of average boiling range.

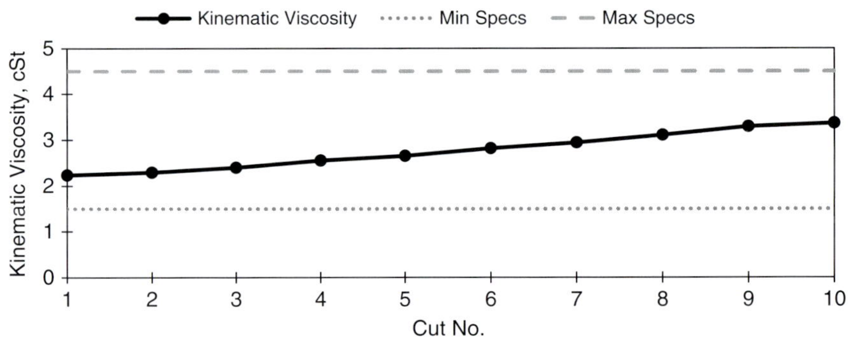

Figure 16.8 Kinematic viscosity of heavy GTL diesel cuts compared with market specification (1.5 to 4.5 cSt).

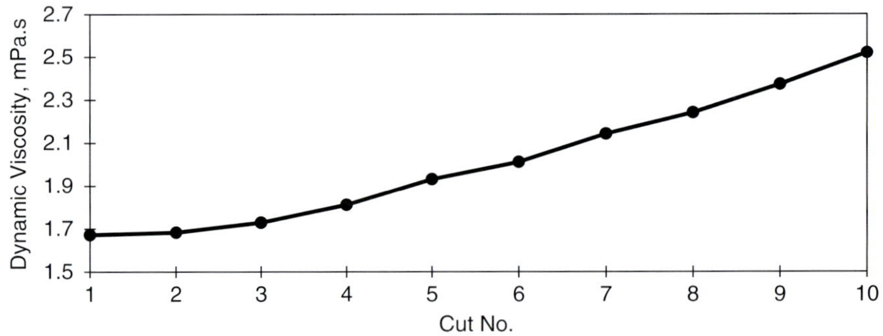

Figure 16.9 Change of dynamic viscosity of GTL diesel heavy cuts with the change of IBP.

cuts falls within the typical marketing specifications limits of 1.5 to 4.5 cSt. Figure 16.9 shows the dynamic viscosity results for the heavy GTL diesel cuts. The recorded readings for kinematic viscosity were bounded by a lower value of 2.24 cSt and an upper value of 3.36 cSt, while the recorded readings for dynamic viscosity were bounded by a lower value of 1.67 mPa.s and an upper value of 2.58 mPa.s.

Both kinematic and dynamic viscosity results showed an increase in their values as the IBP increased, i.e. the heavier the fuel cut, higher is the viscosity.

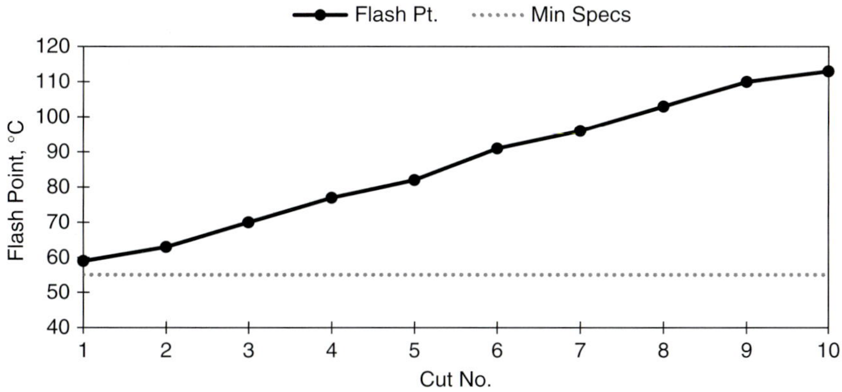

Figure 16.10 Flash point of GTL diesel heavy cuts compared with market specification.

16.3.7 Flash Point Analysis

The fifth test conducted for the GTL diesel heavy cuts was the flash point analysis using Automatic Pensky-Martens Closed Cup Flash Point Tester Model 35000-0, based on ASTM D-93 Procedure A. Figure 16.10 shows that the flash points of all fuel cuts are above the minimum marketing specifications limit of 55 °C. The recorded readings for the flash point were bounded by a lower value of 59 °C and an upper value of 113 °C. A higher GTL diesel flash point is better in terms of product safety storage temperature especially in hot countries like those in the Middle East, as it allows for safe storage in tanks at severe climatic conditions in summer when ambient temperatures may exceed 50 °C. Hence, it is important to ensure safe storage of these fuels in trucks/cars/tanks without vaporizing the volatile hydrocarbon, which may lead to a fire in case these vapors are subject to an ignition source or excessive heat.

16.3.8 Cloud and Pour Point Analysis

The sixth test conducted for the GTL diesel heavy cuts was the cloud point and the pour point analysis using the apparatus Phase Technology 70Xi series equipment, which is based on test MD-5773. Figure 16.11 shows the measured cloud point (CP) for the GTL diesel heavy cuts with respect to typical marketing specifications. The samples with IBP > 130 to > 170 °C were within the maximum allowable limit of −3 °C. However, the rest of the cuts exhibited higher than

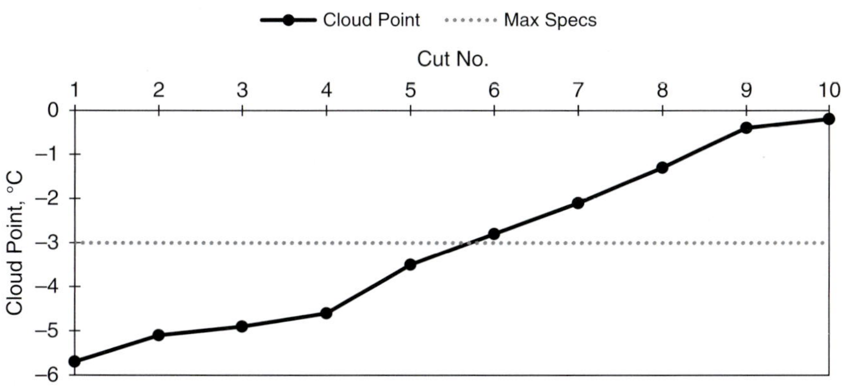

Figure 16.11 Cloud Point of GTL diesel heavy cuts compared with market specification.

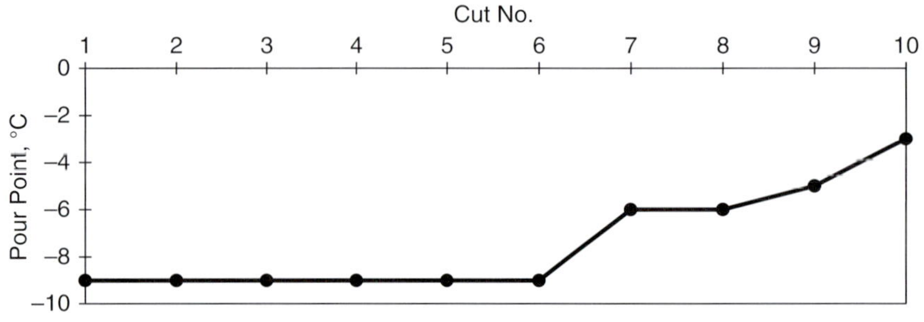

Figure 16.12 Pour Point of GTL diesel heavy cuts.

the market specification. This property is important specifically for the European and North American market since the ambient temperature in the winter may easily reach to −10 °C. The recorded readings for cloud point were bounded by a lower value of −5.7 °C and an upper value of −0.2 °C.

It should be noted here that if the heavier GTL diesel with higher cloud point is considered for Middle Eastern markets, the cloud point parameter will meet the required specification limit. This is because the lowest ambient temperature in winter in most of the region is around 0 °C. Therefore, cloud point and pour point specifications depend on the environment of the consumer market.

The pour point results are plotted in Figure 16.12. Due to possible accuracy issues with the measuring instrument, the pour points of cuts 1–5 are stagnant at −9 °C. The pour point for other cuts shows an increasing trend with increasing IBP.

16.3.9 Cetane Index Calculation

The ease of diesel fuel ignition and the manner in which it burns determine the ignition quality of the diesel fuel. The fuel must then be able to vaporize quickly and ignite without a flame or spark. The ignition quality of a diesel fuel is determined by its Cetane number rating, or Cetane Index [13]. The Cetane Index has been calculated for the GTL diesel heavy cuts as per the ASTM Standard Procedure (D4737-10) as shown in eq. (16.2) below, which is specified for conventional diesel grades as shown earlier in Table 16.2.

$$\begin{aligned} CI = {} & 45.2 + (0.0892)(T_{10N}) + [0.131 + (0.901)(B)][T_{50N}] \\ & + [0.0523 - (0.420)(B)][T_{90N}] + [0.00049][(T_{10N})^2 - (T_{90N})^2] \\ & + (107)(B) + (60)(B)^2 \end{aligned} \tag{16.2}$$

Where:

CI → Calculated Cetane Index by Four Variable Equation

D → Density at 15 °C, g/mL determined by Test Methods D1298 or D4052,

DN → D - 0.85,

B → $[e^{(-3.5)}(DN)] - 1$,

T_{10} → 10% recovery temperature, °C, determined by Test Method D86

T_{10N} → T_{10} - 215,

T_{50} → 50% recovery temperature, °C, determined by Test Method D86

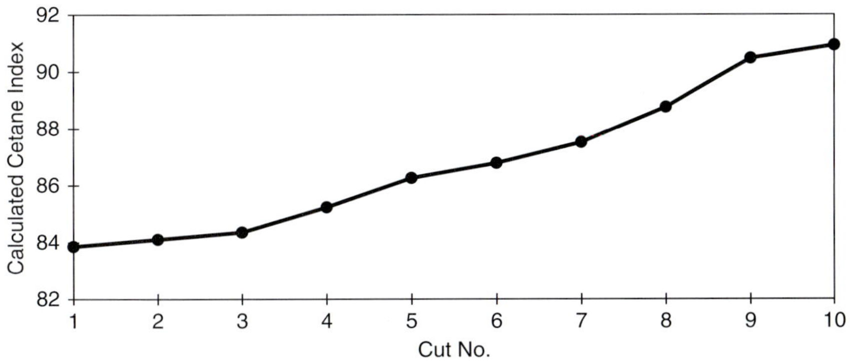

Figure 16.13 The impact of IBP change on GTL diesel heavy cuts CI.

$T_{50N} \rightarrow T_{50}$ - 260,

$T_{90} \rightarrow$ 90% recovery temperature, °C, determined by Test Method D86

$T_{90N} \rightarrow T_{90}$ - 310.

Figure 16.13 shows the results for the CI calculation for each fuel cut. The calculated results were bounded by a lower value of 83.9 and an upper value of 90.9. The CI was found to be increasing with increase in the IBP of GTL diesel cut. This is an expected correlation since CI calculation is a function of GTL diesel density as well as the boiling points data.

It should be noted here that CI is a modeling approach to calculate the Cetane number, which is used to study the ignition quality of fuel in compression ignition engines [14]. Cetane number (CN) is evaluated experimentally in standard test conditions on a standardized diesel engine. The models for CI are a quick way to estimate the CN of fuel. However, since the models for CI (e.g., equation (16.2)) are usually calculated by established correlations based on density and distillation data, these models cannot be used for diesel fuels with additives.

16.4 Mathematical Models for Properties-Composition Relationship

This section describes the second part of the study, which uses the experimental data obtained in the previous section to develop mathematical correlations for GTL diesel properties at different boiling ranges and their carbon distribution. Most of the developed correlations in the literature [15, 16] are specific to conventional diesel obtained from crude oil while very limited studies addressed similar correlations for GTL diesel [17].

GTL diesel properties of density, flash point, viscosity, and cloud point have been measured at the different boiling ranges and carbon number distributions as described in the previous section. This data has been used to verify existing models and correlations in the literature that have been developed for conventional diesel fuels. For the properties that failed to fit literature correlations, new correlations and regressions were investigated that can lead to better predictability of the GTL diesel fuels' properties based on their hydrocarbon composition.

Regression analysis is a statistical tool for estimating the relationships among variables. It is used for modeling and analyzing numerical data that can be used to investigate the relationship between two or more variables. Regression analysis aims to determine the strength of the relationship between one dependent variable and a series of other changing variables (known as independent variables). Regression can assist in prediction, estimation, and verification of empirical and calibrating models [18].

The composition-property prediction models for conventional diesel were tested for GTL diesel using the obtained experimental data from the lab analysis. The results of the experiments were used in model formulation based on established modeling methods from literature [19] and are explained in the three-step process:

1) Model Testing
The first step is to search for an existing model in literature. Once we find a model for the particular property, we test the accuracy of the parameters of the model using our data. This is called "model testing" and is done to check the validity of the model for application to GTL diesel. For instance, for CP, eq. (16.3) gives the formula as well as the regression coefficients for conventional diesel [16].

$$CP = a + bT_{50} + cT_{90} + dT_{90-20} + eT_{50} * T_{90} + fSG + gSG * T_{90} \tag{16.3}$$

The regression coefficients as calculated by Dinkov et al. [16] are listed below:

$a = 40.5188082034628$ $b = 0.352055808730715$ $c = 0.0213385486437754$

$d = -0.11817892070543$ $e = -0.00116227648075031$ $f = -345.341766942041$

$g = 0.91964970145254$

2) Calibrating Model Method
If the first method failed to predict the properties of GTL diesel reasonably, then the "calibrating model" method was used. This refers to the estimation of new coefficients for the model using laboratory data. The model results are then tested with actual experimental data. LINGO was used to estimate the model parameters. Half of the lab results (for cuts number 1, 3, 5, 7, and 9) were used to estimate the regression coefficients, and the rest of experimental results (for cuts 2, 4, 6, 8, and 10) were used to test the accuracy of these coefficients. For instance, for eq. (16.3), new coefficients have been estimated by regression using LINGO, and the coefficients are listed below. The performance of the model with experimental data is shown in Table 16.7.

$a = 1195.917$ $b = -5.382029$ $c = -2.632678$

$d = 0.1358335$ $e = 0.1992243E\text{-}01$ $f = 759.4667$

$g = -4.907416$

3) Empirical Model Method
If both the above methods failed to predict the GTL diesel properties, then the "empirical model" method was used. Since the model from literature for cloud point [16] as well as the newly estimated coefficients (method 2 above) have failed, this necessitates the formulation of a new model. In this study, new models were generated using similar models for conventional diesel but with some changes of the independent variables based on heuristics. The coefficients for this new model were estimated using LINGO. An example for the new model and its coefficients is shown below where the property, P (CP in this case) is estimated using eq. (16.4) and the model performance is shown in Table 16.9:

$$\begin{aligned} &P\text{ (cloud point)} \\ &= a_0 + a_1[\text{Light Kerosene}] + a_2[\text{Heavy Kerosene}] \\ &\quad + a_3[\text{Light Diesel}] + a_4[\text{Heavy Diesel}] \end{aligned} \tag{16.4}$$

Table 16.7 GTL diesel property modeling using distillation data.

<table>
<tr><th>GTL diesel property</th><th>Literature model performance</th><th>Using LINGO regression</th></tr>
<tr><td>Cloud point</td><td>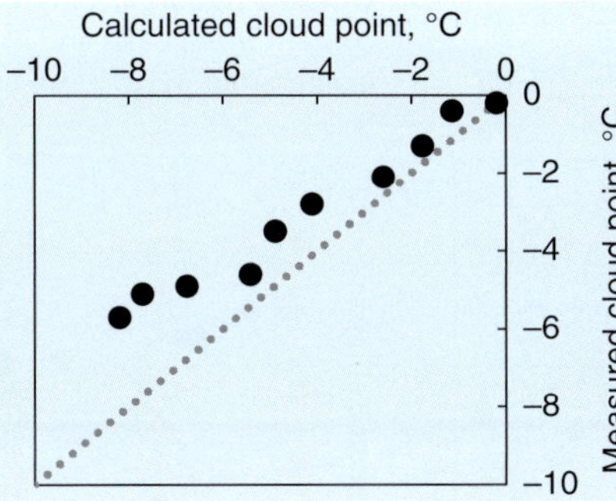
</td><td>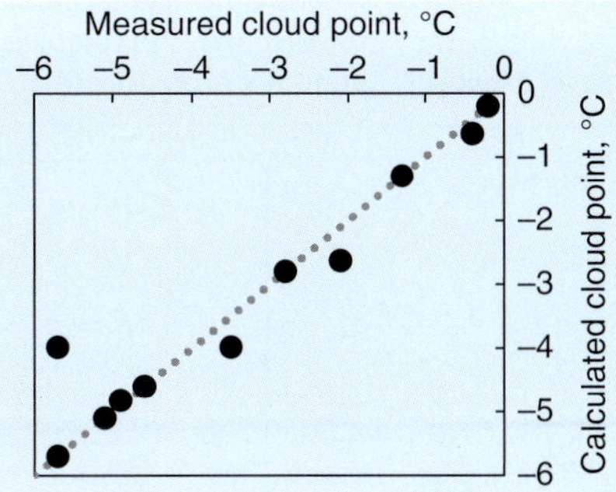
</td></tr>
<tr><td>Kinematic viscosity</td><td>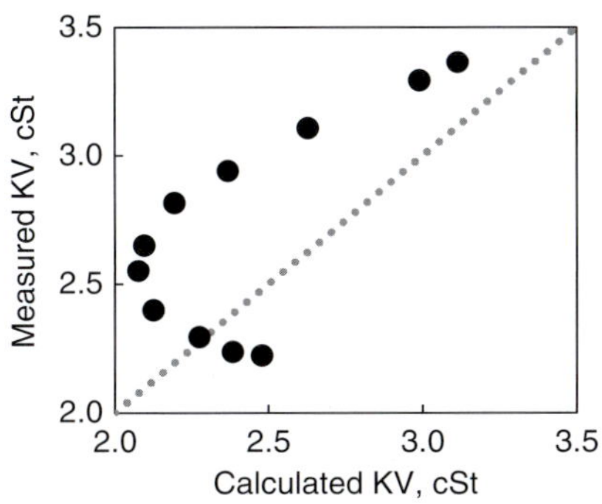
</td><td>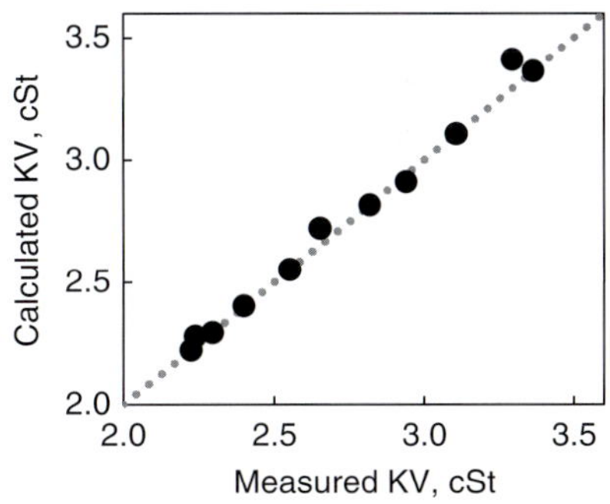
</td></tr>
</table>

where regression coefficients calculated using LINGO are,

$a_0 = 12.6769$ $a_1 = -0.3166$ $a_2 = -0.1909$

$a_3 = 0.1302$ $a_4 = -0.6788$

As can be seen from the above example, some of the regression coefficients have negative values. This is simply because of the process of regression which is a mathematical process and does not always have physical significance.

The figures in Table 16.7 to Table 16.9 show the process of model formulation and fitting as described above for the various GTL diesel properties. The 45-degree dotted line in the figures represents the perfect match (Ideal Model) between the actual laboratory measured properties for each cut and the calculated values obtained from the developed correlation. Some of the figures have 11 data points, and the 11th data point refers to the original GTL diesel sample without any fractionation. Shahin [20] lists all the detailed values of regression coefficients for all of the regression analyses of this study.

LINGO, used in the regression analysis of the work is a powerful, comprehensive software designed to build and solve linear, nonlinear, quadratic constrained, second-order cone, stochastic, and integer optimization models faster, easier, and more efficiently. LINGO provides a completely integrated package that includes a powerful language for expressing optimization models[18].

Table 16.8 GTL diesel property modeling using paraffin content data.

	Literature model performance	Using LINGO regression
GTL diesel property	$P = a_1[n] + a_2[BC] + a_3[Ar]$ where, total normal paraffin [n], total iso-paraffin (branched) [BC], and total Aromatics [Ar]	$P = a_0 + a_1[nP] + a_2[iP]$ where nP is normal paraffins content in wt. basis and iP are iso-paraffins in wt. basis
Cloud point	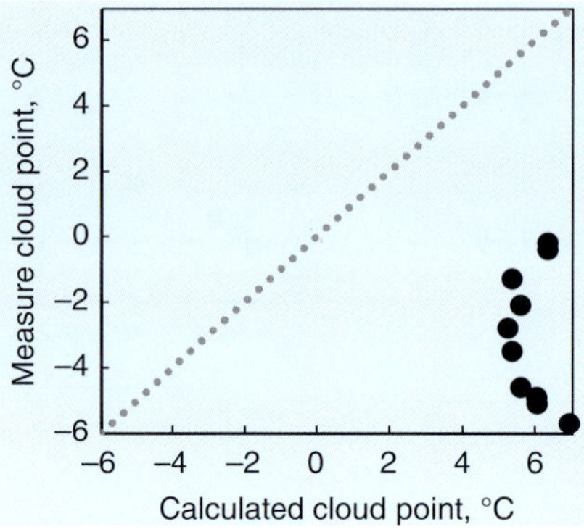	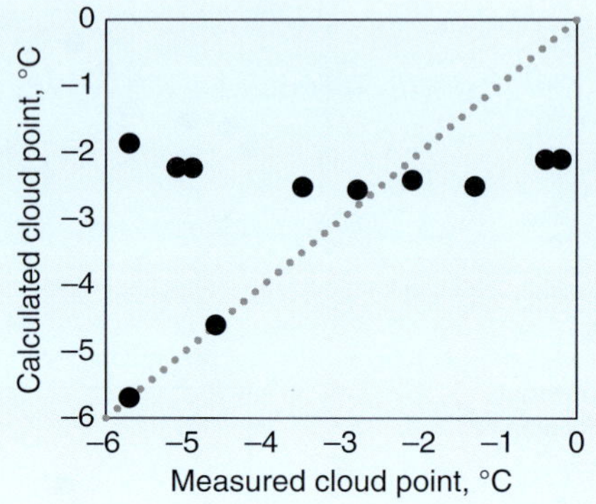
Density	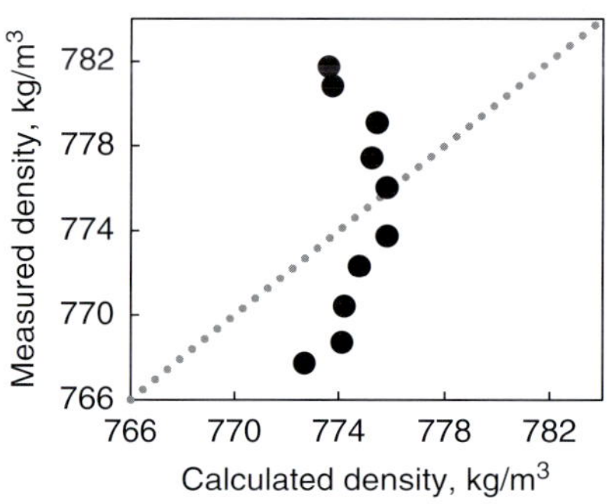	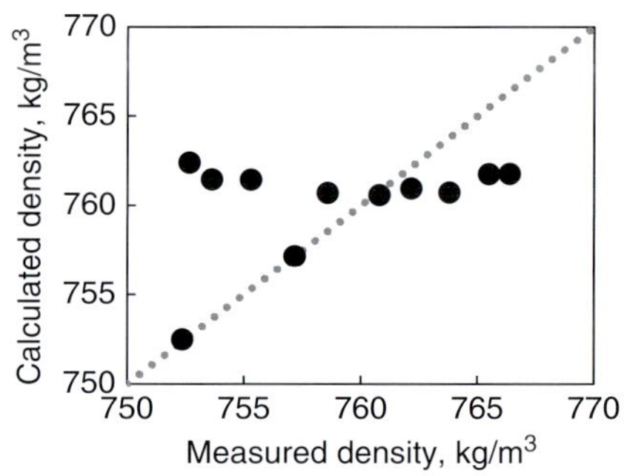
Dynamic viscosity		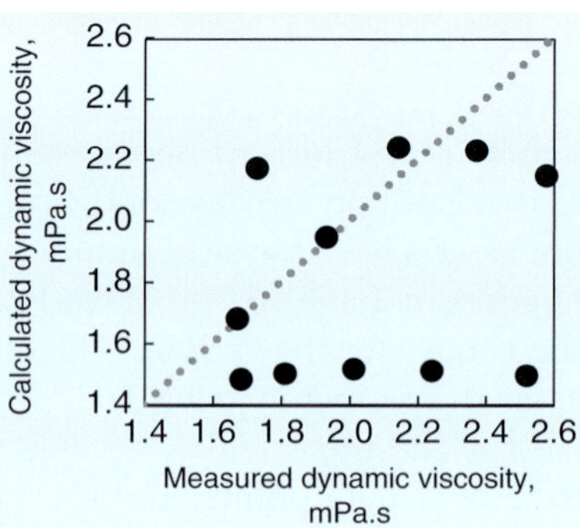

Table 16.8 (Continued)

GTL diesel property	Literature model performance $P = a_1[n] + a_2[BC] + a_3[Ar]$ where, total normal paraffin [n], total iso-paraffin (branched) [BC], and total Aromatics [Ar]	Using LINGO regression $P = a_0 + a_1[nP] + a_2[iP]$ where nP is normal paraffins content in wt. basis and iP are iso-paraffins in wt. basis
Flash point		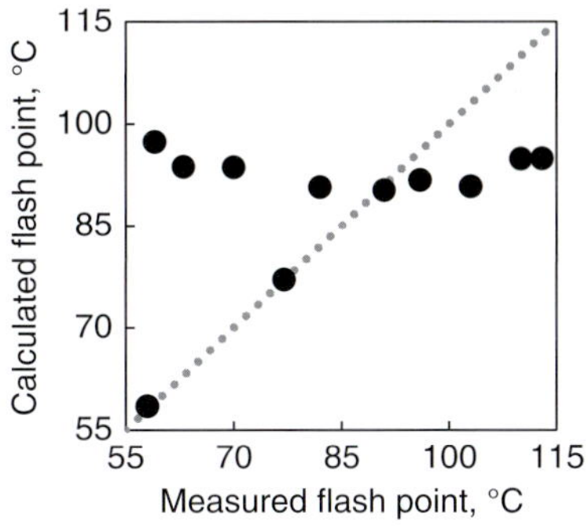

The first GTL diesel properties to be modeled based on Distillation data were CP and Kinematic Viscosity. Models reported by Dinkov et al. [16] for Cloud Point and Karonis et al. [15] for Kinematic Viscosity were used, and the respective plots are shown in Table 16.7. The formulae used in the models are listed below:

$$\text{Cloud Point} = a + bT_{50} + cT_{90} + dT_{90-20} + eT_{50} * T_{90} + fSG + gSG * T_{90} \tag{16.5}$$

Kinematic Viscosity, V_{40}

$$= a.DENS + b.DENS^2 + c.D_{10}^2 + \frac{d}{D_{10}} + e.D_{50}^2 + \frac{f}{D_{50}^2} + g.D_{90}^2 + \frac{h}{D_{90}} + i \tag{16.6}$$

where,

V_{40} : Kinematic Viscosity, cSt at 40 °C

DENS : Fuel Density g/ml at 15 °C

D_{10} : Distillation Temperature for 10% volume of fuel distilled

D_{50} : Distillation Temperature for 50% volume of fuel distilled

D_{90} : Distillation Temperature for 90% volume of fuel distilled

As shown, the literature models of Dinkov et al. [16] for Cloud Point and Karonis et al. [15] for Kinematic Viscosity are not accurate enough to predict these GTL diesel properties reliably. However, the model using LINGO Regression is much more reliable in the prediction of these properties.

Apart from distillation-data-based models, there are other models based on paraffin, iso-paraffin, and aromatics content for conventional diesel. To check their applicability for GTL diesel, a well-studied model developed by Cookson et al. [21] was applied for GTL diesel. The model is based on the normal paraffin, iso-paraffin, and aromatic content in the diesel fuel. The model was not formulated for Dynamic Viscosity and Flash Point and hence, only Cloud Point and Density were calculated. As shown in Table 16.8, the correlation was poor and not reliable.

Table 16.9 GTL diesel property modeling using individual carbon number data and grouped carbon number data.

GTL diesel property	**Individual carbon number based model** $P = a_0 + a_1[C_9] + a_2[C_{10}] + a_3[C_{11}] + a_4[C_{12}] + a_5[C_{13}] + a_6[C_{14}] + a_7[C_{15}] + a_8[C_{16}] + a_9[C_{17}] + a_{10}[C_{18}]$ where, $P \rightarrow$ GTL diesel property $C_i \rightarrow$ weight % of ith carbon number content a_0 to $a_{10} \rightarrow$ regression coefficients	**Grouped carbon number based model** $P = a_0 + a_1$[Light Kerosene] $+ a_2$[Heavy Kerosene] $+ a_3$[Light Diesel] $+ a_4$[Heavy Diesel] where, Light Kerosene $\rightarrow (C_8+C_9+C_{10}+C_{11})$ Heavy Kerosene $\rightarrow (C_{12}+C_{13}+C_{14}+C_{15})$ Light Diesel $\rightarrow (C_{16}+C_{17}+C_{18}+C_{19})$ Heavy Diesel $\rightarrow (C_{20}+C_{21}+C_{22}+C_{23})$
Flash point	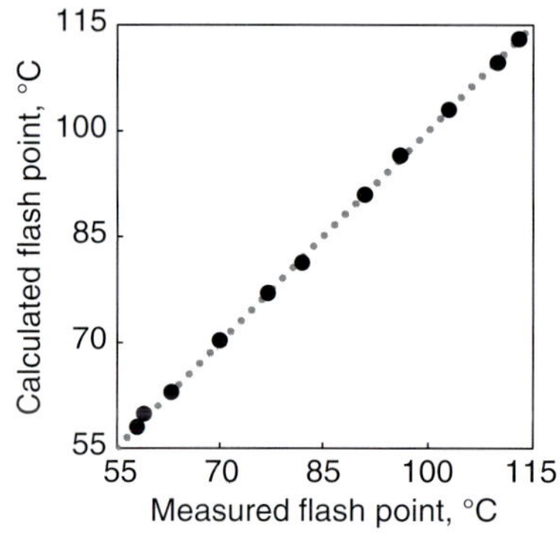	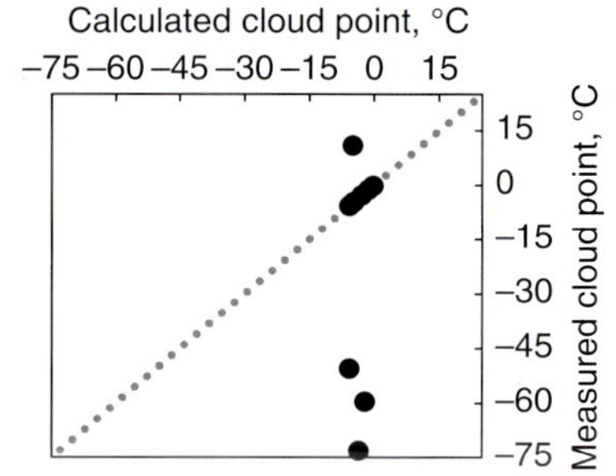
Cloud point	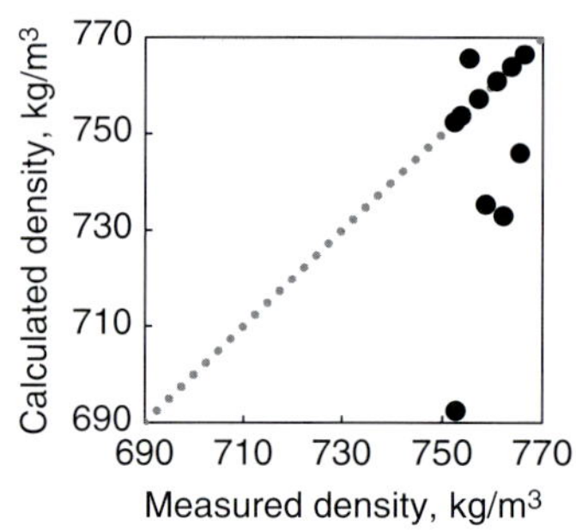	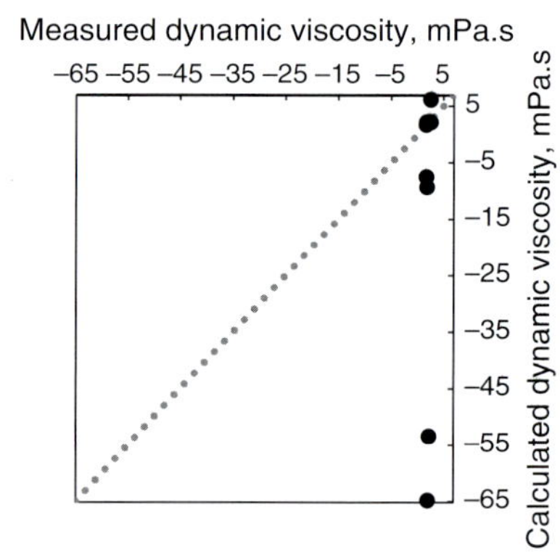
Density	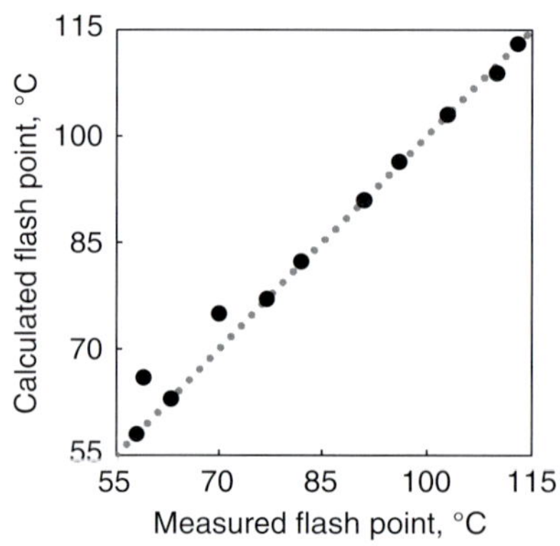	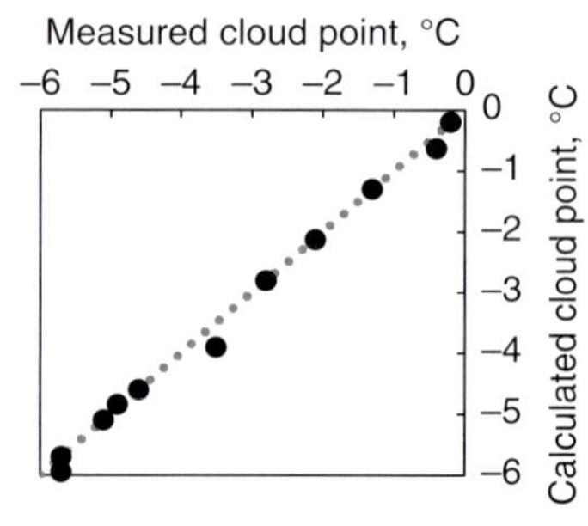

Table 16.9 (Continued)

	Individual carbon number based model	Grouped carbon number based model
GTL diesel property	$P = a_0 + a_1[C_9] + a_2[C_{10}] + a_3[C_{11}] + a_4[C_{12}] + a_5[C_{13}] + a_6[C_{14}] + a_7[C_{15}] + a_8[C_{16}] + a_9[C_{17}] + a_{10}[C_{18}]$ where, $P \rightarrow$ GTL diesel property $C_i \rightarrow$ weight % of ith carbon number content a_0 to $a_{10} \rightarrow$ regression coefficients	$P = a_0 + a_1$[Light Kerosene] $+ a_2$[Heavy Kerosene] $+ a_3$[Light Diesel] $+ a_4$[Heavy Diesel] where, Light Kerosene $\rightarrow (C_8+C_9+C_{10}+C_{11})$ Heavy Kerosene $\rightarrow (C_{12}+C_{13}+C_{14}+C_{15})$ Light Diesel $\rightarrow (C_{16}+C_{17}+C_{18}+C_{19})$ Heavy Diesel $\rightarrow (C_{20}+C_{21}+C_{22}+C_{23})$
Dynamic viscosity	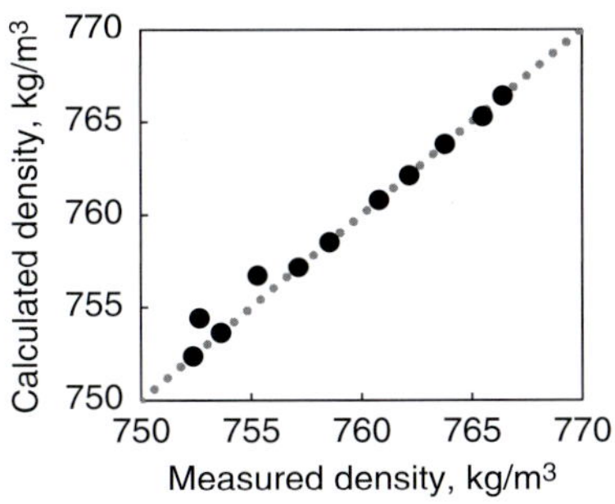	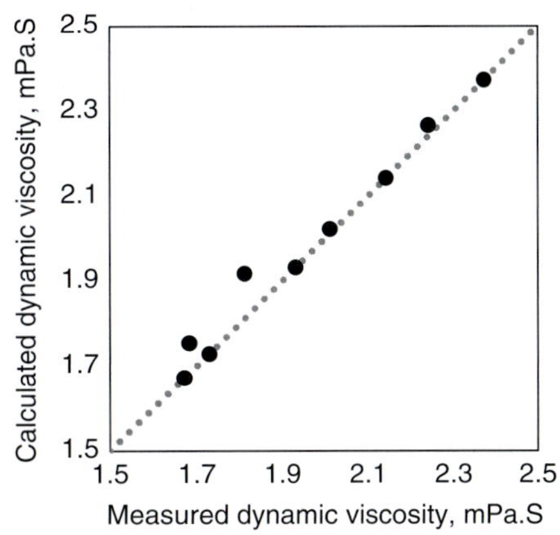

Since the first step did not yield a reliable model, the second method of finding new coefficients was used; however, with a minor change in the model. Since GTL diesel does not contain aromatics, the equation was modified to include n-paraffins and iso-paraffins, both of which are predominant types of molecules in GTL diesel. The modified equation is shown in Table 16.8. By accounting for the major components in the GTL diesel and removing the aromatics term, the aim was to find the right correlation. The properties of Cloud Point, Density, Dynamic Viscosity, and Flash Point were modeled using this modified equation and the results are shown in Table 16.8. These properties were chosen since they are the most important properties used in marketing GTL diesel to customers.

As evident from Table 16.8, neither the literature models nor the LINGO regression with the modified equation gives a good reliable model. Even after the minor modification to the equation, the measured and calculated values do not show a good correlation. This indicates that the segregation of GTL diesel fractions into paraffins and iso-paraffins is not a good classification method for its model formulation.

Since both, the first and second methods failed to predict the properties reasonably, it was decided to test a model based on individual carbon number content. The carbon number content was chosen since it has a major influence on the physical properties of the molecule. The model was formulated as shown in eq. (16.7).

$$P = a_0 + a_1[C_9] + a_2[C_{10}] + a_3[C_{11}] + a_4[C_{12}] + a_5[C_{13}] + a_6[C_{14}] + a_7[C_{15}] + a_8[C_{16}] + a_9[C_{17}] + a_{10}[C_{18}] \quad (16.7)$$

where,

$P \rightarrow$ GTL diesel property

$C_i \rightarrow$ weight percentage of ith carbon number content

a_0 to $a_{10} \rightarrow$ regression coefficients (from LINGO)

When tested, it was found that flash point was very well predicted by this model as shown in Table 16.9. However, the model gave a very poor prediction for the CP, density, and dynamic viscosity. To resolve this, it was decided to group the carbon numbers based on cuts as follows:

$$P = a_0 + a_1[\text{Light Kerosene}] + a_2[\text{Heavy Kerosene}] + a_3[\text{Light Diesel}] + a_4[\text{Heavy Diesel}] \quad (16.8)$$

where,

Light Kerosene $\rightarrow (C_8+C_9+C_{10}+C_{11})$

Heavy Kerosene $\rightarrow (C_{12}+C_{13}+C_{14}+C_{15})$

Light Diesel $\rightarrow (C_{16}+C_{17}+C_{18}+C_{19})$

Heavy Diesel $\rightarrow (C_{20}+C_{21}+C_{22}+C_{23})$

When tested, the performance of this "Grouped Carbon Number" model was excellent, and the properties of cloud point, density, dynamic viscosity, and flash point were very well estimated. This proves that this method of classification combined with LINGO regression works very well in estimating GTL diesel properties.

16.5 Summary and Conclusion

This section presents an overview of this study and some insights into how the results can be used to maximize the utilization of GTL products. The initial part of this work involved an experimental campaign to perform various tests on samples of GTL diesel with different IBPs. The regions where the samples exceed or fall short of the required specifications are shown. The experimental data (either distillation data or carbon number distribution) was used to model some selected important properties using LINGO regression through various literature models as well as newly formulated individual carbon number based and grouped carbon number models. The grouped carbon number model estimated the properties of cloud point, density, dynamic viscosity, and flash point with a good degree of accuracy.

Based on the experimental results of different samples of GTL diesel with different boiling ranges, it was found that the GTL diesel properties were improved by reducing the amount of lighter hydrocarbon content. Higher GTL diesel density and viscosity will support GTL diesel as an independent fuel for diesel engines rather than being used as a blend stock with conventional diesel. The higher flash point will support for more severe storage conditions especially in hot climates like Middle Eastern countries. Higher CI improves the performance of the diesel engines. The only exceptions were the cold properties of CP and PP. For these two specifications, the heavier GTL diesel cuts breached the allowable limits. This is a drawback for the use of GTL diesel in cold countries.

It is important to note here that the GTL diesel specifications are highly dependent on the target market and weather conditions (summer/winter). As mentioned earlier, there is no current ASTM standard for GTL diesel similar to conventional diesel. However, the manufacturing

Table 16.10 Proposed GTL diesel new specification formulation.

Property	Unit	GTL diesel grades		
		Grade A	Grade B	Grade C
Cloud Point	°C, max	0	−3	−6

and distributing companies have formulated a set of specifications for GTL diesel in order to regulate its trading. Based on the results of this study, a new GTL diesel specification can be formulated to classify GTL diesel grades. Table 16.10 proposes the new specification formulation for GTL diesel for the cloud point property. This is similar to the specification of conventional diesel grades in EN-590 European diesel standards, where it specifies six Temperature Climate Grades of diesel fuel (Grade A, B, ... F) which differ in the CFPP (Cold Filter Plugging Point) values (+5 to −20 °C). There are also five Arctic classes of diesel fuel (Class 0, 1, ... 4) characterized by different properties.

The experimental and modeling results from the study can be used to check which of the available composition-properties related prediction models of conventional diesel can be applied to GTL diesel and which ones cannot. New models have been formulated that are able to predict the GTL diesel properties at different boiling ranges and carbon number distributions. In order to test the applicability of each formula, the standard error was calculated for each correlation between the composition and the property; the standard error was calculated as per the below eq. (16.9):

$$\text{Standard Error} = \sqrt{\frac{\sum(\text{Actual Measure} - \text{Model Result})^2}{(\text{Number of points} - 2)}} \tag{16.9}$$

Table 16.11 shows the standard error calculated for each GTL diesel property where the lowest standard error (gray highlighted) is the best model for that GTL diesel specification.

Best Model tested

Cloud Point → Grouped carbon number model with regression coefficients calculated from LINGO

Table 16.11 GTL diesel properties prediction equations, standard errors.

Model based on →	Distillation data		Paraffin content		Individual Carbon number	Grouped Carbon number
Equation source	Literature		Literature		This study	This study
Regression coefficients source	Literature	LINGO	Literature	LINGO	LINGO	LINGO
Cloud point	1.67	0.66	10.28	2.09	37.3	0.19
Kinematic viscosity	0.44	0.05	–	–	–	–
Density	–	–	5.16	5.28	26.23	0.05
Dynamic viscosity	–	–	–	–	31.19	3.06
Flash point	–	–	–	–	0.47	0.80

Kinematic viscosity → Literature model (Karonis et al. [15]) utilizing the distillation data with regression coefficients calculated from LINGO

Density → Grouped carbon number model with regression coefficients calculated from LINGO

Dynamic viscosity → Grouped carbon number model with regression coefficients calculated from LINGO

Flash Point → Individual carbon number model with regression coefficients calculated from LINGO

As evident from Table 16.11 below, the grouped carbon number model was the most successful model to predict most of the GTL diesel properties by knowing its carbon number distribution.

This work shows the potential to improve GTL diesel quality by changing its composition, wherein the higher the carbon number content in GTL diesel, the better the properties. This information can be utilized in a GTL diesel production facility where the GTL diesel composition can be modified to meet certain specifications. GTL diesel production is always challenging due to the dynamic nature of diesel markets. The price of GTL diesel compared to GTL naphtha could fluctuate frequently. The same applies to kerosene (jet fuel) prices. Hence, the GTL diesel plant should be designed to be as flexible as possible in terms of operation. This will give the operators much-needed flexibility to be able to produce the product slate that maximizes profits based on market demand.

For a GTL plant in operation that only produces three products of LPG, naphtha and diesel; there is a good opportunity to modify the facility to produce an intermediate product between naphtha and diesel, i.e. kerosene. Accordingly, the plant will have good flexibility to change the product slates based on market demand. Modifying an existing GTL plant is much more complicated and difficult than modifying the plant during design phase. Hence, GTL plants should be designed for operational flexibility to alter product slates which will enhance profits in the long run.

For an existing GTL plant designed to produce GTL diesel, simple modifications can be implemented to enhance the product slate; e.g., by adding additional draw-off trays for the existing fractionation column, or by adding a Diesel splitter downstream of the fractionation column as shown in Figure 16.14. A technical and economic feasibility study for the specific case will

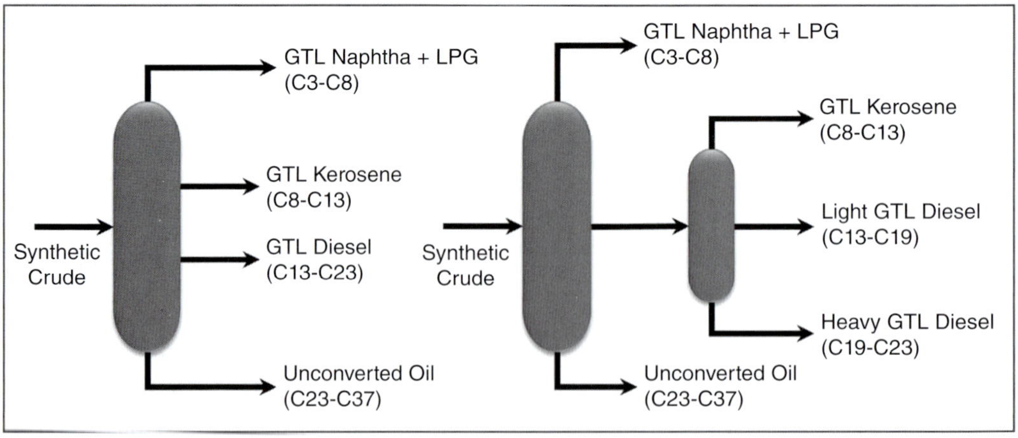

Figure 16.14 Two Possible Modifications to Existing GTL Facility to Produce Additional Products.

determine the benefits and challenges of each of these options. The original design of Shell's SMDS (Shell Middle Distillate Synthesis) [22] process, already has a fractionation column that is designed to produce more products.

As shown in Figure 16.2, for the >215 °C IBP sample of GTL diesel, the percentage of the light fraction was about 30% of the whole sample. If the goal is to improve the GTL diesel quality by removing the lighter ends, then the lighter ends should result in an economic benefit elsewhere to justify the modification. This Light GTL diesel cut as shown in Figure 16.5 consists of C7 to C12 range. This carbon distribution is almost in the kerosene region, which is separated as GTL Jet fuel (SPK) in Shell's SMDS [22] process. Elmalik et al. [23] and Al-Nuaimi et al. [24] have conducted extensive investigations in the area of GTL Jet fuels to understand the contribution of each hydrocarbon building block in satisfying the required specifications, and also in the field of GTL diesel on designing surrogate fuels [17, 25].

The C9-C14 cut is the most suitable feedstock for the production of Linear Alkyl Benzene (LAB). An alternative GTL product separation configuration could separate out this cut and process it to produce the economically valuable LAB used in the manufacture of detergents. However, this cut should be normal paraffin [12] and not be mixed between iso- and normal paraffins as is the current case of SPK. All iso- components should be converted to normal paraffins or separated out. A feasibility study will ascertain if this route is more attractive than the existing GTL Jet fuel configuration.

Another potential area of study to expand the application of GTL diesel is the blending with conventional diesel. Blending with conventional fuels is a much simpler way to enhance the properties of GTL-derived fuels than modifications in the process [26]. Blending of GTL Kerosene with conventional jet fuel has been studied by Al-Nuaimi et al. [24] where different blend ratios were studied to verify the regions where all important jet fuel specifications are met. A similar study performed on GTL diesel will help identify the possibility of blending GTL diesel thereby enhancing its applicability in diesel markets.

Acknowledgment:

This publication was made possible by NPRP award [NPRP 5-066-2-023] from the Qatar National Research Fund, which is a member of the Qatar Foundation. The statements made herein are solely the responsibility of the authors. Also, we are very grateful for ORYX GTL for supporting this project through the ORYX GTL Excellence Program.

References

1 EIA. International Energy Outlook 2016; 2016.
2 OPEC Secretariat. 2014 World Oil Outlook; 2014.
3 International Agency for Research on Cancer (IARC), Diesel Fuels, Chemical and Physical Data. In: IARC Monographs on the Evaluation of Carcinogenic Risks to Humans. Lyon:IARC; 1988, p. 220–221.
4 Shell GTL Fuel & Shell GTL Fuel Marine, Product Information Version 2.0. 2014.
5 Bays J, King D. A NMR-Based Carbon-Type Analysis of Diesel Fuel Blends from Various Sources, prepared for the U.S. Department of Energy under Contract DE-AC05-76RL01830. Pacific Northwest National Laboratory; 2013.
6 Chevron. Diesel Fuels Technical Review. Available rom: https://www.chevron.com/-/media/chevron/operations/documents/diesel-fuel-tech-review.pdf.
7 NiQuan ENERGY. GTL Diesel Specifications. Available from: http://www.niquanenergy.com/sites/default/files/DieselSpecssheet.pdf.

8 ASTM International. ASTM D86: Standard Test Method for Distillation of Petroleum Products at Atmospheric Pressure. 2014.

9 ASTM International. ASTM D4052: Standard Test Method for Meter Density, Relative Density, and API Gravity of Liquids by Digital Density Meter. 2011.

10 ASTM International. ASTM D7042: Standard Test Method for Dynamic Viscosity and density of Liquids by Stabinger Viscometer (and the Calculation of Kinematic viscosity). 2012.

11 ASTM International. ASTM D93: Standard Test Methods for An American National Standard Flash Point by Pensky-Martens Closed Cup Tester. 2002.

12 UOP, UOP Linear Alkylbenzene (LAB) Complex Datasheet. Available from: https://www.honeywell-uop.cn/wp-content/uploads/2011/02/UOP-LAB-Complex-Data-Sheet.pdf.

13 Norman A, Corinchock JA, Scharff R. Diesel Fuels. In: Diesel Technology. Illinois: Goodheart-Wilcox. 1988, p. 303–314.

14 Bezaire N, Wadumesthrige K, Simon Ng KY, and Salley SO. Limitations of the use of cetane index for alternative compression ignition engine fuels. Fuel, 89 (12) (2010): 3807–3813.

15 Karonis D, Lois E, Stournas S, Zannikos F. Use of Mathematical Expressions for the Estimation of Selected Diesel Fuel Properties. In: Preprints Division of Fuel Chemistry, American Chemical Society, 215th ACS National Meeting, Dallas. Volume 43. Dallas: American Chemical Society; 1988; p. 32–37.

16 Dinkov R, Stratiev D, Penev D, and Cholakov G. Investigation on Diesel Cold Flow Properties. University of Chemical Technology and Metallurgy - Sofia, Bulgaria. Available from: https://www.academia.edu/17065610/INVESTIGATION_ON_DIESEL_COLD_FLOW_PROPERTIES.

17 Choudhury HA, Intikhab S, Kalakul S, et al. Designing a surrogate fuel for gas-to-liquid (GTL) derived diesel. Energ Fuels, 31 (10) (2017): 11266–11279.

18 LINGO 15.0. Optimization modeling software for linear, nonlinear, and integer programming. Available from: http://www.lindo.com/.

19 Vas L. Empirical models. Available from: https://liavas.net/courses/math422/files/Empirical_models.pdf.

20 Shahin MMMI. GTL Diesel way forward to improve properties by separating light ends and formulating the specifications. Texas A&M University at Qatar, 2016. Available from: http://oaktrust.library.tamu.edu/bitstream/handle/1969.1/156970/SHAHIN-THESIS-2016.pdf?sequence=1.

21 Cookson DJ, Smith BE. Observed and predicted properties of jet and diesel fuels formulated from coal liquefaction and Fischer-Tropsch. Energ Fuels, 6, (1992): 581–585.

22 Klerk A. De. Shell Middle Distillate Synthesis (SMDS) Facilities. In: Fischer-Tropsch Refining. Weinheim: Wiley-VCH Verlag GmbH & Co. KGaA; 2011, p. 231–240.

23 Elmalik EE, Raza B, Warrag S, Ramadhan H, Alborzi E, Elbashir NO. Role of hydrocarbon building blocks on gas-to-liquid derived synthetic jet fuel characteristics. Ind Eng Chem Res, 53 (5) (2014): 1856–1865.

24 Al-Nuaimi IA, Bohra M, Selam M, Choudhury HA, El-Halwagi MM, Elbashir NO. Optimization of the aromatic/paraffinic composition of synthetic jet fuels. Chem Eng Technol, 39 (12) (2016): 2217–2228.

25 Choudhury HA, Intikhab S, Kalakul S, Gani R, Elbashir NO. Integration of computational modeling and experimental techniques to design fuel surrogates. J Nat Gas Sci Eng, 55 (2018): 585–594.

26 Klerk A. De. Diesel Fuel Refining. In: Fischer-Tropsch Refining. Weinheim: Wiley-VCH Verlag GmbH & Co. KGaA; 2011, p. 559–579.

17

An Energy Integrated Approach to Design a Supercritical Fischer-Tropsch Synthesis Products Separation and Solvent Recovery System

Tala Katbeh[2], Nimir O. Elbashir[1,3], and Mahmoud El-Halwagi[4]

[1] *Petroleum Engineering Program, Texas A&M University at Qatar, Doha, Qatar*
[2] *Chemical Engineering Program, Texas A&M University at Qatar, Doha, Qatar*
[3] *TEES Gas and Fuels Research Center, Texas A&M University, Texas, USA*
[4] *Artie McFerrin Department of Chemical Engineering, Texas A&M University, College Station, Texas, 77843, USA*

CHAPTER MENU

17.1 Introduction

For several decades, pivotal research in FT technology has attracted attention as a result of its provision of a relatively new source of clean fuels as well as value-added chemicals from abundant natural resources. The FT process involves an exothermic heterogeneous reaction over a cobalt-based or iron-based catalyst in which syngas is converted into hydrocarbons with a range of variable chain length (Dry 2002; Steynberg 2004).

The development of GTL technology using natural gas as a feedstock through the FT synthesis evolved from coal-to-liquid (CTL) technology developed by Franz Fischer and Hans Tropsch to provide transportation fuels for German military machines during World War II (Fischer and Tropsch 1926). The GTL technology then gained traction as an attractive new source of ultraclean fuels that met the stringent environmental regulations of the developed nations (Dancuart and Steynberg 2007; Stranges 2007).

The GTL technology consists mainly of the three processes as shown in Figure 17.1. Block 1 represents a reformer unit that converts natural gas into syngas; block 2 denotes the FT reactor that converts the syngas into syncrude (long-chain hydrocarbons, mostly paraffinic C_nH_{2n+2}); and block 3 indicates the last process, which is the refinery section for the separation and upgrading of the hydrocarbon products to fuels and chemicals.

17.1.1 Block 1: Syngas Generation (Natural Gas Reformer)

The common syngas-generation technologies implemented commercially include steam methane reforming (SMR), partial oxidation (POX), autothermal reforming (ATR), and heat exchange reforming. SMR is an endothermic catalytic reaction in which methane and steam react to form syngas. POX technology involves an exothermic reaction involving

Natural Gas Processing from Midstream to Downstream, First Edition.
Edited by Nimir O. Elbashir, Mahmoud M. El-Halwagi, Ioannis G. Economou, and Kenneth R. Hall.

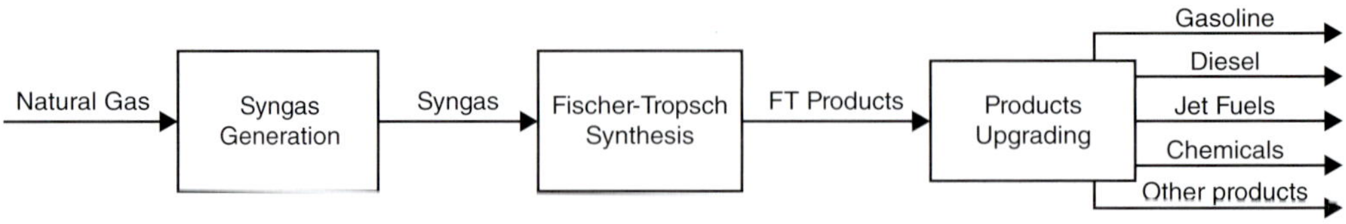

Figure 17.1 Schematic of the overall GTL process.

the conversion of methane and oxygen to form syngas. ATR combines the previous two technologies such that oxygen and steam react in an endothermic reaction with methane to produce syngas. Heat exchange reforming utilizes high reforming temperature that can be obtained from the reformed gas through heat recovery thereby leading to a more compact size with reduced emissions (Wilhelm et al. 2001).

In addition to the technologies mentioned above, extensive research is dedicated to reducing carbon footprints from a GTL process by utilizing the dry reforming of methane (DRM) process. DRM involves an endothermic reaction that converts methane and carbon dioxide into syngas. The process limitations associated with DRM (carbon formation, high endothermicity, and low syngas ratio) are some of the areas that attract numerous research avenues. One such process that synergistically combines the benefits of the three reforming processes is the combined reforming of methane (CRM). CRM is shown to provide significant carbon dioxide conversion at low energy requirements when integrated into a GTL plant (Gabriel et al. 2014; Noureldin, Elbashir, and El-Halwagi 2014; Noureldin et al. 2015; Challiwala et al. 2017).

17.1.2 Block 2: Fischer-Tropsch Synthesis

In 1926, Franz Fischer (1877–1947) and Hans Tropsch (1889–1935) at the Kaiser Wilhelm Institute for Coal Research (KWI) in Mülheim invented the process for converting syngas produced from coal into other hydrocarbons (Stranges 2007). The chemistry of the reaction that converts natural gas to syngas followed by the FT reaction can be simplified in reactions (17.1–17.3) (Khodakov, Chu, and Fongarland 2007; Speight 2007).

$$CH_4 + {}^1\!/_2O_2 \rightarrow 2H_2 + CO \tag{17.1}$$

$$2nH_2 + nCO \rightarrow C_nH_{2n} + nH_2O \tag{17.2}$$

$$(2n+1)H_2 + nCO \rightarrow C_nH_{2n+2} + nH_2O \tag{17.3}$$

Conducting the FT reaction over a cobalt-based catalyst produces water as the main byproduct. However, conducting the FT reaction over an iron-based catalyst produces mainly CO_2 as a byproduct due to its high activity for the WGS reaction (17.4) (Choi et al. 1997).

$$CO + H_2O \rightleftarrows CO_2 + H_2 \tag{17.4}$$

The FT product distribution model follows the Anderson-Schulz-Flory (ASF) distribution model (Sie and Krishna 1999). The hydrocarbon chain growth is represented by the chain growth probability factor (α-value), which is the fraction of the rate of propagation, $r_{P,n}$, to the sum of the rates of propagation and termination of the hydrocarbon chains, $r_{T,n}$, as described in eq. 17.5. The chain length is represented by the carbon number n (Förtsch, Pabst, and Groß-Hardt 2015).

$$\alpha_n = \frac{r_{P,n}}{r_{P,n} + r_{T,n}} \tag{17.5}$$

The mass fraction of the hydrocarbon products is represented by W_n and is determined by the ASF equation (eq. 17.6) below:

$$W_n = n\alpha^{n-1}(1-\alpha)^2 \quad (17.6)$$

This model portrays that as the α-value increases, the heavy hydrocarbons production increases. It is necessary to have information on the FT products' composition in order to determine the product fractions from the process.

The FT process is classified primarily into two operating modes: the low-temperature Fischer-Tropsch (LTFT) process and the high-temperature Fischer-Tropsch (HTFT) process. The LTFT ranges at temperatures between 200 and 240°C and mainly use cobalt-based catalysts, while the HTFT operates at temperatures between 300 and 350°C and mainly uses iron-based catalysts. They are selected depending on the desired final products, as they differ in carbon number distribution. The HTFT process produces gasoline and alpha-olefins while the LTFT process is used for the production of diesel and wax (de Klerk 2007).

17.1.2.1 Conventional FT Reactors

There are four types of conventional FT reactors currently used (referring to Figure 17.2 (a–d)): (a) circulating fluidized bed reactor, (b) fixed fluidized bed reactor, (c) slurry bubble reactor, and (d) multitubular fixed bed reactor. The fluidized bed reactors operate as HTFT reactors while the multitubular and slurry phase reactors operate for LTFT processes.

The LTFT multitubular reactor requires a large number of tubes to remove the heat released by the highly exothermic FT reaction. Furthermore, the multitubular reactor consists of many tubes; this leads to a nonuniform temperature profile. The challenge that faces this design is the problem of economies of scale that can be addressed through building a mega scale facility such as the Pearl GTL plant. To avoid the issues mentioned above, Sasol developed a slurry bubble FT reactor that uses a liquid phase media (heavy wax). The design results in a uniform temperature profile inside the reactor bed. The limitations faced in this type of reactor include the separation of the solid catalyst from the liquid products and the slow diffusion of the syngas in the liquid media (Yokota and Fujimoto 1989; Elbashir et al. 2010; Bao 2012).

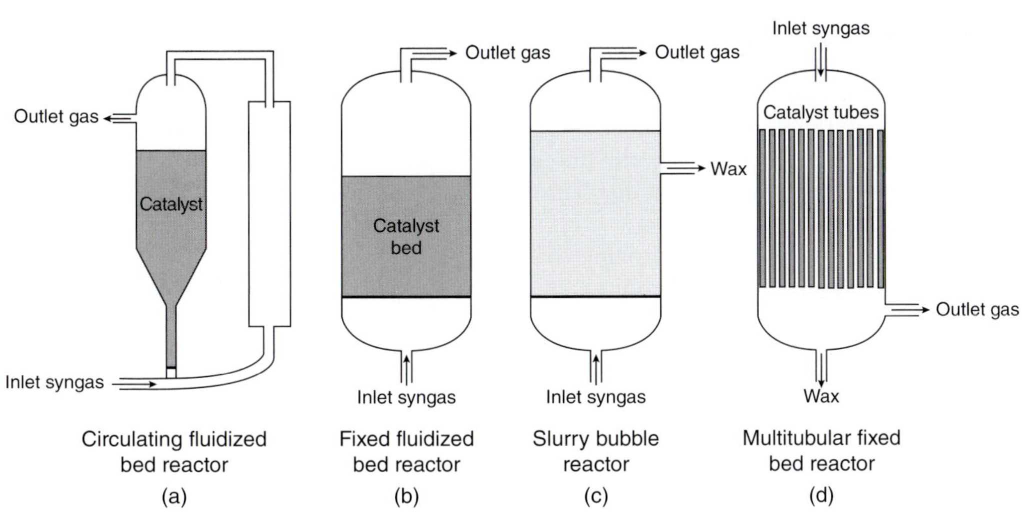

Figure 17.2 Types of reactors.

17.1.3 Introduction on the Utilization of Supercritical Fluids in the FT Synthesis

The operation of FT with SCF conditions was introduced to avert the limitations posed by the fixed-bed reactor and the slurry reactor while combining the simplicity of the former and the improved heat transfer characteristics of the latter (Yokota and Fujimoto 1989; Elbashir et al. 2010 (and references therein)). The SCF-FT process combines the advantages of the conventional FT reactors due to the unique characteristics of gas-like diffusivity as well as liquid-like heat transfer and solubility offered by the SCF solvent (Elbashir, Bao and El-Halwagi 2009). The application of SCF-FT is aimed mainly at LTFT reactor which operates at temperatures between 220 and 250°C (Elbashir et al. 2010).

The SCF-FT process comes with the following considerations that would affect the economics of the process. The sizing of the reactor and auxiliary equipment needs to be taken into account to accommodate the large amount of solvent required by the process. Using the SCF as a medium reduces coke formation on catalyst, and it prevents the buildup of wax on the catalyst surface. Hence, this type of media extends the lifetime of the catalyst and is more cost effective and economical (Elbashir et al. 2005). The SCF-FT process mainly produces diesel while other conventional processes produce heavy hydrocarbons. Consequently, the SCF-FT process reduces the load on the hydrocracker, which in turn reduces its associated catalyst costs (Elbashir and Roberts 2005).

Common SCFs for the FT reaction are pentane (C_5), hexane (C_6), or a C5-C6 blend. These SCFs are most suitable because their critical properties are within the required reaction operating conditions (e.g., for the common SCF solvents their critical properties are within the range of the FT reaction conditions, i.e., temperature 200–240°C and pressure 45–60 bar). On the other hand, heptane (C_7) is a challenging cut to upgrade, and octane (C_8) is more complicated to recover than C_6 due to the presence of concentrated oxygenates. Although, Naphtha has the advantage of available separation techniques within existing GTL plants; however, it results in significant loss of products such as gasoline and jet fuel during the upgrading of the products. Diesel can be used alone or blended as fuel, but the downside is that it has a very high critical temperature (Elmalik et al. 2011). In this study, the separation sequence for solvent recovery and recycle has been developed for n-hexane as the SCF solvent.

17.1.3.1 Block 3: Products Upgrading

The Fischer-Tropsch products possess the benefits of having zero sulfur content and very low aromaticity in the fuel fractions. The product fuel types vary from diesel, naphtha, gasoline, jet fuel, and LPG. The compositions of each type are shown in Table 17.1. To obtain the desired product fuel types, the FT products need to undergo an upgrading process to attain the desired saleable product cuts. Distillation is the first step to separate the syncrude into fractions in a refinery. Therefore, this work utilizes distillation based separation units to recover the products as well as the solvent.

Bao et al. (2012) focused on improving the SCF-FT separation sequence by removing the heavy components (heavy hydrocarbons C20+) first followed by the supplemental separation of the products into light components, water, and the solvent as fraction cuts in a supercritical Fischer-Tropsch process. A series of flash columns with a decreasing pressure drop was used to separate the products to replace the utilization of a distillation column. Also, Bao et al. (2012) reported an energy optimization technique for the process presented in Figure 17.3 by performing sensitivity analysis to determine the configuration of the products separation units.

Four different optimization scenarios were studied including (1) optimizing the design of the flash column sequence, (2) optimizing the heavy components recovery, (3) effects of replacing the Radfrac column with a flash column in separating the solvent, and (4) effects of adding

Table 17.1 Components of each type of fuel.

Fuel name	Components	
Fuel gas	C_1-C_2	Light hydrocarbons
LPG	C_3-C_4	
Gasoline	C_5-C_{12}	
Naphtha	C_8-C_{12}	Heavy hydrocarbons
Kerosene (jet fuel)	C_{11}-C_{13}	
Diesel (fuel oil)	C_{13}-C_{17}	
Middle distillates (light gas oil)	C_{10}-C_{20}	
Wax	C_{19+}	

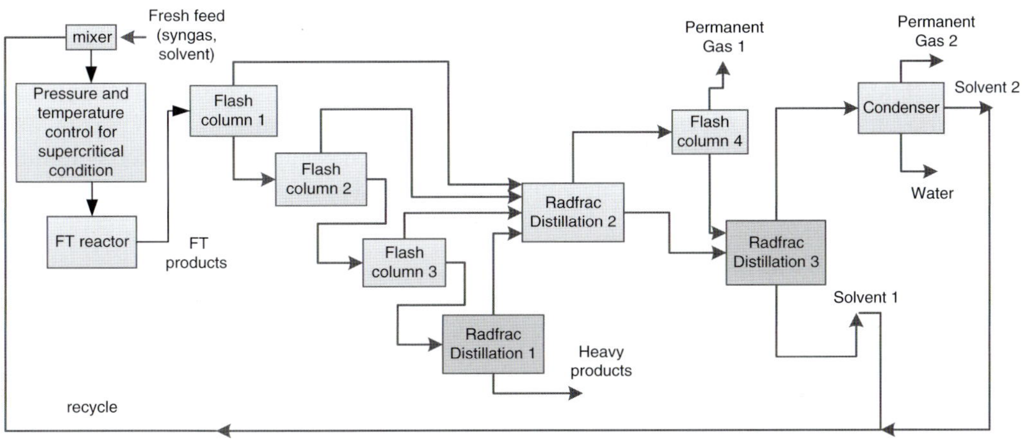

Figure 17.3 Flowsheet for supercritical solvent separation process.

a condenser to increase the permanent gas purity (refer to Figure 17.3). After simulating the different scenarios, an economic analysis was performed. It showed which design had the highest return on investment, the highest recoverability and purity, the highest sale of production, and the highest energy saved. The information obtained from the economic analysis is useful as it allows the designer to choose the best case depending on the objective of the design.

Previous studies focused mainly on syngas generation and FT reaction (Hao et al. 2008; Elbashir, Bao, and El-Halwagi 2009; Kim et al. 2009; Malek Abbaslou, Soltan Mohammadzadeh, and Dalai 2009; Bao, El-Halwagi, and Elbashir 2010; Elbashir et al. 2010; Panahi et al. 2012; Gabriel et al. 2014) with very few studies covered the development of techniques for FT products separation.

As an extension of the relevant research previously performed for SCF-FT product separation (Elbashir, Bao, and El-Halwagi 2009; Bao, El-Halwagi, and Elbashir 2010; Elbashir et al. 2010; Bao et al. 2012), the present work proposes an alternate method for addressing the challenges associated with SCF-FT product separation. The main focus of this work is to design a separation sequence for the syncrude obtained from SCF-FT synthesis while separating the typical FT products and recovering the supercritical solvent that would be recycled back to the reactor. The separation sequence needs to be operationally feasible, and the products yield, solvent recovery, and energy consumption need to be considered when selecting the designs.

The separation sequences are to be compared to existing conventional processes implemented commercially.

In this work, process design was typically accompanied by the methodical construction of multicomponent separation sequences. The steps to accomplishing that involve determining separation methods and sequences and then selecting the best separation sequence to achieve the optimal target. Establishing separation flowsheets requires the use of heuristic rules to "narrow down" the potential sequences (Malek and Glavic 1994).

The work covers alternative separation methods for the SCF-FT products. An initial separation sequence design was simulated in which the heavy components were initially separated with the gases, middle distillate, solvent, and water separated in the final stages. An alternative approach is to tackle the products separation by using three-phase separation units in order to remove the majority of the water produced for it not to be carried on until the end. Separating the water in the early stages would eliminate the need for it to act as an inert component through the later separation stages and would reduce the need of having larger equipment sizes and larger flowrates along the process.

The following three primary scenarios are considered: Scenario 1: Separation of the heavy components first; Scenario 2: Separation of the water first; and Scenario 3: Separation of the vapor and liquid components and use of a three-phase separator to recover water, solvent, and syngas. Different approaches to utilizing pressure and temperature in inducing separation were considered. Additionally, various technologies (e.g., flash units, distillation columns, three-phase separators) in different arrangements were considered.

17.2 Approach and Methodology

The use of SCF in FT reaction complicates the product separation process because it introduces a significant volumetric fraction of the solvents within the reaction mixture. In this work, the solvent-to-syngas molar ratio was 3:1 as it results in high CO conversions as well as good methane selectivity and chain growth probability (Elbashir et al. 2005). The products of the SCF-FT can be aggregated into five fractions: the permanent gases (H_2, CO, and CO_2), the light hydrocarbons (C_1-C_4), the solvent fraction (C_5-C_7), the water fraction, and the medium distillate and heavy components (C_{8+}). However, the products are typically separated into three cuts: the permanent gases and light hydrocarbons, the solvent and water, and the heavy hydrocarbons (see Figure 17.4). Since the solvent and water fraction is the middle cut, there are

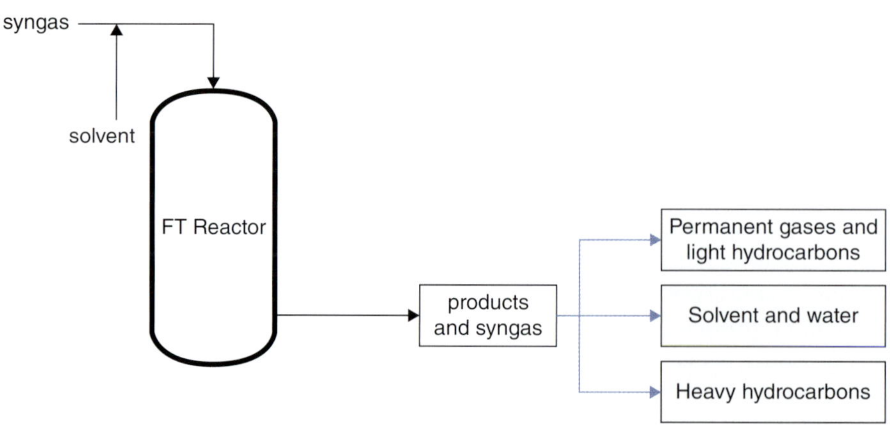

Figure 17.4 SCF-FT synthesis product cuts.

significant complications and challenges for separating the solvent. As shown in Figure 17.4, the required large volume of solvent is co-fed into the reactor alongside the syngas. Having this significant amount of solvent means that the equipment sizes would be larger and thus, the size of the entire process would be larger than that of a conventional FT technology (e.g., gas phase FT). Another complication, is that the solvent fraction shares the same constituents as a portion of the reaction products (e.g., light naphtha), and this adds to the amount of solvent available in the product stream that must be separated. Therefore, the main challenge facing SCF-FT design will be to establish an optimized separation sequence for the SCF-FT products and solvent recovery system.

The goal of establishing a separation sequence that recovers the solvent while separating the water from the products was achieved through the following tasks: (1) an initial separation design was constructed, (2) various separation sequences were proposed and analyzed, and (3) the equipment parameters were optimized to determine the sequence with the least energy requirements.

The study was aided by utilizing Aspen Plus® process simulator version 8.8, which enables the user to construct flowsheets, perform product analysis, and carry out what-if scenarios to optimize the process and the equipment design. The methods used to achieve that are through the prediction of the energy consumption and costs by using the economic analysis, sensitivity analysis and optimization tools that are built in the software (Aspen Technology 2017).

17.2.1 The FT Reactor Conditions

For the process simulation, the selected thermodynamic model was the Non-Random Two-Liquid Redlich-Kwong (NRTL-RK) property method. It was selected to consider the polar nature of water. The simulation takes into account the material and energy balances of the process. The NRTL-RK method utilizes the NRTL liquid phase activity coefficient model and the Redlich-Kwong equation of state, which calculates the thermodynamic properties of the vapor phase. It also uses the Rackett model to compute the liquid molar volumes and Henry's law for supercritical components (Aspen Technology 2013).

The SCF-FT reactor conditions used were at a temperature of 240°C and a pressure of 45 bar (Fan and Fujimoto 1999). This high pressure is beneficial to be utilized for the separation. The reaction takes place on a cobalt catalyst with the syngas feed molar ratio H_2/CO of 2.15 and 75% of the CO is reacted with the CO (Dry 2002; Bao, El-Halwagi, and Elbashir 2010). The solvent selected for this was hexane, C_6, due to its supercritical properties as its critical temperature and pressure are 234.5°C and 30.2 bar, respectively. This solvent was fed to the reactor along with the syngas at a molar ratio of 3:1 (Elbashir et al. 2005). For the products distribution, the ASF products distribution model was used, and the chain growth probability factor (α-value) was 0.96. The SCF-FT products stream data (named FTPROD) was obtained from Elbashir et al. (2009). The hydrocarbon products range from C_1 up to C_{100}; however, for ease of computation in the simulation, all hydrocarbons with carbon number larger than 30 were accumulated as C_{30+} components. The molar composition of the FTPROD stream consists of approximately 7.2% syngas, 7% water, 85.4% solvent, and 0.4% hydrocarbon products (with the molar flowrates represented in Table 17.2). It can be noted that the solvent constitutes the majority of the reactor outlet stream composition.

17.2.2 The Process Design Approach

Through the utilization of engineering heuristics, the viable options to approach the separation of the SCF-FT products are (1) remove the heavy first, (2) remove the most volatile first, (3) remove the water first, or (4) remove the fraction with the highest weight percent.

Table 17.2 SCF-FT reactor outlet stream composition from the Aspen Plus® simulation results.

Component	Mole flow (kmol/hr)	Mole fraction (%)
Syngas	4,599	7.20
H_2O	4,536	7.09
C_1-C_5	33.5	0.05
C_6	54,628	85.4
C_{7+}	187	0.29
Total	63,983	100

In the SCF-FT process, in order to start the separation with the removal of the fraction with the highest weight percent (which is the solvent fraction), this option would, in essence, simultaneously occur with options (2) and (3) in which the gases and the water fractions would be separated as well. The C_6 solvent would have to be recovered by separating it with another fraction since it is a middle cut. Removing the solvent with the water would not be the best option since water is available by a significant amount (since the water is produced at a ratio of 1:1 with the CO reacted) and comprises 7% of the SCF-FT products and it needs to be separated as well. Therefore, the most suitable method would be to use a three-phase separator, as compared to what is used in industry, in which the permanent gases, the solvent and hydrocarbon products, and the water would be separated.

The successive steps taken in order to achieve the desired outcome are illustrated in Figure 17.5. First, a base case flowsheet was generated using an existing GTL plant design as a framework while bearing in mind the common heuristic rules. Next, the flowsheet was simulated using Aspen Plus® to determine the suitable unit conditions and the stream compositions around each equipment. Each local section in the flowsheet was optimized by conducting a sensitivity analysis on the different equipment. The results enabled the selection of the promising operating conditions and allowed the avoidance of unsuitable conditions. The sensitivity analysis resulted in the process conditions optimization. If the desired outcome of achieving the products separation while recovering the water and solvent was not met, the previous steps were repeated until a valid flowsheet was obtained. Once each piece of equipment was optimized, and the desired outcomes were achieved, it was possible to identify the feasibility of the process and determine its comparability to an existing plant.

The separation sequences were optimized by performing a sensitivity analysis for each separation unit in the flowsheet. It is especially essential for optimizing the distillation columns in the process such that the energy consumption is reduced while maintaining maximum product separation. The optimization process mainly consists of the following: the optimization function, the design variables, and the constraints. In the case for optimizing the distillation column, the optimization function was to determine the reboiler and condenser duty and select the conditions at which the duties were minimal. The design variables were the molar reflux ratio, the feed stage, and the feed temperature. The constraints would be the recovery of the desired output products from the distillate and bottoms streams. The sensitivity analysis performed on the three-phase separators and the flash columns, on the other hand, involved determining the effect of changing the feed-stream conditions on the separation such that maximum recovery and purity are reached.

Figure 17.5 The Process Design Approach.

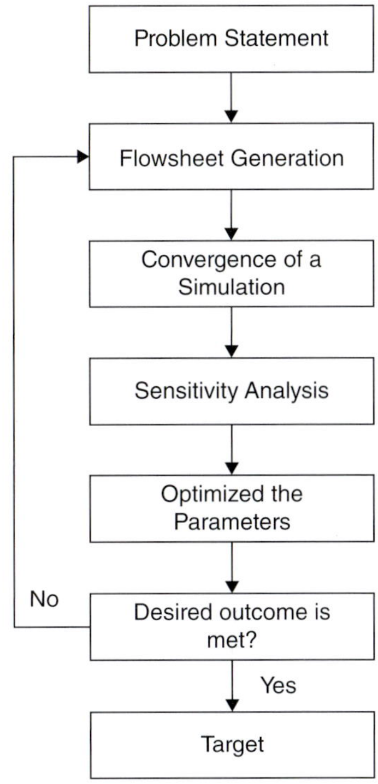

17.3 Results and Discussion

17.3.1 Scenario 1: Separation of the Heavy Components First

The initial design for the SCF-FT synthesis product separation was simulated as shown in Figure 17.6 (Bao et al. 2012). The design exploits that ease of separation using gradual reduction in pressure of the product stream.

The separation starts with a flash column sequence, FC1, FC2, and FC3, which separates the middle distillate and the heavy components (C_{8+}) from the lighter components (this includes the permanent gases, the solvent, and the water). The flash columns are operated in this arrangement as it results in a successive drop in the columns' pressures. That is followed by a distillation column (DIST1) that separates the heavier products (C_{8+}) from the vapors. The vapor streams are then mixed and input into a series of two-phase flash drums, which further separate the lighter components: the permanent gases. The distillation column, DIST2, recovers most of the solvent while the condensers, C1 and C2, separate the permanent gases from the remaining water and solvent. The streams' compositions and the recovery of the components in the exit streams can be found in Table 17.3 and Table 17.4, respectively.

The flash columns operating conditions were as illustrated in Table 17.5. The temperature was kept constant while the pressures gradually decreased along the series of flash separators in order to maximize the separation. These conditions were selected based on a previous sensitivity analysis (Bao 2012), which showed that at these specific temperatures and pressures the heavy components recovery met the target of this separation sequence.

The distillation columns (DIST1 and DIST2) have 30 stages and are operating at a pressure of 5 bar. The bottoms rate was specified for each column in order to recover the desired

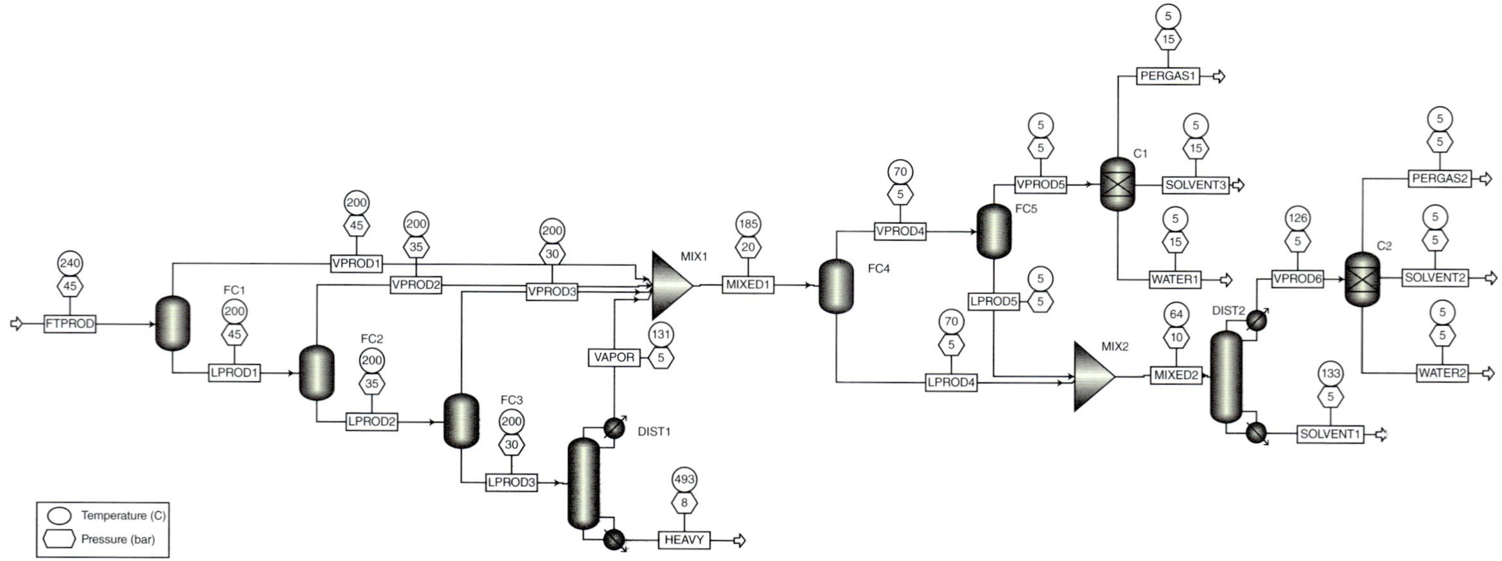

Figure 17.6 Simulation Flowsheet for Scenario 1.

Table 17.3 Scenario 1 stream compositions.

Mole frac (%)	FTPROD	VPROD1	LPROD1	VPROD2	LPROD2	VPROD3	LPROD3	VAPOR	HEAVY	PERGAS1	SOLVENT3	WATER1	SOLVENT1	PERGAS2	SOLVENT2	WATER2
Syngas	7.19	12.9	0.09	7.79	0.09	4.85	0.07	0.07	0	99.2	1.47	1.43	0	97.3	2.17	2.11
H_2O	7.09	6.96	0.29	15.2	0.29	14.3	0.24	0.24	0	0.06	0.17	98.5	0	0.18	0.20	97.7
C1-C5	0.05	0.03	0.04	0.07	0.04	0.08	0.04	0.04	0	0.10	0.12	0.06	0	0.80	0.34	0.24
C6	85.4	41.4	99.0	76.9	99.0	80.7	99.0	99.6	1.4	0.60	98.2	0	99.9	1.70	97.3	0
C7+	0.29	0.02	0.62	0.03	0.62	0.03	0.62	0	98.6	0	0	0	0.02	0	0	0

Table 17.4 Recovery of Scenario 1 components in exit streams.

Recovery (%)	HEAVY	PERGAS1	SOLVENT3	WATER1	SOLVENT1	PERGAS2	SOLVENT2	WATER2
Syngas	0	94.8	0.03	0	0	0.81	2.24	2.12
H_2O	0	0.06	0	0.23	0	0	0.21	99.5
C_2-C_5	0	13.5	0.36	0.02	4.63	0.91	48.0	32.6
C_6	0	0.05	0.19	0	91.3	0	8.46	0
C_{7+}	93.5	0	0	0	6.50	0	0	0

Table 17.5 Flash columns' operating conditions.

Flash column	Temperature (°C)	Pressure (bar)
FC1	200	45
FC2	200	35
FC3	200	30

components (the heavy components for DIST1 and the solvent for DIST2). It was found that as the bottoms rate increases, the recoverability of the heavier components increases. The feed stage was determined by studying the effect of the reboiler duty and the recoverability of the heavy components on the inlet stage. The flash columns, FC4 and FC5, were both at a pressure of 5 bar but at decreasing temperatures of 70°C and 5°C, respectively. The low temperature of FC5 enabled the condensation of the liquids such that the majority of the permanent gases can be recovered. From the results obtained after optimizing the equipment conditions, 93.5% of the middle distillate and heavy components were recovered in the HEAVY stream, 95.6% of the syngas was recovered in the PERGAS streams, 99.95% of the solvent was recovered in the SOLVENT streams, and 99.7% of the water was recovered in the WATER streams. As for the energy consumption, the heating utilities amounted to 574.6 MW while the cooling utilities were 1227 MW where the total energy consumed was 1802 MW. The details are shown in Table 17.6.

17.3.2 Alternate Separation Design for Scenario 1

An alternative approach to separating the products from the SCF-FT synthesis is to start by removing the water at the initial stages of separation. The main purpose is to avoid carrying the water on throughout the process as it would lead to an unnecessary larger treatment capacity as it acts as an inert component along the process.

A typical commercial GTL plant is illustrated in Figure 17.7 (de Klerk 1994), and it shows a simplified description to the Fischer-Tropsch unit and the separation/refining units. The syngas enters the FT reactor that is a low-temperature slurry bed reactor for this case. Water is initially knocked out using three-phase separators, and the light gases are further separated in which the tail gas is sent for further hydrocarbon recovery, while the syngas is recycled as an external recycle stream back to the reformer in the syngas production unit. Meanwhile, the heavy ends recovered (HER) and the hydrocarbon condensate is further separated through the hydrocracker, followed by a series of hot and cold high-pressure and low-pressure flash separation

Table 17.6 Total utilities for Scenario 1.

	Heating utilities (MW)	Cooling utilities (MW)
FC1	–	–220.0
FC2	2.155	–
FC3	1.485	–
FC4	–	–610.2
FC5	–	–73.91
DIST1 Condenser	–	–146.5
DIST1 Reboiler	201.7	–
DIST2 Condenser	–	–55.91
DIST2 Reboiler	369.3	–
C1	–	–1.055
C2	–	–119.3
Total	574.6	–1227

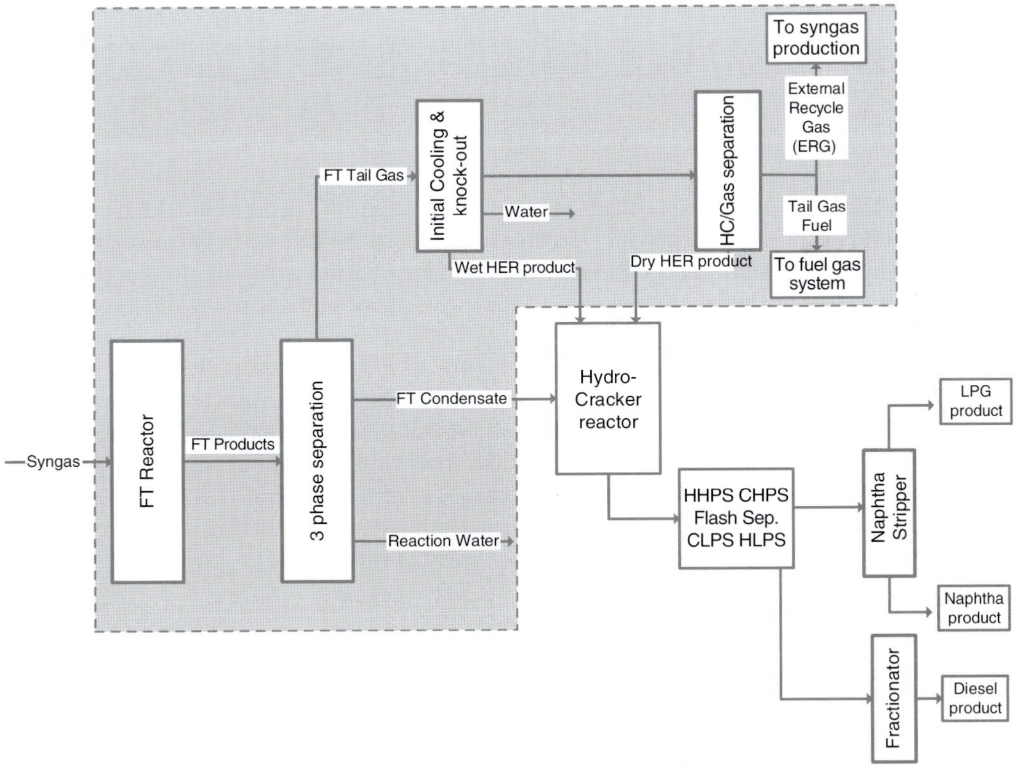

Figure 17.7 An example of a conventional GTL plant setup.

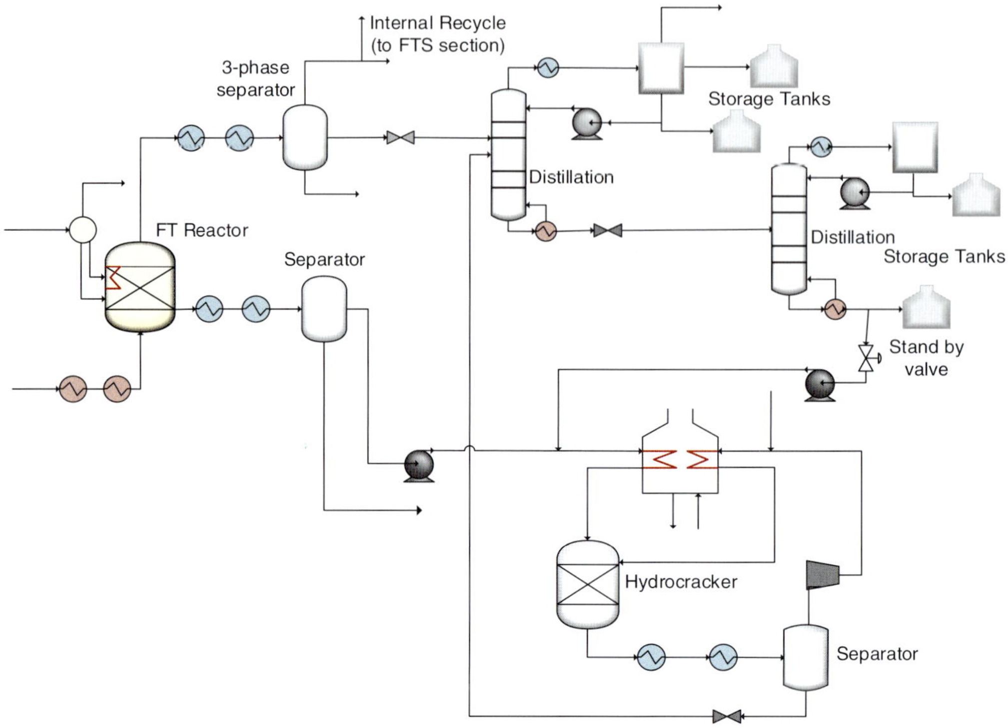

Figure 17.8 Another example of a conventional GTL plant setup.

units. They are then taken to a stripper column and a fractionator to produce the final cuts: LPG product, light and heavy naphtha product, and diesel product. For this work, however, the focus would only be on the separation occurring before the hydrocracker as highlighted in the setup. It should be noted that the hydrocracking of the heavy hydrocarbons was not taken into account for the SCF-FT separation sequence simulations.

Another conventional setup can be shown in Figure 17.8. In this setup, it shows the vapor and liquid components leaving in a separate stream out of the FT reactor, whereas the gaseous components first undergo a three-phase separation for water knockout and the middle fraction is then inputted into a series of distillation columns that would further separate the products into the desired fractions.

The following sections describe the simulated separation sequences using the two examples of the conventional GTL processes described previously as references.

17.3.3 Scenario 2: Separation of the Water First

The second separation sequence (shown in Figure 17.9) flowsheet was constructed based on an existing GTL plant that utilizes a similar sequence to the FT products. It starts with cooling the FT products stream from a temperature of 240°C to a temperature of 125°C since at that temperature and pressure, the three-phase separator would be able to separate the water effectively based on the properties of water.

In order to improve the recovery of the major components to be separated from 3-PHASE1, a sensitivity analysis was implemented. The design variables were the pressure and temperature at which the FTPROD stream needs to be fed into the three-phase separator. The pressure was varied from 10 bar to 45 bar, and the temperature was varied from 50°C to 125°C (since

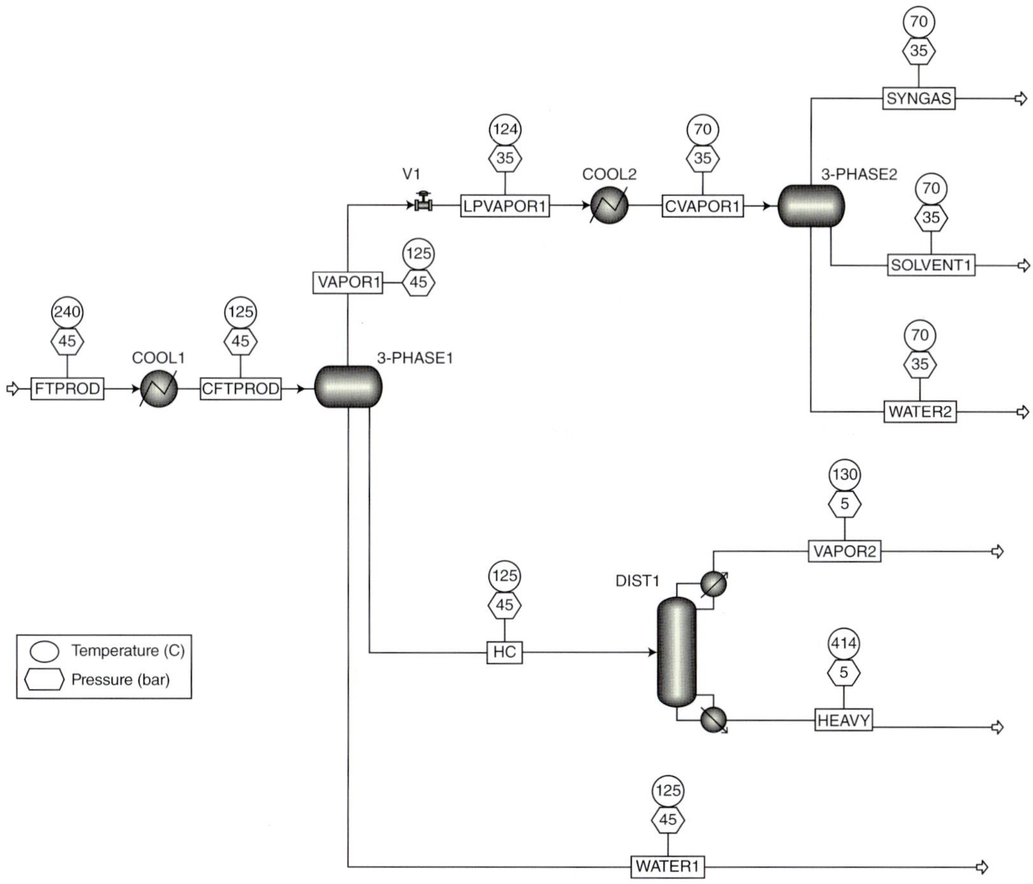

Figure 17.9 Scenario 2 Simulation Flowsheet.

at higher temperatures, the water recovery would be <80% until the three-phase separator would cease to separate the water as it would not be liquefied at the combination of higher temperatures and the inputted pressure range). The constraints taken into account were the following: the recovery of water, the solvent, and the syngas. The sensitivity analysis showed that there needs to be a trade-off between the recovery of water and that of CO since their relationship is inversely proportional. With increasing temperature, CO recovery increases while water recovery decreases. And with increasing pressure, CO recovery decreases while the water recovery increases. The solvent recovery shows a similar trend to that of water; however, the difference is minimal and could be neglected for this purpose. From the sensitivity analysis results (presented in Figures 17.10–17.12), the 3-phase separator's operating conditions for this case were set to 125°C and 45 bar. This lead to 95.16% of the syngas to be separated to the VAPOR1 stream, 85.51% of the water separated into the WATER1 stream, and 98.37% of the solvent C_6 separated into the HC stream.

The HC and VAPOR1 streams were then further separated to recover the rest of the products. The VAPOR1 stream's temperature and pressure were lowered to 70°C and 35 bar before being inputted to another 3-phase separator, 3-PHASE2. This resulted in the separation of the remaining water and solvent in the gas stream.

The HC stream with the hydrocarbon products and most of the solvent were separated using a distillation column (DIST1) for efficient and maximum separation. Using a flash drum resulted

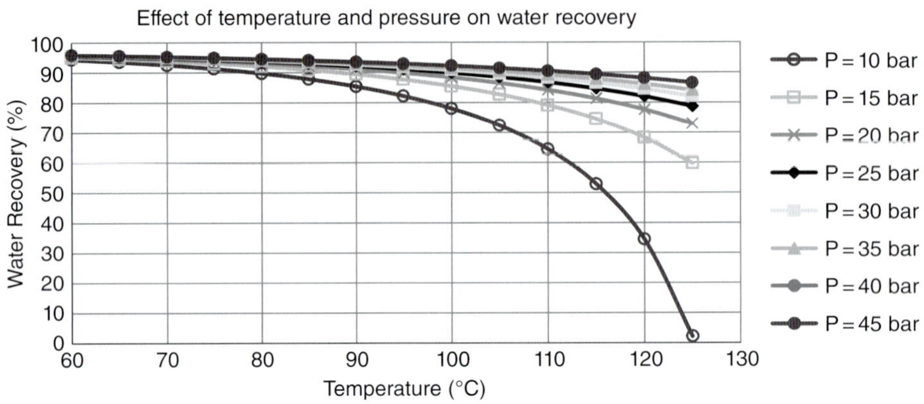

Figure 17.10 Effect of Temperature and Pressure on Water Recovery.

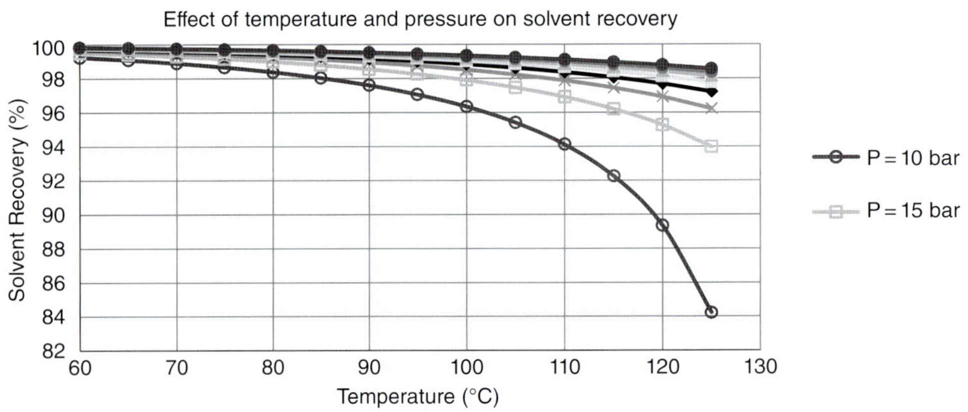

Figure 17.11 Effect of Temperature and Pressure on Solvent Recovery.

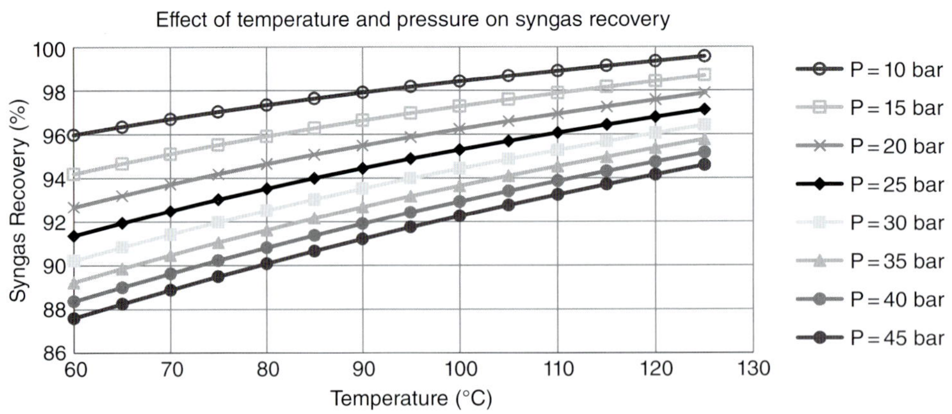

Figure 17.12 Effect of Temperature and Pressure on Syngas Recovery.

Table 17.7 Scenario 2 stream compositions.

Mole frac (%)	FTPROD	VAPOR1	HC	WATER1	SYNGAS	SOLVENT1	WATER2	VAPOR2	HEAVY
Syngas	7.19	77.3	0.38	0.37	95.0	0.88	0.85	0.38	0
H_2O	7.09	6.88	0.49	99.4	1.00	0.32	99.1	0.49	0
C_1-C_5	0.05	0.09	0.05	0.04	0.09	0.08	0.06	0.05	0
C_6	85.4	15.7	98.8	0.01	3.88	98.7	0	99.1	0.09
C_{7+}	0.29	0	0.33	0.24	0	0.01	0	0	99.9

Table 17.8 Recovery of Scenario 2 components in the exit streams.

Recovery (%)	FTPROD	VAPOR1	HC	WATER1	SYNGAS	SOLVENT1	WATER2	VAPOR2	HEAVY
Syngas	–	–	–	0.31	95.0	0.14	0.06	4.53	0
H_2O	–	–	–	85.5	1.02	0.05	7.52	5.90	0
C_1-C_5	–	–	–	4.33	12.9	1.69	0.62	80.5	0
C_6	–	–	–	0	0.33	1.30	0	98.4	0
C_{7+}	–	–	–	4.93	0	0.05	0	0.17	94.9

in an inefficient separation of the solvent, and there was no condition in which the flash separation would lead to an adequate separation.

The distillation column optimized by performing a sensitivity analysis on the column. The design variables selected were the molar reflux ratio and the feed stage. The constraints were the solvent (C_6) recovery in the VAPOR2 stream as well as the middle distillate and heavier components recovery (in this case, C_{7+} recovery was examined) in the HEAVY stream. The optimization functions, on the other hand, were the reboiler and condenser duties.

The distillation column, DIST1, was set to have 33 stages, with a pressure of 5 bar, and the stage pressure drop for the column was set at 0.01 bar according to the allowable pressure drop in Parkash's book *Refining Processes Handbook* (Parkash 2003). The column's operating specifications were set such that the feed stage was at 10, the molar reflux ratio was 1.2, and the bottoms rate was 177 kmol/hr (since the C_{7+} products made up this quantity in the feed HC stream).

The second three-phase separator, 3-PHASE2, recovered the syngas from the other components in the VAPOR1 stream. The stream's pressure and temperature were lowered to 35 bar and 70°C, respectively.

This separation sequence resulted in the recovery of 95% of the syngas, 94.85% of the heavy components (C_{7+}), 99.67% of the solvent, and 93.03% of the water. Table 17.7 and Table 17.8 tabulate the stream compositions and the components recovery, respectively. The net duties used in the process added up to be 1919 MW. The values are illustrated in Table 17.9.

17.3.4 Scenario 3: Separation of the Vapor and Liquid Components and Use of 3-phase Separator to Recover Water, Solvent, and Syngas

As for the third separation sequence, the setup shown in Figure 17.8 was used as a reference for this design. The simulated flowsheet is as shown in Figure 17.13. The FT products streams

Table 17.9 Total Utilities for Scenario 2.

	Heating Utilities (MW)	Cooling Utilities (MW)
COOL1	–	−606.2
COOL2	–	−8.824
3-PHASE1	–	−31.40
3-PHASE2	–	−3.190
DIST1 Condenser	–	−437.2
DIST1 Reboiler	832.3	–
Total	832.3	−1087

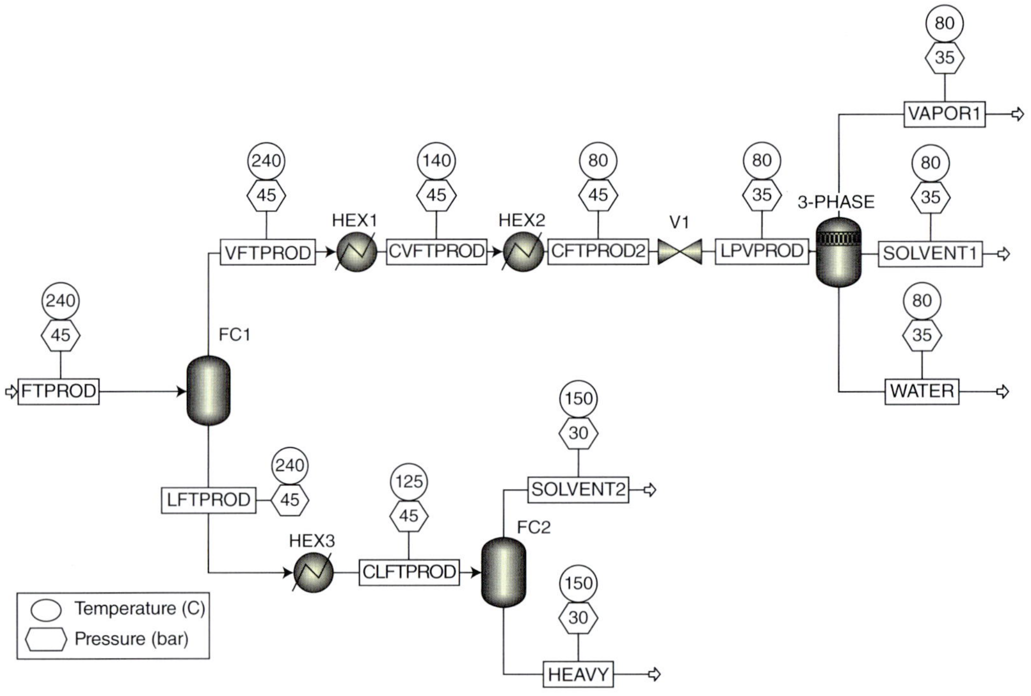

Figure 17.13 Scenario 3 Initial Simulation Flowsheet.

enter a flash column (FC1) that would separate the vapors from the liquids under the conditions of 240°C and 45 bar. However, with this initial setup, it was found that 51.2% of the heavy components were recovered in the vapor stream (VFTPROD) that was separated by FC1. Under those conditions, that is not typically the case, and the heavy components would exist in the liquid phase. Moreover, using FC2 resulted in an inefficient separation of the heavy components from the remaining solvent. Thus, the flowsheet was modified (Figure 17.14).

In order to separate the liquid and vapor components from the FT products stream, a component separator (SEP1) was used. The separator would separate the components based on split

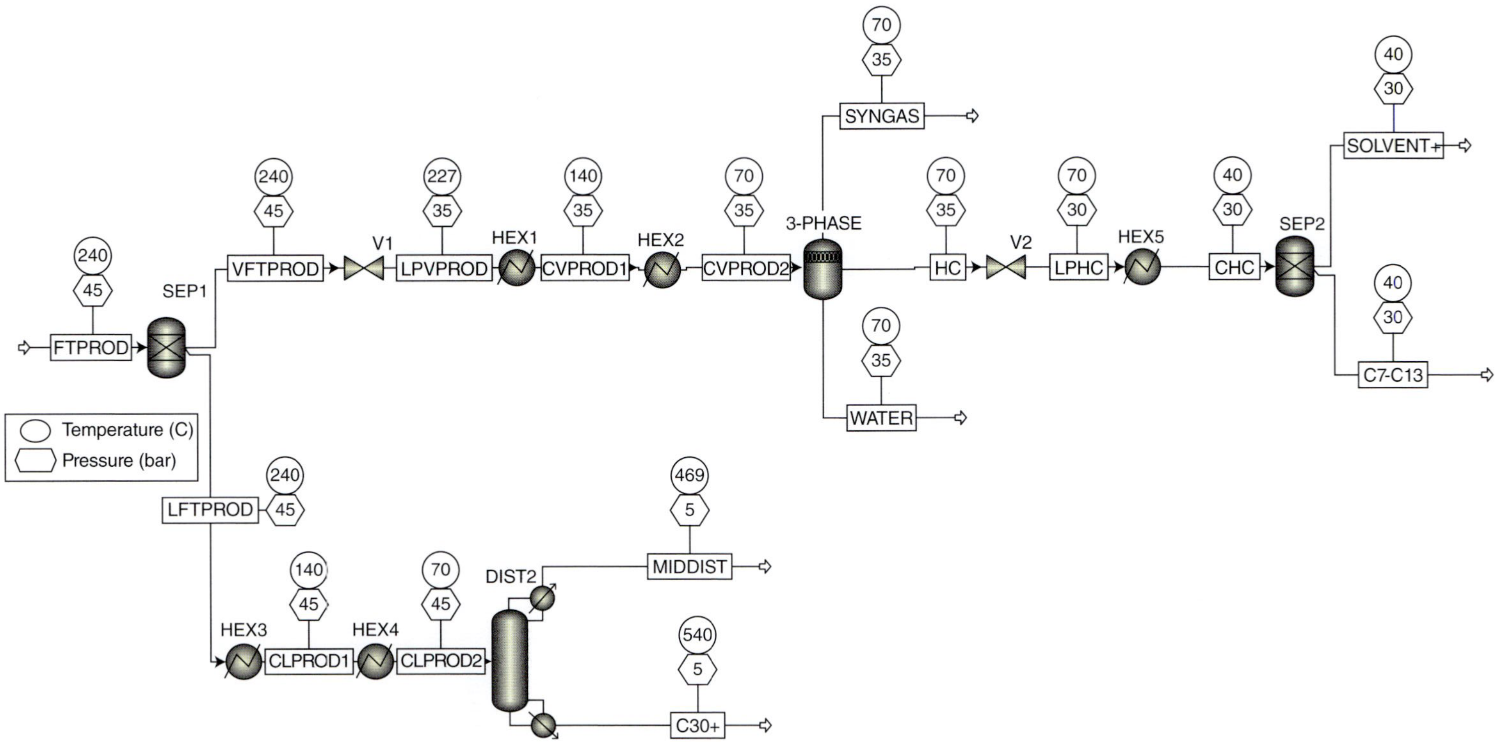

Figure 17.14 Scenario 3 Modified Simulation Flowsheet.

Table 17.10 Scenario 3 stream compositions.

Mole frac (%)	FTPROD	VFTPROD	LFTPROD	SYNGAS	HC	WATER	SOLVENT+	C7-C13	MIDDIST	C30+
Syngas	7.19	7.21	0	95.1	0.74	0.71	0.74	0	0	0
H_2O	7.09	7.11	0	1.00	0.32	99.2	0.32	0	0	0
C_1-C_5	0.05	0.05	0	0.06	0.05	0.04	0.05	0	0	0
C_6	85.4	85.6	0	3.85	98.8	0	98.9	0	0	0
C_7-C_{13}	0.06	0.06	0	0	0.06	0	0	100	0	0
C_{14}-C_{29}	0.08	0	33.9	0	0	0	0	0	91.0	4.62
C_{30+}	0.16	0	66.1	0	0	0	0	0	8.96	95.4

Table 17.11 Recovery of Scenario 3 components in exit streams.

Mole frac (%)	FTPROD	VFTPROD	LFTPROD	SYNGAS	HC	WATER	SOLVENT+	C7-C13	MIDDIST	C30+
Syngas	–	–	–	90.5	–	0.67	8.83	0	0	0
H_2O	–	–	–	0.97	–	95.2	3.83	0	0	0
C_1-C_5	–	–	–	7.33	–	5.29	87.4	0	0	0
C_6	–	–	–	0.31	–	0	99.7	0	0	0
C_7-C_{13}	–	–	–	0.03	–	0	0	99.97	0	0
C_{14}-C_{29}	–	–	–	0	–	0	0	0	91.0	9.02
C_{30+}	–	–	–	0	–	0	0	0	4.59	95.4

fractions. Therefore, it was assumed that the liquid stream composed of all components C_{14+} and their split fractions were set to 1 since that is typically what the liquid stream consists of in the conventional process. This resulted in the recovery of 99.75% of the solvent, 99.3% of the water, and 99.2% of the syngas, and 99.9% of the heavy components. The stream compositions and the components recovered in the exit streams' data are shown in Table 17.10 and Table 17.11, respectively.

The heating utilities amounted to 23.96 MW while the cooling utilities were −947.3 MW. The net duties used in the process added up to be 971.26 MW. The values are illustrated in Table 17.12.

Table 17.13 shows the percent recovered for each of the major components: the syngas, solvent, water, and heavy components in each of the simulated scenarios. It can be seen that Scenario 3 resulted in the highest recovery. Comparing the energy consumption of the three scenarios (shown in Table 17.14), it can be noted that Scenario 3 had the least energy consumption of the three separation sequences.

Considering the design simplicity, it was shown that separating the water at the earlier stages of the separation (Scenarios 2 and 3) resulted in fewer pieces of separation equipment as compared to Scenario 1 where the water was carried on to the end. While Scenario 2 showed the highest energy consumption and a reasonable amount of products recovered, it should be noted that this design is the closest to what is implemented commercially. Scenario 1 lacks equipment that would control the streams' conditions prior to the flash columns and distillation column,

Table 17.12 Total utilities for Scenario 3.

	Heating utilities (MW)	Cooling utilities (MW)
3-PHASE	–	−8.54
DIST1 Condenser	–	−1.29
DIST1 Reboiler	24.0	–
HEX1	–	−523.3
HEX2	–	−310.5
HEX3	–	−4.33
HEX4	–	−2.66
HEX5	–	−96.7
Total	24.0	−947.3

Table 17.13 Recovery of the components for each design.

Recovery (%)	Scenario 1	Scenario 2	Scenario 3
Syngas	95.6%	95%	99.2%
Solvent	99.95%	99.67%	99.75%
Water	99.7%	93.03%	99.3%
Heavy components	93.5%	94.85%	99.9%

Table 17.14 Comparison of the energy consumed for each scenario.

	Scenario 1	Scenario 2	Scenario 3
Heating utilities (MW)	574.6	832.3	23.96
Cooling utilities (MW)	−1227	−1087	−947.3
Total utilities (MW)	1802	1919	971.3

which would affect the energy requirements data obtained. Scenario 3 resulted in the highest recovery of the separated products, the lowest energy consumption, and is open to suggestions to replace the separators that are used in the simulation when it is to be selected as the method of separation.

A comparison was then made to determine the energy consumption of Scenario 3 with and without the use of the solvent. Using the same amount of syngas that was used for the reaction, the ASF product distribution calculations were then made. The FT products stream data was inputted into the separation sequence with the same setup as Scenario 3 but at an inlet temperature of 240°C and 24 bar, respectively.

The data tabulated in Table 17.15 show that the energy consumption is higher with the SCF-FT synthesis since with the addition of the solvent at a solvent-to-syngas molar ratio of 3:1, the molar flow rate would be four times that without the solvent. Therefore, the larger flow rate results in greater energy requirements since the heat exchanger duties are a function of

Table 17.15 Comparing Scenario 3 energy consumption with and without solvent.

	Scenario 3 with solvent	Scenario 3 without solvent
Heating utilities (MW)	23.9	17.5
Cooling utilities (MW)	947.4	78.5
Total utilities (MW)	971.3	96.9
Molar flowrate (kmol/hr)	63,983	15,996
Mass flowrate (kg/hr)	4,894,570	289,000

the flowrates. Further studies need to be conducted to determine potential for heat integration and the economic evaluation of an SCF-FT products separation as it is already known that the application of an SCF-FT process leads to improved thermal management of the reactor, longer life of the catalyst, and products selectivity control.

17.4 Conclusion

The application of supercritical fluid (SCF) solvents has emerged from extensive research due to the limitations faced by the media used in conventional Fischer-Tropsch synthesis. The benefits of using an SCF are mainly due to its properties. SCF possess high diffusivity (relative to the liquid phase in the slurry reactor), better heat transfer control (relevant to the gas phase reaction in multitubular fixed bed reactor), and improved solubility of the solvent in the hydrocarbon products (compared to both technologies). While previous studies focused mainly on the first two sections of the GTL process: the syngas production and the FT process, fewer studies have looked at the advancements in the FT products separation when utilizing the supercritical solvents in the reaction.

Hence the work presented here focuses on the less-studied field of SCF-FT product separation. The work proposed an alternative method for separating water from the solvent and products at the earlier stages of the separation before the medium and heavy hydrocarbons. Previous studies have shown that removing the water at later stages would result in a larger treatment capacity and larger flowrates along the process. The work focus was given to three alternatives, by synthesizing several separation sequences for the SCF-FT products separation and assessing their performance. The key differences involve the gradual manipulation of temperature and pressure and the decisions on the sequence of separating specific cuts. The three scenarios are synthesized, simulated, designed, and assessed. These scenarios are Scenario 1: Separation of the heavy components first; Scenario 2: Separation of the water first; and Scenario 3: Separation of the vapor and liquid components and use of 3-phase separator to recover water, solvent, and syngas. Computer-aided simulation of the three scenarios was carried out using Aspen Plus®. For each separation sequence, the results from the analysis of energy consumption, product recovery, and the design simplicity were used as parameters for comparison. The results showed that Scenario 1 was a more complicated design amongst the alternatives and it had more equipment and an extra distillation column, which would add to the costs of the process. The Scenario 2 flowsheet sequence is the closest when compared to that of a commercial one; yet Scenario 3 had the lowest energy intensity and the highest products recovery. This Scenario also offers the designer the flexibility to select the most suitable equipment in place of the separators used (e.g., flash separation, 3-phase separators, and distillation columns).

Acknowledgements

This book chapter was made possible by a NPRP award [NPRP 7-843-2-312] from the Qatar National Research Fund (a member of the Qatar Foundation). The statements made herein are solely the responsibility of the authors.

References

Aspen Technology. 2013. Aspen Physical Property System. Available from: *aspentech*.

Aspen Technology. 2017. Design and optimize chemical processes with aspen plus®, aspentech. Available from: http://www.aspentech.com/products/aspen-plus/.

Bao B, et al. 2012. Elevated pressure supercritical fischer-tropsch synthesis: product separation and energy integration. In: The proceedings of the 2012 International Symposium of Supercritical Fluids (ISSF), San Fransisco, California, P0605.

Bao B. 2012. process design and optimization of biorefining pathways. Texas A&M University, PhD dissertation.

Bao B, El-Halwagi MM, Elbashir NO. 2010. Simulation, integration, and economic analysis of gas-to-liquid processes. Fuel Proc Technol, 91(7): 703–713. doi: 10.1016/j.fuproc.2010.02.001.

Challiwala MS, et al. 2017. A combined thermo-kinetic analysis of various methane reforming technologies: comparison with dry reforming. J CO2 Utiliz, 17, p. 99–111. doi: 10.1016/j.jcou.2016.11.008.

Choi GN, et al. 1997. Design and economics of a Fischer-Tropsch plant for converting natural gas to liquid transportation fuels. ACS Division of Fuel Chemistry, Preprints, 42(2), p. 667–670. doi: 10.1016/S0140-6701(97)84578-X.

Dancuart LP, Steynberg AP. 2007. Fischer-Tropsch based GTL technology: a new process? Studies Surface Sci Catal . Elsevier B.V. doi: 10.1016/S0167-2991(07)80490-3.

Dry ME. 2002.The Fischer-Tropsch process: 1950–2000. Catal Today, 71(3–4), p. 227–241. doi: 10.1016/S0920-5861(01)00453-9.

Elbashir NO, et al. 2005. Impact of cobalt-based catalyst characteristics on the performance of conventional gas-phase and supercritical-phase Fischer-Tropsch synthesis. Appl Catal A: General, 285, p. 169–180. doi: 10.1016/j.apcata.2005.02.023.

Elbashir NO, et al. 2010. Advancement of Fischer-Tropsch synthesis via utilization of supercritical fluid reaction media. AIChE J., 56(4), p. 997–1015. doi: 10.1002/aic.

Elbashir NO, Bao B, El-Halwagi MM. 2009. An approach to the design of advanced fischer-tropsch reactor for operation in near-critical and supercritical phase media. Advances Gas Process, 1 (October 2015), p. 423–433. doi: 10.1016/B978-0-444-53292-3.50051-1.

Elbashir NO, Roberts CB. 2005. Catalyst in near-critical and supercritical hexane media. Indus Engg Chem Res, 44, p. 505–521.

Elmalik EE, et al. 2011. Solvent selection for commercial supercritical Fischer-Tropsch synthesis process. Fuel Proc Technol, 92(8), p. 1525–1530. doi: 10.1016/j.fuproc.2011.03.014.

Fan L, Fujimoto K. 1999. Fischer-Tropsch synthesis in supercritical fluid : characteristics and application. App Catal A: General, 186, p. 343–354.

Fischer F, Tropsch H. 1926. The synthesis of petroleum at atmospheric pressures from gasification products of coal. Brennstoff-Chemie, p. 97–104.

Förtsch D, Pabst K, Groß-Hardt E. 2015. The product distribution in Fischer-Tropsch synthesis: an extension of the ASF model to describe common deviations. Chem Engg Sci, 138, p. 333–346. doi: 10.1016/j.ces.2015.07.005.

Gabriel KJ, et al. 2014. Targeting of the water-energy nexus in gas-to-liquid processes: a comparison of syngas technologies. Indus Engg Chem Res, 53 (17), p. 7087–7102. doi: 10.1021/ie4042998.

Hao, X, et al. 2008. Simulation analysis of a gas-to-liquid process using aspen plus. Chem Engg Technol, 31(2), p. 188–196. doi: 10.1002/ceat.200700336.

Khodakov A Y, Chu W, and Fongarland P. 2007. Advances in the development of novel cobalt fischer-tropsch catalysts for synthesis of long-chain hydrocarbons and clean fuels. Chem Rev, 107(5) p. 1692–1744.

Kim YH, et al. 2009. A simulation study on gas-to-liquid (natural gas to Fischer-Tropsch synthetic fuel) process optimization. Chem Engg J, 155 (1–2): 427–432. doi: 10.1016/j.cej.2009.08.018.

de Klerk, A. 2007. Environmentally friendly refining: Fischer-Tropsch versus crude oil. Green Chem, 9(6): 560. doi: 10.1039/b614187k.

de Klerk, A. 2011. Oryx and Escravos Gas-to-Liquids Facilities. in Fischer-Tropsch Refining, Wiley-VCH Verlag GmbH & Co. KGaA p. 241–248. Malek, N. and Glavic, P. 1994. Theoretical bases of separation sequence heuristics. Computers and Chemical Engineering, 18(SUPPL), p. 0–4. doi: 10.1016/0098-1354(94)80024-3.

Malek Abbaslou RM, Soltan Mohammadzadeh JS, Dalai AK. 2009. Review on Fischer-Tropsch synthesis in supercritical media. Fuel Proc Technol, 90 (7–8): 849–856. doi: 10.1016/j.fuproc.2009.03.018.

Noureldin MMB, et al. 2015. A process integration approach to the assessment of CO2 fixation through dry reforming. ACS Sustain Chem Engg, 3(4): 625–636. doi: 10.1021/sc5007736.

Noureldin MMB, Elbashir NO, El-Halwagi MM. 2014. Optimization and selection of reforming approaches for syngas generation from natural/shale gas. Indus Engg Chem Res, 53(5): 1841–1855. doi: 10.1021/ie402382w.

Panahi M, et al. 2012. A natural gas to liquids process model for optimal operation. Ind Eng Chem Res, 51, p. 425–433. doi: 10.1021/ie2014058.

Parkash, S. 2003. Refining processes handbook. Elsevier.

Sie ST, Krishna R. 1999. Fundamentals and selection of advanced Fischer–Tropsch reactors. App Catal A: General, 186 (1–2), p. 55–70. doi: 10.1016/S0926-860X(99)00164-7.

Speight JG. 2007. Natural gas: a basic handbook. CRC Press, Taylor and Francis Group.

Steynberg AP. 2004. Introduction to Fischer-Tropsch technology, studies in surface science and catalysis. Studies in Surface Science and Catalysis: Fischer-Tropsch Technology, Elsevier B.V., Amsterdam, 2004, pp. 1-63. Stranges AN. 2007. A history of the Fischer-Tropsch synthesis in Germany 1926–45. Studies in Surface science and catalysis. Elsevier B.V. doi: 10.1016/S0167-2991(07)80469-1. Amsterdam.

Wilhelm D J, et al. 2001. Syngas production for gas-to-liquids applications: technologies, issues and outlook. Fuel Proc Technol, 71(1–3): 139–148. doi: 10.1016/S0378-3820(01)00140-0.

Yokota K, Fujimoto K. 1989. Supercritical phase synthesis reaction. Fuel, 68, p. 255–256.

18

Multi-Scale Models for the Prediction of Microscopic Structure and Physical Properties of Chemical Systems Related to Natural Gas Technology

Konstantinos D. Papavasileiou[1*], *Manolis Vasileiadis*[1], *Vasileios K. Michalis*[1], *Loukas D. Peristeras*[1], *and Ioannis G. Economou*[1,2,3*]

[1] *Molecular Thermodynamics and Modeling of Materials Laboratory, Institute of Nanoscience and Nanotechnology, National Center for Scientific Research "Demokritos", Greece*
[2] *Chemical Engineering Program, Texas A&M University at Qatar, Qatar*
[3] *TEES Gas and Fuels Research Center, Texas A&M Engineering Experiment Station, USA*

CHAPTER MENU

18.1 Introduction

Natural gas (NG) production, monetization, and conversion processing technologies have been under constant evolution and development for decades. The continuous efforts of researchers in academia and industry are inherently tied to optimizing the individual processes involved, by coping with challenges related to sustainable economics and high-quality end products. These implicate treatments of natural gas that is recovered either from conventional or unconventional (e.g., shale formations) reservoirs at the midstream level, which involves, among others, sweetening, sulfur recovery, and removal of water. Engineers face challenges not only during the midstream processing of natural gas but also at the subsequent steps in downstream processes such as different physicochemical treatments, storage, end user distribution etc. Using various molecular simulation techniques, a large number of processes involved in natural gas life cycle have been investigated, improving their understanding and resulting in their optimization. In particular, molecular modeling can contribute to advancement of our comprehension of the material science fundamentals, since it is possible to reveal the underlying molecular mechanisms and their connection with the observed macroscopic properties and behavior. This is essential for the optimization of processes and the design of new materials meeting desired specifications and requirements, and makes up the subject of this chapter (Figure 18.1a). In this context, a walkthrough on selected simulation methods relevant to the time scales and systems

*Corresponding authors:
Konstantinos D. Papavasileiou: k.papavasiliou@inn.demokritos.gr
Ioannis G. Economou: ioannis.economou@qatar.tamu.edu

Natural Gas Processing from Midstream to Downstream, First Edition.
Edited by Nimir O. Elbashir, Mahmoud M. El-Halwagi, Ioannis G. Economou, and Kenneth R. Hall.

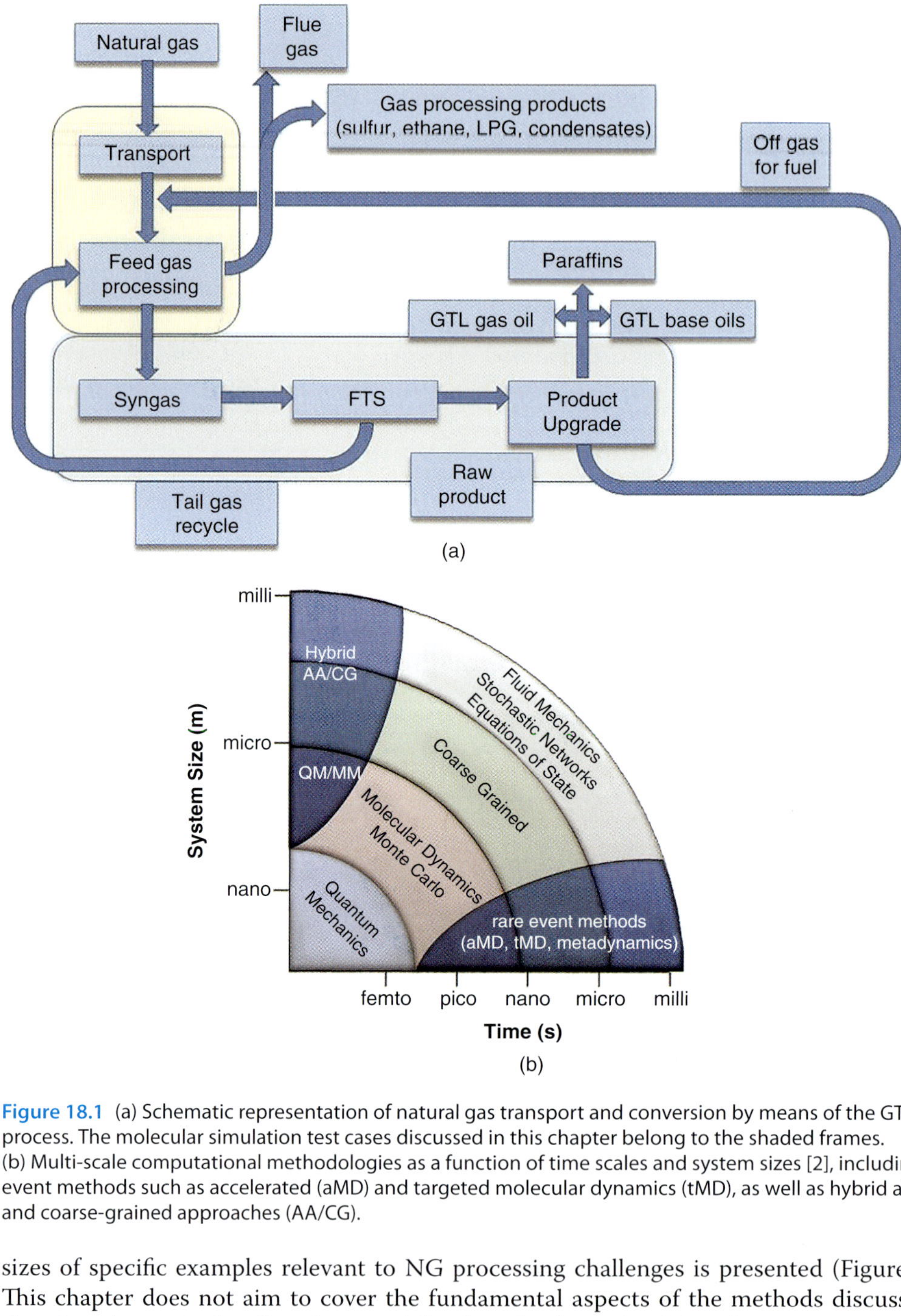

Figure 18.1 (a) Schematic representation of natural gas transport and conversion by means of the GTL process. The molecular simulation test cases discussed in this chapter belong to the shaded frames. (b) Multi-scale computational methodologies as a function of time scales and system sizes [2], including rare event methods such as accelerated (aMD) and targeted molecular dynamics (tMD), as well as hybrid all-atom and coarse-grained approaches (AA/CG).

sizes of specific examples relevant to NG processing challenges is presented (Figure 18.1b). This chapter does not aim to cover the fundamental aspects of the methods discussed, and therefore the reader is referred to specialized textbooks and articles where appropriate.

Midstream processing, and particularly the blockage of transportation pipes by the formation of natural gas hydrates, constitutes a prime example of the aforementioned challenges addressed with the aid of simulation. Hydrates are aqueous solutions of low-molecular-weight species like methane (CH_4), ethane (C_2H_6), propane (C_3H_8), carbon dioxide (CO_2), hydrogen sulfide (H_2S), nitrogen (N_2), and many more, which, under suitable conditions of pressure, temperature, and composition, can form a solid crystalline phase [1]. In these crystalline inclusion compounds, which deceptively resemble ice, the gas molecules, commonly termed hydrate formers or guests, are enclathrated (encaged) inside cages of hydrogen-bonded water (H_2O) molecules. The size of the guest molecules and their interactions with H_2O define the type of cages, which together with their respective ratio inside a crystal unit define the crystal structure. There are three common crystal structures: namely sI, sII, and sH. CH_4 and CO_2 hydrate, as well as their respective mixture, form the sI cubic structure with a lattice constant of about 1.2 nm, where each unit cell consists of 2 pentagonal dodecahedron cages (5^{12}) and 6 tetrakaidecahedron cages ($5^{12}6^2$), having a stoichiometric composition of 46 H_2O molecules with 8 guest molecules for a fully occupied cell.

There are more than 130 known guests that form hydrates, and many of these have high technological importance. Given that most of the natural gas components are hydrate formers, it is not surprising that it was the gas and oil industry that first seriously invested in the study of their properties [3], as gas hydrates pose a formidable flow assurance and safety risk. Under relatively high pressure and low temperature, the combination of free H_2O and natural gas leads to hydrate formation inside pipelines and other equipment, which can have disastrous consequences [1].

For example, during transport through subsea pipelines at depths greater than 1200 m, the nearly constant ocean floor temperature at around 4 °C provides constant cooling, which, at some point along the way, reduces the fluid temperature below the hydrate formation temperature that can be as low as 10 °C at 100 bar. It is thus necessary to know the thermodynamic, as well kinetic behavior of these systems, in order to implement the appropriate mitigation or prevention strategy. Such actions can consist of H_2O removal or the addition to the stream of hydrate inhibitors [4], which can either act thermodynamically, by shifting the equilibrium line away from the operational conditions, or kinetically, by preventing hydrate agglomeration for a sufficient residence time inside the pipeline.

Furthermore, molecular simulations have been used extensively to calculate the bulk volumetric, transport and phase equilibrium properties of CH_4-based mixtures, which are necessary for the design and optimization of natural gas processes and transportation. This is particularly important in cases where the experimental measurements are costly, potentially dangerous, and time consuming. These difficulties restrict the use of thermodynamic models, since their parametrization in order to access the systems properties in a wide range of conditions with certain accuracy depends on experimental data. This is particularly true in the case of polar substances and their mixtures. Gas sweetening for the removal for H_2S is a characteristic example of the use of molecular dynamics (MD) simulations in a range of conditions for the calculation of the volumetric properties of methane-based ternary mixtures ($CH_4/H_2S/CO_2$), while employing the Gibbs ensemble Monte Carlo (MC) simulation technique for phase diagram predictions of CH_4 and other hydrocarbon binary mixtures with H_2S [5].

Natural gas is not only a valuable combustible gaseous fuel, but it also serves as raw material for the production of transportation fuels and value-added chemicals. Natural gas reserves are much larger compared to oil, having a high hydrogen-to-carbon ratio with a source-dependent composition [6]. The development of effective routes for CH_4 conversion to produce clean hydrocarbon fuels and fine chemicals has been a field of great interest for both industrial and academic researchers. Conversion routes are either direct or indirect. Even though several techniques have been developed for direct conversion (e.g., pyrolysis, oxidative coupling, selective and partial oxidation, as well as steam or CO_2 reforming) [7], in most cases, industrial applications have not been fully developed. Commercialized natural gas conversion proceeds primarily via indirect routes, which heavily utilize heterogeneous catalysis. The gas-to-liquids (GTL) process is an established, robust and economically viable technology in harnessing natural gas resources for its efficient conversion in the production of environmentally benign and high-quality transportation fuels by the petrochemical industry.

The GTL process initiates by manufacturing synthesis gas (syngas), a crucial step for the procedural steps that follow, which is responsible for 60% of the capital cost of the entire process [8, 9]. Syngas is a mixture of hydrogen and carbon monoxide (CO) [10, 11]. It is normally derived from coal gasification or natural gas reforming using either partial oxidation or steam-reforming processes [10–12] and serves as an intermediate gas feed for many different petrochemical processes, including GTL [10].

Upon removal of impurities, syngas is in turn converted to liquid hydrocarbons, using a catalyst. The liquid formed resembles wax at room temperature, having a similar look and feel. Then, the final stage of cracking and isomerization follows, where the hydrocarbon "wax" chains are converted into high-quality fuels such as diesel, kerosene, and lubricant oil products. In particular, diesel, a fuel that is significantly more energy efficient compared to gasoline, has an important contribution to the global efforts in reducing CO_2 emissions in the transportation sector, by simultaneously retaining the existing vehicle technology. The GTL process is heavily utilized in the production of clean transportation fuel, as the diesel produced is a much higher-quality fuel compared to its crude oil refined counterpart. It has a high cetane number (at least 70), low sulphur (less than 5 ppm) and aromatics (less than 1%) content [10], while its usage generates lower emissions (lower NO_x formation) [13].

The Fischer-Tropsch (FTS) synthesis is a highly exothermic polymerization reaction fundamental to the GTL production process, as it converts syngas into fractions of synthetic liquid hydrocarbons such as paraffins (from CH_4 to waxes) and olefins. Other products include water and relatively small quantities of oxygenated compounds, such as alcohols and carboxylic acids [14, 15]. The principal equation that describes the conversion to n-alkanes is [16]:

$$nCO + (2n + 1)H_2 \rightarrow H(CH_2)_nH + nH_2O \tag{18.1}$$

where the water and wax products form a mixture through which the dissolved reactants (CO and H_2) diffuse, reach the catalytic nanoparticles, and react.

As with any synthetic route, optimizing the final product outcome by maximizing catalyst activity, selectivity, and yield towards synthetic fuels is of primary importance. Liquid hydrocarbon production is controlled by both the choice of the metal catalyst and its support material. The former is usually a transition metal such as Fe, Co, Ni, and Ru [17], with Co and Fe being an established choice in terms of their performance in industrial FTS applications [10, 16, 18]. Even though Co is more expensive than Fe, it exhibits improved selectivity towards paraffins, producing olefins and oxygenates in smaller fractions [19]. In addition, operating conditions categorize FTS processes into high-temperature (HTFT, 573–623 K) and low-temperature (LTFT, 473–513 K), respectively [20]. Co is almost exclusively used in LTFT processes [16, 19], due to its high yield towards linear C_{5+} paraffin production, increased stability, and high resistance to deactivation [20] (Figure 18.2a).

The physical and morphological properties of the catalyst support is a key feature in controlling and ensuring FTS activity and stability. For example, the pore size and surface structure of the support influences the catalyst dispersion, particle size, their arising strong metal-support interactions (SMSI) [21], as well as catalyst reducibility and diffusion coefficient of reactants and products [20]. Typical catalyst carriers include titanium dioxide (TiO_2), silicon carbide (SiC), silicon dioxide (SiO_2), and aluminum oxide (Al_2O_3) [22–24]. Unfortunately, FTS catalysts deactivate over time; contributing factors have been extensively reviewed in the literature [23, 24]. Among them, reaction conditions with H_2O in excess amounts are of particular interest, since—under this regime—sintering of catalytic nanoparticles is observed, which is responsible for FTS reactors' reduced lifetimes and increased operational costs [25] (Figure 18.2b). The degree of catalyst deactivation increases with increasing H_2O partial pressure caused by high H_2O loads [19, 26–29]. Furthermore, the effects of catalyst support features such as the pore

(a) (b)

Figure 18.2 (a) Illustration of Co nanoparticles supported on a TiO_2 anatase surface {101}. (b) Schematic representation of catalyst nanoparticle sintering.

structure and pore size distribution, as well as the surface dispersion are directly correlated with C_{5+} selectivity and productivity [16] are significant, extremely complicated, and not fully understood. The extent and exact manner that these factors affect the performance of the FTS also depend on the reactor design and operating conditions [16].

It becomes apparent that ensuring stable catalyst performance is essential to GTL process economics, as the GTL industry is exposed to significant oil and gas price risks. Recent studies have shown that for the commercial viability of the FTS GTL with oil prices of less than about $40/barrel, there needs to be a significant reduction in plant capital costs, as well as operating and feed gas [10].

Therefore, tackling the numerous and diverse challenges faced at the individual natural gas processes and developing ways of improving existing technologies is vital for the economic sustainability of the natural gas processing industry. In the following sections, three examples highlighting how molecular modeling can be used to predict microscopic structure and physical properties of model systems relevant to natural gas midstream and downstream processes are presented.

18.2 Natural Gas Pipeline Transportation: Modeling Gas Hydrates

The design of more efficient hydrate inhibitors is an active field of study, in which the prominent role is naturally held by the experimental approach and to a certain extent by the macroscopic theoretical approach that is based on the van der Waals-Platteeuw theory [30, 31]. Nevertheless, molecular simulation is playing an increasingly important role as it is gradually maturing to become a reliable tool for the determination of hydrate properties, also having the inherent capability of providing unique understanding of formation, growth, and hydrate dissociation processes at a molecular level. Using quantum mechanical (QM) theoretical methods, the interactions between formers, water molecules of the hydrate cages [32], and potential inhibitors [33] have been investigated, giving a detailed picture of the cage structure. Performing atoms-in-molecules (AIM) analysis [34] allows the study of chemical bonding nature and strength, in terms of the electronic charge density topological features [34, 35]. These features provide information on hydrate hydrogen bonding patterns, through the determination of bond critical points (BCP): i.e., points where the gradient of the electronic charge density vanishes. In the context of classical modelling, both MC and MD simulation methods have been used to investigate a number of hydrate properties [36, 37] such as lattice constants, guest occupancies, heat capacity, thermal expansivity and conductivity, nucleation, growth and dissociation. In the following section, we offer an example of the determination of P-T diagrams of hydrates with the use of MD simulation, where the combination of appropriate force fields and calculation methodology can provide accurate predictions of the three-phase coexistence conditions for the CH_4 and CO_2 hydrates, as well as their respective mixture. Such studies lay the ground for the systematic use of molecular simulation for the study of hydrate inhibitors.

The determination of the thermodynamic coexistence conditions of gas hydrates has been approached [38–40] with the direct phase coexistence methodology. The initial configuration of the system consists of all three phases, namely the solid hydrate phase, the liquid aqueous phase, and the guest rich phase, which can be either gas or liquid depending on the type of the guest and the conditions. By simulating with MD, the time evolution of such a system under various regimes of pressure and temperature, either hydrate growth or hydrate dissociation will be observed. Thus, for a pure hydrate, by fixing, for example, the pressure and carrying out simulations at various temperatures, the coexistence temperature can be identified as the average value between the highest temperature for which the hydrate growth is observed and the lowest temperature for which hydrate dissociation takes place. Michalis et al. [38] have shown that both hydrate growth and dissociation in the regime very close to the three-phase coexistence temperature T_3 are highly stochastic, which effectively means that for a few degrees around T_3, the same system can exhibit either hydrate growth or dissociation, and thus a simple temperature scan is not enough. In order to be able to provide accurate results, it is necessary to carry out multiple independent simulations at the same conditions and extract a statistically meaningful average. For a relatively small system size, where a $2 \times 2 \times 2$ sI supercell must be grown by a couple of unit cells in the direction normal to the hydrate-water interface, simulations that are several hundred nanoseconds long are required. It is thus clear that such calculations are computationally demanding. They can nevertheless offer high-quality predictions, comparable to the experimentally determined values.

The quality of the force fields used in these simulations is of paramount importance. Conde and Vega [41] and Michalis et al. [38] have shown the importance of the H_2O model in this type of calculations, which is necessary for the correct description of the H_2O–H_2O [38] interactions. In general, it has been found that a model such as TIP4P/ice [42], that can offer

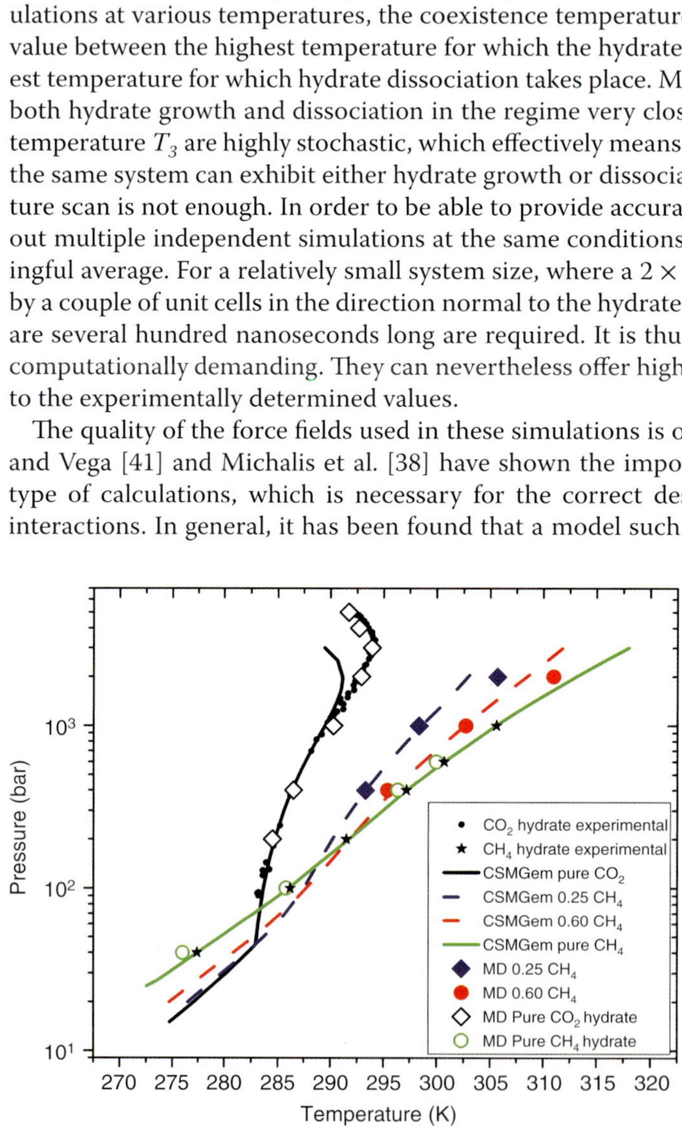

Figure 18.3 MD predictions of the three-phase coexistence conditions of the binaries $CH_4 - H_2O$, $CO_2 - H_2O$ and the ternary $CH_4 - CO_2 - H_2O$ hydrate systems. Comparison with CSMGem predictions and for the case of CO_2 with experimental data as well [1, 43, 44].

good predictions for the melting temperature of ice, captures accurately the thermodynamic behavior of hydrates as well. Furthermore, the H_2O–guest interactions must be adequately described. Costandy et al. [39] have shown that for the case of CO_2 hydrate, a force field combination that can correctly describe the solubility of the guest molecule in the aqueous phase offers the best predictions of the hydrate coexistence temperature for a wide range of pressures. Having correctly described the H_2O–H_2O and H_2O–guest interactions for the pure CH_4 and CO_2 hydrates, Michalis et al. [40] showed that the use of their combination can be extended for the prediction of the coexistence line of the respective ternary mixed hydrate system for various compositions.

As an example, MD results [40] are presented in Figure 18.3 for the three-phase equilibrium conditions of the ternary mixture hydrate system for two CH_4 bulk compositions, as well as results of the pure binary systems and the respective experimental data. In both cases presented, i.e., pure methane and pure carbon dioxide hydrates, the agreement between the experimental data and MD predictions is excellent. For the ternary system, the MD-predicted values are compared with the respective predictions by CSMGem [1] (an implementation of the van der Waals-Platteeuw approach) with good agreement up to 1000 bar. The deviations observed at 2000 bar and above, are attributed to the difficulty of CSMGem to correctly describe the equilibrium behavior of the pure CO_2 hydrate in this pressure region, whereas for the same case it has been shown [39] that the MD predictions have good agreement with the available experimental data for the whole pressure range. Furthermore, given the intrinsic capability of MD to provide molecular level insights, these simulations can be used in order to extract additional information about the system. For the CH_4–CO_2–H_2O hydrate mixture, the analysis of the MD trajectories can reveal the ratio of the guest molecules in the hydrate structure, as well as their distribution in the different types of cages. In Figure 18.4, two snapshots of an MD trajectory [40] are presented (initial and final configuration), where the grown mixed hydrate can be analyzed in order to determine the selectivity of the system.

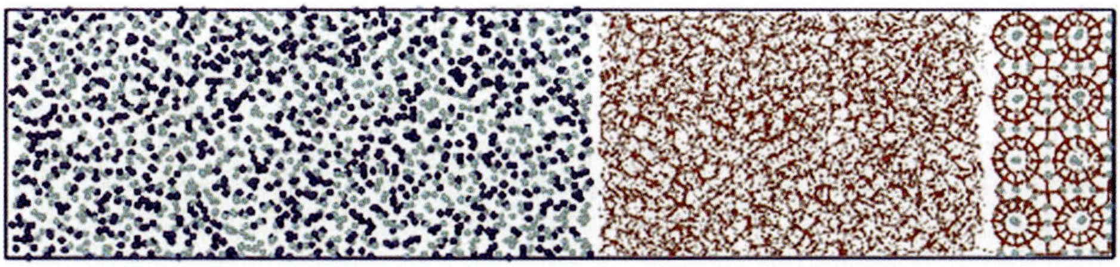

(a)

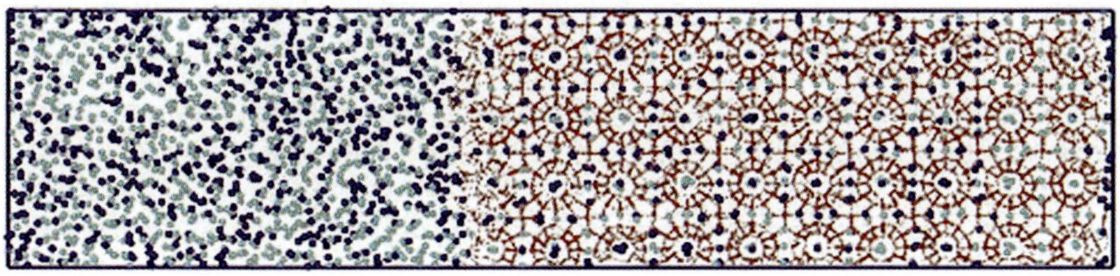

(b)

Figure 18.4 Initial (a) and final (b) configurations of an MD trajectory at 2000 bar and 295 K, where growth has taken place of a mixed CH_4 – CO_2 hydrate. Dark blue spheres correspond to the carbon atom of the CO_2 molecule, light blue spheres correspond to CH_4 molecules, and red lines represent the hydrogen bonds between the H_2O molecules.

The successful predictions of the three-phase coexistence line of CH_4 and CO_2 hydrates serve as test cases that exhibit the capabilities of molecular simulation to play a more active role in the study of gas hydrates. At present, there is a lack of atomistic simulation studies regarding multicomponent gas mixtures. However, given the continuous increase of available computational power and the existence of efficient methodologies, it is expected that more complex systems involving combinations of fluids with real-gas compositions and inhibitors will be simulated with increased accuracy in the near future. This can lead to a paradigm shift in our approach for the determination of hydrate phase behavior.

18.3 Modeling Porous Media in Separation and Storage Procedures

Porous media are present in every step of natural gas exploitation activity, from its extraction to its monetization to value-added products. Their transport, sorption, swelling, and mechanical properties define the viability of natural gas extraction from both conventional and unconventional reservoirs, and control the efficiency of separation midstream and downstream processes crucial for natural gas cleaning, transportation, and refinement [45]. As discussed previously, porous media are also used in catalytic processes either as substrates or as catalysts for the production of synthetic fuels and other valuable chemicals like methanol, formaldehyde, dimethyl ether, heavier hydrocarbons, hydrogen, and aromatics, which contributes to the increased capitalization of natural gas economy. Last but not least, novel highly porous materials are gaining increasing interest due to their potential use as on-board storage media for automotive vehicles, contributing to the replacement of the conventional petroleum fuels with CH_4 due to its environmental and economic advantages [46].

Seeking high-performance adsorbents that can be used in separations or for on-board storage meeting the existing specifications and requirements, a wide range of classes of materials is being considered by the industry and research communities. This set includes, among others, zeolites, metal–organic frameworks (MOFs), zeolitic imidazolate frameworks (ZIFs), porous polymer networks (PPNs), carbon nanotubes (CNTs), functionalized graphene, polymers, and hybrid composite materials (mixed matrix membranes) consisting of inorganic particles placed inside a polymer matrix. Molecular modeling offers a powerful means to predict porous media properties, understand the underlying molecular mechanisms, and contribute to the design of novel materials, complementary to the experiment.

The essential step to model porous media is the complete characterization of their porosity. Consequently, a significant effort has been invested to devise schemes able to analyze efficiently the void space that exists in atomistic configurations. These schemes are based on either grid [47, 48] or other non-uniform space tessellation algorithms [49–51]. The free and accessible volume and surface, the pore size distribution (PSD), the maximum pore diameter, the limiting pore diameter (LPD, defined as the largest penetrant that can freely diffuse in the material), the character of the porosity (percolated or not), and the developed channels or pore networks are some of the properties that can be calculated. These basic properties are needed to explore the relation between the chemistry and microscopic structure of porous materials and the exhibited adsorption and transport efficiency, as it will be indicated in the next paragraphs.

Adsorption efficiency of the aforementioned materials has been extensively studied using grand canonical Monte Carlo (GCMC) simulations, in combination with schemes for the calculation of the chemical potential [52, 53] or Gibbs ensemble MC simulations [54–60]. In addition, *ab initio* density functional theory (DFT) methods have been utilized to explore how a material's adsorption efficiency is affected by changes at the molecular level. For example, Dutta et al. [61] used van der Waals-corrected *ab initio* DFT calculations to explore the possibility to improve CH_4 and CO_2 uptake for gas storage (and sensing), in high-surface-area graphene-derived materials with modified local structure and morphology. In another study, Wang et al. [62] using *ab initio* DFT and classical MD simulations investigated the performance of nitrogen and fluorine functionalized nonporous graphene for the separation of CO_2, N_2, and CH_4 mixtures. Focusing on the development of novel adsorbents, the majority of the examined materials have been created *in silico* to construct a large number of structures and realistically sample the chemical space of the considered material class. Calculation workflows included "virtual" characterization and calculation of the adsorption properties. An impressive example is the screening of a large database containing over 650,000 computer-generated materials including zeolites, MOFs, and PPNs, in an attempt to identify suitable materials for CH_4 storage [63]. It is worth noting that both simulation studies and experimental findings confirm that the requirements for on-board storing materials set from the Advanced Research Project Agency–Energy (ARPA–E) are difficult to meet [46]. Apart from the actual calculation of permselectivity, as well as the uptake and the deliverable capacity in a range of conditions of interest, an additional desired outcome of screening studies is to reveal structure-property relationships that may be used as new screening descriptors in the quest of designing and synthesizing new materials.

MD simulations, along with calculations of free energy profiles and jump barriers, are the most important tools used for calculating diffusivities (Fick, Onsager, and Maxwell-Stefan) of fluids, gases, or liquids, confined inside crystallographic porous structures. Regarding narrow micro-pores, where molecules move passing windows and channels of small diameter compared to molecular dimensions, transition state theory (TST) is the suitable method for the study of transport properties. In this case, the identification of the channels spanning the structures is important for the efficient performance of the calculations. The diffusion of CH_4 and its mixtures in meso- and microporous crystallographic structures is well studied, and the molecular mechanisms together with the various factors affecting the observed behavior have been identified and classified to a large extent [64–66]. This is particularly important in the design of membranes used in separations and the optimization of catalytic reactors. Examples of classical simulation studies of adsorbents (zeolites) used for separations in the context of the FTS process are presented in section 18.4.3.

Polymeric membrane technology is considered a mature option for hydrocarbon and acid gas separations, and new polymer families have been proposed as suitable to be used in these processes [67–70] overcoming performance problems including plasticization, compaction and aging, also caused by the extreme operation environment [71]. The gas transport mechanism through a rubbery or glassy membrane generally follows the solution-diffusion model [72]. Studying the permeability of these polymeric matrices presents extra levels of difficulty. Besides the need to create atomistic configurations with relaxed configurational properties, the diffusion at low temperatures near or below the T_g is the result of infrequent penetrant jumps inside the matrix. Also in this case, TST methods need to be used for the calculation of the jump rates and the estimation of the diffusion coefficients by means of kinetic MC schemes (KMC), in a way similar to diffusion studies in narrow micropore crystallographic structures [72]. The identification of jumps is performed with a detailed analysis of the matrix accessible volume, since a penetrant will most probably move from its current position to a nearby location characterized by a "lower" density (compared to the bulk). A general implementation of these calculation workflows is demanding compared to the conventional MD, but their use is essential for attaining qualitatively reasonable results.

18.3.1 Modeling Kerogens Porosity: A Case Study

Shale gas is natural gas hosted in shale formations. It is an increasingly important unconventional source of energy that has attracted a lot of interest over the past two decades [73]. In common organic solvents, kerogens are the insoluble part of the organic matter found in shale formations that are capable of generating oil or gas. Based on the hydrogen-to-carbon (H/C) and oxygen-to-carbon (O/C) ratios of their elemental analysis, kerogens are classified in three basic types: namely, I, II, and III [74]. Adsorption experiments have shown that in many cases, their maximum CH_4 uptake is linearly correlated to their total organic content (TOC), indicating that the organic matter's contribution to gas adsorption is the most significant. Better understanding of kerogen properties is important for predicting or improving the productivity of a well, a challenging task due to the unconventional (ultralow permeability) nature of the shales, while limiting the environmental impact associated with shale gas extraction [73]. Molecular simulation can play a significant role towards this direction, given the difficulties associated with the isolation of kerogen from the shale.

The molecular (chemistry/architecture) and porous structure of a kerogen sample is of a statistical nature, since it is essentially a mixture of molecules that resulted from the degradation of large organic biomolecules under extreme conditions of temperature and/or pressure. Both structures depend, among other factors, on their origin and level of maturity and can therefore vary between different shale formations. This fact poses a challenge for modeling kerogens at the atomistic level, i.e., the creation of realistic kerogen structures. One way to construct representative kerogen models is to use reverse MC (RMC) based methods, where the connectivity of the system for a given element constitution is changed using MC moves, in such a way that experimental information like the radial distribution function (RDF) is reproduced. MD hybrid reverse Monte Carlo method (MD-HRMC) is a variation of RMC that has been used to study kerogens [75]. In MD-HRMC, MD steps as elementary MC moves are included in order to evolve and relax system configuration. This approach requires the use of a reactive force fields (e.g., ReaxFF [76], REBO [77]), which are of higher computation cost compared to their classical counterparts and require demanding parameterization to correctly reproduce the chemistry of the targeted system.

An alternative method that has been used in the literature is to construct representative molecular models based on available experimental data (such as elemental analysis or scattering data even of a specific sample) and prior knowledge on molecular fragments that likely exist in kerogen. Using these molecular models, bulk kerogen configurations are constructed, applying a staged cooling protocol [78]. Starting from a high temperature and small density configuration, which can be easily created, the temperature is reduced to the desired value (usually room temperature) in a stepwise fashion using cycles consisting of a linear cooling stage followed by equilibration in the isothermal-isobaric (*NPT*) ensemble. Due to the geometry of the type II-D molecular model, the resulting kerogen structure from this cooling procedure consists of blocks of molecules stacked together in "parallel". The relative orientation of these blocks is random, and the free volume is created mostly at their boundaries [78]. A way to impose porosity is to include a number of Lennard-Jones (LJ) dummy particle(s) in the initial configuration and remove them at the end of the staged cooling, performing an extra equilibration *NPT* simulation [79]. In this case, several factors affect the quality of the resulted configuration: i.e., the force field, the size of the system, and the number of LJ particle(s) (LJP) used to induce porosity. We adopted this later procedure for the construction of overmature type II kerogen, which is commonly found in industrially important shale formation.

Using a scheme developed by Vasileiadis et al. [51] for the characterization of the void space in both amorphous and crystalline materials, we explored the effect of the aforementioned factors that may control the porosity of the constructed kerogens. The scheme is based on

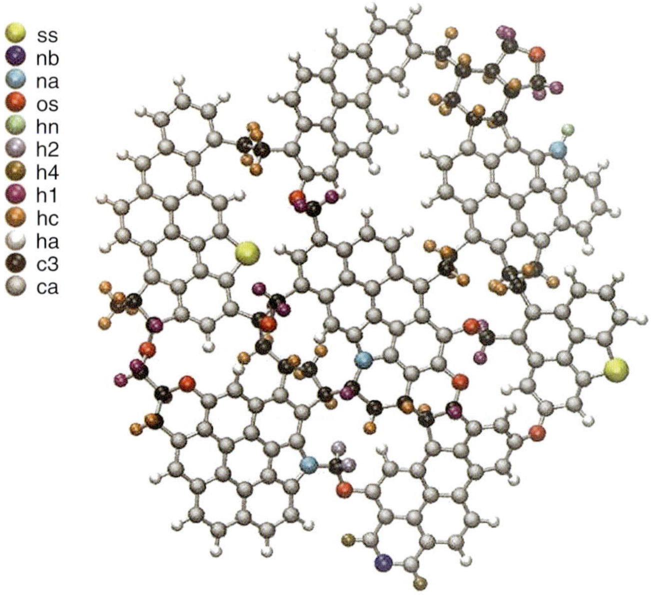

Figure 18.5 The II-D kerogen model consists of five- and six-membered fused rings forming a disk-like molecule with chemical formula $C_{175}H_{102}N_4O_9S_2$. Color map for the GAFF assigned types [82] is: yellow (sp^3 S, ss), blue (inner sp^2 N in a conjugated ring, nb), cyan (sp^2 N, na), red (sp^3 O, os), lime (N-bonded H, hn), light blue (H on aliphatic C with 2 electron-withdrawal groups, h2), ochre (H on aromatic C with 1 electron-withdrawal group, h4), purple (H on aliphatic C with 1 electron-withdrawal group, h1), light orange (aliphatic C-bonded H, hc), white (aromatic C-bonded H, ha), black (sp^3 C, c3) and gray (aromatic C, ca).

Voronoi tessellation analysis of the structure of interest, using algorithms from graph theory in a computationally efficient manner.

A model of type II kerogen available in the literature [78], namely, the II-D molecular model (illustrated in Figure 18.5), was used, as it is representative of many productive shale fields such as the Duvernay and Barnett. Applying a cooling protocol, structures with 15, 50, and 100 II-D molecules were constructed. Two force fields were examined: 1) the Dreiding force field [80] with electrostatics modeled with Gasteiger [81] point charges and 2) the general AMBER force field (GAFF) [82] with charges calculated with a standard protocol [83]. The number of LJ particles ranged from 0 to 3 and the examined sized where 15, 20, 30 and 40 Å.

By comparing the two force fields, minor differences were observed in the calculated densities and the porosity characteristics (both PSD and LPD distributions), suggesting that even the simple, widely available and easy-to-use Dreiding is suitable for a first approach to characterize kerogen structures of similar type. Nevertheless, our study was focused on structures generated using GAFF, which is expected to be more accurate for the prediction of the adsorption and transport properties. On the other hand, system size seems to affect both the observed densities and pore size distributions. Therefore, it is in general desirable to use as many molecules as possible. In all cases, ultralow size pores of below 10 Å are present. The use of LJ particle(s) increases the free volume available in the kerogen model with simultaneous decrease of the density of the system. This trend stops if one particle is used and its diameter exceeds the characteristic II-D molecule diameter (~28 Å). Going from 30 to 40 Å LJ particle diameter the density increases, an indication that the initially imposed free volume collapses after particle removal in the last step of the staged cooling procedure. The use of LJ particles allows to access the remaining spectrum of micropores (up to 20 Å), since increasing either their size or number results in broadening the pore size distribution [Figure 18.6].

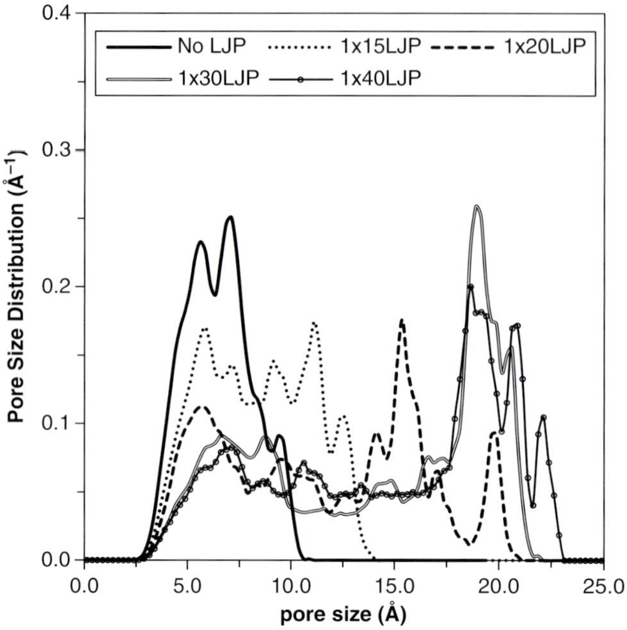

Figure 18.6 Pore size distributions of the 50 II-D bulk kerogen structures modeled using GAFF, as function of the diameter (15 to 40 Å) of a single LJ dummy particle.

The LPD is an important property, since transport properties are expected to be more sensitive to it. The effect of the LJ dummy particle(s) on the LPD is complicated. While force field and system size do not affect the observed LPD, the presence of LJ dummy particles can be used to alter it. Increasing the number of randomly placed LJ particles of small size (15 Å) does not effectively alter the LPD distribution. Instead, using larger particles randomly placed in the initial configuration, allows the construction of pores large enough to study the diffusion of several molecules of considerable size. Certain combinations of number and size of dummy particle result in the kerogen structures whose CH_4-accessible volume is percolated. This is indicated by LPDs greater than the effective diameter of CH_4 (3.73 Å in TraPPE model [84]), as shown in Figure 18.7.

The use of LJPs is necessary for the construction of structures with percolated pores accessible to CH_4 molecules (i.e., their LPD is larger than the effective diameter of CH_4). According to data shown in Figure 18.7, percolation was achieved to a smaller extend (1–2%) when a single small-size particle was used (less or equal to 30 Å). On the contrary, over 20% of the configuration pores were percolated for the larger size (40 Å) and especially when the number of particles was increased, reaching 70% when 2 particles with diameter of 30 Å were used.

Performing the analysis with a probe of smaller diameter (2 Å), 99% of the produced configuration pores were found percolated. Using this probe radius, the elemental analysis of the traced diffusion paths surfaces reveals a clear tendency of specific atom types to lie predominantly at the pore surface instead of the bulk. For example, more than half of the pyrrolic hydrogens (atom type hn) are found at the surface of pores. Figure 18.8 presents the percentage of the atom of each type that constitute the surface of the pores. The same analysis for the regions of diffusion channels with diameters comparable with the LPD shows that these regions are dominated by the hydrogens (atom type ha) and carbons (atom type ca) of the phenyl rings.

GCMC simulations were then performed to study the adsorption behavior of pure CH_4, C_2H_6, and butane (C_4H_{10}) in the generated kerogen structures. The TraPPE force field [84] was

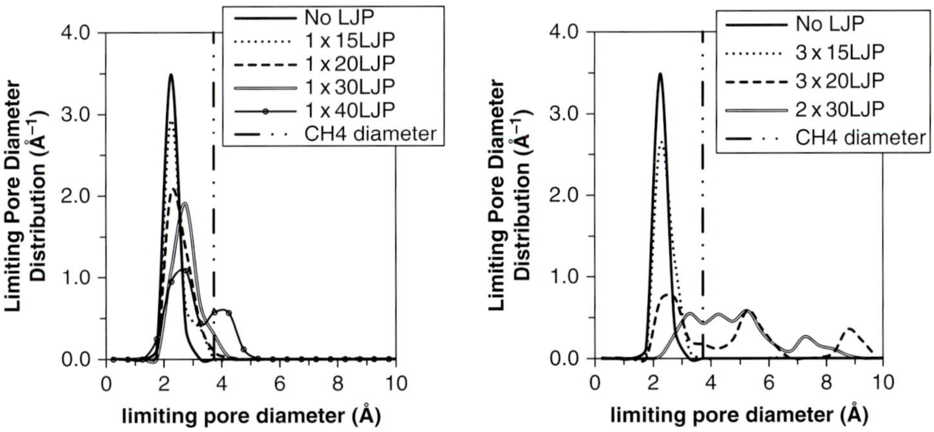

Figure 18.7 Limiting pore diameter distribution for 50 II-D kerogen structures constructed using GAFF, with the aid of (a) 1 × 15 Å, 1 × 20 Å, 1 × 30 Å, 1 × 40 Å, and (b) 3 × 15 Å, 3 × 20 Å, 2 × 30 Å LJ dummy particles set ups.

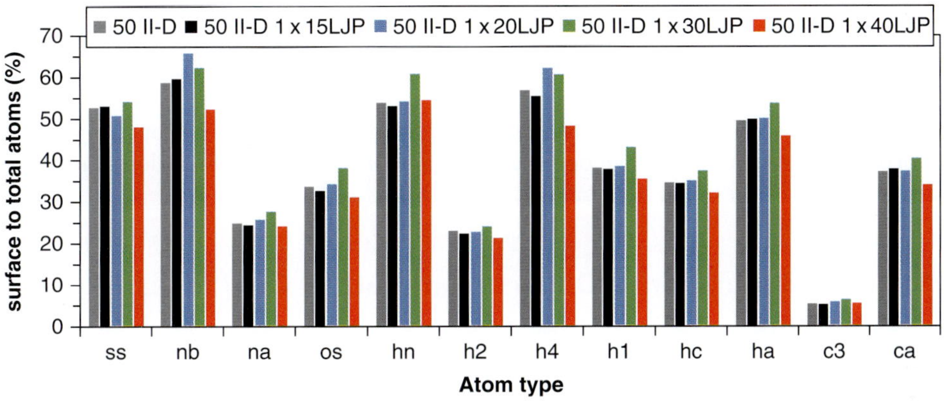

Figure 18.8 Percentage of each GAFF atom type that is located at the pore surface with respect to the total number of atoms of each type.

used to model the adsorbents, and the achieved uptake was correlated linearly with porosity as a function of the pressure and the molecular size. According to simulation results shown in Figure 18.9, the slope of the correlation line for each molecule increases with increasing pressure, especially for CH_4. The small negative slope observed at 1 atm for CH_4 is a consequence of the reduced gas-wall interactions manifested in the case of larger pores. This corresponds to higher porosity, leading to decreased loading compared with the one achieved in the case of smaller micropores [85]. Moreover, an increased statistical uncertainty contributes to this behavior, since the value of the loading for this pressure ranges from 0.72 mmol g^{-1} (for the low porosity structures) up to 0.62 mmol g^{-1} (for the large porosity configurations), with errors in the order of ± 0.1 mmol g^{-1}. For the case of C_2H_6, saturation is achieved at 20 atm, as it is indicated by the fact that the data for this pressure almost coincide with the data at 100 atm. Finally, for C_4H_{10} saturation is achieved at 1 atm, and no other data are reported for the sake of clarity. Note that the adsorbed amount of CH_4 at 1 atm has negligible effect with the accessible volume.

With the boost of the shale gas industry in the USA, industrial and academic research interest in shale gas has been revived. Several methods have been investigated for the construction of

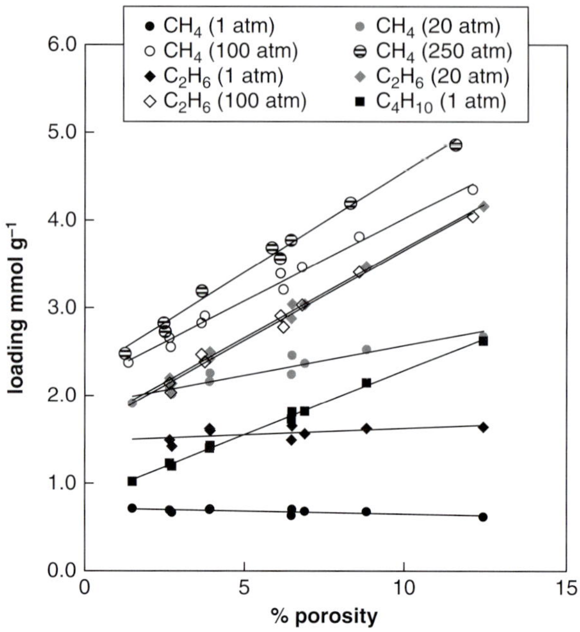

Figure 18.9 Adsorbed amount of CH_4, C_2H_6 and C_4H_{10} as function of porosity for different pressures.

realistic kerogen models. These models have been used for the prediction of properties (volumetric, adsorption, transport) and have been proven valuable in qualitative understanding of the underlying physics of these phenomena [75, 78, 86–88].

18.4 Molecular Simulation of Downstream Natural Gas Processing: The GTL Technology

18.4.1 Investigations at the Quantum Level

18.4.1.1 Methods and Models

The GTL technology is a very complex industrial process, having a significant number of individual steps and parameters controlling its yield and performance (Figure 18.1a). Molecular modeling techniques can be used to obtain the microscopic information that aids in understanding the fundamental mechanisms controlling individual steps of the process, which may otherwise be unattainable by current experimental methods. These techniques cover a wide spectrum of these steps, both in size and time scales. The following section is concerned with presenting the relevant molecular simulation methodologies employed at selected GTL technology stages as we move up the scale (Figure 18.1b), the current status and challenges that simulation studies face.

The theoretical investigation of the properties of interest to GTL relies on molecular simulation. The core of GTL catalytic processes involves metal catalysts and the successful design of reactors is inevitably tied to the knowledge of fundamental principles governing their chemistry. Ideally, all elementary reaction steps involved could be modeled from first principles (*ab initio*) QM methods [89]. These can be subdivided into three different approaches, namely: 1) wavefunction methods, 2) DFT, and 3) semiempirical methods, the theoretical framework of which can be found elsewhere [90–93].

Ab initio wavefunction methods are very accurate if a high level of theory is employed, but are usually limited to treating a few tenths of atoms. Even though modern computers have vastly promoted their efficiency and speed, this poses serious limitations in the case of transition metal surfaces, since the sizes of the systems of interest are typically too big to handle at any level above Møller–Plesset second order perturbation theory (MP2) [94]. This generally makes these methods unattractive for routine treatments of the complex catalyst systems.

On the other hand, semiempirical methods are very efficient computationally, since they possess several approximations compared to their *ab initio* counterparts. They are derived from the Hartree-Fock (HF) formalism by omitting or approximating contributions to the Fock matrix, such as two-electron integrals, by fitting to experimental or *ab initio* calculated data (i.e., ionization energies, dipole moments, etc.) [91]. However, these methods are bound to the parameterization set used and lack the desired accuracy needed for the energetic description of systems concerning heterogeneous catalytic surfaces [94]. DFT methods provide the ideal trade-off between efficiency and accuracy in describing reactions of atoms and small molecules with metal surfaces and have therefore been established in the studies of heterogeneous catalysis relevant to the GTL process [95]. For a broad and concise overview of the DFT functionals used in surface catalysis studies, the reader is referred to the work of Cramer and Truhlar [96]. DFT functionals are commonly combined with two basic types of basis sets, namely: 1) localized functions (Gaussian-Type Orbitals, etc.) and 2) plane waves [97].

In general, the overall cycle of heterogeneous catalysis reactions is thought to consist of four basic elementary steps: 1) chemisorption, 2) dissociation, 3) recombination, and 4) molecular desorption [89]. Chemisorption, in particular, is a definitive step and has been a topic receiving great attention by both experimental and theoretical researchers. An inherent limitation for the theoretical study of these elementary processes is that describing the catalyst systems in their entirety is infeasible. Therefore, there is a need to limit the model system size to the absolute necessary number of atoms that require explicit treatment. In addition, molecular simulation provides accurate scientific results and interpretations only under the condition that representative model systems are selected, capturing the essential features of the actual system [90]. This is usually performed in two ways, as shown schematically in Figure 18.10. The first is by constructing a cluster, which comprises only a small representative number of the catalyst surface atoms [97, 98]. The second approach is by setting up a periodic slab [94, 99, 100], which contains the active surface region within a unit cell of appropriate size so that neighboring unit cells do not interact, repeated in three dimensions. In both cases, system size controls the computational effort required.

18.4.1.2 Methane Conversion to Syngas

A crucial first step in natural gas GTL conversion is the reforming of CH_4 to syngas. Syngas production controls the FTS process, and it is not surprising that this reaction has drawn a lot of

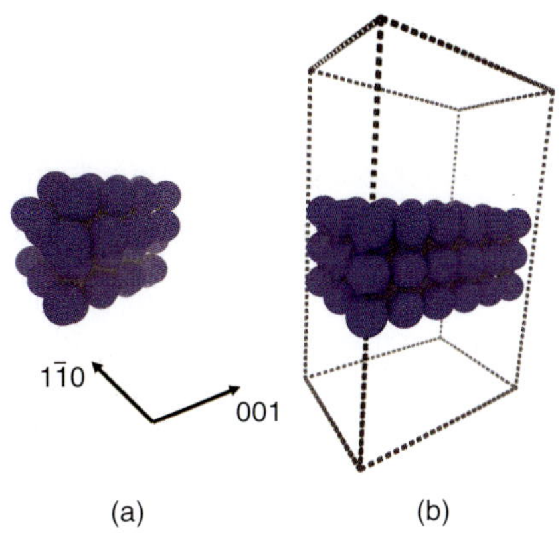

Figure 18.10 Illustration of representative (a) cluster and (b) slab models for face centered cubic (fcc) Co {110} surfaces.

attention from the simulation world. However, as CH_4 is a molecule with characteristic stability and symmetry, this reaction is unfavorable and proceeds in the presence of metal catalysts [101], among which, Ni-based catalysts have been extensively used [95]. Regarding simulation, Siegbahn and Panas [102] performed a theoretical study of the CH_x chemisorption on the Ni {100} and Ni {111} surface clusters of varying size by means of the complete active space self-consistent field (CASSCF) multiconfigurational method, which is used to generate qualitatively correct reference states of molecular systems [103]. Their calculations showed that CH radicals chemisorb more favorably on Ni {100} clusters, while CH_3 was the most stable species on the Ni {111} surface cluster [102]. Watwe et al. [104] performed Generalized Gradient Approximation (GGA) DFT calculations using the Perdew-Wang exchange and correlation functional (PW91) [105, 106], expanding the Kohn-Sham one-electron states on a plane wave basis to study CH_x species on a two-layer periodic Ni {111} slab. Their study revealed the preference of CH_x to Ni {111} threefold sites and that CH is the most abundant species on the surface [104]. Wang et al. [107] performed a study of CH_x species on Ni surfaces, using a periodic slab model by means of GGA DFT calculations using the Perdew-Burke-Ernzerhof (PBE) [108, 109] functional on plane waves. Their calculations showed that Ni {100} was the most preferred surface for CH_4 dissociation, compared to Ni {111} and Ni {110}, while CH was the most abundant CH_x species on Ni {111} [107]. These studies serve as a prime example of the effect of level of theory, model type (cluster or slab), and size upon the calculated properties, and more have followed in the same spirit [110, 111].

Unfortunately, Ni-based catalysts have been found to be susceptible to deactivation from carbon deposition [112, 113], as well as sintering [114]. Therefore, there was a need to explore other alternatives and, to this end, several other transition metals have been studied theoretically in terms of their performance in CH_4-to-syngas conversion. For example, PBE [108, 109] plane wave basis set simulations performed by Kokalj et al. [115] on a Rh {111} slab consisting of four layers, showed the significant enhancement for the first dehydrogenation of CH_4, compared to Ni {111} [104], Ru {0001} [116], and Pd {100} [117]. This was further illustrated by Jones et al. [118], who showed by means of the Revised Perdew-Burke-Ernzerhof functional (RPBE) [119] that Ru and Rh had higher activity compared to Ni, Pd, Pt at the same particle size and dispersion in CH_4 conversion [118]. This finding has important implications in promoting Ni catalysts with these metals, as they enhance catalyst stability and performance [120]. Recently, in the work of Li et al. [121], the thermal conversion of CH_4 to syngas mediated by a heteronuclear metal oxide cluster ($RhAl_3O_4^+$) has been identified for the first time by both mass spectrometry and DFT calculations using the hybrid Becke's Three Parameter Hybrid Perdew 86 functional (B3P86) [122, 123]. Their results showed that trace amounts of Rh can promote the direct participation of a chemically inert support's (e.g., Al_2O_3) lattice oxygen to oxidize CH_4 to CO [121].

18.4.1.3 Syngas Conversion to Hydrocarbons

Moving deeper into the GTL process and with regard to the FTS synthesis, several DFT studies have been devoted to CO chemisorption and activation on different metal surfaces, which is considered as the first reaction step of FTS syngas conversion, under both gas and liquid phase conditions [124]. This step remains to this day an ongoing subject of debate [125]. It is generally thought to proceed via three main mechanisms, namely the 1) carbide, 2) CO insertion, and 3) hydroxycarbene (enol) mechanisms, respectively [18, 125–127]. While the former two conflicting mechanisms are considered to be the most dominant [128], none is capable of fully describing the mechanistic details of FTS or the entire product range, as the process reactions are subject to the various reaction conditions and catalysts [125]. Therefore, it is not surprising that a substantial body of computational research effort has been devoted to this critical step of the process.

Ge et al., investigated by means of the PW91 functional [129, 130] the adsorption and activation of CO over a series of Co surfaces, namely close-packed {0001}, corrugated $\{11\bar{2}0\}$, and stepped $\{10\bar{1}2\}$ and $\{11\bar{2}4\}$ [131]. The calculated CO adsorption was exothermic, without being severely affected by the chosen surface structure (between −160.4 and −164.6 kJ mol^{-1} on all the surfaces examined) [131]. Dissociation of CO was found to proceed through a characteristic transition state with a pronounced C-O bond stretch reaching 2.2 Å, while stepped surfaces exhibited low-energy pathways of spontaneous CO dissociation in the gas phase [131], in agreement with experimental observations [132, 133]. The favorable CO dissociation on stepped Co and Ni surfaces was also shown in other DFT studies [134, 135]. Cusinato et al. [136] investigated the adsorption of H_2 or CO at the surface of Ru nanoparticles for several coverage values, as well as the H/CO co-adsorption according to a thermodynamic model fed with zero-point corrected energies calculated with the PBE functional. Their results showed that surface hydrides are replaced by CO ligands under a syngas flow, in agreement with gas-phase chromatography measurements [136]. Recently, Petersen et al. [137] investigated CO at step and kink sites on face-centered cubic (fcc) Co {221} and Co {321} periodic slab surfaces with the PW91 functional and a plane wave basis set. Their results showed that the direct CO dissociation is most preferable energetically compared to other hydrogenation-assisted routes, also confirming that the step and kink defect sites are more active in direct CO dissociation [137].

18.4.1.4 Solvation Effects

Unfortunately, even though QM calculations provide unique insight in the mechanistical aspects of the FTS process, they are mostly limited in equilibrium model systems in the gas phase. Solvent effects are important, as they may bear significant impact on a reaction's direction and efficiency [138], and, as discussed previously, H_2O presence in FTS process has substantial negative effects, particularly on Co/TiO_2-based catalysis [139, 140]. This is tackled by introducing solvation effects, either explicitly or implicitly, which in either case are accompanied by various difficulties [141].

Adding explicit H_2O molecules in the simulations not only increases the number of atoms, hence rendering computations very expensive, but also the conformational states of the system. In order to ensure proper sampling for calculating properties, multiple configurations are needed, which becomes a nontrivial task for systems possessing multiple degrees of freedom [142]. Adequate sampling is commonly achieved by means of *ab initio* molecular dynamics (AIMD), quantum mechanics/molecular mechanics (QM/MM), and coupling quantum mechanical calculations with classical MD simulations [142]. The AIMD approach [143] has been extensively followed for modeling H_2O solvated systems; however, it is limited to a few hundreds of atoms and simulation times in the picosecond regime [144]. Alternatively, it is often preferred to either use a small H_2O region (i.e. monolayer or less) or fix the entire aqueous phase molecules, with the exception of those lying in proximity of the active region [141].

Faheem and Heyden have suggested an alternative solvation scheme for computing the free energies of reactions at metal–H_2O interfaces by combining QM calculations with molecular mechanics (MM) [144], an approach widely applicable in biomolecular systems [145], which they employed to model the C–C cleavage reaction at the Pt {111}/H_2O interface [144]. Another approach is to couple DFT with classical MD simulations. Recently, Asiaee et al. [124] studied the adsorption of CO on periodic Co {0001} surface slab systems in the presence of supercritical *n*-hexane, by initially conducting DFT calculations for evaluating adsorption energies coverages *in vacuo*, which also served as input for a Morse potential fitting parametrization procedure that allowed establishing a classic description of the CO carbon-cobalt interaction. This potential function was coupled with the universal force field (UFF) [146] for Co and CO and the united-atom transferable potentials for phase equilibria (TraPPE-UA) [84] for *n*-hexane. Then,

MD simulations in the canonical ensemble (*NVT*) were executed for the subsequent calculation of the Potential of Mean Force (PMF) landscapes for CO chemisorption at conditions ranging from ideal gas to supercritical hexane densities. Their results indicated that supercritical *n*-hexane increases CO stability in the adsorbed state in a density-dependent manner, consistent with experimental observations for general fluid property behavior under near- and super-critical fluid conditions [124].

On the other hand, implicit solvation with continuum methods [147] offer an inexpensive alternative to introduce a water medium, by treating the aqueous phase as a dielectric material. Recently, Montemore et al. [141] compared explicit and implicit solvation in their AIMD study of Cu {111} supported alkyl chain growth in H_2O, which are important intermediates in aqueous FTS. According to their results, the description of alkyl–H_2O interactions is in general agreement with either solvation model for low alkyl coverage on the surface [141].

18.4.2 Moving Upscale: Modeling FTS Kinetics, Kinetic Monte Carlo

Apart from the solvent effects, possible kinetic effects are also absent in QM calculations, along with the multisite nature of the catalyst and the complete set of the competing elementary surface steps involved [89]. FTS kinetics are extremely complicated, and these steps such as adsorption, surface reaction, surface diffusion, and desorption of molecular or atomic species are approached as a two-fold process, involving 1) their accurate description at a high level QM level or with an accurate empirical approach and 2) establishing their actual occurrence with regards to the different reaction mechanisms [148]. Kinetic Monte Carlo (KMC) methods treat the changes in numbers and positions of the adsorbed molecular or atomic species onto the catalyst as a lattice of sites, namely a lattice gas [148]. The reactions steps are modeled as a series of discrete transformations of the lattice, hence depicting system evolution. These are formulated in the Master Equation (ME), which captures the dynamic behavior of the system as the rate change of microstate probabilities, which depends on QM derived rate constants [148–150]. Solving the ME yields ensemble averages of properties such as turnover frequencies (TOF) and adsorbate coverages and is straightforward for systems with a small number of states. However, for cases like the FTS process reactions it cannot be solved analytically, and this has led to the development of several algorithms and computational approaches. A recent example of such an algorithm was published by Dybeck et al. and was utilized to model the initial CO activation, formation of C_2 intermediates, and the termination of C_1 and C_2 species FTS reactions over Ru nanoparticles, using results from previous DFT simulations of this process [151].

KMC has been particularly helpful in evaluating the sensitivity of the FTS process to the metal catalyst surface size and morphology. Van Santen et al. [148] employed KMC techniques to study the overall kinetics of the Fischer-Tropsch process. DFT calculations were performed for computing reaction and activation energies using the PW91 functional for CO dissociation and CH–CH bond formation on the Ru and Co surfaces. These calculations served as input for the KMC simulation model analysis of which showed that the dissociation energy of CO is strongly structure sensitive and stepped surface sites are required for dissociation. For this reason, metal nanoparticles should be at least 10 and 6 nm for Ru and Co, respectively, otherwise apart from CO dissociation, hydrocarbon chain growth is also hindered [148].

In another example, Tian et al. [152] performed systematic KMC simulations at different temperatures, total pressures, and H_2/CO ratios on the catalytic performance of Fe as a function of FTS operating conditions. Instead of performing DFT calculations, the energy barriers of elementary steps were estimated by means of the unity bond index quadratic exponential potential (UBI-QEP) method [153–155]. Their results showed a strong influence of the catalyst's activity and selectivity by the reaction conditions, in agreement with experimental trends. They concluded that the TOF for CH_4 production increases with increasing temperature, declines with increasing total pressure, and reaches a maximum with increasing H_2/CO ratio. TOF for C_2 hydrocarbon decreases with increasing temperature and H_2/CO ratio, while it increases as total pressure increases. Simulations also showed that catalyst performance is determined by the surface H/CO ratio, while surface coverage of all intermediates is controlled by process conditions. As process conditions changed, increasing hydrogen coverage and decreasing CO coverage was found to favor the production of lighter hydrocarbons [152].

18.4.3 Classical Approaches: Molecular Simulation at Larger Size and Time Scales of the FTS Process

Even though the QM and AIMD approaches provide useful mechanistic insights and quantitative data of the FTS elementary reaction step energy barriers, they are still limited to a small number of atoms and short timescales. On the other hand, molecular simulation methods such as MD, allow for treatment of millions of atoms in the micro [156, 157] or even the millisecond time scale [158]. These methods close the gap between QM calculations and experiments, by describing molecular interactions using a parametrized set of potential energy functions, namely a force field. The majority of these force fields are classical, in the sense that the individual contributions are in the form of simple harmonic potentials, hence incapable of describing bond breaking and formation. Therefore, their use is restricted to phenomena that do not involve chemical change.

One way of tackling this deficiency was to develop potentials that indirectly include electronic effects, namely reactive force fields [159]. These force fields succeed in describing chemical change by employing concepts like bond order and electronegativity [160]. Their use in molecular simulations can provide a much more detailed molecular description of the complex FTS process compared to KMC, at least in principle. Among the different existing types of reactive force fields [160], the ReaxFF potential is frequently used for the study of FTS-related systems, as it includes parameters developed for hydrocarbons, transition metal catalysts, metal oxides, and hydrides, as well as other FTS intermediate species [76, 161]. In a recent example, Zou et al. [162] developed a ReaxFF parameter set describing interactions of C/H/Fe systems, which they employed in MD simulations examining hydrogen dissociation and desorption pathways on iron and iron carbide (CH_4Fe_3) at low and elevated temperatures. Their results showed that even though the barrier associated with hydrogen dissociation is energetically higher in CH_4Fe_3 than pure iron, the corresponding dissociated state in the former is more stable. The surface of CH_4Fe_3 was found to possess more hydride species at high temperatures compared to pure iron, which directly affects activity of these catalysts in FTS [162]. Unfortunately, reactive force fields are computationally more expensive than their classical counterparts, and their development is a nontrivial and elaborate procedure. In addition, even though the availability in parameter sets is growing [161], their use is hampered by the lack of transferability [162].

On the other hand, classical MD and MC simulations have been established as important tools in the study of several FTS process aspects. For example, controlling tail gas expulsion, which is typically a mixture of CO_2, CO, CH_4, N_2, H_2, and hydrocarbons larger than CH_4 (C_{2+}), is very important in reducing the hydrocarbon matter consumption and pollutant emissions, with simultaneous increase of the FTS overall yield. Currently, pure silica zeolites are evaluated in their potential to separate tail gas mixture components. This has led Perez-Carbajo et al. [163] to perform MC simulations in the Grand Canonical ensemble (μVT) for the adsorption behavior study of selected tail mixture adsorbents for a range of pressures at 303.15 K, considering four pure silica zeolites of different aluminum concentrations and locations. From the computed pure component and mixture isotherms, they proposed an operating procedure to achieve a complete five-component separation for improving the targeted industrial process, by successive utilization of the different zeolites examined [163].

Another important aspect of the FTS process involves the study of transport properties of the various components involved, namely CO, H_2, H_2O, and *n*-alkanes. This is of considerable practical and theoretical interest, as the ability to predict and control the viscosity, and thus the diffusion in systems of pure *n*-alkanes and their mixtures is crucial for product design and process optimization in the petrochemical industry. Makrodimitri et al. performed extensive MD simulations for the study of *n*-alkanes mixture viscosities, as well as the self-diffusivity of H_2, CO, and H_2O in these mixtures, at high temperature and pressure conditions [164]. For the latter, a newly developed macroscopic model was proposed, based entirely on simulation data [164]. The calculated viscosities of all mixtures examined were found to be in good agreement with available experimental data, appropriately reproducing the observed trends with varying conditions. Similarly, predicted self-diffusion values of the aforementioned species were also in good agreement with their experimental counterparts, as shown in Figure 18.11 [164].

Apart from investigating the properties of bulk FTS product mixtures, it is very important to elucidate their behavior in the environment of catalyst confinement. The interfaces of the catalyst support control the interactions with the product wax and H_2O mixture, hence the flow and efficient reactor performance [167]. Therefore, the transport properties of mixtures under confinement and particularly inside catalyst nanopores is a topic of significant interest for sustaining the performance of the GTL process. Molecular simulation and particularly MD approaches are a reliable tool, because of their ability to access nanoscale regimes and buried interfaces, that are experimentally challenging to characterize [167], especially under reaction conditions. For example, it is generally assumed that the overall mole fraction of H_2O inside the catalyst nanopores varies from less than 0.01 up to 0.20, corresponding to the inlet and the outlet, respectively, of the FT fixed-bed reactor. This assumption may not hold in the outlet of the FT reactor, where the water concentration can be significantly higher than 0.20.

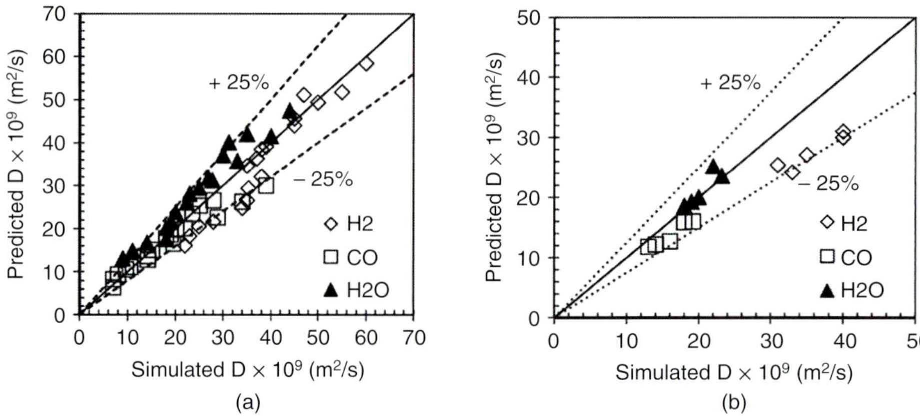

Figure 18.11 Comparison between simulated and predicted self-diffusion coefficients of H_2 (open diamonds), CO (open squares) and H_2O (solid triangles) in (a) pure *n*-alkanes (*n*-C_8, *n*-C_{12}, *n*-C_{16}, *n*-C_{20}, *n*-C_{64}, and *n*-C_{96}) and (b) *n*-alkane mixtures (mixtures **H** to **L** in [164], namely: **H.** 7% *n*-C_7 + 21% *n*-C_{12} + 19.3% *n*-C_{16} + 17.6% *n*-C_{20} + 19.3% *n*-C_{25} + 15.8% *n*-C_{30}, **I.** 3.5% *n*-C_6 + 5.3% *n*-C_8 + 28.1% *n*-C_{14} + 19.3% *n*-C_{18} + 24.5% *n*-C_{24} + 19.3% *n*-C_{30}, **J.** 2.5% *n*-C_8 + 25% *n*-C_{16} + 52.5% *n*-C_{28} + 12.5% *n*-C_{36} + 5% *n*-C_{48} + 2.5% *n*-C_{80}, **K.** 4.4% *n*-C_{28} + 34.8% *n*-C_{38} + 26.1% *n*-C_{45} + 21.7% *n*-C_{55} + 8.7% *n*-C_{70} + 4.3% *n*-C_{90}, **L.** 4.4% *n*-C_{25} + 17.4% *n*-C_{32} + 39.1% *n*-C_{42} + 21.7% *n*-C_{50} + 13% *n*-C_{60} + 4.4% *n*-C_{85}). Predicted diffusivities were evaluated by a macroscopic model [164], while simulated diffusivities were calculated by long *NVT* MD simulations [164–166]. In both cases (a and b), predicted values fall within 25% of their simulated counterparts. For pure *n*-alkanes (a), the absolute average deviations between computed and predicted data equal 10%, 9.3% and 14.5% for H_2, CO and H_2O respectively. In the case of (b) these deviations are 24.3% 14.8% and 3.7% for H_2, CO and H_2O respectively. Predictions of H_2 diffusivities in *n*-alkane mixtures (b) are systematically biased toward lower values; this is attributed to quantum effects, which were included in analysis [164]. Figure is reproduced from [164] with permission from Elsevier.

As discussed previously, H_2O presence in excess amounts can lead to phase segregation inside the catalyst nanopores, water-assisted sintering of catalytic nanoparticles, and catalyst deactivation. Reduced catalyst lifetimes increase GTL operational cost, and it is thus important to understand the wax–water liquid–liquid equilibrium (LLE) at reaction conditions, as it determines the maximum allowable amount of water in the FT wax. To this end, we focused on simulating the phase behavior of the *n*-octacosane (*n*-C_{28})–water mixture inside TiO_2 nanopores by means of atomistic modeling [168]. Titania was selected as it is a typical material for Co catalyst support, which is almost exclusively used in commercial LTFT processes [16, 19], especially in its anatase polymorph that is more efficient and more widely used in catalysis [21]. A series of MD simulations was performed using the LAMMPS simulation package [169], examining different pore sizes and mixture concentrations, by means of the TraPPE-UA [84] and SPC/E [170] force fields for *n*-C_{28} and H_2O, respectively. All systems were prepared in the Scienomics MAPS software package [171]. Herein, only the mixture containing excess amount of water ($x_{H_2O} = 0.9655$) is discussed. All calculations refer to 473 K.

Molecular simulation results showed that even though phase-segregated mixtures retain their structural properties compared to their bulk counterparts [168], significant deviations arise in terms of density profiles and transport properties inside the nanopore. Specifically, water molecules organize into two discrete layers on the TiO_2 {101} anatase surface, shielding *n*-C_{28} from the nanopore walls (Figure 18.12a). From the axial density profile along the z direction of *n*-C_{28} and H_2O confined in the pore (Figure 18.12b), it is apparent that the density of *n*-C_{28} in the center of the pore corresponds to the density of *n*-C_{28} in the bulk. Two H_2O peaks separated by around 3 Å appear and correspond to the first and second H_2O monolayers. A third much less structured layer, 3 Å away from the second one, is also observed (Figure 18.12b). These discrepancies in H_2O density delimit three regions inside the pore, in agreement with the reported findings of H_2O in cylindrical rutile pore confinement by Solveyra et al. [172], bearing important implications on its diffusion properties. This behavior is attributed to the hydrophilicity of TiO_2, as the presence of the surface promotes phase separation even in lower H_2O concentrations.

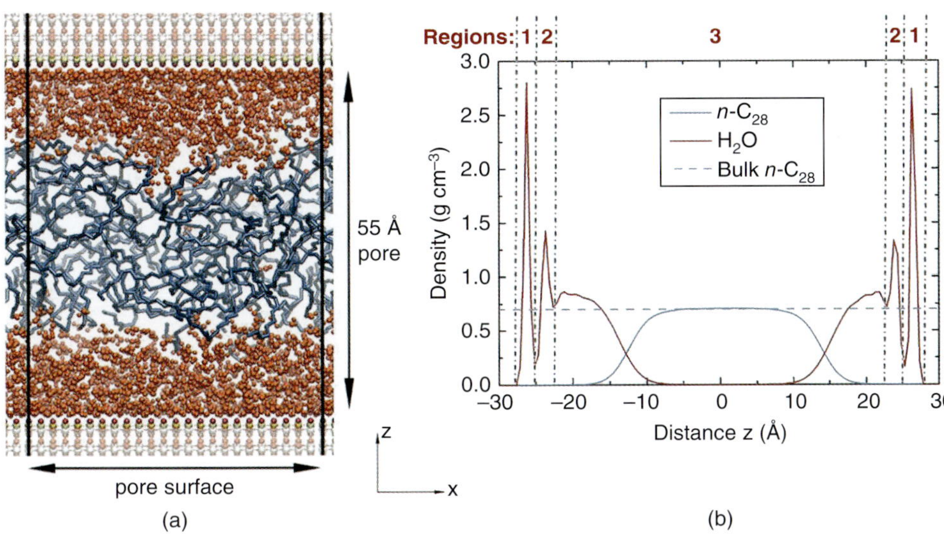

Figure 18.12 (a) Snapshot of the equilibrated *n*-C_{28} – H_2O mixture with $x_{H_2O} = 0.9655$ inside the 55 Å TiO_2 {101} anatase nanopore. (b) Density profile of the mixture, clearly depicting the partitioning of its components inside the nanopore. The preference of H_2O towards the TiO_2 surface is evident, as it is organized into two monolayers, while a third bulk-like region appears closer to the pore center [168].

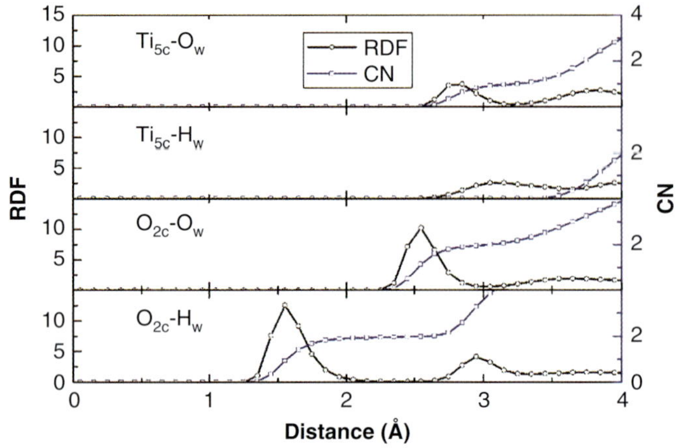

Figure 18.13 RDFs clearly illustrating the ordered H_2O structure close to the TiO_2 {101} anatase surface.

To further elucidate the H_2O monolayer at the TiO_2 interface and gain better insight into H_2O coordination on the unsaturated O_{2c} and Ti_{5c} atomic centers, RDF analysis was performed. Specifically, the intermolecular RDFs between individual TiO_2 sites and H_2O atoms revealed that H_2O oxygen (O_w) is very well structured around O_{2c} and Ti_{5c}, as shown in Figure 18.13, in agreement with previous studies [173, 174].

This highly ordered H_2O structure on the TiO_2 surface is in agreement with temperature-programmed desorption (TPD) studies, which revealed three H_2O desorption peaks at 160, 190, and 250 K, assigned to a) multilayer H_2O, b) H_2O adsorbed on O_{2c}, and c) H_2O adsorbed on Ti_{5c} [175], respectively. In addition, water's organized surface structure and its loss when moving towards the bulk is in agreement with findings in previous calculations [174]. This behavior observed in our simulations is also in agreement with DFT MD simulations of a TiO_2 {101} anatase slab in bulk H_2O, indicating molecular adsorption of H_2O at the Ti_{5c} active site [173]. In addition, it has been shown in the literature that H_2O molecules adsorb via the formation of hydrogen bonds (HB) onto O_{2c} (hydrogen acceptor) [176]. Our geometric HB analysis (donor–acceptor distance: 3 Å, donor–hydrogen–acceptor angle: 30°) yielded approximately 1.65 hydrogen bonds per O_{2c} atom, thus affirming the formation of approximately two hydrogen bonds between O_{2c} and water (Figure 18.14).

The calculated lateral self-diffusion coefficients (D_{xy}) for n-C_{28} and H_2O were obtained as a function of the TiO_2 surface distance, by calculating the 1/4 of the slope of the mean square displacement (MSD) of each component along the x and y directions at long simulation times according to the following equation:

$$D_{xy} = \lim_{t \to \infty} \frac{1}{4} t \langle (x(t) - x(0))^2 + (y(t) - y(0))^2 \rangle \tag{18.2}$$

n-Octacosane's self-diffusion is not influenced by confinement; water on the other hand is severely hindered by the TiO_2 nanopore surface, with its diffusivity bearing a strong dependence on the distance from the nanopore center. The diffusivity of n-C_{28} and H_2O in the pore is considered as the result of a two-dimensional move in the xy plane since the movement along the z direction is constrained by the confining pore surfaces. From the results presented in Figure 18.15, it is evident that the discrepancies in D_{xy} coincide with the three regions delimited by the water density profile of Figure 18.12b. The water monolayer at the TiO_2 interface with the titania surface shows no exchange of molecules with the outer region, which is reflected by

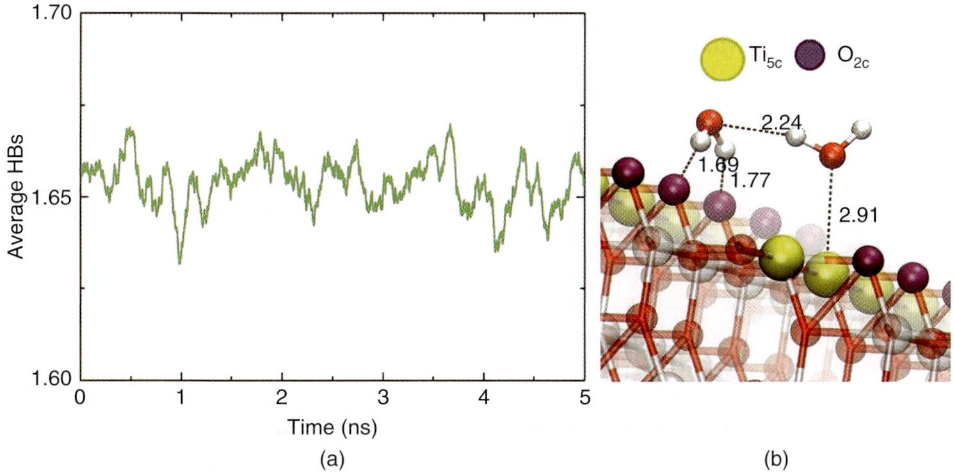

Figure 18.14 (a) Calculated number of hydrogen bonds between water and O_{2c} ($O_w{-}H_w{\cdots}O_{2c}$) in the excess H_2O mixture. (b) Characteristic snapshot of the hydrogen bonded network of H_2O with the unsaturated TiO_2 surface atoms.

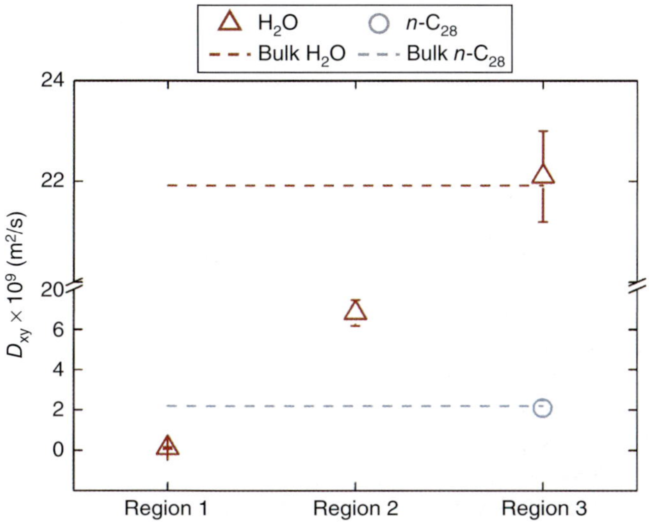

Figure 18.15 MD simulation results for the diffusion coefficients of H_2O and n-C_{28} in the regions delimited inside the TiO_2 {101} anatase nanopore for $x_{H_2O} = 0.9655$. Regions 1, 2 and 3 are shown in Figure 18.12b.

the very low diffusion coefficient in this regime, characteristic of a non-Fickian behavior (i.e., $D_{xy} \propto t^{0.8}$, indicating subdiffusive behavior).

18.5 Future Outlook

From the examples presented in this chapter, it becomes apparent that molecular simulation is a valuable ally in the investigation of the phenomena governing natural gas processes at the atomistic scale. However, as already discussed, reactivity studies by means of first principle methodologies are currently limited to only a few tens to hundreds of atoms. Furthermore, solvent effects are difficult to include, and dynamics at this level are restricted to a few picoseconds. Molecular simulation methods such as MD and MC achieve the migration to larger systems and timescales, but still face inherent obstacles for studying complex systems. Among them, the lack of force field parameters specialized for model systems relevant to natural gas processing is prominent. The accuracy of the molecular models, describing the intra- and inter-molecular interactions, employed in simulations are of outmost importance for attaining accurate results. Even though there is a wealth of developed models, which are, to some extent, transferable to studies involving natural gas processes, there is always the need for potentials that better capture the elementary processes in question. For example, even though several force fields have been developed to either study the behavior of pure H_2S or its binary mixtures with small n–alkanes [177–180], a model that can accurately describe the phase behavior of gas hydrates is still lacking. The same is true for porous media and FTS systems, where, for example, classical parameters are not always available for catalyst nanoparticles and their supports. This is even more profound in the case for reactive force fields.

In general, development of atomistic force fields is a challenging and nontrivial task, due to the computational demands required for accurately capturing the multidimensional potential energy surface of the system. In addition, potential function parameterization of such complex systems is not trivial, as nonlinear optimization methods involving a large number of parameters are prone to entrapment in local minima far away from the desired global minimum. With relation to classical atomistic force fields, even though they are very efficient and reliable for most chemical systems, they are not suitable for polar systems such as water, which are better described by explicit polarizable models. Unfortunately, their use is not straightforward and is not yet supported in all atomistic computational codes and platforms. Therefore, a lot of effort is expected to be dedicated by researchers in rendering these models as accessible and as efficient as possible.

Moreover, these phenomena are usually observed in timescales of hundreds of nanoseconds (ns) and are unattainable by means of conventional MD simulations. This caveat can be addressed by resorting to enhanced sampling methods such as accelerated molecular dynamics (aMD) [181], which can greatly accelerate simulations to reach timescales relevant to natural gas processing at the mesoscale and above. An alternative way to overcome these limitations is by lowering the chemical accuracy of the modeled system, i.e., treating molecular species in a "coarse" manner by sacrificing the degrees of freedom for the sake of timescale. Several coarse-grained simulation methods have been established to understand the structure–property relationships of material interfaces, which include Brownian dynamics (BD) [182–184], dissipative particle dynamics (DPD) [185–187], and coarse-grained molecular dynamics (CGMD) [188, 189]. CGMD can be used to study the effects of interfaces at the mesoscale level on the transport properties of wax–water mixtures. However, these approaches have yet to be fully implemented in the context of natural gas processing, as they face the inherent drawback of the lack of force fields discussed previously, which have not been developed to capture all the essential process components, and most importantly metal-based materials. CGMD methods have been enriched recently with force fields developed with the SAFT-γ Mie EoS top-down approach, which has been successfully applied in MD simulations for the prediction of several water/oil mixture properties [32], but no parameters exist for more complex setups.

Lastly, performing simulations in NG systems such as gas hydrates is highly demanding computationally, and often inefficient to perform on CPUs (central processing unit). The advancement of hardware and specifically the exploitation of existing resources with excellent computing capabilities such as GPU (graphics processing unit) has given a new boost to the field of molecular simulation [190]. The significant progress made in terms of migrating classical molecular simulation codes to GPUs has also been a contributing factor in porting QM codes as well, which is expected to significantly increase the current limits in terms of system sizes.

List of Abbreviations

AA/CG	all-atom and coarse-grained
AIM	atoms-in-molecules
AIMD	*ab initio* molecular dynamics
aMD	accelerated molecular dynamics
ARPA–E	Advanced Research Project Agency–Energy
B3P86	Becke's Three Parameter Hybrid Perdew 86 functional
BCP	bond critical point
BD	Brownian dynamics
CASSCF	complete active space self-consistent field
CGMD	coarse-grained molecular dynamics
CNT	carbon nano tube
DFT	density functional theory
DPD	dissipative particle dynamics
EoS	equations of state
FTS	Fischer-Tropsch synthesis
GAFF	general AMBER force field
GCMC	grand canonical Monte Carlo
GGA	Generalized Gradient Approximation
GTL	gas-to-liquids
HF	Hartree-Fock
HTFT	high temperature Fischer-Tropsch
KMC	kinetic Monte Carlo
LAMMPS	Large-scale Atomic/Molecular Massively Parallel Simulator
LLE	liquid-liquid equilibrium
LJ	Lennard-Jones
LJP	Lennard-Jones particle
LPD	limiting pore diameter
LTFT	low temperature Fischer-Tropsch
MC	Monte Carlo
MD	Molecular Dynamics
MD-HRMC	Molecular Dynamics Hybrid Reverse Monte Carlo
ME	Master Equation
MM	molecular mechanics
MOF	metal–organic frameworks
MP2	Møller–Plesset second order perturbation theory
MSD	mean square displacement
NG	natural gas
PBE	Perdew–Burke–Ernzerhof functional
PMF	Potential of Mean Force
PPN	Porous Polymer Networks
PW91	Perdew–Wang exchange and correlation functional

QM	quantum mechanics
RDF	radial distribution function
REBO	Reactive Empirical Bond-Order
RMC	reverse Monte Carlo
RPBE	revised Perdew–Burke–Ernzerhof functional
tMD	targeted molecular dynamics
TOC	total organic content
TOF	turnover frequency
TPD	temperature-programmed desorption
TraPPE	Transferable Potentials for Phase Equilibria
TST	transition state theory
UBI-QEP	unity bond index quadratic exponential potential
UFF	universal force field
ZIF	zeolitic imidazolate frameworks

Acknowledgements

We are grateful to the European Commission for funding our research under Horizon 2020 Programme for Research and Innovation, grant agreement No. 640979. In addition, financial support from Shell Global Solutions International BV through a contracted research agreement is gratefully acknowledged. The Niarchos Foundation and Scienomics SARL are acknowledged for generous post-doctoral fellowship to Dr. Konstantinos D. Papavasileiou. This publication was made possible by NPRP Grant No. 6-1547-2-632 from the Qatar National Research fund (a member of the Qatar Foundation). The statements made herein are solely the responsibility of the authors. This work was supported by computational time granted from the Greek Research & Technology Network (GRNET) in the National High Performance Computing facility—ARIS—under project ID002043. We are also thankful to the High Performance Computing Center of Texas A&M University at Qatar for generous resource allocation.

References

1 Sloan ED, Koh CA. 2008. Clathrate Hydrates of Natural Gases. 3rd ed. Boca Raton, FL: CRC Press.

2 Nielsen SO, Bulo RE, Moore PB, Ensing B. 2010. Recent progress in adaptive multi-scale molecular dynamics simulations of soft matter. Phys Chem Chem Phys, 12 (39): 12401–12414.

3 Zerpa LE, Salager JL, Koh CA, et al. 2011. Surface chemistry and gas hydrates in flow assurance. Ind Eng Chem Res, 50 (1): 188–197.

4 Kelland MA. 2006. History of the development of low dosage hydrate inhibitors. Energy Fuels, 20 (3): 825–847.

5 Ungerer P, Nieto-Draghi C, Lachet V, et al. 2007. Molecular simulation applied to fluid properties in the oil and gas industry. Mol Simul, 33 (4–5): 287–304.

6 Olah GA, Molnár Á. 2003. Hydrocarbon chemistry. 2nd ed. Hoboken, NJ: Wiley-Interscience.

7 Iglesia E. 2004. Challenges and progress in the conversion of natural gas to fuels and chemicals. Preprint Papers - American Chemical Society, Division of Fuel Chemistry, 49 (1): 49–50.

8 Nahreen S, Praserthdam S, Beltran SP, et al. 2016. Catalytic upgrading of methane to higher hydrocarbon in a nonoxidative chemical conversion. Energy Fuels, 30 (4): 2584–2593.

9 Holmen A. 2009. Direct conversion of methane to fuels and chemicals. Catal Today, 142 (1–2): 2–8.

10 Wood DA, Nwaoha C, Towler BF. 2012. Gas-to-Liquids (GTL): A review of an industry offering several routes for monetizing natural gas. J Nat Gas Sci Eng, 9, 196–208.

11 Wilhelm DJ, Simbeck DR, Karp AD, Dickenson RL. 2001. Syngas production for gas-to-liquids applications: technologies, issues and outlook. Fuel Process Technol, 71 (1–3): 139–148.

12 Rahmim II. 2003. Gas-to-Liquid technologies: recent advances, economics, prospects. 26th IAEE Annual International Conference. Prague, p. 1–36.

13 Gill SS, Tsolakis A, Dearn KD, Rodriguez-Fernandez J. 2011. Combustion characteristics and emissions of Fischer-Tropsch diesel fuels in IC engines. Prog Energy Combust Sci, 37 (4): 503–523.

14 Di Sanzo FP. 1981. Characterization of high boiling Fischer-Tropsch liquids by liquid and gas-chromatography. Anal Chem, 53 (12): 1911–1914.

15 Steynberg A, Dry M. 2004. Fischer-Tropsch technology. Amsterdam: Elsevier.

16 van de Loosdrecht J, Botes FG, Ciobica IM, et al. (2013). Fischer-Tropsch synthesis: catalysts and chemistry. In: Poeppelmeier K, ed., Comprehensive inorganic chemistry II. 2nd ed. Amsterdam: Elsevier, p. 525–557.

17 Dry ME. 2002. The Fischer-Tropsch process: 1950–2000. Catal Today, 71 (3–4): 227–241.

18 Schulz H. 1999. Short history and present trends of Fischer-Tropsch synthesis. App Catal A: General, 186 (1–2): 3–12.

19 Jahangiri H, Bennett J, Mahjoubi P, et al. 2014. A review of advanced catalyst development for Fischer-Tropsch synthesis of hydrocarbons from biomass derived syn-gas. Catal Sci Technol, 4 (8): 2210–2229.

20 Khodakov AY, Chu W, Fongarland P. 2007. Advances in the development of novel cobalt Fischer-Tropsch catalysts for synthesis of long-chain hydrocarbons and clean fuels. Chem Rev, 107 (5): 1692–1744.

21 Hadjiivanov KI, Klissurski DG. 1996. Surface chemistry of titania (anatase) and titania-supported catalysts. Chem Soc Rev, 25 (1): 61–69.

22 Oh J-H, Bae JW, Park S-J, et al. 2009. Slurry-phase Fischer–Tropsch synthesis using Co/γ-Al_2O_3, Co/SiO_2 and Co/TiO_2: Effect of support on catalyst aggregation. Catal Lett, 130 (3): 403–409.

23 Tsakoumis NE, Ronning M, Borg O, et al. 2010. Deactivation of cobalt based Fischer-Tropsch catalysts: A review. Catal Today, 154 (3–4): 162–182.

24 van de Loosdrecht J, Bazhinimaev B, Dalmon JA, et al. 2007. Cobalt Fischer-Tropsch synthesis: deactivation by oxidation? Catal Today, 123 (1–4): 293–302.

25 Qin Q, Ramkrishna D. 2004. Modeling of surface structure of supported catalysts. Application to deactivation of FT reactors. Ind Eng Chem Res, 43 (12): 2912–2921.

26 Bezemer GL, Remans TJ, van Bavel AP, Dugulan AI. 2010. Direct evidence of water-assisted sintering of cobalt on carbon nanofiber catalysts during simulated Fischer–Tropsch conditions revealed with in situ Mössbauer spectroscopy. J Am Chem Soc, 132 (25): 8540–8541.

27 Eschemann TO, de Jong KP. 2015. Deactivation behavior of Co/TiO_2 catalysts during Fischer-Tropsch synthesis. ACS Catal, 5 (6): 3181–3188.

28 Kiss G, Kliewer CE, DeMartin GJ, et al. 2003. Hydrothermal deactivation of silica- supported cobalt catalysts in Fischer–Tropsch synthesis. J Catal, 217 (1): 127–140.

29 Das TK, Conner WA, Li JL, et al. 2005. Fischer-Tropsch synthesis: Kinetics and effect of water for a Co/SiO_2 catalyst. Energy Fuels, 19 (4): 1430–1439.

30 van der Waals JH, Platteeuw JC. 1959. Clathrate solutions In: Prigogine I, ed., Advances in chemical physics. Volume 2. New York: Wiley-Interscience, p. 1–57.

31 Holder GD, Zetts SP, Pradhan N. 1988. Phase-behavior in systems containing clathrate hydrates: a review. Rev Chem Eng, 5 (1–4): 1–70.

32 Kumar P, Sathyamurthy N. 2011. Theoretical studies of host-guest interaction in gas hydrates. J Phys Chem A, 115 (50): 14276–14281.

33 Tariq M, Atilhan M, Khraisheh M, et al. 2016. Experimental and DFT approach on the determination of natural gas hydrate equilibrium with the use of excess N_2 and choline chloride ionic liquid as an inhibitor. Energy Fuels, 30 (4): 2821–2832.

34 Bader RFW. 1990. Atoms in molecules: a quantum theory. Oxford: Clarendon Press.

35 Bader RFW. 1998. A bond path: a universal indicator of bonded interactions. J Phys Chem A, 102 (37): 7314–7323.

36 Barnes BC, Sum AK. 2013. Advances in molecular simulations of clathrate hydrates. Curr Opin Chem Eng, 2 (2): 184–190.

37 English NJ, MacElroy JMD. 2015. Perspectives on molecular simulation of clathrate hydrates: progress, prospects and challenges. Chem Eng Sci, 121: 133–156.

38 Michalis VK, Costandy J, Tsimpanogiannis IN, et al. 2015. Prediction of the phase equilibria of methane hydrates using the direct phase coexistence methodology. J Chem Phys, 142 (4): 044501.

39 Costandy J, Michalis VK, Tsimpanogiannis IN, et al. 2015. The role of intermolecular interactions in the prediction of the phase equilibria of carbon dioxide hydrates. J Chem Phys, 143 (9): 094506.

40 Michalis VK, Tsimpanogiannis IN, Stubos AK, Economou IG. 2016. Direct phase coexistence molecular dynamics study of the phase equilibria of the ternary methane-carbon dioxide-water hydrate system. Phys Chem Chem Phys, 18 (34): 23538–23548.

41 Conde MM, Vega C. 2010. Determining the three-phase coexistence line in methane hydrates using computer simulations. J Chem Phys, 133 (6): 064507.

42 Abascal JLF, Sanz E, Fernández RG, Vega C. 2005. A potential model for the study of ices and amorphous water: TIP4P/Ice. J Chem Phys, 122 (23): 234511.

43 Nakano S, Moritoki M, Ohgaki K. 1999. High-pressure phase equilibrium and Raman microprobe spectroscopic studies on the methane hydrate system. J Chem Eng Data, 44 (2): 254–257.

44 Bollengier O, Choukroun M, Grasset O, et al. 2013. Phase equilibria in the H_2O–CO_2 system between 250–330 K and 0–1.7 GPa: stability of the CO_2 hydrates and H_2O-ice VI at CO_2 saturation. Geochim Cosmochim Acta, 119, 322–339.

45 Saha D, Grappe HA, Chakraborty A, Orkoulas G. 2016. Postextraction separation, on-board storage, and catalytic conversion of methane in natural gas: A review. Chem Rev, 116 (19): 11436–11499.

46 Kumar KV, Preuss K, Titirici MM, Rodriguez-Reinoso F. 2017. Nanoporous materials for the onboard storage of natural gas. Chem Rev, 117 (3): 1796–1825.

47 Sarkisov L, Harrison A. 2011. Computational structure characterisation tools in application to ordered and disordered porous materials. Mol Simul, 37 (15): 1248–1257.

48 Bhattacharya S, Gubbins KE. 2006. Fast method for computing pore size distributions of model materials. Langmuir, 22 (18): 7726–7731.

49 Greenfield ML, Theodorou DN. 1993. Geometric analysis of diffusion pathways in glassy and melt atactic polypropylene. Macromolecules, 26 (20): 5461–5472.

50 Willems TF, Rycroft CH, Kazi M, et al. 2012. Algorithms and tools for high-throughput geometry-based analysis of crystalline porous materials. Micropor Mesopor Mat, 149 (1): 134–141.
51 Vasileiadis M, Peristeras LD, Papavasileiou KD, Economou I.G. 2017. Modeling of bulk kerogen porosity: methods for control and characterization. Energy Fuels, 31 (6): 6004–6018.
52 Boulougouris GC, Economou IG, Theodorou DN. 2001. Calculation of the chemical potential of chain molecules using the staged particle deletion scheme. J Chem Phys, 115 (17): 8231–8237.
53 Widom B. 1963. Some topics in the theory of fluids. J Chem Phys, 39 (11): 2808–2812.
54 Lee SJ, Bae YS. 2016. Review of molecular simulations of methane storage in metal-organic frameworks. J Nanosci Nanotechnol, 16 (5): 4284–4290.
55 Schrier J. 2012. Carbon dioxide separation with a two-dimensional polymer membrane. ACS Appl Mater Interfaces, 4 (7): 3745–3752.
56 Shah MS, Tsapatsis M, Siepmann JI. 2015. Monte Carlo simulations probing the adsorptive separation of hydrogen sulfide/methane mixtures using all-silica zeolites. Langmuir, 31 (44): 12268–12278.
57 Sun C, Wen B, Bai B. 2015. Application of nanoporous graphene membranes in natural gas processing: Molecular simulations of CH_4/CO_2, CH_4/H_2S and CH_4/N_2 separation. Chem Eng Sci, 138, 616–621.
58 Tian Z, Dai S, Jiang DE. 2015. Expanded porphyrins as two-dimensional porous membranes for CO_2 separation. ACS Appl Mater Interfaces, 7 (23): 13073–13079.
59 Liu H, Cooper VR, Dai S, Jiang D-E. 2012. Windowed carbon nanotubes for efficient CO_2 removal from natural gas J Phys Chem Lett, 3 (22): 3343–3347.
60 He J, To JWF, Psarras PC, et al. 2016. Tunable polyaniline-based porous carbon with ultrahigh surface area for CO_2 capture at elevated pressure. Adv Energy Mater, 6 (14): 1502491.
61 Dutta D, Wood BC, Bhide SY, et al. 2014. Enhanced gas adsorption on graphitic substrates via defects and local curvature: a density functional theory study. J Phys Chem C, 118 (15): 7741–7750.
62 Wang Y, Yang Q, Zhong C, Li J. 2017. Theoretical investigation of gas separation in functionalized nanoporous graphene membranes. Appl Surf Sci, 407, 532–539.
63 Simon CM, Kim J, Gomez-Gualdron DA, et al. 2015. The materials genome in action: identifying the performance limits for methane storage. Energy Environ Sci, 8 (4): 1190–1199.
64 Krishna R, van Baten JM. 2008. Onsager coefficients for binary mixture diffusion in nanopores. Chem Eng Sci, 63 (12): 3120–3140.
65 Phan A, Cole DR, Weiss RG, et al. 2016. Confined water determines transport properties of guest molecules in narrow pores. ACS Nano, 10 (8): 7646–7656.
66 Skoulidas AI, Sholl DS. 2005. Self-diffusion and transport diffusion of light gases in metal-organic framework materials assessed using molecular dynamics simulations. J Phys Chem B, 109 (33): 15760–15768.
67 George G, Bhoria N, AlHallaq S, et al. 2016. Polymer membranes for acid gas removal from natural gas. Sep Purif Technol, 158, 333–356.
68 Yampolskii Y, Starannikova L, Belov N, et al. 2014. Solubility controlled permeation of hydrocarbons: new membrane materials and results. J Memb Sci, 453, 532–545.
69 Sridhar S, Smitha B, Aminabhavi TM. 2007. Separation of carbon dioxide from natural gas mixtures through polymeric membranes: a review. Sep Purif Rev, 36 (2): 113–174.

70 Chen G, Koros WJ, Jones CW. 2016. Hybrid polymer/UiO-66(Zr) and polymer/NaY fiber sorbents for mercaptan removal from natural gas. ACS Appl Mater Interfaces, 8 (15): 9700–9709.

71 Jusoh N, Yeong YF, Chew TL, et al. 2016. Current development and challenges of mixed matrix membranes for CO_2/CH_4 separation. Sep Purif Rev, 45 (4): 321–344.

72 Theodorou DN. 2006. Principles of molecular simulation of gas transport in polymers. In: Yampolskii Y, Pinnau I, Freeman B, eds, Materials science of membranes for gas and vapor separation. Hoboken, NJ: John Wiley & Sons, Ltd, p. 49–94.

73 Striolo A, Cole DR. 2017. Understanding shale gas: Recent progress and remaining challenges. Energy Fuels, 31 (10): 10300–10310.

74 Vandenbroucke M. 2006. Kerogen: from types to models of chemical structure. Oil Gas Sci Technol, 58 (2): 243–269.

75 Bousige C, Ghimbeu CM, Vix-Guterl C, et al. 2016. Realistic molecular model of kerogen's nanostructure. Nat Mater, 15 (5): 576–582.

76 van Duin ACT, Dasgupta S, Lorant F, Goddard WA. 2001. ReaxFF: a reactive force field for hydrocarbons. J Phys Chem A, 105 (41): 9396–9409.

77 Ni B, Lee K-H, Sinnott SB. 2004. A reactive empirical bond order (REBO) potential for hydrocarbon–oxygen interactions. J Phys Condens Matter, 16 (41): 7261–7275.

78 Ungerer P, Collell J, Yiannourakou M. 2015. Molecular modeling of the volumetric and thermodynamic properties of kerogen: Influence of organic type and maturity. Energy Fuels, 29 (1): 91–105.

79 Collell J, Galliero G, Gouth F, et al. 2014. Molecular simulation and modelisation of methane/ethane mixtures adsorption onto a microporous molecular model of kerogen under typical reservoir conditions. Micropor Mesopor Mat, 197, 194–203.

80 Mayo SL, Olafson BD, Goddard WA. 1990. Dreiding: a generic force-field for molecular simulations. J Phys Chem, 94 (26): 8897–8909.

81 Gasteiger J, Marsili M. 1980. Iterative partial equalization of orbital electronegativity: a rapid access to atomic charges. Tetrahedron, 36 (22): 3219–3228.

82 Wang JM, Wolf RM, Caldwell JW, et al. 2004. Development and testing of a general Amber force field. J Comput Chem, 25 (9): 1157–1174.

83 Bayly CI, Cieplak P, Cornell WD, Kollman PA. 1993. A well-behaved electrostatic potential based method using charge restraints for deriving atomic charges: the RESP model. J Phys Chem, 97 (40): 10269–10280.

84 Martin MG, Siepmann JI. 1998. Transferable potentials for phase equilibria. 1. united-atom description of *n*-alkanes. J Phys Chem B, 102 (14): 2569–2577.

85 Mosher K, He JJ, Liu YY, et al. 2013. Molecular simulation of methane adsorption in micro- and mesoporous carbons with applications to coal and gas shale systems. Int J Coal Geol, 109, 36–44.

86 Collell J, Galliero G, Vermorel R, et al. 2015. Transport of multicomponent hydrocarbon mixtures in shale organic matter by molecular simulations. J Phys Chem C, 119 (39): 22587–22595.

87 Michalec L, Lísal M. 2016. Molecular simulation of shale gas adsorption onto overmature type II model kerogen with control microporosity. Molec Phys, 115 (9–12): 1086–1103.

88 Obliger A, Pellenq R, Ulm FJ, Coasne B. 2016. Free volume theory of hydrocarbon mixture transport in nanoporous materials. J Phys Chem Letters, 7 (19): 3712–3717.

89 van Santen RA, Neurock M. 1995. Concepts in theoretical heterogeneous catalytic reactivity. Catal Rev Sci. Eng, 37 (4): 557–698.

90 Leach AR. 2001. Molecular modelling: principles and applications. 2nd ed. Harlow: Prentice Hall.

91 Jensen F. 2007. Introduction to computational chemistry 2nd ed. Chichester: John Wiley & Sons.

92 Levine IN. 2009. Quantum chemistry. 6th ed. Upper Saddle River, NJ: Pearson Prentice Hall.

93 Szabó A, Ostlund NS. 1996. Modern quantum chemistry: introduction to advanced electronic structure theory. Mineola, NY: Dover Publications.

94 van Santen RA, Neurock M. 2006. Molecular Heterogeneous Catalysis: A Conceptual and Computational Approach. Weinheim, Germany: Wiley-VCH.

95 Inderwildi OR, Jenkins SJ. 2008. In-silico investigations in heterogeneous catalysis-combustion and synthesis of small alkanes. Chem Soc Rev, 37 (10): 2274–2309.

96 Cramer CJ, Truhlar DG. 2009. Density functional theory for transition metals and transition metal chemistry. Phys Chem Chem Phys, 11 (46): 10757–10816.

97 Gates BC, Knözinger H. 2000. Impact of surface science on catalysis. San Diego: Academic Press.

98 Siegbahn PEM, Nygren MA, Wahlgren U. 1992. The use of the cluster model for the calculation of chemisorption energetics In: Pacchioni G, Bagus PS, Parmigiani F, eds, Cluster models for surface and bulk phenomena. NATO ASI Series, Series B: Physics. Boston: Springer US, p. 267–279.

99 Keith JA, Anton J, Kaghazchi P, Jacob T. 2011. Modeling catalytic reactions on surfaces with density functional theory In: Deutschmann O, ed., Modeling and simulation of heterogeneous catalytic reactions. Weinheim: Wiley-VCH, p. 1–38.

100 Sautet P. 2012. Quantum chemistry methods. In: Che M, Védrine JC, eds, Characterization of solid materials and heterogeneous catalysts. Weinheim: Wiley-VCH, p. 1119–1145.

101 Gharibi M, Zangeneh FT, Yaripour F, Sahebdelfar S. 2012. Nanocatalysts for conversion of natural gas to liquid fuels and petrochemical feedstocks. App Catal A: General, 443: 8–26.

102 Siegbahn PEM, Panas I. 1990. A theoretical study of CH_x chemisorption on the Ni(100) and Ni(111) surfaces. Surface Sci, 240 (1–3): 37–49.

103 Olsen J 2011. The CASSCF method: a perspective and commentary. Int J Quantum Chem, 111 (13): 3267–3272.

104 Watwe RM, Bengaard HS, Rostrup-Nielsen JR, et al. 2000. Theoretical studies of stability and reactivity of CH_x species on Ni(111). J Catal, 189 (1): 16–30.

105 Perdew JP, Chevary JA, Vosko SH, et al. 1992. Atoms, molecules, solids, and surfaces: applications of the generalized gradient approximation for exchange and correlation. Phys Rev B, 46 (11): 6671–6687.

106 White JA, Bird DM. 1994. Implementation of gradient-corrected exchange- correlation potentials in Car-Parrinello total-energy calculations. Phys Rev B, 50 (7): 4954–4957.

107 Wang SG, Cao DB, Li YW, et al. 2006. CH_4 dissociation on Ni surfaces: density functional theory study. Surf Sci, 600 (16): 3226–3234.

108 Perdew JP, Burke K, Ernzerhof M. 1996. Generalized gradient approximation made simple. Phys Rev Lett, 77 (18): 3865–3868.

109 Perdew JP, Burke K, Ernzerhof M. 1997. Generalized gradient approximation made simple. Phys Rev Lett, 78 (7): 1396–1396.

110 Roy G, Chattopadhyay AP. 2017. Dissociation of methane on Ni_4 cluster: a DFT study. Comput Theor Chem, 1106, 7–14.

111 Li JD, Croiset E, Ricardez-Sandoval L. 2012. Methane dissociation on Ni (100): Ni (111): and Ni (553): a comparative density functional theory study. J Mol Catal A: Chem, 365, 103–114.

112 Rostrup-Nielsen JR. 1984. Sulfur-passivated nickel-catalysts for carbon-free steam reforming of methane. J Catal, 85 (1): 31–43.

113 Ashik UPM, Daud WMAW, Hayashi J. 2017. A review on methane transformation to hydrogen and nanocarbon: relevance of catalyst characteristics and experimental parameters on yield. Renew Sustain Energ Revs, 76, 743–767.

114 Mei DH, Glezakou VA, Lebarbier V, et al. 2014. Highly active and stable $MgAl_2O_4^-$ supported Rh and Ir catalysts for methane steam reforming: a combined experimental and theoretical study. J Catal, 316, 11–23.

115 Kokalj A, Bonini N, Sbraccia C, et al. 2004. Engineering the reactivity of metal catalysts: a model study of methane dehydrogenation on Rh(111). J Am Chem Soc, 126 (51): 16732–16733.

116 Ciobîcă IM, Frechard F, van Santen RA, et al. 2000. A DFT study of transition states for C–H activation on the Ru(0001). surface. J Phys Chem B, 104 (14): 3364–3369.

117 Zhang CJ, Hu P. 2002. Methane transformation to carbon and hydrogen on Pd(100): pathways and energetics from density functional theory calculations. J Chem Phys, 116 (1): 322–327.

118 Jones G, Jakobsen JG, Shim SS, et al. 2008. First principles calculations and experimental insight into methane steam reforming over transition metal catalysts. J Catal, 259 (1): 147–160.

119 Hammer B, Hansen LB, Norskov JK. 1999. Improved adsorption energetics within density-functional theory using revised Perdew-Burke-Ernzerhof functionals. Phys Rev B, 59 (11): 7413–7421.

120 Pakhare D, J. 2014. A review of dry (CO_2) reforming of methane over noble metal catalysts. Chem Soc Rev, 43 (22): 7813–7837.

121 Li YK, Yuan Z, Zhao YX, et al. 2016. Thermal methane conversion to syngas mediated by Rh_1-doped aluminum oxide cluster cations $RhAl_3O_4^+$. J Am Chem Soc, 138 (39): 12854–12860.

122 Becke AD. 1993. Density-functional thermochemistry. III. The role of exact exchange. J Chem Phys, 98 (7): 5648–5652.

123 Perdew JP. 1986. Density-functional approximation for the correlation-energy of the inhomogeneous electron-gas. Phys Rev B, 33 (12): 8822–8824.

124 Asiaee A, Benjamin K.M. 2016. Molecular simulation of CO chemisorption on Co(0001) in presence of supercritical fluid solvent: a potential of mean force study. J Chem Phys, 145 (8): 084506.

125 Qi YY, Yang J, Chen D, Holmen A. 2015. Recent progresses in understanding of Co-based Fischer-Tropsch catalysis by means of transient kinetic studies and theoretical analysis. Catal Lett, 145 (1): 145–161.

126 Shao M, Li Y, Chen J, Zhang Y. 2015. Mesoscale effects on product distribution of Fischer-Tropsch Synthesis. In: Guy BM, Jinghai L, eds, Advances in chemical engineering. Volume 47. New York: Academic Press, p. 337–387.

127 Hindermann JP, Hutchings GJ, Kiennemann A. 1993. Mechanistic aspects of the formation of hydrocarbons and alcohols from CO hydrogenation. Cat Rev Sci Eng, 35 (1): 1–127.

128 van Santen RA, Markvoort AJ, Filot IAW, et al. 2013. Mechanism and microkinetics of the Fischer-Tropsch reaction. Phys Chem Chem Phys, 15 (40): 17038–17063.

129 Perdew JP, Burke K, Wang Y. 1996. Generalized gradient approximation for the exchange-correlation hole of a many-electron system. Phys Rev B, 54 (23): 16533–16539.

130 Perdew JP, Burke K, Wang Y. 1998. Erratum: Generalized gradient approximation for the exchange-correlation hole of a many-electron system (vol 54, pg 16 533, 1996). Phys Rev B, 57 (23): 14999–14999.

131 Ge QF, Neurock M. 2006. Adsorption and activation of CO over flat and stepped Co surfaces: a first principles analysis. J Phys Chem B, 110 (31): 15368–15380.

132 Papp H. 1983. The chemisorption of carbon monoxide on a Co(0001) single crystal surface; studied by LEED, UPS, EELS, AES and work function measurements. Surf Sci, 129 (1): 205–218.

133 Weststrate CJ, van Helden P, de Loosdrecht JV, Niemantsverdriet JW. 2016 Elementary steps in Fischer-Tropsch synthesis: CO bond scission, CO oxidation and surface carbiding on Co(0001). Surf Sci, 648, 60–66.

134 Gong XQ, Raval R, Hu P. 2004. CO dissociation and O removal on Co(0001): a density functional theory study. Surf Sci, 562 (1–3): 247–256.

135 Li T, Bhatia B, Sholl DS. 2004. First-principles study of C adsorption, O adsorption, and CO dissociation on flat and stepped Ni surfaces. J Chem Phys, 121 (20): 10241–10249.

136 Cusinato L, Martinez-Prieto LM, Chaudret B, et al. 2016. Theoretical characterization of the surface composition of ruthenium nanoparticles in equilibrium with syngas. Nanoscale, 8 (21): 10974–10992.

137 Petersen MA, van den Berg JA, Ciobîcă IM, van Helden P. 2017. Revisiting CO activation on Co catalysts: Impact sites from DFT of step and kink. ACS Catal, 7 (3): 1984–1992.

138 Zhao S, Liu Y, Chen X, et al. 2015. Unified framework of multiscale density functional theories and its recent applications. In: Guy BM, Jinghai L, eds., Advances in chemical engineering. Volume 47 New York: Academic Press, p. 1–83.

139 Davies PR. 2016. On the role of water in heterogeneous catalysis: a tribute to Professor M. Wyn Roberts. Top Catal, 59 (8–9): 671–677.

140 Iglesia E. 1997. Design, synthesis, and use of cobalt-based Fischer-Tropsch synthesis catalysts. App Catal A, 161 (1–2): 59–78.

141 Montemore MM, Andreussi O, Medlin JW. 2016. Hydrocarbon adsorption in an aqueous environment: a computational study of alkyls on Cu(111). J Chem Phys, 145 (7): 074702.

142 Zhang XH, Sewell TE, Glatz B, et al. 2017. On the water structure at hydrophobic interfaces and the roles of water on transition-metal catalyzed reactions: A short review. Catal Today, 285, 57–64.

143 Kühne TD. 2014. Second generation Car-Parrinello molecular dynamics. Wiley Interdiscip Rev Comput Mol Sci, 4 (4): 391–406.

144 Faheem M, Heyden A. 2014. Hybrid quantum mechanics/molecular mechanics solvation scheme for computing free energies of reactions at metal-water interfaces. J Chem Theory Comput, 10 (8): 3354–3368.

145 Senn HM, Thiel W. 2009. QM/MM Methods for biomolecular systems. Angew Chem Int Ed , 48 (7): 1198–1229.

146 Rappé AK, Casewit CJ, Colwell KS, et al. 1992. UFF, a full periodic-table force-field for molecular mechanics and molecular-dynamics simulations. J Am Chem Soc, 114 (25): 10024–10035.

147 Cramer CJ, Truhlar DG. 1999. Implicit solvation models: equilibria, structure, spectra, and dynamics. Chem Rev, 99 (8): 2161–2200.

148 van Santen RA, Ghouri MM, Shetty S, Hensen EMH. 2011. Structure sensitivity of the Fischer-Tropsch reaction: molecular kinetics simulations. Catal Sci Technol, 1 (6): 891–911.

149 Nieminen RM, Jansen APJ. 1997. Monte Carlo simulations of surface reactions. App Catal A, 160 (1): 99–123.

150 Stamatakis M. 2015. Kinetic modelling of heterogeneous catalytic systems. J Phys Condens Matter, 27 (1): 013001.

151 Dybeck EC, Plaisance CP, Neurock M. 2017. Generalized temporal acceleration scheme for kinetic Monte Carlo simulations of surface catalytic processes by scaling the rates of fast reactions. J Chem Theory Comput, 13 (4): 1525–1538.

152 Tian L, Huo CF, Cao DB, et al. 2010. Effects of reaction conditions on iron-catalyzed Fischer-Tropsch synthesis: a kinetic Monte Carlo study. Comput Theor Chem, 941 (1–3): 30–35.

153 Akimov AV, Prezhdo OV. 2015. Large-scale computations in chemistry: a bird's eye view of a vibrant field. Chem Rev, 115 (12): 5797–5890.

154 Shustorovich E. 1984. Activation barrier for adsorbate surface-diffusion, heat of chemisorption, and adsorbate registry: theoretical interrelations. J Am Chem Soc, 106 (22): 6479–6481.

155 Shustorovich E, Sellers H. 1998. The UBI-QEP method: a practical theoretical approach to understanding chemistry on transition metal surfaces. Surf Sci Rep, 31 (1–3): 5–119.

156 Dror RO, Mildorf TJ, Hilger D, et al. 2015. Structural basis for nucleotide exchange in heterotrimeric G proteins. Science, 348 (6241): 1361–1365.

157 Galindo-Murillo R, Roe DR, Cheatham TE. 2014. On the absence of intrahelical DNA dynamics on the μs to ms timescale. Nat Commun, 5, 5152.

158 Shaw DE, Maragakis P, Lindorff-Larsen K, et al 2010. Atomic-level characterization of the structural dynamics of proteins. Science, 330 (6002): 341–346.

159 Farah K, Müller-Plathe F, Böhm MC. 2012. Classical reactive molecular dynamics implementations: state of the art. Chem Phys Chem, 13 (5): 1127–1151.

160 Liang T, Shin YK, Cheng YT, et al. 2013. Reactive potentials for advanced atomistic simulations. Ann Rev Materials Res, 43, 109–129.

161 Senftle TP, Hong S, Islam MM, et al. 2016. The ReaxFF reactive force-field: development, applications and future directions. Nat Comput Mater 2, 15011.

162 Zou CY, van Duin ACT, Sorescu DC. 2012. Theoretical investigation of hydrogen adsorption and dissociation on iron and iron carbide surfaces using the ReaxFF reactive force field method. Top Catal, 55 (5–6): 391–401.

163 Perez-Carbajo J, Gomez-Alvarez P, Bueno-Perez R, et al. 2014. Optimisation of the Fischer-Tropsch process using zeolites for tail gas separation. Phys Chem Chem Phys, 16 (12): 5678–5688.

164 Makrodimitri ZF, Heller A, Koller TM, et al. 2015. Viscosity of heavy *n*-alkanes and diffusion of gases therein based on molecular dynamics simulations and empirical correlations. J Chem Thermodyn, 91, 101–107.

165 Makrodimitri ZA, Unruh DJM, Economou IG. 2011. Molecular simulation of diffusion of hydrogen, carbon monoxide, and water in heavy *n*-alkanes. J Phys Chem B, 115 (6): 1429–1439.

166 Makrodimitri ZA, Unruh DJM, Economou IG. 2012. Molecular simulation and macroscopic modeling of the diffusion of hydrogen, carbon monoxide and water in heavy *n*-alkane mixtures. Phys Chem Chem Phys, 14 (12): 4133–4141.

167 Jha KC, Singh V, Tsige M. 2016. Interfacial engineering for oil and gas applications: role of modeling and simulation. In: Jin C, Cusatis G, eds. New frontiers in oil and gas exploration. Basel: Springer International Publishing, p. 257–283.

168 Papavasileiou KD, Makrodimitri ZA, Peristeras LD, et al. 2016. Molecular simulation of n-octacosane-water mixture in titania nanopores at elevated temperature and pressure. J Phys Chem C, 120 (43): 24743–24753.

169 LAMMPS Molecular Dynamics Simulator, version (10 Jan 2015). Available from: http://lammps.sandia.gov.

170 Berendsen HJC, Grigera JR, Straatsma TP. 1987. The missing term in effective pair potentials. J Phys Chem, 91 (24): 6269–6271.

171 Scienomics, MAPS Platform. 3.4.2, Paris, France, 2015. Available from: http://www.scienomics.com/.

172 Solveyra EG, de la Llave E, Molinero V, et al. 2013. Structure, dynamics, and phase behavior of water in TiO_2 nanopores. J Phys Chem C, 117 (7): 3330–3342.

173 Sumita M, Hu CP, Tateyama Y. 2010. Interface water on TiO_2 Anatase (101) and (1) surfaces: First-principles study with TiO_2 slabs dipped in bulk water. J Phys Chem C, 114 (43): 18529–18537.

174 Zhao ZY, Li ZS, Zou ZG. 2012. Structure and properties of water on the anatase TiO_2(101) surface: from single-molecule adsorption to interface formation. J Phys Chem C, 116 (20): 11054–11061.

175 Herman GS, Dohnalek Z, Ruzycki N, Diebold U. 2003. Experimental investigation of the interaction of water and methanol with anatase-TiO_2(101). J Phys Chem B, 107 (12): 2788–2795.

176 Mattioli G, Filippone F, Caminiti R, Bonapasta AA. 2008. Short hydrogen bonds at the water/TiO_2 (anatase). interface. J Phys Chem C, 112 (35): 13579–13586.

177 Delhommelle J, Millie P, Fuchs AH. 2000. On the role of the definition of potential models in Gibbs ensemble phase equilibria simulations of the H_2S- pentane mixture. Mol Phys, 98 (22): 1895–1905.

178 Jorgensen WL. 1986. Intermolecular potential functions and Monte-Carlo simulations for liquid sulfur-compounds. J Phys Chem, 90 (23): 6379–6388.

179 Kamath G, Lubna N, Potoff JJ. 2005. Effect of partial charge parametrization on the fluid phase behavior of hydrogen sulfide. J Chem Phys, 123 (12): 124505.

180 Nath SK. 2003. Molecular simulation of vapor-liquid phase equilibria of hydrogen sulfide and its mixtures with alkanes. J Phys Chem B, 107 (35): 9498–9504.

181 Pierce LCT, Salomon-Ferrer R, de Oliveira CAF, et al. 2012. Routine access to millisecond time scale events with accelerated molecular dynamics. J Chem Theory Comput, 8 (9): 2997–3002.

182 Dunweg B, Paul W. 1991. Brownian dynamics simulations without Gaussian random numbers. Int J Mod Phys C, 2 (3): 817–827.

183 Plimpton S. 1995. Fast parallel algorithms for short-range molecular-dynamics. J Comput Phys, 117 (1): 1–19.

184 Schneider T.E. 1978. Molecular-dynamics study of a 3-dimensional one-component model for distortive phase-transitions. Phys Rev B, 17 (3): 1302–1322.

185 Espanol P, Warren P. 1995. Statistical-mechanics of dissipative particle dynamics. Europhys Lett, 30 (4): 191–196.

186 Groot RD, Warren PB. 1997. Dissipative particle dynamics: bridging the gap between atomistic and mesoscopic simulation. J Chem Phys, 107 (11): 4423–4435.

187 Hoogerbrugge PJ, Koelman JMVA. 1992. Simulating microscopic hydrodynamic phenomena with dissipative particle dynamics. Europhys Lett, 19 (3): 155–160.

188 Muller-Plathe F. 2002. Coarse-graining in polymer simulation: from the atomistic to the mesoscopic scale and back. Chemphyschem, 3 (9): 754–769.

189 Marrink SJ, Risselada HJ, Yefimov S, et al. 2007. The MARTINI force field: coarse grained model for biomolecular simulations. J Phys Chem B, 111 (27): 7812–7824.

190 Xu D, Williamson MJ, Walker RC. 2010. Advancements in molecular dynamics simulations of biomolecules on Graphical Processing Units. Ann Rep Comput Chem, 6 2–19.

19

Natural Gas to Acetylene (GTA)/Ethylene (GTE)/Liquid Fuels (GTL) The Synfuels International, Inc. Process

Kenneth R. Hall[1], *Joel G. Cantrell*[1], *and Ben R. Weber, Jr*[2]

[1] *Bryan Research & Engineering, LLC, Texas, USA*
[2] *Synfuels International, INC, Parkland Hall, Texas, USA*

CHAPTER MENU

19.1 Introduction

Ever since the discovery of producible fields of petroleum, humankind has used its properties as a fuel and a feedstock to advance its needs and desires. A historical description of the advance of petroleum and its products is beyond the scope of this chapter, but suffice it to say that life today would be vastly different, and doubtless more primitive, without petroleum (natural gas and oil).

Chemical engineering and chemistry have played a vital role in advancing use of petroleum over the past century. Among the early advances pertinent to the story in this chapter, is the development of the Fischer-Tropsch process. Originally it was a means for Germany to turn its ample coal supplies into liquid fuels for its war machine in World War I. Even today, it remains a useful process to convert natural gas into liquid fuels. However, it produces many undesired products that require extensive recycling. In addition, it is an extremely expensive process because it requires processing vast amounts of feedstock to improve economics. A normal plant might process on the order of a billion standard cubic feet per day (BSCFD) of natural gas and cost \$10–20 billion. That said several plants exist in the world, notably two massive plants at the Ras Laffan industrial complex in Qatar.

In the years since World War I, the petrochemical industry has grown enormously and has greatly enhanced the quality of life for mankind. World War II provided another major impetus to its growth, which included obtaining data and theories to propel the industry into the gigantic operation it is today.

Along the way, some interesting things happened. Natural gas was considered a nuisance in production unless a ready market was available very close to the production site. As a result, producers usually flared the gas to be rid of it. Yet, natural gas was also a valuable fuel

Natural Gas Processing from Midstream to Downstream, First Edition.
Edited by Nimir O. Elbashir, Mahmoud M. El-Halwagi, Ioannis G. Economou, and Kenneth R. Hall.

without the distribution problem. Furthermore, it became a vital feedstock for major chemical industries. Then came the idea that "peak oil" had passed, and natural gas could help get the world through that "crisis" if only processes existed to convert it into liquid fuels. Enter cryogenic processing to liquefy natural gas, which could relieve the load on liquid fuels to provide for home heating. Then came the idea to use the Fischer-Tropsch process to convert natural gas into liquid fuels. This was and is a viable, although expensive, option. Another option would have been to provide a means to "activate" methane that would allow easy transformation into liquid fuels (this concept fizzled for technical reasons).

Over the past few years, it has become obvious that "peak oil" has not passed after all. In fact, it may never have existed. Given modern production technology, IHS Markit estimates that the Permian Basin by itself contains ~75 billion barrels of petroleum. We have passed from speaking of "peak oil" to dreaming of "infinite oil" in less than a decade. So why do we need natural gas under these circumstances?

Well, as a fuel it burns much cleaner than oil products. Also, converting it into a liquid greatly reduces transportation costs, thereby providing a greater supply from remote production facilities assuming a low-cost option exists for liquefaction.

In 1996, some professors at Texas A&M University sought a way to destroy toxic wastes using microwaves. This idea was short-lived because of costs. However, it gave rise to a new way to convert natural gas into liquid products (or to activate methane). This became known as the Synfuels process (named for the company that licensed the technology). The rest of this chapter details the Synfuels process, and demonstrates some of its advantages.

19.2 Additive and Subtractive Processes

Recently, additive technology has demonstrated many advantages compared to subtractive technology. Subtractive technologies have been the normal means to produce products. If a sculptor wants a statue, he carves away the marble that is not the statue. If a refiner wants to produce gasoline from oil, it devises a process that separates the gasoline molecules from all the other molecules in the oil. "Heavy" oils, such as those from tar sands, can have a large fraction of components heavier than those in diesel fuel. To the extent possible, it is necessary to remove these compounds and to devise uses for them other than normal fuels. The Fischer-Tropsch process produces a product distribution with approximately 50% of the compounds heavier than diesel fuel even when using natural gas as the feedstock. Recovering these molecules in a more useful form is a major expense. However, another undesirable characteristic is that is it impossible to remove all of these compounds, and some long-chain molecules remain in the products.

However, recently another manufacturing process has begun to emerge. In general, it is additive technology (commonly referred to as 3-D printing). Now the sculptor could simply write a computer code and "print" the statue from treated marble powder. Likewise, the refiner could construct its gasoline from small molecules, and stop when it reaches the desired properties. So what is the advantage of "printing" a liquid fuel? Well, think about jet fuel. Petroleum-derived jet fuel still retains some long-chain molecules that were present in the original oil. These molecules greatly influence the freezing point of the jet fuel, making it freeze at temperatures such as those encountered at high altitudes. Why go to such altitudes? To achieve more efficient use of fuel commercially, or to achieve a tactical advantage militarily. The Synfuels process is an additive technology to produce jet fuel from small molecules that remains free of the long-chain molecules that provide undesirable properties. For example, if jet fuel is the desired product, the product would have essentially no molecules longer than about C_{18}.

19.3 The Synfuels Process

The concept behind the Synfuels process is to use molecules available from natural gas conversion to build liquid fuel molecules. If this sounds like methane activation, it is (in a practical process that works). The feed gas should be sweet natural gas (passed through an amine unit if necessary) with minimal LPG, e.g., cut at butane.

The trick is to subject methane to temperatures in the range of 4,000 °F and convert it into acetylene. The reaction in its basic form is

$$2CH_4 \rightarrow C_2H_6 + H_2 \rightarrow C_2H_4 + H_2 \rightarrow C_2H_2 + H_2 \rightarrow 2C + H_2$$

These reactions require only heat to occur, but the Synfuels reactor does have a proprietary design to utilize the heat efficiently. The first three reactions (methane to ethane to acetylene) occur very fast, on the order of microseconds. However, the final reaction (acetylene to carbon) is relatively "slow" on the order of 0.5 millisecond. In addition, although acetylene is a very reactive molecule, unlike most hydrocarbons it is stable at these temperatures. Thus, with careful attention to residence time (<1 ms), it is possible to design a reactor that produces acetylene and hydrogen with very little carbon or coke formation. In effect, this is a *natural gas to acetylene* process at this stage.

However, the equation above shows that hydrogen forms in each step. It is therefore possible to recombine the hydrogen with acetylene to make ethylene for further processing. In other words, we now have a *natural gas to ethylene* process. Now all ethylene chemistry becomes available, but more germane to this discussion is the ability to convert ethylene into liquid fuels affording us a *natural gas to liquid fuels* process.

Figure 19.1 presents one possible arrangement for the Synfuels process. The process starts by importing sweet natural gas (if only sour gas is available, an amine unit would be necessary to remove the hydrogen sulfide and carbon dioxide) and commercial-grade oxygen. The oxygen goes directly to the burner/cracker/quench unit, while the natural gas first mixes with recycle gas coming from the product separation step and then proceeds to the burner/cracker/quench unit. Removing any LPG fraction of the natural gas is not necessary, but it would reduce the cost of the process somewhat while recovering the LPG fractions for direct sale. The burner/cracker/quench unit operates at approximately 4,000 °F (2 200 °C), which is sufficient to crack 50–65% of the methane and essentially 100% of all other hydrocarbons in the feed. The unreacted methane composes a large fraction of the recycle gas that mixes with the inlet natural gas stream. Introducing steam into the burner/cracker/quench unit greatly reduces coke formation in the unit to less than 1% of the inlet carbon. The exit stream from this unit contains acetylene, hydrogen, unconverted methane, carbon oxides, and nitrogen/nitrogen oxides. Commercial absorbents are readily available to separate the acetylene from this gas mixture.

The spray tower washes the exit stream from the burner/cracker/quench unit to remove carbon granules/particles and any green oil produced in the converter. Should acetylene be the desired product, the stream would proceed to a unit to purify the acetylene. Should ethylene or liquids be the desired product, this unit then cools the stream to ambient temperature before compressing it from about 100 psia to 400 psia (7 to 27 bar) before it enters the reactor to hydrogenate the acetylene to ethylene using hydrogen already in the stream produced in the burner/cracker/quench unit.

Normally, hydrogenation of acetylene into ethylene is a gas-phase reaction. However, it is possible to use a proprietary Synfuels process to perform the hydrogenation in a liquid phase. Liquid-phase hydrogenation has safety advantages over the gas-phase hydrogenation. Obviously, the reactor would be smaller in a liquid-phase process, which reduces the cost. However, the liquid-phase process also produces a better product for continued processing.

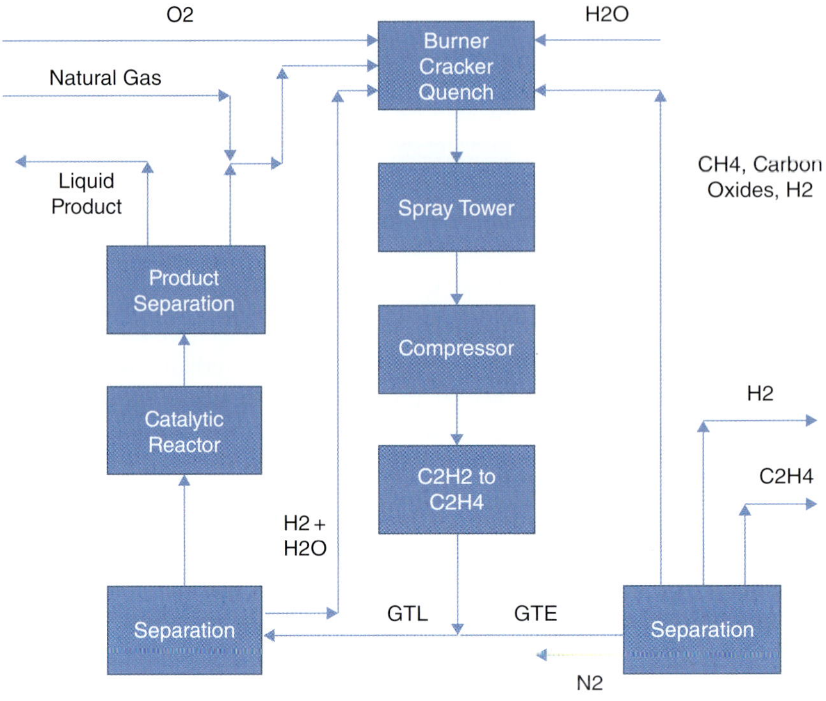

Figure 19.1 Possible arrangement for gas to acetylene/ethylene/liquids process.

At this point, the operator can elect to produce ethylene, liquid fuels, or both. Producing both probably would be an unlikely choice from an economic viewpoint, but the possibility exists if it does make sense. It is noteworthy that production of acetylene or ethylene in this process does not require recycle of the product.

Assuming ethylene is the desired product, the stream from the hydrogenation unit would pass to a cryogenic separation, which would purify the ethylene and could split out the hydrogen should that be desirable. The remainder of the stream (composed of unreacted methane, carbon oxides, possibly hydrogen) would recycle to the burner/cracker/quench unit primarily as fuel. Removal of nitrogen and nitrogen oxides would be beneficial in this unit before recycling to the burner/cracker/quench unit. This choice presumes the proximity of chemical-processing plants that utilize ethylene and hydrogen. Transporting either of these products long distances to a user would not be an economical enterprise in almost any circumstances, even though they are usually more profitable than liquid fuels. Normally, it would be less expensive to transport the natural gas directly to the proximity of the end user and convert it into ethylene and hydrogen there.

Assuming liquid fuel is the desired product, the stream from the hydrogenation unit would pass to a catalytic reactor that would oligomerize the ethylene (and possibly a small amount of the hydrogen) into a liquid stream, which could be primarily gasoline, jet fuel, or diesel fuel, and a recycle stream composed of light hydrocarbons (C_5 and lighter) that passes to the burner/cracker/quench unit.

Operation of the oligomerization unit at low pressure (near atmospheric) produces gasoline. This product (after stabilization) is a 95-octane, C_8-average fuel that contains about 30% aromatics (but very little benzene), 40% iso-alkanes, and 30% cyclo-alkanes. An important feature of the Synfuels gasoline is its comparison in its product distribution to that of a Fischer-Tropsch process. Figure 19.2 provides the comparison. The obvious advantage of the

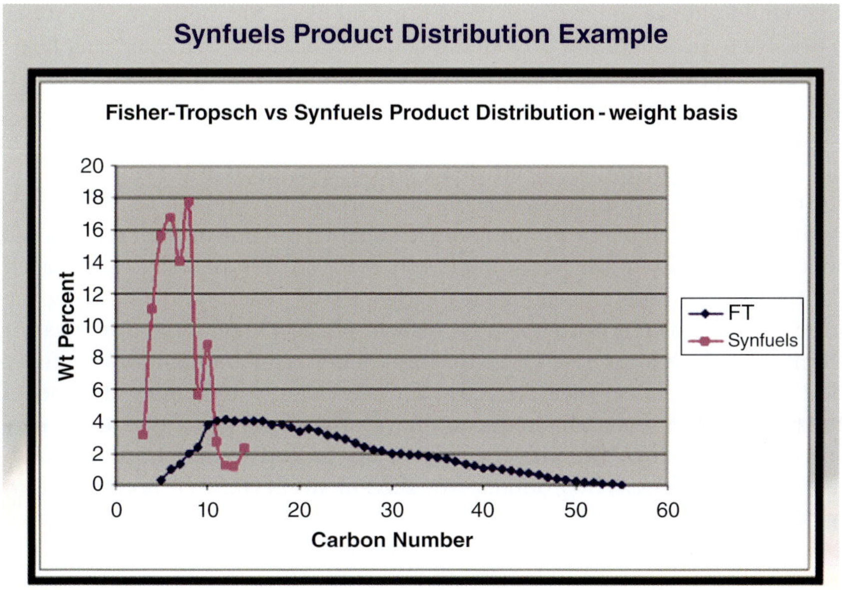

Figure 19.2 Comparison of Synfuels gasoline production to that of Fischer-Tropsch.

Synfuels product is the tight distribution about a carbon number of 8. No long-chain liquids or waxes form. The Fischer-Tropsch products do form a large fraction of very long-chain molecules, which require recycling unless a market exists for these molecules. Regardless, the recycle adds significantly to the cost of operation.

It is possible to operate the Synfuels reactor in a manner that would reduce the aromatic content while the iso- and cyclo-alkanes would remain in essentially the same ratio. This option might be preferable for gasoline production, but if jet fuel is the desired product, the aromatics are essential components. An interesting option is that the Synfuels plant can operate to produce various fractions of aromatics (10–45 wt%), which enables the producer to fine-tune the jet fuel. Jet fuel production requires higher-pressure operation than gasoline production. Operation at still higher pressures can produce diesel fuel.

Another option would be to separate the aromatic compounds as a product. They would be liquid and easily transportable. They would have a high value as feed stock for the chemical industry. Then the alkanes could provide an alkylation stock for the gasoline industry. One other option would be to use a demethanizer on the feed stream. This would allow the process to be more efficient in converting methane into the products described. Then depending upon location, the C_{2+} fraction could feed a smaller, more efficient unit to convert these compounds into acetylene, ethylene or liquid fuels.

19.4 Pilot Plant

Some preliminary scoping of catalysts for the hydrogenation and oligomerization, performed at bench scale, provided high enough yields (~15%) to encourage advancing the project to larger scale. Commercial development would require yields approaching 40–50%. However, the pyrolysis performance depended strongly upon geometry and scale. Development, testing, and optimization of the pyrolysis reactor required pilot tests. Synfuels constructed a 50 MSCFD pilot plant in Robertson County, Texas, for this purpose. Over the course of four years, a design emerged that could endure the extreme conditions required for methane activation, was reasonably inexpensive to produce, and provided the instrumentation and operating flexibility for remote and varied application. In addition, the reactor utilized an internal flame in a hydrocarbon-safe environment, which provided safety and robustness of the operation.

The original design for the hydrogenation reactor was a gas-phase, fixed-bed reactor similar to acetylene hydrogenation found in ethylene production processes. However, the non-isothermal pilot reactor revealed an acetylene concentration high enough that the reaction heat management was unreliable and prohibitively expensive to control with recycle gas dilution. In addition, the propensity of the acetylene to self-react created residues that caused rapid deactivation of the catalyst. Safe management for the heat of reaction and keeping the catalyst clean required a liquid solvent in the process instead of recycle dilution. The mass of the solvent helped limit the temperature rise of the reaction materials and maintain selectivity of the reaction. Additionally, the solvent was selective for acetylene, which allowed removal of acetylene from the bulk pyrolysis gas. This reduced the volume of gas processed, especially when ethylene was the target product. Because the solvent was selective for acetylene, it also favored production of polyacetylenes as byproducts. The solvent swept these compounds from the reactor to a slipstream cleanup process.

The oligomerization unit, when piloted, performed much the same as it had in the lab. The yield, quality, and catalyst deactivation were acceptable.

While the pyrolysis reactor was a success at the pilot scale, it was a design that would be sensitive to geometric scale-up in terms of operation and efficiency. Because energy efficiency is proportional to surface area, as the reactor scales up, fuel and oxygen consumption would decrease. To demonstrate this increase in efficiency, the pyrolysis reactor required scale-up from 50 MSCFD to 1 MMSCFD. The scale was definitely worthwhile. While providing the same burner stability and thermal efficiency as expected, the cracking reaction section revealed that the pyrolysis time estimated from the pilot cracker was insufficient for the mid-size unit. In addition, while the location of the quench was understood in the pilot reactor, the exact quenching temperature profile was not. The mid-size reactor required more reaction time. After increasing the length and optimizing the conditions, the mid-size reactor demonstrated expected performance with improved fuel efficiency. In addition, mid-scale testing allowed refinement of the ignition and quenching systems at realistic reactor volumes.

The 50 MSCFD reactor converted about 33% of carbon in lean gas per pass. The 1 MMSCFD reactor converted about 40% of the carbon in lean gas and about 46% of the carbon in rich gas (associated gas). The pilot plant demonstrated convincingly that the Synfuels process worked.

The Synfuels technology has several advantages over traditional processes from design, construction, and products vantage points. Some of the more visible are:

1) Standard materials of construction: Most of the Synfuels equipment is carbon steel. The pyrolysis reactor is a common grade of stainless steel, and the hydrogenation reactor has an internal lining made of stainless steel.
2) Static catalyst beds: While fluidized catalyst beds can be advantageous, they are more expensive and more difficult to design, build, and operate than fixed-bed units. The Synfuels reactor is a multisection fixed-bed unit with minimal catalyst attrition, and far less chance of downstream damage caused by catalyst infiltration.
3) Very low recycle: Most GTL processes (especially Fischer-Tropsch) have low per pass conversions that require significant recycle to achieve the required conversions. The Synfuels process has very little recycle and yields about 3% light products. The process has lower capital and operating costs, smaller equipment, and less environmental impact than other GTX processes (a Fischer-Tropsch process would be about 3.6x the cost of a Synfuels process per barrel of product).

Pyrolysis yield potential.

	Pilot Plant	Mid-Scale	Commercial
Design pyrolysis gas flowrate[1] MSCFD	50	1000	50000
Actual pyrolysis gas flowrate[1] MSCFD	24.7	454.8	50000
Short-term pyrolysis yield wt% carbon	34.0	41.2	45.3
Long-term pyrolysis yield wt% carbon	N/A	39.9	44.0[2]
Scale-up ratio wrt pilot plant	1	18.5	202.8
Short-term conversion increase wt% carbon	–	7.2	4.1
Heat loss % of nominal burner duty	42.7	17.7	8.4

Notes: 1. Pyrolysis gas carbon number = 1.0.
2. Long-term commercial yield obtained from de-rating short-term yield by 1.3%.

4) Low pressure operation: The maximum operating pressure in a Synfuels plant is about 300 psig (compared to 800–1200 psig for Fischer-Tropsch plants). This reduces cost while improving safety.
5) Safe hydrogenation: The liquid-phase hydrogenation reactor has a temperature rise of about 60 °F compared to 400–800 °F rises in a gas-phase hydrogenation.
6) Minimal by-products: The Synfuels process produces only CO_2 and water as by-products.
7) Scalability: Tests of the Synfuels reactors indicate that they can scale up to any full-size operation.

19.5 Location, Location, Location

Fischer-Tropsch plants are very large and very expensive because the process is relatively dirty and complex. As a result, location of the plants must be near the source of feedstock, coal originally and natural gas recently (the process dates back to World War I when Germany had abundant coal but little oil to fuel its war effort). As a result, without reasonably proximate feedstock sources (think Qatar), transportation costs can render the process uneconomical especially for natural gas. Locating a Fischer-Tropsch plant in remote or cold locations is a prohibitively expensive project. The same is true for many natural gas fields.

The Synfuels process is much smaller and less expensive than Fischer-Tropsch. It is possible to construct an economical Synfuels plant at about 10 million SCFD. It also is possible to construct these plants at sizes required for Fischer-Tropsch plants (BSCFD) – the difference is they would be approximately one-third the size and cost. Thus, it is conceivable to construct Synfuels plants in remote locations with economical gas flows. It is even possible to construct one on a large tanker, which could move when the source field played out. Even smaller plants might have military applications in which mobility needs might offset additional expense.

19.6 Biofuels

Interestingly, the Synfuels process is amenable to use in biofuel production. This process does not use corn nor any other food crop (food to fuel) to make ethanol for mixing with gasoline. Byogy Renewables, a company separate from Synfuels, has developed a process, based upon the Synfuels process, to convert biomass into ethanol and ethanol into fuels.

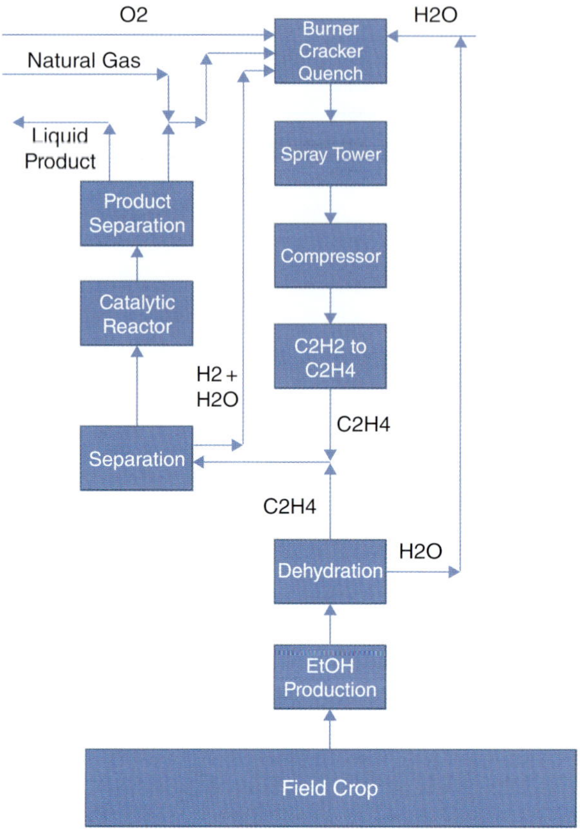

Introduction of Byogy Renewables process into Synfuels process.

Consider Figure 19.2, which is very similar to Figure 19.1. The additional reactions are:

$$\text{Biomass} \rightarrow C_2H_5OH \quad \text{(using any of a variety of processes)}$$

$$C_2H_5OH \rightarrow C_2H_4 + H_2O$$

Both process produce ethylene from a feedstock: natural gas in the case of Synfuels and biomass in the case of Byogy Renewables. Clearly, the ethylene streams can mix and proceed to common processing. However, the location issue rears its head again. What if no biomass source is near the Synfuels process? Well, ethanol is relatively inexpensive to transport as one option. Figure 19.3 presents another option.

This option could have a location near the biomass source. In addition, the greatly simplified Synfuels process would cost significantly less than the one to convert natural gas into fuels. Conversely should a Synfuels plant location abut a biomass source, the addition of a Byogy Renewables unit would add little additional expense.

Pilot testing of the Byogy process primarily targeted achieving volumes of product that would meet customer requirements. Bench scale tests verified the catalyst sequence to generate jet fuel from ethanol (which is actually superior to petroleum –derived jet fuel). The main challenge was to move from an isothermal to a non-isothermal reactor environment, particularly for a highly exothermic reaction in which yield and selectivity were sensitive to temperature. The pilot reactors represented the expected length and diameter of commercial reactor tubes. The reactors proved to be challenging to control, but ultimately stable and effective. The product quality exceeded the customer specification.

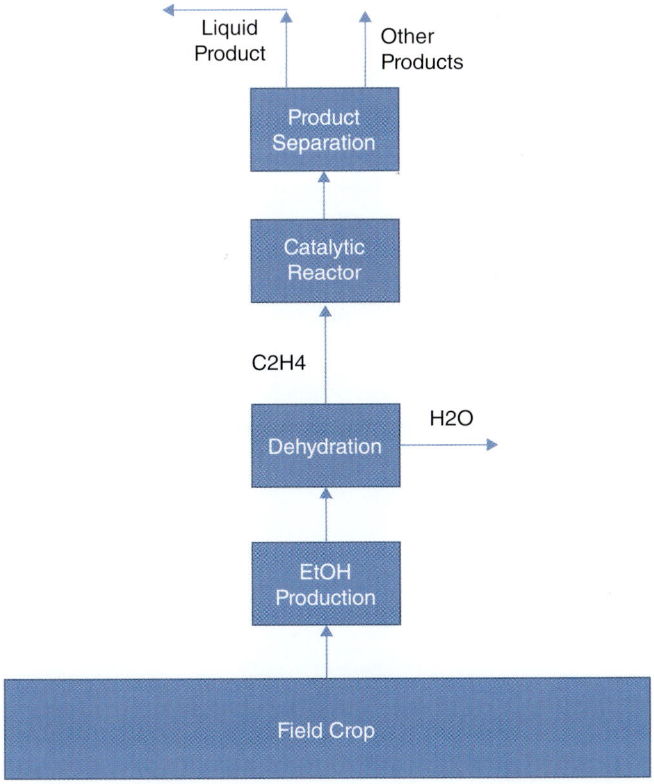

Figure 19.3 Byogy Renewables process utilizing only Synfuels conversion technology.

It is important to note that Byogy Renewables has identified a nonfood biomass source that provides much more ethanol per mass than corn. The synergistic combination of these technologies offers an intriguing source for transportation fuels derived from nonconventional sources.

19.7 Conclusion

Synfuels International has conceived and tested through pilot scale, and partially through mid-range, a unique process to convert natural gas to acetylene, ethylene, or liquid fuels including gasoline, jet fuel, and diesel fuel. The process has the advantage of using additive technology to produce its products, which greatly reduces plant size and cost compared to Fischer-Tropsch technology. Commercial plants as small as 10 MMSCFD can be practical, but scale-up to essentially any throughput is feasible. The pilot and mid-scale testing proved that the products would meet or exceed customer specifications.

An added bonus is that the process can accept non–food-derived ethanol as the feedstock. In contrast to use as a fuel diluent, the ethanol transforms into ethylene, gasoline, jet fuel, or diesel fuel. The jet fuel is particularly interesting in that it has a freezing point lower than −80 °C. It also is possible to produce the jet fuel with any desired amount of aromatic concentration while maintaining a very low concentration of benzene.

The process is ready for commercialization. It has received in-depth scrutiny at multiple levels from potential users. None have found any technical flaws.

20

Natural-Gas-Based SOFC in Distributed Electricity Generation: Modeling and Control

Gerald S. Ogumerem [1,2], *Nikolaos A. Diangelakis* [1,2], *and Efstratios N. Pistikopoulos* [1,2]

[1] *Artie McFerrin Department of Chemical Engineering, Texas A&M University, USA*
[2] *Texas A&M Energy Institute, Texas A&M University, USA*

CHAPTER MENU

20.1 Introduction

The recent global natural gas production increase and its low price in the United States has led to an increase in the utilization of natural-gas-based power generation (U.S. Energy Information Administration and U.S. Department of Energy 2011). Some studies have estimated the Ultimately Recoverable Resources (URR) of natural gas to be between 10,700 and 18,300 EJ, for conventional sources and 4,250 to 11,000 EJ for unconventional sources while the peak yearly natural gas recovery period is estimated between 2025 and 2066 at 140 to 217 EJ/y (Al-Fattah and Startzman n.d.; U.S. Energy Information Administration and U.S. Department of Energy 2011; Mohr and Evans 2011). Natural gas is a feedstock for several industrial processes. Some of them convert natural gas to more valuable chemicals. Catalytic steam reforming is a popular industrial process for converting natural gas to syngas, and other valuable chemicals for other industrial processes. The main routes for converting natural gas to power includes converting its chemical energy (directly or indirectly) to thermal, kinetic or electrical energy. Internal combustion engines (ICE) and gas turbines are the most popular equipment for converting natural gas to kinetic and/or thermal energy, through combustion. The kinetic energy is subsequently converted to electrical energy, and the thermal energy is typically harnessed through an integrated system to improve the overall efficiency of the system through steam cycles. The use of natural gas for (distributed) electricity generation is gaining traction, currently propelled by low natural gas prices (U.S. Energy Information Administration and U.S. Department of Energy 2011). Figure 20.1 shows a projected increase in the use of natural gas (NG) for various activities including electricity generation (U.S. Energy Information Administration and U.S. Department of Energy 2011).

Over the years, there have been reports of increasing greenhouse gas (GHG) emissions, which have been correlated with the increasing fossil fuel usage. This has driven recent interest

Natural Gas Processing from Midstream to Downstream, First Edition.
Edited by Nimir O. Elbashir, Mahmoud M. El-Halwagi, Ioannis G. Economou, and Kenneth R. Hall.

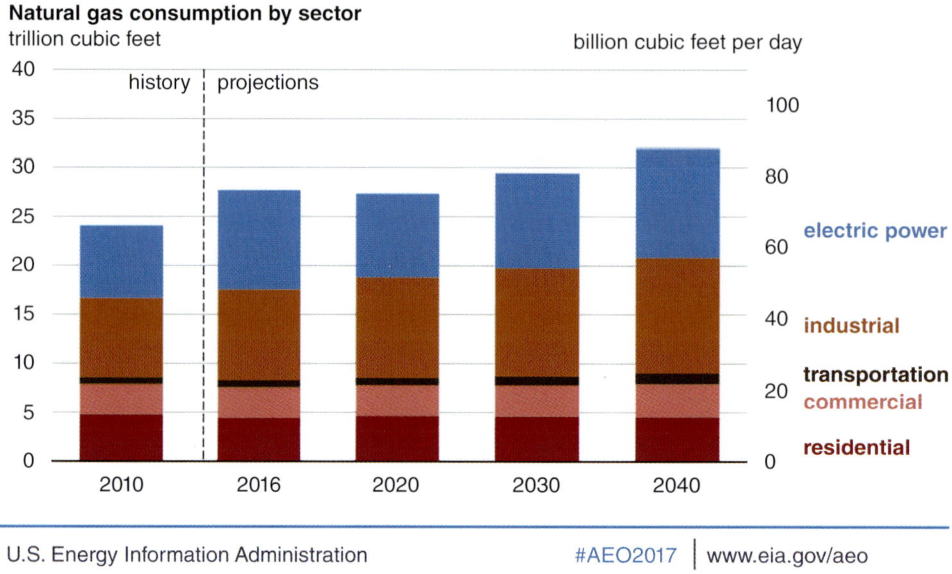

Figure 20.1 Natural Gas Consumption by Sector.

in alternative, environmentally benign energy sources. Comparatively natural gas has less adverse impact to the environment than other fossil fuels, and it is generally accepted as a good transition energy source (Burnham et al. 2011). The composition of natural gas varies significantly based on recovery locations, but methane is its major component, usually more than 80% of the total volume. Methane, is one of the GHGs, the combustion natural gas yields CO_2 which is also a GHG. Emissions from leaks in natural gas production equipment contributes about 40% of estimated national emission of methane (US NETL 2013). Combustion of NG to produce energy generates about 117 pounds of CO_2 per unit of heat (Million Btu equivalent) produced.

Coal produces more than 200 pounds of CO_2 per million Btu while distillate fuel oil produces more than 160 pounds per million Btu equivalent (US EIA 2016). This and its lower impurity content makes NG a cleaner fossil energy source. The energy content of fuels can be categorized by their energy density. The energy density metric can be either gravimetric (energy content per weight) or volumetric (energy content per volume) (Hore-Lacy 2006). Table 20.1 below shows a list some fuel types and their gravimetric energy density.

20.1.1 Distributed Energy Production

Technology developments and environmental concerns are changing electricity generation and transmission. Smaller more distributed generating stations are replacing larger centralized generating facilities (Lasseter 2007). Technologies like fuel cells (FC), internal combustion engines (ICE), wind turbines (WT), gas turbines (GT), and photovoltaics (PV) and a combination of any of these, make up the variety used in most distributed system. These technologies also create avenues for process intensification that would be rather more expensive or unnecessary if applied to centralized generating systems. Distributed electricity-generating systems have numerous advantages over centralized electricity-generating systems. Their ability to be modularized increases their efficiency and flexibility. Cogeneration units like ICE-based combined heat and power (CHP) can produce heat and electricity in a single

Table 20.1 Energy Density of Various Energy Sources.

FuelType	Energy Density(MJ/kg)	Typicaluses
Biodiesel	38	automotive engine
Coal	24	Power plants, Electricity generation
Diesel	45	Diesel engines
Gasoline	46	Gasoline engines
Natural gas	55	Household heating, Electricity generation
Uranium-235	3, 900, 000	Nuclear reactor electricity generation
Wood	16	Space heating, Cooking

Source: (Hore-Lacy 2006).

process without major changes in fuel utilization, thereby increasing the efficiency of the system. Centralized power production units would usually have a minimum required load to be in operation, and the power generated is more than the required amount to compensate for the losses in transmission (Diangelakis and Pistikopoulos 2017). However, in distributed power production units, which are usually modularized units, there is less minimum-required power production, which thus avoids large excess power production and therefore is cost effective and environmentally friendly. Distributed electricity production systems provide the opportunity to adopt renewable energy resources. The use of distributed solar and wind energy systems has gained a lot of interest. NG can have a significant contribution to the development of distributed systems. Technologies like the internal reforming solid oxide fuel cell (IR SOFC) use NG to produce electricity. Most residential units have existing NG supply lines for heating and other domestic activity, which makes the transition from a centralized electricity grid power production to a distributed IR SOFC electricity provision seamless.

20.1.2 Solid Oxide Fuel Cell (SOFC) Overview

Fuel cell technologies have received much attention in the last decade, with the polymer electrolyte membrane (PEMFC) fuel cell receiving the most due to its relevance in the transportation sector. Consequently, SOFC technology is not new; however, its use for distributed power generation in small to medium scale is still under testing and is yet to reach 40,000 hours of operation (the estimated life time of a SOFC unit) (Riensche et al. 1998).

Stand-alone SOFC systems have efficiencies ranging from 40% to 55%. Combining the SOFC with a cogeneration of high-grade thermal energy in a combined heat and power (CHP) system could increase its overall efficiency to approximately 80% (Hawkes and Leach 2005). With its attractive features, one would expect the SOFC to be an important technology for distributed power generation or an auxiliary power unit (APU). However, the commercialization of SOFC has been plagued by its financial implications. Studies by Jin et al. (2004) show that it takes about $8,738 to produce a 5 kW SOFC unit that uses diesel as fuel (Battelle 2014). This estimate is for the Balance of Plant (BOF) of an SOFC with an external reformer (see section 20.1.4) and a modular desulfurizing unit. As shown in Figure 20.2, the major cost-contributing factors are the heat transfer unit and the electronic and controls units.

Like other fuel cells (FC), SOFC uses hydrogen and oxygen to produce electricity. On the contrary to other FC alternative, the SOFC can use NG as fuel, a characteristic that makes this technology very attractive in areas with pre-existing NG availability. This is possible through natural gas reforming, which is discussed next.

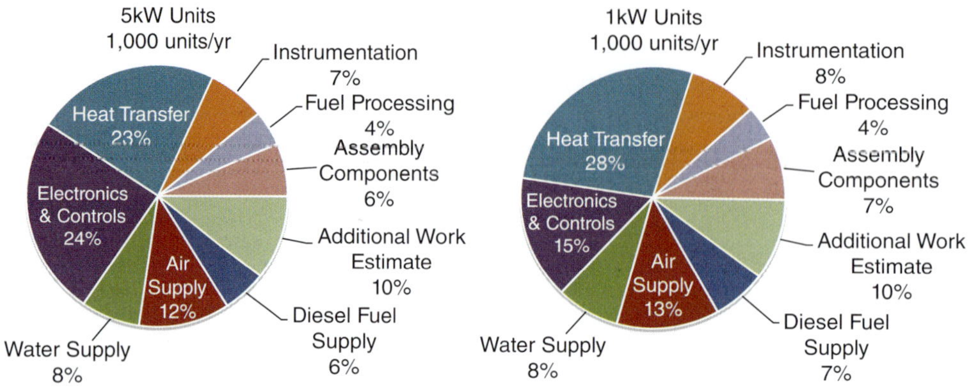

Figure 20.2 Cost Distribution for a 1kW SOFC plant and a 5kW SOFC plant.

20.1.3 Natural Gas Reforming

The main reason developers design a SOFC that uses NG is to thermally integrate the reforming and the electrochemical reaction. The NG undergoes a reforming reaction to produce hydrogen and carbon monoxide, CO. The reforming process could be partial oxidation where oxygen is the oxidant. Partial oxidation is exothermic and produces a substantial amount of steam along with CO_2. NG decomposes to form solid carbon and hydrogen at elevated temperature (1273–1473 K) according to equation 1.

$$CH_4 \overset{\text{Heat}}{\Rightarrow} C + 2H_2 \qquad 74.9\ \text{kJ/mol} \tag{20.1}$$

Studies have shown that regulating the amount of reactant oxygen is an option for controlling the formation of solid carbon (Poppinger and Landes 2001). The reforming process could also be steam reforming, which requires a significant amount of steam and produces relatively more hydrogen per mole of NG reacted. The steam reforming (SR) is well suited for integrated SOFC application since steam produced from the electrochemical reaction can serve as reactant for reforming NG or shifting CO. SR requires a significant amount of thermal energy as show in equation (20.2) since its main reaction is endothermic.

$$\begin{aligned} CH_4 + H_2O &\Leftrightarrow CO + 3H_2 && 224\ \text{kJ/mol} \\ CO + H_2O &\Leftrightarrow CO_2 + H_2 && -37\ \text{kJ/mol} \\ CH_4 + 2H_2O &\Leftrightarrow CO_2 + 4H_2 && 187.5\ \text{kJ/mol} \end{aligned} \tag{20.2}$$

Conversely, SOFC average operating temperature can be (950–1200 K). SOFC developers proposed various way to integrate the reforming process into the SOFC to reduce the amount of energy consumed.

20.1.4 Direct Internal Reforming (DIR) SOFC

Several integration strategies for SOFC have been reviewed (Zhang et al. 2010). One of the design approaches considered is the external reforming SOFC. Here, the reformer is spatially close to the SOFC, and the heat generated for the SOFC is transferred to the reformer through heat integration (Dicks 1998). Contrariwise, in the internal reforming approach the reforming process is within the SOFC. Generally, two design techniques exist for the internal reform of natural gas in SOFC: the Direct internal reforming (DIR) and indirect internal reforming (IIR). In the latter, the stack and the reformer are in thermal contact, and the reformate is fed to

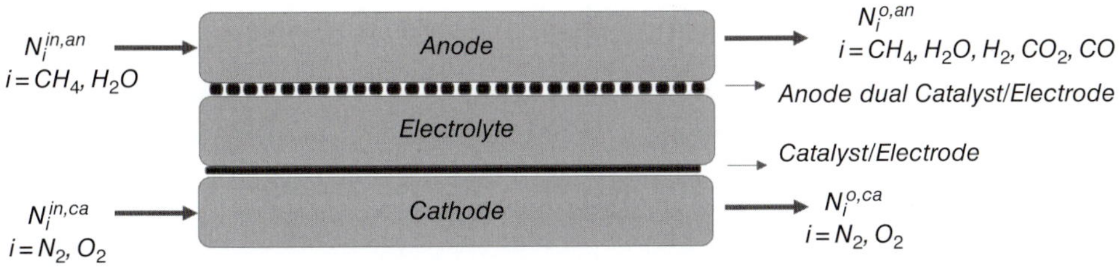

Figure 20.3 Schematic Diagram of a DIR SOFC Cell.

the to the adjacent anode. Also, the electrochemical reaction and the reforming reaction are separated. While in the DIR, the NG reform takes place in the anode. The heat generated by the exothermic electrochemical and shift reaction is utilized by the endothermic reforming reactions. Tubular, planar, and monolithic are the predominant geometric configurations of a SOFC (Ahmed and Foger 2000; Klein et al. 2007). Westinghouse (now Siemens-Westinghouse) developed the tubular SOFC (Kakaç et al. 2007). Planar designs are popular for their compact nature and higher power density. A schematic diagram of a DIR SOFC Cell is shown in Figure 20.3. Due to its high operating temperature, the electrolyte is a nonporous ceramic; Y_2O_3 stabilized ZrO_2 (YSZ). The anode face is made of a porous Nickel/Yttria-stabilized zirconia (Ni/YSZ) cement, and the nickel serves as a catalyst for the reform reaction. The cathode face is porous Sr-doped $LaMnO_3$.

During the operation of the direct internal reforming, natural gas is fed to the anode. The steam from the electrochemical, heat and the nickel catalyst activates the reform reaction producing hydrogen (H_2), carbon monoxide (CO), and carbon dioxide (CO_2). The shift reaction also utilizes the CO and steam to produce more CO_2 and H_2. The H_2 molecule loses its electron as it adsorbs to the electrode. Air is fed to the cathode. The oxygen in the air gains the electron as it adsorbs to the electrode.

$$\begin{aligned} H_2 + O_2^- &\Leftrightarrow H_2O + 2e^- && -285\ \text{kJ/mol} \\ CO + O_2^- &\Leftrightarrow CO_2 + 2e^- && -283\ \text{kJ/mol} \\ O_2 + 4e^- &\Leftrightarrow 2O_2^- && 0\ \text{kJ/mol} \end{aligned} \tag{20.3}$$

The oxygen ion percolates through the electrolyte to the anode, where it reacts with the hydrogen ion to form steam. One of the challenges of the DIR is the formation of carbon. The decomposition of NG to solid carbon is a thermodynamically favored reaction at the nominal operating conditions at the anode. However, a preferred technique is the addition of steam, since it is a coreactant with NG in the reform reaction. The ratio of steam to methane can be used to regulate the formation of solid carbon (Poppinger and Landes 2001). The steam produced from the electrochemical reaction is used for reforming the NG. Steam is also fed into the system to provide a sufficient amount of steam for the reforming reaction. Mathematical models can be used to estimate the required amount of input steam by analyzing the interactions between various components of the system

20.2 Mathematical Model

The interactions in the SOFC are quite complex, cutting across material sciences, electrochemistry, and transport phenomena etc. This makes the mathematical model a requisite tool for studying the static and dynamic behaviors of SOFC. Several models for SOFC have been developed over the years for various applications. These models range from complex high-fidelity dynamic, spatial models to data-driven models. Some researchers have focused on the investigation of the complex interaction in an SOFC using three-dimensional models (Aguiar et al. 2002; Ferguson et al. 1996; Recknagle et al. 2003; Yakabe et al. 2001). In Bhattacharyya and Rengaswamy (2009) a wide range of dynamic models of SOFC with reformers was reviewed. They categorized the various interactions in the SOFC according to their characteristic time constant and evaluated the models accordingly. In Kakaç et al. (2007) a broader spectrum of SOFC considering the effects of the geometry of the unit on the performance of the system was reviewed. A few researchers have developed models for internal reforming SOFC (Klein et al. 2007). Here we develop a dynamic model describing various fundamental principles of a DIR SOFC. The system adopted is a 12 kW SOFC with the following specifications at nominal operation. The following assumption were made in developing the hypothetical model used here;

- Ideal gas behavior
- The pressure correction on the voltage is dependent of hydrogen, oxygen, and steam.
- The kinetics developed through the experiments conducted by Jianguo Xu and Gilbert Foment (Xu and Froment 1989) is the same for a direct internal reforming SOFC since a nickel-base catalyst is used in both cases.
- The anodic and cathodic chambers are well mixed, and the oxidation of hydrogen happens in the anode.

20.2.1 Mass Balance

Anode. The mass balance described here is done according to equations (20.2) and (20.3). The anode takes in NG and a small amount of steam at a molar ratio of 3:1. The NG undergoes a reforming reaction with steam to produce CO and H_2. The CO produced is shifted with steam to produce more H_2. The H_2 produced undergoes an electrochemical reaction with oxygen to produce steam. The steam produced in the electrochemical reaction augments the steam in the inlet stream to make available an abundant steam supply. The material balance in the anode is as follows:

$$\frac{V_{an}}{RT}\frac{dp_i}{dt} = N_i^{i,an} - N_i^{o,an} - r_i^{an} \tag{20.4}$$

Where Van is the volume of the anode, p is the partial pressure of specie i, R and T are gas constant and temperature respectively, N_i is the molar flowrate of specie i, r_i is the sum of reactions involving specie I, and i represents all the species in the anode (CO, CO_2, H_2O, CH_4, H_2).

Cathode. At the cathode, air flows in, and the oxygen from air adsorbs on the electrode. The adsorbed oxygen seeps through the electrolyte and into the anode as oxygen ions where it reacts with hydrogen ions to form water. The composition of the air is assumed to be 21% oxygen and 79% nitrogen. The material balance in the cathode is given as follows

$$\frac{V_{ca}}{RT}\frac{dp_i}{dt} = N_i^{i,ca} - N_i^{o,ca} - r_i^{ca} \tag{20.5}$$

where V_{ca} is the volume of the anode, i represents all the species in the cathode (O_2, N_2).

20.2.2 Energy Balance

The thermal integration that takes place in the DIR SOFC creates a state where the endothermic reaction is complemented by the exothermic reactions. Here the streams are preheated to 700 °C and are fed at 20 bars (Fernandes and Soares 2006). The partial enthalpies of the inlet and outlet streams of the anode and the cathode make up the first two terms in equation (1.6). The combination of the reaction enthalpies and reacted quantities produces a net heat input from the reaction. This is greatly dependent on the kinetics of the reaction with respect to the equilibrium. About 235 kJ of energy generated for every mole of water produced contributes to power, and some of the heat generated is lost due to radiation.

$$\frac{dT}{dt} = \sum_{i=1}^{n_a} N_i^{an}(h_i^{in} - h_i^{st}) + \sum_{i=1}^{n_c} N_i^{ca}(h_i^{in} - h_i^{st}) + \sum_{j=1}^{m} \Delta H_j r_j - P_{st} - H_{rad} \tag{20.6}$$

Where ΔH_j is the enthalpy of reaction j, h_i^{in} and h_i^{st} are the enthalpy of the fluid entering and leaving the stack respectively, na and nc are the number of species in the anode and the cathode respectively. P_{st} and H_{rad} are the power of the stack and the heat loss from radiation respectively.

$$H_{rad} = A\delta\sigma(T^4 - T_{surr}^4) \tag{20.7}$$

Where A is the surface area of the stack, δ and σ are constants and T_{surr} is the temperature of the surrounding.

20.2.3 Kinetics

The Langmuir Hinshelwood kinetics is used to describe the reforming reaction according to Xu and Froment (1989). The rate expressions for the reaction given by.

$$r_1^{an} = \frac{\frac{k_1}{p_{H_2}^{2.5}}\left[p_{CH_4}p_{H_2O} - \frac{p_{H_2}^3 p_{CO}}{K_1}\right]}{DEN^2} \tag{20.8}$$

$$r_2^{an} = \frac{\frac{k_2}{p_{H_2}}\left[p_{CO}p_{H_2O} - \frac{p_{H_2} p_{CO_2}}{K_2}\right]}{DEN^2} \tag{20.9}$$

$$r_3^{an} = \frac{\frac{k_3}{p_{H_2}^{3.5}}\left[p_{CH_4}p_{H_2O}^2 - \frac{p_{H_2}^4 p_{CO}}{K_3}\right]}{DEN^2} \tag{20.10}$$

$$DEN = 1 + K_{CO}^{ads}p_{CO} + K_{H_2}^{ads}p_{H_2} + K_{CH_4}^{ads}p_{CH_4} + K_{H_2O}^{ads}\frac{p_{H_2O}}{p_{H_2}} \tag{20.11}$$

$$k_j = A_j \exp\left[\frac{E_j}{R}\left(\frac{1}{T} - \frac{1}{T_{ref}}\right)\right] \tag{20.12}$$

$$K_i^{ads} = A_i \exp\left[\frac{\Delta H_i^{ads}}{R}\left(\frac{1}{T} - \frac{1}{T_{ref}}\right)\right] \tag{20.13}$$

$$K_j^{equ} = A_j \exp\left[\frac{\Delta H_j^{rxn}}{R}\left(\frac{1}{T} - \frac{1}{T_{ref}}\right)\right] \tag{20.14}$$

where k_j is the rate constant of reaction j; K_j is the equilibrium constant of reaction j and K_i is the adsorption coefficient of component i.

The kinetic model for the electrochemical reaction is based on a Butler Volmer reaction mechanism in which the rate expressions for reactions is given by

$$I = I_o\left[\exp\left(\frac{\alpha_a zF\eta}{RT}\right) - \exp\left(\frac{\alpha_c zF\eta}{RT}\right)\right] \tag{20.15}$$

where I is electrode current density, I_o exchange current density, z number of electrons involved in the electrode reaction, F Faraday constant, α_c cathodic charge transfer coefficient, α_a anodic charge transfer coefficient, η activation overpotential

$$r_4 = \frac{I}{zF} \tag{20.16}$$

The reaction expressions are applied to the various species as follows.

$$r_i = \sum_{j=1}^{4} v_{ij} r_j \tag{20.17}$$

20.2.4 Electrochemistry

The theoretical or reversible voltage can be derived from the standard Gibbs free energy at any defined temperature. However, the actual cell voltage is less than the theoretical voltage due to overpotentials. Thus, the cell requires more energy than thermodynamically determined. The overpotentials are caused by the activation barriers, V_{act} and resistances to electron, ion, and bulk mass transport, through/to the electrolyte, and charge leaks. The actual cell voltage is the open circuit voltage less the activation overpotential, the concentration overpotential, and the ohmic overpotential (Ziogou et al. 2013; (Gebregergis et al. 2009).

$$V = n(E_{oc} - V_{act} - V_{ohm} - V_{conc}) \tag{20.18}$$

Where V is the actual voltage of the stack, V_{oc} is the open circuit voltage, V_{act} is the activation overpotential, V_{ohm} is the ohmic overpotential, and V_{conc} is the concentration overpotential

The open circuit voltage with pressure change considerations can be expressed as follows.

$$E_{oc} = E^o + \frac{RT}{zF} \ln \frac{p_{H_2} p_{O_2}^{0.5}}{p_{H_2O}} \tag{20.19}$$

where E_o is a temperature-dependent theoretical/reversible voltage derived from Gibb's free energy.

Electrochemical reactions must overcome energy barriers for the reaction to proceed.

The activation loss is derived from the ButlerVolmer equation and given as 20.20:

$$V_{act} = \frac{RT}{F} \sinh^{-1} \frac{I}{2I_o} \tag{20.20}$$

Io is the exchange current, and it varies with temperature as follows (Gebregergis et al. 2009):

$$I_o = A\exp(-E/RT) \tag{20.21}$$

Ohmic overpotentials occur because of the resistance to the flow of ions in the electrolyte and the resistance to the flow of electrons through the electrode materials. The characteristic resistance of a fuel cell as a function of cell temperature is given by (Gebregergis et al. 2009):

$$V_{ohm} = \gamma \exp\left[\beta\left(\frac{1}{T_o} - \frac{1}{T}\right)\right] I \tag{20.22}$$

where T_0 is the fuel cell reference temperature; $T_0 = 973$ K, $\gamma = 0.2$, and $\beta = 2870$ K are the constant coefficients of the fuel cell. Concentration overpotentials occur because of the resistance to the transport of material to the triple-phase boundary for adsorption and subsequent ionization. Concentration over potential V_{conc} is higher when the electrochemical reaction is faster than the flow of material to the triple phase boundary. The limiting current density I_L is the current density at which the flow to the triple phase boundary is not sufficient to meet the power demand. V_{conc} related to the current as follows:

$$V_{conc} = \frac{RT}{zF} \ln\left(1 - \frac{I}{I_L}\right) \tag{20.23}$$

20.3 Simulation

The Model was developed and simulated in gPROMS® (Process Systems Enterprise n.d.). The dynamic model is simulated to investigate the interactions between the manipulated and controlled variables, and for the characterization of the effect of disturbances. The results from the simulations show the transient behavior of the materials and temperature changes. The inlet stream is a mixture of methane and steam at 950 K and a composition ratio of 3:2. The reforming reactions, shift reaction, and the electrochemical reactions occur simultaneously. The model is initialized with some hydrogen in the anode. Steam produced from the electrochemical reaction is used by the reforming and the shift reaction. A polarization curve was derived from the model at constant temperature (950 K) as shown in Figure 20.4. The figure shows how the linear combination of the various overpotentials contributes to the cells' actual voltage. The current produced (flow of produced electrons) in the fuel cell depends mostly on the reaction of hydrogen and oxygen (according to equations (20.2) and (20.16)), and the current density is the ratio of the current produced to the total active area of the stack. The active area is a design parameter that is fixed at the design stage. Thus, the polarization curve is used to describe how the voltage changes with the current density. It also shows how the power changes with current density. The limit current density used for the hypothetical model is 2 A/cm^2.

DIR SOFC is usually subjected to varying power demand; thus it should be robust enough to meet varying power demand. To describe this variability, a power demand profile is used to simulate the dynamic model. The power demand profile is developed by varying current. Figure 20.5 shows the response of the species in the anode and cathode to a change in the current demand, which also corresponds to a power change at constant inlet flow of a mixture of natural gas and steam. Hydrogen and steam have more abrupt response to the variability as

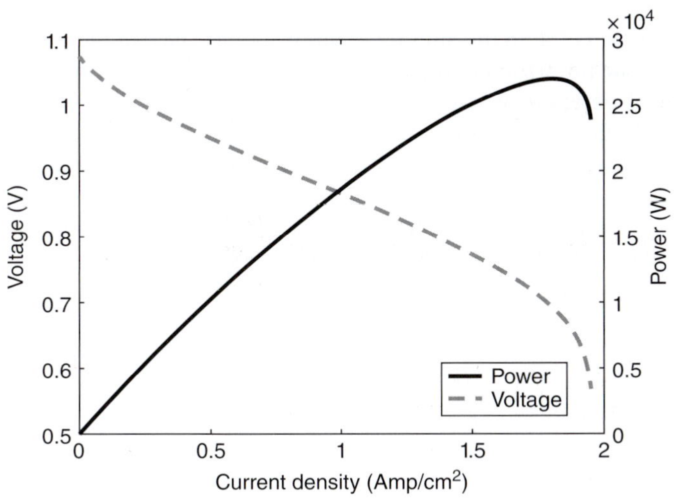

Figure 20.4 Polarization curve.

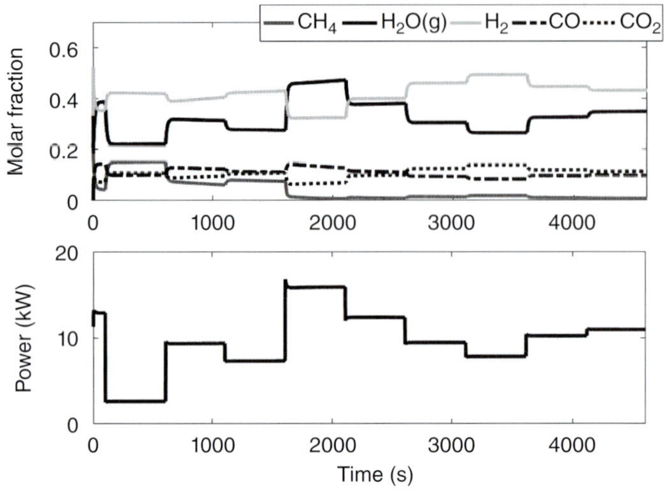

Figure 20.5 Transient change in mole fraction with change in power demand.

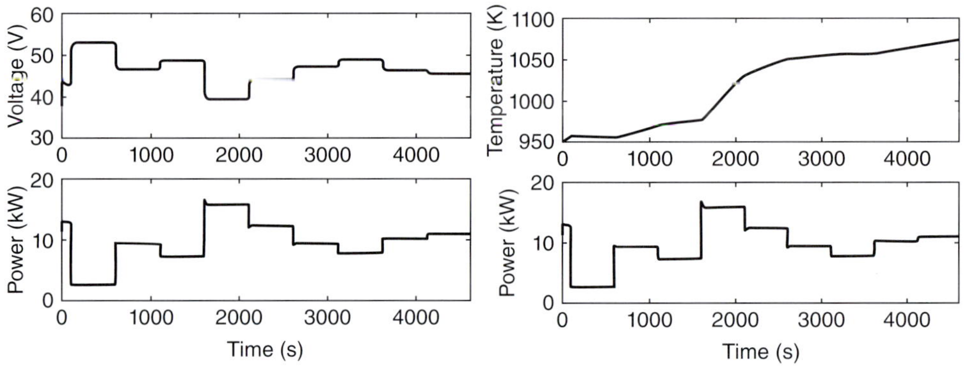

Figure 20.6 Transient change in Voltage and Temperature with change in power demand.

would be expected. An increase in the power demand should result in an increased depletion of the hydrogen and an accumulation of steam, which in turn favors the reforming reaction. Generally, the higher the electrochemical reactant pressure, the higher the open circuit voltage. This interaction could be investigated further since the operating conditions favorable for the reforming reaction will increase the amount of hydrogen generated irrespective of the current demand or the rate of the electrochemical reaction. Moreover, if the reforming reaction is more favorable than the electrochemical reaction (probably due to the current demand), the temperature of the system might start decreasing and will tend to slow down the reforming reaction.

One of the major challenges of fuel cells is the voltage fluctuations that accompany a change in current or power demand. This phenomenon occurs as a result of multiple time constants in the various fuel cell interactions. Figure 20.6 shows how voltage changes with current or power change. An ideal process operation would be to have the voltage remain relatively constant with slight perturbation of the current. A controller can be used to regulate the voltage to a set-point, for slight change in the power demand. A controller, however, cannot be used to effectively circumvent the mass transport limitation of a fuel. The fuel is usually integrated with a battery or a capacitor to make up for its limitations. The thermal coupling of the reforming, the shift and the electrochemical reaction in a direct internal reforming, is a development that needs to be studied critically. It has the potential to greatly improve the energy efficiency and cost of an SOFC. This combination of exothermic and endothermic reactions at varying reaction rates (or time constants) has a significant impact on the temperature of the system. Generally, fuel cell control designs are intended to regulate disturbances caused by changing current demand, track load demand, comply with safety or operational constraints, and improve efficiency. The control developed here will attempt to maintain a voltage set-point while constraining the temperature of the system within a bound.

20.4 Multiparametric Model Predictive Control (mpMPC)

An advanced process control framework is used to design a controller for the DIR SOFC. The PARametric Optimization and Control (PAROC) framework is a framework that enables the representation and solution of demanding model-based operational optimization and control problems following an integrated procedure featuring high-fidelity modeling, approximation techniques, and optimization-based strategies, including multi-parametric programming (Pistikopoulos et al. 2015). The PAROC framework can adapt to several classes of problems through its prototype software platform. PAROC has been used to design controllers for PEMFC (Ziogou et al. 2013); however, the dynamic of an IR SOFC is different and more complex from a control standpoint.

20.4.1 PAROC Framework

The first step of the PAROC framework is to develop a high-fidelity model that correctly describes the various aspects of the system. The dynamic model described above can easily be adapted for controller design. However, the model has some complex nonlinear expressions that make it difficult to be directly adapted for control studies. Consequently, it is necessary to simplify the representation of the model without compromising its accuracy. PSEs gPROMS Model Builder software allows direct communication with MATLAB through gO:MATLAB for further analysis. The simplification of the model is carried out in MATLAB.

20.4.1.1 Linear Model Approximation

A series of simulations of the dynamic model for different input is used to construct a meaningful linear state-space model of the process using statistical methods. The inputs are the inlet flowrate for the anode and the cathode while we measure voltage current and temperature. The current is treated as a disturbance. One of the most widely applied tools within this area is the N4SID algorithm within the System Identification Toolbox from MATLAB. The sampling time for the input-output data is 200ms. A discrete linear state space model is obtained from the input-output data. The mathematical expression of the SS model is of the form of eq. (20.24):

$$\begin{aligned} x_{k+1} &= Ax_k + B_1 u_k + B_2 d_k \\ y_k &= Cx_k + Du_k \\ T_s &= 0.2s \end{aligned} \tag{20.24}$$

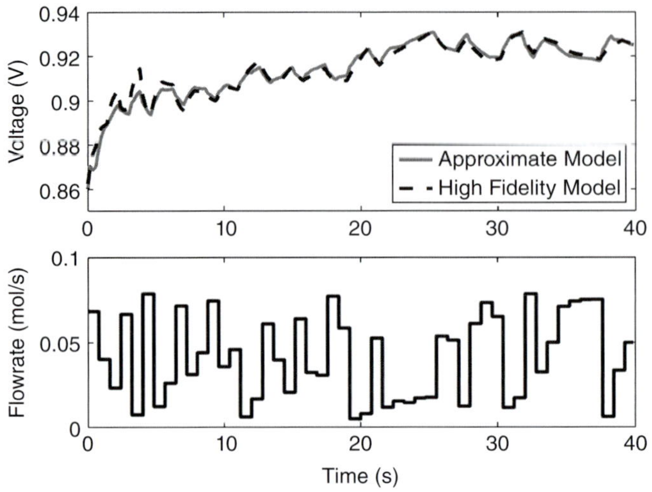

Figure 20.7 Comparison of linear SS model and the original model.

The SS model generated have 4 identified states (x_k), 2 outputs (y_k), 2 inputs (u_k) and 1 disturbance (d_k). The state space model is given in eq. (20.25).

$$
\begin{aligned}
x_{k+1} &= \begin{bmatrix} 1 & 2.75e^{-5} & 2.9e^{-5} & 2.14e^{-6} \\ -0.064 & 0.933 & 0.065 & -0.023 \\ 0.0696 & 0.0526 & 0.887 & -0.223 \\ 9.23e^{-3} & -0.239 & -0.0397 & -0.0472 \end{bmatrix} x_k + \begin{bmatrix} -5e^{-6} & 1.05e^{-4} \\ -1.532 & 0.85 \\ -1.523 & -5.093 \\ -12.12 & -18.07 \end{bmatrix} u_k + \begin{bmatrix} 5e^{-6} \\ 1.5e^{-4} \\ 9.88e^{-5} \\ 2.92e^{-4} \end{bmatrix} d_k \\
y_k &= \begin{bmatrix} -1875 & -0.01 & -0.0336 & 5.9e^{-4} \\ -1.756 & 0.049 & 2.2e^{-3} & 1.41e^{-4} \end{bmatrix} x_k + \begin{bmatrix} 0.236 & 0.114 \\ 0.063 & 2.5e^{-3} \end{bmatrix} u_k \\
T_s &= 0.2s
\end{aligned} \tag{20.25}
$$

The approximate model was compared to the original nonlinear model in Figure 20.7. For brevity, only one input (anode inlet flowrate) and output (voltage) are shown here, however the inlet flow of the cathode was varied too, and the temperature fitting curve was obtained. The temperature has an R^2 value of 99% while that of the voltage (Figure 20.7) is 79%.

There are eight states in the original dynamic model, however; for purposes of simplification, the linear discrete state-space model consists of four states. Not only are the states reduced, but their physical meaning is lost during the system identification process. To maintain the accuracy of the original complex dynamic model, a state observer is employed to estimate the state of the system using the linear model. The Kalman filter is a type of observer that uses a least square estimation approach to characterize the current state of a dynamic system based on the influence of past measurement. Its prime feature is the iterative propagation of the noise measurement risk. For further reading on the Kalman Filter used in this work see Yu et al. (2005).

20.4.1.2 mpMPC Controller Design

The generic expression of the model predictive control is given in eq. (20.26).

$$
\begin{aligned}
\min_{x,y,u} J = x_N^T P x_N &+ \sum_{k=1}^{N-1} (x_k^T Q_k x_k + (y_k - y_k^R)^T QR_k (y_k - y_k^R)) \\
&+ \sum_{k=1}^{N-1} ((u_k - u_k^R)^T R_k (u_k - u_k^R) + \Delta u_k^T R1_k \Delta u_k)
\end{aligned}
$$

$$\mathbf{s.t.}x_{k+1} = Ax_k + Bu_k + Cd_k$$

$$y_k = Dx_k + Eu_k + e$$

$$u_{min} \leq u_k \leq u_{max}$$

$$\Delta u_{min} \leq \Delta u_k \leq \Delta u_{max}$$

$$x_{min} \leq x_k \leq x_{max}$$

$$y_{min} \leq y_k \leq y_{max} \tag{20.26}$$

where x_k are the state variables; u_k and $u^R{}_k$ are the control variables and their respective set-points; Δu_k denotes the difference between two consecutive control actions; y_k and $y^R{}_k$ are the outputs and their respective set-points; d_k are the measured disturbances; Q_k, R_k, $R1_k$, and QR_k are the corresponding weights in the quadratic objective function; P is the stabilizing term derived from the Riccatti equation for discrete systems; N and M are the output horizon and control horizon, respectively; k is the time step; A, B, C, D, and E are the matrices of the discrete linear state-space model; and e denotes the mismatch between the actual system output and the predicted output at initial time.

This quadratic problem in eq. (20.26) can be reformulated into a multiparametric quadratic programming problem (mp-QP) (Bemporad et al. 2002). The uncertain parameters are denoted by θ, and they include the initial states x_0, the set-points $y^R{}_k$, previous control actions u_{-1} for evaluating Δu_k, and the disturbances d_k. The mp-QP is given as follows:

$$\min_{u,\theta} J = u^T Q u + u^T F^T \theta$$

$$\mathbf{s.t.A}u \leq \mathbf{b} + \mathbf{B}\theta$$

$$u \in U = \{u \in \mathbb{R}^n | u_{min} \leq u \leq u_{max}, \forall n\}$$

$$\theta \in \Theta = \{\theta \in \mathbb{R}^q | \theta_{min} \leq \theta \leq \theta_{max}, \forall q\} \tag{20.27}$$

where u is a vector that consists of M control actions, n and q are the size of the control and parameter vector respectively, and consequently the size of the parametric solution and parametric space of the problem.

The multi-parametric model predictive controller problem is formulated and solved using the POP toolbox in MATLAB. The objective function of the mpMPC is to meet the electrical power demand at a constant voltage amid the current charge and to maintain the temperature below 1500 K. The optimal control action is generated as a map of solutions and as an affine function of the aforementioned parameters. Thus, at every time step a set of parameters, realized from measurements and estimations, is used as inputs to evaluate the affine function generated from the mp-QP solution. Table 20.2 shows the mpMPC design specification and bounds on various variables. The bounds on the output variable u_k and y_k have been set according to the complex dynamic model simulation results. However, the bounds on x_k are exaggerated since the states of the linear state space model do not necessarily represent the state of the original system. The choice of bounds on x_k also reflects the usage of an unconstrained state approximation technique: i.e., the Kalman filter. Alternatively, constrained state estimation techniques such as the (multi-parametric) Moving Horizon Estimator ((mp-)MHE) can be utilized (Darby and Nikolaou 2007; Voelker et al. 2013; Nascu et al. 2014; Pistikopoulos et al. 2015). The tuning of the controller is presented in section 20.5.

Table 20.2 Weight tuning for the mpMPC of the DIR SOFC unit.

mpMPC design parameters	Value
N	5
M	2
QR_k, $\forall k \in \{1, \ldots, N\}$	100
$R1_k$, $\forall k \in \{1, \ldots, M\}$	$\begin{bmatrix} 10^{-4} & 0 \\ 0 & 10^{-4} \end{bmatrix}$
x_{min}	$-500[1\ 1\ 1\ 1]$
x_{max}	$500[1\ 1\ 1\ 1]$
u_{min}	[0.002 0.004]
u_{max}	[0.1 0.1]
y_{max}	[2000 1.2]
y_{min}	[900 0.1]
Δu_{min}	−[0.05 0.05]
Δu_{max}	[0.05 0.05]

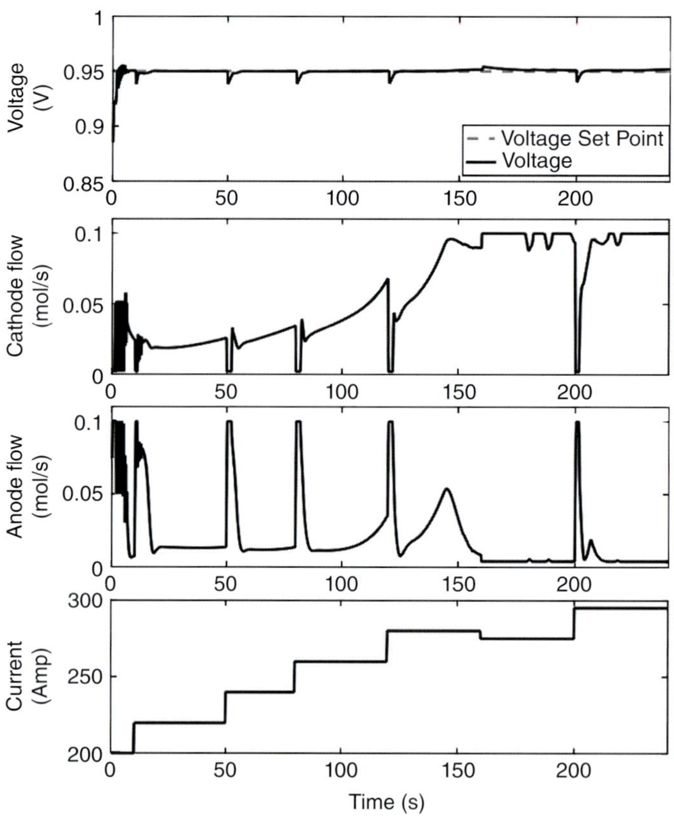

Figure 20.8 Closed-loop validation results of the mpMPC controller showing the voltage set-point tracking (top) the anode (natural gas and steam) and cathode (oxygen) flowrate (middle) and the current profile (bottom).

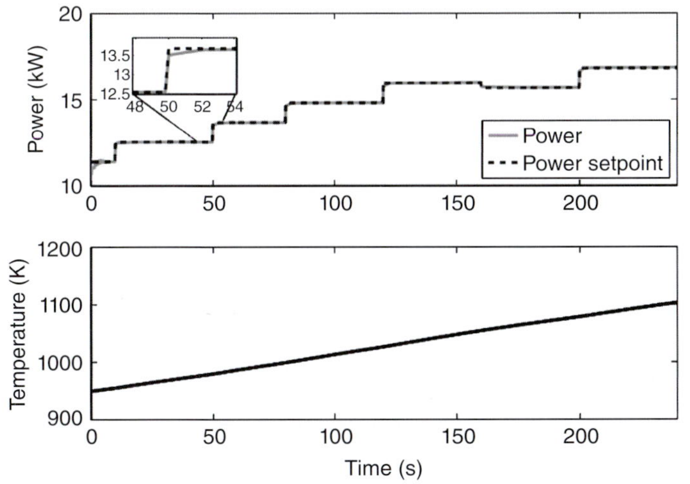

Figure 20.9 Closed-loop validation results showing the power set-point tracking and the temperature.

20.5 Closed-Loop Validation and Results

The controller is validated by closing the loop on the controller design. The controller is used on the original complex dynamic model to check if the controller can actually control the system from which it was designed. The validation process involves calculating the control action from the realized parameters and the mp-QP generated affine function. The control action calculated is used as input to the original dynamic model. The output Voltage is measured to and compared with the set-point value as shown in Figure 20.8. This is enabled through gO:MATLAB, an application that links the gPROMS with MATLAB. Figure 20.8 also shows how the flow of NG in the anode and flow of air in the anode changes with changes in current. The power demand is varied by changing the current. Figure 20.9 shows how the controller-enabled DIR SOFC tracks the power demand (product of the current and voltage). The controller follows the set-point effectively. The temperature is maintained below the bound as also shown in Figure 20.9.

20.6 Conclusion

We demonstrated the use of a natural-gas-based DIR SOFC for distributed electricity generation. In addition, we developed a complex dynamic model of a hypothetical DIR SOFC to describe the various interactions between the reforming reaction, the transport phenomenon, and the electrochemical reaction.

Using the PAROC framework, we designed a multi-parametric model Predictive controller for regulating the voltage of the power generated by the DIR SOFC. The controller was validated with the complex dynamic model. In future work, we intend to explore how the DIR SOFC compares with other intensified distributed electricity generation units. An economic analysis of the DIR SOFC and other auxiliary unit that make up the balance of plant will be essential for benchmarking it with other technologies.

References

Aguiar P, Chadwick D, Kershenbaum L., 2002. Modelling of an indirect internal reforming solid oxide fuel cell. Chemical Engineering Science, 57(10): 1665–1677.

Ahmed K, Foger K. 2000. Kinetics of internal steam reforming of methane on Ni/YSZ-based anodes for solid oxide fuel cells. Catalysis Today, 63(2): 479–487.

Al-Fattah SM, Startzman RA. Forecasting world natural gas supply. Society of Petroleum Engineers. SPE/CERI Gas Technology Symposium, 3–5 April, Calgary, Alberta, Canada.

Battelle 2014. Manufacturing cost analysis of 1 KW and 5 KW Solid oxide fuel cell (SOFC) for auxillary power applications. Available from: https://www.energy.gov/eere/fuelcells/downloads/manufacturing-cost-analysis-1-kw-and-5-kw-solid-oxide-fuel-cell-sofc.

Bemporad A et al. 2002. The explicit linear quadratic regulator for constrained systems. Automatica, 38(1): 3–20.

Bhattacharyya D, Rengaswamy R. 2009. A review of solid oxide fuel cell (SOFC) dynamic models. Ind. Eng Chem Res, 48(13): 6068–6086.

Burnham A et al. 2011. Life-cycle greenhouse gas emissions of shale gas, natural gas, coal, and petroleum. Environ Sci Technol, 46(2): 619–627.

Darby ML, Nikolaou M. 2007. A parametric programming approach to moving-horizon state estimation. Automatica, 43(5): 885–891.

Diangelakis NA, Pistikopoulos EN. 2017. A multi-scale energy systems engineering approach to residential combined heat and power systems. Comput Chem Engg, 102, p.128–138.

Dicks AL. 1998. Advances in catalysts for internal reforming in high temperature fuel cells. J Power Sources, 71(1): 111–122.

Ferguson JR, Fiard JM. Herbin R., 1996. Three-dimensional numerical simulation for various geometries of solid oxide fuel cells. J Power Sources, 58(2): 109–122.

Fernandes FAN, Soares AB. 2006. Methane steam reforming modeling in a palladium membrane reactor. Fuel, 85 (4): 569–573.

Gebregergis A, et al., 2009. Solid oxide fuel cell modeling. IEEE Transactions Indus Electron, 56(1): 139–148.

Hawkes A, Leach M. 2005. Solid oxide fuel cell systems for residential micro-combined heat and power in the UK: Key economic drivers. J Power Sources, 149, p.72–83.

Hore-Lacy I. 2006. Nuclear energy in the 21st century. Elsevier. eBook ISBN: 9780080497532.

Jin W et al. 2004. Low cost fuel cell converter system for residential power generation. IEEE Trans Power Electron, 19 (5): 1315–1322.

Kakaç S, Pramuanjaroenkij A, Zhou XY. 2007. A review of numerical modeling of solid oxide fuel cells. Int. J Hydrogen Energ, 32(7): 761–786. Available from: http://dx.doi.org/10.1016/j.ijhydene.2016.02.089.

Klein JM, et al. 2007. Modeling of a SOFC fuelled by methane: from direct internal reforming to gradual internal reforming. Chem Engg Sci, 62 (6): 1636–1649.

Lasseter RH. 2007. Microgrids and distributed generation. J Energ Engg, 133(3): 144–149.

Mohr SH, Evans GM. 2011. Long term forecasting of natural gas production. Energ Policy, 39 (9): 5550–5560.

Nascu I et al., 2014. Simultaneous multi-parametric model predictive control and state estimation with application to distillation column and intravenous anaesthesia. In: 24th European Symposium on Computer Aided Process Engineering. Computer Aided Chemical Engineering, Budapest, Hungary: Elsevier, pp. 541–546.

Pistikopoulos EN, et al., 2015. PAROC An integrated framework and software platform for the optimisation and advanced model-based control of process systems. Chem Engg Sci, 136: pp.115–138.

Poppinger M, and Landes H. 2001. Aspects of the internal reforming of methane in solid oxide fuel cells. Ionics, 7(1): 7–15.

Process Systems Enterprise. gPROMS. Available from: www.psenterprise.com/gproms.

Recknagle KP et al, 2003. Three-dimensional thermo-fluid electrochemical modeling of planar SOFC stacks. J Power Sources, 113 (1): 109–114.

Riensche E, Stimming U, Unverzagt G. 1998. Optimization of a 200 kW SOFC cogeneration power plant. J Power Sources, 73(2): 251–256.

U.S. Energy Information Administration and U.S. Department of Energy. 2011. Ann Energ Outlook 2011 with Projections to 2035, Available from: http://www.eia.gov/forecasts/aeo/pdf/0383(2011).pdf.

U.S. EIA, 2016. Carbon Dioxide Emissions Coefficients, U.S. Energy Information Adminstration. Available from: https://www.eia.gov/environment/emissions/co2_vol_mass.cfm.

U.S. NETL. 2013. Modern shale gas development in the united states : an update. U.S. Department of Energy, 2013 (September): 1–79.

Voelker A, Kouramas K, Pistikopoulos EN. 2013. Moving horizon estimation: error dynamics and bounding error sets for robust control. Automatica, 49(4): 943–948.

Xu J, Froment GF. 1989. Methane steam reforming, methanation and water-gas shift: I. Intrinsic kinetics. AI Ch E Journal, 35(1): 88–96.

Yakabe H, et al., 2001. 3-D model calculation for planar SOFC. J Power Sources, 102(1): 144–154.

Zhang X, et al., 2010. A review of integration strategies for solid oxide fuel cells. J Power Sources, 195(3): 685–702.

Ziogou C, et al., 2013. Empowering the performance of advanced NMPC by multiparametric programming—an application to a pem fuel cell system. Indus Engg Chem Res, 52(13): 4863–4873.

21

Design of Synthetic Jet Fuel Using Multivariate Statistical Methods

Rajib Mukherjee[1], Noof Abdalla[2], Nasr Mohamed[3], Marwan El Wash[2], Nimir O. Elbashir[1,2,3], and Mahmoud M. El-Halwagi[1,4]*

[1] *TEES Gas and Fuels Research Center, Texas A&M Engineering Experiment Station, USA*
[2] *Chemical Engineering Program, Texas A&M University at Qatar, Qatar*
[3] *Petroleum Engineering Program, Texas A&M University at Qatar, Qatar*
[4] *Department of Chemical Engineering, Texas A&M University, USA*

CHAPTER MENU

21.1 Introduction

Liquid fuels are among the principal forms of energy with forecasts of increasing global demand (Suganthi and Samuel 2012). In addition to the escalating use of gasoline and diesel, there is a noticeable growth in the use of aviation fuels that is unsustainable using only conventional crude oil-derived Jet A-1 fuel(5). With the recent discoveries of abundant shale-gas reserves, gas-to-liquid (GTL) technology (Bao et al. 2010; Martínez et al. 2013; Gabriel et al. 2014a,b; Challiwala et al. 2017) is becoming a promising option to provide a meaningful market share of transportation fuels. These GTL processes yield several fuels including gasoline, diesel, and jet fuels that are complex mixtures of different hydrocarbon components. Such complex mixtures make modeling of their exact physicochemical properties all the more difficult, mainly due to the multivariate nature of the problem. The goal is to find a synthetic fuel that is comprised of only a handful of hydrocarbon components but has characteristics comparable to actual fuel. Even for a small number of components, the real challenge is how to determine the optimal mixture composition that can match a specific fuel of interest (Viana et al. 2012).

Although there are numerous ways to classify jet fuel, such as by hydrogen content, aromaticity, and cetane index (ASTM 2010) a simpler way to match real fuel is to meet industry certification as defined in the American Society for Testing and Materials (ASTM) D1655 specification. A reliable source of alternate jet fuel is the use synthetic paraffinic kerosene (SPK) that can be obtained from syngas using the Fischer-Tropsch synthesis process. SPKs prepared from syngas that is generated from natural gas are known as gas-to-liquid (GTL) fuels. SPKs are often blended with kerosene obtained from petroleum (Liu et al. 2013). ASTM D7566 specification was introduced in 2009 specifically to allow two classes of SPKs to be blended

Natural Gas Processing from Midstream to Downstream, First Edition.
Edited by Nimir O. Elbashir, Mahmoud M. El-Halwagi, Ioannis G. Economou, and Kenneth R. Hall.

with petroleum-derived kerosene. The present trend allows 50% by volume of SPK. The final blend should meet the ASTM D1655 specification. The blend of SPK with petroleum-derived kerosene is required since SPKs have low density and lack aromatics that lead to lower lubricity.

Jet fuel obtained from petroleum is composed of a large number of different hydrocarbon groups with the carbon number ranging between C8 to C16. The proportions of different compounds are not constant and vary from one crude to another, and the refining process (Solash et al. 1978). SPKs, on the other hand, are a mixture of only normal paraffin and iso-paraffin. SPKs obtained from Fischer-Tropsch synthesis may contain cyclo-paraffin with a maximum of 15% as specified by ASTM D7566. In this work, we are conducting a comprehensive study for the development of synthetic fuel composed of different hydrocarbons that are competitive with aviation fuel. Thus, we have not imposed any limitations on the percentage of different building blocks and have made a thorough search over the entire range of different compounds.

In order to systematically identify a right combination of the hydrocarbon components that creates aviation-grade fuel, we have to find the effect of different hydrocarbon groups on properties of the fuel mixture. This will enable us to choose the right percentage of different groups that can meet the properties of the jet fuel. For this purpose, experiments were conducted at the Fuel Characterization Laboratory of Texas A&M Laboratory, Qatar, where blends of hydrocarbons are prepared and their physical properties are tested. The blends were made using four base hydrocarbon components including n-paraffin, iso-paraffins, cyclo-paraffins, and aromatics. These base compounds play a major factor on combustion behavior of jet fuel (Al-Nuami et al. 2016). Four different properties of the blend are monitored, including density, freezing point, flash point, and heat of combustion for the comparison of synthetic fuel with aviation fuel. Other literature on composition property study also used similar compounds for their analysis. For instance, Salem et al. (2014) characterized the role of hydrocarbon building blocks on physical and chemical properties of synthetic jet fuel. For our analysis, we have used limited experimental data on synthetic blends. In order to find the optimal blend composition, experiments have been conducted to obtain blend properties across a broad range of different hydrocarbon components. A plot of the composition property relation in a ternary diagram identifies a region where all properties meet the standard of aviation grade fuel. Data generated by Artificial Neural Network (ANN) were used for complete mapping of the ternary diagram. This process of blend selection and hydrocarbon property correlation has two main limitations. The role of hydrocarbon building blocks on blend properties cannot be uniquely defined. This is because we cannot uniquely vary the composition of single building block by keeping all other compositions constant. An interactive effect of all the hydrocarbon components can play a major role in blend property determination, and a multivariate analysis is essential for this purpose. Selection of the optimal composition from a ternary diagram can also be erroneous, as it requires a visual interpretation. Also, the amount of model-generated data may not be sufficient for complete mapping of the ternary region.

In the present work, we have developed a methodology for the selection and subsequent characterization of synthetic fuel using multivariate statistical methods. Synthetic fuel is characterized based on its composition. In order to fulfill the requirement, we need to find a composition-property correlation. The presence of various hydrocarbon components in the blend leads to a multivariate regression problem where each hydrocarbon component can act as a variable upon which the property of the composition depends. An example of the application of multivariate statistics for the optimal selection of materials and blend composition for polymer blends can be found in Muteki et al. (2006). Partial least squares (PLS) regression is a multivariate regression method that will allow us to correlate the variability of different hydrocarbon components that reflects the most in the composition property (Mukherjee et al. 2013). In this chapter, each property is correlated with the multivariate system. This method of multivariate correlation instead of pairwise correlation helped us to identify the true nature of the properties.

In order to find the most suitable composition, the four base hydrocarbon components that are used to create the blend are varied over the entire range; i.e., each hydrocarbon component can comprise 0 to 100%. The variation of every single hydrocarbon component can affect the resultant properties differently. Even in the range of different hydrocarbon components where the blend shows compliance with ASTM D1655 specifications, different hydrocarbon components may vary differently: i.e., one hydrocarbon component may vary a lot and nevertheless the blend can abide by the specification while a slight variation of another hydrocarbon component may affect the blend property significantly (Elmalik et al. 2014). In order to find the sensitivity of blend properties on different base hydrocarbon components, principal component analysis is performed. A model is also developed to predict the suitability of any composition as aviation fuel based on its principal component score. The score-plot-based analysis can overcome the limitation of the ternary plot, which can only be used to find the results from four hydrocarbon components at maximum.

Section 21.2 (Methodology) describes different statistical methods including partial least squares regression methods and principal component score generation for blend characterization in order to find optimal composition. Section 21.3 shows the results from the application of these methods. It shows the optimal region of blend composition in the ternary plot. The section also gives results from the multivariate correlation analysis, how individual properties are correlated to the composition, characterization of blends, and explains the importance of the proposed methodology and its uniqueness in formulating synthetic aviation fuel. The last section gives the conclusions from the present work.

21.2 Methodology

In order to identify the optimal blend ratio that can meet aviation quality standards, we need to conduct experiments thoroughly over the entire range of different compounds. Estimated properties of the compounds evaluated over an extensive range have enabled us to categorically select optimal compositions. In the process of identifying the right compositions, a prime importance is to know the relation between blend compositions with each and every property of the blend. This leads to a multivariate problem (due to multiple compounds) and is solved using partial least squares (PLS) regression and the support-vector-machine-based regression method. Another important aspect of the multivariate analysis is to identify the contribution of different compounds in defining the properties of the blend. Finding the effect of every compound as well as identifying the quality of a blend without actually conducting the experiment is a challenge that has been successfully answered using principal component analysis (PCA).

The steps involved in the analysis of data for optimal composition selection are given below, followed by a brief description of those methods:

1) The principal component factor loadings for finding individual hydrocarbon component contributions in determining the properties of the blend.
2) Finding the correlation of the blend with their properties using multivariate regression methods.
3) Use of principal component score of qualified blends for blend quality analysis to conduct aviation-grade blend selection.
4) Novel blend identification based on principal component score.
5) Selected blend characterization using already developed multivariate regression models.

21.2.1 Characterization with Principal Component Analysis

Since the hydrocarbons, and their compositions to obtain ASTM D1655 specification, are always chosen prospectively from an understanding of the system under study, it is never certain that redundant, insignificant, or derivative compounds have not been included in the analysis. It is necessary to not include those hydrocarbons in blend preparation if they are identified as redundant or insignificant. Weeding out unnecessary compounds yields those that are essential. That is, we want to reduce the dimensionality of the system. In the matrix of the original composition dataset, each compound is considered a component of a vector. We tacitly assumed when constructing the blend that the chosen compounds are ideally independent of each other, that is, the vector components are orthogonal to each other, in the n-dimensional composition space. Statistically, if some of the hydrocarbon data are in fact correlated, then the compounds are insufficiently orthogonal. Correlated hydrocarbons imply that they are not separately independent contributors of unique features of the blend and there are overlaps in their contribution.

Principal component analysis (PCA) is a statistical method to reduce the dimensionality of systems, giving rise to principal components, which are also known as eigenvectors, as discussed below (Jolliffe 1986). These eigenvectors are by definition orthogonal. The contribution of the hydrocarbons to the principal components will allow us to identify linearly dependent (or independent) compounds. If found redundant or insignificant, the original compound set can be reduced to eliminate derived compounds for optimal composition selection.

The analysis is performed with experimental data on blend composition and its properties. For m different blends, the data can be arranged into the blend composition matrix (**X**) $(m \times n)$ where each observation or blend composition (m) containing the volume percent of different compounds (n) and the blend characteristics matrix (**Y**) $(m \times p)$ containing the (p) blend characteristics as obtained from each different blend compositions (m). The process of PCA creates a transformation matrix of the original data matrix **X**. This transformation matrix is called the correlation matrix **R** $(n \times n)$, such that the following eigenvalue problem can represent the entire dataset:

$$\mathbf{RQ} = \Delta\mathbf{Q} \tag{21.1}$$

where **Q** is an $n \times n$ matrix containing the unit eigenvectors **q**'s, **Δ** is the $n \times 1$ vector of the eigenvalues of the matrix **R**. Thus, PCA designs n-dimensional unit vectors (**q**'s) onto which the n-dimensional input data vectors are projected. The n eigenvalues (λ's, elements of vector **Δ**) are arranged in descending order. The sum of the eigenvalues accounts for the total variability of the data. The percent contribution of an eigenvalue (λ_j) towards total variability can be expressed as:

$$\left(\lambda_j / \sum_{j=1}^{n} \lambda_j \right) \times 100 \tag{21.2}$$

The matrix of eigenvectors, **Q**, gives the principal components. PCA allows selecting a few most important eigenvectors out of the possible maximum of n that need to be used to account for the overwhelming portion of the data variability. Each principal component represents all the input variables (in our case, the hydrocarbon compounds are the variables). Choosing a few of these principal components based on the eigenvalues associated with them is an exercise in looking at the data from several different angles to capture most of the variability of the data in order to reduce the dimensionality of the data.

Mapping square roots of eigenvalues λ_j on to the eigenvector matrix **Q**, we obtain the loading matrix **L**, given in equation (21.3).

$$\sqrt{\Delta}\,\mathbf{Q} = \mathbf{L} \tag{21.3}$$

Each element of the loading matrix **L**, belonging to a principal component, denotes the loading of a variable in a principal component (Sikdar et al. 2017). In a reduced subspace of two principal components, we can obtain a plot that shows loadings of each hydrocarbon component of the blend composition. Hydrocarbon components that overlap each other or form clusters on such a plot are correlated. This process can be repeated with a greater number of principal components until the chosen principal components cover more than 95% of the data variability. For example, in a plot of the loadings of the hydrocarbon components in the first two principal components, if hydrocarbon component A and B overlap each other, then either A or B is redundant in defining the characteristics of the blend. In that case either A or B is sufficient for blend characteristics analysis and can be included in the blend formation, and the other can be discarded. In the present work, we have used loadings to understand the correlation among the variables.

In the present work, principal component (PC) score is used for composition characteristics. The PC score is obtained from the product of the data matrix **X** with the loading matrix **L**; we obtain the score matrix **T**, which is given in equation (21.4).

$$\mathbf{XL} = \mathbf{T} \tag{21.4}$$

The principal component score of each observation (in this case blends) belongs to a principal component that denotes the score of that blend in the direction of that principal component. In the reduced subspace of two principal components, we can obtain a score of each blend. Blends that form clusters in the space of principal components show similar properties, while those that are separated show unique physicochemical features.

21.2.2 Multivariate Regression Model for Blend Property Correlation

A multivariate regression can be achieved either using a linear regression model or a nonlinear one (Gamerman and Moreira 2004). While the PCA or partial least squares (PLS) regression model is useful for linear regression, machine learning with the support-vector-based regression model can be used for nonlinear regression (Cinar et al. 2007; Konishi 1995). The PCA-based principal components regression (PCR) uses the principal component score (**T**) of the blend composition matrix (**X**) and correlates to the blend characteristics (**Y**). PCR uses unsupervised principal components for regression. For the present problem where the input blend composition is highly correlated with the output properties of the mixture, a more accurate multivariate regression can be achieved using the Partial Least Squares model that develops a biased regression model between **X** and **Y**. If the correlation between some of the composite properties is highly nonlinear, SVM-based nonlinear regression is used for our analysis.

21.2.2.1 PLS for Linear Regression

Through PLS, we select latent variables such that the variation in **X** that is most predictive of the characteristics (**Y**) is extracted (Cinar et al. 2007). Unlike PCA, which works on the covariance matrix ($\mathbf{X}^{\mathrm{T}}\mathbf{X}$), PLS works on the covariance matrix ($\mathbf{X}^{\mathrm{T}}\mathbf{Y}$)($\mathbf{Y}^{\mathrm{T}}\mathbf{X}$). The PLS model can be built by the nonlinear iterative partial least squares (NIPALS) algorithm (Wold 1966). The model consists of an outer relation where **X** and **Y** are independent and an inner relation where **X** and **Y** are linked together. The outer relation of **X** and **Y** are represented as:

$$\begin{aligned}\mathbf{X} &= \mathbf{TP}^{T} + \mathbf{E} = \sum_{i=1}^{a} \mathbf{t}_i \mathbf{p}_i^{T} + \mathbf{E}\\ \mathbf{Y} &= \mathbf{UQ}^{T} + \mathbf{F} = \sum_{i=1}^{a} \mathbf{u}_i \mathbf{q}_i^{T} + \mathbf{F}\end{aligned} \tag{21.5}$$

where **E** and **F** are the residual matrices obtained after calculating the data matrix and their characteristics using latent variable scores $\mathbf{t}_i = \mathbf{w}_i^{T}\mathbf{x}$ obtained from **x** vectors and $\mathbf{u}_i = \mathbf{q}_i^{T}\mathbf{y}$

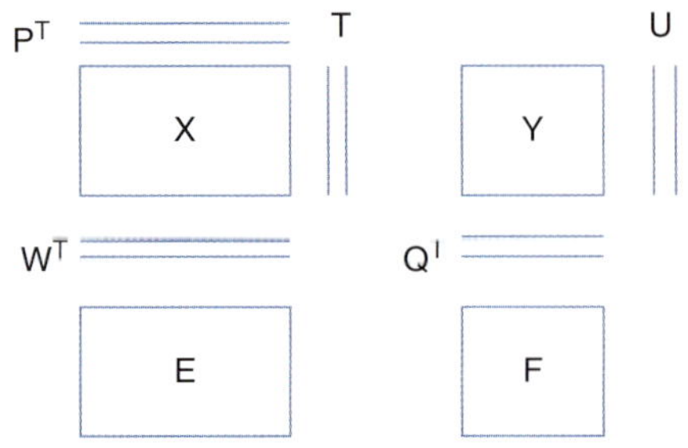

Figure 21.1 Schematic representation of matrix relationship in PLS. T and U are PLS score matrices of X and Y respectively. P is the matrix of loadings of X. W and Q are the weight matrices of X and Y respectively. E and F are the residual matrices of X and Y.

obtained from the **y** vectors. Here **x** vectors and **y** vectors represent blends and are obtained from each row of **X** and **Y** matrix respectively. The $\mathbf{w}_i$ and $\mathbf{q}_i$ are the weight vectors of **X** and **Y** respectively, and $\mathbf{p}_i$is the loading vectors of **X**. The weight vectors $\mathbf{w}_i$ and $\mathbf{q}_i$ are chosen such that they maximize the covariance between **X** and **Y**. The number of latent variables (i) is determined by the cumulative variance as obtained from the eigenvalues of the covariance matrix $(\mathbf{X}^T\mathbf{Y})(\mathbf{Y}^T\mathbf{X})$. A schematic representation of the matrices and their relationship in PLS is shown in Figure 21.1.

To obtain the first latent variable using the NIPALS, the algorithm starts with selecting y_j that is a column of **Y** that has the highest variance as $\mathbf{u}_1$. For the first latent variable, the weights $\mathbf{w}_1^T$ and the score $\mathbf{t}_1$as obtained from **X** matrix is given as:

$$\mathbf{w}_1^T = \frac{\mathbf{u}_1^T\mathbf{X}}{\|\mathbf{u}_1^T\mathbf{u}_1\|}$$
$$\mathbf{t}_1 = \frac{\mathbf{X}\mathbf{w}_1}{\|\mathbf{w}_1^T\mathbf{w}_1\|} \tag{21.6}$$

The corresponding weights $\mathbf{q}_1^T$ and the score $\mathbf{u}_1$as obtained from **Y** matrix are given as:

$$\mathbf{q}_1^T = \frac{\mathbf{t}_1^T\mathbf{Y}}{\|\mathbf{t}_1^T\mathbf{t}_1\|}$$
$$\mathbf{u}_1 = \frac{\mathbf{Y}\mathbf{q}_1}{\|\mathbf{q}_1^T\mathbf{q}_1\|} \tag{21.7}$$

If **Y** is univariate then $\mathbf{q}_1 = 1$. The convergence is checked by comparing $\mathbf{t}_1$ from two successive iterations. Following convergence, the loading vector of **X** matrix is obtained as:

$$\mathbf{p}_1^T = \frac{\mathbf{t}_1^T\mathbf{X}}{\|\mathbf{t}_1^T\mathbf{t}_1\|} \tag{21.8}$$

The loading $\mathbf{p}_1$ is used to rescale the score and weights as:

$$\mathbf{t}_{1n} = \mathbf{t}_{1o}\|\mathbf{p}_{1o}\|$$
$$\mathbf{w}_{1n} = \mathbf{w}_{1o}\|\mathbf{p}_{1o}\| \tag{21.9}$$

where the subscript o refers to old and n refers to new values. With the rescaled score, the regression coefficient b_1 for the inner relation is obtained as:

$$b_1 = \frac{\mathbf{t}_1^T\mathbf{u}_1}{\|\mathbf{t}_1^T\mathbf{t}_1\|} \tag{21.10}$$

For a given latent variable, the scores, weights, loadings, and coefficients are used to adjust **X** and **Y** matrices by excluding the variation explained by that latent variable. The residual matrices $\mathbf{E}_1$ and $\mathbf{F}_1$ obtained after excluding the variation from the first latent variable is shown in equation 21.11.

$$\begin{aligned}\mathbf{E}_1 &= \mathbf{X} - \mathbf{t}_1\mathbf{p}_1^T \\ \mathbf{F}_1 &= \mathbf{Y} - b_1\mathbf{t}_1\mathbf{q}_1^T\end{aligned} \tag{21.11}$$

The entire process is repeated for the next latent variable and weights starting from equation (21.6) by replacing the data matrices from the previous iteration with their residual matrices. For example, after the convergence of the first set of latent variables, **X** and **Y** are replaced with $\mathbf{E}_1$ and $\mathbf{F}_1$ respectively, and all other subscripts are increased by 1 and the procedure is continued.

PLS generates a linear regression model to correlate blend composition with their properties. From our previous study, we have found that some of the properties of the blend, including density and heat content, are linear in nature while other shows high nonlinearity (Elmalik et al. 2014). Property estimation with multivariate regression coefficients obtained using PLS can only be used if the correlation is linear. In the case of a linear relation, the residuals will be randomly distributed. In other cases, data generated by the regression model can be used to check nonlinearity of the system by looking into the nonrandom pattern of the residuals. This will enable us to characterize correlation of different properties of the blend with their compositions.

21.2.2.2 Support Vector Machine (SVM) for Nonlinear Regression

A support vector machine (SVM) is a multivariate statistical method that can be used for regression as well as classification. SVM was developed by Vladimir Vapnik and co-workers (Cortes and Vapnik 1995; Vapnik 2013). SVM is mainly applicable for high-dimensional data and data with nonlinear structures. In SVM, the input vectors **x** of the loading matrix **X** are mapped into a high-dimensional feature space Z through a chosen nonlinear mapping, and in this space a hyperplane is constructed that optimally separates the data. In the case of regression, we minimize a function that optimally reduces the distance between the training data and the regressed data. The regression line can be linear or nonlinear.

The optimally separating hyperplane in the feature space Z can be calculated by the inner product of support vectors and vectors of featured space z. This inner product can be represented in the Hilbert space as well as shown in equation (21.12).

$$(z_i.z) = K(\mathbf{x}, \mathbf{x}_i) \tag{21.12}$$

where z is the image of the input space vector **x** in the feature space Z, and $K(\mathbf{x}, \mathbf{x}_i)$ is known as the kernel function. The kernel function is used to construct the function $f(\mathbf{x})$ for predicting new values as:

$$f(\mathbf{x}) = \sum_{i=1}^{N}(\alpha_n - \alpha_n^*)K(\mathbf{x}, \mathbf{x}_i) - b \tag{21.13}$$

To find the coefficients α_i *and* α_i^*, we need to find maximum of the functional

$$L(\alpha, \alpha^*) = \frac{1}{2}\sum_{i=1}^{N}\sum_{j=1}^{N}(\alpha_i - \alpha_i^*)(\alpha_j - \alpha_j^*)K(\mathbf{x}_i, \mathbf{x}_j) + \varepsilon\sum_{i=1}^{N}(\alpha_i + \alpha_i^*) - \sum_{i=1}^{N}y_i(\alpha_i - \alpha_i^*) \tag{21.14}$$

subject to:

$$\sum_{i=1}^{N}(\alpha_i - \alpha_i^*) = 0 \tag{21.15}$$

$$0 \le \alpha_i \le C, \quad i = 1, \dots \dots ., N \tag{21.16}$$

$$0 \le \alpha_i^* \le C, \quad i = 1, \dots \dots ., N \tag{21.17}$$

Here α_i *and* α_i^* are the Lagrange multipliers. Observations with non-zero Lagrange multipliers constitute the support vectors.

Machine learning that constructs decision functions $f(\mathbf{x})$ as in equation (21.13) are called support vector machines (SVM). In SVMs, different types of nonlinear functions can be used as kernel functions leading to different nonlinear regression lines. The two most popular kernel functions includes a polynomial kernel function and a Gaussian kernel function. The polynomial kernel function is given as:

$$K(\mathbf{x}_i, \mathbf{x}_j) = [1 + (\mathbf{x}_i . \mathbf{x}_j)]^d \textit{ where d is in the set } \{2, 3, \ldots\} \tag{21.18}$$

The Gaussian kernel function is given as:

$$K(\mathbf{x}_i, \mathbf{x}_j) = \exp\left(-\|\mathbf{x}_i - \mathbf{x}_j\|^2\right) \tag{21.19}$$

where || || indicates the second norm of vector $\mathbf{x}_i$ and $\mathbf{x}_j$.

21.3 Results and Discussions

Until now, the process of optimal blend selection was confined to finding regions in a ternary plot (Elmalik et al. 2013). Though this method of visual inspection was very useful, it has its own limitations. The greatest drawback of the present method was the number of hydrocarbon components that can be observed. The current method was also not able to distinguish between hydrocarbon components with respect to their contribution in blend properties. In the present work, multivariate statistics are used to overcome the limitations of ternary plots. A total of 56 experimental data comprising different blends of n-paraffin, cyclo-paraffin, iso-paraffin, and aromatics have been used for our analysis. The initial 56 data points are randomly selected. These data points are collected to create the ternary diagram.

21.3.1 Optimal Blend Selection Using Ternary Diagram

The experimental data plotted in a ternary diagram are shown in Figure 21.2. The figure shows the volume percent of n-paraffin, cyclo-paraffin, and iso-paraffin. The remaining amount is that of aromatics. Four different properties are used that need to be fulfilled to meet the ASTM D1655 specifications. We need to identify the region where properties of the composition match all the required property of the jet fuel. We have used ASTM D1655 limits of density, heat content, flash point, and freezing point to select the regions. Any blend can meet 0 to all 4 properties. With 4 different properties, there will be 15 different combinations (${}^4C_4 + {}^4C_3 + {}^4C_2 + {}^4C_1$). Thus, to mark the region where different criteria are fulfilled, we have used number 1 to 15 to indicate the regions. Here 1 represents where all four properties are fulfilled and 15 represent the region where none of them is matched.

Blend composition for which experiments are conducted is marked with a star in Figure 21.2. A spline interpolation is used to connect between different points. The region plot can be extended with a greater number of sample points where properties are known. With few experimental data, the compositions that meet aviation quality can be achieved provided the experiment is conducted over a wide range of compositions. Since limited experimental data are used for the plot, in order to make the plot more reliable the figure is created by generating data points covering the entire range of different compositions.

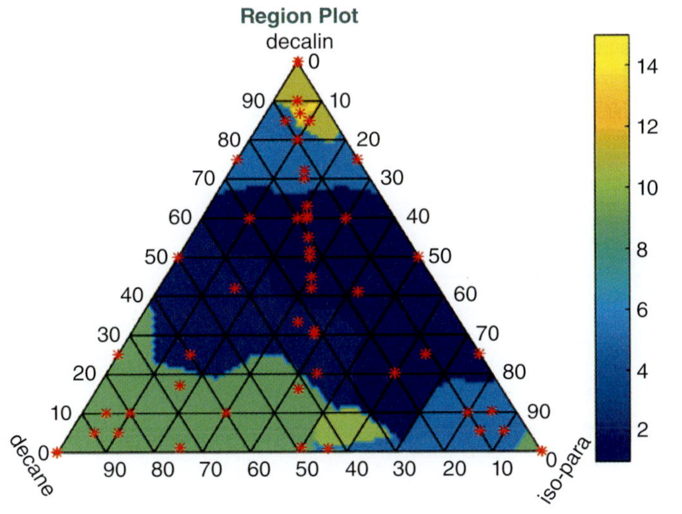

Figure 21.2 Optimal region in a ternary plot with region 1 as shown in the color map has compositions with all four required characteristics.

21.3.2 Optimal Blend Selection Using Multivariate Statistics

The process of blend selection and hydrocarbon property correlation using ternary diagram has limitations in finding the contribution of different hydrocarbon building blocks on blend properties. Since we cannot uniquely vary the composition of a single building block by keeping all other hydrocarbon components constant, unique contributions cannot be evaluated. A multivariate analysis is essential to find the interactive effects of different building blocks. This along with the visual selection of optimal composition from the ternary diagram makes the process of optimal blend selection dependent on human interpretation. In the present work, we have developed multivariate techniques for the selection and subsequent characterization of synthetic fuel that is human-independent and can be implemented for high throughput selection of optimal blends. All the analysis is performed in Matlab®.

21.3.2.1 Contribution of Different Hydrocarbon Components

Four different compounds are used to select the right composition property that fulfills the criteria of ASTM D1655. Though each compound has its own contribution in shaping the properties of the blend, it is useful to identify their relative contribution. Traditionally, to find the sensitivity of each hydrocarbon component of a blend, the individual hydrocarbon component was varied to find their impact in composition properties. In this method, we have to change the proportion of other hydrocarbon components as well. Thus, impact of a single hydrocarbon component cannot be found exclusively. Our goal is to find those criteria that the blend has to maintain in order to be aviation grade and to identify the building hydrocarbon components on which the required quality is most sensitive to.

Principal component analysis is performed on the blend data that meet the ASTM D1655 specifications. In this way, we can measure the variability of the compounds that have contributed in aviation fuel characteristics. The factor loadings of different compounds are used as a measure of their contribution. Figure 21.3 shows the factor loadings in the first two principal

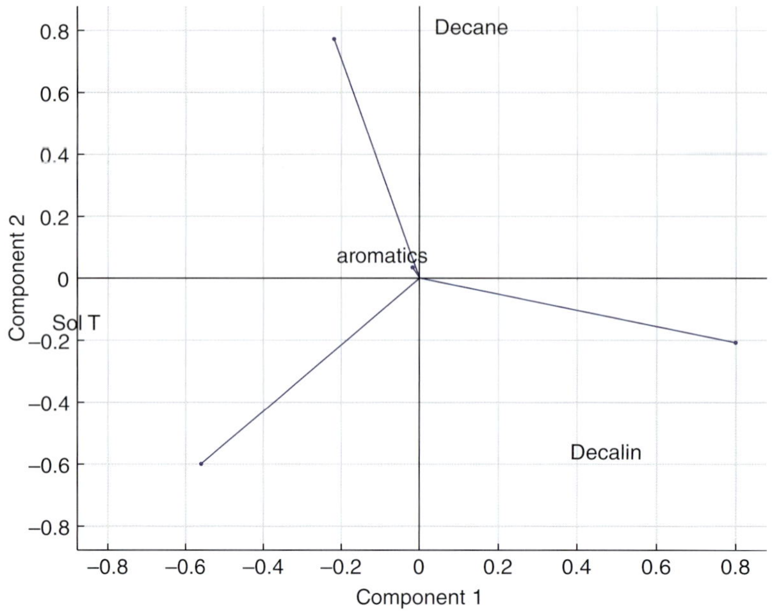

Figure 21.3 Factor loadings as obtained for different hydrocarbon components used for the blend.

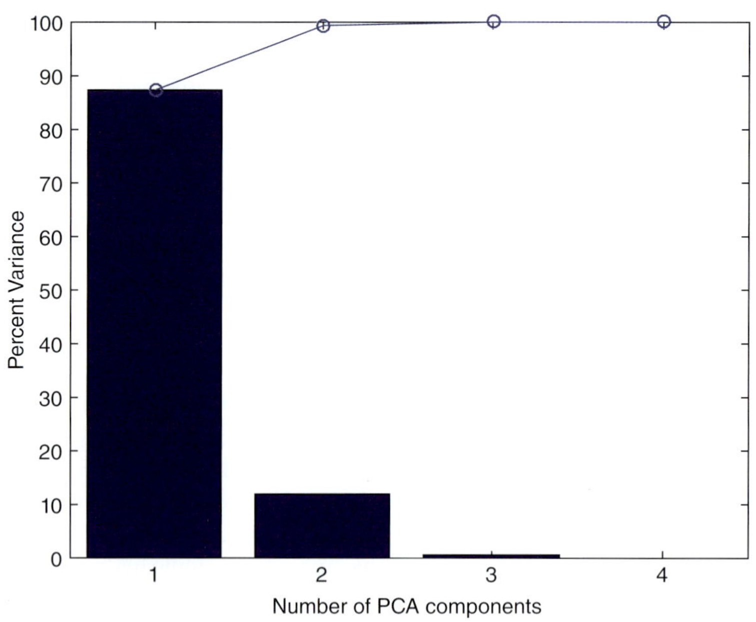

Figure 21.4 Percentage variance for different principal component.

components. Figure 21.4 gives the variability as captured in different principal components (PCs). The first two principal components comprise more that 90% of the variability, indicating our analysis with two PCs is reliable. From the figure, it is clear that paraffinic compounds can undergo a large variation, yet the blend will fulfill ASTM D1655 criteria. On the other hand, aromatics show the least variability. Thus in order to prepare the optimal blend, we may not have a large variation of aromatics. Also, all the compounds show distinct contributions towards blend characteristics, as they do not form clusters.

21.3.2.2 Composition Property Correlation

As mentioned before, selection of the optimal composition is dictated by the correlation of the properties with blends. Though correlation of properties with individual compounds shows a particular trend, the volumetric change of one compound is related to others. Thus, it is useful to find the correlation of the blend composition with properties. This is a multivariate problem, and a multivariate correlation is used to find the nature of correlations between different properties with the blend. The PLS regression method, as described before, is used for this purpose.

The result from multivariate correlation is shown in Figure 21.5. Figure 21.5a–d shows the correlation of the blend with density, heat content, flash point, and freezing point respectively as obtained using a linear regression analysis with PLS regression method. Figure 21.6a–d shows the percentage of variance captured by different latent variables. Since we have only four variables, we have used all four for regression coefficient generation that covers 100 percent of the original variability. In Figure 21.5, the values of the properties are plotted against that obtained using PLS regression coefficients. The coefficient of determination (R^2) for density and heat content is 0.992 and 0.9409 respectively and that for flash point and freezing point are 0.6052 and 0.702 respectively.

From Figure 21.5a,b, it is clear that density and heat content can be regenerated using PLS regression coefficients (data lies in the diagonal). Also from Figure 21.5c,d, we can see that flash point and the freezing point cannot be regenerated by using PLS regression coefficients (data are off-diagonal). The result shows that a linear regression model obtained from the PLS regression method is suitable to predict linear correlations of the blend with density and heat content but it is inadequate in predicting the nonlinear correlation of blend with flash point

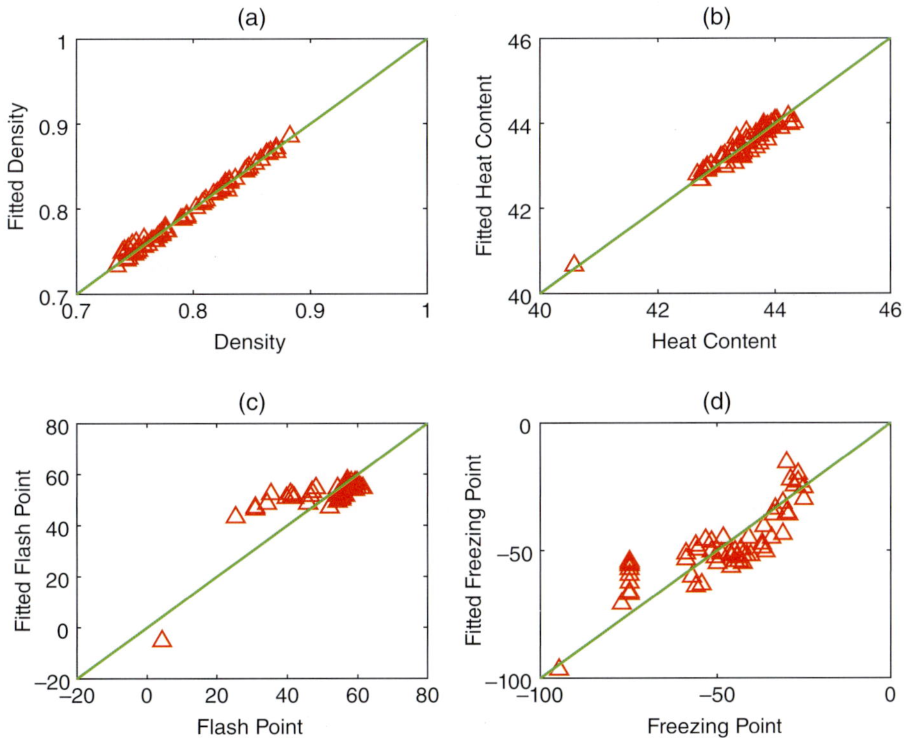

Figure 21.5 Multivariate correlation of different properties with blend composition using PLS regression method.

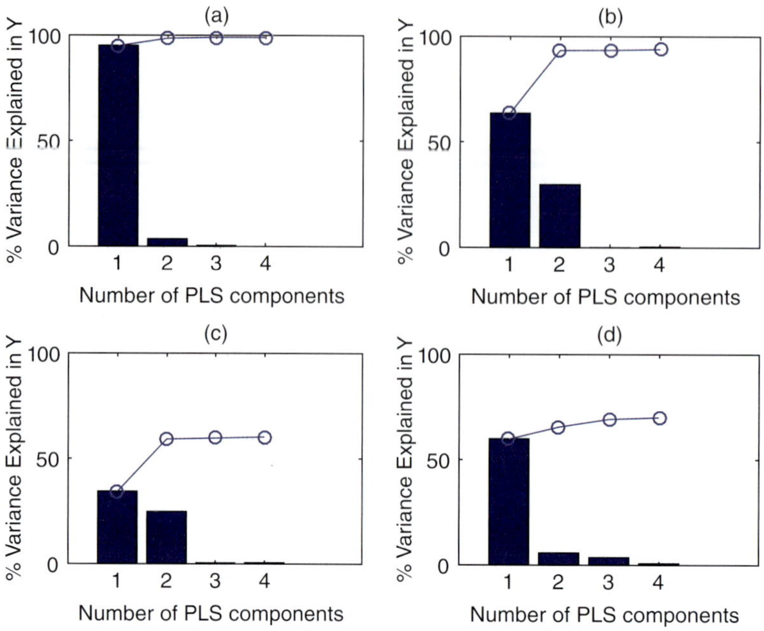

Figure 21.6 Percentage of variance at different latent variables.

and freezing point. The inefficiency of the linear regression method using PLS regression coefficient for flash point and freezing point is also evident from Figure 21.6. While only two latent variables can capture most of the variance in the case of density and heat content as seen from Figure 21.6a and b, even four latent variables are insufficient to capture the variance of flash point and freezing point as seen from Figure 21.6c and d.

The nonlinear correlation of the blend with flash point and freezing point is captured using the support vector machine (SVM). Using the polynomial kernel function for nonlinear regression with SVM, the coefficients of determination (R^2) for flash point and freezing point are found to be 0.9459 and 0.8982 respectively. Among polynomial and Gaussian kernel function, the polynomial kernel function with d=5 gives a higher R^2 value. Figure 21.7a and b shows the values of the properties are plotted against the one obtained using the SVM regression method. From Figure 21.7a and b, it is clear that density and heat content can be regenerated using the SVM regression method (data lies in the diagonal).

21.3.2.3 Reliability Prediction Using Score Plot

Among all the different numerical analyses of hydrocarbon blends for the optimal design of aviation fuel, the most important factor is the ability to predict the quality of a particular composition without conducting experiments. In this way, we can reduce the number of experiments, a reduction that can save resources as well as time for synthetic fuel development. We can measure n characteristic parameters of a blend that need to be fulfilled in order to reach aviation quality. A particular blend can fulfill 0 to n of those characteristics. While a ternary plot can show the region where different characteristics are fulfilled and can thus assist in designing suitable blends, the plot has limitations including the number of compounds that can be used for the diagram. In this work, an alternate approach has been used for blend characteristics using the multivariate statistical method.

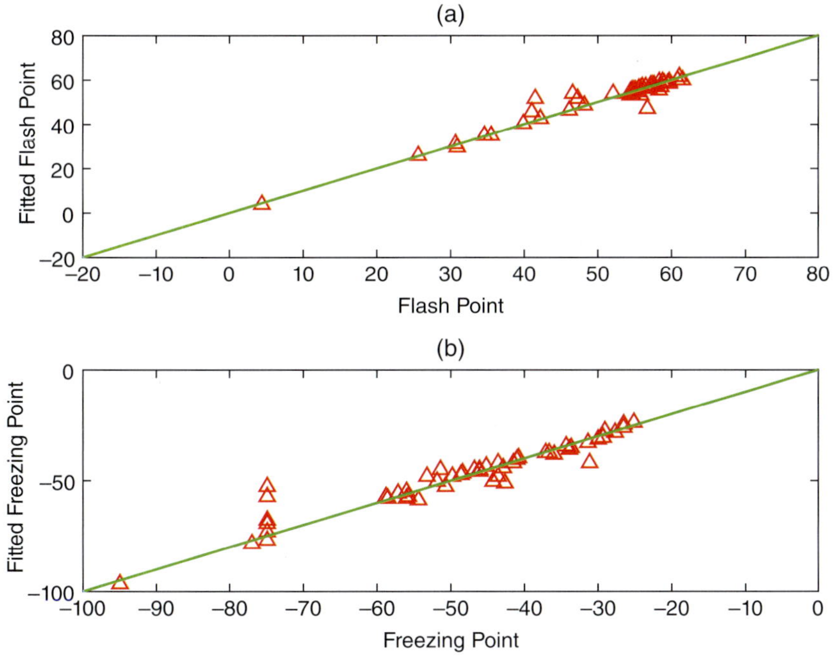

Figure 21.7 Multivariate correlation of flash point and freezing point with blend composition using SVM for regression.

Every single composition needs to be evaluated in order to check if it qualifies for the ASTM D1655 standard. When a particular blend qualifies, the specific composition can be noted as a single observation of success. Similarly, we can have m' such compositions that fulfill ASTM D1655 standards. In the subset of successful observations, every single compound has a range of variability within which we can successfully prepare synthetic fuel. Multivariate analysis of this data using PCA will generate eigenvectors that will give the direction of the maximum possible variability of the compounds. Any composition can be evaluated by finding its contribution in the direction of the principal component (dot product) through the principal component score of the composition using equation 21.4.

For an observation with n variables, we will have principal components with n factors. A dot product of an observation with the principal component will give a single score to evaluate the observation in that particular direction of the eigenvector. The principal component score from the first two eigenvectors (say P1 and P2) can be plotted in the score plot. Scores from observations that fulfill the fuel standard will generate a region (say R) in the score plot. It is to be noted that the observations that fulfill ASTM D1655 standards (m′) are a subset of the total number of possible compositions m. The score of all (m) observations can be evaluated in the direction of P1 and P2. Thus, blends whose composition meets ASTM D1655 criteria will reside inside the region R. Blends that do not meet ASTM D1655 criteria will remain outside the region R and can be treated as outliers.

Initially, the experimental data are separated into five groups. Compositions that meet ASTM D1655 specifications (m′) fulfill all four characteristics and are selected from the experimental data (m′) and assigned a number '4'. Similarly, blends that fulfill 3 characteristics are assigned number '3' and so on. The selected data (m′) are then used to find the eigenvectors through principal component analysis. The score plots as obtained from all the observations (m) in the direction of the first two principal components of the selected data (m′) are shown in

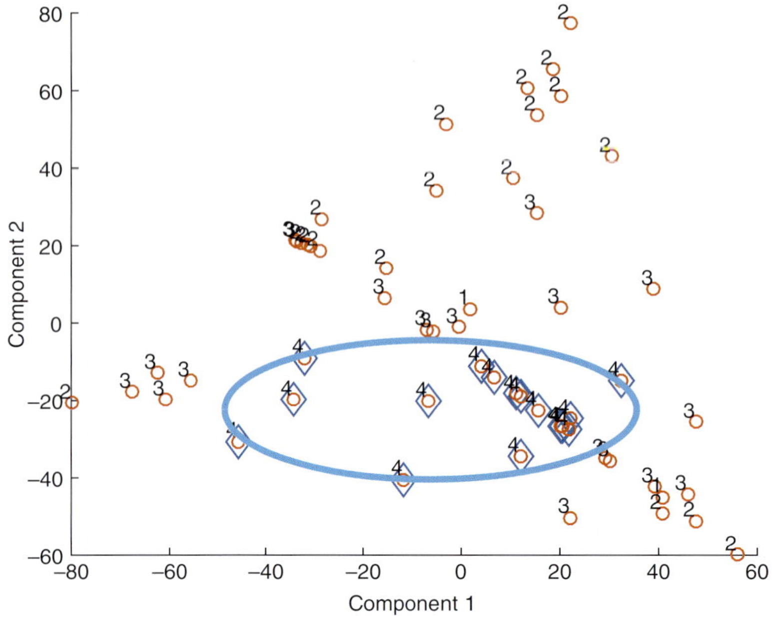

Figure 21.8 Score plot of different observations (blends) with outliers. The optimal region where all four properties are fulfilled is inside the ellipse. The number indicates number of properties that are fulfilled.

Figure 21.8. The scores from different observations are marked with their assigned number. From the figure it is clear that blends that fulfills ASTM D1655 criteria (marked '4') are clustered in a region R. In the present case, we can see that blends that are qualified as aviation grade fuels are scattered more in the direction of the first principal component. An elliptical region in the score plot is identified as the confidence region R. The ellipse is not centered; the original data is scaled but not centered. All other observations that match three or fewer characteristics are outside the region R. It is interesting to note that the distance of an observation located away from region R is inversely proportional to the number of characteristics matched by the blend. Thus, the score plot can be used as testing criteria for any new composition before it can be verified experimentally. The score plot of the observations with the blend number is also shown in Figure 21.9. All the analysis is performed on normalized data.

21.3.3 Experimental Verification of Model Predicted Data

In order to verify the model, eight randomly generated blends are obtained. Using their PC score, the blends are predicted to fulfill ASTM D1655 specifications. These models predicted data points are shown in Table 21.1. The PC score of the selected blends in the score plot is shown in Figure 21.10. They are marked with (*) in the score plot. The adequacy of the PC-score-based selected blends to meet ASTM D1655 specifications are verified in two different ways. Experiments were conducted to find density, heat content, flash point, and freezing of the selected blends. Simultaneously, multivariate regression models developed with m′ compositions that fulfills ASTM D1655 standards are used to predict properties of novel selected blends. PLS regression models are used to predict density and heat content, and SVM-based nonlinear regression models are used to predict flash point and freezing points of the selected blends.

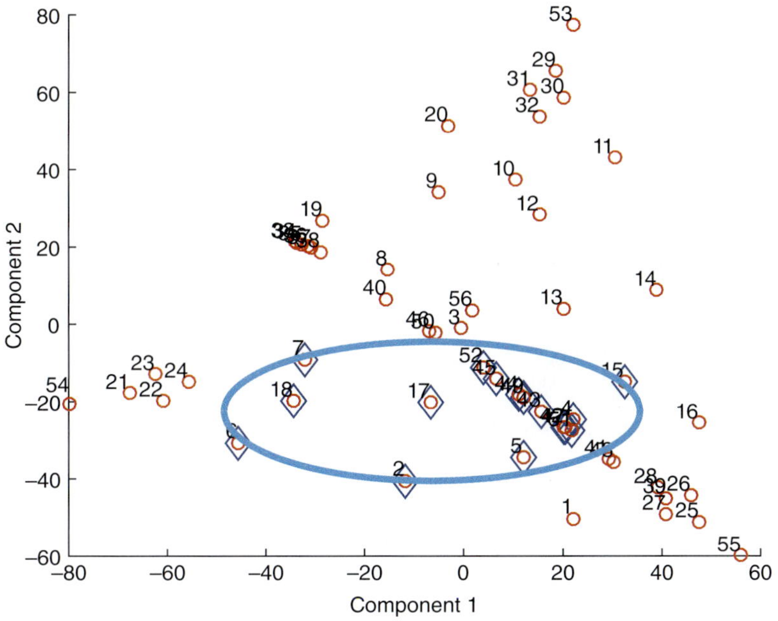

Figure 21.9 Score plot of the blends and outliers with blend numbers.

Table 21.1 Blend compositions for experimental verifications.

Blend ID	Decane (n-para) (vol %)	Shell Sol T (Iso-para) (vol %)	Decalin (Cyclo-para) (vol %)	Ethylbenzene (Aromatics) (vol %)
Blend 1	10	60	30	0
Blend 2	15	42.5	51.5	0
Blend 3	20	35	45	0
Blend 4	10	50	40	0
Blend 5	20	40	40	0
Blend 6	15	40	40	5
Blend 7	20	35	40	5
Blend 8	15	35	45	5

Both experimental results, as well as the regression model predicted properties, verified the qualification of the PC-score-based predicted blends as aviation grade fuel meeting ASTM D1655 criteria. A comparison of the properties predicted by the regression model with those from experimentally determined properties of the selected blends are shown in Table 21.2. Both density and heat content show a deviation of less than 1% while flash point and freezing point shows a maximum deviation of 6% and 16% respectively. From the location of the selected blends in the score plot as seen in Figure 21.10 and the parity of the experimentally determined properties with regression-based models as seen in Table 21.2, it is clear that a PC-score-based model along with the PLS- and SVM-based regression model is sufficient for novel blend selection and subsequent characterization for selection of aviation grade fuel composition. In this process, we can avoid the laborious, expensive, and time-consuming experimental procedure for the same.

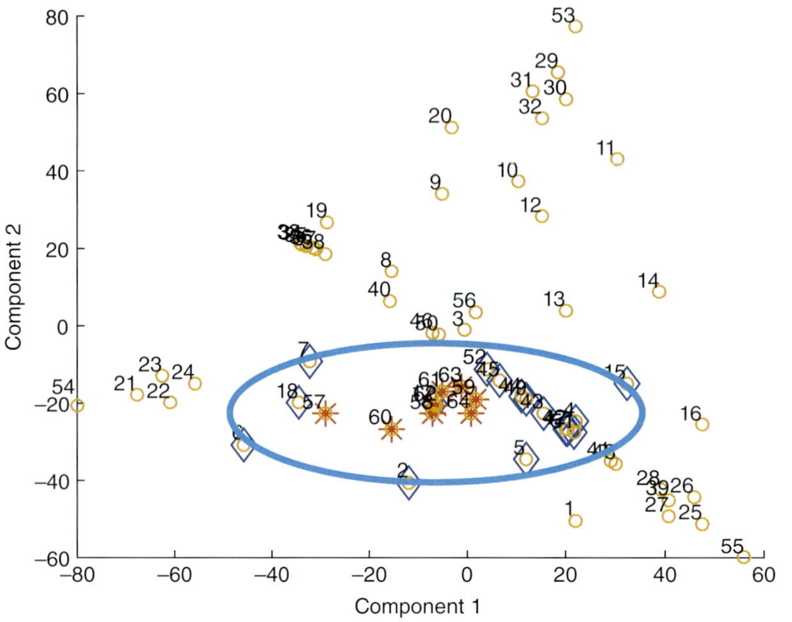

Figure 21.10 Score plot of the blends with outliers. The compositions with predicted properties are marked with "*".

Table 21.2 Comparison of model-generated blend properties with experimental data using PLS regression method for density and net heat content and SVM based regression for flash point and freezing point.

	Density (g/cc)			Heat Content (MJ/kg)		
Blend ID	Experimental	Model Predicted	% deviation	Experimental	Model Predicted	% deviation
Blend 1	0.7949	0.7961	−0.1565	43.5887	43.5191	0.1597
Blend 2	0.8092	0.8099	−0.0804	43.3622	43.4279	−0.1515
Blend 3	0.811	0.8109	0.0169	43.4139	43.4424	−0.0657
Blend 4	0.8097	0.8088	0.1064	43.4734	43.4133	0.1382
Blend 5	0.8048	0.8045	0.0353	43.4422	43.4953	−0.1222
Blend 6	0.8135	0.8121	0.1748	43.3279	43.4014	−0.1695
Blend 7	0.8098	0.8099	−0.0144	43.255	43.4423	−0.4331
Blend 8	0.8175	0.8184	−0.1132	43.1421	43.3485	−0.4784

	Flash Point (°C)			Freezing Point (°C)		
Blend ID	Experimental	Model Predicted	% deviation	Experimental	Model Predicted	% deviation
Blend 1	58.5	59.3925	−1.5257	−63.4	−57.0679	9.9875
Blend 2	58.5	58.3529	0.2514	−59.07	−50.4387	15.5131
Blend 3	57	57.4119	−0.7226	−55.77	−47.2261	15.3199
Blend 4	58.7	58.9659	−0.4529	−64.5	−54.7163	15.1685
Blend 5	57.5	57.6239	−0.2155	−55.63	−47.2837	15.0033
Blend 6	51.33	54.745	−6.6531	−59.77	−50.17	16.0616
Blend 7	51.5	49.014	4.8272	−55.6	−46.2875	16.749
Blend 8	51.5	52.252	−1.4602	−59.77	−50.5114	15.4904

21.4 Conclusions

Our target of developing synthetic fuel that is competitive with aviation fuel is fulfilled using multivariate statistical methods. The first step in this direction is to identify the correlation of different properties with the blend. While density and heat content show a linear relation, flash point and freezing point show highly nonlinear correlation. The different compounds show different impact in blend composition. Thus each of them has different roles in blend characterization. Multivariate statistics allows us to use a unlimited number of variables for analysis. Since there are no limits on the number of variables, unlike a ternary plot, a list of qualified blends can be created for any number of hydrocarbon components. Any new composition can be tested using its score in the direction of principal components obtained from blends with rightful composition. Although the principal components (PC) analysis acts as a linear transformation of the properties in a lower dimensional space, PC scores can be used as an initial selection of qualified blends that have required properties. Selected blends, when followed by property evaluation using regression models, corroborate their qualification as aviation grade fuel.

In this work, eight blends have been identified using their PC score. Experimental verifications have been performed of the physical properties of selected blends that were obtained using the PC score. Properties of the selected blends are also evaluated using multivariate regression models. Both experimental results, as well as regression models, verified the properties of the selected blends as aviation grade fuel. Thus, a unique approach has been performed for blend composition identification and subsequent characterization using multivariate statistical methods that can be used for aviation grade fuel selection without conducting rigorous experiments. In this approach, the effect of the composition of the blend on physical properties is evaluated rather than finding the contribution of individual hydrocarbon component.

Acknowledgements

This paper was made possible by a NPRP award [NPRP 7-843-2-312] from the Qatar National Research Fund (a member of the Qatar Foundation). The statements made herein are solely the responsibility of the authors.

References

Al-Nuaimi I A, Bohra M, Selam M, Choudhury HA, El-Halwagi MM, Elbashir NO. 2016. Optimization of the aromatic/paraffinic composition of synthetic jet fuels. Chem Engg Technol, 39 (12): 2217–2228.

ASTM. 2010. Standard specification for aviation turbine fuels. Annual Book of ASTM Standards, p. 1–16. Available from: https://doi.org/10.1520/D1655-10.2.

Bao B, El-Halwagi MM, Elbashir NO. 2010. Simulation, integration, and economic analysis of gas-to-liquid processes. Fuel Proc Technol, 91(7): 703–713.

Challiwala M, Ghouri M, Linke P, El-Halwagi M, Elbashir N. 2017. A combined thermo-kinetic analysis of various methane reforming technologies: comparison with dry reforming. J CO2 Utiliz, 17, p. 99–111.

Cinar A, Palazoglu A, Kayihan F. 2007. Chemical process performance evaluation. CRC press. Boca Raton.

Cortes C, Vapnik V. 1995. Support-vector networks. Machine Learn, 20 (3): 273–297.

Elmalik E E, Raza B, Warrag S, Ramadhan H, Alborzi E, Elbashir NO. 2014. Role of hydrocarbon building blocks on gas-to-liquid derived synthetic jet fuel characteristics. Indus Engg Chem Res, 53(5): 1856–1865. https://doi.org/10.1021/ie402486c

Gabriel K J, Noureldin M, El-Halwagi MM, Linke P, Jiménez-Gutiérrez A, Martínez DY. 2014a. Gas-to-liquid (GTL) technology: targets for process design and water-energy nexus. Curr Op Chem Engg, 5, 49–54.

Gabriel K J, Linke P, Jiménez-Gutiérrez A, Martínez DY, Noureldin MMB, El-Halwagi MM. 2014b. Targeting of the water-energy nexus in gas-to-liquid processes: a comparison of syngas technologies. Ind Eng Chem Res 53 (17): 7087–7102.

Gamerman D, Moreira ARB. 2004. Multivariate spatial regression models. J Multivariate Analysis, 91 (2): 262–281. Available from: https://doi.org/10.1016/j.jmva.2004.02.016

Huang D, Zhou H, Lin L. 2011. Biodiesel: an alternative to conventional fuel. Energ Procedia, 16 (PART C): 1874–1885. https://doi.org/10.1016/j.egypro.2012.01.287

Jolliffe, IT. 1986. Principal component analysis. 2nd ed. New York: Springer, 37–52.

Konishi, S. 2014. Introduction to multivariate analysis: linear and nonlinear modeling. Boca Raton, FL: CRC Press.

Liu G, Yan B, Chen G. 2013. Technical review on jet fuel production. Renew Sustain Energ Revs, 25, 59–70. Available from: https://doi.org/10.1016/j.rser.2013.03.025.

Martínez DY, Jiménez-Gutiérrez A, Linke P, Gabriel KJ, Noureldin MMB, El-Halwagi MM. 2013. Water and energy issues in gas-to-liquid processes: assessment and integration of different gas reforming alternatives. ACS Sustain Chem Eng, 2, (2): 216–225

Mukherjee R, Sengupta D, Sikdar SK. 2013. Parsimonious use of indicators for evaluating sustainability systems with multivariate statistical analyses. Clean Technol Environ Policy, 15(4): 699–706.

Muteki K, MacGregor JF., Ueda T. 2006. Rapid development of new polymer blends: the optimal selection of materials and blend ratios. Indus Engg Chem Res, 45 (13): 4653–4660.

Selam M, Ahmed W, Bohra M, Mohamed H, Hussain R, Elbashir N. 2014. Role of aromatics in synthetic fuels. Qatar Foundation Annual Research Conference 2014 Nov 18, No. 1, p. EESP0487.

Sikdar SK, Sengupta D, Mukherjee R. 2017. Statistical algorithms for sustainability measurement and decision making. In: Measuring Progress towards sustainability. Switzerland: Springer International Publishing, p. 153–184

Solash J, Hazlett RN, Hall JM, Nowack CJ. 1978. Relation between fuel properties and chemical composition. 1. Jet fuels from coal, oil shale and tar sands. Fuel, 57 (9): 521–528. Available from: https://doi.org/10.1016/0016-2361(78)90036-4.

Suganthi L, Samuel AA. 2012. Energy models for demand forecasting: a review. Renew Sustain Energ Revs, 16 (2); 1223–1240. Available from: https://doi.org/10.1016/j.rser.2011.08.014

Vapnik V. 2013. The nature of statistical learning theory. New York: Springer science & business media.

Viana H, Vega-Nieva DJ, Ortiz Torres L, Lousada, J, Aranha J. 2012. Fuel characterization and biomass combustion properties of selected native woody shrub species from central Portugal and NW Spain. Fuel, 102, 737–745.

Wold, H. (1966) Nonlinear estimation by iterative least square procedures. In: David, F. N., ed., Research Papers in Statistics. Festschrift for J. Neyman, 411–444. London, New York: Wiley.

Index

Natural Gas Processing from Midstream to Downstream, First Edition.
Edited by Nimir O. Elbashir, Mahmoud M. El-Halwagi, Ioannis G. Economou, and Kenneth R. Hall.

o

p